The Arboranan Frogs

Evolution, Biology, and Cytogenetics

Editors

M. Schmid, Würzburg
J.P. Bogart, Guelph, ON
S.B. Hedges, Philadelphia, PA

Authors

Michael Schmid[a] Claus Steinlein[a] Thomas Haaf[a] Wolfgang Feichtinger[a]
Martina Guttenbach[a] James P. Bogart[b] Simone L. Gruber[c] Sanae Kasahara[c]
Wanpen Kakampuy[d] Eugenia M. del Pino[e] Ana-Belén Carrillo[e] Andrés Romero-Carvajal[e]
Michael Mahony[f] Max King[g] William E. Duellman[h] S. Blair Hedges[i]

[a] Department of Human Genetics, University of Würzburg, Würzburg, Germany;
[b] Department of Integrative Biology, University of Guelph, Guelph, ON, Canada;
[c] Department of Biology, Insitute of Biosciences, University Estadual Paulista, Rio Claro, Brazil;
[d] Department of Biological Sciences, Ubon Ratchathani University, Ubon Ratchathani, Thailand;
[e] School of Biological Sciences, Pontificia Universidad Católica del Ecuador, Quito, Ecuador;
[f] School of Environmental and Life Sciences, University of Newcastle, Callaghan, NSW, Australia;
[g] Museums of Arts and Sciences of the Northern Territory, Darwin, NT, Australia;
[h] Biodiversity Institute, University of Kansas, Lawrence, KS, USA;
[i] Center for Biodiversity, Temple University, Philadelphia, PA, USA

956 figures, 117 in color, 10 tables, and 3 appendices, 2018

KARGER

Basel · Freiburg · Paris · London · New York · Chennai · New Delhi ·
Bangkok · Beijing · Shanghai · Tokyo · Kuala Lumpur · Singapore · Sydney

Addresses for correspondence:

Michael Schmid
Department of Human Genetics, University of Würzburg
Biozentrum, Am Hubland
DE–97074 Würzburg (Germany)
Tel. +49 931 318 4077, Fax +49 931 318 4058
E-Mail m.schmid@biozentrum.uni-wuerzburg.de

James P. Bogart
Department of Integrative Biology, University of Guelph
50 Stone Road East
Guelph, ON N1G 2W1 (Canada)
Tel. +1 519 824 4120, ext. 58728, Fax +1 519 767 1656
E-Mail jbogart@uoguelph.ca

S. Blair Hedges
Center for Biodiversity, Science Education & Research Center
Temple University, 1925 N. 12th Street, Suite 502
Philadelphia, PA 19122 (USA)
Tel. +1 215 204 4244, Fax +1 215 204 2121
E-Mail sbh@temple.edu

Reprint of **Cytogenetic and Genome Research** (ISSN 1424–8581)
Vol. 155, No. 1–4, 2018

S. Karger
Medical and Scientific Publishers
Basel · Freiburg · Paris · London ·
New York · Chennai · New Delhi ·
Bangkok · Beijing · Shanghai · Tokyo ·
Kuala Lumpur · Singapore · Sydney

Drug Dosage
The authors and the publisher have exerted every effort to ensure that drug selection and dosage set forth in this text are in accord with current recommendations and practice at the time of publication. However, in view of ongoing research, changes in government regulations, and the constant flow of information relating to drug therapy and drug reactions, the reader is urged to check the package insert for each drug for any change in indications and dosage and for added warnings and precautions. This is particularly important when the recommended agent is a new and/or infrequently employed drug.

Printed on acid-free and non-aging paper (ISO 9706)
ISBN 978–3–318–06404–9
e-ISBN 978–3–318–06405–6

KARGER

E-Mail karger@karger.com
www.karger.com/cgr

Contents

Cytogenet Genome Res 2018;155:6
DOI: 10.1159/000490324

Accepted: February 1, 2018
by J. Smith

Foreword

After a long-standing period of relative taxonomic stability, the popular and charismatic frogs included in the non-ranked group Arboranae (or "true treefrogs") have been the subject of considerable taxonomic turmoil in the current century. These frogs, for many years conforming to the family Hylidae, are now split into three different families: Hylidae, Pelodryadidae, and Phyllomedusidae. Recent taxonomies include also the erection of many new genera and the inclusion in Hylidae of the aquatic and fairly divergent few species formerly placed in the family Pseudidae.

As systematic studies progress, the phylogeny of Arboranae will for sure experience more changes in the future but, whatever the taxonomy adopted, the present book edited by M. Schmid, J.P. Bogart, and S.B. Hedges will stand as a mandatory reference for any study on the group. Like its precedents on terraranan and hemiphractid frogs (2010 and 2012, respectively), this opus represents not only a monumental initiative to put together all available information on the cytogenetics of an enormous monophyletic group of frogs, but also a substantial effort to provide new data, everything properly placed in a phylogenetic, evolutionary, and biogeographic framework.

This book has 994 literature citations, which indicate the extensive literature revision carried out by the authors. It also shows 956 figures, clearly making it a fundamental visual reference, highly useful to anyone interested in cytogenetics in general, and in Arboranae in particular. Many of those figures are devoted to the karyotypes of 204 species belonging to 47 genera, which represent a large coverage of the Arboranae – 20.7% of the species but, more importantly, 78.3% of the genera. Thus, this work offers a unique and comprehensive panorama of chromosome variation in an important component of the world's amphibian fauna. Up to 21 refined methods were used to obtain karyotypes, which are illustrated in an impressively detailed way. Exhaustive data on specimens sampled, localities, techniques used, and diploid number/fundamental number are provided in Appendix 1, which becomes a pivotal part of the work that will doubtlessly serve as a framework for future reference.

Arboranan frogs are almost cosmopolitan, being absent only from sub-Saharan Africa and southeastern Asia (where they are ecologically replaced mostly by the families Hyperoliidae and Rhacophoridae, respectively). In concordance with this fact, this book benefits from being a necessarily cosmopolitan joint venture, bringing together collaborators from Australia, Brazil, Canada, Ecuador, Germany, Thailand, and the USA. These authors contributed with their varied expertise in the introductory chapters on phylogeny, classification, biogeography, life history, and population declines of Arboranae, but the main focus of the book is put on oogenesis, spermatogenesis, early embryogenesis, and, especially, the structure and evolution of chromosomes.

Many years (indeed, decades!) have been necessary to gather and compile the information provided in this book, which together with that of its predecessors on Terraranae and Hemiphractidae amounts to 488 species of frogs already studied from a cytogenetic standpoint. This is an astonishing figure, considering the time and effort involved in obtaining this kind of information.

In summary, this work represents an exhaustive compilation of information by a "dream team" of scientists that have devoted a great part of their careers to study the Arboranae and their chromosomes. After commanding this impressive enterprise, the question is: Which one is the next candidate group of amphibians to be considered for a review like this?

Ignacio De la Riva
Madrid
April 2018

KARGER

E-Mail karger@karger.com
www.karger.com/cgr

Cytogenet Genome Res 2018;155:7
DOI: 10.1159/000490325

Accepted: February 1, 2018
by J. Smith

Preface

The first expedition to South America and chromosome analyses for the present monograph on Arboranae were realized by James P. Bogart 49 years ago in 1969. In 1987, Michael Schmid, Claus Steinlein, Wolfgang Feichtinger, and Thomas Haaf started with their collecting trips and cytogenetic studies in Central and South America, in the Caribbean region, as well as in Australia. In 2010, S. Blair Hedges and William E. Duellman joined this project with their contributions on phylogeny, embryology, and life history of arboranan frogs.

The time required for writing the text of the manuscript, as well as for the preparation of the numerous karyotypes and other illustrations, was approximately 4 years. All correspondence and discussions on arboranan frogs between the authors in the USA, Canada, Ecuador, and Germany took place by electronic mail, courier messages containing photographic negatives, and electronic transmission of huge files with illustrations. Not a single face-to-face meeting of the authors, prior to or during the elaboration of this monograph, was necessary to keep the enthusiasm for this project going.

Although the karyotypes of arboranan frog species are conservative when compared to other groups of neotropical anurans, they raise and answer a wealth of fascinating questions on chromosome evolution, sex determination, meiosis, and cell biology. Furthermore, reproduction, embryology, and life history of the Arboranae belong to challenging aspects in herpetology and developmental biology. It is for this reason that these important chapters are included in the present monograph.

A passable knowledge of phylogenesis is a necessary prerequisite for interpreting chromosome evolution in any group of eukaryotes. Most earlier comparative studies dealing with chromosome evolution in Amphibia were restricted by poor knowledge of their relationships. One of the few exceptions is an extensive analysis of terraranan chromosome evolution [Schmid et al., 2010], which used knowledge of their phylogeny down to the species group and species series level. In the present investigation, the evolution of arboranan karyotypes was also determined by mapping the karyotype characters on their molecular phylogeny to determine the most parsimonious arrangements.

In the present book, the results obtained for the analyzed arboranan frogs are always compared with cytogenetic data published for other amphibian taxa and other vertebrates. The deduced structural and numerical chromosome repatternings in Arboranae are explained on the background of general principles of chromosome evolution. We always aimed for a high quality of chromosome preparations and microphotographs, as well as for precise and explanatory graphics, schematic drawings, and tables.

In spite of the detailed data presented here, we are aware that many questions concerning chromosome evolution in the Arboranae remain unanswered. A careful cytogenetic examination of many additional species in the diverse genera is required in order to obtain a more complete picture of the fascinating chromosomes of these frogs. We also expect our interpretations on arboranan chromosome evolution will likely change in the light of work including molecular cytogenetic techniques, especially those using in situ hybridizations with chromosome-specific DNA probes. Nevertheless, the information presented here provides a good basis and foundation for future research on these frogs.

The present book is the third volume dedicated to amphibian cytogenetics launched by Karger Publishers. We hope it will be useful for providing detailed and archival information on treefrog karyotypes and genomes to recognized specialists in amphibian and vertebrate cytogenetics. We also hope that this monograph will encourage students to participate in future investigations on vertebrate cytogenetics.

Michael Schmid, Würzburg
James P. Bogart, Guelph
S. Blair Hedges, Philadelphia
January 2018

KARGER

E-Mail karger@karger.com
www.karger.com/cgr

Cytogenet Genome Res 2018;155:8–9
DOI: 10.1159/000490326

Accepted: February 1, 2018
by J. Smith

Acknowledgements

The cytogenetic studies preceding the publication of the present monograph on arboranan frogs were carried out for more than 40 years (1969–2012) in several laboratories. In a project of such magnitude, the authors asked for and received the support of many colleagues. The Editors of the monograph (Michael Schmid, James P. Bogart, and S. Blair Hedges) acknowledge with great pleasure all the persons mentioned in the following.

We need to thank all colleagues and students for their support, their immense and untiring engagement, for their intrepidity in coping with all the ventures in, and/or for, their skillful laboratory analyses on arboranan frogs. We thank Christian Lomb, Anna Rohrbacher, Sabine Tauscher, Roland Weimer (Germany), Hinrich Kaiser (USA), José de Aguar, Mariana Albornoz, Marcos Alemán, Raúl Castro, David Diez, Alberto Fernández, Ernesto Fernández†, Marco Gaiani, Manuel Galea†, Jorge Gonzalez, Enrique La Marca, Jesús Manzanilla, Thomas Schmid†, Gregorio Ulloa†, and Richar Visbal (Venezuela), Peter Vogel† (Jamaica), Luis Diaz (Cuba), John Reed, and Steven Donnellan (Australia).

On our expeditions in South and Central America, the Caribbean islands, and Australia, numerous people helped to make the trips successful events. Their kindness, hospitality, and appreciation of our projects on neotropical amphibians are unforgotten.

We very much appreciate the efforts of Gitta Hesse (Department of Human Genetics, University of Würzburg, Germany) for having produced the high-quality photographic prints of chromosomes and cell nuclei of treefrogs in her darkroom. Many students have assisted in cell cultures, chromosome preparations and banding analyses, in situ hybridizations, measurements of genome sizes, and in evaluating cell nuclei parameters of arboranan frogs.

Michael Schmid thanks the former head of the Department of Human Genetics, University of Würzburg, Holger Hoehn, for having supported all expeditions to the Neotropics and for his deep interest in evolutionary cytogenetics. Thea and Gottfried Schmid† (Caracas, Venezuela) rendered possible the expeditions of Michael Schmid to Venezuela. He intensely appreciates the many efforts of Heidi and Wolf Flatzek (Würzburg, Germany), who took substantial burden from his shoulder, allowing him to continue with scientific projects and expeditions.

James P. Bogart thanks his former PhD supervisor, W. Frank Blair† (University of Texas at Austin, USA), for supporting his chromosome research in Texas and in South America even though much of that work was peripheral to his thesis. He also wishes to thank his wife, Jo Ellen, for her capable field assistance in Argentina and Brazil.

S. Blair Hedges thanks his PhD supervisor, Richard Highton (University of Maryland, USA), and postdoctoral supervisor, Linda Maxson (then at Pennsylvania State University, USA), for encouraging and supporting his work on tropical frogs. He also thanks Angela Marion for assistance with the timetree figures and Brooke Gattens, Angela Lu, and Julie Marin for assistance with the distribution figures.

William Duellman thanks Taran Grant, Brian Kubicki, and Ignacio De la Riva for pertinent publications. For responding to his requests for photographs, he is grateful to Leonardo Boquimpani-Freitas, Roy W. McDiarmid, José M. Padial, Gilson Rivas, Mauro Teixeira, Jr., and Victor Hugo Zaracho.

Eugenia del Pino thanks Classius de Oliveira, Harry W. Greene, Renato Nali, and Beata Rozenblut-Kościsty for photographs of frogs and ovaries. She also thanks Maria Ogielska for revision of the oogenesis section.

Andres Romero-Carvajal thanks Classius de Oliveira, Barrie Jamieson, and David Scheltinga for sharing their images on arboranan spermatogenesis and sperm ultrastructure. He also thanks Andrés Merino-Viteri, María José Sala-

KARGER

E-Mail karger@karger.com
www.karger.com/cgr

zar-Nicholls, and Barrie Jamieson for critical reading of the manuscript. The study of *Agalychnis* embryo development was supported by grants from Pontificia Universidad Católica del Ecuador.

Shirlei M. Recco-Pimentel (University of Campinas, São Paulo, Brazil), and Radenka F. Batistic (Universidade de São Paulo, São Paulo, Brazil) helped in obtaining cytogenetic literature on Brazilian arboranan frog species. Ikuo Miura (Amphibian Research Center, Hiroshima University, Higashi-Hiroshima, Japan) and Stanley K. Sessions (Hartwick College, Oneonta, NY, USA) kindly provided the karyotype of *Anotheca spinosa*. Matthias Stöck (Leibnitz-Insitute of Freshwater Ecology, Berlin, Germany) kindly contributed material of *Hyla arborea* and relevant literature on treefrogs, and John Malone (Wildlife and Fisheries Sciences, Texas A&M University, USA) collected the specimens of *Dryophytes femoralis* necessary for this monograph.

Our projects on Arboranae had considerable financial supports of national, academic, and industrial foundations in the USA, Canada, Brazil, and Germany. In addition, the Department of Human Genetics of the University of Würzburg (Germany) generously contributed funds for the production of this monograph. We are grateful to the Governmental Authorities in many countries for providing the permits to collect amphibians within their territories and to bring them to our laboratories in Canada, Germany, and the USA.

Jacqueline Smith (Division of Genetics and Genomics, Roslin Institute, Roslin, UK), Associate Editor of the journal *Cytogenetic and Genome Research*, kindly processed the voluminous manuscript for publication.

The final draft of the manuscript was essentially improved thanks to the efforts of Martina Guttenbach and Karin Schmid (Managing Editors of *Cytogenetic and Genome Research*, Würzburg), who performed a thorough copy editing. We are very thankful to the Publisher, Thomas Karger (Basel, Switzerland), for giving us the opportunity to publish the third large monograph in this series on amphibian biology, phylogeny, and cytogenetics.

Cytogenet Genome Res 2018;155:10
DOI: 10.1159/000490327

Accepted: February 1, 2018
by J. Smith

Remarks

Remarks on the Use of Scientific Names

Throughout this monograph, the scientific Latin names used for the species of Arboranae follow the electronic database of Frost [2017] and Duellman et al. [2016]. Changes of scientific names are regularly updated in the electronic database of Frost [2017].

The scientific names of all other amphibian species referred to in the text and tables do also follow the electronic database of Frost [2017]. The scientific names utilized for all non-amphibian species referred to in the text are the same as in the quoted publications.

Remarks on the Illustrations, Tables, and Appendices

Figures 1–52 are incorporated within the text of chapters 1–3. The legends to these figures are placed beneath or on the page before or after the corresponding illustrations. Figures 53–937 with the arboranan karyotypes are placed one after the other on pages 131–211. They have a common legend on page 130 and are also referred to in the text. Figures 938–956 showing the living arboranan species and the tadpoles are placed on pages 212–221 and are referred to in the legends beneath the figures and in the text. At least 1 species of each of the 59 arboranan genera is depicted. Tables 1–10 are incorporated within the text of chapters 1–3 and, when possible, are placed next to the pages where they are first referred to. The important Appendix 1 which provides essential information is found on pages 250–311.

Remarks on Previous Publications

The results of the cytogenetic studies of the 204 arboranan species presented in this monograph were partially published in previous publications. In some of these species, specific staining techniques and nucleic acid in situ hybridizations have been applied to the chromosomes. The chromosomes of the remaining species were only analyzed using conventional staining.

For the sake of completeness, and for the convenience of the readers, theoretical cytogenetic background, modified figures, schematic drawings, and tables of our previous publications on the Arboranae are included in the present monograph. If necessary for a general understanding, the schematic drawings of other authors were redrawn in a modified form, and the source of the original drawing is acknowledged in the legend.

KARGER

E-Mail karger@karger.com
www.karger.com/cgr

Cytogenet Genome Res 2018;155:11–45
DOI: 10.1159/000489540

Accepted: February 1, 2018
by J. Smith

The Arboranan Frogs:
Introduction

1

Systema Naturae *is one of the major works of the Swedish botanist, zoologist, and physician Carl Linnaeus (1707–1778) and introduced the Linnaean taxonomy. Although the system, now known as binomial nomenclature, was partially developed by Gaspard and Johann Bauhin 200 years earlier, Linnaeus was the first to use it consistently throughout his book. The first edition was published in 1735. The full title of the 10th edition [1758], which was the most important one (left image below), was "Systema naturæ per regna tria naturæ, secundum classes, ordines, genera, species, cum characteribus, differentiis, synonymis, locis" or translated: "System of nature through the three kingdoms of nature, according to classes, orders, genera and species, with characters, differences, synonyms, places". The left illustration below shows the front page of the 10th edition, the right illustration, page 194, introducing the chapter on the class Amphibia. The first amphibian species described by Linnaeus in this 10th edition were* Rana arborea (Hyla arborea), Rana boans (Boana boans), Rana ocellata (Osteopilus ocellatus), Rana paradoxa (Pseudis paradoxa), *and* Rana typhonia (Trachycephalus typhonius).

Unfortunately, and for the despair of many herpetologists, since the publication of the ***Systema Naturae****, the nomenclature has been subjected to many changes, sometimes irrational and unnecessary (see also Appendix 1 of the present monograph).*

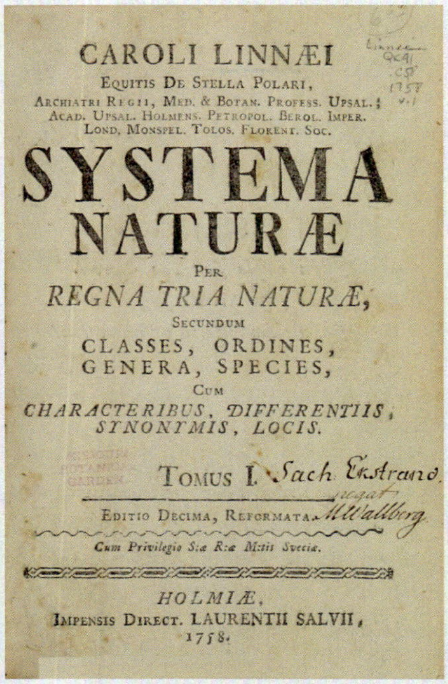

CAROLI LINNÆI
EQUITIS DE STELLA POLARI,
ARCHIATRI REGII, MED. & BOTAN. PROFESS. UPSAL.; ACAD. UPSAL. HOLMENS. PETROPOL. BEROL. IMPER. LOND. MONSPEL. TOLOS. FLORENT. SOC.
SYSTEMA NATURÆ
PER
REGNA TRIA NATURÆ,
SECUNDUM
CLASSES, ORDINES, GENERA, SPECIES,
CUM
CHARACTERIBUS, DIFFERENTIIS, SYNONYMIS, LOCIS.
TOMUS I.
EDITIO DECIMA, REFORMATA.
Cum Privilegio S:æ R:æ M:tis Sveciæ.
HOLMIÆ,
IMPENSIS DIRECT. LAURENTII SALVII,
1758.

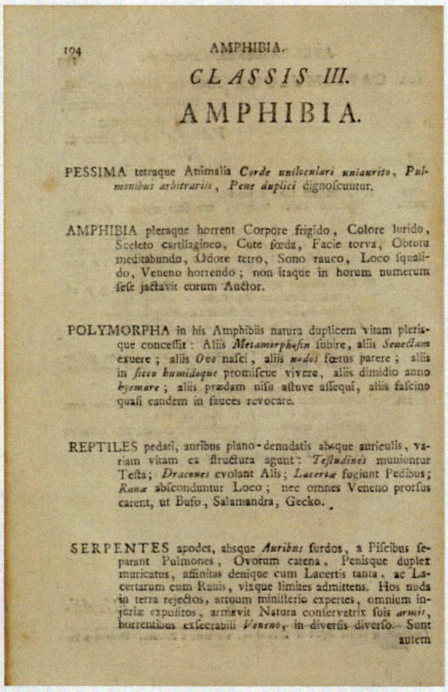

194 AMPHIBIA.

CLASSIS III.

AMPHIBIA.

PESSIMA tetraque Animalia *Corde uniloculari uniaurito*, *Pulmonibus arbitrariis*, *Pene duplici* dignofcuntur.

AMPHIBIA pleraque horrent Corpore frigido, Colore lurido, Sceleto cartilagineo, Cute fœda, Facie torva, Obtutu meditabundo, Odore tetro, Sono rauco, Loco fqualido, Veneno horrendo; non itaque in horum numerum fefe jactavit eorum Auctor.

POLYMORPHA in his Amphibiis natura duplicem vitam plerisque conceffit: Aliis *Metamorphofin* fubire, aliis *Senectam* exuere; aliis *Ovo* nafci, aliis *nudos* fœtus parere; aliis in *ficco humidoque* promifcue vivere, aliis dimidio anno *hyemare*; aliis prædam nifu aftuve affequi, aliis fafcino quafi eandem in fauces revocare.

REPTILES pedati, auribus plano-denudatis abfque auriculis, variam vitam ex ftructura agunt: *Teftudines* muniuntur Tefta; *Dracones* evolant Alis; *Lacertæ* fugiunt Pedibus; *Ranæ* abfconduntur Loco; nec omnes Veneno prorfus carent, ut Bufo, Salamandra, Gecko.

SERPENTES apodes, abfque *Auribus* furdos, a Pifcibus feparant Pulmones, Ovorum catena, Penisque duplex muricatus, affinitas denique cum Lacertis tanta, ac Lacertarum cum Ranis, vixque limites admittens. Hos nuda in terra rejectos, artuum minifterio expertes, omnium injuriæ expofitos, armavit Natura confervatrix fuis *armis*, horrentibus exfecrabili *Veneno*, in diverfis diverfo. Sunt autem

KARGER

E-Mail karger@karger.com
www.karger.com/cgr

1.1 Phylogeny, Classification, and Biogeography of Arboranae

1.1.1 Phylogeny

Diverse types of morphological, cytological, and molecular data have contributed to the understanding of treefrog relationships over the years and have clarified some major groupings and species relationships of individual groups [e.g., Maxson and Wilson, 1975; Hedges, 1986; Duellman, 2001]. Comprehensive DNA sequence phylogenies in the last decade or so have accelerated this discovery of relationships and have built on an increasingly more stable global phylogeny of treefrogs [Faivovich et al., 2005; Frost et al., 2006; Wiens et al., 2010; Duellman et al., 2016]. Here, we review arboranan relationships based on the largest and most recent phylogeny of 503 species (~50% of the described species), 19 genes, and 16,128 aligned sites [Duellman et al., 2016]. That study also produced a timetree useful for inferring the historical biogeography of the Arboranae.

At the highest level (Fig. 1), the tree obtained by Duellman et al. [2016] resolved 3 clades of treefrogs at the family level with strong support: Hylidae (676 species), Phyllomedusidae (59 species), and Pelodryadidae (208 species). The South American phyllomedusids are closest relatives of the Australasian pelodryadids.

Within the Hylidae, Duellman et al. [2016] recognized 7 subfamilies: Acridinae (now Acrisinae, Dubois et al. [2017]), Hylinae, Pseudinae, Dendropsophinae, Lophyohylinae, Scinaxinae, and Cophomantinae. The only subfamily not strongly supported (>95% bootstrap) was Scinaxinae (49%), only because of the weakly associated genus *Sphaenorhynchus*. Otherwise, Scinaxinae was strongly supported (97%).

The tree (Fig. 1) strongly supported (99%) a close relationship between the small North American subfamily Acrisinae (21 species) and the larger northern neotropical and holarctic subfamily Hylinae (165 species). Otherwise, the only other well supported (>70%) nodes in the tree of hyline subfamilies were the placement of Cophomantinae (179 species) as closest relative of all other hyline subfamilies (93%) and the node joining Pseudinae and Dendropsophinae (79%).

For the other 2 families, no subfamilies were recognized for Phyllomedusidae and 2 for Pelodryadidae: Litoriinae [Dubois and Frétey, 2016] and Pelodryadinae. The node joining those 2 subfamilies was strongly supported (100%).

Intergeneric relationships (Fig. 1) also were well supported in general. Of the 38 intergeneric nodes in the tree, within subfamilies, 21 were strongly supported (>95%), 11 more were well supported (>70%), and only 6 were not well supported (<70%), with 5 of those 6 weak nodes in a single subfamily, Lophyohylinae. Turning to the expanded phylogeny of all available species, relationships were generally well supported or strongly supported for many of the 503 species. For example, all but 5 of the 19 species-level nodes in the genus *Pseudacris* (Acrisinae) were strongly supported (>95%). On the other hand, relationships were poorly established for some genera such as *Scinax*, where only 4 of 25 species-level nodes were strongly supported and nearly half of the nodes were not even well supported (>70%). Future studies will undoubtedly add more species and genomes of arboranans, resolving more details of phylogeny, but the major patterns in the current phylogenetic tree are unlikely to change (Fig. 1).

1.1.2 Classification

Based on the foregoing phylogenetic tree, Duellman et al. [2016] proposed a new classification of the frogs in the unranked taxon Arboranae. This taxon includes the genera and species formerly placed in Hylidae, Pelodryadidae, and Phyllomedusidae. In earlier classifications [e.g., Faivovich et al., 2005; Wiens et al., 2010], the latter 2 families were recognized as subfamilies of Hylidae.

Duellman et al. [2016] defined Arboranae as anurans having an arciferal pectoral girdle, 8 procoelous presacral vertebrae, separate calcaneum and astragalus, intercalary elements (usually cartilaginous) between terminal and penultimate phalanges, claw-shaped terminal phalanges, snout-vent lengths of 18–142 mm; eggs deposited in water or on vegetation above water; aquatic larvae in all species; basic labial tooth row formula of 2/3 in tadpoles; and a chromosome complement of 2n = 18–34.

The 3 families contain 935 currently recognized species ranging throughout temperate North America, Central America, the West Indies, and tropical and subtropical South America; arboranans occur in Australia, Tasmania, New Guinea, and the Solomon Islands, as well as in much of temperate Eurasia, Japan, and extreme northern Africa. Three distinct clades are evident in the molecular phylogeny presented by Duellman et al. [2016]. These are recognized as the families Hylidae, Pelodryadidae, and Phyllomedusidae. However, within Hylidae, their phylogenetic analysis shows that the large conglomerate of South American hylids is paraphyletic with respect to Mesoamerican hylids.

1.1.2.1 Family Hylidae Rafinesque, 1815

Hylidae is the largest arboranan family. It is divided into 7 subfamilies containing 49 genera and 676 species. The family occurs throughout most of the tropical and temperate regions in the New World, but 2 genera (one endemic) occur throughout much of temperate Eurasia, Japan, and extreme northern Africa.

There are only 56 species of hylid frogs in the Holarctic Region. Twenty-one of these are members of the subfamily Acrisinae containing 3 genera (*Acris*, *Hyliola*, and *Pseudacris*). All are endemic to temperate North America. These small to medium-sized frogs are primarily terrestrial and semi-aquatic. The discs on the fingers and toes are not, or only slightly, expanded (Fig. 938a–c). Previous to the analysis by Duellman et al. [2016], all Holarctic treefrogs were placed in the genus *Hyla* (subfamily Hylinae). Their mo-

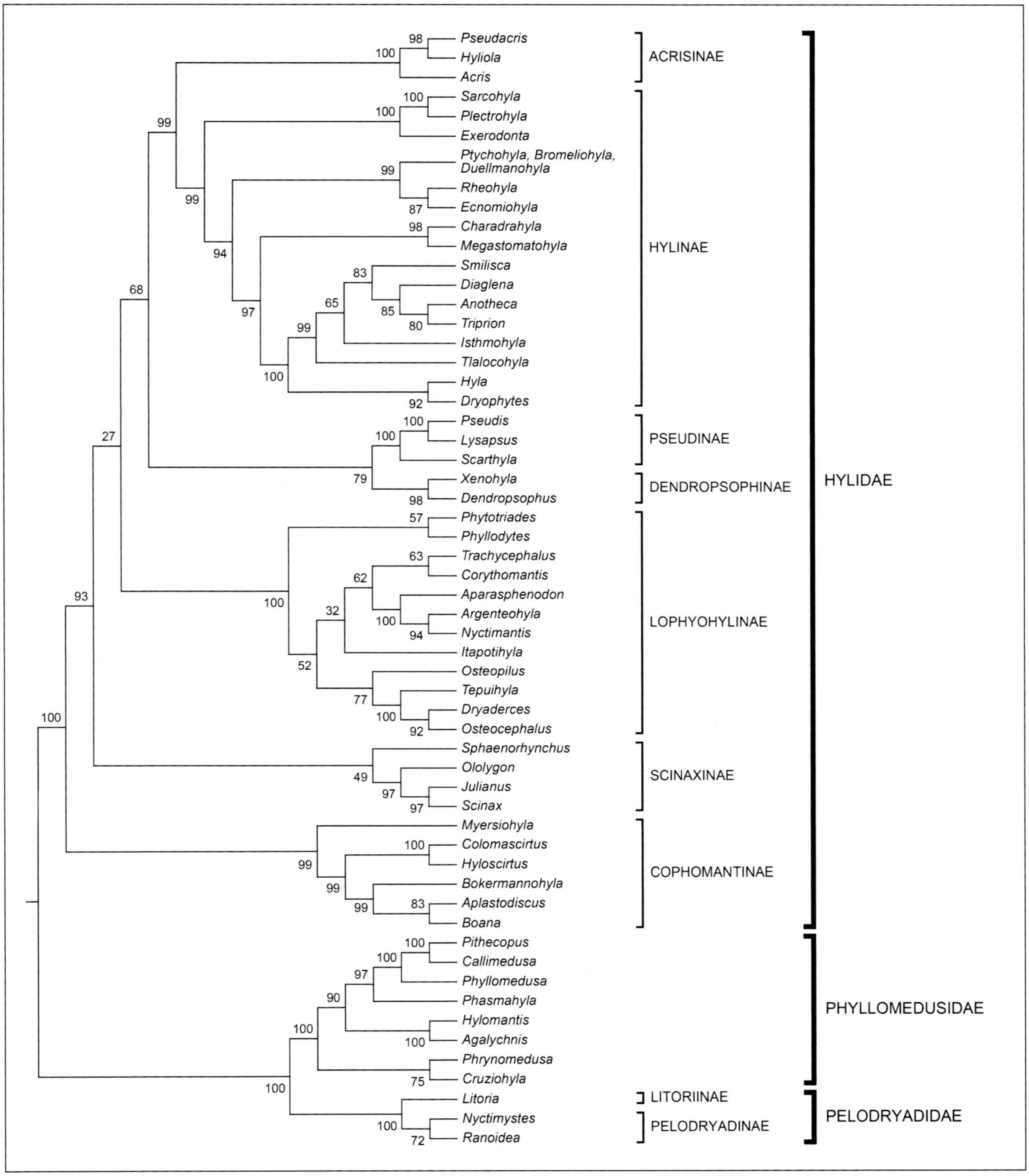

Fig. 1. Maximum likelihood phylogeny of arboranan frogs based on sequence data from 19 genes (16,128 aligned sites) from Duellman et al. [2016]. Bootstrap support values are indicated at the nodes. For further explanations, see chapter 1.1.1.

lecular phylogenetic analysis resulted in the recognition of 2 genera. The 16 species of *Hyla* (Fig. 938d) are endemic to temperate and tropical regions of Eurasia and extreme northern Africa. The genus *Dryophytes* (Fig. 938e, f) has 9 species in temperate North America, 7 species in temperate montane regions in Mexico and Guatemala, and 3 species in temperate eastern Asia, including the Japanese and Ryukyu archipelagos.

The vast majority of frogs of the subfamily Hylinae occur in tropical and subtropical regions of Mexico and Central America [Duellman, 2001]. All species of 14 genera are endemic to Middle America. Two other genera exist mainly in Middle America. *Smilisca baudinii* (Fig. 939a) is the most widespread hyline frog in Middle America. Four species of *Smilisca* live in the lowlands, 2 exist in subhumid northwestern Mexico, and 2 others inhabit moderate elevations in the mountains of Costa Rica and Panama. The ranges of 3 species of *Smilisca* extend into extreme northwestern Colombia. Of the hylines in Middle America, the 4 species of *Tlalocohyla* (Fig. 939d) are restricted to tropical lowlands. Likewise, the bizarre casque-headed frogs of the monotypic genera *Diaglena* and *Triprion* (Fig. 939e, f) occur primarily in subhumid lowlands in Mexico and Guatemala.

Nine genera of hylines inhabit the Middle American highlands and breed in mountain streams. These frogs are most speciose in Mexico; 4 genera (*Charadrahyla*, *Megastomatohyla*, *Rheohyla*, and *Sarcohyla*; Fig. 940a–c, 941a) are endemic to Mexico west of the Isthmus of Tehuantepec. Two other genera (*Exerodonta* and *Ptychohyla*; Fig. 940d, f) are primarily distributed in Mexico but extend into the Central American highlands. *Plectrohyla* (Fig. 941b) is most speciose in northern Central America but also extends into Mexico east of the Isthmus of Tehuantepec. Two genera, *Duellmanohyla* and *Isthmohyla* (Fig. 940e, 941c, d), occur only in the mountains of Central America, where they are most speciose in the highlands of Costa Rica and western Panama. The monotypic *Anotheca* (Fig. 941f) ranges from eastern Mexico to western Panama; it breeds in bromeliads and tree holes. Ten species of the canopy-dwelling frogs of the genus *Ecnomiohyla* (Fig. 939b, c) are known from southern Mexico to Panama, and 1 is known from western Colombia and northwestern Ecuador. For those species whose life history is known, *Ecnomiohyla* also breeds in tree holes. The 2 small species of *Bromeliohyla* (Fig. 941e) in southern Mexico and northern Central America breed in bromeliads.

Hylid frogs are especially diverse in South America, where the number of stream-breeding species is proportionately fewer than in Central America. The 5 genera of South American stream-breeding frogs currently contain 89 recognized species in the subfamily Cophomantinae, which contains the speciose pond-breeding *Boana*. Among the stream-breeding genera, *Aplastodiscus* and *Bokermannohyla* (Fig. 942a, b) inhabit the mountains and foothills in the Atlantic Coastal Forest in eastern Brazil. The 17 species of *Colomascirtus* (Fig. 942c, d) (thoroughly reviewed as the *Hyloscirtus larinopygion* group by Coloma et al. [2012]) inhabit moderately high elevations in the Andes from southern Colombia to Bolivia. *Hyloscirtus* (Fig. 942e) with 19 species inhabits low to moderate elevations in the Andes from western Venezuela to central Peru. Two species of *Hyloscirtus* extend into lower Central America. Each of the 6 species of *Myersiohyla* occurs on an isolated highland (tepui) in Guyana and southern Venezuela; the genus was reviewed by Faivovich et al. [2013].

The ranges of 92 species of the genus *Boana* (Fig. 943a–c) extend throughout tropical South America southward to northern Argentina, and Trinidad and Tobago. Three species of South American *Boana* extend into Central America, and 1 species, *B. rufitela*, is endemic to Central America. These are moderate-sized to large frogs, up to 118 mm in snout-vent length in *B. boans* [Rodríguez and Duellman, 1994]. The genus includes the gladiator frogs, so named because males have sharp prepollical spines that are used in male-male combat [Kluge, 1981]. The genus is represented on the Island of Hispaniola by *B. heilprini* (Fig. 943d).

The large genus *Dendropsophus* with 95 species (Dendropsophinae) occurs throughout tropical and subtropical Middle America and South America. These small, colorful frogs (Fig. 943e, f, 944a, b) are most plentiful in lowland rainforests. Unlike all other hylids, except some *Aplastodiscus*, which have variable chromosome complements, *Dendropsophus* has a diploid chromosome number of 2n = 30. Genomic evidence places the frugivorus *Xenohyla truncata* (Fig. 944c) with a chromosome complement of 2n = 24 in the Dendropsophinae.

The small subfamily Pseudinae with 3 genera is widely distributed in South America east of the Andes. *Lysapsus* (4 species) and *Pseudis* (7 species) are aquatic in lentic waters. These frogs have robust hind limbs and fully webbed feet (Fig. 944d, e). The 2 small species of *Scarthyla* (Fig. 944f) perch on low vegetation in ponds. They are capable of skittering on the surface of the water between emergent plants [Duellman and de Sá, 1988].

The hylid subfamily Scinaxinae contains 4 genera. All are pond breeders except for some *Ololygon* that breed in terrestrial bromeliads. The largest genus, *Scinax* with 69 species, is distributed from southern Mexico to Uruguay and northern Argentina, on Trinidad and Tobago, and St. Lucia in the Lesser Antilles. Most *Scinax* are small (less than 40 mm in snout-vent length) but a few are larger (females to more than 50 mm in snout-vent length) with pointed snouts (Fig. 945a, b). *Ololygon* (Fig. 945c, d) with 44 species ranges throughout the Atlantic Coastal Forests of eastern Brazil and extends southward to northeastern Argentina and westward into gallery forests in the Brazilian Cerrado. The 2 species of *Julianus* (Fig. 945e) occur in Minas Gerais, Brazil, and extreme southern Brazil, Uruguay, and northeastern Argentina. The 16 species of *Sphaenorhynchus* (Fig. 945f) occur in lowland tropical forests east of the Andes in South America. These arboreal frogs are predominately green.

Members of the hylid subfamily Lophyohylinae are morphologically quite diverse. The subfamily contains all of the

South American and West Indian casque-headed treefrogs and related genera. Four of the 5 species of *Aparasphenodon* occur in eastern Brazil, and 1 is in the upper Orinoco Region in Colombia and Venezuela. The 2 known species of *Corythomantis* inhabit subhumid regions in northeastern Brazil. Insofar as known, *Aparasphenodon brunoi* (Fig. 946a) and *Corythomantis greeningi* (Fig. 946c) are unique among all anurans in possessing venom. Both species have bony spines on the periphery of the upper jaw. These spines pierce the skin in areas of glands that produce venom. The spines act as a delivery system of venom that is more potent than that of some pitvipers (genus *Bothrops*) [Jared et al., 2015].

Casqued heads are present in 3 monotypic genera, *Argenteohyla* (Fig. 946b), *Itapotihyla* (Fig. 946d), and *Nyctimantis* (Fig. 947a) occurring in coastal Brazil, northeastern Argentina, and the upper Amazon Basin in Ecuador, Colombia, and adjacent Peru, respectively. Two other genera in South America contain some species that have casque heads. These are *Osteocephalus* (Fig. 947d), a highly variable genus with 24 recognized species in the Guianan Region and Amazon Basin; its sister genus is *Dryaderces* (Fig. 947c) with 2 species in the upper Amazon Basin. The second is *Trachycephalus* that has 14 species ranging throughout tropical America. *Trachycephalus typhonius* (Fig. 947f) (formerly *Phrynohyas venulosa*), well known for its toxic skin secretions, has the greatest distribution of any hylid frog. It ranges from Mexico to Paraguay and northern Argentina. Other South American genera of Lophyohylinae include *Tepuihyla* (Fig. 947b) with 7 species in the mountains of Guyana and southeastern Venezuela, *Phyllodytes* (Fig. 946e) with 11 species in eastern Brazil, and *Phytotriades* (Fig. 946f) in Trinidad and on the adjacent Peninsula de Paria in Venezuela.

One genus, *Osteopilus* (Fig. 947e) with 8 species, is endemic to the Greater Antilles. Three species are endemic to Hispaniola and 4 to Jamaica. *Osteopilus septentrionalis* is native to Cuba, Cayman Islands, and Bahama Islands. It has been introduced into Puerto Rico, various islands in the Lesser Antilles, and peninsular Florida in the United States.

1.1.2.2 Family Phyllomedusidae Günther, 1859

This small family of 62 species was considered to be a subfamily of Hylidae until Duellman et al. [2016] elevated this distinctive group of frogs to familial status. Phyllomedusids differ from hylids by having a chromosome complement of 2n = 26 (2n = 24 in most hylids) and vertically elliptical pupils, as well as distinctive biochemical, morphological, behavioral, and reproductive characteristics [Faivovich et al., 2010]. All of the 8 genera are predominately green dorsally. They exist in tropical and subtropical regions in South America southward to northern Argentina.

Agalychnis and *Cruziohyla* are exceptions in that the former ranges northward to Mexico and the latter to Honduras. Most *Agalychnis* (Fig. 948a) occur in the lowlands of Middle America and the Pacific lowlands of northwestern South America, but 4 small species (Fig. 948b) are capable of changing the dorsal color from green by day to brown at night. These are inhabitants of cloud forests in lower Central America and northwestern South America. The 2 species of *Cruziohyla* are gaudy frogs with dermal ornamentation on their limbs (Fig. 948e). One species occurs in Central America and the Pacific lowlands of Colombia and northwestern Ecuador, and the other is widely distributed in the Amazon Basin.

Callimedusa occurs in the Guiana Region, upper Amazon Basin, and on the Amazonian slopes of the Andes in Ecuador to central Peru. Seven of the 8 species are green dorsally and have yellow, orange, or red flanks with black or purple bars or mottling (Fig. 948c, d). The enigmatic *C. atelopoides* is a small brown terrestrial species. It is unlike any other phyllomedusid in lacking vomerine teeth and clavicles in the pectoral girdle [Duellman et al., 1988].

Three genera (*Hylomantis*, *Phasmahyla*, and *Phrynomedusa*) are restricted to the Atlantic Coast Forest in eastern Brazil. In size and structure, the 2 species of *Hylomantis* (Fig. 948f) resemble the small species of *Agalychnis* in northwestern South America. The 7 species of *Phasmahyla* (Fig. 949a) occur in the forested mountains of southeastern Brazil, where at least some species breed in streams. The 6 species of *Phrynomedusa* (Fig. 949b) in southeastern Brazil have dermal ornamentation on the limbs.

The genus *Phyllomedusa* contains 16 large species (snout-vent length up to 130 mm in *P. bicolor*; Fig. 949c) ranging throughout tropical and subtropical South America from northwestern Colombia, Venezuela, and the Guianas southward to Uruguay and northern Argentina. Species inhabiting subhumid areas (*P. boliviana*, *P. iheringii*, and *P. sauvagii*; Fig. 949d) produce dermal secretions that reduce evaporative water loss while aestivating in the dry season. The frogs spread the secretion over their entire bodies by wiping motions with their hands and feet [Shoemaker and McClanahan, 1975]. *Pithecopus* with 9 species (Fig. 949e, f) ranges throughout tropical South America east of the Andes. These moderate-sized (up to ±50 mm snout-vent length) are agile branch-walkers.

1.1.2.3 Family Pelodryadidae Günther, 1859

Pelodryadidae is the sister taxon to Phyllomedusidae. Both have a chromosome complement of 2n = 26. Duellman et al. [2016] recognized 2 subfamilies, 3 genera, and 208 species occurring principally in Australia, Tasmania, New Guinea, and associated islands – Solomon Islands, Bismarck Archipelago, New Britain, Vanuatu, and Molucca Islands and Seram Island in Indonesia. Nomenclature of some pelodryadid frogs was changed by Dubois and Frétey [2016], who revealed that Pelobiinae was preoccupied by its earlier use in Coleoptera. Thus, they proposed a new subfamily Litoriinae. They also argued that *Ranoidea* had priority over *Dryopsophus* as used by Duellman et al. [2016]. The recognition of subfamilies and genera is based on the analysis of genomic data; morphological data are needed to define these groups.

Cytogenet Genome Res 2018;155:11–45
DOI: 10.1159/000489540

The subfamily Litoriinae contains 1 genus, *Litoria* (Fig. 950a, b), with 92 species in northern and eastern Australia, New Guinea, Solomon Islands, and other archipelagos in the Australo-Papuan Region. Most *Litoria* are arboreal, but some are terrestrial. All have horizontally elliptical pupils and unpigmented palpebral membranes.

The subfamily Pelodryadinae contains 2 genera and 109 species in Australia, Tasmania, and New Guinea and associated islands. Most of the 38 species of *Nyctimystes* (Fig. 950c) occur in the mountains of New Guinea, and 1 is widespread in eastern Australia. *Nyctimystes* is unique among pelodryadids in having vertically elliptical pupils with reticulated palpebral membranes; furthermore, these frogs are stream breeders. One species that is assigned to the genus in the molecular analysis is *N. infrafrenatus* (Fig. 950d), a large frog (snout-vent length up to 135 mm), unlike other members of the genus by having a horizontal pupil, unpigmented palpebral membrane, and breeding in ponds.

The large genus *Ranoidea* (Fig. 950e, f) (*Dryopsophus* in Duellman et al. [2016]) contains at least 71 species; about one-third of the species occur in New Guinea, whereas Australia and Tasmania are home to two-thirds of the species. *Ranoidea* is like *Litoria* in having a horizontal pupil and an unpigmented palpebral membrane. Most *Ranoidea* are pond breeders, but several species in the mountains of New Guinea and Queensland in northern Australia are stream breeders. Most species of *Ranoidea* are arboreal, but some are terrestrial.

1.1.3 Biogeography

Duellman et al. [2016] provided a timetree for treefrog evolution. According to their data analysis, crown Arboranae originated 61.8 (57.5–66.1) million years ago (Mya) in the Paleocene (Fig. 2). From their place of origin in South America, these treefrogs differentiated and dispersed into the Australo-Papuan Region, Central and North America, the West Indies, and Eurasia. The historical events in the evolution of arboranans with respect to time and geography were hypothesized by the calculation of the nodes in the timetree (Fig. 2). Calculated times are given with their confidence intervals as shown by the blue bars on the timetree.

In summary of the extensive biogeographic analysis by Duellman et al. [2016], the treefrog (Arboranae) clade diverged from its nobleobatrachian relatives in South America during the latest Cretaceous or early Cenozoic. Differentiation of arboranans into the 3 families (Hylidae, Pelodryadidae, and Phyllomedusidae) occurred in South America during the early Cenozoic. Ancestral Hylidae diverged from the phyllomedusid-pelodryadid clade in the late Paleocene. Divergence of phyllomedusids and pelodryadids occurred in the mid-Eocene. When South America separated from Antarctica 35.9–30.5 Mya, ancestral pelodryadid frogs were present on Antarctica and Australia before they separated about 35.5 Mya. In the Australo-Papuan Region, the genera of pelodryadids differentiated during the Eocene and Oligocene. In tropical South America, the genera of phyllomedusids differentiated in the Miocene and Oligocene.

In South America, differentiation of the major clades (subfamilies) of Hylidae took place in the Eocene to early Oligocene, and most of their genera differentiated in the Eocene and Oligocene. Middle American hylids diverged from their South American relatives in the early Oligocene. Four major clades of South American arboranans (Phyllomedusidae, Dendropsophinae, Hylinae, Scinaxinae) invaded Central America in the mid-Cenozoic, possibly prior to the connection of the continents in the Miocene and Pliocene. Many South American species entered Central America after the formation of the Panamanian Land Bridge in the mid-Pliocene, but only 3 species of Middle American origin (*Smilisca*) invaded South America. Generic differentiation of hylines in Middle America took place in the Oligocene and Miocene.

Two ancestral clades of hylids reached North America. The first differentiated from the Hylinae in the late Eocene and became the Acrisinae with 3 genera, all restricted to temperate North America. The second was the hyline stock ancestral to *Hyla* in Eurasia and *Dryophytes* that now occupies temperate North America plus 3 species in extreme eastern Asia. As suggested by Borkin [1999], there were 2 invasions of hylids into Eurasia from North America. The first became *Hyla*, the second was an early *Dryophytes* stock. The timing of the first invasion in the Miocene (22.6 Mya; Duellman et al. [2016]) was substantiated by the discovery of fossil *Hyla* from the early-mid Miocene at Lake Baikal in Russia [Syromyatnikova, 2016].

Over-water dispersal by means of flotsam from South America resulted in 2 lineages of Hylidae reaching the Greater Antilles [Hedges et al., 1992; Hedges, 1996; Hass et al., 2001]. One of these, *Osteopilus*, dispersed to the Greater Antilles in the Oligocene or Miocene and then dispersed and differentiated throughout the Greater Antilles in the Miocene (Fig. 2). The other lineage, represented by *Boana heilprini* restricted to Hispaniola, dispersed from South America in the Oligocene or later (Fig. 2).

1.2 Ecology of Arboranae

As their vernacular and unranked taxon names imply, most arboranan frogs are arboreal. There are few places in the New World where hylid frogs do not occur. One species, *Pseudacris triseriata*, exists as far north as Great Bear Lake (65° N) in the Northwest Territories in Canada, where the mean summer temperature is 11.04 °C and the mean winter temperature is –27.22 °C. In Eurasia, *Hyla arborea* reaches 57° N in Denmark, and *Dryophytes japonicus* reaches the same latitude northeast of Lake Baikal in Russia. Obviously survival in these harsh climes necessitates hibernation below frost line. In South America, *Boana pulchella* occurs southward to 39° S in the Andean foothills; there aridity may be the limiting factor.

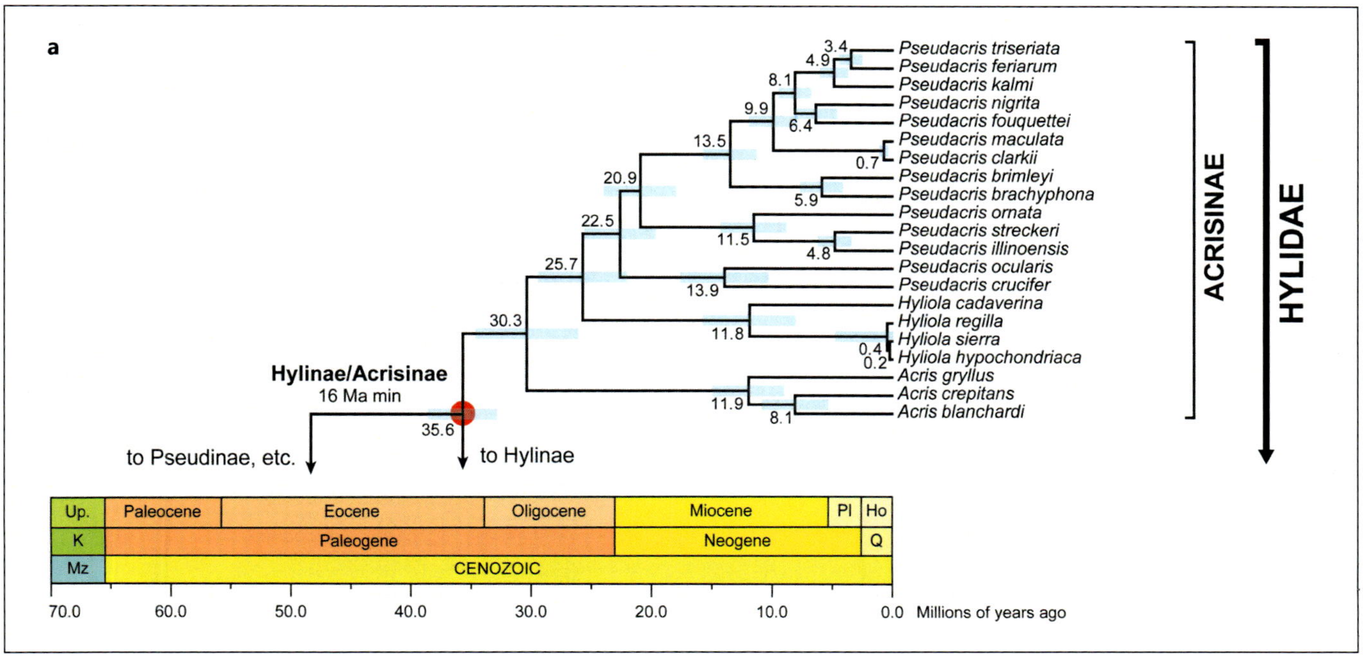

Fig. 2. a–g Timetree of arboranan frogs from Duellman et al. [2016]. Dates in the timetree are estimated with RelTime from DNA sequence data of 19 genes (16,128 aligned sites) based on a maximum likelihood topology. Estimated dates of divergence are indicated at each node (in millions of years), with blue bars representing 95% confidence intervals. Calibrated nodes are marked with red dots. For further explanations, see chapter 1.1.3.

(Figure continued on next pages.)

But, not all arboranans are treefrogs. In forested parts of North America, the small to medium-sized frogs of the genera *Hyliola* and *Pseudacris* are terrestrial, and *Acris* inhabits aquatic margins. Likewise, in the Neotropical Region, some arboranans inhabit open, non-forested areas. For example, in the Villavicencio Region in Colombia, 6 hylids are restricted to forests, whereas 8 hylids and 1 phyllomedusid occur only in the natural grasslands (Llanos) and anthropogenically created pastures; 2 species occur in both habitats [Lynch, 2006]. In the Australo-Papuan Region, 1 clade of the pelodryadid genus *Ranoidea* is terrestrial in northern, western, and central Australia; some species (e.g., *R. platycephala,* the so-called water-holding frog) can survive 2 or more years in a self-constructed burrow. In the hot, arid regions of Australia, tadpoles that develop in shallow ephemeral ponds are subject to high temperatures. Tadpoles of 3 species (*Litoria coplandi, L. rubella*, and *Ranoidea australis*) have been found in water in which the temperature exceeded 40°C [Tyler, 1989].

Geological events during the Cenozoic resulted in the rise of the Middle American highlands and the Andes, as well as an expansion of lowland tropical rainforest in what is now the Amazon Basin. Also, humid montane forests developed on the Andes and on the Brazilian highlands. Likewise, mountains with associated humid forests rose in New Guinea and Australia. Associated with the geological and climate changes, arboranans differentiated and invaded the variety of habitats. This resulted in some geographic regions of high density of species (Fig. 3, 4).

The regions having the highest density of arboranans in South America are much the same as for all amphibians shown by Duellman [1999]. The largest number of arboranans is in the Amazon Basin and in the Atlantic Coastal Forest in Brazil. These are followed by the mountains in Middle America and those in New Guinea (Fig. 3, 4). It is doubtful if these "hot spots" will diminish in importance as the number of species (and densities) continues to grow.

The hylid fauna of South America consists of 511 recognized species in 27 genera. There are 4 large genera of hylid frogs in South America – *Dendropsophus* (95 species), *Boana* (92 species), *Scinax* (69 species), and *Ololygon* (44 species). These 4 genera with 300 species make up 58.7% of the hylid fauna in South America. On that continent, there are 5 genera of stream-breeding hylids. Two of these are in the Andes, *Colomascirtus* (17 species) and *Hyloscirtus* (18 species); two in the Atlantic Coastal Forest in Brazil, *Aplastodiscus* (15 species) and *Bokermannohyla* (42 species); and *Myersiohyla* with 6 species in the Guiana Highlands. These 98 species account for only 19.2% of the hylid fauna in South America. The percentage of stream-breeding hylid frogs in South America is much lower than in Middle America, where 52 of 87 species (59.8%) are stream breeders.

The composition of arboranan communities is highly variable. The numbers of species are greatest in tropical rainforest and humid montane forests (Table 1). Dry scrub forest in Mexico has no more than 5 hylids and 1 phyllomedusid. The semideciduous forests in southern Brazil are

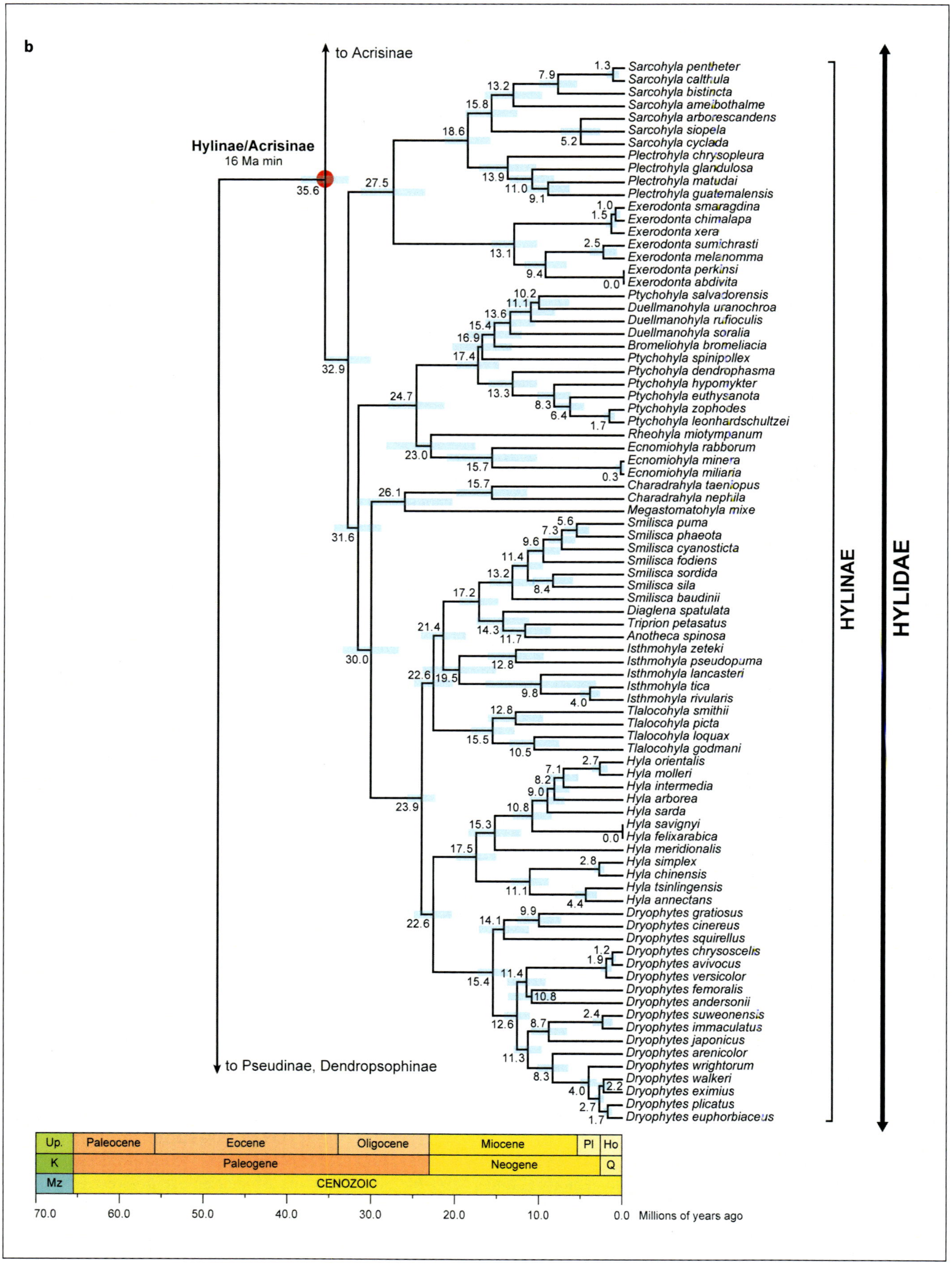
b
to Acrisinae
Hylinae/Acrisinae
16 Ma min
to Pseudinae, Dendropsophinae
Sarcohyla pentheter
Sarcohyla calthula
Sarcohyla bistincta
Sarcohyla ameibothalme
Sarcohyla arborescandens
Sarcohyla siopela
Sarcohyla cyclada
Plectrohyla chrysopleura
Plectrohyla glandulosa
Plectrohyla matudai
Plectrohyla guatemalensis
Exerodonta smaragdina
Exerodonta chimalapa
Exerodonta xera
Exerodonta sumichrasti
Exerodonta melanomma
Exerodonta perkinsi
Exerodonta abdivita
Ptychohyla salvadorensis
Duellmanohyla uranochroa
Duellmanohyla rufioculis
Duellmanohyla soralia
Bromeliohyla bromeliacia
Ptychohyla spinipollex
Ptychohyla dendrophasma
Ptychohyla hypomykter
Ptychohyla euthysanota
Ptychohyla zophodes
Ptychohyla leonhardschultzei
Rheohyla miotympanum
Ecnomiohyla rabborum
Ecnomiohyla minera
Ecnomiohyla miliaria
Charadrahyla taeniopus
Charadrahyla nephila
Megastomatohyla mixe
Smilisca puma
Smilisca phaeota
Smilisca cyanosticta
Smilisca fodiens
Smilisca sordida
Smilisca sila
Smilisca baudinii
Diaglena spatulata
Triprion petasatus
Anotheca spinosa
Isthmohyla zeteki
Isthmohyla pseudopuma
Isthmohyla lancasteri
Isthmohyla tica
Isthmohyla rivularis
Tlalocohyla smithii
Tlalocohyla picta
Tlalocohyla loquax
Tlalocohyla godmani
Hyla orientalis
Hyla molleri
Hyla intermedia
Hyla arborea
Hyla sarda
Hyla savignyi
Hyla felixarabica
Hyla meridionalis
Hyla simplex
Hyla chinensis
Hyla tsinlingensis
Hyla annectans
Dryophytes gratiosus
Dryophytes cinereus
Dryophytes squirellus
Dryophytes chrysoscelis
Dryophytes avivocus
Dryophytes versicolor
Dryophytes femoralis
Dryophytes andersonii
Dryophytes suweonensis
Dryophytes immaculatus
Dryophytes japonicus
Dryophytes arenicolor
Dryophytes wrightorum
Dryophytes walkeri
Dryophytes eximius
Dryophytes plicatus
Dryophytes euphorbiaceus
HYLINAE
HYLIDAE
Up.
Paleocene
Eocene
Oligocene
Miocene
Pl
Ho
K
Paleogene
Neogene
Q
Mz
CENOZOIC
70.0
60.0
50.0
40.0
30.0
20.0
10.0
0.0
Millions of years ago

2

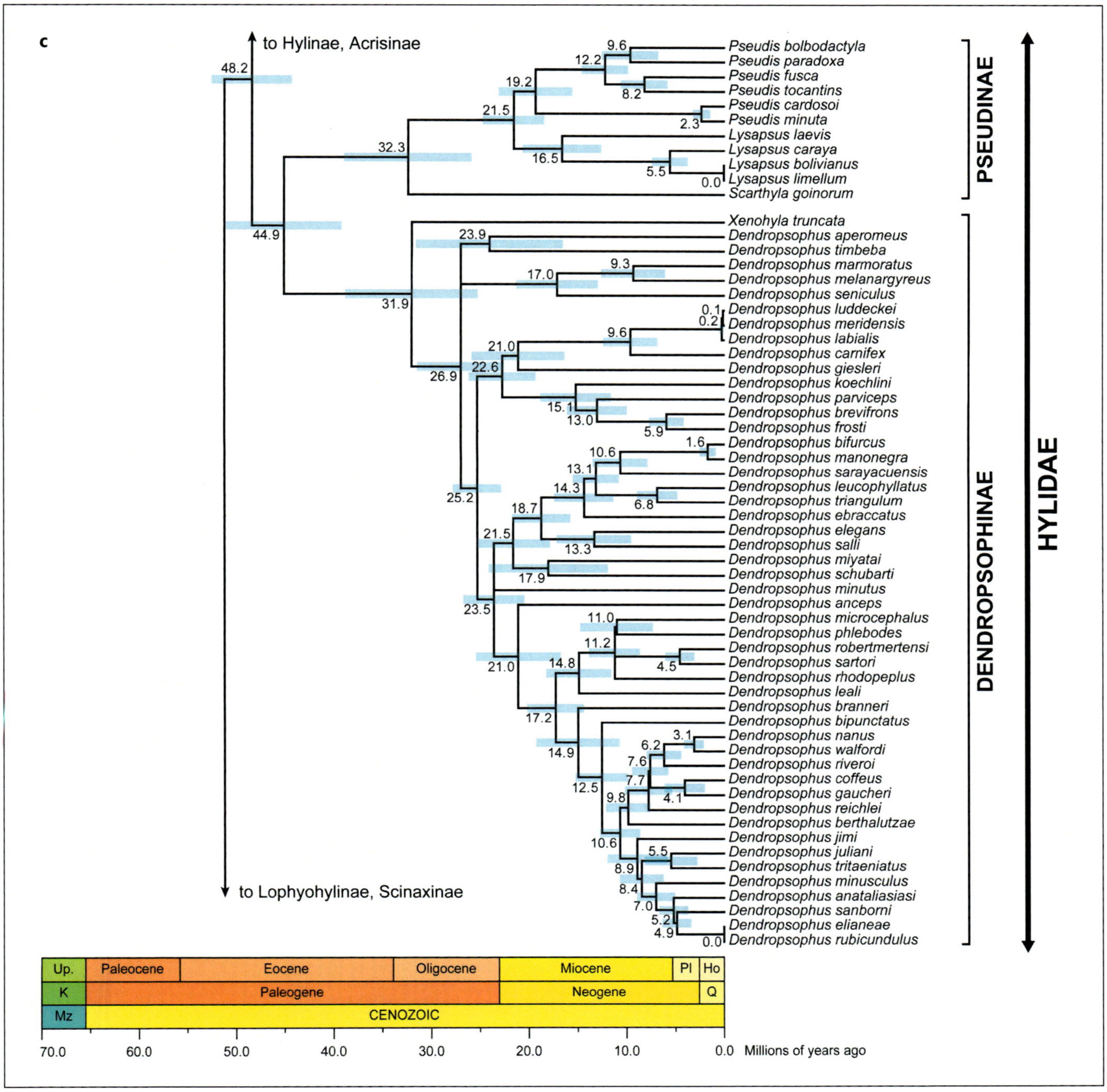

2

inhabited by 8–13 hylids and 0–1 phyllomedusids. Savannas in Bolivia and southern Brazil have 9–17 hylids and 1 or 2 phyllomedusids. Communities in tropical rainforests in southern Mexico and Guatemala contain only 4 hylids and 1 phyllomedusid, whereas in Costa Rica and Panama the numbers are much higher (11–14 hylids and 3 phyllomedusids).

All arboranan frogs are sit-and-wait insectivores, but some large species may eat smaller frogs, small snakes, and small rodents. Analyses of diets of frogs at Santa Cecilia, Ecuador, and Cusco Amazónico, Peru, were provided by Duellman [1978] and Parmelee [1999], respectively. Both studies revealed the consumption of a great variety of insects with beetles, crickets, and grasshoppers being eaten by the majority of species. However, some are specialists. Most commonly, in some large species of Hylidae and Phyllomedusidae (e.g., *Boana boans*, *B. geographica*, *Osteocephalus taurinus*, and *Phyllomedusa tarsius*) orthopterans make up more than 70% of their diet. Species of the hylid genus *Sphaenorhynchus* feed only on ants. Two species of hylids, *Dendropsophus reticulatus* and *D. sarayacuensis*, specialize on moths, whereas *D. minutus* eats caterpillars, and *D. koechlini* feeds almost exclusively on eruciform larvae. In 2 terrestrial species in Australia (*Litoria nasuta* and *Ranoidea*

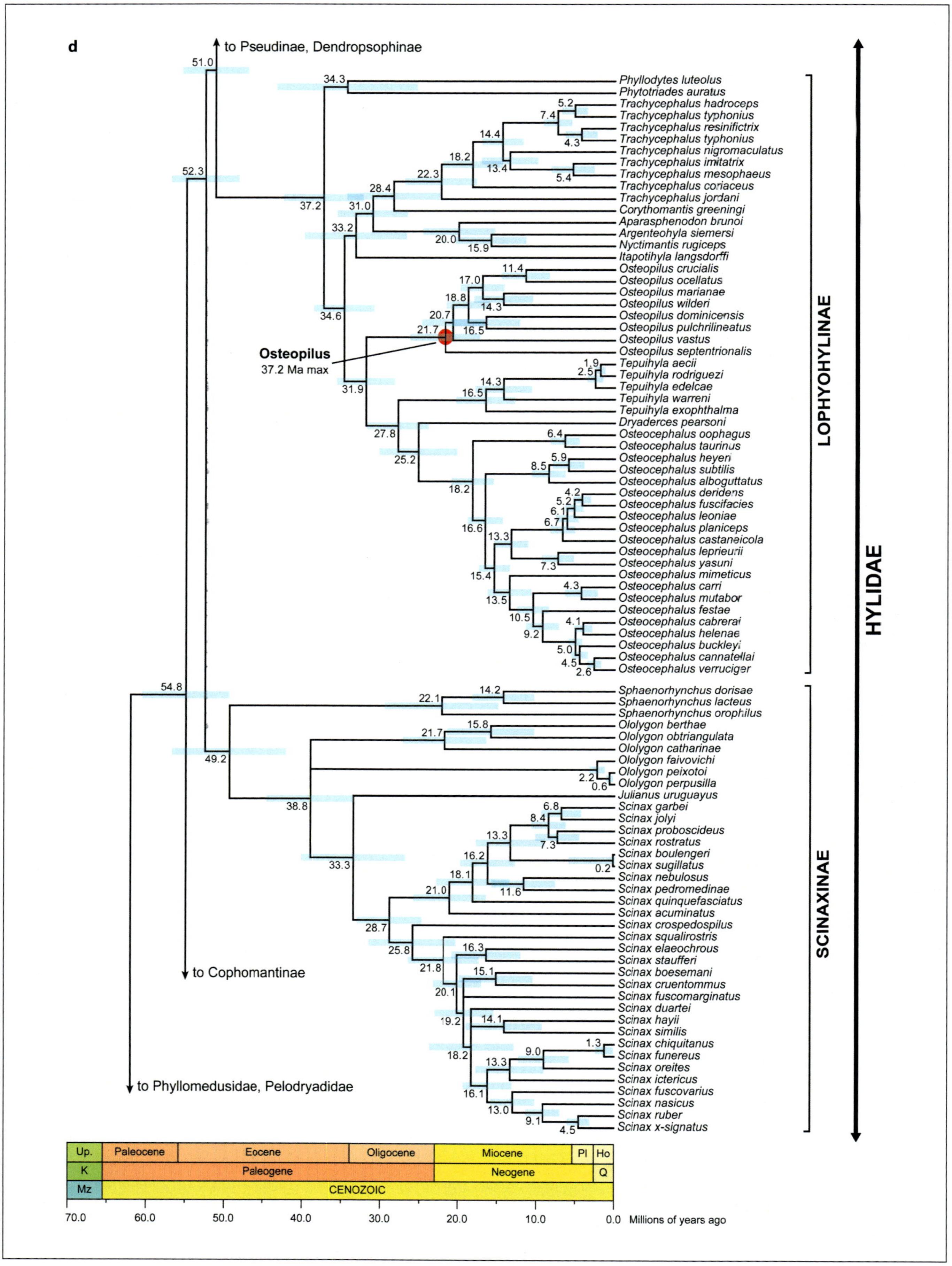
d
to Pseudinae, Dendropsophinae
Phyllodytes luteolus
Phytotriades auratus
Trachycephalus hadroceps
Trachycephalus typhonius
Trachycephalus resinifictrix
Trachycephalus typhonius
Trachycephalus nigromaculatus
Trachycephalus imitatrix
Trachycephalus mesophaeus
Trachycephalus coriaceus
Trachycephalus jordani
Corythomantis greeningi
Aparasphenodon brunoi
Argenteohyla siemersi
Nyctimantis rugiceps
Itapotihyla langsdorffi
Osteopilus crucialis
Osteopilus ocellatus
Osteopilus marianae
Osteopilus wilderi
Osteopilus dominicensis
Osteopilus pulchrilineatus
Osteopilus vastus
Osteopilus septentrionalis
Osteopilus
37.2 Ma max
Tepuihyla aecii
Tepuihyla rodriguezi
Tepuihyla edelcae
Tepuihyla warreni
Tepuihyla exophthalma
Dryaderces pearsoni
Osteocephalus oophagus
Osteocephalus taurinus
Osteocephalus heyeri
Osteocephalus subtilis
Osteocephalus alboguttatus
Osteocephalus deridens
Osteocephalus fuscifacies
Osteocephalus leoniae
Osteocephalus planiceps
Osteocephalus castaneicola
Osteocephalus leprieurii
Osteocephalus yasuni
Osteocephalus mimeticus
Osteocephalus carri
Osteocephalus mutabor
Osteocephalus festae
Osteocephalus cabrerai
Osteocephalus helenae
Osteocephalus buckleyi
Osteocephalus cannatellai
Osteocephalus verruciger
Sphaenorhynchus dorisae
Sphaenorhynchus lacteus
Sphaenorhynchus orophilus
Ololygon berthae
Ololygon obtriangulata
Ololygon catharinae
Ololygon faivovichi
Ololygon peixotoi
Ololygon perpusilla
Julianus uruguayus
Scinax garbei
Scinax jolyi
Scinax proboscideus
Scinax rostratus
Scinax boulengeri
Scinax sugillatus
Scinax nebulosus
Scinax pedromedinae
Scinax quinquefasciatus
Scinax acuminatus
Scinax crospedospilus
Scinax squalirostris
Scinax elaeochrous
Scinax staufferi
Scinax boesemani
Scinax cruentommus
Scinax fuscomarginatus
Scinax duartei
Scinax hayii
Scinax similis
Scinax chiquitanus
Scinax funereus
Scinax oreites
Scinax ictericus
Scinax fuscovarius
Scinax nasicus
Scinax ruber
Scinax x-signatus
to Cophomantinae
to Phyllomedusidae, Pelodryadidae
LOPHYOHYLINAE
SCINAXINAE
HYLIDAE
Up.
K
Mz
Paleocene
Eocene
Oligocene
Miocene
Pl
Ho
Paleogene
Neogene
Q
CENOZOIC
70.0
60.0
50.0
40.0
30.0
20.0
10.0
0.0
Millions of years ago

2

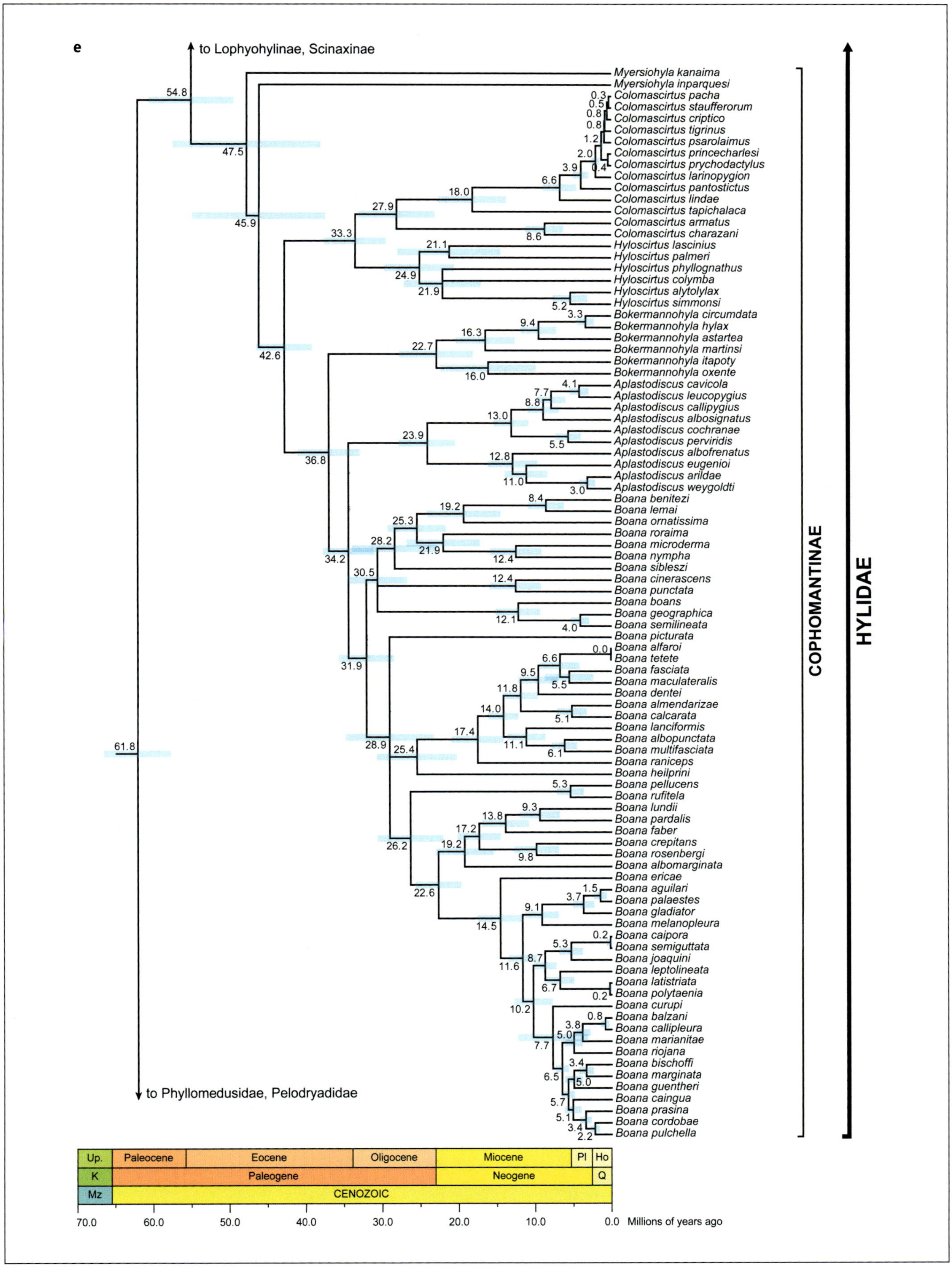
e
to Lophyohylinae, Scinaxinae
to Phyllomedusidae, Pelodryadidae
Myersiohyla kanaima
Myersiohyla inparquesi
Colomascirtus pacha
Colomascirtus staufferorum
Colomascirtus criptico
Colomascirtus tigrinus
Colomascirtus psarolaimus
Colomascirtus princecharlesi
Colomascirtus prychodactylus
Colomascirtus larinopygion
Colomascirtus pantostictus
Colomascirtus lindae
Colomascirtus tapichalaca
Colomascirtus armatus
Colomascirtus charazani
Hyloscirtus lascinius
Hyloscirtus palmeri
Hyloscirtus phyllognathus
Hyloscirtus colymba
Hyloscirtus alytolylax
Hyloscirtus simmonsi
Bokermannohyla circumdata
Bokermannohyla hylax
Bokermannohyla astartea
Bokermannohyla martinsi
Bokermannohyla itapoty
Bokermannohyla oxente
Aplastodiscus cavicola
Aplastodiscus leucopygius
Aplastodiscus callipygius
Aplastodiscus albosignatus
Aplastodiscus cochranae
Aplastodiscus perviridis
Aplastodiscus albofrenatus
Aplastodiscus eugenioi
Aplastodiscus arildae
Aplastodiscus weygoldti
Boana benitezi
Boana lemai
Boana ornatissima
Boana roraima
Boana microderma
Boana nympha
Boana sibleszi
Boana cinerascens
Boana punctata
Boana boans
Boana geographica
Boana semilineata
Boana picturata
Boana alfaroi
Boana tetete
Boana fasciata
Boana maculateralis
Boana dentei
Boana almendarizae
Boana calcarata
Boana lanciformis
Boana albopunctata
Boana multifasciata
Boana raniceps
Boana heilprini
Boana pellucens
Boana rufitela
Boana lundii
Boana pardalis
Boana faber
Boana crepitans
Boana rosenbergi
Boana albomarginata
Boana ericae
Boana aguilari
Boana palaestes
Boana gladiator
Boana melanopleura
Boana caipora
Boana semiguttata
Boana joaquini
Boana leptolineata
Boana latistriata
Boana polytaenia
Boana curupi
Boana balzani
Boana callipleura
Boana marianitae
Boana riojana
Boana bischoffi
Boana marginata
Boana guentheri
Boana caingua
Boana prasina
Boana cordobae
Boana pulchella
COPHOMANTINAE
HYLIDAE
Up.
K
Mz
Paleocene
Eocene
Oligocene
Miocene
Pl
Ho
Paleogene
Neogene
Q
CENOZOIC
70.0
60.0
50.0
40.0
30.0
20.0
10.0
0.0 Millions of years ago

2

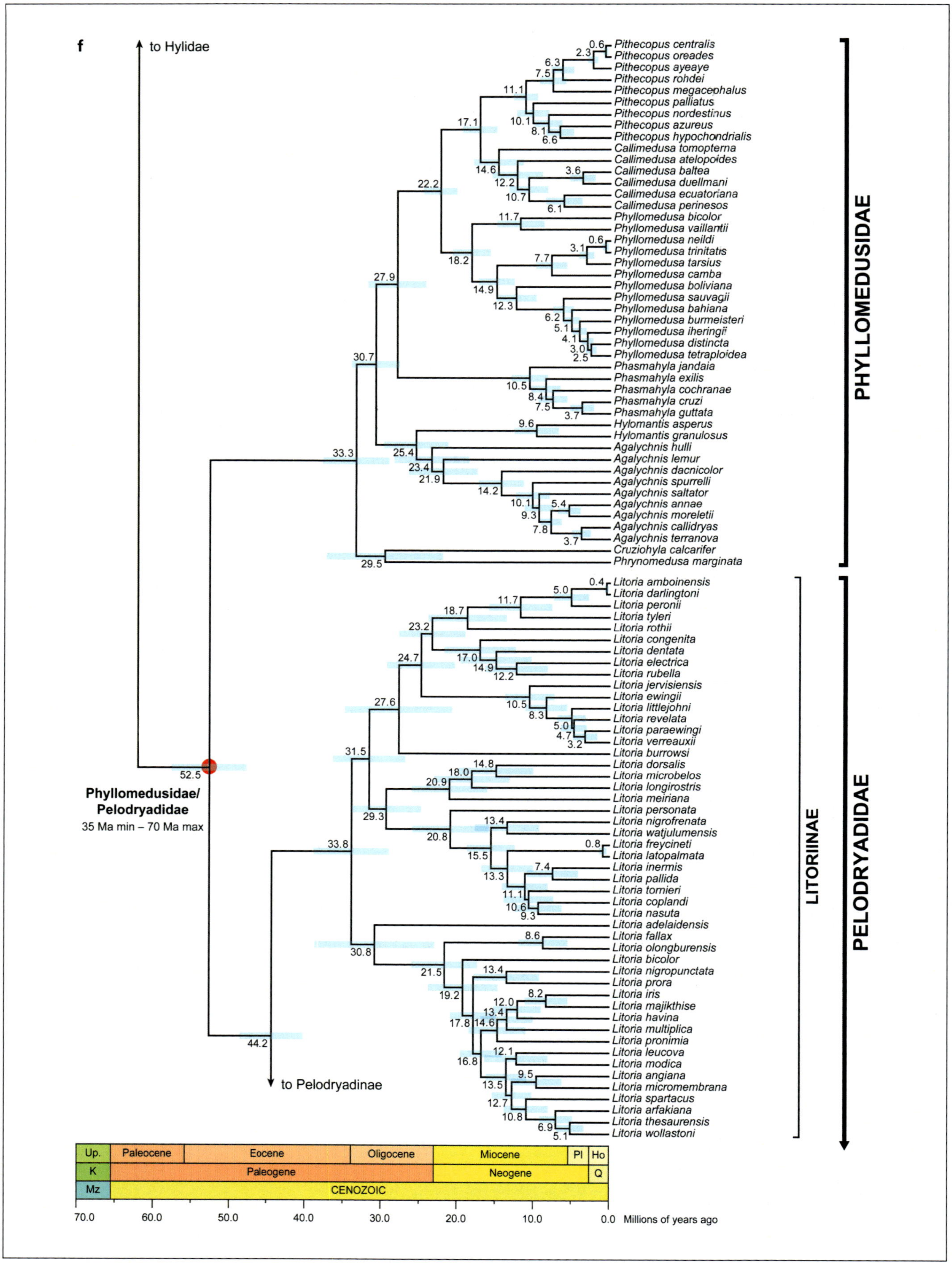
f
to Hylidae
Pithecopus centralis
Pithecopus oreades
Pithecopus ayeaye
Pithecopus rohdei
Pithecopus megacephalus
Pithecopus palliatus
Pithecopus nordestinus
Pithecopus azureus
Pithecopus hypochondrialis
Callimedusa tomopterna
Callimedusa atelopoides
Callimedusa baltea
Callimedusa duellmani
Callimedusa ecuatoriana
Callimedusa perinesos
Phyllomedusa bicolor
Phyllomedusa vaillantii
Phyllomedusa neildi
Phyllomedusa trinitatis
Phyllomedusa tarsius
Phyllomedusa camba
Phyllomedusa boliviana
Phyllomedusa sauvagii
Phyllomedusa bahiana
Phyllomedusa burmeisteri
Phyllomedusa iheringii
Phyllomedusa distincta
Phyllomedusa tetraploidea
Phasmahyla jandaia
Phasmahyla exilis
Phasmahyla cochranae
Phasmahyla cruzi
Phasmahyla guttata
Hylomantis asperus
Hylomantis granulosus
Agalychnis hulli
Agalychnis lemur
Agalychnis dacnicolor
Agalychnis spurrelli
Agalychnis saltator
Agalychnis annae
Agalychnis moreletii
Agalychnis callidryas
Agalychnis terranova
Cruziohyla calcarifer
Phrynomedusa marginata
PHYLLOMEDUSIDAE
Litoria amboinensis
Litoria darlingtoni
Litoria peronii
Litoria tyleri
Litoria rothii
Litoria congenita
Litoria dentata
Litoria electrica
Litoria rubella
Litoria jervisiensis
Litoria ewingii
Litoria littlejohni
Litoria revelata
Litoria paraewingi
Litoria verreauxii
Litoria burrowsi
Litoria dorsalis
Litoria microbelos
Litoria longirostris
Litoria meiriana
Litoria personata
Litoria nigrofrenata
Litoria watjulumensis
Litoria freycineti
Litoria latopalmata
Litoria inermis
Litoria pallida
Litoria tornieri
Litoria coplandi
Litoria nasuta
Litoria adelaidensis
Litoria fallax
Litoria olongburensis
Litoria bicolor
Litoria nigropunctata
Litoria prora
Litoria iris
Litoria majikthise
Litoria havina
Litoria multiplica
Litoria pronimia
Litoria leucova
Litoria modica
Litoria angiana
Litoria micromembrana
Litoria spartacus
Litoria arfakiana
Litoria thesaurensis
Litoria wollastoni
LITORIINAE
PELODRYADIDAE
52.5
Phyllomedusidae/
Pelodryadidae
35 Ma min – 70 Ma max
to Pelodryadinae
Up.
Paleocene
Eocene
Oligocene
Miocene
Pl
Ho
K
Paleogene
Neogene
Q
Mz
CENOZOIC
70.0
60.0
50.0
40.0
30.0
20.0
10.0
0.0
Millions of years ago

2

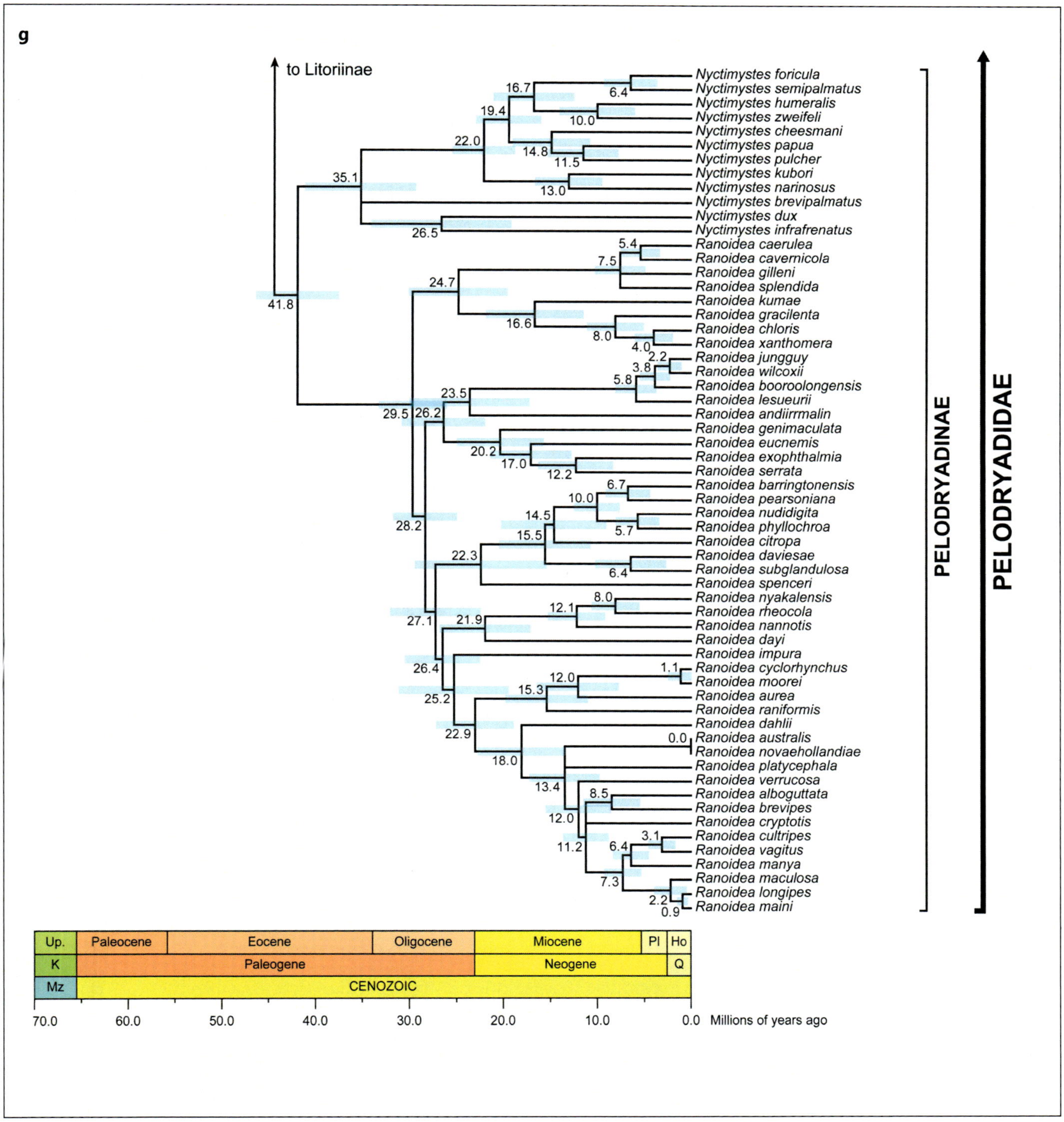

2

australis), beetles and grasshoppers each occurred in more than 35% of the stomachs. Relative abundance of certain kinds of prey may be seasonal. This has been documented in the Australian pelodryadid *Litoria rubella* by Tyler [1989]. At the beginning of the wet season, ants, beetles, and flies were the most common prey, whereas in the latter part of the wet season, ants and beetles were absent and mayflies, hemipterans, and flies were the most abundant prey. The hylid *Xenohyla truncata* is a dietary exception; it is frugivorous. The frog inhabits bromeliads on small trees in the Restinga of southeastern Brazil. It feeds on the small fruits that fall from the trees into bromeliads.

Anurans are consumed by a great variety of predators. Eggs and tadpoles in ponds are ingested by larval arthropods such as those of odonates and predaceous diving beetles. Phyllomedusid eggs on vegetation above water are known to be consumed by various species of the colubrid snake genus *Leptodeira* and at least 1 species of bird, the

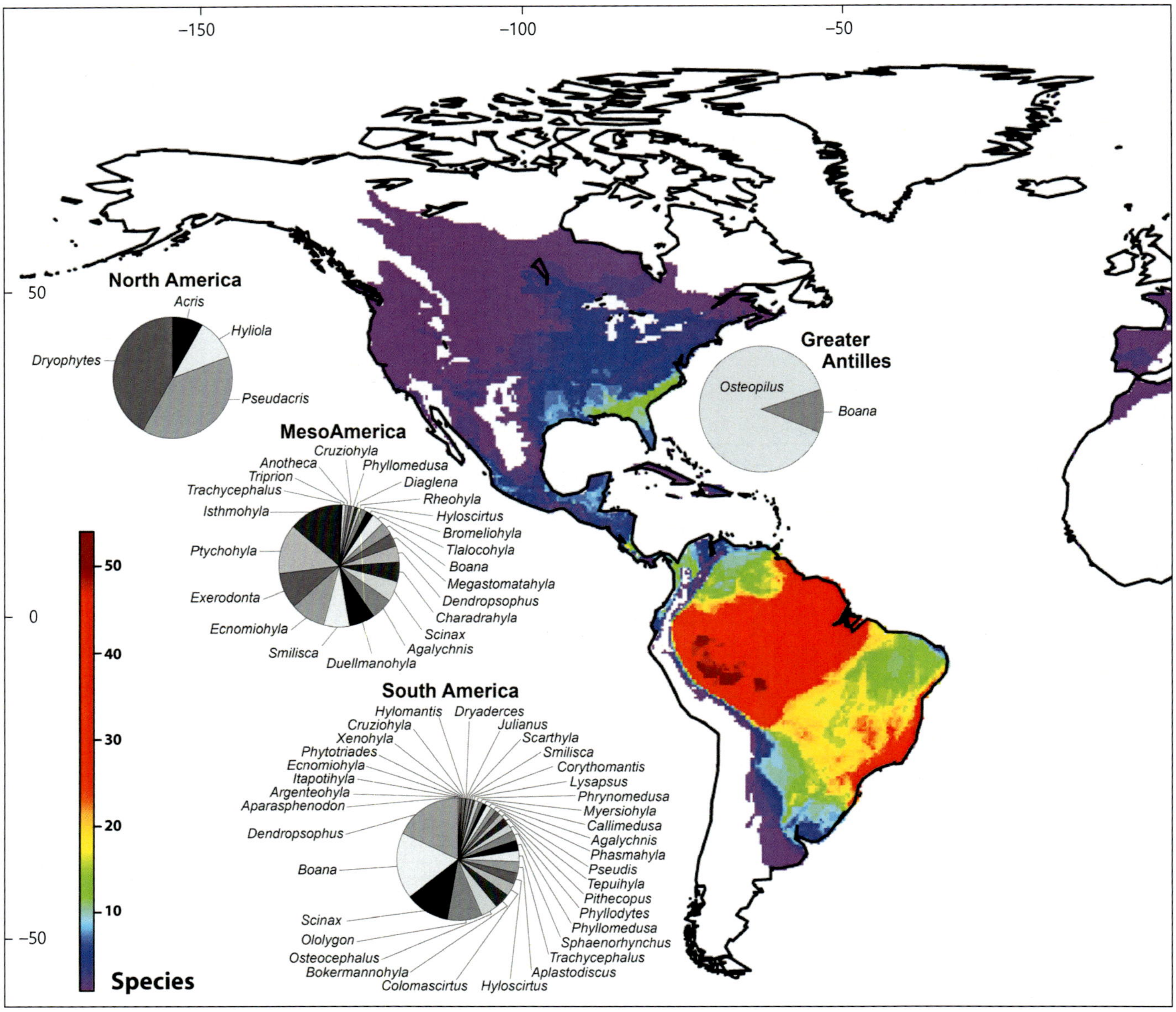

Fig. 3. Map of the New World showing species density of arboranan frogs (color); white areas have 0–1 species. Distributional data are from IUCN [2014]. Pie diagrams show the proportion of species in each genus by geographic regions. The border between North America and Mesomerica is the Isthmus of Tehuantepec in southern Mexico. From Duellman et al. [2016]. For further explanations, see chapters 1.2 and 1.6.

purple gallinule. Water bugs and water scorpions suck body fluids from captured tadpoles. In South America, the large tadpoles of the horned frog *Ceratophrys cornuta* are obligate carnivores that feed primarily on tadpoles. Wading birds (egrets, herons, etc.) are major predators of tadpoles. Some snakes, such as *Leimadophis epinephelus*, include tadpoles in their diets. Metamorphosing young are especially vulnerable as they emerge onto land where they are consumed by mammals (e.g., raccoons and opossums) and many kinds of birds and snakes. Many kinds of spiders prey on anurans. Nocturnal arboreal snakes, such as *Imantodes lentiferus* and all species of *Leptodeira*, especially *L. annulata*, frequent the bushes and other vegetation during frog choruses and consume quantities of frogs, especially hylids. For example, examination of the stomach contents of *L. annulata* from Santa Cecilia, Ecuador, revealed 3 species of *Dendropsophus* and 1 *Boana lanciformis*. *L. annulata* also eats *Trachycephalus coriaceus* and *T. typhonius*, both of which have toxic skin secretions. Anuran vocalizations attract 2 kinds of bats, *Megoderma* in the Old World and *Trachops* in the New World tropics. Diurnal retreats are not always secure. Some birds and diurnal colubrid snakes of the genera *Leptophis* and *Xenoxybelis* actively search diurnal retreats such as bromeliads.

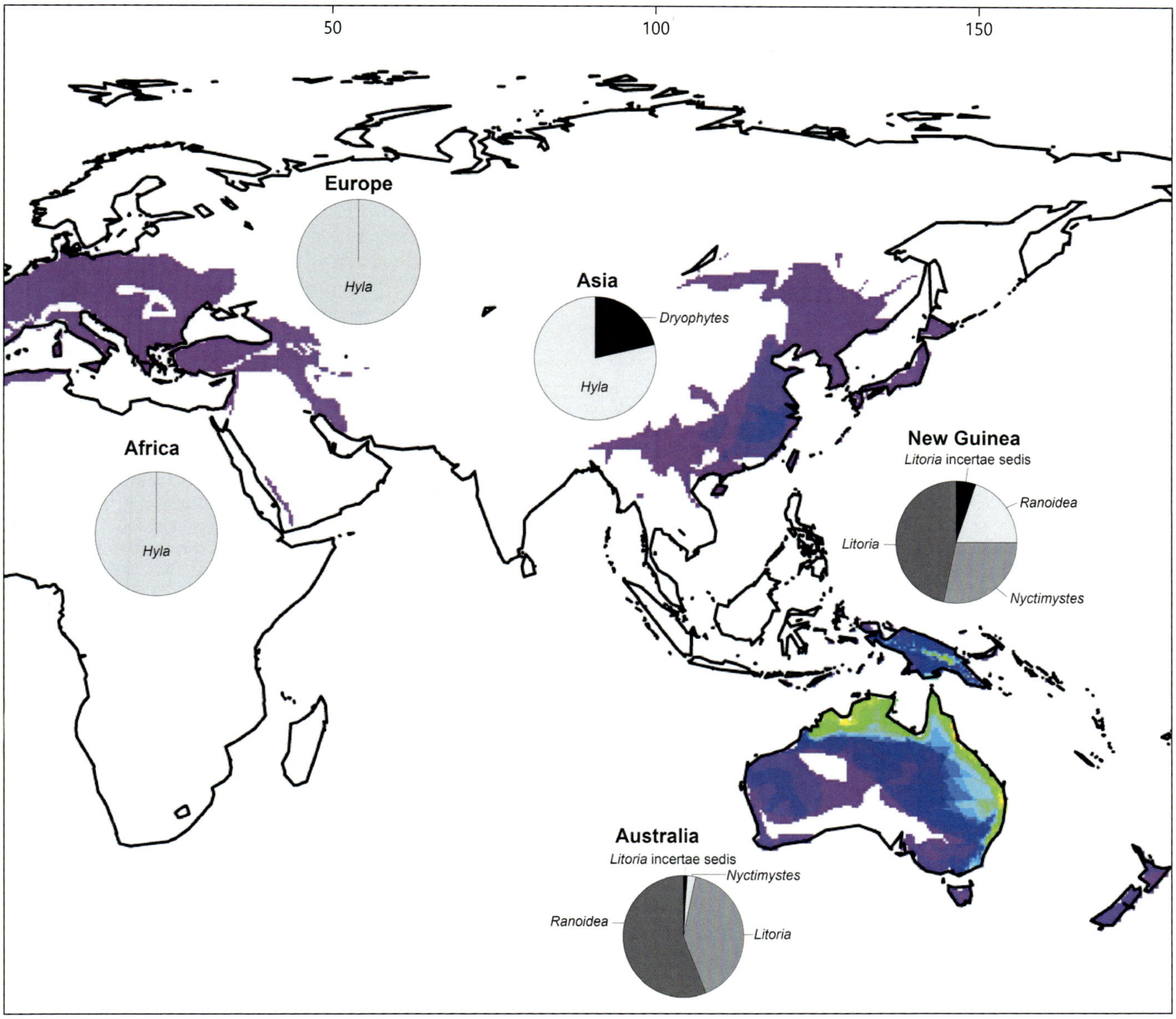

Fig. 4. Map of the Old World showing species density of arboranan frogs (color); white areas have 0–1 species. Distributional data are from IUCN [2014]. Pie diagrams show the proportion of species in each genus by geographic regions. Modified from Duellman et al. [2016]. For further explanations, see chapters 1.2 and 1.6.

1.3 Life Histories of Arboranae

All Arboranae deposit eggs in or above water and have free-living aquatic tadpoles. The terminology used herein follows Altig and McDiarmid [1999], who summarized habitat and structure for all tadpoles known at the time. The labial tooth row formula (LTRF) is the number of anterior rows of teeth/ventral rows of teeth. Numbers in parentheses indicate those rows that are incomplete medially. The ecological features and some of their morphological characters are summarized below within the classification proposed by Duellman et al. [2016]. A community in tropical rainforest in western Ecuador is depauperate (8 hylids and 1 phyllomedusid) in contrast to communities in the Amazon Basin. The largest communities are in the western part of the basin (25–40 hylids, 4–6 phyllomedusids), in contrast to communities in the middle and eastern parts that have 11–20 hylids and 3–5 phyllomedusids. The only habitat that is equal to Amazonia rainforest is humid montane forest in the Atlantic Coastal Forest in Bahia, Brazil, which contains 38 hylids and 4 phyllomedusids.

1.3.1 Family Hylidae

Most hylids deposit their eggs in water. The eggs of many pond breeders may be deposited in single or multiple clutches usually attached to stems of aquatic vegetation (Fig. 951a).

Cytogenet Genome Res 2018;155:11–45
DOI: 10.1159/000489540

Table 1. Sizes of arboranan communities in tropical and subtropical Neotropics

Locality	Coordinates	Habitat	Hylid/ phyllomedusid species, *n*	References
Mazatlán, Sinaloa, Mexico, 10 m	23°13′N, 106°25′W	dry scrub forest	5/1	Hardy and McDiarmid, 1969
Chichén-Itzá, Yucatán, Mexico, 10 m	20°41′N, 88°23′W	dry scrub forest	5/1	Lee, 1996
Nueva Italia, Michoacán, Mexico, 380 m	19°02′N, 102°07′W	dry scrub forest	4/1	Duellman, 1961
Sarabia, Oaxaca, Mexico, 200 m	17°04′N, 94°53′W	tropical rainforest	4/1	Duellman, 1960
Chinajá, Alta Verapaz, Guatemala, 140 m	16°02′N, 90°13′W	tropical rainforest	4/0	Duellman, 1963
La Selva, Heredia, Costa Rica, 30 m	10°26′N, 83°59′W	tropical rainforest	11/3	Guyer, 1990
Barro Colorado Island, Panama, 20 m	09°09′N, 79°51′W	tropical rainforest	14/3	Rand and Myers, 1990
Río Cuyuni, Bolívar, Venezuela, 130 m	06°37′N, 61°33′W	tropical rainforest	11/4	Duellman, 1997
Santa Cecilia, Ecuador, 340 m	00°04′N, 76°59′W	tropical rainforest	33/4	Duellman, 1978
Centro Científico Río Palenque, Pichincha, Ecuador, 220 m	00°35′S, 79°22′W	tropical rainforest	8/1	Duellman, unpublished
APEG Reserve, Pará, Brazil, 10 m	01°27′S, 48°28′W	tropical rainforest	20/3	Crump, 1971
Reserva Adolpho Ducke, Amazonas, Brazil, 140 m	03°08′S, 60°01′W	tropical rainforest	13/4	Lima et al., 2006
10 km N Leticia, Amazonas, Colombia	04°08′S, 69°40′W	tropical rainforest	40/5	Lynch, 2005
Panguana, Ucayala, Peru, 260 m	09°37′S, 74°56′W	tropical rainforest	25/4	Schlüter et al., 2004
Cocha Cashu, Madre de Dios, Peru	11°16′S, 71°00′W	tropical rainforest	30/6	Rodríguez and Cadle, 1990
Cusco Amazónico, Madre de Dios, Peru	12°35′S, 65°05′W	tropical rainforest	27/6	Duellman, 2005
Serra Bonita, Bahia, Brazil, 200–900 m	15°23′S, 39°33′W	humid montane forest	38/4	Dias et al., 2014
Puerto Almacén, Santa Cruz, Bolivia, 288 m	15°37′S, 62°09′W	seasonal tropical forest	23/2	De la Riva, 1993
Fazenda Baía de Pedra, Mato Grosso, Brazil	16°02′S, 58°09′W	savanna	13/1	Pansanota et al., 2011
Santa Cruz del Sara, Santa Cruz, Bolivia	16°52′S, 62°45′W	savanna	9/2	Pinto-Viveros et al., 2015
San Sebastián, Santa Cruz, Bolivia, 500 m	18°22′S, 62°00′W	savanna	17/2	Schultz et al., 2009
Pontinha do Cocho, Mato Grosso, Brazil	19°01′S, 53°53′W	savanna	12/1	Sugai et al., 2014
Corumbá, Mato Grosso do Sul, Brazil, 120 m	19°34′S, 57°00′W	savanna	9/1	Prado et al., 2005
PEMG, Londrina, Paraná, Brazil	21°27′S, 52°12′W	semideciduous forest	9/1	Machado et al., 1999
EEC, São Paulo, Brazil, 500 m	22°24′S, 49°40′W	semideciduous forest	13/1	Brassaloti et al., 2010
BAP/Nupelia, Porto Rico, Paraná, Brazil	22°45′S, 53°15′W	semideciduous forest	8/0	Alfonso et al., 2013
Boracéia, São Paulo, Brazil	23°38′S, 43°52′W	humid montane forest	25/2	Heyer et al., 1990
Parque Estadual Ilha do Cardoso, São Paulo, Brazil	25°03′S, 47°53′W	restinga	10/0	Vilela et al., 2011
Parque Estadual FP, Santa Catarina, Brazil	27°18′S, 52°07′W	semideciduous forest	7/1	Molinari de Bastiani and Lucas, 2013

This is characteristic of *Acris*, *Dryophytes*, *Hyla*, *Pseudacris*, *Tlalocohyla*, and some *Isthmohyla*. Throughout the lowlands of tropical America, many species of hylids deposit eggs in a surface film in temporary ponds; these include many frogs in the genera *Boana*, *Osteocephalus*, *Scinax*, *Smilisca*, *Trachycephalus* (Fig. 951b), and many *Dendropsophus*. However, some *Dendropsophus* deposit small clutches of eggs on vegetation above ponds (Fig. 951c). *Isthmohyla calypsa* and *I. insolita* deposit eggs on vegetation above streams. With the exception of the latter, the tadpoles of these frogs have rounded bodies and high tail fins. The LTRF is 2(1)/3 in most genera but 2/2(1) in *Acris*, 2(2)/4 in some *Boana*, 2(2)/6(1) in *Osteocephalus*, and 3(1, 3)/6(1, 6) in *Trachycephalus*. The tadpoles of *Pseudis paradoxa* (up to 220 mm in total length) are large pond dwellers; they metamorphose into proportionately smaller froglets (45 mm snout-vent length).

The tadpole of *Scarthyla goinorum* (Fig. 953a) has a small oral disc with an LTRF of 2(1)/2; although it is a pond dweller, it has a long muscular tail and low caudal fins. According to Duellman and de Sá [1988], tadpoles swim rapidly just below the duckweed (genus *Lemna*) on the surface of the pond; they are capable of propelling themselves though the duckweed and into the air so as to reenter the water at a distance of 20–30 cm. In the genus *Dendropsophus*, there is a progressive loss of teeth from 1/2 in the *D. labialis* and *D. columbianus* groups, to 1/0 in the *D. minutus* group, and to 0/0 in the *D. leucophyllatus*, *D. marmoratus*, *D. microcephalus*, and *D. parviceps* groups [Duellman and Trueb, 1983].

Small clutches of eggs are attached to leaves of arboreal bromeliads. This is the mode of the 2 species of *Bromeliohyla* in Middle America (Fig. 951d), 3 species of *Osteopilus* in Jamaica, *Isthmohyla zeteki* in Central America, *Phyllotriades auratus* in northern Venezuela and Trinidad, and *Dendropsophus bromeliaceus* in the Atlantic Coastal Forest in Brazil. The tadpoles of these species have long, slender muscular tails with low fins. The oral discs in *Bromeliohyla* and *Phyllotriades* are ventral and have LTRFs of 2(2)/4, 5; these tadpoles feed on vegetal detritus. The tadpoles of *I. zeteki* and the 3 Jamaican species of *Osteopilus* have greatly reduced dentition. The LTRF in *I. zeteki* is 2(2)/2 (Fig. 953c), in *Osteopilus brunneus* 1/0, and in *O. marianae* (Fig. 953b) and *O. wilderi* 0/0; these tadpoles feed on eggs of their own species. Other hylid frogs that are known to deposit eggs in bromeliads or tree holes are *Anotheca spinosa* in Middle America, *Aparasphenodon arapapa* and some species of *Ololygon* in coastal Brazil, and 4 species in the Amazon Basin, *Nyctimantis rugiceps*, *Osteocephalus deridens*, *O. oophagus*, and *Trachycephalus cunauaru*. Clutches are small (up to 311 eggs) in *Anotheca* or large (up to 2,500 eggs) in *Trachycephalus*. In some of these oophagus species, the tadpoles tend to have subterminal mouths, large jaw sheaths, and reduced dentition, 2(2)/2 in *Anotheca* (Fig. 953d). Females of *Anotheca* return to their deposition site and deposit unfertilized eggs for the tadpoles to eat [Jungfer, 1996]. After eggs hatch, males of *Trachycephalus* call to attract another female to deposit eggs for food for the tadpoles [Gordo et al., 2013]. Tadpoles of *Ecnomiohyla* have been found in water-filled cavities in trees in cloud forest in Central America. These tadpoles have short tails with high fins and LTRFs of 2(2)/3(1); dentition is poorly developed in *E. sukia* [Hoffman and Kubicki, 2011]. Those authors suggested *E. sukia* may feed on nutrient eggs, as postulated for *E. salvaje* by Wilson et al. [1985].

Males of a few species are known to construct "nests" for eggs and tadpoles. In the Amazon Basin, *Boana boans* is one of several species that constructs a shallow basin at water level near the edge of a pond or sluggish stream (Fig. 951e). This behavior apparently evolved at least twice, in species in the *Boana faber* and *B. semilineata* groups of Faivovich et al. [2005]. Details of parental behavior in *B. rosenbergi* in Panama were published by Kluge [1981]. Amplexus and oviposition take place in the basin where tadpoles remain until rising water floods the nest. Males of at least some species of *Aplastodiscus* construct subterranean nests in the muddy sides of streams and ponds where tadpoles spend the early stages of development; subsequent to flooding, tadpoles inhabit slow streams or ponds. The pond-dwelling tadpoles of 2 species of *Boana* (*B. geographica* and *B. semilineata*) in the Amazon Basin form large schools of hundreds of individuals.

The tadpoles of many hylid frogs live in streams. Stream-dwelling tadpoles tend to have highly developed caudal musculature and relatively low caudal fins. There is considerable variation in the oral discs of stream-dwelling tadpoles. In some tadpoles, the oral disc is enlarged with many rows of marginal papillae and the development of submarginal papillae but maintenance of an LTRF of 2/3; examples are *Plectrohyla* and *Sarcohyla* in Middle America (Fig. 954a). Also in Middle America, stream-dwelling tadpoles of some species of *Isthmohyla* (e.g., *I. pictipes* and *I. rivularis*) have an LTRF of 2/4; *Exerodonta smaragdina* and *E. sumichrasti* have an LTRF of 3/7. Tadpoles of *Megastomatohyla mixe* and *M. mixomaculata* have an LTRF of 7/11 and a single complete row of small marginal papillae. In the Andes in South America, *Colomascirtus* and *Hyloscirtus* have stream-inhabiting tadpoles that are adapted for slinging to rocks in torrential streams. They have enlarged oral discs with multiple rows of marginal papillae and many rows of labial teeth. Usually the last anterior row of teeth has a median gap, as does the first posterior row in most species. In *Hyloscirtus*, the LTRF varies from 4/5 in *H. bogotensis* and *H. platydactylus* to 8/10 in *H. denticulentus*. In *Colomascirtus*, the LTRF varies from 4/6 in *C. pantostictus* to 15/17 in *C. armatus*. The only stream-dwelling tadpole in the Guiana Highlands is *Myersiohyla*. The tadpoles have enlarged oral discs bordered by a single row of marginal papillae. The LTRF varies from 5/9 in *M. chameleo* to 16/21 in *M. neblinaria* (Fig. 954b). This is the largest known LTRF.

In Middle America, the tadpoles in 2 genera are bottom feeders in pools in streams; they have funnel-like suctorial oral discs completely bordered by a single row of small marginal papillae and having many submarginal papillae. The tadpoles in the *Ptychohyla euthysanota* group have an LTRF of 3/3 short rows of teeth (Fig. 954c). Medial anterior and posterior indentations of the oral disc are characteristic of *Duellmanohyla*, which has an LTRF of 2/2 (Fig. 954d).

1.3.2 Family Phyllomedusidae

Tadpoles of the family Phyllomedusidae are suspension raspers. They usually are just below the surface of the water with their heads up and the body at an angle of about 45° from the surface. With the exception of *Phasmahyla*, which develops in streams in the Atlantic Coastal Forest of Brazil, phyllomedusid tadpoles are pond dwellers with nearly terminal oral discs and LTRFs of 2/3. *Agalychnis*, *Callimedusa*, *Phyllomedusa*, and *Pithecopus* have numerous rows of marginal papillae, especially laterally, but no papillae medially on the anterior lip (Fig. 955a, b). The same is true for *Cruziohyla*, except the marginal papillae are complete across the anterior lip. The oral discs of tadpoles of 2 genera in the Atlantic Coastal Forest of Brazil are remarkably different; both have a complete single row of marginal papillae. The tadpole of *Phrynomedusa* has an LTRF of 2/3 (Fig. 955c), whereas the stream-dwelling tadpole of *Phasmahyla* has an upturned umbelliform oral disc with a median indentation of the anterior lip, numerous submarginal papillae, a small mouth, and an LTRF of 1/1 short rows (Fig. 955d).

1.3.3 Family Pelodryadidae

With the exception of 2 species of *Ranoidea*, all known Pelodryadidae have tadpoles with 2 anterior and 3 posterior rows of teeth [Menzies, 2006; Anstis, 2013]. Commonly, the second anterior row of teeth has a median gap, as does the first posterior row in some species. The oral discs have one or more rows of marginal papillae; these usually are absent on the median part of the anterior lip. The exceptions are *Ranoidea daviesae* and *R. subglandulosa*. The tadpoles of these 2 stream-breeding species have suctorial oral discs with a single complete row of marginal papillae. There are no tooth rows (LTRF 0/0) and no keratinized mouth parts; the entire surface of the oral disc has numerous pointed papillae.

The tadpoles of Australian anurans have been thoroughly documented and beautifully illustrated by Anstis [2013]. In Australia, most species of *Litoria* and *Ranoidea* deposit pigmented eggs in quiet water, lakes, ponds, swamps, and quiet pools in streams. The tadpoles of these species (e.g., *Litoria bicolor*, *L. ewingii*, and *L. peronii*, as well as *Ranoidea australis*, *R. chloris*, and *R. raniformis*) characteristically have moderately high dorsal fins and an LTRF of 2(2)/3(1) (Fig. 956a). In a few species (e.g., *Litoria cooloolensis*, *L. microbelos*, and *L. nasuta*), the LTRF is 2(2)/3, and in *R. lesueuri* the LTRF is 2/3. In nearly all species of pond-dwelling tadpoles, the keratinized jaw sheaths are slender, but the lower sheath is massive in *L. rothii*. Some bottom-feeding stream inhabitants have short tooth rows (e.g., *L. citropa*; Fig. 956b) or no keratinized teeth or beaks (e.g., *R. daviesae*; Fig. 956c); these tadpoles have small papillae in a single marginal row. A few species (e.g., *R. lorica*, *R. rheocola*, and *R. nannotis*; Fig. 956d) have greatly enlarged oral discs, complete rows of marginal papillae, an LTRF of 2/3, and low fins on muscular tails. Their unpigmented eggs are attached to, or under, rocks in fast-flowing streams; the tadpoles adhere to rocks in torrential streams.

Only 1 Australian species, *Litoria longirostris*, is known to deposit eggs above pools in streams. Small clutches of 29–60 pale green eggs are attached to leaves, branches, tree trunks, or the stream bank above water. The tadpoles are bottom feeders in quiet pools. The 2 Australian species assigned to *Nyctimystes* by Duellman et al. [2016] have pond-dwelling tadpoles with anteroventral oral discs with a median gap in the marginal papillae, slender beaks, and an LTRF of 2/3 (*N. brevipalmatus*) or 2(2)/3 (*N. infrafrenatus*).

In New Guinea, most *Litoria* (27 species including *L. bicolor*, *L. dorsalis*, and *L. thesaurensis*) and at least 12 *Ranoidea* (e.g., *R. genimaculata*, *R. graminae*, and *R. impura*) deposit small, pigmented eggs in still water. Their tadpoles tend to have slender tails with expanded dorsal fins and anteroventral oral discs having an LTRF of 2(2)/3. Seven species of *Litoria* (e.g., *L. angiana*, *L. longicrus*, and *L. modica*) and 7 species of *Ranoidea* (e.g., *R. becki*, *R. dorsivena*, and *R. pratti*) are known to deposit small clutches of unpigmented eggs in streams. Likewise, all species of *Nyctimystes* presumably deposit their eggs in streams. Ten species of *Litoria* (e.g., *L. iris*, *L. leucova*, and *L. multiplica*) are known to deposit small clutches of unpigmented eggs on vegetation over streams. All of these stream breeders have tadpoles with muscular tails, low dorsal fins, and slightly enlarged ventral oral discs with 2/3 or 2(2)/3 tooth row formulae.

1.4 Gametogenesis and Embryogenesis of Arboranae

1.4.1 Reproductive Modes, Behavior, and Egg Clutch Plasticity

Reproductive modes are complex phenotypes of a species before, during, and after mating. This concept includes nesting and nesting-associated courtship, nest characteristics, site of oviposition, clutch characteristics, developmental rates, hatching rates, and parental care, among others [Crump, 1974, 2015; Duellman, 1992]. Anuran amphibians, in particular, have the most diverse reproductive modes among tetrapods, from aquatic egg deposition and no parental care, to terrestrial eggs, egg brooding, and viviparity [Duellman and Trueb, 1986; Wells, 2007]. Arboranan species, in particular, spend most of their lifespan on vegetation; yet, breeding is tightly related to water bodies, raising important questions on the reproductive adaptability of these species to different microhabitats in comparison to other anuran clades. In this chapter, it is intended to describe the diversity of reproductive modes in the Arboranae clade and the open questions related to the origin and adaptation of these behaviors. Clearly, data that describe reproductive behaviors in such detail are not thoroughly available for every species given the high diversity of anurans (see below). Even less data are available regarding the types of aquatic environments, types of clutches, and clutch size.

Traditionally, species belonging to the arboranan clade have been described as species with aquatic reproduction. However, the distribution of the distinct reproductive behaviors has not been thoroughly compiled. To address the abundance and distribution of the current reproductive modes, we generated a database for 471 of the 935 species of Arboranae, in all the 59 genera and 10 subfamilies currently included in this clade [Duellman et al., 2016]. Reproductive mode information was manually acquired and curated from the 2 main amphibian species databases: http://www.iucnredlist.org/ and http://www.amphibiaweb.org/. Data quality was ensured by avoiding entries on which the reproductive mode was unknown or presumed, and also by proportionally sampling the most specious genera within Arboranae: *Dendropsophus*, *Litoria*, *Scinax*, and *Boana*. Finally, reproductive mode data were classified according to the currently accepted 39 reproductive modes reviewed by Haddad and Prado [2005].

1.4.1.1 Aquatic Reproduction and Nesting

Reproductive modes can be roughly grouped into aquatic, terrestrial, and arboreal modes, depending on where the eggs are laid [Duellman and Trueb, 1986; Haddad and Prado, 2005; Wells, 2007; Crump, 2015]. From the oviposition data, an unsurprising 80% of the assessed species deposit eggs directly into lentic or lotic water bodies: a trend shared by all anuran species (Fig. 5a) [Duellman and Trueb, 1986]. Indeed, most arboranan frogs deposit films, masses, or scattered eggs on the surface or submerged into these aquatic bodies, where embryos will develop onto exotrophic larvae. Arboranan tadpoles strictly depend on water bodies for the successful completion of metamorphosis. Among aquatic reproductive modes, egg deposition in lentic water is the most common behavior (46%), followed by egg deposition in lotic water (32%) (Fig. 5a). In spite of the large number of species described as lotic and lentic water breeders, there are few actual descriptions of this behavior (see *Pseudis minuta*

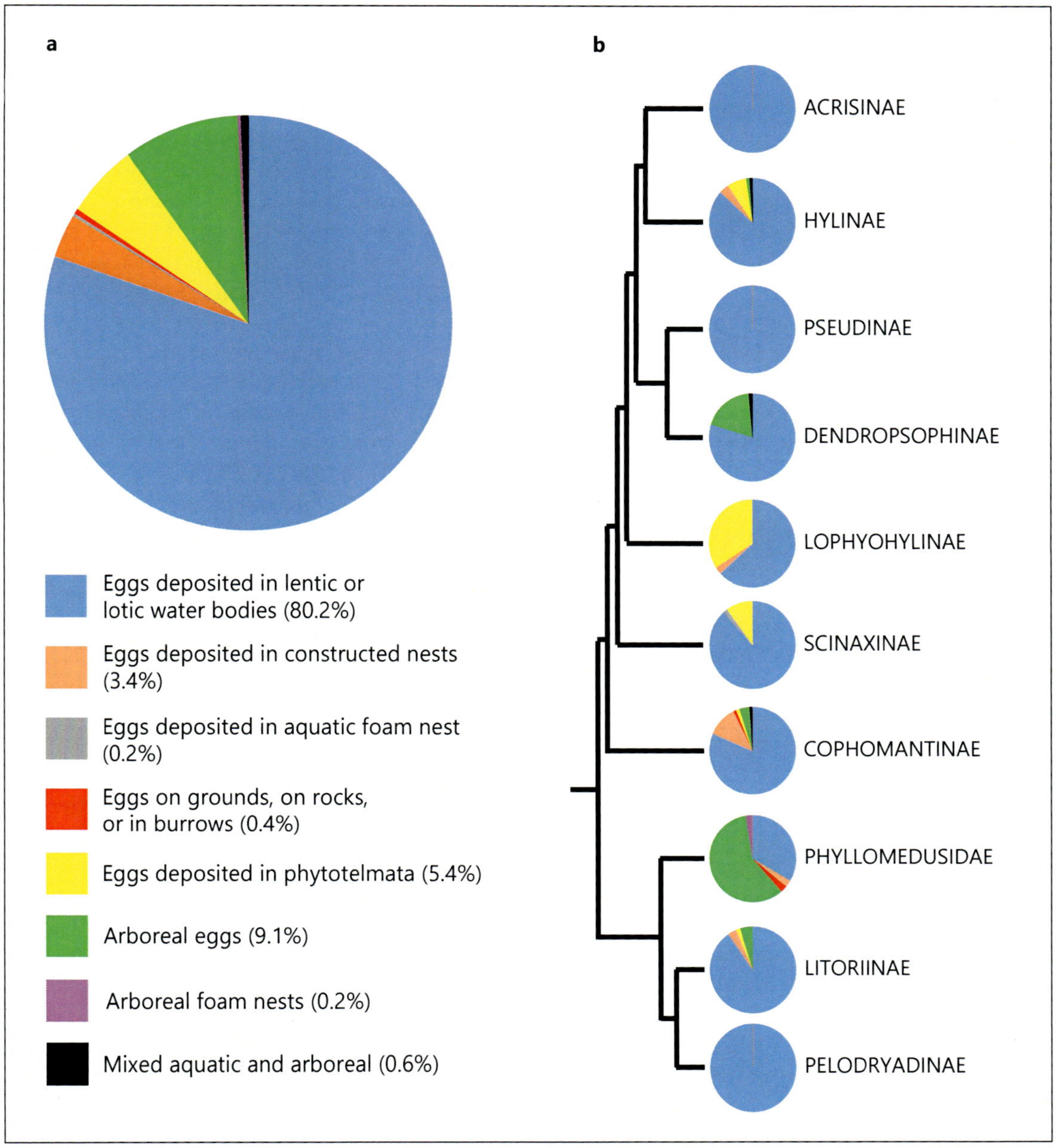

Fig. 5. a Reproductive modes in Arboranae. The data come from 471 of the 934 species of Arboranae, in all the 59 genera and 10 subfamilies currently included in this clade [Duellman et al., 2016]. Data were curated from http://www.iucnredlist.org/ and http://www.amphibiaweb.org/. Reproductive mode data were classified according to Haddad and Prado [2005]. **b** Proportion of reproductive modes present in Arboranae. The data in **a** were subdivided for the 10 subfamilies currently accepted as part of the Arboranae clade [Duellman et al., 2016]. The phylogenetic tree is a reproduction of the distilled maximum likelihood phylogeny presented in Duellman et al. [2016]. For further explanations, see chapter 1.4.

or *Trachycephalus typhonius*) [Huckembeck et al., 2012; Oda et al., 2016]. To compare, some genera show exclusive adaptations to stream breeding like the *Colomascirtus* and *Hyloscirtus* species associated with Andean and Tepui streams [Faivovich et al., 2005]. *Hyloscirtus* tadpoles show particular larval adaptations to dwell in streams, like a suctorial mouth with an increased number of tooth rows and marginal papillae [Cadle and Altig, 1991]. Other adaptations in the egg jelly and embryo might be exclusive to one of these reproductive modes; yet, no descriptions are available.

In anurans, other behavioral-physiological adaptations like the construction of nests in mud, leaves, or by foaming female mucous secretions occur in aquatic, arboreal, and terrestrial nests, adding diversity to the reproductive modes [Caldwell, 1992; Crump, 2015]. A classic example of nest construction in Arboranae are the males of the gladiator frogs (genus *Boana*, including *B. faber*, *B. pardalis*, *B. boans*, *B. rosenbergi*, among others) which form flooded basins out of sand, clay, or mud, next to rivers and pond margins, but also lay eggs in temporary ponds [Caldwell, 1992; Martins, 1993]. Although *Boana* nests are not highly complex, mating success relies on its quality, carefully assessed by the female, and harshly defended by the male, which has evolved a prepollical spine for this purpose [Martins and Haddad, 1988]. Fighting is also a form of parental care during embryo development; however, these behaviors are plastic and might not occur depending on nesting site availability [Höbel, 2008]. Species of *Aplastodiscus*, such as *A. cavicola*, *A. perviridis*, and *A. leucopygius*, traditionally described as stream breeders, can also build subterranean nests subject to flooding, and very little is known about their nesting behavior [Haddad and Sawaya, 2000; Haddad et al., 2005]. There are only 2 Arboranae species that produce foam nests: *Phyllomedusa camba*, an arboreal nester (see below), and *Scinax rizibilis*. *S. rizibilis* is a commonly occurring species in primary and secondary Brazilian forests that builds foam nests in ponds. The nests are of intermediate complexity in comparison to the highly complex foam nests produced by frogs in the leptodactylid subfamily Leiuperinae [Bastos et al., 2010].

1.4.1.2 Arboreal Reproduction and Nesting

The second most common reproductive mode in Arboranae (9%) is the terrestrial adaptation to lay eggs on vegetation, where embryonic development takes place. This behavior is called arboreal reproduction. In arboreal reproductive modes, embryos are not in contact with water until older embryos or exotrophic larvae fall or are washed to a water body. Within arboranan main clades, proper arboreal reproduction occurs only in 5 of the 10 subfamilies, being most common in the phyllomedusid clade (Fig. 5b). The distribution of this character between subfamilies suggests an independent origin of arboreal reproduction. To avoid desiccation, arboreal nesting phyllomedusids have developed particular adaptations such as parental humidification of the egg jelly and the production of thicker jelly coats. In *Phyllomedusa*, nests are intentionally sheltered by the parents using leaf envelopes, and the mother secretes thick eggless jelly plugs rich in fibrous proteinaceous components [Downie et al., 2013]. *Phyllomedusa camba* is an arboreal species with traditional arboreal nesting using leaves [de la Riva, 1999], and it is also capable of producing arboreal foam nests [Wong, 2013]. On the other hand, *Agalychnis* females like *A. callidryas* and *A. dacnicolor* do not form leaf nests but rather travel to ponds to first humidify their skin and bladder. This water will be used to hydrate the jelly coat of the recently laid embryos [Pyburn, 1970; de la Riva, 1999]. Similar adaptations to avoid desiccation are completely unknown for other species and clades where arboreal reproduction also occurs. *Dendropsophus ebraccatus* is the only anuran species in which both arboreal and aquatic reproduction has been reported. This species often lays eggs in ponds; yet, it is capable of nesting in terrestrial surfaces and vegetation, mostly in shaded areas, depending on the environmental conditions [Touchon and Warkentin, 2008]. Such adaptive behavior had been previously unknown for any anuran amphibian and could be present in more species given the lack of data. Indeed, when the availability of data was assessed for all the species of *Dendropsophus*, the largest genus of Arboranae, 35% of the species lack any information regarding their reproductive modes. This percentage is more or less similar for the other large genera: *Litoria*, *Scinax*, and *Boana*. Even though terrestrial nesting in arboreal species is accompanied by higher desiccation rates, egg jelly molecular and structural characteristics might also influence the success in the plasticity of oviposition sites in *D. ebraccatus* [Hödl, 1990; Touchon and Warkentin, 2008; Fleming et al., 2009].

The usage of phytotelmata, like tree holes or bromeliads, that harbor small relatively permanent water bodies, is an aquatic reproductive mode which is fairly common in Arboranae (6%) and is present in 5 of the 10 subfamilies (Fig. 5b). In spite of the fact that the eggs still develop in water, phytotelmata usage could also be considered as a secondary arboreal mode of reproduction, given the discrete groups of phylogenetically related species that show this behavior [Zamudio et al., 2016]. In Scinaxinae, for example, there is no evidence of proper arboreal reproduction; however, an important fraction of the species reproduces using phytotelmata (Fig. 5b). These species, previously classified as part of the genus *Scinax*, apparently belong to a monophyletic clade and have been re-instated in the genus *Ololygon*, more specifically, in the *O. perpusilla* group [Peixoto, 1987; Faivovich et al., 2005; Duellman et al., 2016]. *Ololygon* species are endemic to the tropical Brazilian Atlantic Coastal Forest and exclusively breed and lay eggs in bromeliads where embryos and tadpoles develop [Pederassi et al., 2012; Guimarães et al., 2014]. This adaptation could also be diagnostic in other species groups from the genera *Phyllodytes* and *Osteocephalus*, in the subfamily Lophyohylinae, that also show a bromeligenic tendency [Alves-Silva and da Silva, 2009]. Within this mode, plasticity is also possible, with some species that use both phytotelmata and ground water bodies. For example, *Exerodonta perkinsi* breeds both in water bodies and in the leaf axils of elephant ear plants, and *Boana pardalis* breeds both in ponds and in bromeliads [Acevedo and Smith, 2004; Moura et al., 2011].

1.4.1.3 Outlook: Evolution of Reproductive Modes and Embryo Adaptations

While the tendency to microhabitat specialization and speciation related to the oviposition site does occur, plasticity in reproductive modes might be more widespread, reflecting the high adaptability of this clade to a changing environment. This is of particular interest, since the diversity of reproductive modes might explain the success of anuran amphibians in the colonization of the terrestrial environment [Touchon and Warkentin, 2008; Crump, 2015; Zamudio et al., 2016]. Nevertheless, the mechanisms that drive diversification of reproductive modes are still unknown. Humidity, natural selection, and predation, either alone or in combination, have been proposed as environmental factors that drive reproductive mode diversification, allowing for availability, or forcing the usage of new microhabitats [Magnusson and Hero, 1991; da Silva et al., 2012; Zamudio et al., 2016]. Indeed, 31 of the 39 currently described reproductive modes occur in the neotropical forests where anuran diversity and microhabitat availability is large [Haddad and Prado, 2005]. Zamudio et al. [2016] suggest that sexual selection might have played a major role in the origin of terrestrial reproduction, at least in Leptodactylidae and Arboranae. In these 2 large Anuran clades, terrestrial reproduction is correlated with the presence of hidden amplexus, suggesting male competition is a major driving force [Zamudio et al., 2016].

In all, the success of a reproductive mode seems to be the result of one or multiple parental behaviors, and one or multiple adaptive maternal-embryonic structures. In Arboranae, parental behaviors and the egg jelly coat seem to play a major role in maintaining an aquatic reproductive mode, but also in allowing plasticity and diversification. Other embryonic structures like the cement glands, embryonic gills, and even the prehatching embryo morphology, could be playing major roles in embryo survival. However, very limited information and experimental research are available. All these adaptive characters may have evolved to improve the fitness of a particular reproductive mode. Therefore, by analyzing the genetic architecture of these adaptive characters, it might be possible to uncover the genetic basis of reproductive mode adaptability in amphibians.

1.4.2 Oogenesis

The frog ovary consists of numerous chambers (ovarian lobes) (Fig. 6a, b) that during the breeding season increase in size and occupy a large part of the body cavity [Ecker, 1889]. The ovary contains oogonia, oocytes, and somatic cells. The somatic cells envelop each diplotene oocyte in a follicle (Fig. 6a, b). The terms oocyte and egg represent 2 different cell types: oocytes are immature female germ cells found in the ovary, whereas the eggs are the female gametes that have completed oogenesis, become liberated from the ovary during ovulation, and are released by the female frog [Barth, 1953]. The total pool of oocytes that a female may produce during its life span is established during the juvenile period, as detected in *Rana temporaria* [Ogielska et al., 2013]. Oocytes are arrested in the first meiotic prophase during the long period of amphibian oogenesis. The oocyte nucleus undergoes extrachromosomal amplification of the ribosomal genes, the formation of thousands of extrachromosomal nucleoli, and the formation of transcriptionally active lampbrush chromosomes. These features of oogenesis were previously reviewed [Schmid et al., 2012]. During oogenesis, numerous transcripts, ribosomes, proteins, other gene products, and considerable reserves of nutrients, in the form of yolk platelets, accumulate in the oocyte's cytoplasm [Davidson, 1976]. Comparative analyses revealed that the pattern of ovarian differentiation is universal for anuran amphibians [Ogielska and Kotusz, 2004], and therefore no major deviations of oogenesis are expected in the Arboranae.

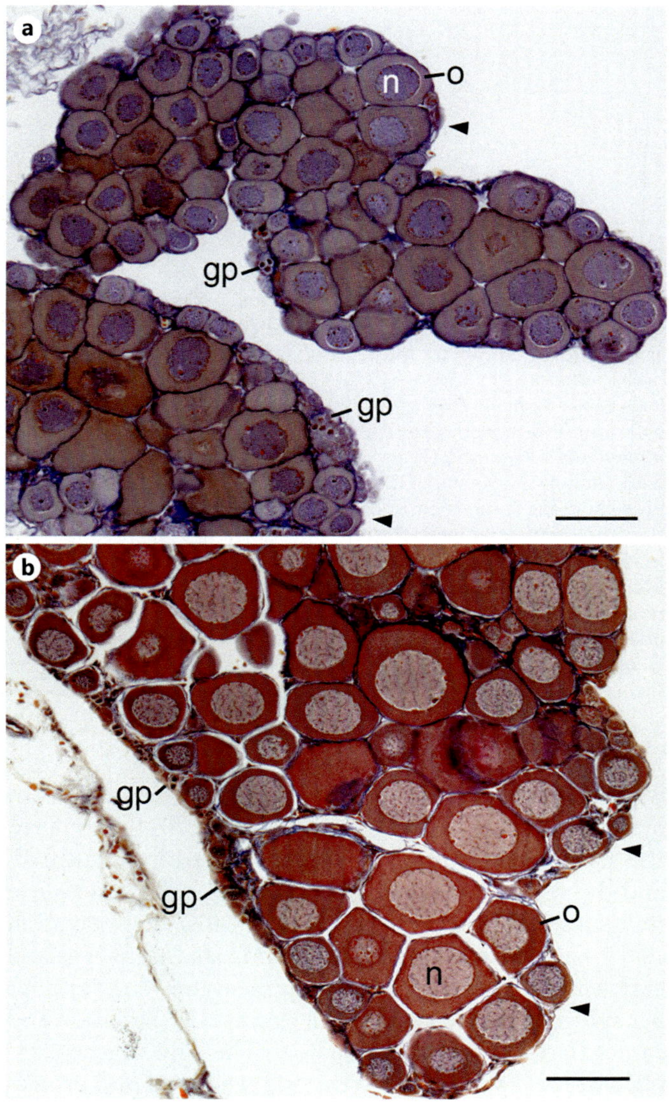

Fig. 6. Histological characterization of the ovary at metamorphosis. **a** *Hyla arborea*. **b** *H. orientalis*. Both sections of the ovaries show ovarian lobes (indicated by arrowheads), germ patches (gp), and numerous mononucleated diplotene oocytes. These oocytes are surrounded by follicle cells. o, previtellogenic diplotene oocyte of stage 1; n, diplotene nucleus. Scale bars, 100 μm. Collection localities: **a** Wrocław, Poland, **b** Urwitałt, Poland. Photographs by B. Rozenblut-Kościsty. For further explanations, see chapter 1.4.2.2.

1.4.2.1 The Ovary in Arboranae

Pigmented and unpigmented oocytes and eggs have been found in Arboranae, as in other frogs (Fig. 7a–d). For example, in Cophomantinae (family Hylidae), the oocytes of *Boana heilprini* are uniformly pale yellow (Fig. 7a, b),

Cytogenet Genome Res 2018;155:11–45
DOI: 10.1159/000489540

Fig. 7. Pigmented and unpigmented oocytes of Arboranae, family Hylidae. **a** *Boana heilprini*. The sex of the specimen is unknown. Collection locality: Haiti. Specimen deposited at the Cornell University Museum of Vertebrates, Ithaca, NY, USA (CUMV 15991). **b** Ovary of *B. heilprini* with vitellogenic oocytes in uniform pale color. **c** Male of *Bokermannohyla ibitiguara*. **d** Ovary of *B. ibitiguara* with vitellogenic oocytes. Oocytes have dark pigment in the animal hemisphere. The white structure in the midline is the intestine. Collection locality: Brazil, State of Minas Gerais, Municipality of Sacramento. Images in **a**, **b** reproduced from Nali et al. [2014] with permission from the authors. Photograph in **a** by H.W. Greene and **b–d** by R.C. Nali. For further explanations, see 1.4.2.1.

whereas the animal hemisphere is darkly pigmented in oocytes of most species of *Boana* [Nali et al., 2014]. Similarly, the animal hemisphere of oocytes is darkly pigmented, and the vegetal pole is beige in color in *Bokermannohyla ibitiguara* (Fig. 7c, d) [Nali and Prado, 2012]. Pigmentation of the animal hemisphere occurs in frogs whose eggs and early embryos develop exposed in the water. The dark upper side may provide protection against UV radiation and camouflage against bird predators. The light vegetal/ventral side looks white against the bright sky and may protect the eggs and embryos against aquatic predators. In contrast, eggs and embryos are unpigmented in frogs that develop in secluded places. It may be that *B. heilprini* eggs develop in sheltered areas [Nali et al., 2014]. The egg deposition sites of *B. ibitiguara* are unknown [Nali and Prado, 2012].

The ovaries of *B. ibitiguara* contain 202 vitellogenic oocytes with a diameter of about 2 mm (Fig. 7d) [Nali and Prado, 2012]. In contrast, the ovaries of *B. heilprini* contain 332 vitellogenic oocytes. These oocytes are unpigmented, and their diameter is about 2 mm (Fig. 7b) [Nali et al., 2014]. The egg diameters of *Nyctimantis rugiceps* are small and range from 1.1 to 1.4 mm. The analyzed egg clutch of this frog consisted of 773 pale eggs, deposited in a bamboo stump [Duellman and Trueb, 1976]. In contrast, the vitellogenic oocytes of *Callimedusa tomopterna* were larger. Oocytes of this frog are pale and reach a diameter of 2.8 mm. The number of eggs in Arboranae ranges from 29 eggs in *Litoria longirostris* to 2,500 eggs in *Trachycephalus* (see 1.3.1). The egg diameter ranges from about 0.8 to 3.5 mm (see 1.4.4.1).

1.4.2.2 The Mode of Oogenesis in Arboranae

In anurans, there are mononucleated and multinucleated modes of oogenesis. In mononucleated oocytes, there is only 1 nucleus per oocyte during the entire period of oogenesis. Mononucleated oogenesis occurs in the majority of frogs. In contrast, in the tailed frog of North America, *Ascaphus truei*, and in 12 of the 33 species of Hemiphractidae that were analyzed, previtellogenic oocytes contain numerous nuclei, ranging from 4 to 3,000 nuclei per oocyte [Macgregor and Kezer, 1970; del Pino and Humphries, 1978; Macgregor and del Pino, 1982]. This type of oogenesis was named multinucleated oogenesis [del Pino and Humphries, 1978]. The oocyte nuclei of a multinucleated oocyte are gradually reduced, and at the end of oogenesis, only 1 nucleus remains in each oocyte [del Pino and Humphries, 1978; Macgregor and del Pino, 1982]. The significance of the many nuclei for accelerating the process of oogenesis was reviewed previously [Schmid et al., 2012].

The study of the juvenile ovary of Arboranae species is of significance to determine whether in frogs of this group, oocytes are mononucleated. The previtellogenic ovary of juvenile *Hyla arborea* and *H. orientalis* at metamorphosis contains oogonia and oocytes at the leptotene, pachytene, and diplotene stages of meiosis [Stöck et al., 2013b; Tamschick et al., 2016b]. Diplotene previtellogenic oocytes of stage 1 are surrounded by follicle cells (Fig. 6). The analysis of published images and oocyte micrographs revealed mononucleated oocytes in the juvenile ovaries of these frogs [Stöck et al., 2013b; Tamschick et al., 2016b] (Fig. 6). Oocytes are also mononucleated in adult females of *Scinax fuscovarius* (Fig. 8) [de Oliveira and de Souza Santos, 2004]. These observations indicate that mononucleated oogenesis occurs in *H. arborea*, *H. orientalis*, and *S. fuscovarius*.

The mode of oogenesis was additionally investigated in 9 species of Arboranae. Pieces of ovary were removed from the female frogs, and oocyte nuclei were observed under a microscope. The oocyte nucleus was released by opening the oocyte with a pair of forceps in a dish filled with a saline solution. In some instances, histological sections of the ovary were prepared and analyzed to detect the mode of oogenesis. The analysis was enhanced by the study of published oocyte images from 7 additional species (Table 2). The subfamilies of Hylidae and the family Phyllomedusidae were represented in the analysis. The family Pelodryadidae was not studied. The oocytes are mononucleated in the 16 species of Arboranae investigated (Table 2). This analysis indicates that the oocytes are mononucleated in Arboranae as in most amphibians and other vertebrates.

1.4.2.3 Stages of Oogenesis in *Scinax fuscovarius*

Detailed studies on chromosomal changes during oogenesis were performed in amphibian oocytes of species belonging to temperate regions of the world [Duryee, 1950; Dumont, 1972; Coggins, 1973; Davidson, 1976]. In contrast, the cytological changes during oogenesis of tropical species of Arboranae are known to a lesser extent [Hunter and Murcia de Valdivieso, 1962; Rodríguez-Gómez et al., 2014]. Oogenesis has been divided into stages to facilitate its analysis. However, the number of the oogenesis stages varies between species. Ovarian differentiation was divided into 10 stages based on the study of 12 species from 6 families [Ogielska and Kotusz, 2004]. Growth and maturation of the *Xenopus laevis* diplotene oocytes was divided into 6 stages based on oocyte size, accumulation of yolk, and cytoplasmic and nuclear characteristics [Dumont, 1972]. Oogenesis of the Brazilian species *Scinax fuscovarius* has been divided into 4 stages [de Oliveira and de Souza Santos, 2004].

The ovary of *S. fuscovarius* consists of nests of oogonia, leptotene and pachytene oocytes, and diplotene oocytes surrounded by follicular cells. The following description of oogenesis in this frog is according to de Oliveira and de Souza Santos [2004], in comparison with the *X. laevis* stages of oogenesis [Dumont, 1972].

Oogonia are large cells with a large and oval nucleus. They are located – together with early leptotene-pachytene oocytes – peripherally in the ovary in areas called "germpatches", comparable to *X. laevis* [Dumont, 1972] and other frogs [Ogielska and Kotusz, 2004]. Diplotene oocytes are categorized in the following stages [Dumont, 1972; de Oliveira and de Souza Santos, 2004]:

Stage 1 oocytes are previtellogenic, and likely transparent, as in *X. laevis*. In *X. laevis*, the nucleus can be observed in the intact transparent oocyte.

Stage 2 oocytes are previtellogenic as in *X. laevis*. The oocytes are larger than in stage 1, and the cytoplasm becomes more basophilic. The nucleus has a central location. Nucleoli and chromosomes are visible (Fig. 8a).

Stage 3 oocytes are early-vitellogenic. Oocytes are larger than in stage 2. Stage 3 oocytes are characterized by intense cytoplasm acidophilia. Yolk platelets are detected in the periphery of the oocyte. The nucleus has a central position. Synthesis of dark pigment has started. Oocytes are surrounded by the vitelline envelope and 2 layers of follicle cells (Fig. 8b). Oocytes of this stage are comparable to stage 3 of *X. laevis* oogenesis.

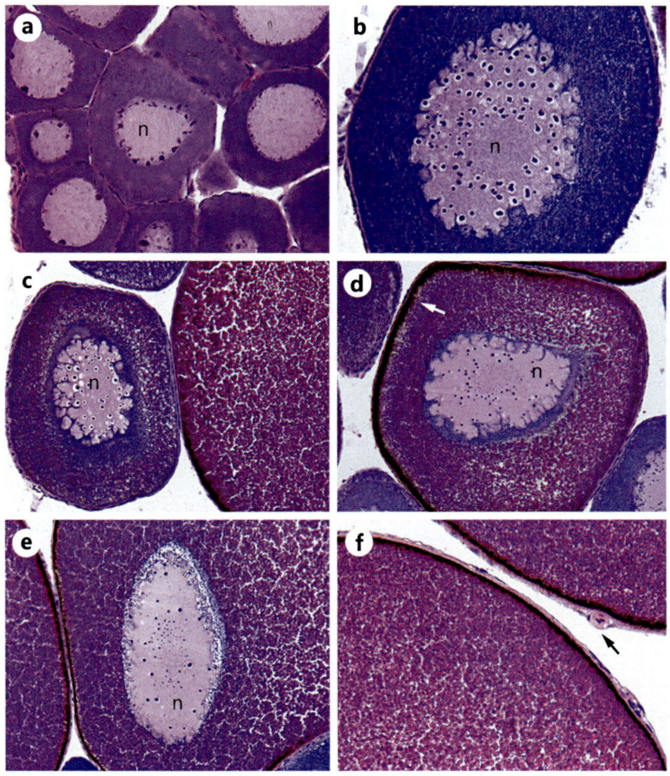

Fig. 8. Histological characterization of *Scinax fuscovarius* oogenesis according to de Oliveira and de Souza Santos [2004]. **a** Previtellogenic oocytes, stage 2. **b** Early vitellogenic oocyte, stage 3. **c** Early to mid-vitellogenic oocyte, stage 4. Cell nuclei are centrally located. Hemispheres are not yet defined by the concentration of dark pigment. **d** Mid-to-late vitellogenic oocyte, stage 4. This oocyte is more advanced than that in **c**. The cell nucleus is centrally located. The 2 hemispheres are clearly defined because the animal hemisphere is darkly pigmented. The white arrow points to the darkly pigmented surface of the animal region. **e** Vitellogenic oocyte, late stage 4. The cell nucleus is distinctly displaced towards the animal pole. Dark pigment is concentrated in the surface of the animal hemisphere. **f** Follicle layers surrounding 2 oocytes. The black arrow points to a blood vessel of the follicle layers. n, nucleus. Collection locality: Botucatu, São Paulo State, Brazil. Photographs by C. de Oliveira. For further explanations, see 1.4.2.3.

Stage 4 oocytes of *S. fuscovarius* are mid-to-late vitellogenic and correspond to stages 4 and 5 of *X. laevis* oogenesis.

Stage 4 mid-vitellogenic oocytes of *S. fuscovarius* are comparable with *X. laevis* stage 4 oocytes. In mid-vitellogenic oocytes, the nucleus is centrally located, and the oocyte hemispheres are not well defined (Fig. 8c). As oogenesis progresses, dark pigment accumulates in the animal hemisphere, and the oocyte hemispheres become clearly defined. The nucleus maintains a central location (Fig. 8d).

Cytogenet Genome Res 2018;155:11–45
DOI: 10.1159/000489540

Table 2. Arboranan species with mononucleated oocytes

Family/subfamily	Species[a]	References	Figure
Hylidae			
Acrisinae	*Acris crepitans*	Reeder et al., 1998	
Hylinae	*Anotheca spinosa*	present sudy	
	Hyla arborea	Tamschick et al., 2016b	6a
	Hyla orientalis	Stöck et al., 2013a	
Pseudinae	*Pseudis paradoxa*	Downie et al., 2009	
Dendropsophinae	*Dendropsophus molitor*	Pinto-Erazo et al., 2016	
Lophyohylinae	*Trachycephalus jordani*	present study	
	Nyctimantis rugiceps	present study	
	Osteopilus septentrionalis	Rodriguez-Gómez et al., 2014	
Scinaxinae	*Scinax fuscovarius*	de Oliveira and de Souza Santos, 2004	8
	Scinax ruber	present study	
	Scinax quinquefasciatus	present study	
Cophomantinae	*Boana pellucens*	present study	
Phyllomedusidae			
	Callimedusa perinesos	present study	
	Callimedusa tomopterna	present study	
	Phyllomedusa tarsius	present study	

[a] Species names according to the electronic database of Frost [2017].

Stage 4 late-vitellogenic oocytes of *S. fuscovarius* are comparable with *X. laevis* stage 5 oocytes. In late-vitellogenic oocytes, the animal and vegetal hemispheres can be identified by the concentration of dark pigment in the animal hemisphere surface. In contrast, the vegetal region is lightly colored. The nucleus of late-vitellogenic oocytes is displaced toward the animal pole (Fig. 8e).

Stage 6 oocytes have not been described for *S. fuscovarius*. In *X. laevis*, stage 6 oocytes are post-vitellogenic. The follicle of *S. fuscovarius* consists of 2 cell layers and contains blood vessels (Fig. 8f).

The description of oogenesis of *S. fuscovarius* and the detailed analysis of gonadogenesis in tadpoles of this species provide a baseline to detect abnormalities during development due to environmental contaminants [de Oliveira and de Souza Santos, 2004; Goldberg, 2015]. Similarly, the description of normal gonadal development in the arboranan *Dendropsophus labialis*, an endemic species from the Eastern Andes of Colombia, is important for the preservation of this species [Pinto-Erazo et al., 2016].

1.4.2.4 Oogenesis and Environmental Contaminants

The presence of endocrine disruptors in the environment, other chemical contaminants, parasites, and the effects of the deadly fungal pathogen, *Batrachochytrium dendrobatidis* disturb frog gonadogenesis [Reeder et al., 1998; Sitjà-Bobadilla, 2009; Goldberg, 2015; Brannelly et al., 2016; Pinto-Erazo et al., 2016; Tamschick et al., 2016a–c]. Hence, the description of normal gonadogenesis is important to provide a baseline for the identification of gonadal abnormalities and sex reversal. The knowledge of normal gonadal development in anurans increases our knowledge about the anuran reproductive biology and contributes to the conservation of the great diversity of Arboranae and other frogs.

1.4.3 Spermatogenesis

Anurans in general have a highly conserved testis and spermatogenic tissue structure [Ferreira et al., 2009; Leite et al., 2015; Yoshida, 2016]. Briefly, testes are mostly ovoid or elongated structures lying ventrally to the kidneys (Fig. 9a). Beneath the peritoneum envelopes (mesorchium), the tunica albuginea is the most external connective tissue that protects the spermatogenic tissue [Lofts, 1974; Lombardi, 1998].

The seminiferous tissue within the testes forms seminiferous lobules that result from the partial invagination of the tunica (Fig. 9b) [Lofts, 1974; Oliveira et al., 2003]. Attached to this basal membrane, each primary spermatogonium is nourished by multiple Sertoli cells (Fig. 9b, c). The descendants of a single spermatogonium produce an independent mass of cells surrounded by Sertoli cells constituting a cyst, where cell amplification and meiosis proceeds synchronously until spermatid differentiation occurs (Fig. 9b, c). This is a characteristic of anamniotes that differs from the seminiferous tubules found in amniotes [Yoshida, 2016].

The histological appearance of primary spermatogonia is heterogeneous depending on their stage of maturation [Oliveira et al., 2003]. These cells often show a multilobular shape, an irregular nuclear shape, and they are located close to the lobular wall (Fig. 9c). In *Dendropsophus minutus*, *Scinax fuscovarius*, *Boana pulchella*, and *Agalychnis dacnicolor*, spermatogonial nuclei are multilobed [Rastogi et al., 1988; Oliveira et al., 2003; Carezzano et al., 2014]. In *Lysapsus limellum*, spermatogonial nuclei are usually bilobed, with relatively loose chromatin [Ferreira et al., 2008]. There are no histological descriptions for pelodryadid species. Within the cyst, the tightly associated Sertoli cells have large nuclei with visible nucleoli (Fig. 9c) [Ferreira et al., 2008, 2009]. Spermatocytes have an irregular shape and a darker asymmetrically located nucleus that shows various degrees of chromatin condensation depending on the meiotic phase (Fig. 9c). As described for *P. limellum*, primary spermatids are rounded, and later spermatids are elongated with long nuclei and a visible proacrosomal vesicle [Ferreira et al., 2008]. The highest degree of chromatin condensation is only appreciable in the mature spermatozoa, which form homogeneous bundles with the acrosomal end tightly associated with Sertoli cells (Fig. 9c) [Oliveira et al., 2003; Santos and de Oliveira, 2008; Curi et al., 2014].

1.4.3.1 Ultrastructure of Spermatozoa and Arboranan Evolution

Multiple subcellular characteristics of spermatozoa have been proposed as characters for phylogenetic reconstruction [Meyer et al., 1997; Garda et al., 2002; Scheltinga et al., 2002; Zeng et al., 2014]. Although the phylogenetic "informativeness" of these structures is still being assessed, the diversity of spermatozoal ultrastructure can provide some cues about the natural history and evolution of anurans given the importance of spermatozoal fitness in reproductive success [Fouquette and Delahoussaye, 1977; Teixeira et al., 1999].

In anurans, the head of a mature spermatozoon is the most prominent part and has an acrosomal complex that consists of a terminal cone-shaped acrosome vesicle and a subacrosomal cone or conical perforatorium (Fig. 9d–g) [Lombardi, 1998]. In multiple species of amphibians, the acrosome, as for all vertebrates, has proteolytic enzymes that allow sperm penetration through the egg jelly [Lofts, 1974]. This reaction, in *Xenopus laevis*, seems to be activated by secretions from the female oviduct [Ueda et al., 2002]. The subacrosomal cone is a mysterious structure, present in all anurans and variably present in animals, that is placed below the acrosome and takes the conical shape of the space

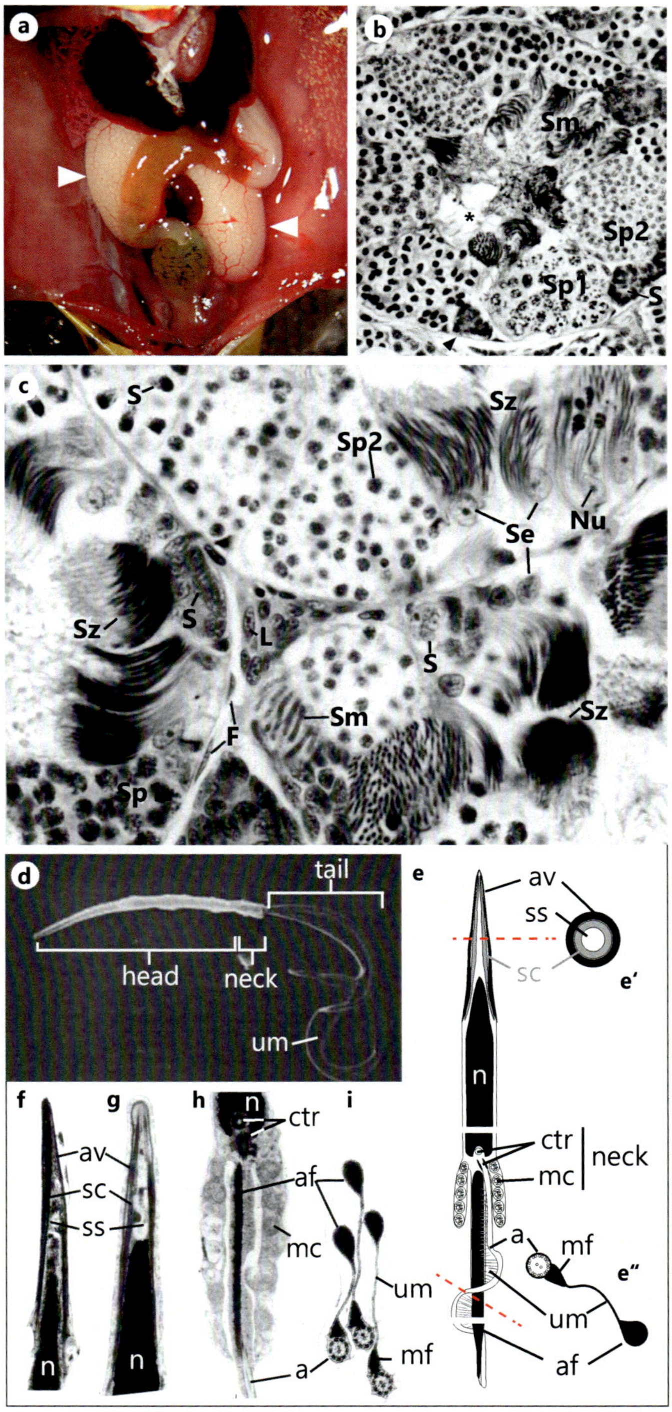

Fig. 9. Spermatogenesis and sperm ultrastructure in Arboranae. **a** Abdominal dissection of a *Scinax fuscovarius* male with mature testes (white arrowheads). **b** Histological section of a seminiferous lobule from *S. fuscovarius* with multiple cysts, each containing cells in a distinct meiotic stage. The asterisk labels the luminal area of the lobule, the black arrowhead points to the tunica albuginea. HE. ×850. **c** Detail of the interstitial area between 3 seminiferous lobules from *S. fuscovarius*. HE. ×1,850. F, possible fibroblasts; L, Leydig cells; Nu, nucleus; S, spermatogonia; Se, Sertoli cells; Sm, spermatids; Sp1, primary spermatocytes; Sp2, secondary spermatocytes; Sz, spermatozoa. **d** Scanning electron micrograph of a *Litoria lesueurii* spermatozoon. **e** Diagram of a bufonoid spermatozoon. The white gaps of the diagram compensate the lack of proportionality. The red dashed lines represent the optical sections from which **e**′ and **e**″ where obtained. **f** Transmission electron micrograph (TEM) of the acrosomal region of a *Dryophytes japonicus* spermatozoon. **g** TEM of the acrosomal region of a *Litoria manya* spermatozoon. **h** TEM of the neck region of a *Cyclorana novaehollandiae* spermatozoon. **i** TEM of 3 cross sections from a *Litoria cryptotis* spermatozoon tail. The diagram in **e**″ is a representation of this image. a, 9 + 2 axoneme; af, axial fiber; av, acrosomal vesicle; ctr, centrioles; mc, mitochondrial collar; mf, minor fiber; n, nucleus; sc, subacrosomal cone; ss, subacrosomal space; um, undulating membrane. **a** Photograph by C. de Oliveira. **b–i** Images modified from Oliveira et al. [2003] (**b**, **c**) and from Scheltinga and Jamieson [2003] (**d–i**) with permission from the authors. For further explanations, see 1.4.3.

between the nucleus and the acrosome [Austin and Bishop, 1958; Lee and Jamieson, 1993; Meyer et al., 1997; Garda et al., 2002]. In mammals, this structure mediates the effectiveness of the acrosomal reaction and is derived from the perinuclear theca [Oko and Sutovsky, 2009]. In amphibians, this function might be maintained, as shown for the newt *Cynops pyrrhogaster* [Yokoe et al., 2014]. In Urodela, Gymnophiona, and basal anurans, like *Ascaphus truei*, there is an additional rod shaped structure termed the axial perforatorium that extends from the apex of the acrosome and becomes endonuclear. This structure could be plesiomorphic to all amphibians [Lee and Jamieson, 1993; Scheltinga et al., 2002]. The role of the axial perforatorium in amphibians is unknown, but in other animals it also contributes to the acrosomal reaction [Baccetti, 1985]. Below the nucleus, the midpiece is composed of a centriolar medial region and a cytoplasmic protrusion or collar that gathers most of the mitochondria and covers the initial part of the tail (Fig. 9d, e, h) [Pitnick et al., 2009]. The tail of the anuran spermatozoon has 2 filaments: a filament of lamellipodial origin termed the axial fiber, which is often not motile, and the axoneme, both connected by a membranous sheath, the undulating membrane (Fig. 9d, e, i) [Lombardi, 1998; Pitnick et al., 2009].

The sperm structure of Arboranae can be grouped with most anurans of the Hyloidea taxon, which includes the families Bufonidae, Eleutherodactylidae, Craugastoridae, Centrolenidae, and Leptodactylidae, among others [Pyron and Wiens, 2011], mostly due to the presence of the subacrosomal cone [Lee and Jamieson, 1993; Garda et al., 2002]. This subacrosomal cone is not present or less conspicuous in the Ranoidea taxon, which includes Microhylidae, Ranidae, and Rhacophoridae [Pyron and Wiens, 2011]. The midpiece in the sperm of the Hyloidea is prominent compared to other groups; however, the mitochondrial collar and the tail are variable within Arboranae (Fig. 9h) [Garda et al., 2002]. Briefly, most Arboranae have a prominent mitochondrial collar. The tail has a large axial fiber widely separated from the axoneme but linked through the undulating membrane. In *Scinax rostratus*, *S. squalirostris* (Scinaxinae), *Bokermannohyla pseudopseudis* (Cophomantinae), and the pelodryadids *Litoria fallax*, *L. longirostris*, *Ranoidea gracilenta*, *R. lesueuri*, *R. alboguttata*, *R. brevipes*, *R. novaehollandiae*, and *R. cryptotis*, the axial fiber has approximately the same diameter as the axoneme [Lee and Jamieson, 1993; Meyer et al., 1997; Scheltinga et al., 2002; Costa et al., 2004]. In *Ranoidea* species, the undulating membrane appears to be thicker [Meyer et al., 1997]. In *Dryophytes japonicus*, the axial fiber and the axoneme are attached without an undulating membrane [Lee and Kwon, 1992]. In *Dendropsophus minutus*, there is no axial fiber and no undulating membrane [Ferreira and Mehanna, 2012]. To compare, pseudids seem to have a more derived pattern. In *Pseudis paradoxa*, *P. bolbodactyla*, *P. tocantins*, *P. minuta*, and *Lysapsus laevis*, the mitochondrial collar is less prominent, and the axial fiber and undulating membrane are absent, similar to *D. minutus* [Garda et al., 2004]. Regarding these variations within Arboranae, Scheltinga et al. [2002] have hypothesized that differences in certain structures, such as the subacrosomal cone, are the result of natural selection. In this case, a seemingly larger structure might improve the ability to trespass the gelatinous layer that covers the egg mass in *L. longirostris*. Accordingly, other species that deposit additional gelatinous layers, like *Agalychnis dacnicolor* and other phyllomedusids, should show similar adaptations [Rastogi et al., 1988]. However comparative morphometry from transmission electron microscopy would require simultaneous preparation. Hence, further experiments using new molecular, cellular, and histological tools are required to determine the adaptive value of the sperm ultrastructure in anuran amphibians.

1.4.3.2 Patterns of Testis Maturation

The process of testis differentiation occurs during tadpole stages. In *Hyla arborea*, the most distal metameric parts of the undifferentiated gonads degenerate, while the most proximal parts differentiate into the adult testes [Haczkiewicz and Ogielska, 2013]. While the beginning of gonad differentiation seems to be a constant across anuran species, the timing of differentiation is highly variable, and the biotic and abiotic factors that influence this variability remain unknown. For most anurans, ovary differentiation occurs earlier than testis differentiation compared to somatic development [Ogielska and Kotusz, 2004; Haczkiewicz and Ogielska, 2013]. In *Scinax fuscovarius*, oogenesis occurs simultaneously to the differentiation of the follicular tissue; though, the seminiferous tubules only differentiate after metamorphosis, at juvenile stages [Goldberg, 2015]. This pattern is considered to be a delayed pattern of testes maturation compared to somatic development. Other Arboranae, however, show distinct maturation rates or heterochronic changes. Multiple pseudid species, like *Pseudis minuta*, *P. paradoxa*, *P. platensis*, and even *Dendropsophus labialis* (Dendropsophinae), show an accelerated pattern of testis differentiation with highly differentiated seminiferous tubules and advanced spermiogenic cysts present during tadpole stages [Downie et al., 2009; Fabrezi et al., 2010; Goldberg et al., 2016; Pinto-Erazo et al., 2016]. Metamorphosis is another developmental feature subject to environmentally induced variability. The newly metamorphosed *P. minuta* can be seen in ponds alongside larger males, producing mating calls with their still developing vocal sacs [Goldberg et al., 2016]. For Phyllomedusidae, a few studies show that there is some variability in the speed of testis differentiation: *Phyllomedusa boliviana* has an accelerated pattern of testis differentiation [Rengel et al., 1995]. *Pithecopus azureus* and *Phyllomedusa sauvagii*, on the other hand, have a standard to delayed pattern, on which seminiferous tubule differentiation occurs during metamorphosis [Rodrigues et al., 2007; Goldberg, 2015]. Patterns of gonad differentiation in Pelodryadidae are yet to be described.

The hormone regulatory mechanisms that control metamorphosis might influence the decoupling of gonad maturation, since *Xenopus laevis* tadpoles that lack the thyroid gland do not undergo metamorphosis, but increase in body size and have completely differentiated testes [Rot-Nikcevic, 2004]. To compare, *Pseudis platensis* and a few other species of *Pseudis* produce giant tadpoles that take longer than other sympatric tadpoles to undergo metamorphosis and also show early testis maturation during tadpole stages [Fabrezi et al., 2010]. Remarkably, *Pseudis minuta* does not produce giant tadpoles in contrast to *P. platensis*. The endocrine mechanisms that regulate testis maturation in *P. minuta* and *P. platensis* are unknown.

1.4.3.3 Environmental Kinetics of Spermatogenesis

In species found in tropical regions such as *Dendropsophus nanus*, *Lysapsus limellum*, *Pseudis paradoxa*, and *Scinax acuminatus*, from the Brazilian Chaco and Cerrado, it is possible to find active meiotic cysts in all the stages as well as mature spermatozoa in a single tissue sample [Ferreira et al., 2008, 2009]. Testicular structure and histology are also constant when sampled throughout the year, which correlates to the reproductive behavior by which mating and nesting occur at any time [Ferreira et al., 2008]. In *L. limellum*, cysts with secondary spermatocytes are rarely found, and there is a higher amount of spermatozoa, a character possibly related with the transient timing of the second meiosis in this species [Ferreira et al., 2009]. Another particular temporal change in spermatogenesis kinetics of tropical species are variations in the number of spermatozoa and the volume occupied by late spermiogenic cells, which is larger during rainy seasons, as shown for *Dendropsophus minutus* [Santos et al., 2011].

Species that dwell in latitudes with well-marked seasonal changes, like the *Pelophylax esculentus* complex, increase testis volume during the summer only when temperature and water are appropriate for breeding [Chieffi et al., 1980]. In these species, testis volume is related to the amount of spermatogenesis occurring and the accumulation of steroid lipids; for example, during winter time, the number of spermatophorous cysts is reduced, and spermatogonia degenerate as a result of modifications in testosterone levels [Lofts, 1964]. A more drastic effect occurs in the aestivating frogs *Ranoidea platycephala* and *R. maini* from Southern Australia. These burrowing pelodryadids form cocoons at least 1 meter below the surface to avoid desiccation and remain there for almost a year [Tracy et al., 2007]. In these frogs, once aestivation starts, formed spermatogonia undergo at least 1 cycle of meiosis before sperm number and lobule diameter in the testis decay. Later, meiosis reactivates independently from environmental conditions between 4 to 7 months of aestivation without spermiogenesis, which apparently occurs only when rain starts [Shalan et al., 2004].

1.4.4. Prehatching Development in Agalychnis spurrelli

The reproductive biology of phyllomedusids is distinctive because of the oviposition site and clutch structure, as discussed above. Embryos within arboreal nests are threatened by desiccation and predation, for which multiple adaptive behaviors of the embryo occur to ensure survival [Warkentin, 1995, 2007]. The ability of hatching before embryonic development completion and the adaptive sensing of the external environment during the embryonic stages have been analyzed and reviewed for *Agalychnis callidryas* [Warkentin, 1999, 2000; Warkentin et al., 2005, 2006; Gomez-Mestre et al., 2010]. In particular, these adaptive behaviors require the physiological activation of embryonic structures that are still developing, like the external gills, eyes, mechanosensory organs, and the central nervous system [Gomez-Mestre and Warkentin, 2007]. Yet, very little is known about the cellular and molecular mechanisms of development in *Agalychnis*. This is also true for most species of Arboranae. Although early embryo development has been described for several species of Arboranae, most of these descriptions are incomplete or do not analyze internal embryo characteristics [Fox, 1984]. Interestingly, some of the first experimental approaches to understand gastrulation in anurans were performed in *Hyliola regilla*; however, the amenability and historic usage of *Xenopus laevis* relegated the experimental usage of other amphibians [Schechtman, 1942; Eakin, 1946; Rugh, 1962]. To review and discuss current aspects of embryo development in Arboranae, here as reference, the morphological description of embryo development in the species *Agalychnis spurrelli* is utilized.

1.4.4.1 Morphology of the Embryos of *Agalychnis spurrelli*

The adults of *A. spurrelli*, similar to other species of *Agalychnis*, lay their eggs on terrestrial vegetation, and the egg clutch is covered by a dense jelly. On natural sites of oviposition, development from fertilization to tadpole hatching occurs within approximately 8 days [Vargas-Salinas et al., 2000]. Under laboratory conditions, this process is slightly delayed, with most tadpoles hatching 11 days post fertilization at 20°C. The fertilized eggs are mostly white with a light green appearance and dim dark coloration in the animal pole, similar to other species of *Agalychnis* (Fig. 10a). The light green coloration apparently acts as a form of camouflage, given the use of exposed leaves as nesting sites [Altig and McDiarmid, 2007]. The color is the result of lutein (blue) and biliverdin (yellow) pigment deposition in the yolk platelets of *Agalychnis* embryos [Marinetti and Bagnara, 1983]. Apart from the phyllomedusids, the only other Arboranae species in which this pattern of pigmentation has been reported is *Pseudacris clarkii* [Eaton and Imagawa, 1948].

The eggs of *A. spurrelli* are large (2.9 mm) in comparison to those of *Xenopus laevis*. Phyllomedusids in general seem to have large eggs, as seen in *Phyllomedusa trinitatis* (2.5 mm) and *P. sauvagii* (2.4 mm) [Salica et al., 2011; Downie et al., 2013]. The vitellogenic oocytes of *Callimedusa tomopterna* reach a diameter of 2.8 mm (see 1.4.2.1). In other Ar-

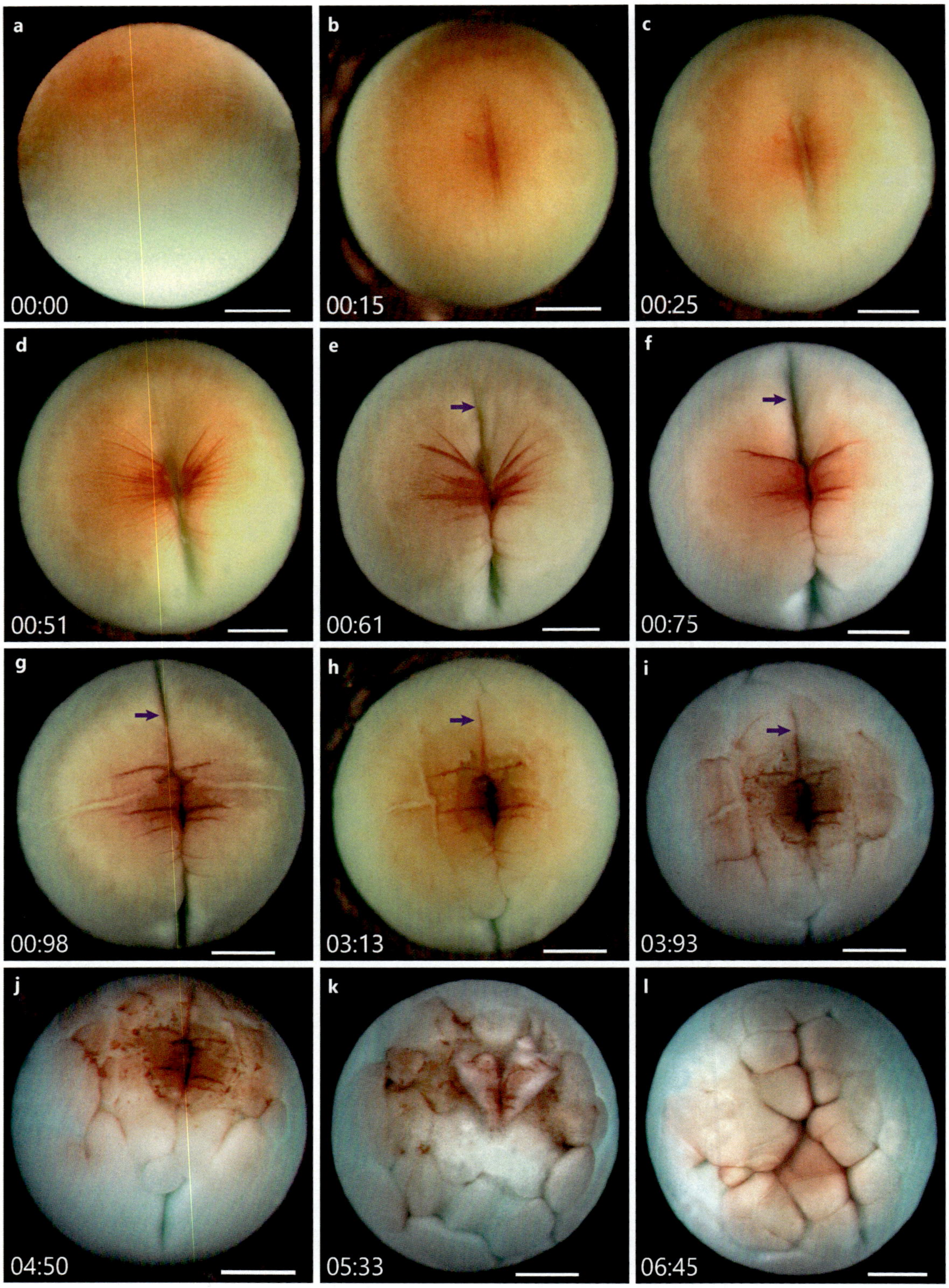

Fig. 10. Early cleavage in *Agalychnis spurrelli*. The descriptions of these stages are the results from observations of a single embryo. **a** Lateral view of the fertilized egg. **b–i**, **k**, **l** Views from the animal pole. **b–f** Formation of the initial cleavage furrow. **f** The second cleavage event consists of multiple incomplete furrows perpendicular to the first furrow. **h**, **i** The third cleavage event consists of furrows parallel to the first one. **j** Animal-lateral view of the embryo, now with approximately 10 cells. The first furrow has not reached the vegetal pole yet (not shown). **k** Embryo with approximately 20 cells. **l** Embryo with approximately 30 cells. Arrows indicate the initial cleavage furrow. The time, indicated in minutes and hours, represents the time elapsed between each cleavage event. Scale bars, 500 μm. For further explanations, see chapters 1.4.4.1 and 1.4.4.2.

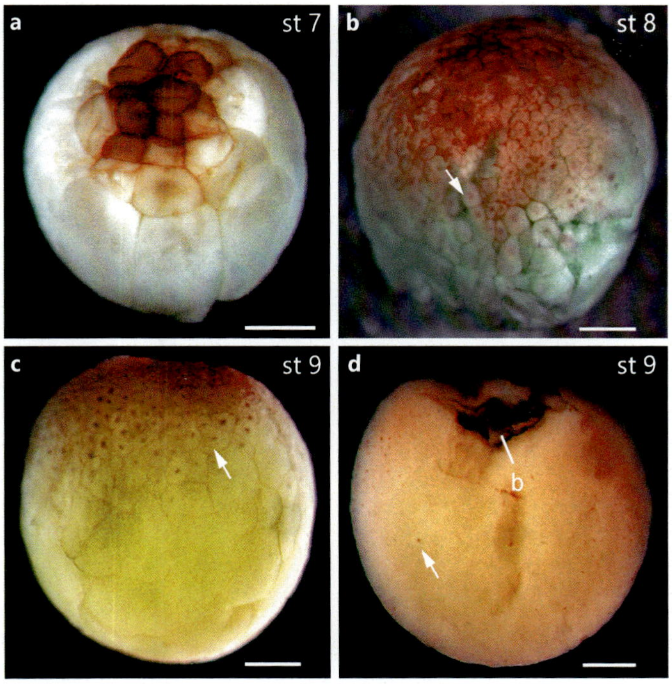

Fig. 11. Blastula stages in *Agalychnis spurrelli* embryos. **a**, **b** Animal pole lateral views. **a** Embryo in large-cell blastula stage with approximately 64 cells. **b** Medium-staged cell blastula. **c** Lateral view of a late blastula. **d** Sagittal section of a late blastula embryo showing a small blastocoel (b). In **b–d**, the arrows indicate the pigment around cell nuclei. The stage numbers (upper right corners) have been assigned according to Gosner [1960]. Scale bars, 500 μm. For further explanations, see chapter 1.4.4.2.

boranae, egg size is highly variable and does not correlate with their reproductive modes. For example, hylid egg size seems to be constant, from 1 to 1.3 mm [Eaton and Imagawa, 1948; Volpe et al., 1961; Iwasawa and Futagami, 1992; Sayim and Kaya, 2008; Sarria, 2010]. In Pelodryadidae, even though most species of *Litoria* and *Ranoidea* deposit eggs in water, egg diameter varies widely, from 0.8–0.9 mm in *L. microbelos* to 2.7–3.6 mm in *R. nannotis* [Tyler, 1998].

1.4.4.2 Early Development of *Agalychnis spurrelli*

A. spurrelli embryo development follows a general pattern, previously described for most anurans with biphasic life cycle. Simplified developmental tables can be used for field staging purposes [Gosner, 1960]. In the present study, all the stages were assigned using this table. However, there are a few distinctive features in the development of this species that are worth reviewing.

Perhaps the most divergent event in *A. spurrelli* embryo development is early cleavage. During cleavage stages, the first cell division starts with a single polar furrow that appears 15 minutes after oviposition and extends, during the next 80 minutes, through the animal-vegetal plane without reaching the vegetal pole (Fig. 10a–f). The following cleavage events appear superficially in the animal pole without a clear position in relation to the first furrow (Fig. 10g). A latitudinal furrow is not observed either. Such events will generate dissimilar animal blastomeres, while blastomere cytokinesis has not been completed at the vegetal pole (Fig. 10i–l).

The cleavage pattern in *A. spurrelli* is defined as atypical holoblastic, due to the formation of irregular blastomeres in an asynchronous fashion [Nelsen, 1953]. In the few species of Arboranae from which a detailed description of cleavage is available, none have this cleavage pattern. *Dryophytes japonicus*, *Hyla orientalis*, *H. arborea*, and *Litoria rubella* show typical holoblastic cleavage patterns with synchronous stereotypical cell divisions: the first 2 cell divisions occur in the animal-vegetal plane, one perpendicular to the other, and the third cleavage plane is latitudinal supraequatorial, which results in 8 blastomeres, 4 small animal blastomeres and 4 large vegetal blastomeres radially symmetric [Iwasawa and Futagami, 1992; Tyler, 1998; Sayim and Kaya, 2008]. Although there is a clear tendency towards a meroblastic pattern of early cleavage, *A. spurrelli* embryos will still form complete vegetal blastomeres during the following cleavages. Meroblastic cleavage has been associated with increased egg size due to larger amounts of yolk, like the amniote egg, in which cell divisions only occur in a yolk-free place of the cytoplasm, where an embryonic disc forms [Elinson and Beckham, 2002].

Atypical holoblastic cleavage patterns can be seen in other amphibians with larger eggs like the direct-developing frog *Eleutherodactylus portoricensis* (Terrarana), the egg-breeding frog *Gastrotheca riobambae* (Hemiphractidae), and the salamander *Ensatina eschscholtzii* [Gitlin, 1944; del Pino and Loor-Vela, 1990; Collazo and Keller, 2010]. In these cases, the amount of yolk and the large egg size might be important factors that cause drastic changes in early development [Arendt and Nübler-Jung, 1999]. Yolk density or other properties of the cytoplasm of large eggs might not allow latitudinal cleavages during the first cellular divisions. In this context, irregular cleavages might be the result of the inability of the mitotic spindles to orient themselves at right angles, as hypothesized by del Pino and Loor-Vela [1990]. However, there might be also other factors, not related to the egg size, that influence the cleavage pattern. For example, dendrobatids with similarly large eggs still preserve the symmetrical holoblastic pattern of cleavage [del Pino et al., 2007; Hervas et al., 2015].

Approximately at the 64-cell stage, vegetal furrows appear, and blastomeres become more homogeneously arranged, with small cells in the animal pole and large cells in the vegetal pole (Fig. 11a). This stage is called large-cell blastula and occurs 8 hours post fertilization. At this stage, cells in the animal pole become more pigmented and smaller than the cells of the vegetal pole. Despite the potentially higher exposure to UV radiation in *Agalychnis* nests, compared to eggs deposited on land or water bodies, the dark pigmentation in the animal pole is reduced in comparison to other anuran embryos. Later, in blastula stages, this pig-

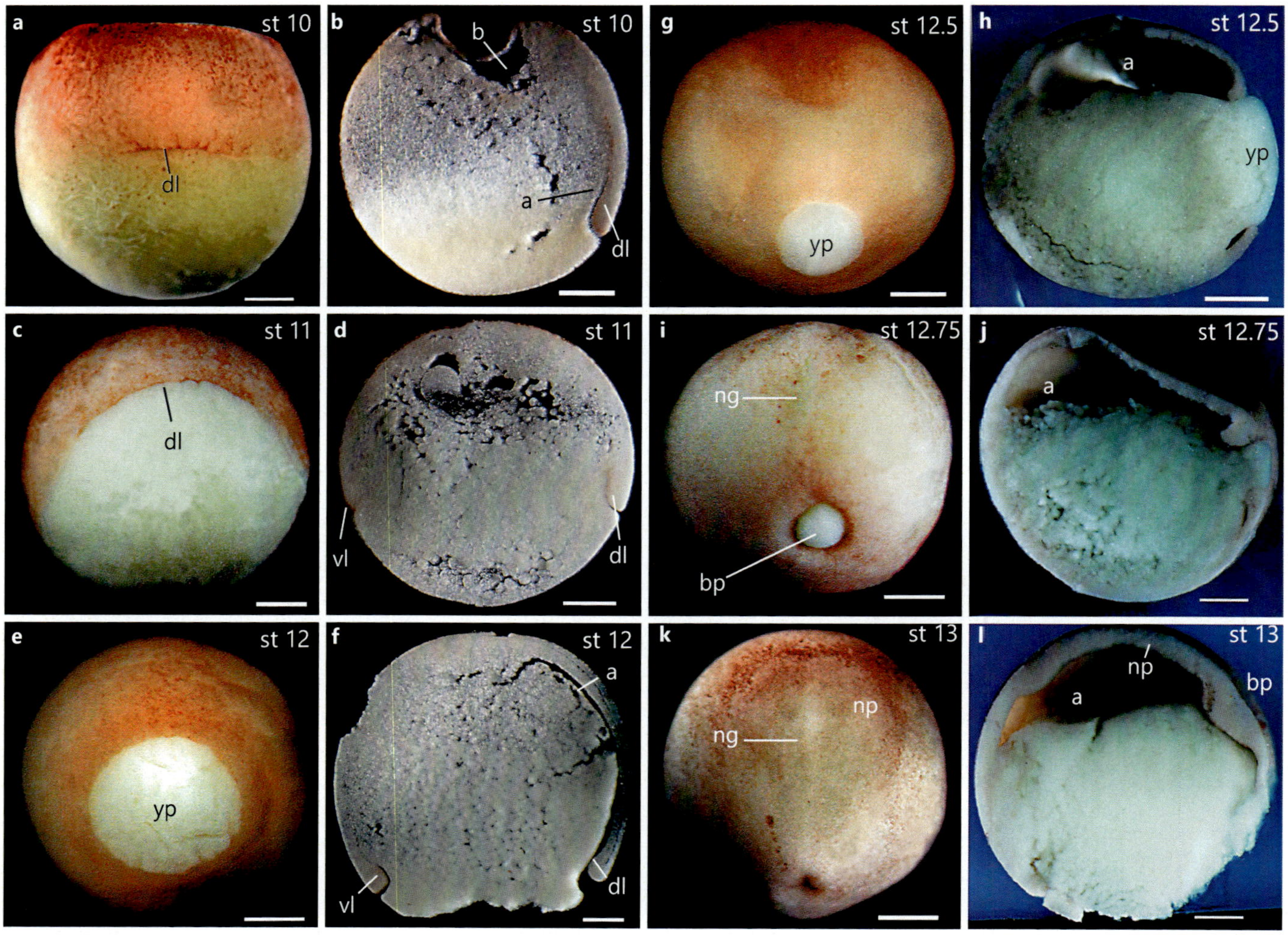

Fig. 12. Gastrula stages in *Agalychnis spurrelli*. **a**, **c**, **e**, **g**, **i**, **k** Dorsal views. **b**, **d**, **f** Images of sagittal sections. **h**, **j**, **l** Images of sagittal bisections. **a** During the early gastrula, the blastopore dorsal lip forms in the equatorial region. **b** Internally, the early gastrula shows a small blastocoel and an elongating archenteron. **c** Mid-gastrula embryos have large yolk plugs. **d** Mid-gastrula. During this stage, the ventral lip is formed and the blastocoel collapses. **e** Late gastrula embryo with a mid-sized yolk plug. **f** Late gastrula. At this stage, the archenteron reaches its maximum extension without expansion. **g–j** Small yolk plug gastrula and archenteron expansion. **g** External view of an embryo with a small yolk plug. **h** Bisection showing the expansion of the archenteron. **i** External view of an embryo with a yolk plug smaller than in **g**. **j** Late gastrula embryo bisection; during late gastrula, the neural groove and neural plate are distinguishable. **k** Early neural plate stage. **l** Early neural plate bisection; the yolk plug is completely internalized at this stage. The stage numbers (upper right corners) have been assigned according to Gosner [1960]. a, archenteron; b, blastocoel; bp, blastopore; dl, dorsal blastopore lip; ng, neural groove; np, neural plate; vl, ventral blastopore lip; yp, yolk plug. Scale bars, 500 μm (**a–i**, **k**) and 200 μm (**j**, **l**). For further explanations, see chapter 1.4.4.2.

ment shows a perinuclear localization (Fig. 11a–c). Finally, the blastocoel formed during these stages appears to be relatively smaller when compared to other species (Fig. 11d) [del Pino et al., 2007]. One possible reason for the small blastocoel could be the amount of yolk of the embryos, as seen in the spotted reed frog *Hyperolius puncticulatus*, which has a small egg diameter (1.5–1.8 mm), but large amounts of yolk [Chipman et al., 1999].

Gastrulation begins 24 hours after fertilization with the formation of the blastopore dorsal lip close to the equatorial region of the embryo (Fig. 12a). In the early gastrula stage, *A. spurrelli* embryos maintain a small blastocoel with a thick roof (Fig. 12b). In most anurans studied, the blastocoel increases its volume in the animal region, and the embryo volume also expands uniformly [Keller, 1986; Keller and Shook, 2004; del Pino et al., 2007]. Conversely, this structure is transient and disappears in the early gastrula stages of *A. spurrelli* (Fig. 12b, d, f). In *G. riobambae*, *E. portoricensis*, and *E. coqui*, the blastocoel roof will reduce its thickness down to a 1-cell layer, possibly due to the movements from the outer ectodermal prospective layer [Gitlin, 1944; Elinson and del Pino, 1985, 2012; Collazo and Keller, 2010]. Through gastrulation, the blastopore lip extends to the ventral region, and the pigmented external cell layer engulfs the vegetal cells that form the yolk plug (Fig. 12c–h). In the late gastrula stage, the archenteron reaches its maximum extension without expansion, which will only occur by the end of gastrulation (Fig. 12f). In general, the amphibian archenteron forms because of multiple cellular movements that extend, internally and anteriorly, the initial invagination of the dorsal lip [Keller and Shook, 2004]. The expansion of the archenteron is the result of the extension movements plus the active transport of fluid from the blastocoel to the archenteron, and the timing of this event has been shown to be variable [Ewald et al., 2004]. The delay in expansion of the archenteron might be the result of the small blastocoel in *A. spurrelli* which completely disappears by the mid-gastrula. Nevertheless, in the late gastrula stage of *A. spurrelli*, the yolk plug is reduced, and the archenteron is fully expanded (Fig. 12g–j). Variability in the timing of expansion has been documented in *Xenopus laevis* [Ewald et al., 2004]. The process of gastrulation, from the formation of the blastopore lip to the closure of the blastopore, lasts approximately 19 hours. Other species with aquatic embryos, like *X. laevis*, develop faster, and multiple developmental events occur simultaneously [Moya et al., 2007; Elinson and del Pino, 2012]. Neurulation in *A. spurrelli* starts during the late gastrula stages, and the thickening of the neural plate is visible from this stage onwards (Fig. 12i–l). Species like dendrobatids and the hemiphractid *G. riobambae*, which do not deposit eggs in water bodies, have longer developmental periods, and gastrulation can last from 4 days up to 15 days. In these species, archenteron expansion and mesoderm movements are delayed and decoupled [del Pino et al., 2007; Moya et al., 2007].

The early neurula stages of *A. spurrelli*, similar to other anurans, can be detected by the thickening of the most dorsal ectoderm that forms the neural plate, and the formation of the neural groove in the dorsal mid line (Fig. 12i–l, 13a). The most anterior part of the neural plate starts to differentiate in the anterior brain and the branchial arches simultaneously (Fig. 13a). During neurula stages, mesoderm internalization also occurs. Internally, the notochord is exposed to the archenteron cavity, called gastrocoel, while the endoderm is visible as lateral crests (Fig. 13b). The exposed mesoderm is a conserved structure in all amphibians analyzed which acts as a signaling tissue called gastrocoel roof plate. In *X. laevis*, it acts in the determination of left-right asymmetries [Shook et al., 2002, 2004; Blum et al., 2009; Saenz-Ponce et al., 2012]. During mid-neurula stages, the neural folds begin to elevate forming a more pronounced groove (Fig. 13c). The notochord is completely covered by endoderm at this stage (Fig. 13d). When the gastrocoel is completely lined with endoderm, it receives the name archenteron. Neural tube closure occurs by the end of the third day post fertilization (Fig. 13e, f).

It is worth to mention that *A. spurrelli* embryo elongation does not occur during the neurula stages, and the rounded shape is maintained until later stages. Indeed, tail bud and head differentiation, together with embryo elongation, occur independently from the large ventral yolk mass (Fig. 14). The embryo shape at these stages is similar to other phyllomedusids and drastically different from the hydrodynamic shape of Arboranae aquatic embryos like *Dryophytes japonicus* and *Hyla arborea* [Iwasawa and Futagami, 1992; Sayim and Kaya, 2008]. Although most detectable landmarks of the stages proposed by Gosner [1960] could be identified in these embryos, several structures like the eyes, dorsal fin, embryonic gills, and yolk mass follow independent and distinctive developmental programs; hence, more detailed staging methods might be required to study the particularities of each structure [Warkentin, 1999]. The largest and more distinctive structures in the prehatching embryo are the external embryonic gills. These structures develop from a bud in the conspicuous branchial arches (Fig. 14a–c). In anurans, generally, the 2 independent external gills originate in the third and fourth branchial arches. The secondary filaments reach their largest extension at stage 23, with 12 filaments in the anterior gill and 6 filaments in the posterior gill (Fig. 14c–g). The development of large external gills seems to ensure the capture of oxygen that is not homogeneously distributed in the aquatic environment within the egg layers of arboreal embryos [Warkentin, 2002; Warkentin et al., 2005]. Similar observations have been made for embryos of the gladiator frog genus *Boana* [Martins, 1993]. In *Agalychnis callidryas*, in addition to gas exchange, the external gills seem to be sensorial structures exclusively adapted to the egg environment. If hatching occurs before normal gill regression, gills will rapidly regress once embryos touch the water [Warkentin, 2002]. Without predation stress, *A. spurrelli* embryos hatch after the external gills retract within the newly formed opercular aperture (Fig. 14h).

X. laevis is the current model organism for the study of vertebrate embryo development, and there are multiple histological, molecular, and genetic tools available [Harland and Grainger, 2011]. This species is an aquatic breeder that lays relatively small eggs on the water surface. Importantly, species that differ in their reproductive modes show important changes in developmental processes, compared to *X. laevis*. Dendrobatid frogs and *G. riobambae*, for example, produce large terrestrial eggs that develop slower than those of *X. laevis*. These terrestrial breeders seem to have distinct cellular and molecular mechanisms to establish the germ layers and are able to characteristically dissociate morphogenic events during gastrulation [del Pino et al., 2004, 2007; Elinson and del Pino, 2012]. The underlying genetic and molecular mechanisms of these developmental changes are still elusive and represent an exciting endeavor to understand how behavioral and physiological adaptations influence embryo development.

1.5 Population Declines and Conservation of Arboranae

The recognition of amphibian populations in decline as a global phenomenon was first enunciated in 1989 at the First Congress of Herpetology in Canterbury, England [Blaustein and Wake, 1990]. It was established beyond rea-

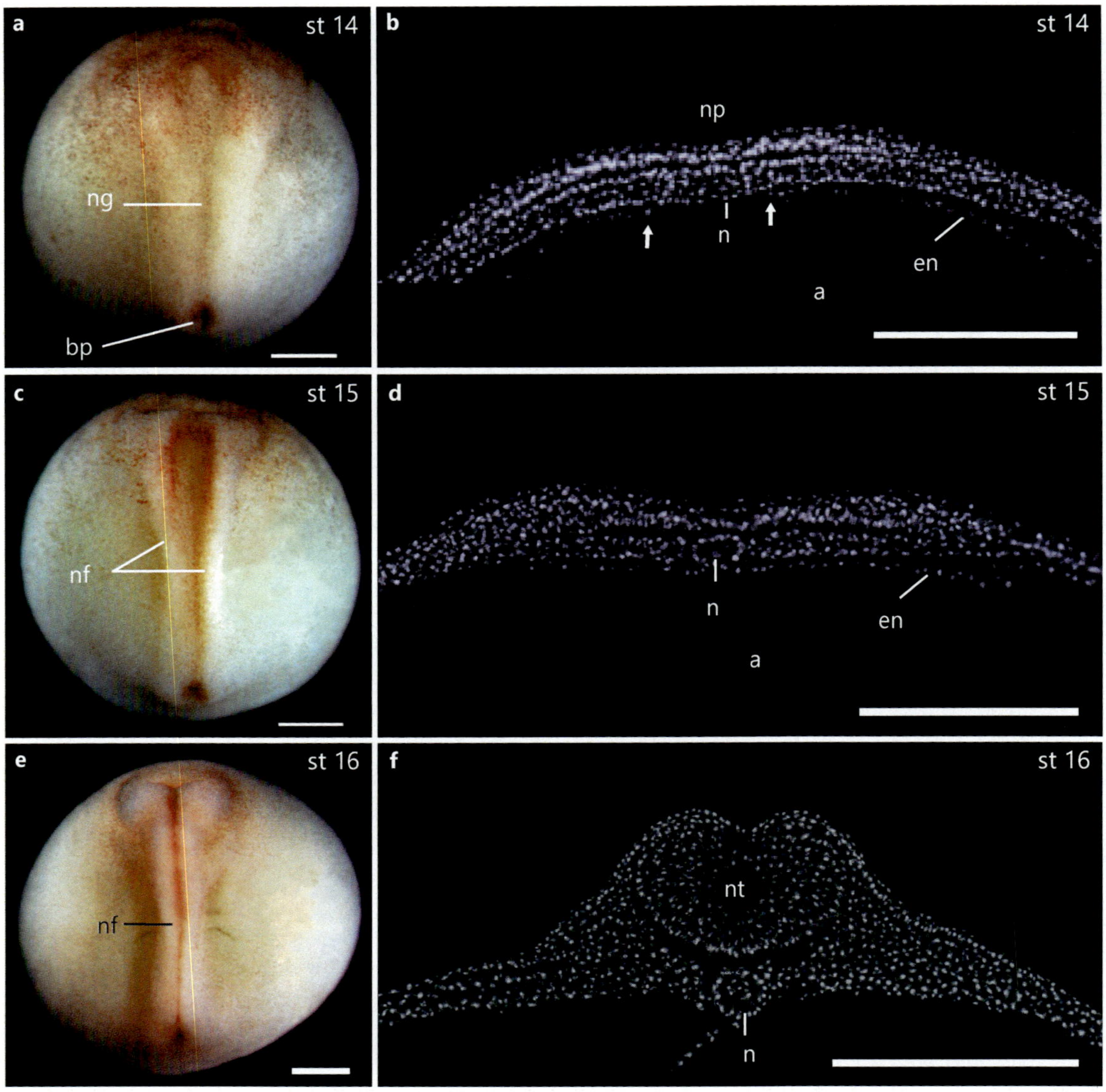

Fig. 13. Neurulation stages in *Agalychnis spurrelli*. **a**, **c**, **e** External dorsal views. **b**, **d**, **f** Cross-sections stained for DNA using Hoechst 33258. **a** Neural folds stage. Formation of the neural plate and the neural groove. **b** Rostral region of an early neurula embryo. The arrows point to the border of the lateral endodermal crests. The notochord is exposed to the gastrocoel. **c** Mid-neurula stage. **d** Rostral region of a mid neurula. The endoderm covers the archenteron roof. **e** Late neurula stage. **f** Medial region of a late neurula embryo. The neural folds are completely closed forming the neural tube. The stage numbers (upper right corners) have been assigned according to Gosner [1960]. a, archenteron; bp, blastopore; en, endoderm; n, notochord; nf, neural folds; ng, neural groove; np, neural plate; nt, neural tube. Scale bars, 500 μm (**a**, **c**, **e**) and 200 μm (**b**, **d**, **f**). For further explanations, see chapter 1.4.4.2.

sonable doubt that amphibians belong to the most threatened vertebrates. According to the Tree of Life database [http://tolweb.org/Salientia/14938], there are 172 extant species of Gymnophiona, 557 species of Urodela, and 5,424 species of Anura. One-third of these amphibians are considered to be on decline with up to 5 species becoming extinct each year [Stuart et al., 2004]; and there is proof that amphibian declines are occurring at higher pace than would be expected by chance [Houlahan et al., 2000]. Extensive amphibian monitoring and inventory efforts documented population declines in South- and Central America [e.g., Heyer et al., 1988; Pounds et al., 1997; Lips, 1998, 1999; Bonaccorso et al., 2003; Ron et al., 2003; Lips et al., 2006]. However, compared to Australia, Europe, and the USA, relatively little efforts have been made to consult herpetologists in Latin America with respect to detailed knowledge concerning the declines of amphibian populations [Wake, 1998; Young et al., 2001].

Die-offs of neotropical amphibians have been reported to occur in Honduras, Costa Rica, Panama, Venezuela, Ecuador, and Brazil [Heyer et al., 1988; Weygoldt, 1989; La Marca and Reinthaler, 1991; Pounds and Crump, 1994; Coloma, 1995; Pounds et al., 1997, 1999; La Marca and Lötters, 1997; Lips, 1998, 1999; Wilson and McCranie, 1998; La

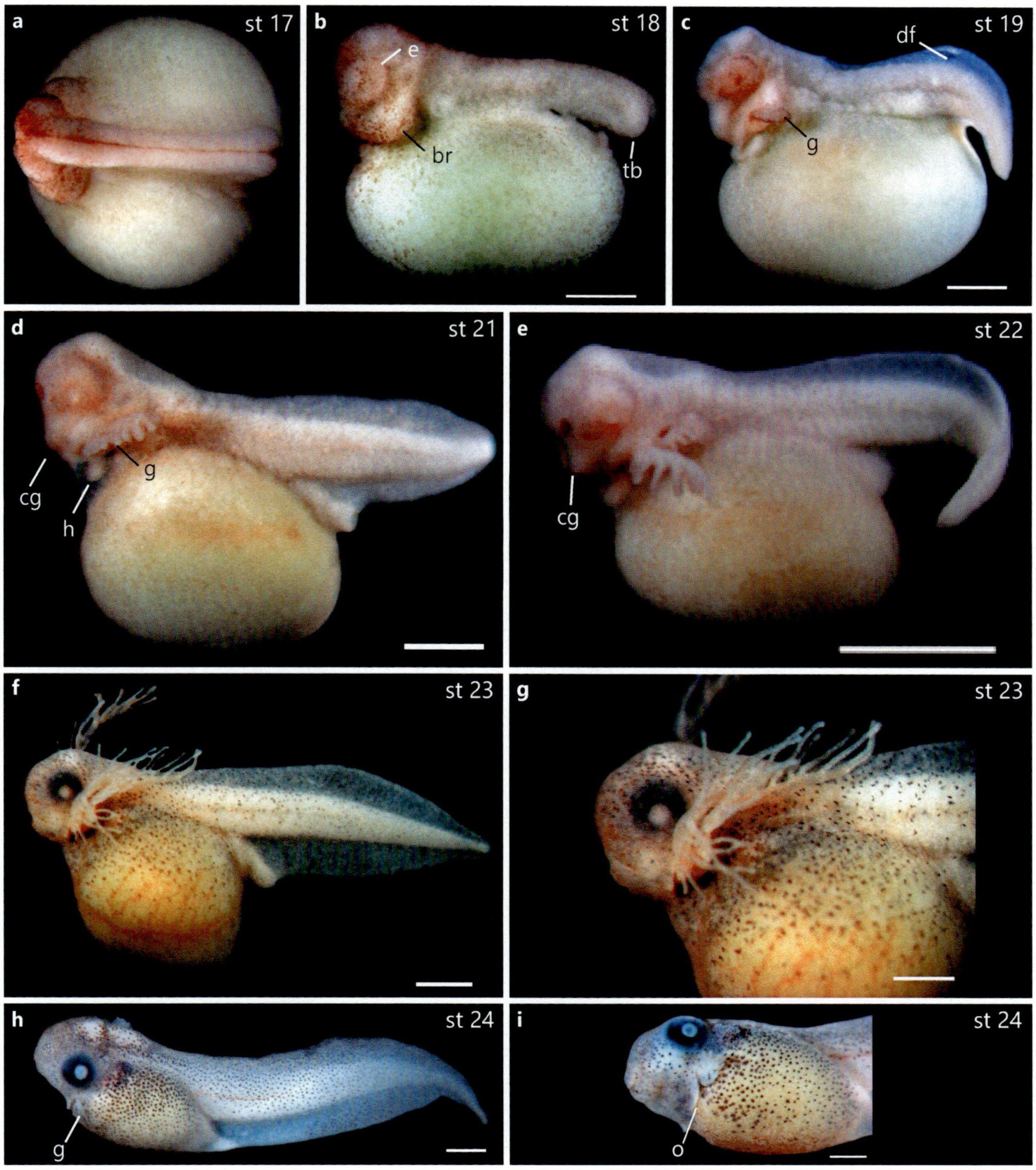

Fig. 14. Gill development and hatching in *Agalychnis spurrelli*. **a** Dorsal view of a tail bud stage embryo. **b–i** Lateral views. **b** Eye differentiation at the stage of muscular response. **c** Gill bud stage. The first bud develops from one of the branchial arches. The dorsal fin is translucent as soon as it differentiates from the elongating tail. **d** Transparent cornea stage. The pericardial ectoderm is translucent at this stage too. The gills have now 10 secondary filaments. The cement glands and mouth start to form at this stage. **e** Tail fin circulation stage. All gill secondary filaments are present and will only elongate in the following stages. **f** Maximum gill extension. Neural crest and retina pigmentation appear during this stage. **g** Close-up image of the embryo in **f**. **h** Tadpole at hatching. **i** At hatching, the gills have been reabsorbed, and only a little part of the gills emerges from the operculum. The stage numbers (upper right corners) have been assigned according to Gosner [1960]. br, branchial arch; cg, cement glands; df, dorsal fin; e, eye; g, gills; h, heart; o, operculum; tb, tail bud. Scale bars, 500 μm (**a**, **i**) and 1 mm (**b–h**). For further explanations, see 1.4.4.2.

Marca et al., 2005]. Further reports of amphibian population declines in the Neotropics have been compiled by Young et al. [2001]. These authors listed 13 countries in Latin America where notable declines of anuran populations were registered over 20 years, which include species representing 30 genera and 9 families of amphibians where declines or even extinctions have occurred. From the data available at that time, they note that most amphibian declines were reported in remote highlands (above 500 m altitude in Central America and above 1,000 m in the Andes), and that stream-associated species were more affected than terrestrial species. In contrast, lowland regions with declines of amphibian populations were most likely to be affected by human activities.

Young et al. [2001] listed 10 probable or confirmed factors which cause the worldwide declines of amphibian populations: climate change, habitat modification, habitat fragmentation, introduced species, solar UV-B radiation, chemical contaminants, acid precipitation, contaminated soil, disease, and the pet trade. Two or more of these factors can act synergistically and cause mortality or sublethal effects [Kiesecker et al., 2001; Blaustein and Kiesecker, 2002]. There are only a few instances where specific causes of amphibian declines have been identified with certainty. The human-spread fungal disease chytridiomycosis, caused by *Batrachochytrium dendrobatidis* [Berger et al., 1998; Longcore et al., 1999; Piotrowski et al., 2004], has received great attention as possibly being the primary agent of amphibian decline [Blaustein et al., 1994a; Berger et al., 1998, 2000; Kriger and Hero, 2007a, b]. One persuasive case for this is the mortality of amphibians associated with the appearance of the fungus at specific localities [Beebee, 1995; Bonaccorso et al., 2003; Lips et al., 2003, 2006; Ryan et al., 2008]. Other direct and indirect evidence has implicated this emerging disease as a major factor [Collins and Crump, 2009], and climate change is exacerbating the negative effect of the chytrid fungus on neotropical amphibians [Pounds et al., 2006]. Nonetheless, some amphibian populations living with the chytrid fungus appear to be unaffected [Daszak et al., 2005]. Further, some frog populations have declined at the same time and to the same degree as co-occurring lizard populations suggesting that the primary cause is not the fungus, but perhaps global warming [Whitfield et al., 2007]. The occurrence of chytridiomycosis in South America was first reported by Ron and Merino [2000].

Climate change is considered to be, at least partially, responsible for declines of amphibian populations [Pounds et al., 1999; Alexander and Eischeid, 2001; Pounds, 2001; Reading, 2007]. Several studies found a relationship between increased solar UV-B irradiation and amphibian declines [Blaustein et al., 1994b; Middleton et al., 2001; Pahkala et al., 2002]. Furthermore, many habitats that appear to be pristine may be overrun with introduced predators which could also be involved in amphibian declines [Hedges, 1993].

In the second half of the last century, neotropical amphibians have been collected from their natural habitats in large quantities and traded internationally for pet markets. Of particular concern are brightly colored species, like dendrobatid frogs, that are highly sought after by the pet trade. Over-collecting also occurred for neotropical and some Australian species of Arboranae which are highly prized as pets by hobby herpetologists in Europe and North America. In certain years of the 1980/1990 decades, the European market was saturated with imports of exotic frog species. Fortunately, in the last 25 years such mass exports were counterbalanced by international efforts against animal trade. Blaustein and Kiesecker [2002] argue that it is unlikely that amphibian declines are monocausal. Although there is undoubtedly no single primary cause for amphibian declines, the ultimate cause is human activity and habitat loss of many amphibian species [Laurance, 1998, 2001; Peres, 2001; Brooks et al., 2002; Dodd and Smith, 2003]. The small distributions of many amphibians make them especially vulnerable for extinction [Sodhi et al., 2008], and the destruction of their natural habitats is one of the most significant threats to biodiversity [Vitousek et al., 1997; Mittermeier et al., 1998; Pimm and Raven, 2000; Young et al., 2001; Brooks et al., 2002]. In South America, habitat loss is primarily caused by deforestation, followed by intensive cattle grazing and/or agriculture. Nearly 6 million hectares of the world tropical forests are destroyed each year by logging [Whitmore, 1997]. Many arboranans now appear to be missing from areas of their formerly known ranges, and these populations, which had low densities historically, may have significantly declined in more recent time. Unfortunately, time-based population density estimates have not been performed for any of these species [Young et al., 2001].

In an effort to develop guidelines for amphibian protection, a series of 3 regional workshops were held in November 1999 in Chetumal (Mexico), Panama City (Panama), and Quito (Ecuador). These workshops were supplemented with published data and information solicited from herpetologists who were unable to attend these workshops [Young et al., 2001]. As a result of their presentations and in-depth discussions, the workshop participants elaborated an exemplary memorandum of conclusions, as well as priority recommendations for research and training in Latin America for amphibian conservation. Reading the conclusions and recommendations [Young et al., 2001] is highly recommended.

1.6 Early Cytogenetic Studies on Arboranae (1930–2004)

Arboranae belong to the anurans whose chromosomes were frequently examined in the past (for reviews, see Morescalchi [1973]; King [1990]; Green and Sessions [1991a, 2007] and Appendix 1). This is not surprising because arboranan frogs occur world-wide (Fig. 3, 4), and many species are, unlike other anurans, not confined to remote and sometimes barely accessible regions. Some arboranan species lived in large populations and in regions where an easy access was possible even before the middle of the last century. Furthermore, many Arboranae can be kept in captivity where they survive and breed without difficulties. One reason for the many cytogenetic publications on the families Hylidae and Phyllomedusidae is that a plethora of species occur in the neotropical forests of Brazil and Argentina (Fig. 3, 4), i.e., in countries where, historically, many cytogenetic groups have been active during the last 60 years (Appendix 1). The nearctic hylid genera were traditionally examined by cytogeneticists in the USA and Canada. The karyotypes of the few hylids inhabiting the palearctic realm were repeatedly and thoroughly analyzed by pioneering European and Japanese cytogeneticists. Finally, in Australasia, experts conducted detailed cytogenetic investigations in species of the family Pelodryadidae (Appendix 1).

Fortunately, despite the immense array of extant arboranan species and in contrast to other amphibian taxa, where incorrect species identifications frequently occurred in cytogenetic publications, relatively few misidentifications were made in the early cytogenetic studies on Arboranae. This was due to the fact that the specimens used for those early cytogenetic analyses were identified by professional herpetologists who were recognized experts in their respective countries or by international authorities. Opportunely, the same was the case during the various cytogenetic projects leading to the present monograph. However, what seriously hampered cytogenetic (and other) investigations in Arboranae were the frequent historical changes of species names that occasionally led to frustrating entanglements (chapter 3.1, Appendix 1).

In the last decades, our knowledge of arboranan systematics has significantly improved. Species identifications are not solely dependent on traditional morphological, anatomical, acoustical, biogeographical, etc. characters, but are incremented by present-day molecular techniques. New species of Arboranae are constantly being discovered and described, mainly in the neotropical areas. It is likely that future chromosome studies on arboranans will have the enormous advantage in relying on specimens which have been preidentified by species-specific DNA sequences. Molecular methods have become an important tool for the identification of new species [Blaxter, 2003; Marshall, 2005; Duryea et al., 2009]. Recently performed studies using molecular markers have significantly increased the number of known species, including amphibians [e.g., Fouquet et al., 2007], and have contributed to assess genetic diversity, population genetic structure, and phylogeography [e.g., Velo-Antón et al., 2007]. Fine-scale genetic studies on arboranan species can be accompanied by analyses of DNA microsatellites. These are short sequences of noncoding DNA characterized by tandem repeats of several nucleotides (e.g., GATA, GACA). DNA microsatellites show rapid rates of mutations. They have proven to be appropriate markers for population genetic studies because they are co-dominant, often highly polymorphic, comparable among studies, and can reliably estimate gene flow [Sunnucks, 2000].

The very first cytogenetic analyses on arboranan species date back to Iriki [1930, 1932a, b] who examined *Dryophytes japonicus* (reported as *Hyla arborea japonica*), Galgano [1933] who analyzed *Hyla arborea*, and Bushnell et al. [1939] who studied the chromosomes of *Dryophytes chrysoscelis* and *Acris crepitans*. These early cytogenetic studies were performed using the classical squash technique of tissues, mainly testes. Following these pioneering studies, numerous chromosome analyses on multiple arboranan species using more advanced techniques were published (Appendix 1).

The first chromosome banding analyses in hylids were carried out by Schmid [1978a], Wiley [1982], and Anderson [1986]. In further investigations, starting in the 1990ies, complete arrays of banding techniques were applied to mitotic and meiotic chromosomes of arboranans (Appendix 1). BrdU replication banding analyses were initiated with the chromosomes of *D. chrysoscelis* and *D. versicolor* [Wiley et al., 1989] and later continued with *D. japonicus* [Miura, 1995a] as well as several neotropical arboranan species [e.g., Kasahara et al., 2003; Gruber et al., 2007; Catroli, 2008]. The first restriction endonuclease bandings (digests) in Arboranae chromosomes were done in various neartic and neotropical species by Anderson [1986].

The original in situ hybridizations on arboranan chromosomes used a radioactively labeled 18S + 28S ribosomal DNA (rDNA) probe in several *Litoria* and *Ranoidea* species [King et al., 1990]. Later, the nonradioactive fluorescence in situ hybridization (FISH) technique with biotinylated 18S + 28S rDNA probes was first applied to the chromosomes of many individuals of *D. chrysoscelis* and *D. versicolor* collected in numerous populations [Wiley et al., 1989]. This was followed by 18S + 28S rDNA FISH experiments by Brazilian cytogeneticists to the metaphases of many neotropical hylids [e.g., Medeiros et al., 2003, 2006; Catroli, 2008; Nunes and Fagundes, 2008a; Carvalho et al., 2009a, b] and phyllomedusids [e.g., Barth et al., 2013; Bruschi et al., 2013; Gruber et al., 2013]. FISH with biotinylated telomeric DNA probes on arboranan chromosomes was started by Anderson [1996] and continued again by South American cytogeneticists [e.g., Carvalho et al., 2009b; Ferro et al., 2012; Gruber et al., 2012a, b; Suárez et al., 2013].

More recently, FISH with DNA microsatellites [Peixoto et al., 2015] and specific repetitive DNA sequences [Gatto et al., 2016], chromosome microdissection and subsequent chromosome painting [Gruber et al., 2014], comparative genomic hybridization [Gatto et al., 2016], and immunofluorescence of hypermethylated repetitive DNA sequences [Schmid and Steinlein, 2016b; present study] were commenced in Arboranae.

The present monograph aims to show that, although most Arboranae exhibit evolutionarily conserved karyotypes, they offer solid data for amphibian cytogenetics. Unfortunately, because many anuran species are currently in strong decline (chapter 1.5), all efforts must be made to investigate the many species whose karyotypes remain unknown. New arboranan species, especially those in the family Hylidae are being discovered in the Neotropics. The sampling of these species, some of them living in distant and isolated habitats, and the preparation of high-quality chromosomes under primitive field conditions or in well-equipped conventional cytogenetic laboratories, is still a challenging and rewarding research task.

Cytogenet Genome Res 2018;155:47–53
DOI: 10.1159/000489541

Accepted: February 1, 2018
by J. Smith

The Arboranan Frogs:
Materials and Methods

2

The course of discovery of arboranan frogs. The curve shows the cumulative number of valid species plotted against time. The rate of species discovery was approximately 1–5 species per year during the 19th century, but then rose sharply during the second half of the 20th century to 15–20 species per year. Species continue to be described at an increasing rate especially in the Neotropics.

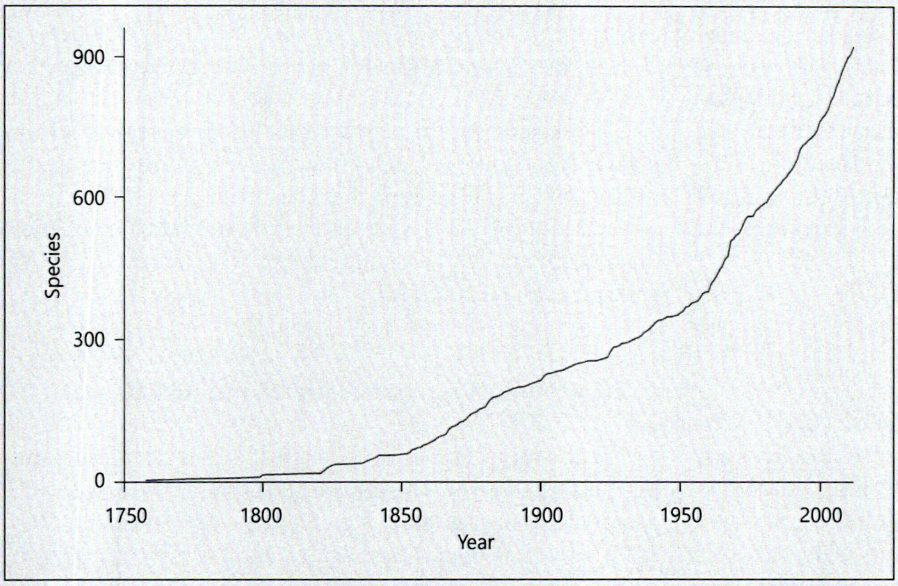

KARGER

E-Mail karger@karger.com
www.karger.com/cgr

2.1 Collecting Expeditions, Taxon Sampling, and Breeding Experiments

For the present monograph, a total of 204 arboranan species belonging to 47 genera were either freshly collected for the cytogenetic analyses or reevaluated from studies published earlier (Appendix 1). The chromosome data and figures of 177 previously published species were included once more in this monograph for necessary comparative purposes (Appendix 1). Furthermore, a vast and long-lasting attempt was made to assemble all original and review cytogenetic reports on the Arboranae published from 1930 to 2016. These reports are also compiled in Appendix 1. It shows that countless errors, inconsistencies, and inaccuracies regarding correct citations of species names, diploid chromosome numbers, and fundamental numbers occurred in the past. The Appendix 1 tries to disentangle and rectify these historical inaccuracies. The expeditions took place over a period of 37 years (1967–2003) during numerous journeys in America (Argentina, Brazil, Colombia, Costa Rica, Cuba, Jamaica, Mexico, Panama, Peru, Venezuela, USA), Australia, Europe (Germany, Serbia), and Japan. Some few specimens were obtained by specialized animal dealers in Germany.

The frogs were collected mostly at night from the ground, branches of bushes, bromeliads, and trees along trails, from stream banks and cliffs of waterfalls. Captured frogs were kept in humid cloth bags, air-filled plastic bags, or plastic boxes with wet leaves. They were processed for chromosomes either in the field, in permanent or temporary cytogenetic laboratories in South America, Australia, and Japan, or transported to our laboratories in Germany, Canada, and the USA. For the transfer to Germany, the fixed chromosome material obtained was stored at room temperature, 4°C or –20°C and transported either packed in dry ice or at room temperature.

2.2 Species Examined

The present monograph represents the most comprehensive comparative cytogenetic study on arboranan frogs (families Hylidae, Phyllomedusidae, and Pelodryadidae). It comprises a total of 204 taxa of Arboranae. The Appendix 1 is a compilation of all original studies and review articles containing cytogenetic data on Arboranae. The species examined and reexamined in the present study are included. The mitotic chromosomes of 32 species are presented for the first time; their male meiotic chromosomes (from the stage of diakinesis) were possibly known from very early examinations. Furthermore, for completeness and for comparative purposes, all cytogenetic data on Arboranae that we previously published have been included. This latter material was reevaluated in the following manner: (1) considerably more conventionally stained karyotypes were examined, (2) whenever possible chromosome banding and in situ hybridizations were performed, and (3) male meiosis was examined. Details of the chromosome analyses of the 204 cytogenetically analyzed arboranan species are provided in chapters 3.1.1–3.4.10.

2.3 Collection Localities

As far as possible, the exact sampling sites are provided in Appendix 1. During the more recent expeditions, the exact geographic coordinates were determined directly in the field using Global Positioning Systems (GPS). For the earlier expeditions, some geographic coordinates were calculated from different geographical maps. Unfortunately, the exact collection sites for some of the specimens provided by colleagues or purchased from specialized animal dealers were not available. However, most of these latter species are considered to be common in the respective countries (Appendix 1).

2.4 Gonadal Sex Identification

Whenever possible, the cytogenetic analyses were preceded by microscopic determination of the gonadal sex of the frogs using the technique described in detail by Schmid et al. [2010, 2012]. Regrettably, there are no available data on the sex for numerous of the specimens examined (Appendix 1).

2.5 Mitotic and Meiotic Chromosome Preparations

The various techniques used for the preparation of mitotic chromosomes from somatic tissues (bone marrow, intestine, cornea of the eyes) and cell cultures (blood lymphocytes, kidney fibroblasts), or for the preparation of male meiotic chromosomes from testes, were summarized by Schmid et al. [2010, 2012].

2.6 Chromosomal Replication Banding Patterns

For the induction of BrdU or BrdU/dT replication banding patterns in the chromosomes of cell cultures several protocols were available [e.g., Schmid and Steinlein, 1991, 2015; Miura, 1995a; Kasahara et al., 1998; Schmid et al., 2003a, 2004, 2012; Gruber et al., 2012a, 2013, 2017].

2.7 Conventional Staining and Regional Banding of Chromosomes

Ordinary Giemsa staining, C-banding of constitutive heterochromatin, silver (Ag)-staining of nucleolus organizer regions (NORs), direct fluorescence staining with the fluorochromes quinacrine mustard, DAPI, Hoechst 33258, or mithramycin, and fluorescence counterstaining with distamycin A/mithramycin or distamycin A/DAPI were all performed according to methods described by Schmid et al. [2010].

2.8 Fluorescence in situ Hybridization of Telomeric DNA Repeats

The oligonucleotide probe $(GGGTTA)_7$ and its complementary sequence $(TAACCC)_7$, synthesized using an automated DNA synthesizer, were used for the demonstration of the canonical telomeric sequences in the chromosomes. The FISH technique described by Schmid et al. [2010, 2012] was applied. As a very easy and efficient option for detecting telomeric repeats, a fluorescein-conjugated peptide nucleic acid (PNA) probe (Telomere PNA FISH kit/FITC; Dako Cytomation, Denmark: K5325) was used. The detailed procedure of in situ hybridization and signal detection is provided by the manufacturer. Alternative protocols were followed for several Brazilian species [Gruber et al., 2017].

2.9 Fluorescence in situ Hybridization of 18S + 28S rDNA

The hybrid plasmid pXlr 101A, which contains inserted ribosomal RNA gene sequences from *Xenopus laevis*, was used as a source of ribosomal DNA (rDNA) [Morgan et al., 1980]. Processing of this probe and FISH were performed according to Schmid et al. [2010, 2012]. Alternative protocols were followed for several Brazilian species [Gruber et al., 2017].

2.10 Detection of Hypermethylated Chromosome Regions

Repetitive hypermethylated DNA sequences in the chromosomes were detected by indirect immunofluorescence using a monoclonal antibody against 5-methylcytosine (Imprint® monoclonal anti-5-methylcytosine [33D3] antibody, Sigma-Aldrich). The technique of indirect immunofluorescence elaborated by Schmid and Steinlein [2016b] was employed.

2.11 Measurements of Chromosome Sizes

Length measurements of chromosomes were only performed in few instances. The cells were in mid-metaphase stage. The chromosome lengths were determined by comparing the chromosome photographs to a 10-μm objective micrometer photographed at the same magnification. However, because of the differing stages of chromosome contraction, such measurements yield data that are not helpful for general interspecies comparisons.

2.12 Interphase Cell Nuclei Preparations

In addition to examining interphase nuclei from bone marrow, lymphocytes, fibroblasts, and gonads, cell nuclei suspensions were prepared from intestine, liver, spleen, lung, kidney, heart, and muscle. Cell nuclei suspensions were obtained according to Schmid et al. [2010, 2012].

2.13 Microscopic Analysis and Microphotography

Microscopic analyses were conducted either on Zeiss photomicroscopes III equipped with an ordinary bright-field system or phase-contrast optics, or Zeiss fluorescence microscopes and Zeiss Axiophot microscopes equipped with incident HBO 50W mercury lamp illumination. The various filter combinations necessary for the analyses of metaphases stained with the different fluorochromes or for FISH were as described by Schmid et al. [2010, 2012]. Conventional photographs were taken on Ortho 25 films (Rollei, Germany) or films with similar properties. Digital images of the metaphase plates were acquired using a bright-field microscope coupled with a CCD camera and appropriate digital imaging software (Applied Spectral Imaging).

2.14 Karyotype Recording and Chromosome Analysis

Metaphase plates were photographed for each of the frog specimens analyzed. For detailed analysis, the chromosomes were cut out of the photographic prints and affixed to double-sided tape in parallel rows. Chromosomes from bright-field microscopy were affixed to a background of white cardboard and chromosomes from fluorescence microscopy to black cardboard. This karyotype record system permits (1) the determination of the conformity of chromosome morphologies, banding patterns, and in situ hybridization signals, (2) facilitates the identification of sex chromosomes, (3) allows the recognition of inter- and intraindividual variations of specific chromosome bands or labeling signals, and (4) reveals the presence of spontaneous chromosome aberrations.

The number of staining techniques applied on the chromosomes of the 204 species of Arboranae examined was limited by the number of sampled specimens per species and the amount of cytological material available. With some exceptions, the karyotypes of all species were characterized by ordinary Giemsa staining (Appendix 1). In addition, all efforts were made to obtain at least the C-band patterns of the constitutive heterochromatin and Ag-staining of the NORs in the chromosomes. Provided that sufficient cytological material was available, staining with the various fluorochromes and in situ hybridizations with the telomeric DNA probe and the 18S + 28S rDNA probe were

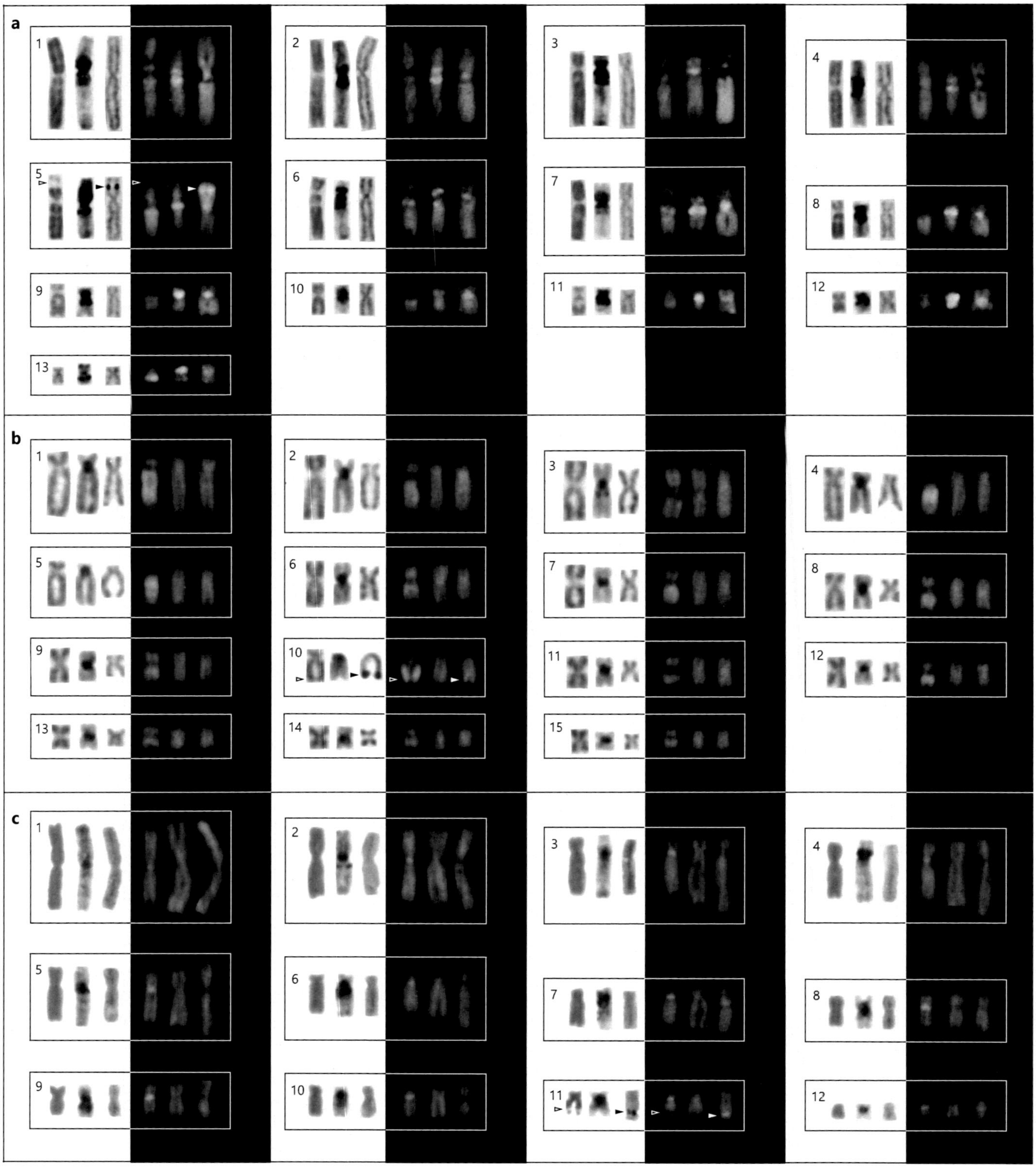

Fig. 15. Karyotypes of *Agalychnis callidryas* (**a**), *Dendropsophus ebraccatus* (**b**), and *Boana faber* (**c**) displaying the different staining techniques applied in the present study. The framed chromosomes show conventional Giemsa staining, C-banding of constitutive heterochromatin, Ag-staining of NORs, quinacrine staining, distamycin A/DAPI counterstaining, and distamycin A/mithramycin counterstaining (from left to right). Note the complementarity of the fluorochrome stainings. Open arrowheads indicate the nucleolar constrictions, filled arrowheads indicate the NORs. For further explanations, see chapter 2.14.

DOI: 10.1159/000489541

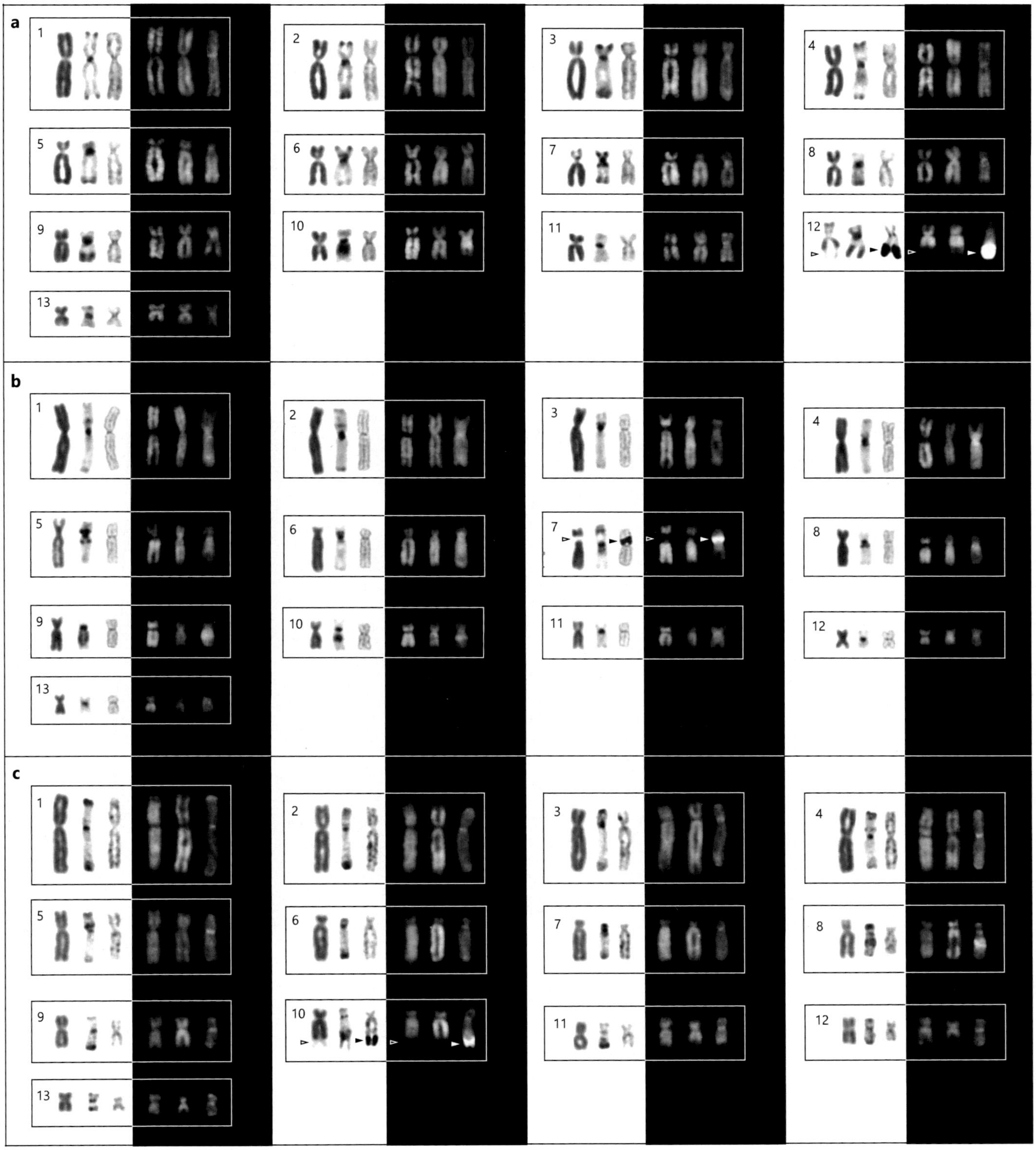

Fig. 16. Karyotypes of *Ranoidea cyclorhynchus* (**a**), *Ranoidea genimaculata* (**b**), and *Litoria rothii* (**c**) displaying the different staining techniques applied in the present study. The framed chromosomes show conventional Giemsa staining, C-banding of constitutive heterochromatin, Ag-staining of NORs, quinacrine staining, distamycin A/DAPI counterstaining, and distamycin A/mithramycin counterstaining (from left to right). Note the complementarity of the fluorochrome stainings. Open arrowheads indicate the nucleolar constrictions, filled arrowheads indicate the NORs. For further explanations, see chapter 2.14.

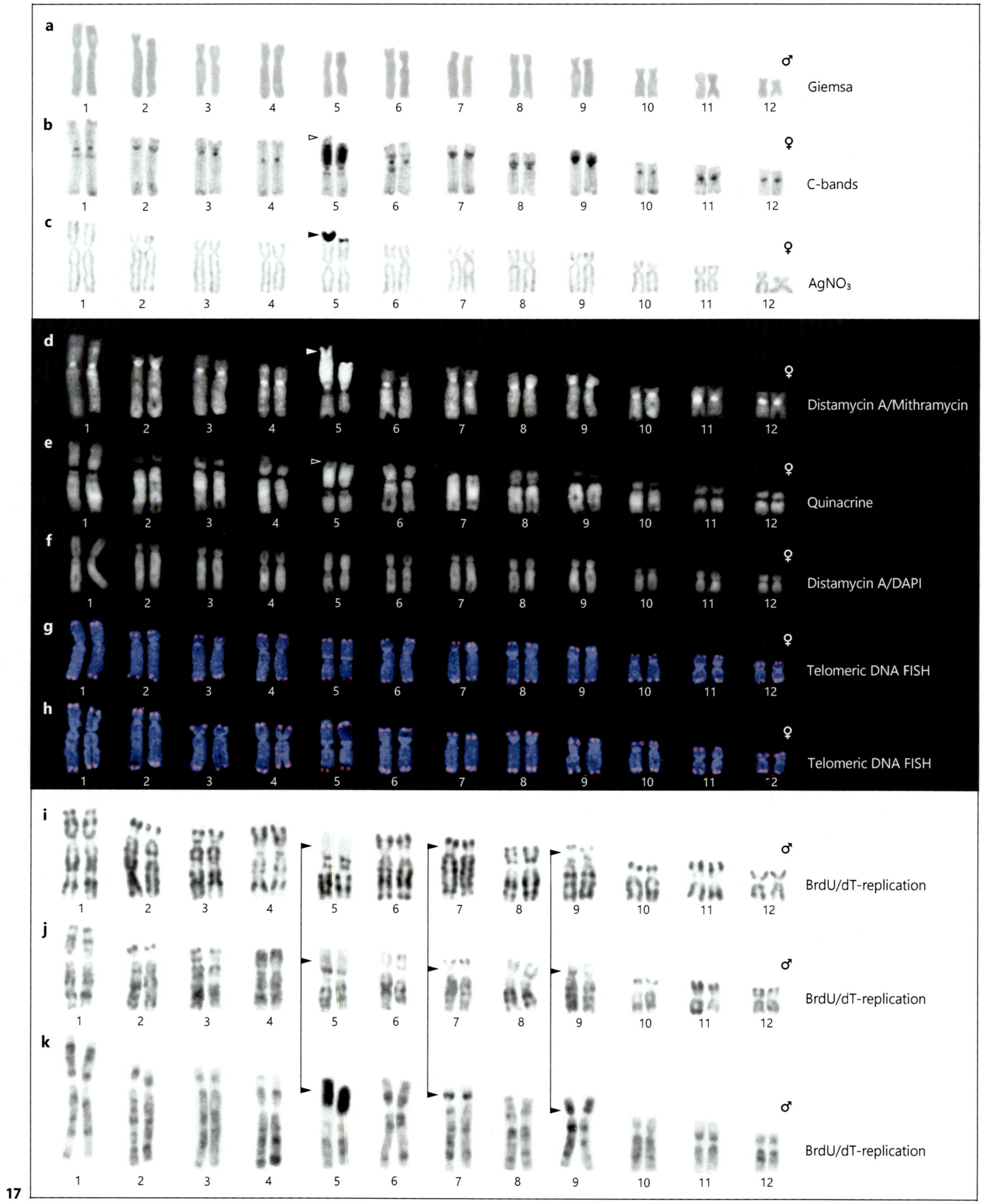

a
Giemsa
b
C-bands
c
AgNO3
d
Distamycin A/Mithramycin
e
Quinacrine
f
Distamycin A/DAPI
g
Telomeric DNA FISH
h
Telomeric DNA FISH
i
BrdU/dT-replication
j
BrdU/dT-replication
k
BrdU/dT-replication
17

also performed. In optimal cases, 6 different staining techniques and in situ hybridizations with the 2 probes could be applied to the chromosomes of a species (Appendix 1). This is exemplified for *Agalychnis callidryas*, *Ranoidea cyclorhynchus*, and *Nyctimystes infrafrenatus* in Figures 15–17. When less material was available, the in situ hybridization experiments had to be omitted. In several species, when living cell cultures (lymphocytes, kidney fibroblasts) were available and could be maintained in sterile conditions, it was possible to carry out BrdU replication banding analyses.

The chromosomes in the stages of male meiosis I (diakinesis) and meiosis II were recorded using the same system as for the mitotic chromosomes (see above). Male meioses of 73 arboranan species were examined. The chromosomes were mostly stained with Giemsa or C-banded; in some instances, they were also submitted to fluorochrome staining. When necessary, in situ hybridization with the telomeric DNA probe as described in chapter 2.8 was performed.

Fig. 17. Karyotypes of *Nyctimystes infrafrenatus* displaying the different staining techniques (**a–f**), in situ hybridizations (**g**, **h**), and BrdU/dT replication bandings (**i–k**) used in the present study. Open arrowheads in **b** and **e** indicate the nucleolar constriction, filled arrowheads in **c** and **d** indicate the NOR. **i–k** Time sequence of BrdU/dT replication banding patterns in each of the 12 chromosome pairs from mid S-phase (**i**) to late S-phase (**k**). The connected filled arrowheads mark the prominent late replicating regions of constitutive heterochromatin in chromosomes 5, 7, and 9. The order of the chromosomes in the karyotypes differs slightly from that previously published [Schmid et al., 2003a]. For further explanations, see chapter 2.14.

Cytogenet Genome Res 2018;155:55–221
DOI: 10.1159/000489839

Accepted: February 1, 2018
by J. Smith

The Arboranan Frogs:
Results and Discussion

3

"The great geneticists of the recent past (e.g., R.A. Fisher, J.B.S. Haldane, H.J. Muller) hardly considered chromosomal rearrangements as an important or significant factor in evolution. For Fisher [1930] 'the evolutionary possibilities of these kinds of change are evidently extremely limited compared to those of the type of change to which the term gene-mutation is applied, ... only in very special cases could they contribute appreciably to the genetic diversity of an interbreeding population'. Sewall Wright [1941] calculated that selection against translocations must be so severe that they would hardly ever manage to establish themselves. H.J. Muller seems to have regarded inversion polymorphism as a kind of temporary imperfection of the evolutionary process, destined to be soon superseded by improvements based on point mutations. And even Haldane, who probably had a deeper knowledge of all aspects of the evolutionary process than any of his contemporaries, never seems to have quite appreciated the evolutionary role of chromosomal rearrangements, in all its complexity. With this neglect by geneticists, it is hardly surprising that chromosomal rearrangements have been left out of account by many non-genetical evolutionists."

Michael J.D. White
Animal Cytology and Evolution [1973a]

KARGER

E-Mail karger@karger.com
www.karger.com/cgr

3.1 General Properties of Arboranan Karyotypes and Chromosomes

For the understanding of the present monograph, a careful and recurrent consulting of the voluminous Appendix 1 is necessary and highly recommended. It represents a central element of this work with a compilation of most previous original studies and review articles containing the published cytogenetic data on Arboranae thus far. Furthermore, all species examined and reexamined in the present study are included. The relevant information concerning the current and former scientific names of the species and genera, the countries and localities where the frogs were collected, the number and sex of the individuals examined, the cytogenetic methods applied, and the diploid and fundamental chromosome numbers are listed. Finally, and most importantly, the fourth column in this Appendix 1 discloses the many historical changes of species names, erroneous annotations, and inconsistencies that have occurred in the literature and in many cases have perpetuated. This has interfered with an easy study of the karyophylogeny in these anurans. The Appendix 1 is an attempt to rectify these factual errors and imprecise data.

A total of 204 species of Arboranae have been cytogenetically examined for this monograph. The karyotypes of these species are displayed in Figures 53–937. In addition, the karyotype data of 177 arboranan species previously published in the literature, but not available for the present study, have been considered for this analysis as well (Appendix 1). The mitotic chromosomes of 32 species are presented for the first time; their male meiotic chromosomes (from the stage of diakinesis) were possibly known from very early examinations. Thus, this monograph examines a total of 381 karyotypes belonging to the anuran families Hylidae (266 species), Pelodryadidae (86 species), and Phyllomedusidae (29 species). Hence, this represents the broadest comparative cytogenetic study on these 3 taxa published up to date.

In general terms, it can be clearly stated that the karyotypes of the 3 Arboranae families resemble those of most other anuran families. The approximate lengths, the morphology, and the structure of arboranan mitotic chromosomes are not exceptional. Some interesting peculiarities noticed in arboranan chromosomes during the present study always have parallels in other anurans and are treated in detail below (chapters 3.1.1–3.3.6). Furthermore, the staining and banding properties of arboranan chromosomes, in particular their reactivity with AT- and GC-specific fluorochromes, are identical to those found in the other amphibians. The properties and functions of such DNA base-specific fluorochromes were explained at length in the 2 previous large monographs of this series on amphibian chromosomes [Schmid et al., 2010, 2012]. The specific DNA replication kinetics in the chromosomes of Arboranae, as determined by 5-bromodeoxyuridine (5-BrdU) incorporation in living cells, is known from other frogs as well. The results obtained with fluorescence in situ hybridizations (FISH) using repetitive DNA sequences (telomeric DNA, 18S + 28S rDNA) in arboranan chromosomes are, in principle, indistinguishable from those achieved in other frogs [for review, see Schmid et al., 2010, 2012].

Throughout the present monograph, the nomenclature system for chromosome morphologies suggested by Green and Sessions [1991b] was applied. Thus, chromosomes are classified as "meta-," "submeta-," "subtelo-," and "telocentric," but the term "acrocentric" is not used. It is very important to recognize that the telocentric chromosomes of the arboranan species examined for this monograph are in fact subtelocentric. Thus, the existence of tiny small arms in the telocentric chromosomes of Arboranae is demonstrated by the presence of telomeric DNA sequences at the centromeric termini of these chromosomes (Fig. 872–937). These telomeric $(TTAGGG)_n$ repeats are spatially separated from the kinetochore of the centromeric region and must therefore reside at the end of tiny short arms. From a historical point of view, the existence of truly telocentric chromosomes, without a minute short arm, has been controversial and is still debated. Although there is no doubt that telocentric chromosomes can be obtained experimentally, and that under certain circumstances they are stable and can be transmitted through several subsequent cell cycles, all evidence leads to the conclusion that naturally occurring, apparently telocentric chromosomes are indeed subtelocentric [Prokofieva, 1935a, b; Prokofieva-Belgovskaya, 1938; White, 1957; White et al., 1967]. An excellent review of the classical debates about the existence of truly telocentric chromosomes and a compilation of many references was provided by White [1973a]. For reasons of simplicity, the tiny short arms of the subtelocentric chromosomes are omitted in the schematic drawings of the Figures 18–21.

Multiple and high-contrast (G-, R-, Q-) bands in euchromatic chromosome segments are a property of mammalian and, in somewhat reduced quality, of avian chromosomes [reviewed by Sumner, 1990]. Additionally, G-bands, and to a lesser extent Q-bands, which are complementary to R-bands, have also been induced in the chromosomes of some reptiles, including turtles, snakes, and lizards [e.g., Stock and Mengden, 1975; Bickham and Baker, 1976; King and Rofe, 1976; Mengden and Stock, 1980; Moritz, 1984]. The metaphase chromosomes of all amphibians and fishes (Osteichthyes and Chondrichthyes) do not exhibit any characteristic multiple (G-, R-, Q-) bands. The absence of characteristic multiple chromosome bands also holds true, without any exception, for the species of Arboranae analyzed in the present study. A careful revision of all karyotypes of the species that have been stained with quinacrine and mithramycin shows that the chromosomal euchromatin fluoresces with a uniform, but moderate, intensity. All the chromosome regions fluorescing either distinctly brighter or distinctly weaker than the euchromatin consist of constitutive

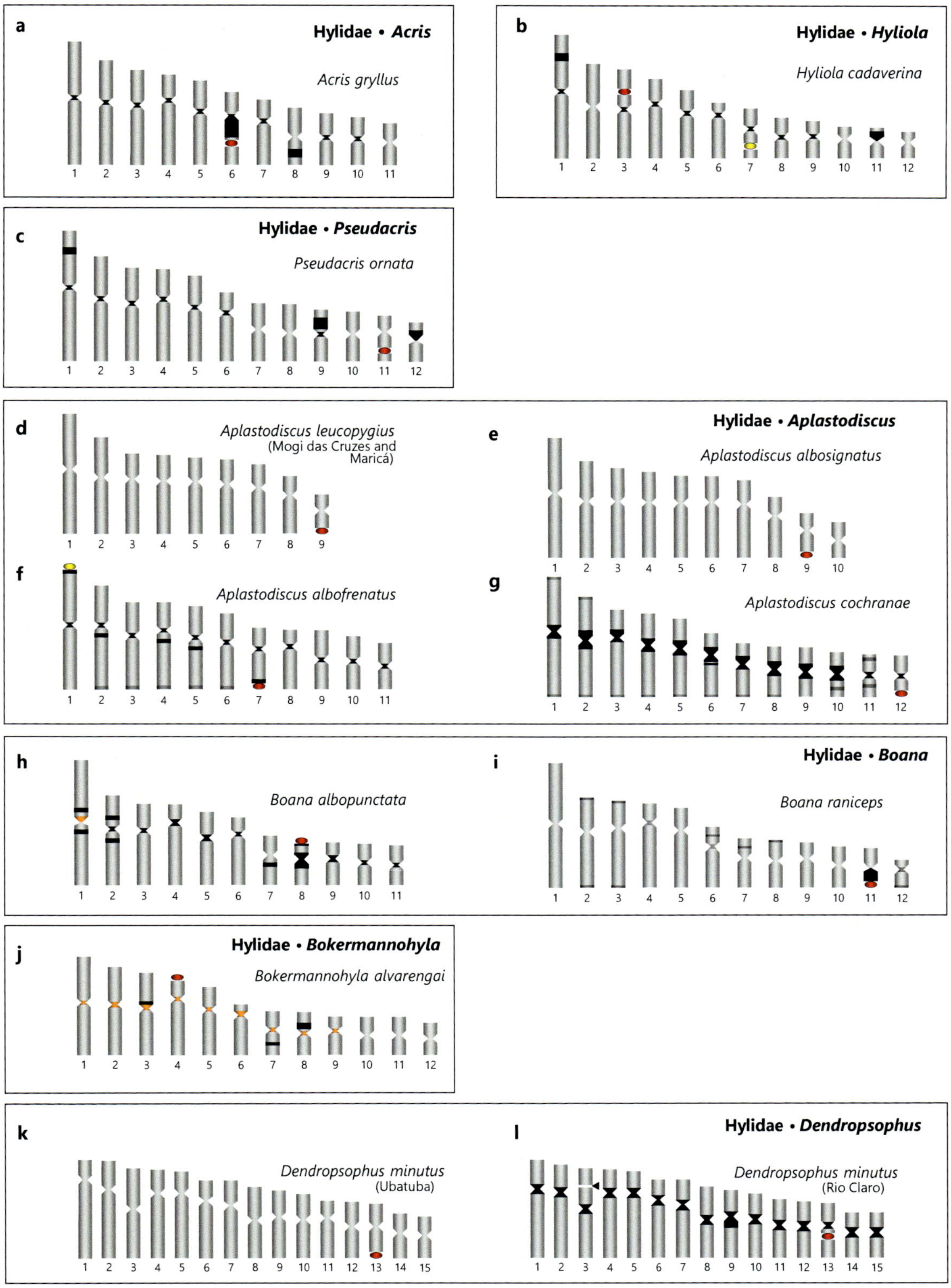

Fig. 18–21. Idiograms of the haploid chromosome set of at least 1 species of all hylid, phyllomedusid, and pelodryadid genera in which mitotic chromosomes have been described up to date. C-banding and Ag-staining were not performed in all karyotypes. Black areas, constitutive heterochromatin, no further characterization with fluorochromes; orange areas, mithramycin-positive constitutive heterochromatin; red ovals, nucleolus organizer regions; arrowheads, visible nucleolar constrictions. Idiograms according to results obtained in the present study and/or in the former publications of Cole [1974], Sessions [1978], Batistic [1989], Anderson [1991, 1996], Kasahara et al. [2003], Gruber et al. [2005, 2007, 2012b], Busin et al. [2001, 2006, 2008], Carvalho et al. [2009a, b], Paiva et al. [2010], Cardozo et al. [2011], Catroli et al. [2011], and Suárez et al. [2013]. For further explanations, see chapter 3.1.

Cytogenet Genome Res 2018;155:55–221
DOI: 10.1159/000489839

19

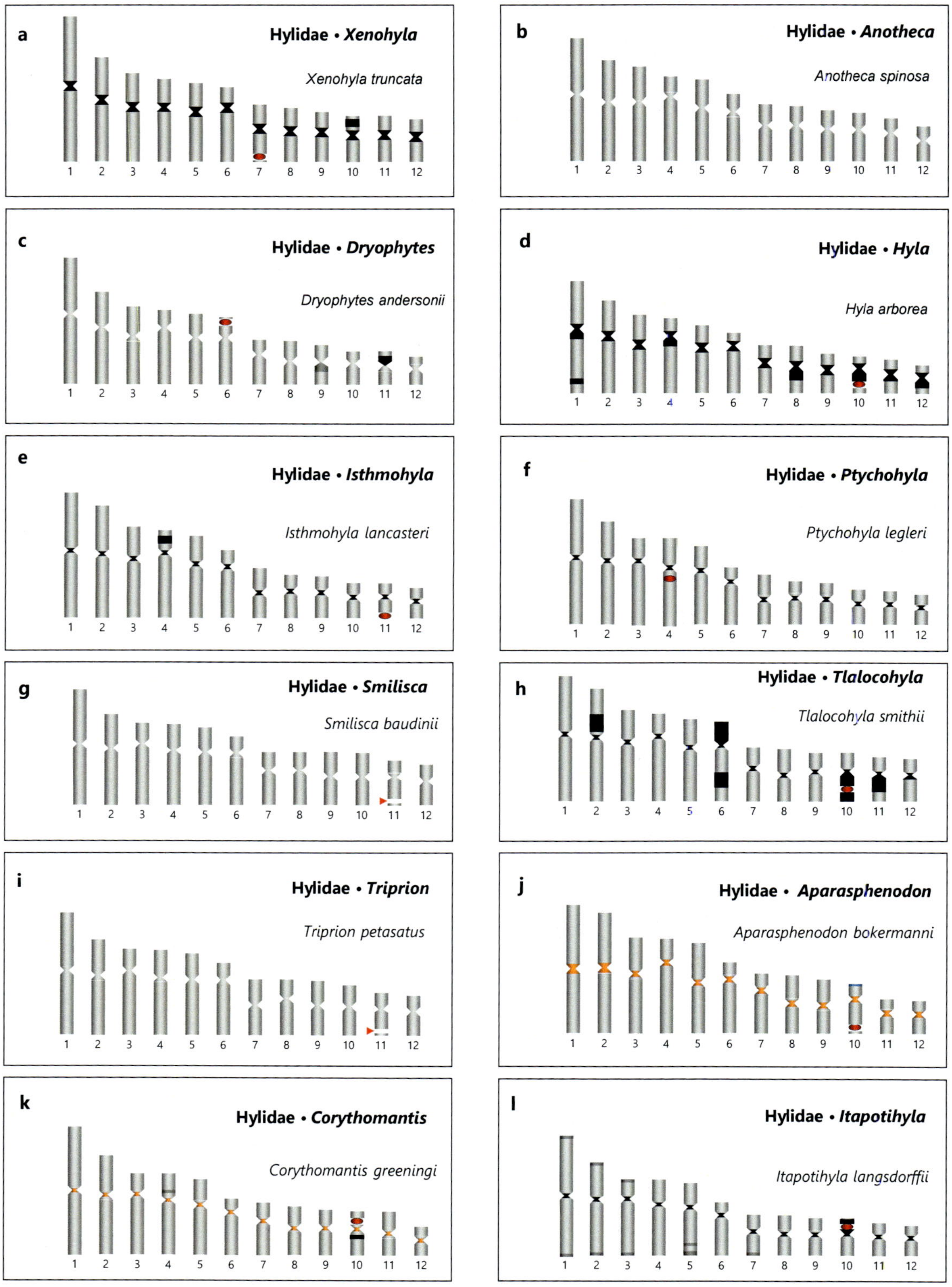

heterochromatin or nucleolus organizer regions (NORs) as confirmed by C-banding or Ag-staining. In the euchromatic chromosome segments of some of the fluorescent arboranan karyotypes analyzed, subtle differences in staining intensities along both chromatids of the chromosomes become visible. These bands do not, however, resemble the high-quality Q-bands obtained in mammalian metaphase chromosomes. In keeping with other amphibian species [Schmid, 1978a, b], it has not been possible to induce G-bands in the euchromatic regions of arboranan chromosomes by prolonged enzymatic treatment with either trypsin [Seabright, 1972] or pronase [Dutrillaux et al., 1971]. At-

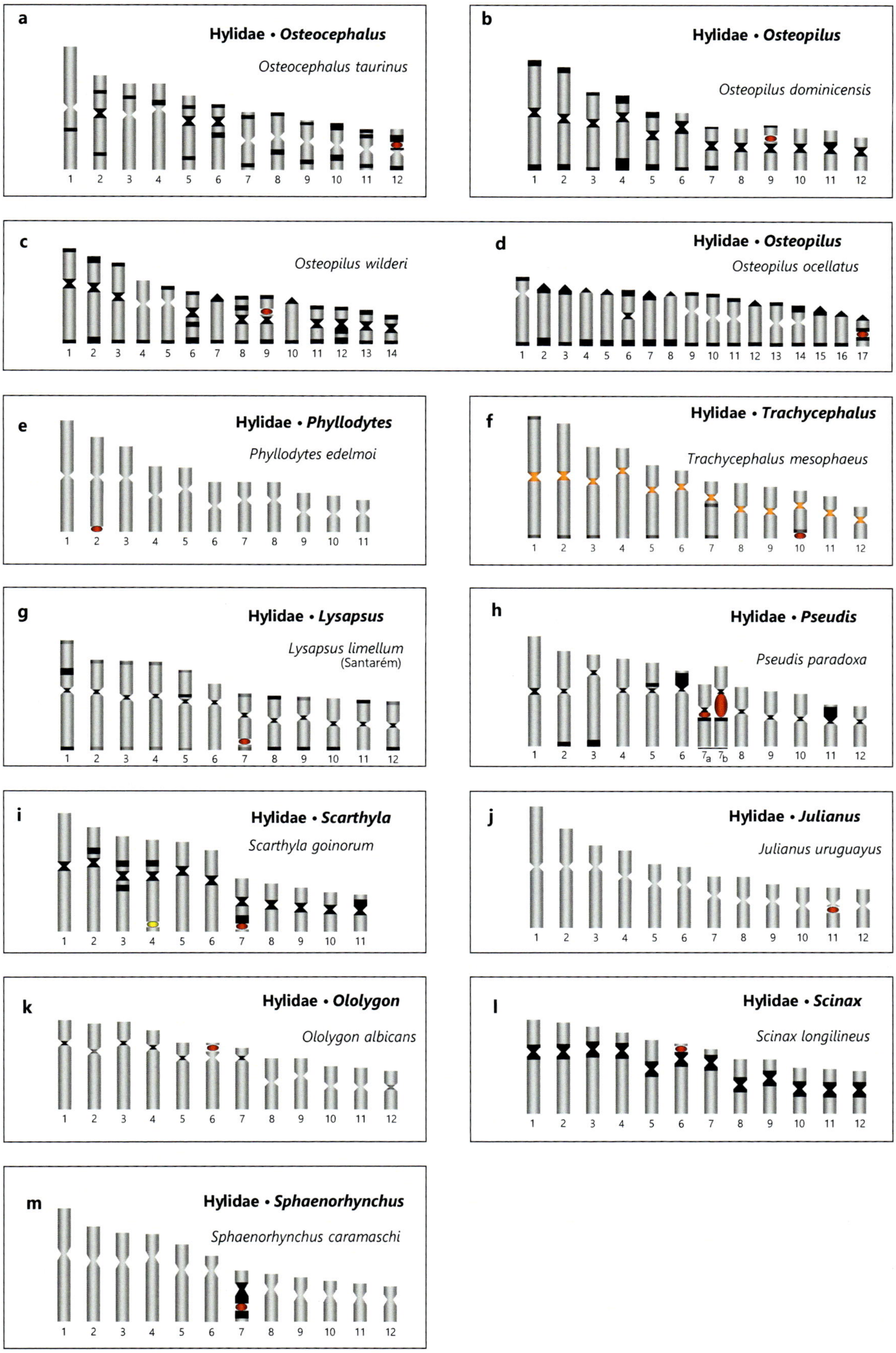
a
Hylidae • Osteocephalus
Osteocephalus taurinus
b
Hylidae • Osteopilus
Osteopilus dominicensis
c
Osteopilus wilderi
d
Hylidae • Osteopilus
Osteopilus ocellatus
e
Hylidae • Phyllodytes
Phyllodytes edelmoi
f
Hylidae • Trachycephalus
Trachycephalus mesophaeus
g
Hylidae • Lysapsus
Lysapsus limellum
(Santarém)
h
Hylidae • Pseudis
Pseudis paradoxa
7a 7b
i
Hylidae • Scarthyla
Scarthyla goinorum
j
Hylidae • Julianus
Julianus uruguayus
k
Hylidae • Ololygon
Ololygon albicans
l
Hylidae • Scinax
Scinax longilineus
m
Hylidae • Sphaenorhynchus
Sphaenorhynchus caramaschi

20

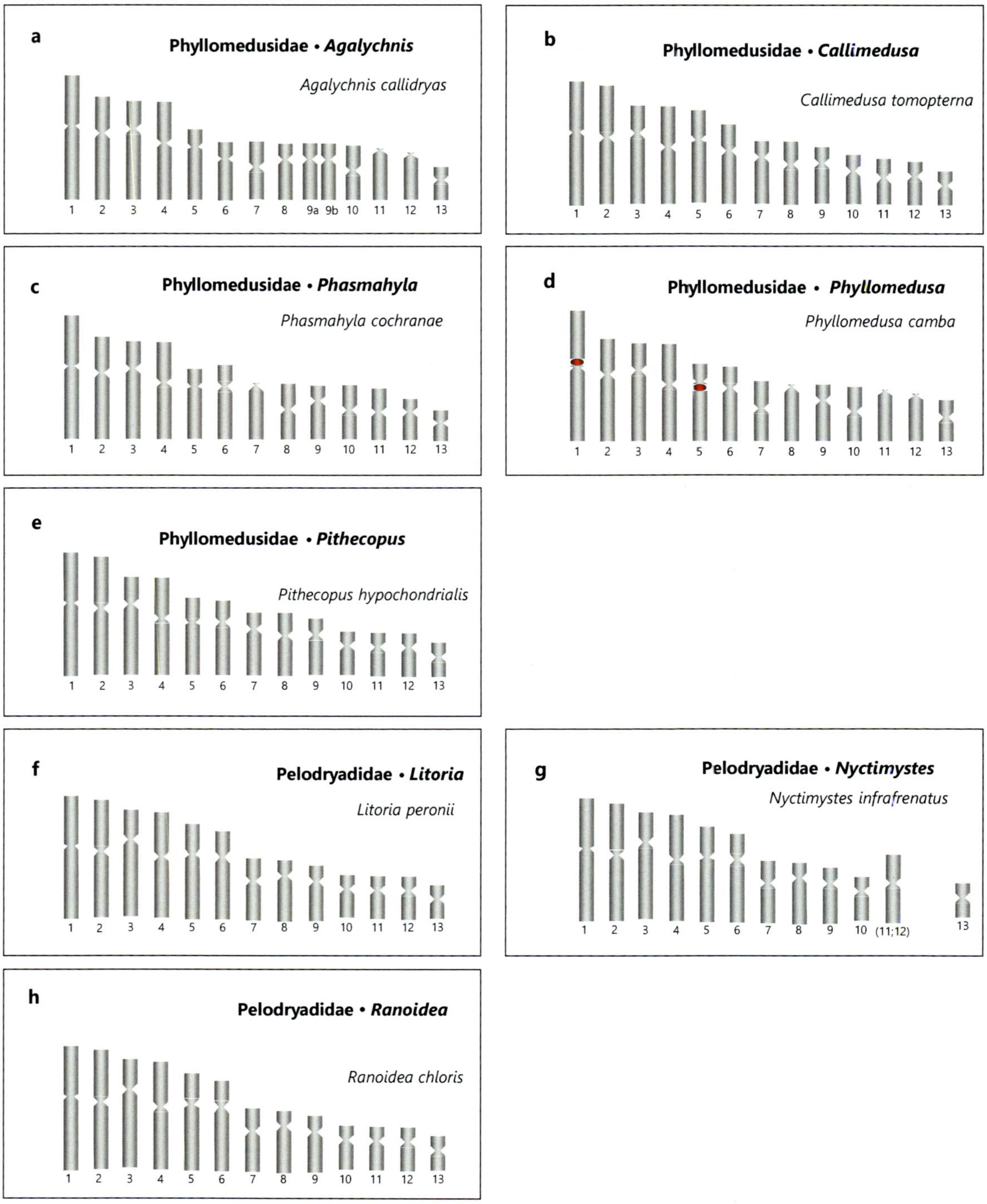

21

tempts to produce R-bands in arboranan chromosomes by heating chromosome preparations in phosphate buffer at 87°C followed by staining with either Giemsa [Dutrillaux et al., 1972] or the fluorochrome acridine orange [Bobrow et al., 1972] have also not been successful.

The difficulty in obtaining multiple banding patterns along fish and amphibian chromosomes is a well-known fact and has led to contradictory published data. For example, it has not been possible to induce Q-bands in the chromosomes of the eel *Anguilla anguilla* [Park and Grimm, 1981], but the experiments of Wiberg [1983] yielded distinct G-bands, and Medrano et al. [1988] observed faintly fluorescing R-bands with acridine orange after partial denaturation of eel chromosomes. A further example are the chromosomes of the frog genus *Xenopus* which fluoresce uniformly following quinacrine and mithramycin staining [Schmid et al., 1987], whereas the treatment with trypsin or a trypsin-urea combination resulted in a G-band-like pattern [Stock and Mengden, 1975; Sekiya and Nakagawa, 1983]. The quality of multiple bands induced in the chromosomes of 12 species of North American hylid frogs by a G-band technique [Wiley, 1982] cannot be considered as satisfactory. Most authors agree that none of the G-, Q-, and R-banding patterns that were induced in the chromosomes of a few fishes and amphibians are of the same quality as those obtained under identical conditions in warm-blooded vertebrates.

The absence of multiple banding patterns in the euchromatin of amphibian chromosomes was originally attributed to the strong spiralization of their chromosomes at the metaphase stage [Schmid, 1978a, 1980a, c]. Comparative measurements between human chromosomes and those of 10 toad species in the family Bufonidae showed that the average amount of DNA per micrometer of chromosome length was higher in the amphibian metaphase chromosomes by a factor of 1.5–3.0 [Schmid, 1978a]. Thus, it was concluded that the distances between individual bands in amphibian chromosomes are probably too small to be resolved with a microscope. Later studies on other amphibian species, however, showed that the DNA density in their metaphase chromosomes is as low or even 1.3–2.0 times lower as in mammals (i.e., 0.04 pg DNA/μm chromosome) [Schmid et al., 1987]. Nevertheless, it was still not possible to induce multiple banding patterns in any of these species. To compound matters, distinct BrdU replication banding patterns have been obtained in the chromosomes of a number of frog and salamander species, regardless of their considerably different degree of condensation [Schempp and Schmid, 1981; Kuro-o, 1986; Kuro-o et al., 1986, 1987; Kohno et al., 1990, 1991; Schmid and Steinlein, 1991; Schmid et al., 1993b; Gruber et al., 2007]. All these data show that there must be other factors than the degree of chromosome condensation at metaphase that are responsible for the absence of euchromatin banding in amphibians.

A generally accepted and plausible explanation for the presence of high-quality G-, Q-, and R-band patterns in fixed metaphase chromosomes of warm-blooded vertebrates and their almost complete absence in the chromosomes of cold-blooded vertebrates is summarized in the following [for review, see also Bickmore and Craig, 1997]. DNA equilibrium density gradient centrifugation in the presence of certain DNA sequence-specific ligands partitions the main-band DNA of warm-blooded vertebrates into light (AT) and heavy (GC) major components [Filipski et al., 1973; Macaya et al., 1976; Thiery et al., 1976]. The DNA fragments of these major components originate during preparation by breaking much longer DNA segments (>>300 kb) called "isochores" [Cuny et al., 1981], which have a rather homogeneous base pair composition and are distinguished by different GC levels. The evolutionary conservation of chromosome banding patterns in vertebrates is paralleled by conservation of the base pair composition and the relative amounts of the major components in the genomes [Thiery et al., 1976; Cuny et al., 1981; Olofsson and Bernardi, 1983]. The average amount of DNA per band in mammalian chromosomes, at the limit of light microscopic resolution, is 1,250 kb [Holmquist, 1988] and is thus in the range of the isochore sizes. Finally, there are strong indications that G-bands correspond to AT-rich DNA and R-bands to GC-rich DNA [Comings, 1975, 1978; Schweizer, 1976; Latt et al., 1977]. These results demonstrate that the genomes of the homeothermic mammals and birds are strongly compartmentalized by DNA base composition and clearly suggest that the isochores could correspond to the DNA sequences in Q-, G-, and R-bands [Cuny et al., 1981]. In contrast, isochores are either absent or rare in the genomes of the poikilothermic fishes, amphibians, and, with reservations, reptiles [Thiery et al., 1976; Hudson et al., 1980; Pizon et al., 1984; Medrano et al., 1988]. This parallels the difficulties in inducing multiple banding patterns in euchromatic regions of their chromosomes. A complete investigation of the correlation between DNA base pair composition heterogeneity and chromosomal banding would require supplementary density gradient centrifugations of the DNAs in those poikilothermic vertebrates whose chromosomes show different degrees of euchromatin banding quality.

Considerable information on the DNA base pair composition of the constitutive heterochromatin in amphibian chromosomes was obtained by analyses with AT- and GC-specific fluorochromes or by combining these fluorochromes with nonfluorescent reagents (counterstaining). These results are comprehensively discussed in the 2 preceding monographs on terraranan and hemiphractid cytogenetics [Schmid et al., 2010, 2012]. The AT-specific fluorochromes DAPI, Hoechst 33258, and quinacrine produce a fluorescence pattern complementary to that of the GC-specific fluorochromes mithramycin and chromomycin A_3. AT-rich heterochromatin is characterized by DAPI-, Hoechst 33258-, and frequently by quinacrine-positive fluorescence, but by mithramycin- and chromomycin A_3-negative fluorescence. Conversely, GC-rich heterochromatin shows mithramycin- and chromomycin A_3-positive fluorescence, but DAPI-, Hoechst 33258-, and quinacrine-negative fluorescence. The various chromatin categories that have been identified in amphibian chromosomes along with the staining techniques and the deduced DNA base pair compositions are summarized in Table 3. Supplementary information concerning the molecular composition of the various categories of constitutive heterochromatin can be derived from analytical density gradient centrifugation and in situ hybridization of isolated highly repetitive satellite DNA. Such data should show how precise the correlation is between the chromosomal distribution of AT-, GC-, and AT + GC-rich satellite DNA and the fluorescence patterns obtained with DAPI, Hoechst 33258, quinacrine, mithramycin, and chromomycin A_3. Such correlations are especially needed for GC-rich satellite DNA and mithramycin- and chromomycin A_3-positive heterochromatin because this category of heterochromatin occurs much more frequently and with greater heterogeneity in anuran genomes than does the DAPI-, Hoechst 33258-, and quinacrine-positive heterochromatin.

Table 3. Differential stainability of the various chromatin categories in Anura observed using different banding methods and the deduced DNA base pair composition

Chromatin category	Differential stainability[a]						Base pair composition of highly repetitive DNA
	C	DAPI	Hoechst 33258	Q	MM	CMA_3	
Euchromatin	–	–	–	–	–	–	
Heterochromatin	+	+	+	(+)[b]	–	–	AT-rich[c]
	+	–	–	–	+	+	GC-rich
	±[d]	–	–	–	+	+	GC-rich
	+	–	–	–	–	–	GC + AT-equality
Nucleolus organizer regions (NORs)	–	–	–	–	++	++	GC-rich[e]
	–	–	–	–	±[f]	±[f]	GC-rich[f]
	–	–	–	–	–[f]	–[f]	GC-rich[f]

[a] Differential stainability indicates a positive staining reaction of the chromatin in metaphase chromosomes and interphase nuclei. C, C-bands; DAPI, 4′,6-diamidino-2-phenylindole fluorescence; Hoechst 33258, 2′-(4-hydroxyphenyl)-5-(4-methyl-1-piperazinyl)-2,5′-bi-1*H*-benzimidazole trihydrochloride hydrate, bisbenzimide fluorescence; Q, quinacrine mustard fluorescence; MM, mithramycin fluorescence; CMA_3, chromomycin A_3 fluorescence. Euchromatin shows a uniform staining without banding patterns in the metaphase chromosomes with all fluorochromes.

[b] Positive quinacrine fluorescence is an optional but not obligatory property of AT-rich heterochromatin [Schmid et al., 2010].

[c] Two AT-rich satellite DNAs have been characterized for *Xenopus muelleri* in the CsCl density gradient [Stern, 1972], and one of these ($\rho = 1.683$) has been localized by in situ hybridization in the Q^+ heterochromatin [Pardue, 1974].

[d] Weakly stained, mainly telomeric constitutive heterochromatin in most metaphase chromosomes.

[e] The overall GC content of *Xenopus laevis* 18S + 28S rDNA is 67% (spacer = 73%) and of *X. muelleri* 65% (spacer = 69%). The number of 18S + 28S rDNA repeating units in *X. laevis* is 400–500 per NOR [Brown et al., 1972].

[f] NORs are only moderately fluorescent or nonfluorescent with MM and CMA_3. They are considered to be CG-rich but contain a low number of 18S + 28S rDNA repeating units.

An explanation is given at this point on the system that was applied with respect to how the karyotypes were organized and how the chromosomes were marked in this monograph. The mitotic karyotypes are arranged in parallel rows in Figures 53–937. For many species, several karyotypes of different staining techniques are included in the figures. This was done to show the consistency of the karyotypes and banding patterns. In all karyotypes, the chromosomes are arranged from left to right according to decreasing size. Male and female karyotypes are marked with the symbols ♂ and ♀, respectively. Karyotypes of some juvenile specimens are labeled as "juv". If no indication is given behind the species name, the sex of the corresponding specimen is unknown. Heteromorphic XY♂/XX♀ chromosome pairs are framed in the figures. The position of the heteromorphic sex chromosomes within the karyotypes is determined by the size of the X chromosome. The B chromosomes are placed between square brackets behind the A chromosomes and are labeled by the capital letter B. In the Ag-stained karyotypes in Figures 546–663 and distamycin A/mithramycin-counterstained karyotypes in Figures 801–871, the NORs are labeled with arrowheads.

With few exceptions, length measurements of arboranan chromosomes were not performed in the present study. Because of the different stages of chromosome contraction during the stage of metaphase as well as varying amounts of constitutive heterochromatin present in homoeologous chromosomes of different species, such measurements can yield results that generally are not very useful for interspecies karyotype comparisons.

In the following chapters 3.1.1–3.1.3, the karyotype formulae of the Arboranae are described, and the different mechanisms of chromosome mutations that played a role during karyophylogeny of the different taxa are discussed. For the convenience of the readers, it must be accentuated that constant checking through Appendix 1 is essential to keep an overview on the variety (or constancy) of the diploid chromosome numbers (2n) and fundamental numbers (FN), as well as to account for the complex nomenclature remarks and the numerous former cytogenetic literature on these frogs.

Furthermore, from each one of the arboranan genera, 1 species has been taken as example, and an idiogram of its haploid chromosome complement is presented in the Figures 18–21. This example stands as representative for the genus. For those genera where more than 1 diploid chromosome number occurs, several idiograms are depicted.

3.1.1 Karyotypes of the Hylidae

So far, the chromosomes of 266 species of the family Hylidae have been analyzed (Appendix 1). These species belong to the subfamilies Acrisinae (genera *Acris*, 3 species; *Hyliola*, 2 species; *Pseudacris*, 10 species), Cophomantinae (genera *Aplastodiscus*, 9 species; *Boana*, 33 identified species, 2 unidentified species; *Bokermannohyla*, 9 identified species, 4 unidentified species; *Hyloscirtus*, 1 species), Dendropsophinae (genera *Dendropsophus*, 38 identified species, 6 unidentified species; *Xenohyla*, 1 species), Hylinae (genera *Anotheca*, 1 species; *Charadrahyla*, 1 species; *Diaglena*, 1 species; *Dryophytes*, 15 species; *Duellmanohyla*, 3 species; *Exerodonta*, 1 species; *Hyla*, 6 species; *Isthmohyla*, 7 species; *Plectrohyla*, 2 species; *Ptychohyla*, 3 species; *Sarcohyla*, 3 species; *Smilisca*, 7 species; *Tlalocohyla*, 2 species; *Triprion*, 1 species), Lophyohylinae (genera *Aparasphenodon*, 2 species; *Argenteohyla*, 1 species; *Corythomantis*, 1 species; *Itapotihyla*, 1 species; *Osteocephalus*, 1 species; *Osteopilus*, 5 species; *Phyllodytes*, 2 species; *Trachycephalus*, 2 identified species, 2 unidentified species), Pseudinae (genera *Lysapsus*, 4 species; *Pseudis*, 7 species; *Scarthyla*, 1 species), and Scinaxinae (genera *Julianus*, 1 species; *Ololygon*, 22 identified species, 3 unidentified species; *Scinax*, 32 identified species, 3 unidentified species; *Sphaenorhynchus*, 5 species). These 266 species correspond to 39.3% of the 676 known extant species of the Hylidae (chapter 1.1.1; Duellman et al. [2016]). There are still 410 (60.7%) hylid species with unknown cytogenetic data. Moreover, it must be emphasized that not a single species has been karyotyped for 10 genera. These genera, which are all in Central and South America, are in the subfamilies Cophomantinae (genera *Colomascirtus* and *Myersiohyla*), Hylinae (genera *Bromeliohyla*, *Ecnomiohyla*, *Megastomatohyla*, and *Rheohyla*), and Lophyohylinae (genera *Dryaderces*, *Nyctimantis*, *Phytotriades*, and *Tepuihyla*) (Appendix 1). Certainly, among these 410 hylid species, there are more karyotypes awaiting description which will contribute to the understanding of chromosome evolution in these frogs.

The 266 hylid species so far examined have diploid chromosome numbers of 2n = 18 (1 species), 2n = 20 (2 species), 2n = 22 (12 species), 2n = 24 (202 species), 2n = 26 (1 species), 2n = 28 (2 species), 2n = 30 (44 species), 2n = 34 (1 species), and 4n = 48 (1 species) (Appendix 1).

3.1.1.1 Hylidae with the Karyotype Formula 2n = 24

In the earlier periods of amphibian genetics, there was practically no other taxon in the order Anura with so many chromosome studies as for the hylids [Morescalchi, 1973, 1979; Bogart, 1991; Green and Sessions, 2007]. In chapter 1.6 and Appendix 1, the efforts in the recording of karyological data of the Hylidae are compiled. Consequently, a large amount of our knowledge on amphibian cytogenetics rests on these pioneering efforts.

The vast majority of hylids (75.9%) have a diploid chromosome number of 2n = 24 and fundamental number of FN = 48 (Fig. 18–20). Consequently, and as the most parsimonious hypothesis, this modal number or type number can be considered as the basal (ancestral) karyotype of the entire family Hylidae. Nonetheless, as cautioned by White [1973a] in the earlier cytogenetic literature, one cannot assume that the modal number is necessarily ancestral for the taxon in question and that all others have been derived from it by whatsoever mutational events. While this might be true for some or even many animal groups, it is an unproven, and perhaps unprovable, hypothesis. White [1973a] convincingly stated that, in many cases, the modal number of a group is simply the chromosome number of a subgroup that is numerically dominant in the sample of species under analysis. In the case of the Hylidae, the modal number is 2n = 24. This predominance, although unlikely, may well be the result of overrepresentation of species belonging to the genera listed in Appendix 1.

The structure of this 2n = 24, FN = 48 hylid karyotype exhibits a very remarkable degree of evolutionary conservation. Figures 18–20 show at least 1 idiogram of each hylid genus in which mitotic chromosomes are known and confirm this striking similarity. The karyotype consists of 12 pairs of mainly meta- to submetacentric chromosomes. In general terms, the chromosome pairs 1–6 are larger than the remaining pairs 7–12. In some karyotypes, the middle-sized chromosome 6 is the most submetacentric element, showing a very short p arm whose length does not exceed the diameter of 1 chromatid (e.g., Fig. 66, 79, 80). In all complements, there is only a slight decrease in chromosome length from pair 1 to pair 6 and from pair 7 to pair 12. It must be emphasized that in the majority of hylid karyotypes, the chromosomes 3 are characterized by a very distinct conserved and typical submetacentric morphology. The p:q ratio in these chromosomes is approximately 1:4 (e.g., Fig. 79, 80, 83). In the karyotypes of some hylid taxa, and as a result of small pericentric inversions or varying amounts of pericentromerically located constitutive heterochromatin, this chromosome 3 has acquired a longer p arm and a p:q ratio of approximately 1.5:4 (e.g., Fig. 58, 60). The morphological appearance of the latter chromosomes 3 clearly differs from the former category of chromosomes 3. In the genomes of the family Hylidae, the nucleolar constrictions are not located in conserved (homoeologous) positions in the karyotypes but occupy a variety of chromosomal sites. This is in conformity with the detection of the Ag-stained NORs (see chapter 3.3.2). Multiple NOR sites are rather rare in hylid karyotypes. It must be emphasized that in the mitotic karyotypes of all hylids investigated and reinvestigated in the present monograph and in the numerous former studies, there are no conspicuous cytological characteristics which distinguish these from the chromosome complements of other anurans. Neither their chromatin contractions and chromatin stainabilities nor their banding affinities, FISH accessibilities, restriction enzyme digestions, and detectabilities of hypermethylated chromosome regions are unusual. The low amount of constitutive heterochromatin noted in a variety of species (Fig. 383–545) does not necessarily represent a feature of the chromosomes in this anuran group but can also be the result of insufficient experimental design of the C-banding technique [Schmid, 1978a]. Fur-

Fig. 22. Males of *Dryophytes chrysoscelis* (**a**) and *D. versicolor* (**b**). Photographs by James P. Bogart. For further explanations, see chapter 3.1.1.1.

thermore, the male meiotic chromosomes of the Hylidae do not possess exceptional properties (chapter 3.2.1), and in the literature, there exist no data on lampbrush chromosomes in female meiosis that show distinctive features.

Deviating genome sizes are also not present in hylid frogs and, in turn, not responsible for any particular cytological observation. The genome sizes measured for more than 400 amphibian species have been compiled by Gregory [2005a, b]. In the order Anura, the genome sizes range from 1.90 to 19.71 pg DNA/nucleus [for details, see Schmid et al., 2010]. As comparisons show, the anurans with the lowest average genome size are the Pelobatidae and Scaphiopodidae (4.71 pg DNA/nucleus), whereas the largest average is found in the Bombinatoridae (19.71 pg DNA/nucleus). The anuran families with the largest genome size variations are the Myobatrachidae (1.90–17.50 pg DNA/nucleus) and the Ranidae (2.80–22.72 pg DNA/nucleus). The Hylidae analyzed are characterized by diploid values of 3.08–15.58 pg DNA/nucleus. The frequencies of the anuran genome sizes show a clear unimodal distribution, with a modal value between 8 and 10 pg DNA/nucleus [Schmid et al., 2010]. The genome sizes of the hylids fit quite well in this frequency distribution. It should be noted, however, that the hylid species whose genome sizes have been determined represent only a very small fraction. No genome size measurements have been published for most neotropical taxa. Therefore, until the size of the nuclear genome of at least 1 representative species in these genera has been measured, the picture of the variability of genome sizes in the Hylidae remains seriously incomplete. Quantification of nuclear genome sizes using DAPI- or ethidiumbromide-stained cell nuclei and DNA flow cytometrical measurements is a noninvasive technique that circumvents major intervention or even sacrification of animals. In this regard, the relative quantities of unique, middle repetitive, and highly repetitive DNA sequences have not yet been determined in any species of Hylidae [for examples, see Straus, 1971; Davidson et al., 1975; Baldari and Amaldi, 1976; Morescalchi and Olmo, 1982], but it is anticipated that such DNA reassociation kinetics experiments will not unravel unexpected qualities of hylid chromosomes.

Dryophytes versicolor – Dryophytes chrysoscelis Complex

Diploid *Dryophytes chrysoscelis* and tetraploid *D. versicolor* (Fig. 22a, b) are cryptic species that are widely distributed in eastern North America [Bogart, 1980]. The 2 species cannot be distinguished by morphology, but they have distinctly different vocalizations that were originally used to differentiate these species [Johnson, 1966] prior to their identification as a diploid (*D. chrysoscelis*) and tetraploid (*D. versicolor*) species pair [Wasserman, 1970; Bogart and Wasserman, 1972; Bachmann and Bogart, 1975]. Furthermore, *D. versicolor* and *D. chrysoscelis* can reliably and noninvasively be differentiated by simply analyzing fixed tissues. The tetraploid *D. versicolor* cells have large nuclei and more nucleoli than the diploid *D. chrysoscelis* cells [Pritts and Miller, 1976; Cash and Bogart, 1978].

Based on mitochondrial sequences [Ptacek et al., 1994], *D. versicolor* arose at least 3 times, twice from *D. chrysoscelis* and once from an unknown maternal ancestor. Using mitochondrial and nuclear markers, Holloway et al. [2006] also suggested that tetraploids originated multiple times from *D. chrysoscelis* and from 2 other extinct lineages. Rather than tetraploidy arising de novo from diploid ancestors, the available mitochondrial data would also support a "triploid bridge hypothesis" [Ralin and Selander, 1979; Bogart and Bi, 2013] where new tetraploids derive from triploid hybrids. The occurrence of polyploid anuran species is summarized in Table 4.

3.1.1.2 Hylidae with 2n = 24 and Heteromorphic XY♂ Sex Chromosomes

The very first indications of heteromorphic sex chromosomes in the large treefrog taxon were obtained by a comprehensive study of Anderson [1991] (Appendix 1) on 21 Holarctic species. In 35 male specimens of *Dryophytes femoralis*, a NOR was observed on only 1 homolog of chromosomes 2, and the interphase nuclei had only 1 nucleolus each. In 17 female individuals examined, NORs were present on both homologs of pair 2, and the interphase nuclei had 2 nucleoli each. The absence of a NOR on one homolog 2 in the male specimens marked the sex of the individual and represented a morphological differentiation of heteromorphic XY♂/XX♀ sex chromosomes. The present monograph definitively confirms the existence of these primitive XY♂/XX♀ sex chromosomes in *D. femoralis*, adds more

Table 4. Occurrence of polyploidy in Anura

Family Species[a]	Ploidy level	References
Alsodidae		
Eupsophus vertebralis	3n = 42	Formas, 1994
Arthroleptidae		
Astylosternus diadematus	4n = 54	Bogart and Tandy, 1976, 1981
Bombinatoridae		
Bombina orientalis[b]	3n = 36	Litvinchuk and Rosanov, 2016
Bufonidae		
Sclerophrys poweri	3n = 30	Schmid, 1978a
Bufotes baturae	3n = 33	Stöck et al., 2002
Strauchbufo raddei[b]	3n = 33	Litvinchuk et al., 2011, 2012
Strauchbufo raddei	3n = 33	Stöck et al., 1999, 2001; Borkin et al., 2001; Cavallo et al., 2002
Bufotes viridis	4n = 40	Bogart and Tandy, 1976; Bogart, 1980; Tandy et al., 1980
Sclerophrys asmarae	4n = 44	Bogart, 1972; Mazik et al., 1976; Pisanets, 1978; Toktosunov, 1984; Borkin et al., 1986a, b, 2001; Orlova and Uteshev, 1986; Roth and Ráb, 1986, 1987; Wu and Zhao, 1987; Borkin and Kuzmin, 1988; Stöck, 1998; Stöck et al., 2001
Bufotes oblongus	4n = 44	Stöck et al., 2005
Bufotes pewzowi	4n = 44	Stöck et al., 2005
Ceratophryidae		
Ceratophrys aurita	8n = 104	Beçak, 1967; Beçak et al., 1967
Ceratophrys ornata	8n = 104	Bogart, 1967; Bogart and Wasserman, 1972
Ceratophrys joazeirensis	8n = 104	Mercadal de Barrio, 1986
Dicroglossidae		
Hoplobatrachus occipitalis	4n = 52	Bogart and Tandy, 1976, 1981
Hylidae		
Dryophytes versicolor	4n = 48	Wasserman, 1970; Bogart and Wasserman, 1972; Bachmann and Bogart, 1975; Cash and Bogart, 1978; Wiley, 1982; Anderson, 1986
Phyllomedusa tetraploidea	4n = 52	Beçak et al., 1970a; Batistic et al., 1975; Haddad et al., 1994
Leptodactylidae		
Pleurodema bibroni	4n = 44	Kuramoto, 1972; Bogart and Wassermann, 1972; Veloso et al., 1973
Pleurodema kriegi	4n = 44	Kuramoto, 1972; Bogart and Wassermann, 1972; Veloso et al., 1973
Limnodynastidae		
Neobatrachus aquilonius	4n = 48	Mahony and Roberts, 1986
Neobatrachus centralis	4n = 48	Mahony and Roberts, 1986; Schmid et al., unpublished
Neobatrachus kunapalari	4n = 48	Mahony and Roberts, 1986; Schmid et al., unpublished
Neobatrachus sudelli	4n = 48	Mahony and Robinson, 1980; Schmid et al., 1990b; Schmid et al., unpublished
Odontophrynidae		
Odontophrynus americanus	4n = 44	Beçak et al., 1966, 1967, 1970a, b; Bogart, 1967
Pipidae[c]		
Xenopus borealis	4n = 36	Tymowska, 1976, 1977, 1991; Tymowska and Fischberg, 1973
Xenopus clivii	4n = 36	Tymowska, 1977; Tymowska and Fischberg, 1973
Xenopus fraseri	4n = 36	Tymowska, 1977; Tymowska and Fischberg, 1973
Xenopus gilli	4n = 36	Tymowska, 1977, 1991; Tymowska and Fischberg, 1973
Xenopus laevis laevis	4n = 36	Wickbom, 1945; Weiler and Ohno, 1962; Morescalchi, 1963; Tymowska and Kobel, 1972; Tymowska, 1977
Xenopus laevis bunyoniensis	4n = 36	Tymowska, 1991
Xenopus laevis petersi	4n = 36	Tymowska, 1977, 1991; Tymowska and Fischberg, 1973
Xenopus laevis poweri	4n = 36	Tymowska, 1991
Xenopus laevis sudanensis	4n = 36	Tymowska, 1991
Xenopus laevis victorianus	4n = 36	Tymowska, 1977, 1991; Tymowska and Fischberg, 1973
Xenopus muelleri	4n = 36	Tymowska and Kobel, 1972; Tymowska and Fischberg, 1973; Tymowska, 1991
Xenopus pygmaeus	4n = 36	Loumont, 1986
Xenopus sp. nov. III	4n = 36	Tymowska, 1991
Xenopus sp. nov. VI	4n = 36	Tymowska, 1991

Table 4 (continued)

Family Species[a]	Ploidy level	References
Xenopus sp. nov. IX	4n = 36	Tymowska, 1991
Xenopus epitropicalis	4n = 40	Tymowska and Fischberg, 1982; Tymowska, 1991
Xenopus sp. nov. VII	4n = 40	Tymowska, 1991
Xenopus amieti	8n = 72	Kobel et al., 1980
Xenopus andrei	8n = 72	Loumont, 1983
Xenopus boumbaensis	8n = 72	Loumont, 1983
Xenopus vestitus	8n = 72	Tymowska, 1976, Tymowska et al., 1977
Xenopus wittei	8n = 72	Tymowska, 1976; Tymowska and Fischberg, 1980
Xenopus sp. nov. X	8n = 72	Tymowska, 1991
Xenopus ruwenzoriensis	12n = 108	Tymowska and Fischberg, 1973; Tymowska, 1991
Xenopus sp. nov. VIII	12n = 108	Tymowska, 1991
Xenopus itombwensis	12n = 108	Evans et al., 2008
Pyxicephalidae		
Pyxicephalus delalandii	4n = 52	Bogart and Tandy, 1976, 1981; Channing and Bogart, 1996
Ranidae		
Pelophylax cerigensis[b]	3n = 39	Litvinchuk et al., 2015
Pelophylax cretensis[b]	3n = 39	Litvinchuk et al., 2015
Pelophylax nigromaculatus[d]	3n = 39	Kawamura, 1951
Pelophylax esculentus	3n = 39	Heppich, 1978; Heppich and Tunner, 1979
Strabomantidae		
Holoaden luederwaldti	3n = 27	Campos et al., 2012

Included are cases of occasional polyploid individuals, as well as pure polyploid species. Polyploids so far found in the order Caudata are compiled by Schmid et al. [2010]. No polyploids have been detected in the order Gymnophiona.
[a] The name of the species is the same as used in the references.
[b] Determined by DNA flow cytometry.
[c] Four more tetraploid and 2 more dodecaploid *Xenopus* species were more recently described by Evans et al. [2015].
[d] Experimentally obtained polyploid individuals.

data obtained by differential chromosome banding and FISH analyses, and discusses the origin of these sex chromosomes and the consequences of the NOR deletion in the Y chromosome.

D. femoralis (pine woods treefrog, Fig. 23) is a slender, long-legged species with a body length of about 40 mm and a medium to dark brown dorsal coloration. These frogs occur in southeastern USA (Atlantic and Gulf Coastal Plains from southeastern Virginia to southeastern Louisiana and southwestern Mississippi). They are locally common in most of their range. Adults migrate from arboreal retreats to breeding sites (depressional wetlands in pine flatwoods, vernal pools, ditches and other shallow water-filled depressions, and swamps). Pine woods treefrogs are early colonizers and often call from pools in clearcuts and fields. Calling sites are on vegetation around breeding sites, as high as several meters above the ground. Large choruses are heard after heavy rains. Pine woods treefrogs climb to the tops of longleaf pines and seek shelter in cabbage palms, bromeliads, and pitcher plants [AmphibiaWeb, 2017].

In 5 previous studies, the conventionally stained and banded karyotypes of *D. femoralis* were investigated, and some BrdU replication bandings and FISH (18S + 28S rDNA) analyses were carried out [Anderson, 1986, 1991; Anderson and Moler, 1986; Schmid and Steinlein, 2003; Wiley, 2003]. After conventional staining, all male and female specimens of *D. femoralis* exhibited a diploid chromosome number of 2n = 24 and a fundamental number of FN = 48 (Fig. 24a–d, 26a, 98, 99). In agreement with these former analyses, the *D. femoralis* metaphases analyzed in the present study can be arranged in 12 pairs which progressively decrease in size (Fig. 24a–d, 26a, 98, 99). Chromosome pairs 1–6 are considerably larger than pairs 7–12. Chromosomes 1 and 2 are nearly metacentric, 3 and 5 submetacentric, and 6 metacentric. Chromosomes 4 show the typical, submetacentric morphology of the common 2n = 24 hylid chromosome complement. Of the remaining 6 pairs, chromosomes 8 and 9 are nearly metacentric, and chromosomes 7 and 10–12 are submetacentric. There is only a slight decrease in chromosome length from pair 1 to pair 6 and from pair 7 to pair 12. As pointed out by Anderson [1991], the karyotype of *D. femoralis* differs in several aspects from those of all other Holarctic hylid species: (1) chromosome 6 which in all other species has a telocentric morphology, is metacentric in *D.*

Fig. 23. Male *Dryophytes femoralis* from West of Southport, NC, USA. Photograph by James P. Bogart. For further explanations, see chapter 3.1.1.2.

femoralis, (2) it is not possible to detect constitutive heterochromatin in any chromosome pair by the classical C-banding technique, and (3) the NOR is sexually dimorphic. In all *D. femoralis* males analyzed in the present and former studies, a single nucleolar constriction is located in the proximal third of the short arm of only one of the chromosomes 2. The homologous chromosome 2 lacks this constriction (Fig. 24b, d, 98). In contrast, the females exhibit a nucleolar constriction in both chromosomes 2 (Fig. 24a, c, 99). Since the NOR heteromorphism is present only in males but not in females, it can be confirmed that the XY♂/XX♀ mechanism of chromosomal sex determination operates in *D. femoralis*. The chromosome 2 carrying the nucleolar constriction is the X chromosome, and its homolog without the constriction is the Y chromosome.

The X chromosome of *D. femoralis* is distinctly larger than the Y chromosome (Fig. 24a–d, 26a, 98, 99). Five high-quality metaphases in the middle stage of metaphase without chromosome overlaps were selected for chromosome length measurements [Schmid and Steinlein, 2003]. With a length of 8.2 ± 0.1 μm, the X chromosome is in fact one of the longest chromosomes in the karyotype, nearly the size of the autosome 1 (8.4 ± 0.4 μm). The X constitutes 11.1% of the haploid karyotype (11 autosomes + X + X). In contrast, the total length of the Y chromosome is only 6.4 ± 0.02 μm (11 autosomes + X + Y).

The constriction in the X chromosome short arm is the NOR (Fig. 25a–f, 26b, c, e, g). As has already been found in many other anuran species [Schmid, 1980a, 1982; Schmid et al., 2010, 2012], the nucleolar constrictions can vary considerably in length between homologous chromosomes and between different individuals. The lengths of the nucleolar constrictions represent a constitutive property in the karyotypes of individual frogs.

Specific Ag-staining of the NORs in the metaphase chromosomes of *D. femoralis* indicates that the sole NOR is located in the X chromosome short arm (Fig. 25a–c, 26b, g). This confirms the earlier findings of Anderson [1991], Schmid and Steinlein [2003], and Wiley [2003]. In metaphases prestained with quinacrine and subsequently stained with silver, the dark Ag-label is clearly located within the nucleolar constriction. In the chromosomes stained with mithramycin and subsequently with silver, the bright mithramycin fluorescence of the NORs and the Ag-deposit colocalize. The size differences between the nucleolar constrictions in different X chromosomes determined with conventional staining are found with Ag-staining as well. Thus, X chromosomes with very long nucleolar constrictions exhibit considerably larger Ag-blocks than X chromosomes with short constrictions. In female animals with heteromorphic nucleolar constrictions in their X chromosomes, the 2 Ag-stained NORs have a distinctly different size. Moreover, 2 unequal-sized nucleoli are always observed in the interphase nuclei of these females. Ag-stained NORs can neither be detected in the Y chromosome nor in the autosomes of *D. femoralis* (Fig. 25a–c, 26b).

As expected, the fluorochromes mithramycin and quinacrine mustard induce no multiple banding patterns along the metaphase chromosomes of *D. femoralis* such as those obtained in mammals and birds. As in all other Amphibia examined, the euchromatic segments of the *D. femoralis* chromosomes fluoresce with uniform intensity (Fig. 24e, f, 25d–f, 26c, 675, 676, 807–809). All those regions fluorescing either brighter or fainter than the euchromatin are constitutively heterochromatic or represent the NORs. Following mithramycin staining, the centromeric regions of all 12 chromosome pairs and the short arm telomeric regions of chromosomes 8 show a distinctly brighter fluorescence intensity than the euchromatin (Fig. 25d–f, 26c). These regions must be composed of constitutive heterochromatin, although, as the study of Anderson [1991] and the present monograph show, it was not yet possible to label them using the C-banding technique. A faint, interstitial mithramycin-positive heterochromatic band is also located in the middle of the Y chromosome short arm (Fig. 26c, e). This band is positioned in exactly the same site as is the NOR in the X chromosome short arm. As shown for many other amphibian species [Schmid, 1980a, 1982], the NOR of *D. femoralis* exhibits the brightest fluorescence after distamycin A/mithramycin counterstaining (Fig. 25d–f, 26c, e). The fluorescence signal of the NOR is characterized by a much sharper contrast than the enhanced fluorescence of the constitutive heterochromatin. In the quinacrine mustard-stained preparations, the heterochromatic regions and the NORs appear as nonfluorescent gaps (Fig. 24e, f). This is to be expected from the general rule of complementarity of mithramycin and quinacrine banding [Schmid, 1980a; Schmid et al., 1990b].

Whereas female (XX) individuals have 2 NORs in diploid cells, male (XY) individuals only have 1 NOR. Accordingly, one would expect a dose ratio of 2:1 between female and male frogs with regard to the number of 18S + 28S rRNA genes. It should be noted, however, that positive Ag-staining is restricted to transcriptionally active NORs as demonstrated by studies on interspecific cell hybrids of

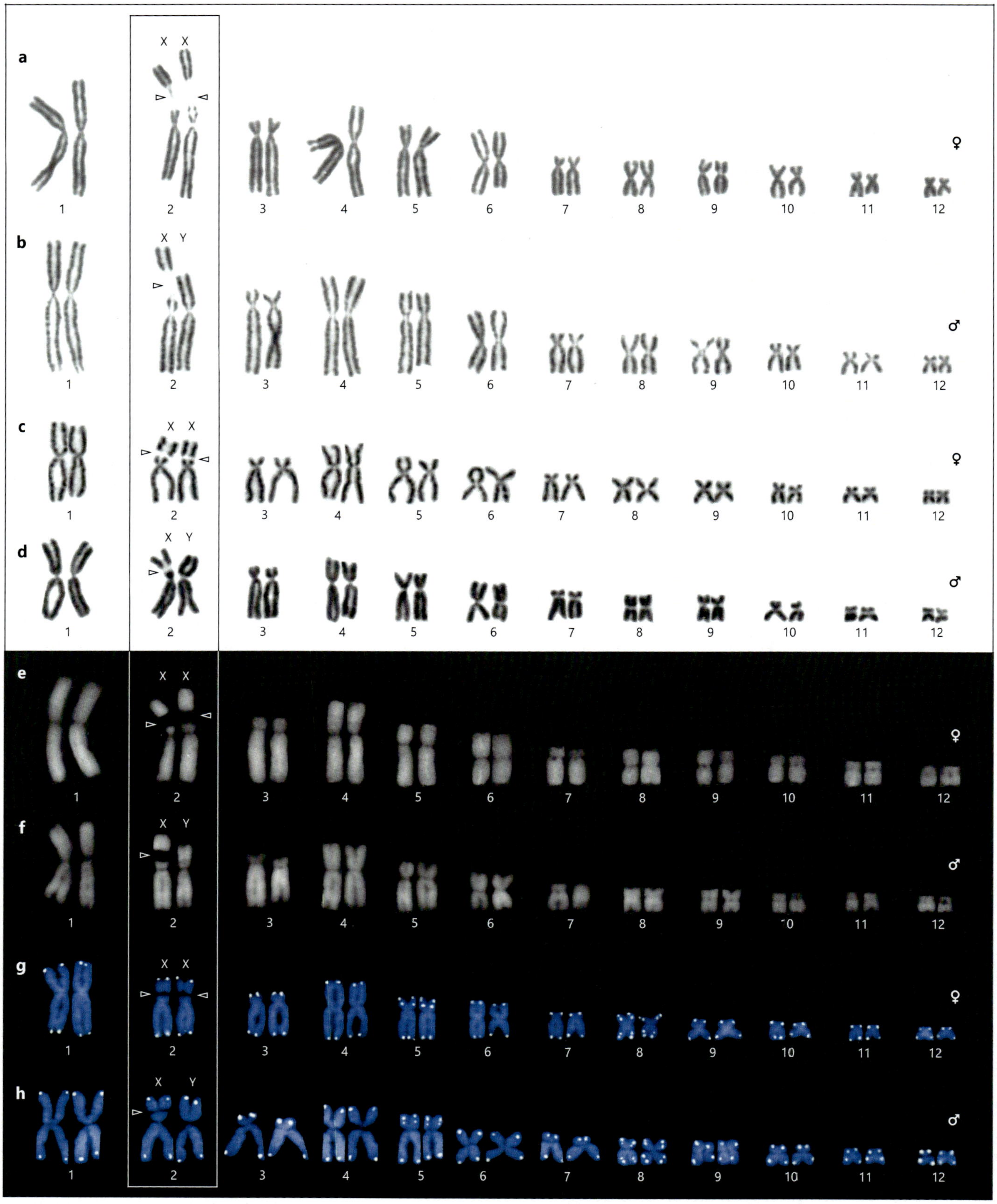

Fig. 24. Karyotypes of female (**a**, **c**, **e**, **g**) and male (**b**, **d**, **f**, **h**) *Dryophytes femoralis* showing conventional Giemsa staining (**a–d**), quinacrine fluorescence (**e**, **f**), and FISH of telomeric DNA sequences (**g**, **h**). The XX♀/XY♂ sex chromosomes are framed. Open arrowheads indicate nucleolar constrictions in the X chromosomes. For further explanations, see chapter 3.1.1.2.

mammals [Miller DA et al., 1976; Miller OJ et al., 1976], embryogenesis of the mouse [Engel et al., 1977], spermatogenesis of a wide variety of vertebrates [Schmid et al., 1977, 1982; Hofgärtner et al., 1979a], and actinomycin D treatment of human fibroblast cultures [Hofgärtner et al., 1979b; Schmid and Guttenbach, 1988]. Extensive cytochemical analyses prove that the Ag-stained material consists of non-histone protein(s) rich in sulfhydryl- and disulfide groups, which rapidly associate with the newly transcribed rRNA [Howell, 1977; Schwarzacher et al., 1978; Varley and Morgan, 1978; Buys and Osinga, 1984]. These proteins have been isolated from nucleoli and designated as B23 and C23 proteins [Busch et al., 1982; Ochs et al., 1983; Ochs and Busch, 1984]. Residues of these proteins remain associated with the NORs in metaphase and can be demonstrated with Ag-staining, although at this stage the NORs themselves are transcriptionally inactive [Fan and Penman, 1971]. Inactive NORs will not be detected by Ag-staining because the Ag-stainable proteins are absent [Buys and Osinga, 1984]. Lack of Ag-staining in metaphase indicates that either a particular NOR was not transcribed in the preceding interphase nucleus and thus does not possess the associated Ag-stainable protein or that the NOR was completely lost by deletion. The black blocks of the Ag-stained NORs presumably consist of precipitated metallic silver which derives from the reduction of the Ag^+ ions by the sulfhydryl- and disulfide groups of the NOR-associated protein. In an electron microscopic study, the spatial distribution of Ag-stainable NOR proteins and chromatin fibers was examined in a human cell line [Hernandez-Verdun et al., 1982]. It was shown that there is a constant overlapping of the Ag-deposit sites with dispersed intranucleolar chromatin fibers, suggesting that the Ag-stainable proteins adhere to the dispersed nucleolar DNA.

The lack of an Ag-stained NOR in the Y chromosome in the metaphase stage of *D. femoralis* makes it impossible to decide whether there is no NOR in this chromosome or whether the NOR is only transcriptionally inactive in the preceding interphase nucleus. Therefore, it was essential to examine the chromosomes of *D. femoralis* with methods which label the NOR and the 18S + 28S rDNA without regard to their state of transcriptional activity. Thus, FISH experiments using a specific probe (pXlr 101A) annealing to the evolutionarily conserved 18S + 28S rDNA sequences were performed on metaphase chromosomes from bone marrow of several frogs. The plasmid pXlr 101A is a derivative of pMB9 (4,900 bp) into which one complete rRNA gene (10,550 bp) of *Xenopus laevis* has been inserted. The inserted DNA sequence, obtained by digestion of whole *X. laevis* rDNA with the restriction endonuclease *Hin*dIII, contains all of the non-transcribed spacer and the 40S precursor coding sequence [Morgan et al., 1980]. When applying FISH with the pXlr 101A probe, a bright and distinct hybridization signal can be identified at the NOR in the X chromosome short arm (Fig. 25g–j). No other 18S + 28S rDNA FISH signals are visible in the karyotype of *D. femoralis*. Although the chromosomes were not preidentified by banding, the X and Y can be unequivocally differentiated from the autosomes on the basis of their morphology. In conclusion, the experiments with silver staining, distamycin A/mithramycin counterstaining, and FISH definitely demonstrate that the NOR located in the X chromosome short arm actually is the only NOR present in the genome of *D. femoralis*.

Sex-specific differences in the number of NORs, such as those found in *D. femoralis*, with a ratio of 2 NORs (♀):1 NOR (♂), are uncommon in vertebrates. In mammals, there are only 3 known cases: the Australian Kangaroo Island wallaby *Macropus eugenii*, the Australian kangaroo rat *Potorous tridactylus*, and the American bat *Carollia perspicillata*. The female karyotype of *M. eugenii* contains 2 X-linked NORs, whereas there is a single X-linked NOR in the male chromosome complement [Haaf et al., 1991]. Also, in *C. perspicillata*, a single NOR is located in the X chromosome with no other NOR in the Y chromosome or in the autosomes. An additional feature is that this X chromosome is translocated onto an autosome, which gives the male an odd diploid chromosome number (2n = ♀20, ♂21). In situ hybridization with tritium-labeled rRNA and Ag-staining showed that the NOR is located at the junction of the translocation between the X chromosome and the autosome [Hsu et al., 1975a; Goodpasture and Bloom, 1975]. In the related bat, *Carollia castanea*, which still possesses the ancestral karyotype before the X-autosome translocation took place [Patton and Gardner, 1971; Stock, 1975], the NORs are located in both the X and the Y sex chromosomes [Goodpasture and Bloom, 1975]. In *P. tridactylus* (2n = ♀12, ♂13), the X chromosome is also translocated onto an autosome [Hsu et al., 1967] and contains the sole NOR in the karyotype [Hsu et al., 1975a; Goodpasture and Bloom, 1975]. The X chromosome also has a NOR in all members of the Australian marsupials of the superfamily Phalangeroidea that have been studied [Hayman and Martin, 1974; Hayman and Rofe, 1977; Hayman and Sharp, 1981], and in the South American marsupials *Monodelphis domestica* and *M. brevicaudata* [Merry et al., 1983; Schmid et al., 2002c]. Because NORs are also located in the Y chromosome and/or autosomes in these species, a clear-cut sex-specific difference in the NOR number with a ratio of 2♀:1♂, as found in *D. femoralis*, is not achieved. For the sake of completeness, it is important to add that there are mammalian species with a Y chromosome-linked NOR which is absent in the X chromosome. This is observed in the antelope squirrel *Ammospermophilus harrisii*, the domestic dog [Pathak et al., 1982], the raccoon dog *Nyctereutes procyonoides* [Makinen and Fredga, 1980], and the primate *Hylobates (Symphalangus) syndactylus* [Ledbetter, 1981]. In all these mammalian species, in which the Y chromosome has a NOR and the X chromosome does not, additional NORs are located in one or more autosome pairs. Finally, in the karyotype of the brush-tailed possum *Trichosurus vulpecula*, NORs are present in the X and Y sex chromosome pair and in 1 autosomal pair [Stock and Mengden, 1982]. Additional mammalian species with

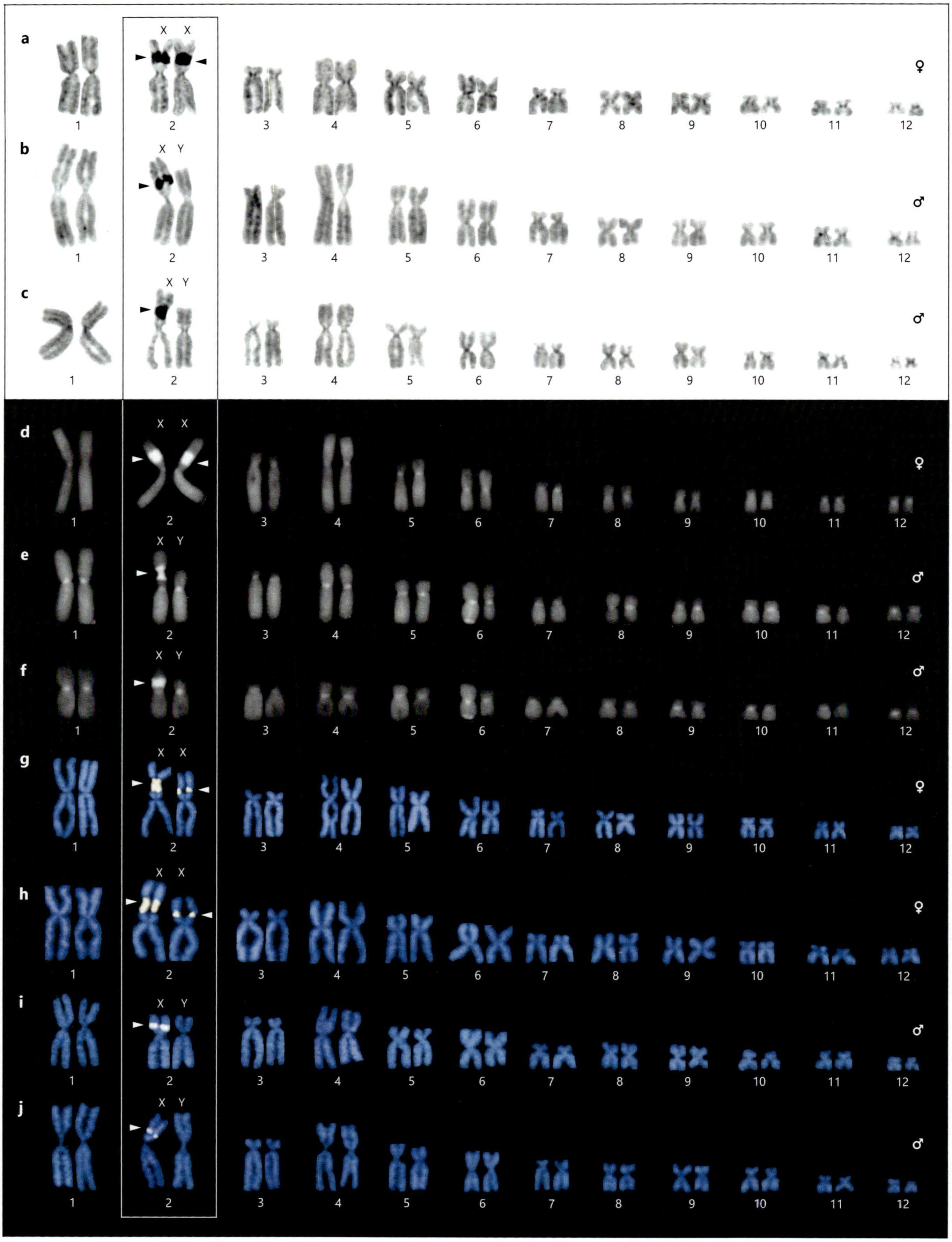

25

NOR-bearing sex chromosomes have been compiled by Merry et al. [1983].

It is interesting to note that among the cytogenetically studied Amphibia, 10 more anuran species and 3 urodelan species with a sex chromosome-linked NOR have been discovered in addition to *D. femoralis* (Table 5). Those with a NOR in only one of the sex chromosomes (X or Z) are the marsupial frog *Gastrotheca riobambae* (family Hemiphractidae), the common Japanese bell-ring frog *Buergeria buergeri* (family Rhacophoridae), and an undescribed terraranan frog *Pristimantis* sp. n. H of the family Strabomantidae. All these karyotypes and the logical cytological implications have been treated with most details in a former monograph of this series [Schmid et al., 2012].

D. femoralis, *G. riobambae*, *B. buergeri*, and *Pristimantis* sp. n. H are comparable to the anucleolar Oxford mutation of *X. laevis* that was described by Elsdale et al. [1958] and Fischberg and Wallace [1960] and which has been thor-

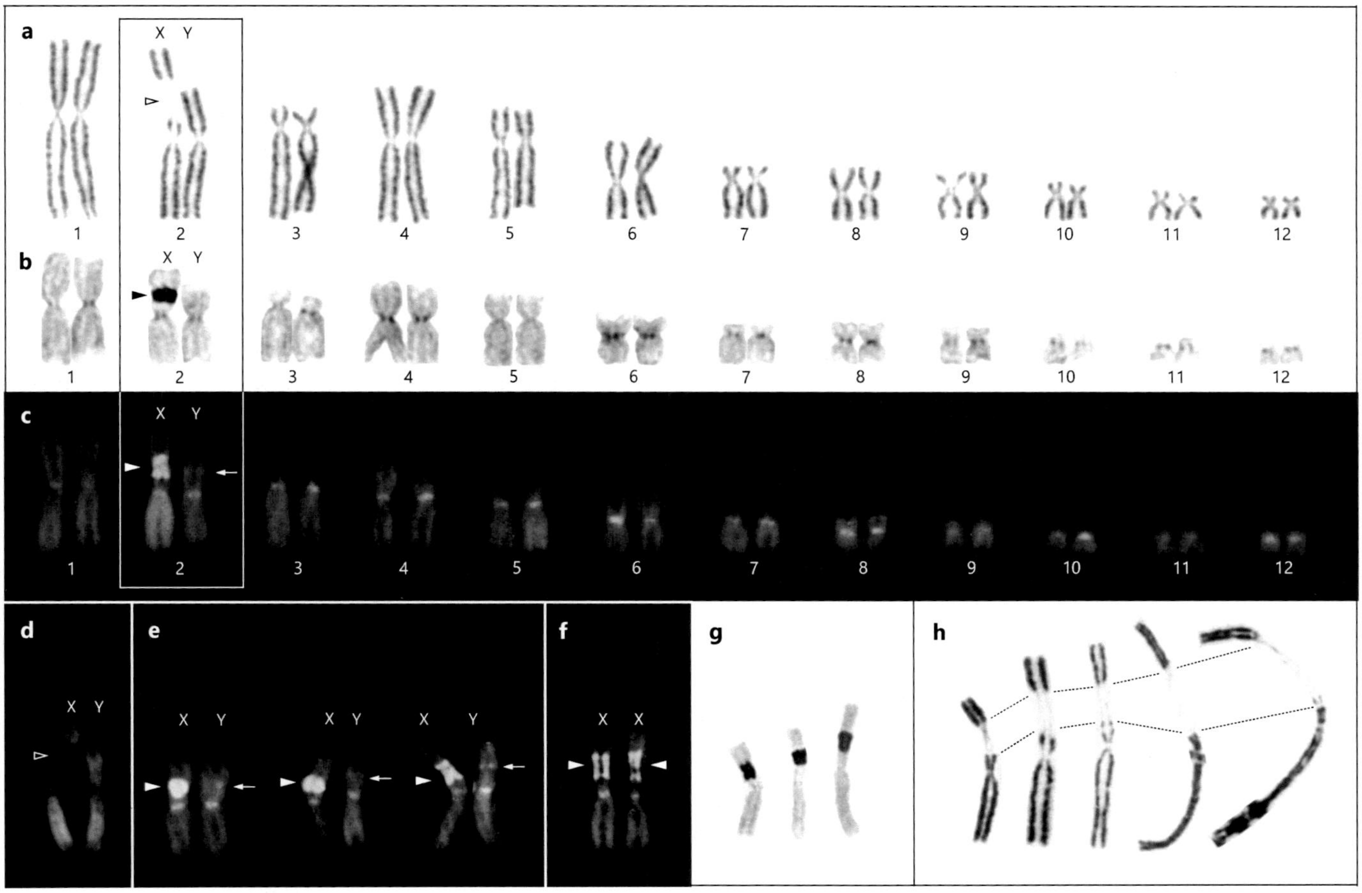

Fig. 26. a–c Karyotypes of male *Dryophytes femoralis* showing conventional Giemsa staining (**a**), Ag-staining of NORs (**b**), and distamycin A/mithramycin counterstaining (**c**). The heteromorphic XY sex chromosomes (pairs 2) are framed. The constitutive heterochromatin at the centromeric regions of all chromosomes and in the chromosome 8 short arm telomeric region is characterized by enhanced mithramycin fluorescence. **d** Selected XY sex chromosome pair after quinacrine staining showing the nonfluorescent nucleolar constriction in the X chromosome. **e, f** Selected XY (**e**) and XX (**f**) sex chromosome pairs exhibiting distamycin A/mithramycin fluorescence. **c, e, f** Note the extremely bright and sharp mithramycin fluorescence of the NOR located in the middle of the X short arm. In the corresponding region of the Y chromosome, only a faint mithramycin-positive heterochromatic band is present. **g** Three selected X sex chromosomes after Ag-staining with a prominent dark labeling of the NOR. **h** Five selected X sex chromosomes from BrdU-treated kidney cell cultures after conventional Giemsa staining showing distinct elongations of the nucleolar constrictions in their short arms. The proximal and distal ends of the nucleolar constrictions are connected by dotted lines. The open arrowheads indicate the nucleolar constriction, the filled arrowheads mark the NOR in the X chromosome. The small white arrows point to the very small mithramycin-positive heterochromatic region located in the middle of the Y short arm. For further explanations, see chapter 3.1.1.2.

Fig. 25. Karyotypes of female (**a, d, g, h**) and male (**b, c, e, f, i, j**) *Dryophytes femoralis* showing Ag-staining (**a–c**), distamycin A/mithramycin counterstaining (**d–f**), and FISH with an 18S + 28S rDNA probe and DAPI staining (**g–j**). The XX♀/XY♂ sex chromosomes are framed. Arrowheads indicate NORs in the X chromosomes. Note the absence of NORs in the Y chromosomes. For further explanations, see chapter 3.1.1.2.

Cytogenet Genome Res 2018;155:55–221
DOI: 10.1159/000489839

Table 5. Nucleolus organizer regions (NOR) located in autosomes and sex chromosomes of amphibians

Species (order, family)[a]	NOR-carrying chromosomes[b]					References
	Autosomes	X	Y	Z	W	
Gastrotheca riobambae (Anura, Hemiphractidae)	–	+	–			Schmid et al., 1983a, b, 1986, 2012; Schmid and Klett, 1994
Dryophytes femoralis (Anura, Hylidae)	–	+	–			Anderson, 1991; Schmid and Steinlein, 2003; Wiley, 2003; present report
Pristimantis sp. n. H (Anura, Strabomantidae)	–	+	–[c]			Schmid et al., 2010
Pseudopaludicola saltica (Anura, Leptodactylidae)	–	+	+			Duarte et al., 2010
Eleutherodactylus johnstonei (Anura, Eleutherodactylidae)	+[d]	+	+			Schmid et al., 2010
Buergeria buergeri (Anura, Rhacophoridae)	–			+	–	Ohta, 1986, 1987; Ohta et al., 1999; Schmid et al., 1993a
Pseudis tocantins (Anura, Hylidae)	–			+	+	Busin et al., 2008
Physalaemus ephippifer (Anura, Leptodactylidae)	–			+	+[e]	Nascimento et al., 2010
Leiopelma archeyi (Anura, Leiopelmatidae)	–			+	+	Green, 2002; Green and Sharbel, 1988
Leiopelma hamiltoni (now *L. pakeka*) (Anura, Leiopelmatidae)	–			+	+	Green, 1988a
Pyxicephalus adspersus (Anura, Pyxicephalidae)	+			–	+[f]	Schmid, 1980b
Hynobius hidamontanus (Caudata, Hynobiidae)	+			–	+	Ikebe et al., 2000
Hynobius quelpaertensis (Caudata, Hynobiidae)	+[g]			–	+[h]	Ikebe et al., 2005
Hynobius tokyoensis (Caudata, Hynobiidae)	+			–	+[h]	Kuro-o et al., 2002

[a] The name of the species is the same as used in the original publications. For further information, see Frost [2017].
[b] +, presence of NOR; –, absence of NOR.
[c] NORs have been deleted in ♂$_{a,b,c}$ type males because of 3 different Y-autosome centric fusions.
[d] Only specimens on the islands of Montserrat and Bermuda carry a single ectopic autosomal NOR.
[e] One NOR in the Z chromosome, 2 NORs in the W chromosome.
[f] Only some of the female specimens have a NOR-carrying W chromosome.
[g] In some specimens, a single ectopic NOR is present in an autosome in addition to the standard autosomal NOR pair.
[h] Three distinct types of W chromosomes (W^A, W^B, and W^C) have been detected in different populations.

oughly analyzed in subsequent studies. In this mutant, 1 of the 2 standard NORs located in the short arms of chromosome pair 12 is deleted. Heterozygous animals carrying the Oxford mutation are fully viable, show a single nucleolar constriction in the chromosomes, a single nucleolus in the cell nuclei, and contain about half the normal amount of rRNA-hybridizable 18S + 28S rDNA [Kahn, 1962; Barr and Esper, 1963; Birnstiel et al., 1966; Rafferty and Sherwin, 1969]. The homozygous anucleolar condition is lethal, and the embryos die at the tail bud stage showing a consistent syndrome of abnormalities [Elsdale et al., 1958; Wallace, 1960; Gurdon, 1977]. These embryos have no nucleoli and do not synthesize rRNA or high molecular weight rRNA precursor molecules [Brown and Gurdon, 1964]. Death occurs at this embryological stage because the maternal supply of rRNA in the embryonic cells is exhausted and their genome contains no genes for independent rRNA transcription.

The most likely explanation for the exclusive location of the NOR in 1 of the 2 sex chromosomes of *D. femoralis, G. riobambae, B. buergeri*, and *Pristimantis* sp. n. H is a deletion of the homologous NOR in the ancestral W and Y chromosomes. It would be most favorable for a stochastic fixation of this mutation by genetic drift if such a spontaneous deletion occurred in an individual of a small population or during an evolutionary bottleneck situation. The concept of genetic drift can be nicely applied to these cases of NOR deletions in Y and W chromosomes. Once the deletion has occurred in a founder individual, the random fixation or elimination of the deletion may occur in subsequent generations. The probabilities for fixation or elimination of the NOR deletion increase with decreasing size of the population. Random extinction and fixation are only extreme cases. In less extreme cases, both alleles (presence of NOR, deletion of NOR) are maintained, but their frequencies show a random fluctuation (genetic drift).

The presence of Ag-positive NORs in each of the 2 X chromosomes of *D. femoralis, G. riobambae*, and *Pristimantis* sp. n. H, as well as in each of the 2 Z chromosomes of *B. buergeri*, indicates that the X or Z chromosomes are not subject to an inactivation mechanism in the homogametic sex. Of course, this observation alone would not be enough to base general conclusions. Thus, in the case of the bat *C. perspicillata* and the marsupial species mentioned above, in which genetic inactivation of 1 of the 2 X chromosomes occurs in the female sex, the X-linked NORs escape inactivation [Goodpasture and Bloom, 1975; Hayman et al., 1982].

NORs are also located in both ZW sex chromosomes of the frog *Pseudis tocantins* (Fig. 33) [Busin et al., 2008] and in both XY sex chromosomes of the frog *Pseudopaludicola saltica* [Duarte et al., 2010] (Table 5). The New Zealand frogs *Leiopelma archeyi* and *L. hamiltoni* (now recognized as *L.*

Fig. 27. Five metaphases from BrdU-treated kidney cell cultures of a male *Dryophytes femoralis* after conventional Giemsa staining. **a** Metaphase showing early replication bands. **b–d** Metaphases with extreme BrdU-induced elongations of the nucleolar constriction in the X sex chromosome (arrowheads connected by lines) and chromatin fragments (arrows). **e** Metaphase containing a chromosome with completely decondensed chromatin (dc). For further explanations, see chapter 3.1.1.2.

pakeka, see Bell et al. [1998]) possess heteromorphic ZZ♂/ZW♀ sex chromosomes. NORs are located in the telomeric regions [Green, 1988b, 2002; Green and Sharbel, 1988]. Finally, the South African bull frog *Pyxicephalus adspersus* exhibits heteromorphic ZZ♂/ZW♀ sex chromosomes. The standard pair of NORs is normally located in the autosomes 6 [Schmid, 1980b]. In *Eleutherodactylus johnstonei* (family Eleutherodactylidae), the standard pair of NORs is located in both of the XX♀ and XY♂ sex chromosomes 14 [Schmid et al., 2010] (Table 5). The NORs are located in the W sex chromosomes of the 3 urodelan species listed in Table 5. *Hynobius hidamontanus*, *H. quelpaertensis*, and *H. tokyoensis* have been discussed in detail in the preceding monographs of Schmid et al. [2010, 2012].

Treatment of the *D. femoralis* cell cultures with BrdU for 24 h resulted in a very distinct elongation of the nucleolar constriction (Fig. 26h, 27). The thin chromatin threads within the constriction have a uniform appearance. No further constrictions are induced by BrdU in any other chromosome. It is interesting to note that such impressive elongations of the nucleolar constriction after in vitro BrdU treatment of fibroblast cultures were also observed in the X chromosome of the hemiphractid frog *G. riobambae* [Schmid et al., 2010].

In Ag-stained interphase nuclei from various somatic tissues of *D. femoralis*, the nucleoli are labeled as black blocks. Since the rest of the chromatin in the nuclei has only a bright yellow color, the nucleoli appear very highly contrasted. This allows exact determination of the number of nucleoli in male and female frog cells. As expected, the majority of male cells have a single nucleolus (Fig. 28b, e, f). However, the quantitative analysis reveals that male cells with 2 Ag-stained nucleoli also occur with a frequency of about 20%. One of the 2 blocks is always distinctly smaller

Cytogenet Genome Res 2018;155:55–221
DOI: 10.1159/000489839

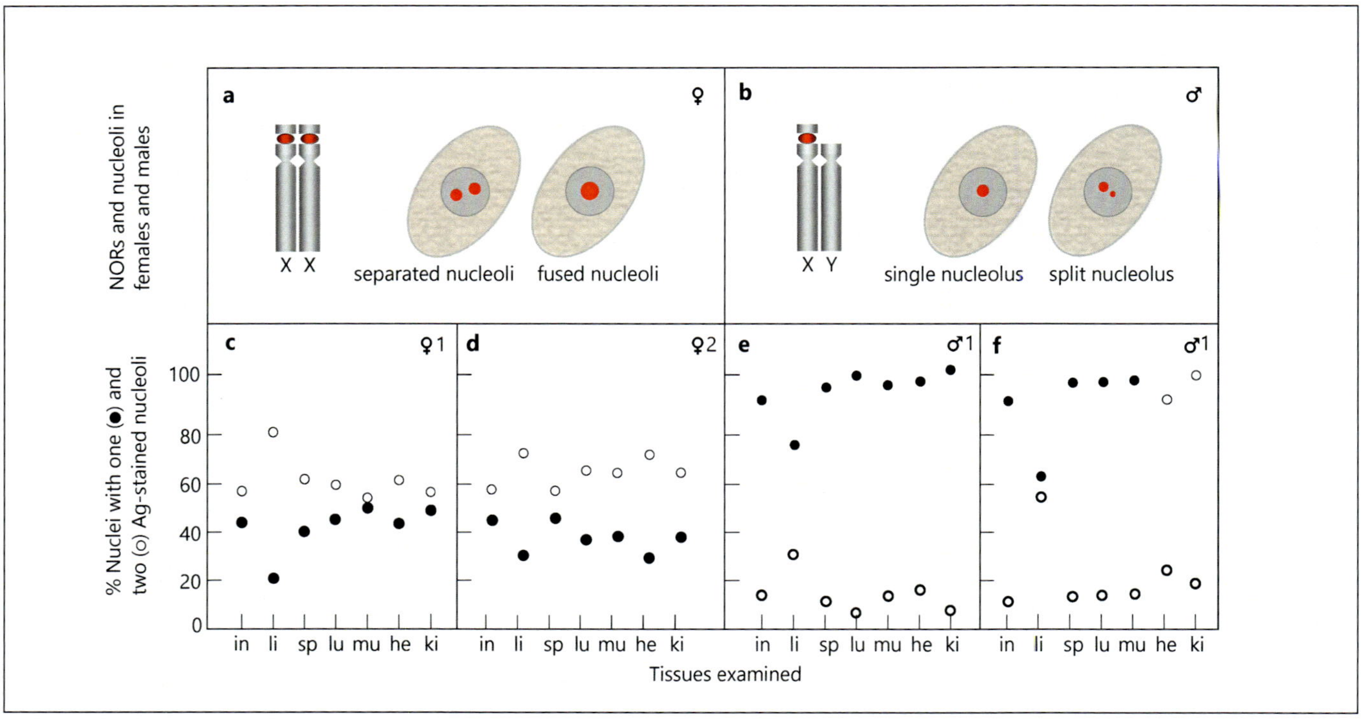

Fig. 28. a, b Schematic drawings of the XX♀/XY♂ sex chromosomes and interphase cells of female (**a**) and male (**b**) *Dryophytes femoralis*. **a** Female cells contain 2 X chromosomes and 2 X-linked nucleolus organizer regions (NORs; red ellipses in the short arms of the chromosomes). In interphase nuclei, the 2 NORs form 2 nucleoli (red circles), which often fuse to produce 1 larger nucleolus. **b** Male cells contain only 1 X chromosome and 1 NOR (red ellipse). In interphase nuclei, the single NOR forms 1 nucleolus (red circle), from which distinctly smaller nucleoli sometimes split off. **c–f** Frequencies of interphase nuclei with 1 and 2 Ag-stained nucleoli in various somatic tissues from 2 female (**c**, **d**) and 1 male (**e**, **f**) individual of *D. femoralis*. For all tissues, 100 nuclei each of the cell types were analyzed. in, intestine; li, liver; sp, spleen; lu, lungs; mu, muscle; he, heart; ki, kidney. For further explanations, see chapter 3.1.1.2.

than the other and is interpreted as a splitting of the large nucleolus. A maximum of 2 Ag-stained nucleoli is demonstrated in the cell nuclei of female animals (Fig. 28a, c, d). Of the nuclei from the various tissues examined, 60–80% have distinctly separated nucleoli. The remaining 20–40% of the nuclei show 1 single and large Ag-block which apparently originates by fusion of the 2 NORs of the homologous X chromosomes. The exact rates of fused NORs in the interphase nuclei were determined for 7 different tissues from 2 females. Figure 28c, d shows that these frequencies are in good agreement for both specimens.

The presence of a single, X chromosome-linked NOR in the *D. femoralis* male genome raises an interesting and fundamental question concerning the supply of the haploid male gamete cells with a sufficient amount of 18S + 5.8S + 28S rRNA required for their differentiation. In the following, this question is addressed in detail [see also Schmid et al., 2012].

The process of spermatogenesis in *D. femoralis* develops like it does in other arboranans and anurans. After a series of mitotic divisions, the primary spermatogonium produces a large number of secondary spermatogonia which enter into meiosis I as primary spermatocytes and then into meiosis II as secondary spermatocytes. The diakineses contain 1 X chromosome with 1 active X-linked NOR (with a complete set of rRNA genes) and 1 Y chromosome (Fig. 29).

In the primary spermatocytes, all chromosomes and the NOR have doubled. They now contain 2 X chromosomes, 2 active NORs, and 2 Y chromosomes (4n stage). After the first meiotic division, 50% of the secondary spermatocytes contain 2 X chromosomes and 2 inactive X-linked NORs, the remaining 50% contain 2 Y chromosomes and no NOR (2n stage). Following completion of the second meiotic division, the early and simple-structured spherical spermatids originate. 50% of these spermatids contain 1 X chromosome with its NOR, the other 50% contain 1 Y chromosome and no NOR. In the X chromosome-carrying spherical spermatids, the NOR is reactivated, reaching the maximum of its postmeiotic activity. This can be recognized by the very large size of the Ag-stainable nucleoli. During early spermiohistogenesis, the spherical spermatids rapidly attain the morphology of elongated spermatids. Finally, the elongated spermatids differentiate into very thin mature sperm cells with their typical cytoplasm-derived structures, the acrosome, midpiece, and flagellum. In the X chromosome-carrying spermatids, the X-linked NOR continuously decreases activity until the late stages of spermiohistogenesis and again becomes inactivated in nearly mature sperm cells. Thus, 50% of the mature sperm cells contain an X chromo-

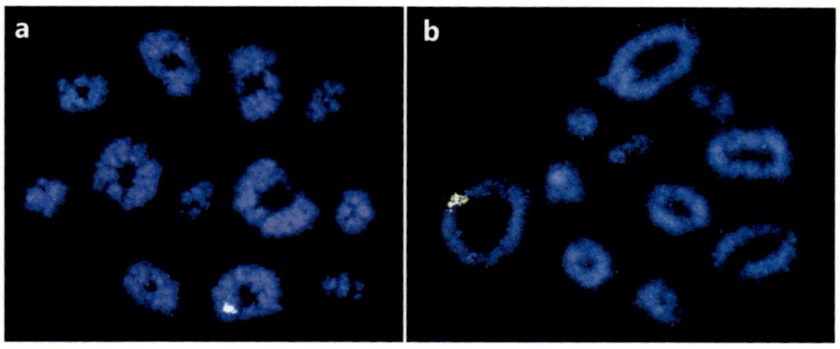

Fig. 29. Two diakineses of a male *Dryophytes femoralis* after FISH with an 18S + 28S rDNA probe and DAPI staining. The XY sex bivalents are still in a ring-like pairing configuration. The only site of rDNA hybridization is the NOR in the short arm of the X sex chromosome. No hybridization signal is present in the Y sex chromosome at the opposite site of the sex bivalent. For further explanations, see chapter 3.1.1.2.

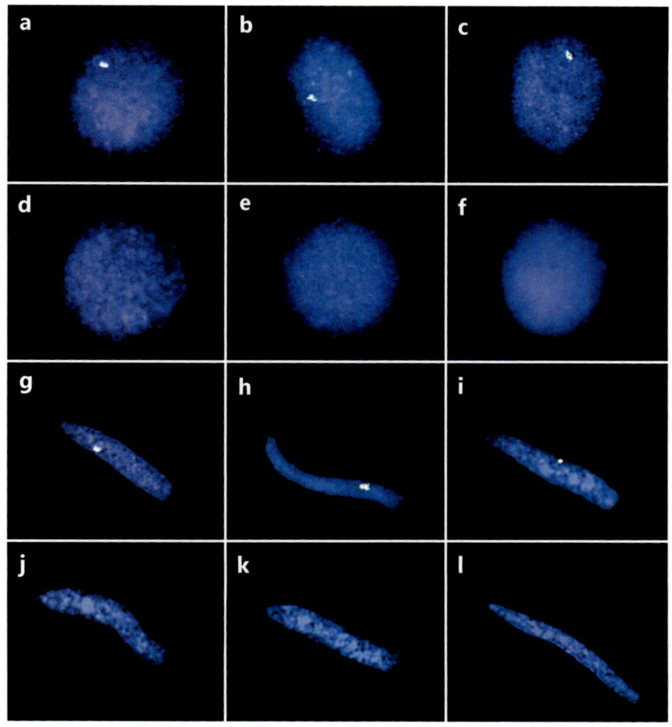

Fig. 30. Early spherical (**a–f**) and elongating (**g–l**) spermatid nuclei of *Dryophytes femoralis* after FISH with an 18S + 28S rDNA probe and DAPI staining. **a–c**, **g–i** Nuclei containing the X sex chromosome with its NOR (hybridization signals). **d–f**, **j–l** Nuclei carrying the Y sex chromosome which lacks a NOR. For further explanations, see chapter 3.1.1.2.

some and a completely inactivated NOR. The other 50% of the sperm cells contain the Y chromosome and lack a NOR.

In the spermatogenesis of all vertebrates, the NORs are postmeiotically reactivated in the early spherical spermatids where they form large nucleoli. This holds also true for the spermatogenesis of *D. femoralis*. After specific Ag-staining, in about 50% of the early spherical spermatids 1 single Ag-stained nucleolus is observed. These must contain the X chromosome and the active X-linked NOR. In the remaining 50% of the spherical spermatids, which must contain the Y chromosome, Ag-labeled nucleoli are absent. Analogously, in approximately 50% of the early spherical spermatids, a very distinct hybridization signal is present after FISH with an 18S + 28S rDNA probe (Fig. 30a–c), whereas in the other 50% of the spermatid nuclei there is no hybridization signal (Fig. 30d–f). The same situation holds true for the elongating spermatids and mature sperm nuclei (Fig. 30g–l).

The NOR is the specific chromosome region which causes the formation of a nucleolus in the interphase stage of eukaryote cell nuclei and which contains the main rRNA genes (18S + 5.8S + 28S rDNA). The rRNA is synthesized and processed into pre-ribosomes in the nucleoli, and then, together with a large number of different proteins, becomes part of the mature ribosomes in the endoplasmic reticulum. A further component of the nucleolus is the 5S rRNA that is transcribed from multiple genes that are not located in the NOR. During spermiohistogenesis, a high amount of specific protein synthesis occurs which requires an intensive synthesis of sufficient ribosomes. The Y chromosome-carrying early spherical spermatids of *D. femoralis* do not contain any 18S + 5.8S + 28S rRNA genes. Therefore, the only ribosomes they contain at the endoplasmic reticulum come from the primary spermatocytes from which they originate by meiotic division. If this supply is exhausted, the Y chromosome-carrying spermatids will fail to differentiate further and die. As a consequence, only X chromosome-carrying mature sperm cells and female offspring will be produced, which is clearly not the case. There are 2 possible explanations that, in spite of their anucleolar condition, the Y chromosome-carrying spermatids complete spermiohistogenesis successfully. (1) The supply of ribosomes coming from the preceding primary spermatocytes must be sufficient to complete spermiohistogenesis. Because of the duration of amphibian spermiohistogenesis (approximately 16 days from meiotic telophase I to mature sperm cells in *Rana esculenta*) and the half-life of ribosomes at the endoplasmic reticulum (approximately 4.5 days; data from mammalian liver cells) [Nordgren and Stenram, 1972; Tsurugi et al., 1974; Nikolov et al., 1983], this seems rather unlikely unless *D. femoralis* has a considerably shorter spermiohistogenesis and/or a distinctly longer half-life of ribosomes at the endoplasmic reticulum in their spermatids. (2) The 4 spermatids resulting from a secondary spermatocyte do not divide completely from each other, but remain together forming a syncytial complex. Owing to an incomplete cytokinesis, the spermatids are connected by intercellular cytoplasmic bridges that break up during late spermiohistogenesis. The existence of such postmeiotic syncytial complexes overcomes the potential lethality of the Y chromosome-carrying and NOR-deficient spermatids by metabolic cooperation. The X-linked NOR in the X chromosome-carrying spermatids can support differentiation of the neighboring Y chromosome-carrying and NOR-deficient spermatids by providing 18S + 5.8S + 28S rRNA.

A similar conclusion can be reached for the marsupial frog *G. riobambae* and the terraranan frog *Pristimantis* sp. n. H (family Strabomantidae) in which the males also possess a single NOR located in the X chromosome (Table 5) [Anderson, 1991; Schmid and Steinlein, 2003; Wiley, 2003; Schmid et al., 2010]. A further and classical example is the well-known case of the anucleolar Oxford mutation of *X. laevis*. Crosses between heterozygous individuals carrying the mutation yield 25% embryos that are homozygous for the mutation and die at the young tail bud stage. This demonstrates that anucleolate spermatids differentiate into mature and fully viable sperm cells, which fertilize without any problem. In the Australian marsupial *M. eugenii*, the sole NOR in the karyotype is X-linked [Haaf et al., 1991]. In spermiohistogenesis, the anucleolate Y-carrying spermatids differentiate into mature sperm cells in the same way as do the spermatids containing the X and the NOR. Further examples are the American bat *C. perspicillata* and the Australian kangaroo rat *P. tridactylus* in which the X chromosomes are translocated onto an autosome and contain the sole NOR in the karyotype [Hsu et al., 1967, 1975a; Goodpasture and Bloom, 1975]. Also in these 2 species, the anucleolate Y-carrying spermatids have no apparent developmental disadvantages. Finally, the cytogenetic analysis of thousands of amphibians collected in the field and belonging to a variety of taxa revealed a number of male individuals with a deletion of one of the 2 NORs [Schmid, 1982; Schmid et al., 2010]. Examination of the spermiohistogenesis of these males always shows no reduction in the number of mature spermatozoa.

3.1.1.3 Hylidae with 2n = 24 and Homomorphic XY♂ Sex Chromosomes

In the last years, the group of Perrin accumulated convincing experimental evidence that the XY♂/XX♀ system is operating in taxa of the genus *Hyla* (*H. arborea* [Fig. 31], *H. intermedia*, *H. meridionalis*, *H. molleri*, *H. orientalis*, *H. sarda*) (for references, see Table 6). The sex chromosomes of these *Hyla* species are absolutely homomorphic. These studies relied on recent developments of DNA markers [Stöck et al., 2001; Berset-Brändli et al., 2006; Yoshimoto et al., 2008; Cano et al., 2011; Guerrero et al., 2012]. Their analyses demonstrated that these species share the same pair of sex chromosomes with identical patterns of male heterogamety and occasional XY recombination in XY females during rare sex reversal events [Perrin, 2009]. However, because of so many particular presuppositions, a non-consideration of female meiosis at the lampbrush phase, and the lack of mathematical models, this so-called "fountain-of-youth" model for homomorphic XY sex chromosomes is poorly conceived and not applicable to other forms of vertebrate sex chromosomes. Nonetheless, these authors showed that homoeology, or at least partial homoeology still exists between the sex chromosomes of *H. arborea*, *Bufotes*

Fig. 31. Male *Hyla arborea* from France. Photograph by James P. Bogart. For further explanations, see chapter 3.1.1.3.

siculus, *Rana temporaria*, the clawed frog *Xenopus tropicalis*, and the zebra finch *Taeniopygia guttata* [Brelsford et al., 2013]. And in a more recent study using a transcriptome of the green toad *Bufotes viridis*, they could demonstrate that, depending on the phylogenetic distance, several markers which are sex-linked in *B. viridis* show high cross-amplification success across the anuran phylogeny, involving 9 systematic anuran families [Gerchen et al., 2016].

3.1.1.4 Hylidae with 2n = 24 and Heteromorphic ZW♀ Sex Chromosomes

Dryophytes squirellus (squirrel treefrog, Fig. 32a) is a small frog native to the United States. It is found along the Atlantic coastal plain from southeastern Virginia to the Florida Keys, and west along the gulf plain to southeastern Texas. It has been introduced to Grand Bahama Island and Little Bahama Bank. Adult frogs are about 4 cm in length. They have a range of coloration, from brown to yellow to green, with a variety of markings. They are nocturnal and can be found hunting insects around lights at night. Between March and October, they migrate to breeding areas, usually shallow temporary water bodies where fish are absent. They especially prefer open wetlands, marshes, and flooded ditches.

In a cytogenetic examination on a limited number of *D. squirellus*, Anderson [1991] recognized an initial morphological differentiation of heteromorphic ZW♀ sex chromosomes. Of the 6 female treefrogs analyzed, all presented a distinct heterochromatic band in the chromosome 1 short arm immediately adjacent to the centromeric constriction. This C-band was absent in both chromosomes 1 of the male animals. Furthermore, restriction enzyme digestions revealed that the W chromosome-specific band has a unique DNA sequence within *D. squirellus*. None of the restriction enzymes was able to digest this C-band [Anderson, 1991].

Several of the initial stages of sex chromosome differentiation were conserved in vertebrates. It is possible to create evolutionary series in which the increasing structural complexity of the sex chromosomes can be reconstructed. In a study on sex chromosomes and W interphase chromatin in snakes, Ray-Chaudhuri et al. [1971] concluded, for the first time, that heterochromatinization of the W chromosome

Table 6. Sex-determining mechanisms and sex-specific chromosomes found in Anura

Family Species[a]	Genetic sex[b]	Sex chromosomes		References
		type[c]	morphology[d]	
Alsodidae				
Eupsophus insularis		XX/XY	X = Y (ci, ch)	Cuevas and Formas, 1996
Eupsophus migueli		XX/XY	X = Y (ci, ch)	Iturra and Veloso, 1981, 1989
Eupsophus roseus		XX/XY	X = Y (ch)	Iturra and Veloso, 1989
Bombinatoridae				
Bombina orientalis	XX♀/XY♂			Kawamura and Nishioka, 1977; Ueda, 1980
Bufonidae				
Bufo bufo	ZZ♂/ZW♀			Harms, 1926; Ponse 1941, 1942, 1950; Kawamura and Nishioka, 1977; Engel and Schmid, 1981
Duttphrynus melanostictus		XX/XY	X = Y (ch, rbp)	Siripiyasing et al., 2008
Bufotes viridis	XX♀/XY♂		X = Y	Kawamura et al., 1980; Ueda, 1985, 1990; Stöck et al., 2013b
Bufotes viridis (population-specific)	ZZ♂/ZW♀	ZZ/ZW	Z = W (ch, rbp)	Odierna et al., 2007
Centrolenidae				
Vitreorana antisthenesi		XX/XY	X = Y (ci, ch)	Schmid et al., 1989
Dicroglossidae				
Quasipaa boulengeri	XX♀/XY♂	XX/XY	X = Y (ci)	Qing et al., 2012, 2013
Fejervarya limnocharis	XX♀/XY♂	XX/XY	X > Y (ci)	Patawang et al., 2014
Hemiphractidae				
Gastrotheca ovifera		XX/XY	X > Y	Schmid et al., 1988, 2002a, 2012
Gastrotheca peruana		XX/XY	X = Y (ci, ch)	Schmid et al., 2012
Gastrotheca pseustes		XX/XY	X = Y (ch)	Schmid et al., 1990a, 2012
Gastrotheca riobambae		XX/XY	X < Y (nor)	Schmid et al., 1983a, b, 1986, 2012; Schmid and de Almeida, 1988; Schmid and Klett, 1994
Gastrotheca walkeri		XX/XY	X > Y	Schmid et al., 1988, 2002a, 2012
Hylidae				
Hyla arborea	XX♀/XY♂			Berset-Brändli et al., 2006
Hyla intermedia	XX♀/XY♂			Berset-Brändli et al., 2006
Hyla meridionalis	XX♀/XY♂			Berset-Brändli et al., 2006
Hyla molleri	XX♀/XY♂			Stöck et al., 2011
Hyla orientalis	XX♀/XY♂			Stöck et al., 2013a
Hyla sarda	XX♀/XY♂			Berset-Brändli et al., 2006
Dryophyters japonicus	XX♀/XY♂			Kawamura and Nishioka, 1977
Dryophytes femoralis		XX/XY	X > Y (nor)	Anderson, 1986; Anderson and Moler, 1986; Anderson, 1991; Schmid and Steinlein, 2003; Wiley, 2003; present monograph
Dryophytes squirellus		ZZ/ZW	Z < W (ch)	Anderson, 1991
Dryophytes suweonensis		ZZ/ZW	Z = W (ch)	Yu and Lee, 1990
Pseudis tocantins		ZZ/ZW	Z < W	Busin et al., 2008; Gatto et al., 2016
Leiopelmatidae				
Leiopelma archeyi		ZZ/ZW	Z < W	Green, 2002
Leiopelma hamiltoni		ZZ/ZW	Z < W	Green, 1988a, 2002
Leiopelma hochstetteri		00/0W		Green, 1988b
Leptodactylidae				
Physalaemus ephippifer		ZZ/ZW	Z = W (ch, nor)	Nascimento et al., 2010
Engymostops petersi		XX/XY	X > Y	Lourenço et al., 1999
Pseudopaludicola saltica		XX/XY	X = Y (ci)	Duarte et al., 2010
Myobatrachidae				
Crinia bilingua		ZZ/ZW	Z < W	Mahony, 1991
Odontophrynidae				
Proceratophrys boiei		ZZ/ZW	Z = W (ch)	Ananias et al., 2007
Pelodytidae				
Pelodytes punctatus	XX♀/XY♂			Zaborski, 1979
Pipidae				
Xenopus gilli	ZZ♂/ZW♀			Engel and Kobel, 1983
Xenopus laevis	ZZ♂/ZW♀			Gallien, 1953; Chang and Witschi, 1955, 1956; Wachtel et al., 1975
Xenopus tropicalis	ZZ♂/ZW♀			Kochan et al., 2003
Pyxicephalidae				
Pyxicephalus adspersus	ZZ♂/ZW♀	ZZ/ZW	Z > W	Schmid, 1980b; Engel and Schmid, 1981; Schmid and Bachmann, 1981
Pyxicephalus edulis	ZZ♂/ZW♀	ZZ/ZW	Z > W	Schmid, unpublished
Tomopterna delalandii	ZZ♂/ZW♀	ZZ/ZW	Z = W (ci, ch)	Schmid, 1980b; Schmid and Bachmann, 1981
Ranidae				
Amolops mantzorum	XX♀/XY♂	XX/XY	X = Y (ci)	Liu et al., 2017
Lithobates berlandieri	XX♀/XY♂			Wright et al., 1983
Lithobates blairi	XX♀/XY♂			Wright et al., 1983

Table 6 (continued)

Family Species[a]	Genetic sex[b]	Sex chromosomes type[c]	Sex chromosomes morphology[d]	References
Lithobates catesbeiana	XX♀/XY♂			Elinson, 1981
Lithobates clamitans	XX♀/XY♂			Elinson, 1981, 1983
Lithobates pipiens	XX♀/XY♂			Wachtel et al., 1975; Richards and Nace, 1978; Wright and Richards, 1983
Lithobates sphenocephalus	XX♀/XY♂			Wright et al., 1983
Pelophylax esculentus	XX♀/XY♂	XX/XY	X = Y (rbp)	Witschi, 1923; Schempp and Schmid, 1981
Pelophylax porosus	XX♀/XY♂			Kawamura and Nishioka, 1977
Pelophylax ridibundus	XX♀/XY♂			Kawamura and Yokota, 1959; Zaborski, 1979
Rana japonica	XX♀/XY♂	XX/XY	X = Y (rbp, ch)	Moriwaki, 1959; Kawamura and Yokota, 1959; Kawamura and Nishioka, 1977; Miura, 1994
Rana nigromaculata	XX♀/XY♂		X = Y (rbp)	Kawamura, 1939; Kawamura and Nishioka, 1977; Wu and Zhang, 1985
Rana sakuraii	XX♀/XY♂	XX/XY	X = Y (rbp, ch)	Ryuzaki et al., 1999
Rana tagoi	XX♀/XY♂	XX/XY	X = Y (rbp, ch)	Ryuzaki et al., 1999, 2006
Rana temporaria	XX♀/XY♂			Witschi, 1929
Rana tsushimensis	XX♀/XY♂			Kawamura and Nishioka, 1977
Glandurana rugosa	XX♀/XY♂	XX/XY	X = Y (ci, rbp)	Kawamura and Nishioka, 1977; Nishioka et al., 1993a, b, 1994; Nishioka and Hanada, 1994; Miura et al., 1997
Glandurana rugosa	ZZ♂/ZW♀	ZZ/ZW	Z = W (ci, rbp)	Nishioka et al., 1993a, b, 1994; Nishioka and Hanada, 1994; Miura et al., 1997
Rhacophoridae				
Buergeria buergeri	ZZ♂/ZW♀	ZZ/ZW	Z = W (ch, nor)	Ohta, 1986, 1987; Schmid et al., 1993a; Ohta et al., 1999; Hanada, 2002
Eleutherodactylidae				
Eleutherodacytylus albipes		ZZ/ZW	Z < W	Schmid et al., 2010
Eleutherodacytylus casparii		ZZ/ZW	Z < W	Schmid et al., 2010
Eleutherodactylus cuneatus		ZZ/ZW	Z < W	Schmid et al., 2010
Eleutherodactylus emiliae		ZZ/ZW	Z < W	Schmid et al., 2010
Eleutherodactylus glamyrus		ZZ/ZW	Z < W	Schmid et al., 2010
Eleutherodactylus turquinensis		ZZ/ZW	Z < W	Schmid et al., 2010
Eleutherodactylus sp. n. G		ZZ/ZW	Z < W	Schmid et al., 2010
Eleutherodactylus johnstonei		XX/XY	X = Y (ci)	Schmid et al., 2010
Eleutherodactylus oxyrhyncus		XX/XY	X < Y	Schmid et al., 2010
Eleutherodactylus cavernicola		XX/XY	A^Y fusion	Schmid et al., 2010
Strabomantidae				
Pristimantis euphronides		ZZ/ZW	Z < W	Schmid et al., 2002d, 2010
Pristimantis shrevei		ZZ/ZW	Z < W	Schmid et al., 2002d, 2010
Pristimantis pulvinatus		XX/XY	A^Y fusion	Schmid et al., 2010
Pristimantis riveroi		XX/XY	A^Y fusion	Schmid et al., 2003b, 2010
Pristimantis sp. n. H		XX/XY	several A^Y fusions	Schmid et al., 2010
Strabomantis biporcatus		XX/XY	A^Y fusion	Schmid et al., 1992, 2002e, 2010

[a] For valid species names, see Frost [2017]. Individuals of *Leiopelma hamiltoni* with ZZ/ZW sex chromosomes are now considered to represent the cryptic species *Leiopelma pakeka* [Bell et al., 1998]. The reports on heteromorphic ZW sex chromosomes in *Rana cyanophlyctus* [Yadav and Pillai, 1975] and *R. tigrina* [Yadav and Pillai, 1975; Chakrabarti et al., 1983] as well as XY sex chromosomes in *Hyla arborea japonica* [Yosida, 1957] are not listed because of the insufficient quality of the chromosomes presented in these reports. The observation of heteromorphic ZW sex chromosomes in *Xenopus laevis* [Weiler and Ohno, 1962; Morescalchi, 1963] and *Discoglossus pictus* [Morescalchi, 1964, 1973; Olmo et al., 1982] have been refuted by Tymowska and Kobel [1972], Schmid and Steinlein [1991], and Odierna et al. [1999]. The ZZ/ZW sex chromosomes found in 3 male and female individuals of a single population of *Rana clamitans* [Mengden, 1981] must be confirmed in this and other populations. The finding of ZZ/ZW or XX/XY sex chromosomes in *Bufo bufo gargarizans* [Shang and Deng, 1983; Wen et al., 1983], *B. raddei* [Deng and Shan, 1984], *B. marinus* [Abramyan et al., 2009], *B. melanostictus* [Siripiyasing et al., 2008], *Dryophytes suweonensis* [Yu and Lee, 1990] as well as XX/XY sex chromosomes in *Fejervarya limnocharis* [Patawang et al., 2014] still needs corroboration.

[b] Determined by the analysis of the sex ratio in the progeny of sex-reversed animals, parthenogenetically bred individuals, triploid animals, patterns of inheritance, and expression of isozymes encoded by the sex chromosomes, HY antigen typing, analysis of DNA microsatellite or AFLP markers.

[c] Determined by cytogenetic analyses.

[d] X = Y or Z = W, both sex chromosomes have the same size or nearly the same size but differ in their replication banding patterns (rbp), centromeric index (ci), amount of constitutive heterochromatin (ch), presence of a nucleolus organizer region in only one of the 2 sex chromosomes (nor), or the Y is involved in a centric (robertsonian) fusion with an autosome (A^Y). X < Y, the X is smaller than the Y. X > Y, the X is larger than the Y. Z < W, the Z is smaller than the W. Z > W, the Z is larger than the W. In these cases one of the 2 sex chromosomes contains distinctly more constitutive heterochromatin than its partner. In *Leiopelma hochstetteri*, no Z chromosomes are present. The XY sex chromosomes of *Physalaemus petersi* are present only in male individuals with the karyotype I in the populations examined. The sex chromosomes of *Rana rugosa*, *Proceratophrys boiei*, *Leiopelma archeyi*, and *Eleutherodactylus johnstonei* show geographic variability. In *Eleutherodactylus cavernicola*, *Pristimantis pulvinatus*, *P. riveroi*, *P.* sp. n. H, and *Strabomantis biporcatus*, male individuals with a Y-autosome fusion coexist with males without the fusion. In *P.* sp. n. H, 3 different Y-autosome fusions occur in males [Schmid et al., 2010].

rather than an inversion in the W chromosome can be the primary step in the evolution of heteromorphic ZW sex chromosomes. This was later confirmed in investigations on constitutive heterochromatin and a W chromosome-associated repetitive satellite DNA in snakes [Singh et al., 1976; Jones and Singh, 1981; Jones, 1983a, b]. Thus, analytical density gradient centrifugation of DNA and in situ hybridization on metaphase chromosomes revealed a specific satellite DNA that is located in the highly differentiated W chromosomes of advanced snake families. The same W-associated DNA was also demonstrated in those highly evolved snake species that still possess undifferentiated (homomorphic) ZW sex chromosomes. A number of the known sex chromosomes of amphibians support the assertion that one of the initial steps in the evolution of sex chromosomes was an accumulation of repetitive DNA in the original W and Y chromosomes. In the 2 former comprehensive monographs in this series (The Chromosomes of Terrararan Frogs [Schmid et al., 2010]; The Hemiphractid Frogs [Schmid et al., 2012]), these mechanisms were explained in detail.

It must be pointed out here that whereas in *Dryophytes femoralis* male heterogamety (XY♂/XX♀) plays the decisive role in sex determination, in *D. squirellus* female heterogamety (ZZ♂/ZW♀) operates. Hillis and Green [1990] used the knowledge of amphibian phylogeny to examine the evolution of sex-determining mechanisms in Amphibia. After the phylogenetic tree was constructed, the heterogametic character states of 63 amphibian species were mapped to the phylogeny in the most parsimonious arrangements (i.e., in the fewest number of possible changes [Fitch, 1971]). The analysis suggested that the ancestral state for amphibians was female heterogamety (ZZ♂/ZW♀). During evolution, male heterogamety (XY♂/XX♀) evolved independently at least 7 times, whereas in only 1 case, male heterogamety reversed to female heterogamety. Based on this phylogenetic analysis, Hillis and Green [1990] proposed a recessive mutation to be involved in the eventual acquisition of male heterogamety. A reversal from male to female heterogamety would require a de novo evolution of a functional control gene. This model would explain the observed bias in the evolution from ZZ♂/ZW♀ to XY♂/XX♀ sex-determining systems.

a

Fig. 32. a *Dryophytes squirellus* from Tampa, FL, USA. Photograph by James P. Bogart. **b** *Pseudis tocantins* from Formoso do Araguaia, Tocantins State, Brazil. Photograph by Adrian Garda. For further explanations, see chapter 3.1.1.4.

The second hylid frog species with female heterogamety and heteromorphic ZW♀ sex chromosomes is *Pseudis tocantins* (Fig. 32b). It is a common and colorful aquatic frog, about 4 cm in length endemic to central Brazil in the States of Goias, Tocantins, and Mato Grosso, along the Tocantins and Araguaia River. The species occurs at low altitudes and is found in moist savanna, subtropical or tropical moist shrubland, swamps, freshwater lakes, freshwater marshes, and ponds. It has several adaptations to aquatic life, such as protuberant eyes, robust hind limbs, and fully webbed feet.

Busin et al. [2008] and, more recently, Gatto et al. [2016] conducted chromosome analyses on 4 males and 7 females of *P. tocantins* collected in Porto Nacional, Tocantins State, Brazil. The Z chromosome of this hylid is metacentric and presents an interstitial NOR in the long arm that is adjacent to a pericentromeric heterochromatic band (Fig. 33; Table 6). In contrast, the submetacentric W chromosome carries a pericentromeric NOR in the long arm, which is adjacent to a clearly evident heterochromatic band that is larger than the band found in the Z chromosome. This explains the size difference observed between these chromosomes. Using molecular and molecular cytogenetic techniques, Gatto et al. [2016] provided evidence that the noncentromeric heterochromatic bands in the long arms of the Z and W chromosomes differ not only in size and location but also in composition, based on comparative genomic hybridization and an analysis of the anuran PcP190 satellite DNA. Seven groups of PcP190 sequences were recognized in the genome of *P. tocantins*. PcP190 probes mapped to the heterochromatic band in the long W arm, and a Southern blot analysis indicated the accumulation of PcP190 in the female genome of *P. tocantins*, which suggests the involvement of this satellite DNA sequence during the evolution of the sex chromosomes of this species.

Busin et al. [2001, 2008] compared the karyotypes of 7 *Pseudis* species using conventional Giemsa staining, C-banding, and Ag-NOR staining. With the exception of *P. cardosoi* (*P.* sp. (aff. *minuta*)), all species have 2n = 24 chromosomes and FN = 48 (Fig. 33, 34a; Appendix 1). Interestingly, the chromosomes 7 which constitute the heteromorphic ZZ♂/ZW♀ sex chromosomes in *P. tocantins* also exhibit structural heteromorphisms in *P. paradoxa*, *P. platensis*, *P. bolbodactyla*, *P. fusca*, and *P. minuta* from the São Jerônimo population (Fig. 33). Further studies on more male and female individuals should clarify whether these heteromorphisms are early stages of sex chromosome differentiation.

P. cardosoi differs from all other *Pseudis* species in its diploid chromosome number (2n = 28), although its fundamental number is the same (FN = 48). Contrary to the other *Pseudis* species, *P. cardosoi* has 4 telocentric chromosomes which obviously originated through 2 fission cycles (Fig. 34a).

Yu and Lee [1990] claimed to have found evidence for the existence of heteromorphic ZW♀ sex chromosomes in the Asiatic hylid species *Dryophytes immaculatus* (formerly *D. suweonensis*). The Z chromosome is the third chromosome in the karyotype, and it is supposed to contain more constitutive heterochromatin than the W chromosome (Table 6). However, judging from the chromosome illustrations that

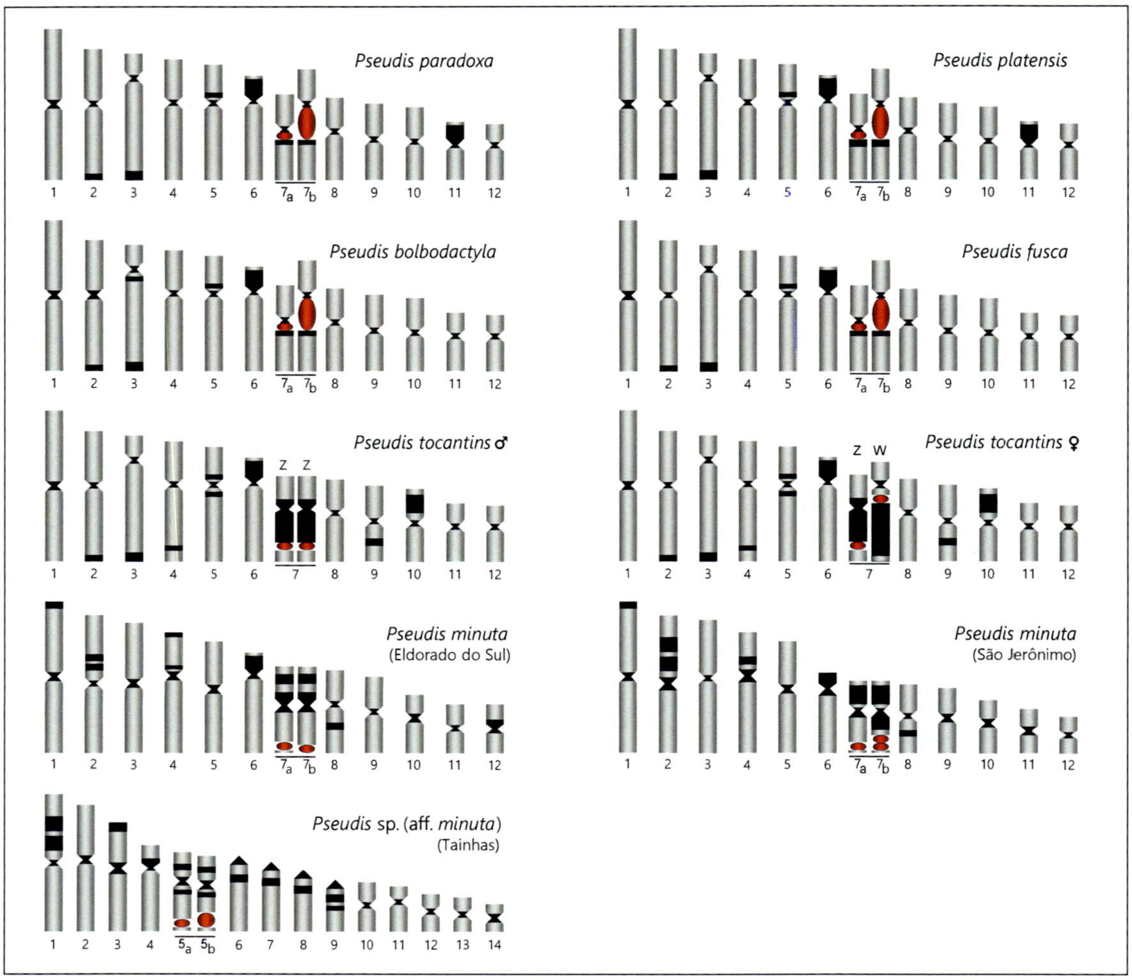

Fig. 33. Schematic drawings of the chromosomes of 9 species of the genus *Pseudis*. Note the heteromorphic ZZ♂/ZW♀ sex chromosomes in *Pseudis tocantins*. Black areas, constitutive heterochromatin; red circles and ovals, NORs. The NORs in the chromosomes 7_a and 7_b of *P. platensis*, *P. bolbodactyla*, *P. fusca*, and *P. minuta*, and in chromosomes 5_a and 5_b of *P.* sp. (aff. *minuta*) are distinctly heteromorphic. According to Busin et al. [2001, 2008]. For further explanations, see chapter 3.1.1.4.

Yu and Lee [1990] presented, these results can only be considered as preliminary.

So far, in hylid species with a diploid chromosome number of 2n = 24 and NF = 48, only 4 species with microscopically distinguishable and 4 species with undifferentiated sex chromosomes have been documented (Table 6). At a first glance, this frequency appears to be rather low, but it is comparable with the number of those species recognized for the anuran order [Schmid et al., 2010, 2012]. It can be predicted that with application of modern chromosome banding techniques, assisted by molecular methods, the number of sex chromosome pairs discovered in the Hylidae will rapidly expand.

3.1.1.5 Hylidae with the Karyotype Formulae 2n = 18, 20, 22, 26, 28, 34, and 48

Deviations from the typical hylid karyotype formula of 2n = 24 are found in the genera *Acris*, *Aplastodiscus*, *Boana (albopunctata)*, *Dendropsophus*, *Dryophytes* (*versicolor*), *Osteopilus*, *Scinax* (*constrictus*), and *Phyllodytes* (Appendix 1).

All 3 *Acris* species possess 2n = 22 meta- to submetacentric chromosomes (Fig. 18a, 69–73, Appendix 1). The pairs 1–6 are larger than the pairs 7–10. The last chromosome pair is again a large chromosome and can be interpreted to be the result of a nonreciprocal translocation or a tandem fusion between chromosome 11 and 12 [tfu(11;12)] in the ancestral hylid karyotype. This mutation, after becoming homozygous in the population, reduced the diploid chromosome number from 2n = 24 to 2n = 22. In the Giemsa-stained karyotypes of *Acris crepitans* and *A. gryllus* examined, a distinct nucleolar constriction is located in the proximal third of the long arm of one each of the tfu(11;12) chromosomes (Fig. 69–73). This constriction could not be demonstrated in *Acris blanchardii* [Cole, 1966], but the metaphase chromosomes shown in the publication of Cole [1966] appear overstained by propiono-orcein.

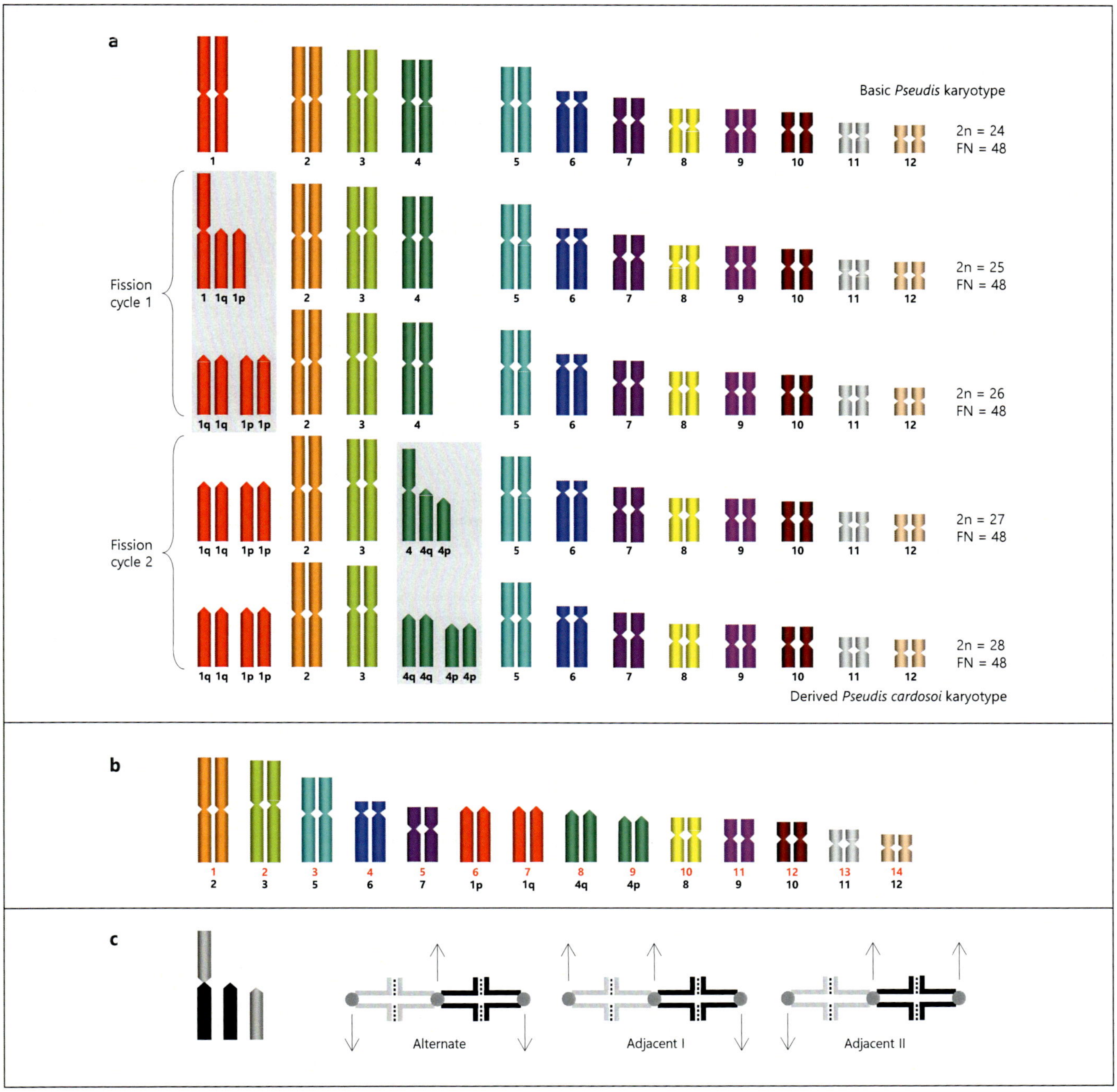

Fig. 34. a Series of 2 subsequent fission cycles that convert the theoretical basic (ancestral) *Pseudis* karyotype consisting of 24 biarmed chromosomes (top) into the derived *Pseudis cardosoi* karyotype with 28 biarmed and uniarmed chromosomes (bottom). After each fission cycle, 2 newly formed fission chromosomes are present in homozygous condition in some individuals and may be fixed in the evolving population. The fundamental number of the basic karyotype (FN_b = 48) remains the same, but with each new centric fission the diploid chromosome number (2n) increases by 1, and with each completed fission cycle by 2. Karyotypes with a new fission chromosome always have an odd diploid chromosome number. p, short arm; q, long arm. **b** The chromosomes of the last row in **a** are arranged according to their decreasing lengths. Black numbers indicate the original position of the chromosomes or chromosome arms in the last row of **a**, the red numbers their position in **b**. **c** Diagrams demonstrating the meiotic pairing configurations and the possible segregation types of the original biarmed chromosome and its free telocentric homologues (fission products). In the meiotic pairings, the large grey dots represent the centromeres, the small dotted lines between the chromosome arms symbolize the regions of synapsis and crossing-over, and the arrows mark the direction in which the paired chromosomes segregate into the 2 daughter nuclei. For further explanations, see chapter 3.1.1.4.

In the genus *Aplastodiscus*, a continuous series of diploid chromosome numbers of 2n = 24, 22, 20, and 18 was observed (Fig. 18d–g, 251–269, Appendix 1). Again, in the 2n = 24 karyotypes of *Aplastodiscus,* all chromosomes are meta- to submetacentric, with the pairs 1–6 larger than the pairs 7–12 (Fig. 18g, 268, 269). One possibility to explain the karyotypes with the reduced chromosome numbers of 2n = 22, 20, and 18 is to suppose nonreciprocal translocations or tandem fusions between chromosomes 6 and 9 [tfu(6;9)], 7 and 10 [tfu(7;10)] as well as between the smallest chromosome 12 and another chromosome in the ancestral hylid karyotype with 2n = 24 chromosomes. Interestingly, in 1 single individual of *A. leucopygius* collected in Teresópolis (Brazil), the 2 submetacentric chromosomes 8 apparently are involved in a reciprocal translocation between their short and long arms resulting in a large metacentric rt(8q;8q) plus a small metacentric rt(8p;8p) chromosome (Fig. 259–261). No more specimens with this rare structural aberration were found in this or other *A. leucopygius* populations. In order to obtain conclusive results on the karyotype phylogeny in this genus, the karyotypes of the 9 further extant *Aplastodiscus* species, which all live in central and southeastern Brazil to adjacent Argentina [Frost, 2017], must be examined.

At this point, it is important to emphasize that in 1973 a far-reaching misidentification of an *Aplastodiscus* species (or individual) occurred, and that this error was procrastinated for several years (Appendix 1). The exact course and the implications of this error are best summarized by Gruber et al. [2012a]. According to these authors, distinct diploid chromosome numbers in the genus *Aplastodiscus* were originally reported [Bogart, 1973]. More recently, the sample of karyotyped species was enlarged [Carvalho et al., 2009a, b]. Now a total of 12 representatives of the genus have described karyotypes: *A. cochranae* and *A. perviridis* with 2n = 24; *A. albofrenatus, A. arildae, A. ehrhardti, A. eugenioi,* and *A. weygoldti* with 2n = 22; *A. albosignatus* and *A. callipygius* with 2n = 20; and *A. cavicola, A. ibirapitanga,* and *A. leucopygius* with 2n = 18. Several individuals of these species were collected from more than 1 locality, and no karyotypic intraspecific variation in the chromosome number has been found except, at the first sight, in the sample of *A. albofrenatus* (2n = 24 in the Floresta da Tijuca, Rio de Janeiro, Rio de Janeiro State, and 2n = 22 in Boraceia, São Paulo State), and *A. albosignatus* (2n = 20 in Boraceia, São Paulo State, and 2n = 18 in Teresópolis, Rio de Janeiro State) [Bogart, 1973], although this variation was probably the consequence of misidentification according to the author. Later, Carvalho et al. [2009a, b], based on the geographical distribution of the genus *Aplastodiscus* and on the cytogenetic data collected in the same or near the localities screened by Bogart [1973], concluded that the formerly karyotyped specimens were actually misidentifications and suggested that the frogs with 2n = 22, 2n = 20, and 2n = 18 corresponded to *A. arildae, A. albosignatus,* and *A. leucopygius*, respectively. Additionally, considering that the range of *A. perviridis* includes the state of Rio de Janeiro [Frost, 2017] and that the diploid chromosome number for this species is 2n = 24 [Gruber et al., 2012a, b], the *A. albofrenatus* from the Tijuca forest [Bogart, 1973], could be *A. perviridis* or some other species of *Aplastodiscus*. These misidentifications and corrections are considered in Appendix 1.

A total of 92 *Boana* species have been described by now [Frost, 2017]. Of these species, 35 (38.0%) have been comprehensively karyotyped (Appendix 1). Thirty-four species exhibit the classical hylid karyotype with 2n = 24 chromosomes of meta- to submetacentric morphology (Fig. 18i, 274–302). As in most other hylid species, the chromosome pairs 1–6 of *Boana* are larger than the remaining pairs 7–12. The middle-sized chromosome 6 is the most submetacentric element in the karyotype, and the chromosome 3 is characterized by the conserved typical submetacentric morphology (see above). However, 2 species, *B. albopunctata* and *B. lanciformis*, possess a chromosome number of 2n = 22 (Fig. 18h, 270–273, Appendix 1). The decreased chromosome number in *B. albopunctata* was detected independently by various cytogeneticists during the last 50 years in several geographically separated populations existing in Brazil (São Paulo State, Goiás State) and Argentina (Misiones Province, Corrientes Province) (Appendix 1) [Beçak, 1968; Bogart, 1973; Feitosa et al., 1995; Gruber et al., 2007; Ferro et al., 2012, 2016; present study]. The smallest chromosome pair 12 is missing in the karyotypes (Fig. 18h, 270–273). This again suggests that a tandem fusion or nonreciprocal translocation between chromosome 12 and another chromosome in the ancestral hylid karyotype with 2n = 24 chromosomes has led to the current karyotype with 2n = 22 chromosomes. This assumption was already made by Bogart [1973] who recognized that the reduced chromosome number of *B. albopunctata* may have originated by an end-to-end fusion (tandem fusion) that involved 2 chromosomes of the ancestral karyotype or by multiple translocations from one of the smallest chromosomes (11 or 12) to different chromosomes of the original complement. The same arguments are valid for *B. lanciformis* [de Mattos et al., 2014].

A total of 71 extant species of *Scinax* have been described [Frost, 2017]. Of these taxa, 35 (49.0%) are karyotyped, and 34 show the monotonous 2n = 24 karyotype (Fig. 20l), but *S. constrictus* clearly has 2n = 22 chromosomes [Sousa, 2007; de Oliveira et al., 2012] (Appendix 1). The best explanation is that this reduced number of 2n = 22 chromosomes originates by a tandem fusion, centric fusion, or nonreciprocal translocation (chapters 3.4.2–3.4.8).

Eight species of the genus *Osteopilus* were discovered up to date [Frost, 2017]. They live on the Greater Antilles, Jamaica, Bahama Islands, and southern Florida (USA). Five of these species have been karyotyped (Fig. 175–182, Appendix 1). *Osteopilus dominicensis, O. marianae,* and *O. septentrionalis* are characterized by the basic hylid karyotype of 2n = 24 (Fig. 20b, 175, 180, 181), *O. ocellatus* has 2n = 34 chromosomes (Fig. 20d, 176–179), whereas *O. wilderi* displays 2n = 28 chromosomes (Fig. 20c, 182). All species possess a fundamental number of FN = 48 and a differing number of telocentric and biarmed chromosomes. This is a clear indication that the karyotypes of *O. ocellatus* and *O. wilderi* were derived from the basic *Osteopilus* karyotype by a series of centric fissions as schematically drawn in Figure 35a, b.

3.1.1.6 Hylidae with the Karyotype Formula 2n = 30

Of the large genus *Dendropsophus* with 95 known species, 44 have been cytogenetically described by now (Appendix 1). Some of these karyotypes are shown in the Figures 18k, l, 36, and 105–163. Without exceptions, all species of *Dendropsophus* are characterized by a diploid chromosome number of 2n = 30. Their fundamental numbers are highly variable and range from FN = 60 (all biarmed chromosomes) like in *D. leali* (Fig. 120, 121) to FN = 50 (mixture of bi- and monoarmed elements) like in *D. marmoratus* (Fig. 125, 126) or *D. labialis* (Fig. 116–119).

The *Dendropsophus* karyotype must have originated from the basic ancestral 2n = 24 hylid karyotype very early in the phylogenesis and conserved since then. The mechanism of origin is straightforward explained by assuming 3 fission cycles which yielded the current 2n = 30 karyotype (Fig. 37a). The exact mechanism of centromeric fission, including the different steps until reaching homozygosity of the various chromosomes, has been explained in detail in the first volume of these 3 monographs on amphibian cytogenetics [Schmid et al., 2010] and, for the convenience of the readers, is also partially repeated in chapter 3.4.1 of the present volume. At the end of these fission cycles, 1 or several rounds of heterochromatin amplifications were possible in some taxa (Fig. 37a, last row). In Figure 36, the schematic drawings of the chromosomes of 13 species of the genus *Dendropsophus* are shown. C-banding was not performed in all karyotypes. The diagrams demonstrate the high variability of the size and structure of the short arms in the chromosomes.

The successful persistence of the *Dendropsophus* karyotype demonstrates that even in a huge anuran group, whose most members still have conserved for millions of years its basic ancestral karyotype (2n = 24), unpredictable exceptions can spontaneously originate in phylogenesis and win.

To complete matters, when considering the *Dendropsophus* systematics, it must be considered that recent investigations showed that widespread single species actually comprise species complexes, meaning that species richness may be significantly higher than current estimates [e.g., Caminer et al., 2017].

3.1.1.7 Hylidae with B Chromosomes

In *Boana albopunctata*, middle-sized supernumerary and mitotically stable B chromosomes were demonstrated (Fig. 271). Ferro et al. [2012] conducted a systematic analysis on 106 males and 2 females collected from 8 natural populations in northeastern Argentina (Misiones Province and Corrientes Province). Four different chromosome numbers with 0, 1, 2, and 3 B chromosomes were found. Of the 108 individuals examined, 74 (68.5%) had the karyotype with 2n = 22 A chromosomes. Twenty-nine frogs (26.85%) had 1 B chromosome and a karyotype formula of 2n = 22A + 1B. Two frogs possessed 2 B chromosomes and the karyotype formula of 2n = 22A + 2B. Finally, only a single specimen (0.9%) was found having 3 B chromosomes and, consequently, the karyotype of 2n = 22A + 3B (Appendix 1). The frequency of supernumerary B chromosomes in the study of Ferro et al. [2012] differed significantly between the Argentinian populations investigated. Northern populations had a higher frequency of B chromosomes than the southern populations. As correctly stated by the authors, the apparent nonappearance of B chromosomes in some populations could be simply explained because of the small size of the collected sample and not because of their real absence.

The supernumerary B chromosomes revealed mostly a euchromatic structure, with little pericentromerically located AT-rich heterochromatin. According to the authors, more detailed analyses using a C-banding pretreatment plus DAPI fluorescence revealed the presence of 2 types of B chromosomes. One of them (B1) presented limited amounts of constitutive heterochromatin in the short arm proximal region, whereas the other one (B2) had heterochromatic bands in the proximal regions of both arms [Ferro et al., 2012]. A mandatory caveat has to be made, however, that the unconventional technique of demonstration of constitutive heterochromatin performed by these authors, namely C-banding pretreatment followed by staining with AT- or GC-specific fluorochromes, does not necessarily reveal the actual chemical composition and molecular structure of the repetitive DNA in the heterochromatin. As demonstrated by Ag-staining and 18S rDNA FISH, active or inactive NORs are not located in the B chromosomes of *B. albopunctata*. FISH with telomeric DNA probes showed the presence of brightly fluorescing signals at both telomeric ends of the B1 chromosome. The B2 chromosome was not studied with the telomeric DNA FISH.

Gruber et al. [2007] analyzed the chromosomes of 17 *B. albopunctata* males from a population in Rio Claro (São Paulo State, Brazil) and also demonstrated the existence of stable metacentric B chromosomes. The frequency of individuals with 1 supernumerary B chromosome was 41.2% (Appendix 1). However, in contrast to the study of Ferro et al. [2012], the B chromosome in their specimens proved to be almost completely heterochromatic except for its telomeric regions. A similar observation is depicted in the present study (Fig. 459). Although Gruber et al. [2007] speculated about the origin of the supernumerary B chromosome from a partial or complete duplication of the 8th chromosome in the karyotype of *B. albopunctata*, there is no experimental evidence for this event. More novel high-resolution techniques like microdissection and painting of chromosomes are required to uncover the origin of the B chromosomes [e.g., Sharbel et al., 1998; Teruel et al., 2009]. Gruber et al. [2014] successfully isolated a DNA probe from

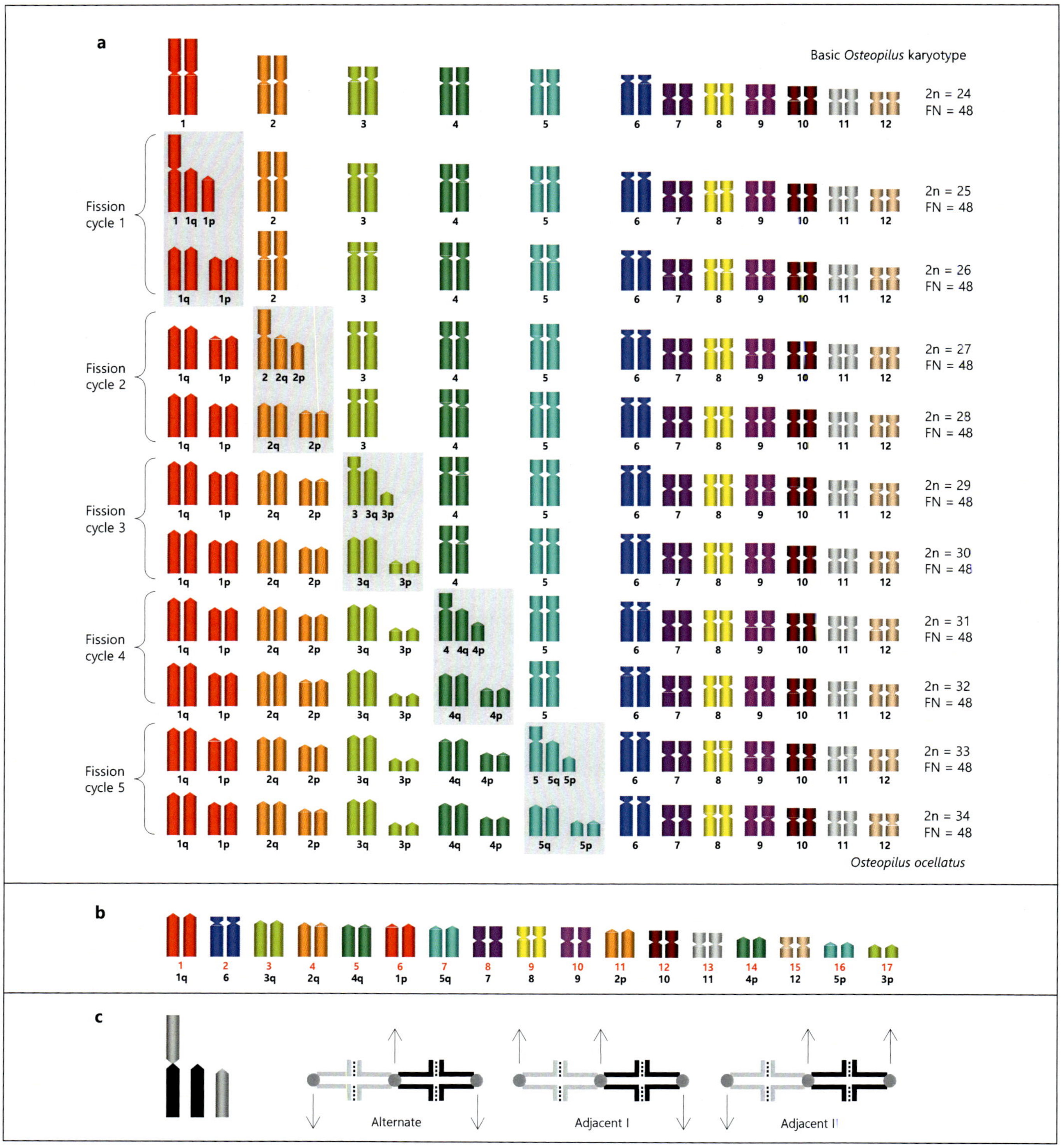

Fig. 35. a Series of 5 subsequent fission cycles that are supposed to have converted the theoretical basic (ancestral) *Osteopilus* karyotype consisting of 24 biarmed chromosomes (top) into the derived *Osteopilus ocellatus* karyotype with 34 biarmed and uniarmed chromosomes (bottom). After each fission cycle, 2 newly formed fission chromosomes are present in homozygous condition in some individuals and may be fixed in the evolving population. The fundamental number of the basic karyotype ($FN_b = 48$) remains the same, but with each new centric fission, the diploid chromosome number (2n) increases by 1, and with each completed fission cycle by 2. Karyotypes with a new fission chromosome always have an odd diploid chromosome number. p, short arm; q, long arm. **b** The chromosomes of the last row in **a** are arranged according to their decreasing lengths. Black numbers indicate the original position of the chromosomes or chromosome arms in the last row of **a**, the red numbers their position in **b**. **c** Diagrams demonstrating the meiotic pairing configurations and the possible segregation types of the original biarmed chromosome and its free telocentric homologues (fission products). In the meiotic pairings, the large grey dots represent the centromeres, the small dotted lines between the chromosome arms symbolize the regions of synapsis and crossing-over, and the arrows mark the direction in which the paired chromosomes segregate into the 2 daughter nuclei. For further explanations, see chapter 3.1.1.5.

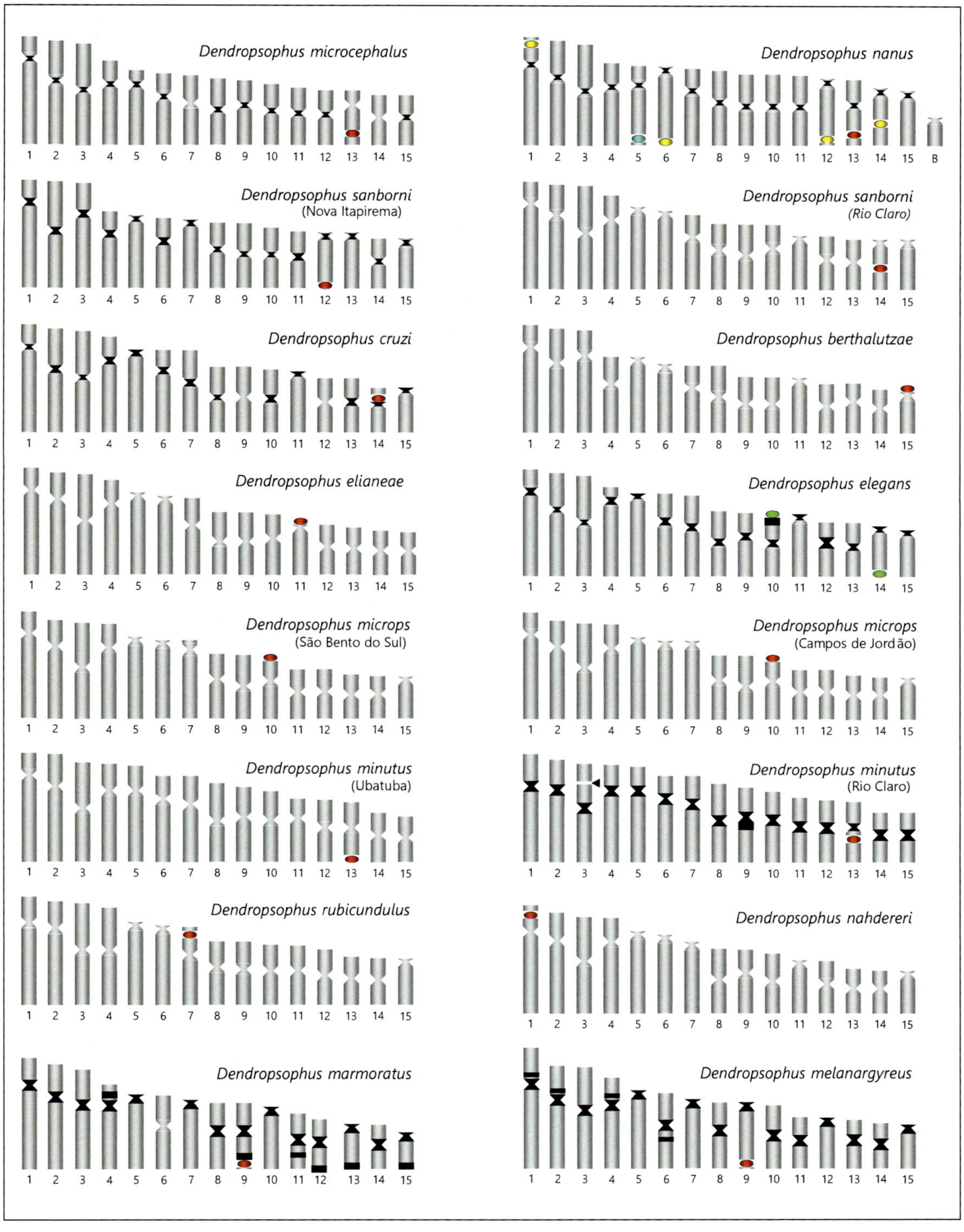

Fig. 36. Schematic drawings of the haploid chromosome set of 13 species of the genus *Dendropsophus*. Note the variable structure and size of the short arms. Black areas, constitutive heterochromatin (C-banding was not performed in all karyotypes); red ovals, nucleolus organizer regions. According to Anderson [1991], Medeiros et al. [2003, 2006], Gruber et al. [2005], and Suárez et al. [2013]. For further explanations, see 3.1.1.6.

the B chromosome in the diakineses of 2 male specimens of *B. albopunctata* with microdissection and DOP-PCR amplification. They used the probe to search for homologies between the B and regular A chromosomes of *B. albopunctata* and the closely related *B. raniceps*. The B DNA probe exclusively painted the supernumerary chromosome but did not hybridize to any other chromosome or chromosome region in *B. albopunctata*. However, the B DNA probe hybridized at multiple interspersed sites in all *B. raniceps* chromosomes. This result may indicate differences between the

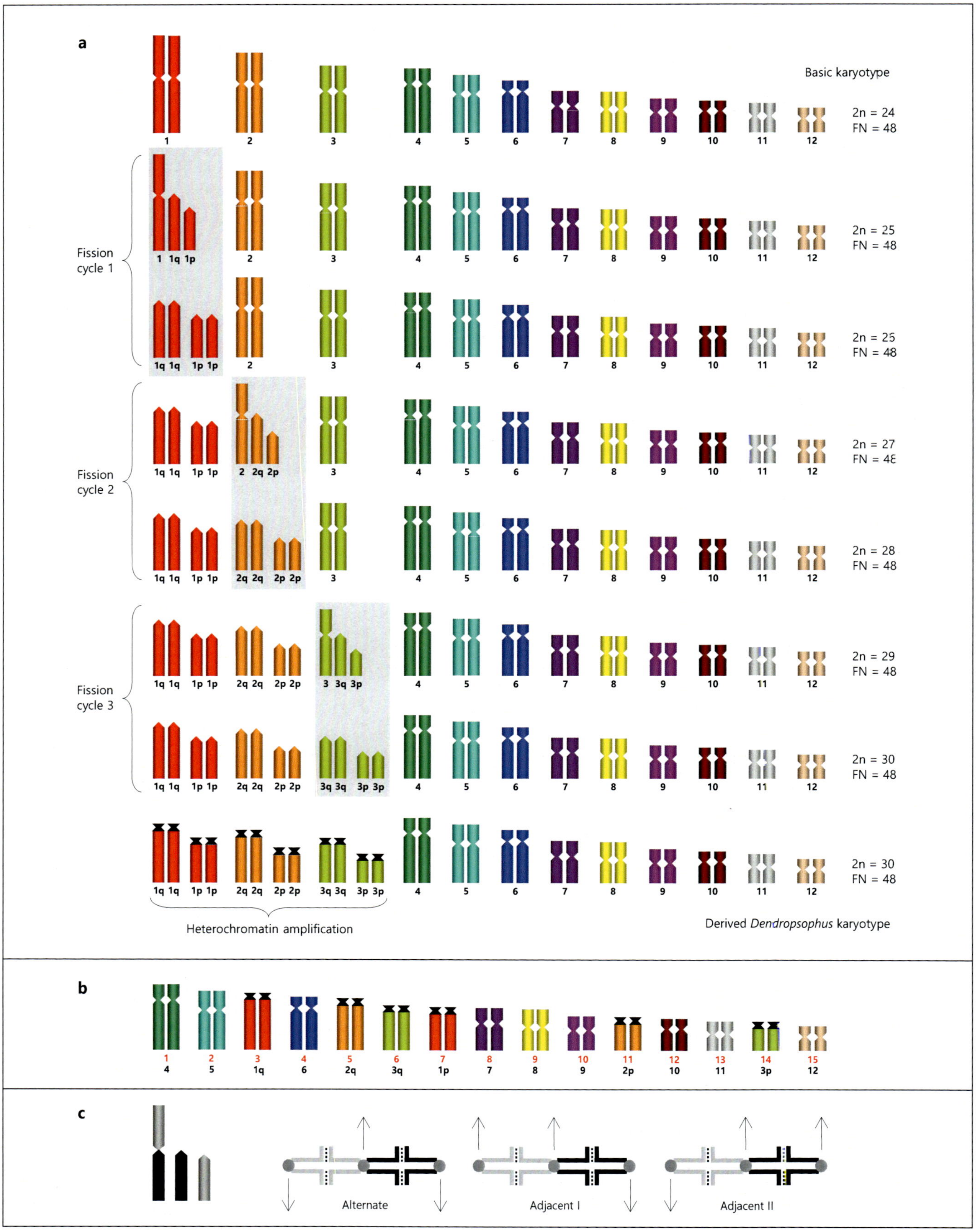
a
Basic karyotype
1 2 3 4 5 6 7 8 9 10 11 12
2n = 24
FN = 48
Fission cycle 1
1 1q 1p
2n = 25
FN = 48
1q 1q 1p 1p
2n = 25
FN = 48
Fission cycle 2
2 2q 2p
2n = 27
FN = 48
2q 2q 2p 2p
2n = 28
FN = 48
Fission cycle 3
3 3q 3p
2n = 29
FN = 48
3q 3q 3p 3p
2n = 30
FN = 48
1q 1q 1p 1p 2q 2q 2p 2p 3q 3q 3p 3p 4 5 6 7 8 9 10 11 12
2n = 30
FN = 48
Heterochromatin amplification
Derived Dendropsophus karyotype
b
1 2 3 4 5 6 7 8 9 10 11 12 13 14 15
4 5 1q 6 2q 3q 1p 7 8 9 2p 10 11 3p 12
c
Alternate
Adjacent I
Adjacent II

37

repetitive sequences in the A and B chromosomes of *B. albopunctata* and similarities between the repetitive DNA sequences in the *B. albopunctata* B chromosome and the *B. raniceps* A chromosomes [Gruber et al., 2014]. According to these authors, the repetitive DNA sequence shared by the *B. albopunctata* B chromosome and the *B. raniceps* A chromosomes seems to correspond to a transposable element. Based on the painting results, the authors concluded that it is very tempting to suggest an interspecific origin of the *B. albopunctata* B chromosome. However, they admitted that this may be a premature conclusion, since the distribution of reiterated DNA sequences in the A and B chromosomes can be unequal [Montiel et al., 2012]. Therefore, the origin of the *B. albopunctata* B chromosome still remains unsolved.

B. albopunctata has a very ample geographical distribution. It occurs from Eastern Bolivia to central and southern Brazil, eastern Paraguay, northern Uruguay, and the northeastern part of Argentina [Frost, 2017]. Many further studies are required to obtain a broader knowledge on the frequencies and banding properties as well as an insight on the origin of the B chromosomes in this anuran species.

Although the properties of B chromosomes in general, as well as in amphibians in particular, were explained at length in the 2 previous monographs of this series on amphibian chromosomes [Schmid et al., 2010, 2012], it seems convenient to recapitulate the most important facts on these chromosomes at this point. The existence of B chromosomes in animals was first described by Wilson [1906] in the karyotype of a hemipteran insect belonging to the genus *Metapodius*. B chromosomes are accessory genomic elements that are now known to occur in about 1,300 species of plants and 500 species of animals. This corresponds to approximately 15% of the extant species that have been karyotyped. A remarkable exception are birds where, so far, only 1 species, the zebra finch *Taeniopygia guttata*, with a B chromosome (that surprisingly enough is restricted to the germline of both sexes) has been discovered [Pigozzi and Solari, 1998]. It is difficult to generalize on B chromosomes, but the consensus is that they are dispensable elements, present in variable numbers in some individuals in some populations of a given species that coevolve with the normal A genome to maximize their representation in the next generation [for reviews, see White, 1973a; Jones and Rees, 1982; Jones, 1985; Lyttle, 1991; Beukeboom, 1994; Camacho et al., 2000; Camacho, 2004]. Generally, B chromosomes are morphologically distinct from the regular A chromosomes. In most cases they are smaller than A chromosomes. In some exceptional species, however, the B chromosomes can be very large and may even surpass the length of the largest autosome in the karyotype comprising approximately 10% of the genome [Ziegler et al., 2003; Schmid et al., 2006]. B chromosomes do not pair meiotically with A chromosomes and are inherited in a non-mendelian fashion. B chromosomes can effectively pair with one another at the pachytene stage of meiosis, but their synaptonemal complex axes are distinctly denser than those of the autosomes. Both bivalent and monovalent meiotic B chromosomes have been observed. When 3 B chromosomes are present in a meiotic cell, a parallel alignment of 3 synaptonemal complexes can develop [Switonski et al., 1987; Shi et al., 1988]. According to the classic meiotic concept, only 2 homologous chromosomes are supposed to synapse in meiosis. Hence, the presence of triple pairings between B chromosomes is enigmatic. Many of the B chromosomes are characterized by a high transmission rate in meiosis and/or mitosis that contributes to their accumulation in natural populations. There are reports on the existence of genes in the standard A chromosomes that are capable of suppressing the meiotic drive of B chromosomes [Henriques-Gil et al., 1982; Nur and Brett, 1985, 1987, 1988; Romera et al., 1991]. Furthermore, B chromosomes may also be mitotically unstable which leads to an intraindividual variability in their number. In this case, the variation probably results from anaphase lagging with subsequent elimination of B chromosomes from some cells, or, alternatively, it can be caused by mitotic nondisjunction with sister chromatids segregating to the same pole. In some vertebrates, such as the frogs *Leiopelma hochstetteri* and *Gastrotheca espeletia* as well as the fish *Astyanax scabripinnis*, B chromosomes are distinctly more frequent in females than they are in males [Green, 1988a; Vicente et al., 1996; Schmid et al., 2002b]. On the other hand, in the fish *Moenkhausia sanctaefilomenae*, the B chromosomes are confined to the male sex [Portela-Castro et al., 2001]. A revision on possible relationships between B chromosomes and sex in animals has been provided by Camacho et al. [2011].

B chromosomes range from those that are almost devoid of constitutive heterochromatin to those that are completely heterochromatic. In the latter case, the highly repetitive DNA sequences located in the B chromosomes replicate in the very late S-phase of the cell cycle [Pauls and Bertollo, 1983; Schmid et al., 2006]. Some of the reiterated DNA se-

Fig. 37. a Series of 3 subsequent fission cycles that are supposed to have converted the theoretical basic (ancestral) karyotype consisting of 24 biarmed chromosomes (top) into a derived *Dendropsophus* karyotype with 30 bi- and uniarmed chromosomes (bottom). After each fission cycle, 2 newly formed fission chromosomes are present in homozygous condition. The fundamental number of the basic karyotype ($FN_b = 48$) remains the same, but with each new centric fission, the diploid chromosome number (2n) increases by 1, and with each completed fission cycle by 2. Karyotypes with a new fission chromosome always have an odd diploid chromosome number. Once the fission cycle 3 was completed, heterochromatin amplification formed the tiny short arms of the uniarmed chromosomes (last row). p, short arm; q, long arm. **b** The chromosomes of the last row in **a** are arranged according to their decreasing lengths. Black numbers indicate the original position of the chromosomes or chromosome arms in the last row of **a**, the red numbers their position in **b**. **c** Diagrams demonstrating the meiotic pairing configurations and the possible segregation types of the original biarmed chromosome and its free telocentric homologues (fission products). In the meiotic pairings, the large grey dots represent the centromeres, the small dotted lines between the chromosome arms symbolize the regions of synapsis and crossing-over, and the arrows mark the direction in which the paired chromosomes segregate into the 2 daughter nuclei. For further explanations, see chapter 3.1.1.6.

quences in constitutively heterochromatic B chromosomes are specific to B chromosomes [Ziegler et al., 2003], whereas others are shared with the A chromosomes [Beukeboom, 1994]. In some cases, the repetitive DNA sequences have provided a useful tool to ascertain the intraspecific or interspecific origin of B chromosomes [Lopez-Leon et al., 1994; McAllister and Werren, 1997; Camacho et al., 2000; Mestriner et al., 2000]. Interestingly, in the fish *Alburnus alburnus*, a B chromosome-specific DNA probe obtained by the PCR-based amplified fragment length polymorphism (AFLP) technique [Vos et al., 1995] revealed a strong homology to a Gypsy/Ty3 LTR retrotransposon of *Drosophila* and the medaka fish (*Oryzias latipes*) [Ziegler et al., 2003].

Active NORs have been visualized in the B chromosomes of 12 plant species and 10 animal species [Green, 1990; Schmid et al., 2002b, 2010, 2012]. Furthermore, a latent NOR is present in a B chromosome of the grasshopper *Eyprepocnemis plorans* which can be activated by a fusion of the B chromosome with an autosome [Cabrero et al., 1987]. It has been suggested, although without experimental evidence, that NORs in the standard A chromosome set are prone to chromosome breakage, and this may provide a mechanism for the generation of B chromosomes.

Different primary sources for B chromosomes have been considered. An intragenomic fragment may acquire the characteristic of a B chromosome (from duplicated or fragmented pieces within a genome) as demonstrated in the plant *Crepis capillaris* [Jamilena et al., 1994, 1995] and in *Drosophila subsilvestris* [Gutknecht et al., 1995]. The first experimental evidence in favor of this hypothesis has been the finding of similar polytene band patterns on the B chromosome and the paracentromeric region of the A chromosome IV in the fly *Chironomus plumosus* [Keyl and Hägele, 1971]. Since pairing and crossing-over between B and A chromosomes have not been observed in meiosis, it seems obvious that the original homology between them would be rapidly lost.

Another pathway for the evolution of B chromosomes within genomes is by derivation from the Y or W sex chromosomes, as has been postulated for the frog *L. hochstetteri* [Green et al., 1993; Sharbel et al., 1998] and the collared lemming *Dicrostonyx groenlandicus* [Berend et al., 2001]. Finally, B chromosomes can originate by interspecies hybridization, when foreign DNA from a closely related species transforms into a B chromosome. Interspecific B chromosome origin by hybridization is postulated for the gynogenetic Amazon molly *Poecilia formosa* [Schartl et al., 1995] and for the wasp *Nasonia vitripennis* [McAllister and Werren, 1997]. Owing to the absence of selection pressure, mutations accumulate and recombination is prevented in B chromosomes. This presumably dictates the loss of functional loci in the B chromosomes. Such a transition to neutrality (frequently termed "parasitic"), often accompanied by heterochromatinization of the element, may be a common pathway in the evolution of B chromosomes. Their subsequent accumulation in some populations is often based on a drive-like mechanism that can occur before, during, or after meiosis [Jones, 1991].

Presently, there is little consensus among cytogeneticists on a possible functional significance of B chromosomes. Individuals carrying B chromosomes are, in general, indistinguishable from those without these chromosomes. Therefore, the conclusion can be offered that B chromosomes are depauperate in functional genes. The fact that B chromosomes have been maintained during innumerable generations in certain populations indicates that they play a hitherto unknown role in the genetic variability of species. Several investigations show that some genes located in B chromosomes can be expressed [Green, 1990, 2004; Jones, 1995; Covert, 1998; Camacho et al., 2000; Camacho, 2004]. When present in high numbers, B chromosomes may influence fertility [Hewitt et al., 1987], developmental rates, or the meiotic process [Parker et al., 1981; Bell and Burt, 1990]. Such effects can develop either by their mere presence in the nuclei (causing meiotic disturbances) or because of transcription of B chromosome-linked genes that somehow interfere with normal cellular activities.

According to White [1973a, b] and Hewitt [1974], interchanges between B chromosomes and A chromosomes would be the most likely way for the B chromosomes to become integrated into the regular genome and provide some possible advantageous effects. Indeed, spontaneous translocations between B chromosomes and either an autosome or the X sex chromosome have been observed in several individuals of 2 Spanish populations of the grasshopper *E. plorans* [Henriques-Gil et al., 1983]. These interchanges do not seem to have major effects on the cytological behavior and may be the beginning of an integration of B chromosomes into the genome of this species.

So far, B chromosomes have been demonstrated in 20 species and 11 families of Anura (Table 7), and further taxa with supernumerary chromosomes are being discovered at regular intervals. As with most reports on B chromosomes in other eukaryotes, amphibian B chromosomes are always restricted to particular individuals within certain populations and do not represent a general feature of a whole taxon. B chromosomes have been frequently found in the neotropical salamander family Plethodontidae. Most amphibian B chromosomes have only been studied by conventional staining or by some banding techniques. As outlined by Green [2004], however, in the frog *L. hochstetteri* from the North Island and the Great Barrier Island of New Zealand, as well as in the coastal giant salamander *Dicamptodon tenebrosus* from the Pacific coast of North America, B chromosomes are not rare and have been studied down to the molecular level (for references see Table 7).

The largest B chromosomes found so far in anurans are those of the North American spadefoot toad *Spea hammondii* [Green, 1991, 2004] and of the South American leptodactylid frog *Megaelosia massarti* [Rosa et al., 2003]. In *S. hammondii*, the B chromosome is nearly as large as the largest A chromosome, mitotically stable, and almost completely het-

Table 7. B chromosomes discovered in species of Anura

Family Species[a]	2n[b]	Maximum number of Bs	B morphs, *n*	C[c]	NORs[d]	References
Odontophrynidae						
Odontophrynus americanus	44	2	2	n.e.	n.e.	Rosset et al., 2006; Lanzone et al., 2008
Discoglossidae						
Discoglossus pictus	28	2	1	n.e.	–	Schmid et al., 1987
Hemiphractidae						
Gastrotheca espeletia	26	9	3	+	+	Schmid et al., 2002b, 2010
Hylidae						
Acris crepitans	22	5	1	n.e.	n.e.	Nur and Nevo, 1969
Boana albopunctata	22	1	1	(+)	–	Cardoso, 1986; Gruber et al., 2007; Ferro et al., 2012
Bokermannohyla luctuosa	24	1	1	+	–	Baldissera et al., 1993
Dendropsophus nanus	30	3	1	–	–	Medeiros et al., 2006
Smilisca baudinii	24	1	1	–	–	Hernández-Guzmán et al., 2011
Hylodidae						
Megaelosia massarti	28	1	2	+, (+)	–	Rosa et al., 2003
Leiopelmatidae						
Leiopelma hochstetteri	22♂ 23♀	15	3	(+)	–	Morescalchi, 1968b; Stephenson et al., 1972; Green, 1988b, 1994; Green et al., 1987, 1993; Zeyl and Green, 1992; Sharbel et al., 1998
Leptodactylidae						
Physalaemus olfersii	22	1	1	(+)	–	Milani et al., 2011
Ranidae						
Amolops liangshanensis	26	1	1	(+)	–	Wu and Zhao, 1985
Sanguirana everetti	26	1	1	(+)	–	Kuramoto, 1989
Rana temporaria	26	4	1	(+)	–	Ullerich, 1967; Schmid, 1978b; Belcheva and Sofianidou, 1990; Voitovich et al., 2006
Scaphiopodidae						
Spea hammondii	28	1	1	+	+	Green, 1991
Craugastoridae						
Craugastor sp. n. B	18	2	1	+	–	Schmid et al., 2010; Ferro et al., 2016
Oreobates discoidalis	22	1	2	n.e.	n.e.	Schmid et al., 2010
Oreobates barituensis	22	3	2	(+)	+	Ferro et al., 2016
Oreobates berdemos	22	1	1	–	–	Ferro et al., 2016
Eleutherodactylidae						
Eleutherodactylus gundlachi	32	10	4	+	(+)	Schmid et al., 2010

Supernumerary chromosomes in the order Caudata have been compiled by Schmid et al. [2010]. So far, no B chromosomes have been discovered in species of the order Gymnophiona.

[a] The name of the species is the same as used in the original publications. For further information, see Frost [2017].

[b] *Odontophrynus americanus* is a tetraploid (4n) species.

[c] C-banding properties; +, B chromosome completely heterochromatic; (+), B chromosome partially heterochromatic; –, B chromosome euchromatic; n.e., not examined.

[d] NORs, nucleolus organizer regions; +, NOR present in B chromosome; –, NOR not present in B chromosome; n.e., not examined.

erochromatic except for a small pericentromeric region where an ectopic NOR can be detected by Ag-staining. The situation in *M. massarti* is more complex. Two out of 4 individuals karyotyped showed B chromosomes in all cells. In 1 specimen, the B chromosome belonged to the smallest elements of the karyotype (~1 μm), had a metacentric morphology, and was completely heterochromatic. The B chromosome of the other individual was as large as the largest chromosome of the A complement (~4.5 μm), was of submetacentric shape, and had C-bands only in the telomeric regions of both arms. All other B chromosomes in Anura are considerably smaller than most of the A chromosomes in the corresponding karyotypes. Another anuran species with completely heterochromatic B chromosomes, and with NORs located in some of them, is the Ecuadorian marsupial frog *G. espeletia* [Schmid et al., 2002b, 2012]. In this species, 1–9 mitotically stable B chromosomes with 3 different morphologies have been found in all 47 individuals examined, indicating a substantial variation and frequency of B chromosomes. Among all amphibians, the B chromosomes of the endemic New Zealand frog *L. hochstetteri* are the most diverse and the most thoroughly analyzed ones (for references, see Table 7). This species carries 0–15 B chromosomes per individual, up to a mean of 11.4 B chromosomes per individual in one of the examined populations, which is the highest number of B chromosomes known for any wild population of animals. Three different morphologies of mitotically stable and partially heterochromatic B chromosomes have been found. They remain as univalents in male meiosis and exhibit a variety of lampbrush structures and lateral transcription loops in female meiosis that indicate the presence of actively transcribed genes. So far nothing is known about these transcripts or whether they are translated into functional proteins. In some populations, the lampbrush chromosomes of *L. hochstetteri* form conspicuous multichromosome associations of various B chromosomes that clump together at their telomeres. Most interestingly, cytogenetic and molecular studies suggest that the B chromosomes derived one or more times from the univalent W sex chromosome (for references, see Table 7) that is present in only some populations of this species. This in turn implies a very recent derivation (approximately 10,000 years) of the univalent W chromosome and the B chromosomes after the post-glacial isolation of those populations.

A bizarre B isochromosome occurs in the salamanders *Oedipina poelzi* from Costa Rica as well as in *Chiropterotriton lavae* from Mexico where 2 identical B chromosomes are centromerically fused and pair with each other forming an isobivalent in meiosis. Callan [1986] demonstrated the presence of lateral transcription loops in the lampbrush B chromosome in a species of the genus *Thorius* from Mexico. Individuals of the coastal giant salamander *D. tenebrosus* have 0–10 telocentric B chromosomes per individual. These B chromosomes are devoid of constitutive heterochromatin but can contain active NORs.

In both, the frog *L. hochstetteri* and the salamander *D. tenebrosus*, highly reiterated DNA sequences have been detected in the B chromosomes [Zeyl and Green, 1992; Sharbel et al., 1998; Brinkman et al., 2000]. In *L. hochstetteri,* the B chromosomes are at least partly composed of repetitive DNA sequences that also exist in numerous copies in the standard A genome. The largest similarities have been found between the B chromosomes and the univalent W sex chromosome. The B chromosomes of *D. tenebrosus* are composed of about 39% repetitive DNA sequences that are also located in the standard A chromosomes. The repetitive DNA in these B chromosomes contains many sequences that are 98–100% identical for much of their length.

In conclusion and as already recognized by Green [2004], amphibian B chromosomes are highly variable elements. Some of them are very large, others are small. They can be either completely euchromatic, whereas others contain high amounts of reiterated DNA sequences. Some of them occur in very low numbers, whereas others are numerous. In most species they are mitotically stable. They are exceptionally found in single species of a genus, or only in certain populations.

3.1.2 Karyotypes of the Phyllomedusidae

Thus far, the chromosome sets of 29 species of the family Phyllomedusidae have been studied in detail (Appendix 1). These species are included in the genera *Agalychnis* (4 species), *Callimedusa* (1 species), *Cruziohyla* (1 species), *Phasmahyla* (3 species), *Phyllomedusa* (12 species), and *Pithecopus* (7 identified species, 1 unidentified species). This corresponds to nearly half (49.2%) of the 59 known species of the family Phyllomedusidae (chapter 1.1.1; Duellman et al. [2016]). Thirty of these species (50.8%) have not been previously karyotyped. Furthermore, no chromosome data have been published so far for species in the neotropical genera *Hylomantis* and *Phrynomedusa* (Appendix 1).

Twenty-seven of the 29 phyllomedusid species so far examined have diploid chromosome numbers of 2n = 26 (26 species) and 2n = 52 (1 species, tetraploid) (Fig. 21, Appendix 1). The fundamental numbers are FN = 52 and 104, respectively. In 2 species, very exceptional karyotypes were observed by Nunes and Fagundes [2008b] (Appendix 1). In 3 male individuals of *Phasmahyla exilis* collected in the Reserva Biológica de Duas Bocas, Cariacica, Espirito Santo State, Brazil, highly heteromorphic XY♂ sex chromosomes were found. The X chromosome is submetacentric and medium-sized, whereas the Y chromosome is a small and submetacentric element. Female specimens of this species could not be examined. In a single male specimen of *Phyllomedusa rohdei* sampled in Linhares, Espírito Santo State, Brazil, an unexpected diploid chromosome number of 2n = 40 (FN = 80) was observed. Again, highly heteromorphic XY♂ sex chromosomes were detected. The X is submetacentric and medium-sized, whereas the Y is a small submetacentric chromosome. Female specimens of this species were also not available. Nunes and Fagundes [2008b] admit that before major conclusions can be drawn, the females of both taxa must be included in the cytogenetic examinations, and that the presumed XY♂ sex chromosomes in the male specimens should be carefully analyzed by C-banding.

As the most likely parsimonious hypothesis, the modal diploid chromosome number of 2n = 26 and fundamental number of FN = 52 can be considered as the basal (ancestral) karyotype of the entire family Phyllomedusidae. The structure of this 2n = 26/FN = 52 phyllomedusid karyotype exhibits a high degree of conservation (Fig. 303–316). Figures 21a–e show 1 idiogram of each phyllomedusid genus in which mitotic chromosomes are known and confirm this striking similarity. The karyotype consists of 13 pairs of mainly meta- to submetacentric chromosomes. In general terms, the chromosome pairs 1–6 are larger than the remaining pairs 7–13. In some species, the middle-sized chromosome 7 is the most submetacentric element in the karyotype, showing a very short p arm whose length does not exceed the diameter of one chromatid (e.g., Fig. 306, 310). In

Fig. 38. Adult individuals of *Phyllomedusa distincta* (**a**), *P. tetraploidea* (**b**), and their triploid hybrid (**c**). From Gruber et al. [2013]. Photographs by Célio Fernando Baptista Haddad. For further explanations, see chapter 3.1.2.

all complements, there is only a slight decrease in chromosome length from pair 1 to pair 6 and from pair 7 to pair 13.

It must be stressed that like in the arboranan family Hylidae (see chapter 3.1.1), in the majority of phyllomedusid karyotypes, the chromosomes 3 are characterized by a very distinct conserved and typical submetacentric morphology. The p:q ratio in these chromosomes is approximately 1:3 to 1:4 (e.g., Fig. 21, 303–316). It is tempting to speculate that the chromosomes 3 in the karyotypes of the Phyllomedusidae and Hylidae have evolutionarily conserved a structural and genetic homoeology, but this can only be confirmed or refuted by experiments using the appropriate chromosome paints and FISH.

Phyllomedusa tetraploidea

This species (Fig. 38b) can be found in the interior regions of the States of São Paulo and Paraná (Brazil), the Province Misiones (Argentina), and the southern Oriental Region of Paraguay. It occurs at an elevation of 400–1,000 m. It lives in the rainforest on herbaceous vegetation near swamps. The egg clutch is deposited in leaf nests above the water in permanent pools in forests. Larvae develop in pools. It occurs in secondary forests and pastures.

The tetraploid species *P. tetraploidea* (4n = 52; Fig. 38b, 309) and the diploid *P. distincta* (2n = 26; Fig. 38a, 306) occur in syntopy in some localities of southeastern Brazil where they produce triploid hybrids (3n = 39; Fig. 38c) [Pombal and Haddad, 1992; Gruber et al., 2013]. Hybrid individuals are numerous and vigorous, and the advertisement call is indistinguishable from those of the diploid and tetraploid species [Haddad et al., 1994]. Based on bioacoustics and ecology, Pombal and Haddad [1992] suggested an autopolyploid origin of *P. tetraploidea* from *P. distincta* or *P. iheringii*. However, the possibility of allopolyploidy by crossing *P. distincta* with *P. iheringii* cannot be excluded [Gruber et al., 2013]. Recent analyses of mitochondrial DNA and nuclear genes showed that both hypotheses are feasible, but in the molecular phylogeny, *P. distincta* and *P. tetraploidea* were sister groups, whereas *P. iheringii* appeared in a separate clade [Brunes et al., 2010].

Following these molecular studies, Gruber et al. [2013] examined in detail the mitotic and meiotic chromosomes of *P. distincta*, *P. tetraploidea*, and their triploid hybrids with classical, banding, and molecular cytogenetic techniques in order to obtain some evidence concerning the origin of the polyploidy in *P. tetraploidea*.

In conventionally Giemsa-stained preparations, the chromosomes can be arranged in 13 pairs, quartets or tercets, respectively, and are of metacentric (1, 4, 8, 11, 13), submetacentric (2, 3, 5, 6, 10, 12), and subtelocentric (7, 9) morphology (Fig. 306, 309). In some of the tetraploid *P. tetraploidea* specimens, the quartet 3 and the quartet 8, or both, can be subdivided into 2 pairs of homologues: chromosomes 3a have short arms that are longer than those of chromosomes 3b, and chromosomes 8a have short arms that are slightly longer than those of chromosomes 8b. Secondary constrictions can occasionally be observed in the proximal short arms of chromosome 1 in the 3 cytotypes and in addition in the proximal short arms of chromosome 8 in *P. tetraploidea*, and in 8p and 9q in the triploid hybrids [Gruber et al., 2013].

C-banding of the constitutive heterochromatin in the chromosome pairs, quartets, and tercets is approximately equivalent. Chromosomes 2–5, 8, 9, 12, and 13 exhibit centromeric C-bands, but some chromosomes show only a tiny or almost absent C-band. The remaining chromosomes 1, 6, 7, 10, and 11 have large pericentromeric heterochromatic blocks, and depending on the chromatin condensation, these blocks continuously extend to the centromeric band. In the 3 cytotypes, the chromosomes 2 and 3 exhibit additional interstitial C-bands in the short arms that are very prominent in some metaphases [Gruber et al., 2013]. The authors realized that the C-bands are not always identical among the homologous chromosomes within each pair, quartet, or tercet. Nevertheless, for some tetraploid specimens, quartet 3 can be subdivided presenting a conspicuous interstitial C-band in the short arms of chromosome 3a that is absent in chromosome 3b, and quartet 8 exhibits slightly different C-banding and can also be subdivided into 2 pairs, with a pericentromeric block in the short arms of chromosome 8a which is larger than the block present in 8b (Fig. 470/471). Interestingly, the chromosomes in each pair, quartet, and tercet are characterized by a particularly distinctive C-banding pattern, so the homologues within each of these groups can be precisely identified. These observations indicate an early diploidization in the polyploid karyotype.

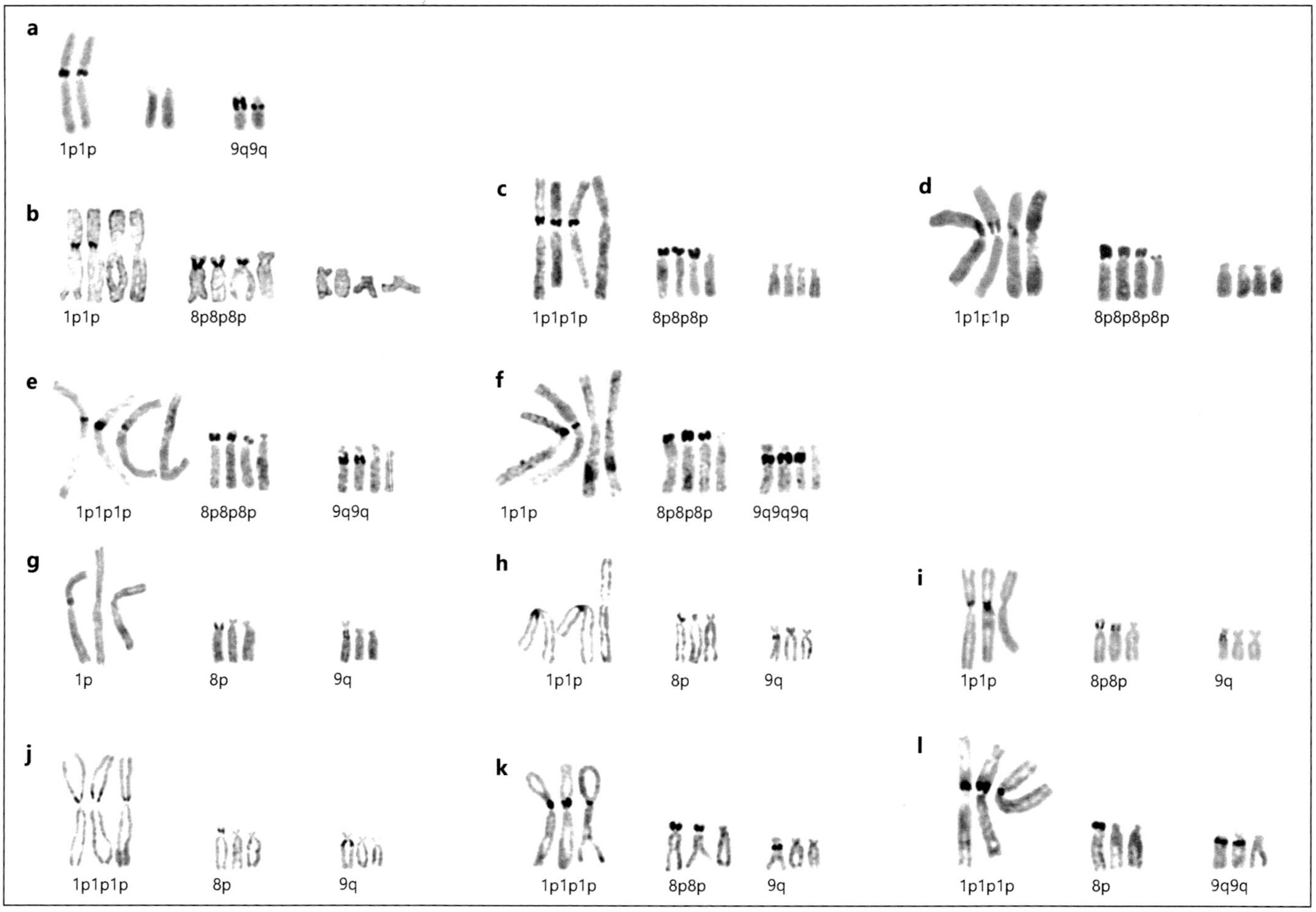

Fig. 39. Chromosomes 1, 8, and 9 of *Phyllomedusa distincta* (**a**), *P. tetraploidea* (**b–f**), and the triploid hybrid of *P. distincta* × *P. tetraploidea* (**g–l**) showing the different labeling patterns obtained after specific Ag-staining. For further explanations, see chapter 3.1.2.

Following specific Ag-staining, Ag-NORs in *P. distincta* are visible on both homologues 1 (1p) and both homologues 9 (9q) according to the 1p1p9q9q pattern (Fig. 39a) [Gruber et al., 2013]. In *P. tetraploidea*, Ag-NORs are present in chromosome 1p and chromosome 8p. In 6 of the individuals examined, the total number of Ag-NORs ranged from 4 to 8 per metaphase (Fig. 39b–d). In 2 further frogs, Ag-NORs were also observed in chromosome 9 (9q) according to the 1p1p1p8p8p8p9q9q pattern in one, and the 1p1p8p8p8p9q9q9q distribution in the other specimen (Fig. 39e, f). In the triploid hybrids, Ag-NORs are present at 1p, 8p, and 9q, and the total number ranges from 3 to 6 per metaphase (Fig. 39g–l). One triploid hybrid exhibited the 1p1p1p8p9q9q Ag-NOR pattern with 2 Ag-impregnated chromosomes 9 (Fig. 39l).

FISH with a specific rDNA probe was applied to the chromosomes of 3 individuals of each cytotype (not shown) [Gruber et al., 2013]. In *P. distincta*, 4 hybridization sites per metaphase were always observed and followed the 1p1p9q9q pattern. The specimen of *P. tetraploidea* that showed intra-individual Ag-NOR variation exhibited 1p1p8p8p8p8p-type hybridization in all metaphases analyzed. The specimen which had the 1p1p8p8p8p Ag-NOR pattern had a 1p1p1p1p8p8p8p hybridization pattern. And the frogs which had 1p1p1p8p8p8p9q9q and 1p1p8p8p8p9q9q9q Ag-NOR patterns had 1p1p1p8p8p8p9q9q9q-type hybridization patterns. The triploid hybrid that exhibited the 1p8p9q Ag-NOR pattern showed a 1p1p1p8p9q-type hybridization. The frog that had the 1p1p8p8p9q Ag-NOR pattern displayed the 1p1p8p8p9q-type hybridization pattern. And 1 specimen which had the 1p1p1p8p9q Ag-NOR pattern showed the 1p1p1p8p9q9q-type hybridization.

FISH with the canonical telomeric sequences labels the terminal regions of all chromosomes in the 3 cytotypes (not shown) [Gruber et al., 2013]. In the metaphases of *P. distincta*, the chromosomes 7 and 11 exhibit additional hybridization signals in their centromeric regions, whereas in *P. tetraploidea*, 2 chromosomes 7 and 2 chromosomes 11 possess interstitial centromeric signals. In the triploid hybrid, the centromere of 1 chromosome 7 and 1 chromosome 11

Cytogenet Genome Res 2018;155:55–221
DOI: 10.1159/000489839

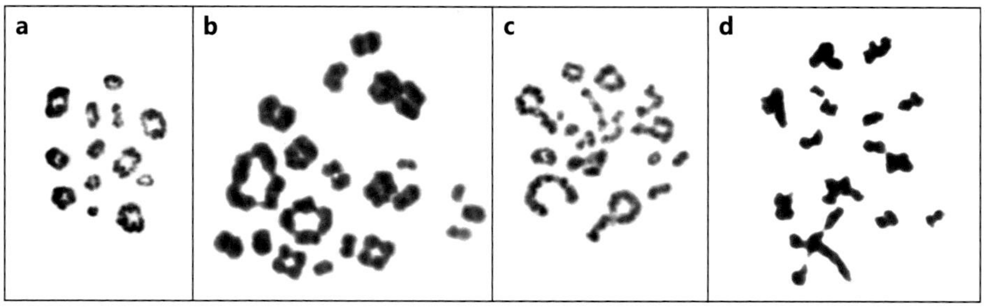

Fig. 40. Giemsa-stained meiotic cells from male *Phyllomedusa distincta* (**a**), *P. tetraploidea* (**b**), and their triploid hybrid (**c**, **d**). **a–c** Diakineses. **d** Metaphase II. From Gruber et al. [2013]. For further explanations, see chapter 3.1.2.

show additional hybridization signals with the telomeric DNA probe. In the 3 cytotypes, chromosome 6 also exhibits a small hybridization signal in the centromeric region.

The replication bands confirm the high degree of homoeology between the chromosomes of *P. distincta*, *P. tetraploidea*, and their triploid hybrids. However, the multiple banding patterns of the homoeologues within each pair, tercet, or quartet are not absolutely identical. In some metaphases, the subdivision into 2 chromosome pairs reinforces the hypothesis of diploidization as revealed by the C-banding patterns.

Diploidization is a very important process in the evolution of polyploid karyotypes. Details are summarized in the first part of this trilogy on amphibian chromosomes [Schmid et al., 2010]. It stabilizes the selective advantages produced by polyploidization, and it has been described as a natural evolutionary pathway in which a disomic status is reestablished. In *P. tetraploidea*, it is possible to ascertain diploidization by accurate cytogenetic analysis (see above) which revealed differences in the morphology of standard-stained chromosomes, in the quantity and distribution of C-bands, in the 18S + 28S rDNA sequences, and in the BrdU replication banding patterns. Although diploidization in *P. tetraploidea* is observed in the quartets 3 and 8, this process is supposed to continue in other quartets.

Meiotic analyses performed in males of *P. distincta* reveal 13 bivalents during diplotene, diakinesis (Fig. 40a), and metaphase I cells, and 13 chromosomes in metaphase II [Gruber et al., 2013]. Male specimens of *P. tetraploidea* possess a variable number of tetravalents and bivalents in the diplotene, diakinesis, and metaphase cells (Fig. 40b). In the testes preparations of the triploid hybrid males, early meiotic stages with univalent, bivalent, and trivalent configurations are observed (Fig. 40c). The tetravalent formation in *P. tetraploidea* could be evidence for autopolyploidy, but considering the high degree of evolutionary chromosomal conservation in species of the genus *Phyllomedusa*, allopolyploidy cannot be excluded as possibility. The triploid hybrids show initial meiotic stages with degenerated chromosomes, but some few cells have relatively normal appearance. Some spermatids, perhaps those with nearly balanced chromosomal constitutions, are able to undergo differentiation into spermatozoa, which explains the fertility of the hybrid individuals. According to field observations that were confirmed in laboratory experiments [Haddad et al., 1994], the triploid males mate with diploids, triploids, or tetraploids, but the egg clutches have a low number of eggs or a low fertilization rate. Some of these eggs produce small, unpigmented, and severely deformed tadpoles unable to complete further development.

To conclude, and for convenience of the readers, the occurrence of polyploid anuran species is summarized in Table 4.

According to Duellman et al. [2016], the Arboranae clade diverged from its nobleobatrachian relatives in South America during the latest Cretaceous or early Cenozoic (Fig. 1, 2). Differentiation of arboranans into the 3 families Hylidae, Pelodryadidae, and Phyllomedusidae occurred in South America during the early Cenozoic. Ancestral Hylidae diverged from the phyllomedusid-pelodryadid clade in late Paleocene. Divergence of phyllomedusids and pelodryadids occurred in the mid-Eocene. When South America separated from Antarctica 35.9–30.5 Mya, ancestral pelodryadid frogs were present on Antarctica and Australia before they separated about 35.5 Mya. In the Australo-Papuan Region, the Pelodryadidae differentiated during the Eocene and Oligocene. In tropical South America, the Phyllomedusidae differentiated in the Miocene and Oligocene (see chapter 1.1.1; Fig. 1, 2). With these data in mind, it becomes compelling evidence that the basal (ancestral) karyotype of the Arboranae must have had a diploid chromosome number of 2n = 26 and a fundamental number of FN = 52. This formula was maintained by the extant species of the phyllomedusid-pelodryadid clade, whereas in the ancestors of the hylid clade, the diploid chromosome number was reduced to 2n = 24 and the fundamental number to FN = 48. This important change implicated only 1 chromosome mutation, probably a nonreciprocal translocation, tandem fusion, or inversion-fusion cycle (for details, see chapters 3.4.2–3.4.4). Further changes in the chromosome number and structure within the 3 families Hylidae, Phyllomedusidae, and Pelodryadidae occurred independently after separation of the 3 clades.

3.1.3 Karyotypes of the Pelodryadidae

The karyotypes of 86 species of the family Pelodryadidae have been examined (Appendix 1). These species belong to the subfamilies Litoriinae (genus *Litoria*, 39 identified spe-

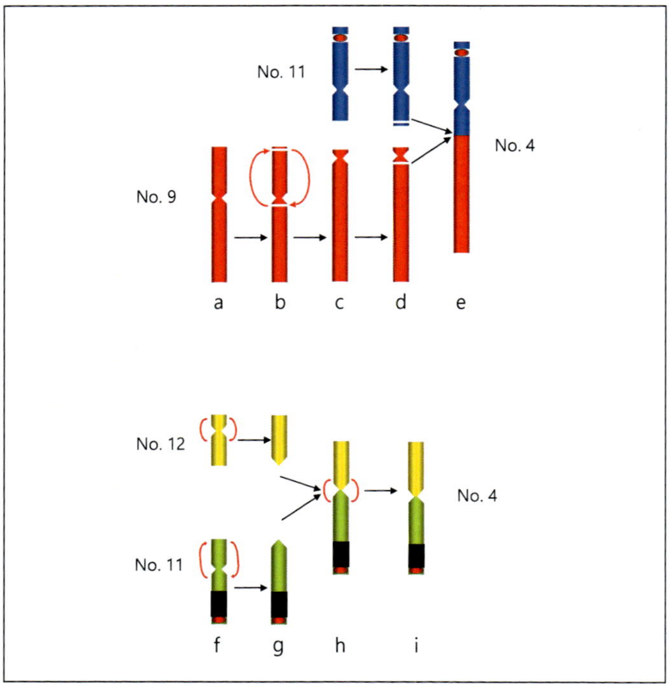

Fig. 41. Two different models for the origin of chromosome 4 in *Nyctimystes infrafrenatus* from the original chromosomes 9 and 11 (a–e) or 11 and 12 (f–i). Curved arrows indicate pericentric inversions. For further explanations, see chapter 3.1.3.

cies, 1 unidentified species) and Pelodryadinae (genera *Nyctimystes*, 10 identified species, 1 unidentified species; *Ranoidea*, 35 identified species). These 86 species are equivalent to 41.3% of the extant species (208) of the family Pelodryadidae (chapter 1.1.1, Duellman et al. [2016]), and there are still 122 (58.7%) pelodryadid species with unknown cytogenetic data.

Eighty-five of the 86 pelodryadid species so far examined have conserved diploid chromosome numbers of 2n = 26 (Appendix 1; Fig. 21f, h, 317–382). The fundamental number is FN = 52. Only a single species, *Nyctimystes infrafrenatus* (see below), has a reduced karyotype with 2n = 24 and FN = 48 (Fig. 21g, 353, 354). In all other species, the karyotype consists of 13 pairs of mainly meta- to submetacentric chromosomes. Again, like in the arboranan families Hylidae and Phyllomedusidae (chapters 3.1.1, 3.1.2), in many karyotypes of the Pelodryadidae, the chromosomes 3 are characterized by a very distinct conserved and typical submetacentric morphology. The p:q ratio in these chromosomes is approximately 1:4 (e.g., Fig. 21f–h, 317–382). It is tempting to speculate that the chromosomes 3 in the karyotypes of the Hylidae, Phyllomedusidae, and Pelodryadidae are derived from the same ancestral chromosome pair. Suitable chromosome paints and the FISH technique will help to solve this assumption.

The chromosome pairs 1–5 are larger than the remaining pairs 6–13. In some species, the middle-sized chromosome 6 is the most submetacentric element in the karyotype showing a very short p arm whose length does not exceed the diameter of one chromatid (e.g., Fig. 21, 320, 323, 346). In all complements, there is only a slight decrease in chromosome length from pair 1 to pair 5 and from pair 6 to pair 13. There are very subtle differences between the karyotypes of species within the genus *Litoria* and within the genus *Ranoidea*. These were recorded in the present study by classifying the *Litoria* species in 3 species groups (Fig. 477–503) and those of the *Ranoidea* species in 5 species groups (Fig. 506–545). Otherwise, the karyotypes are highly conserved. Figures 21f–h show 1 idiogram of each pelodryadid genus and confirm this striking similarity.

As already mentioned above (chapters 3.1.1, 3.1.2), the most likely parsimonious hypothesis is that the modal diploid chromosome number of 2n = 26 and fundamental number of FN = 52 can be considered as the basal (ancestral) karyotype of the taxon Arboranae. It cannot be decided with certainty whether the initial chromosome repatterning that caused the reduction of the chromosome number from 2n = 26 to 2n = 24 in hylids was one and the same or if these events occurred independently from one another. By comparing the karyotypes of the extant phyllomedusids and pelodryadids (Fig. 303–316 and 317–382), it becomes apparent that after the separation of the 2 clades a number of different structural rearrangements must have occurred.

Nyctimystes infrafrenatus

The white-lipped treefrog (Fig. 950d) occurs in the coastal and adjacent areas of northeastern Queensland, north of Townsville and the lowlands of New Guinea, westwards to the islands of Timor, Halmahera, Seram, Buru, and Talaud Island; eastwards to New Ireland, Bismarck Archipelago, and south to the Aru Islands. The area of occurrence of the species is approximately 123,800 km^2. The frog lives in a variety of habitats including rainforests, wet sclerophyll forests, cultivated and suburban habitats. It is active on warm and humid nights and breeds in spring and summer. About 4,100 eggs are laid in clumps, and tadpoles develop in about 8 weeks [AmphibiaWeb, 2017; Frost, 2017].

Up to date, *N. infrafrenatus* is the only karyotyped species of the family Pelodryadidae with a reduced karyotype of 2n = 24 chromosomes and FN = 48 (Fig. 353, 354). Menzies and Tippet [1976], who first described this karyotype (Appendix 1), also proposed the origin of this derived karyotype. According to their model, a chromosome 11 and a chromosome 9 (the latter one with a large pericentric inversion) fused to the large chromosome 4 of the current karyotype (Fig. 41a–e). However, other models are also conceivable (Fig. 41f–i).

Donnellan and Mahony [2004] screened 36 enzyme loci in 179 individuals of the *Ranoidea lesueuri* species group from 33 locations in eastern Australia. They identified 3 genetic groups that they considered separate species. The northernmost group, *R. jungguy* had a distinctive standard and C-band karyotype (Fig. 42c, d, 511–513), while the

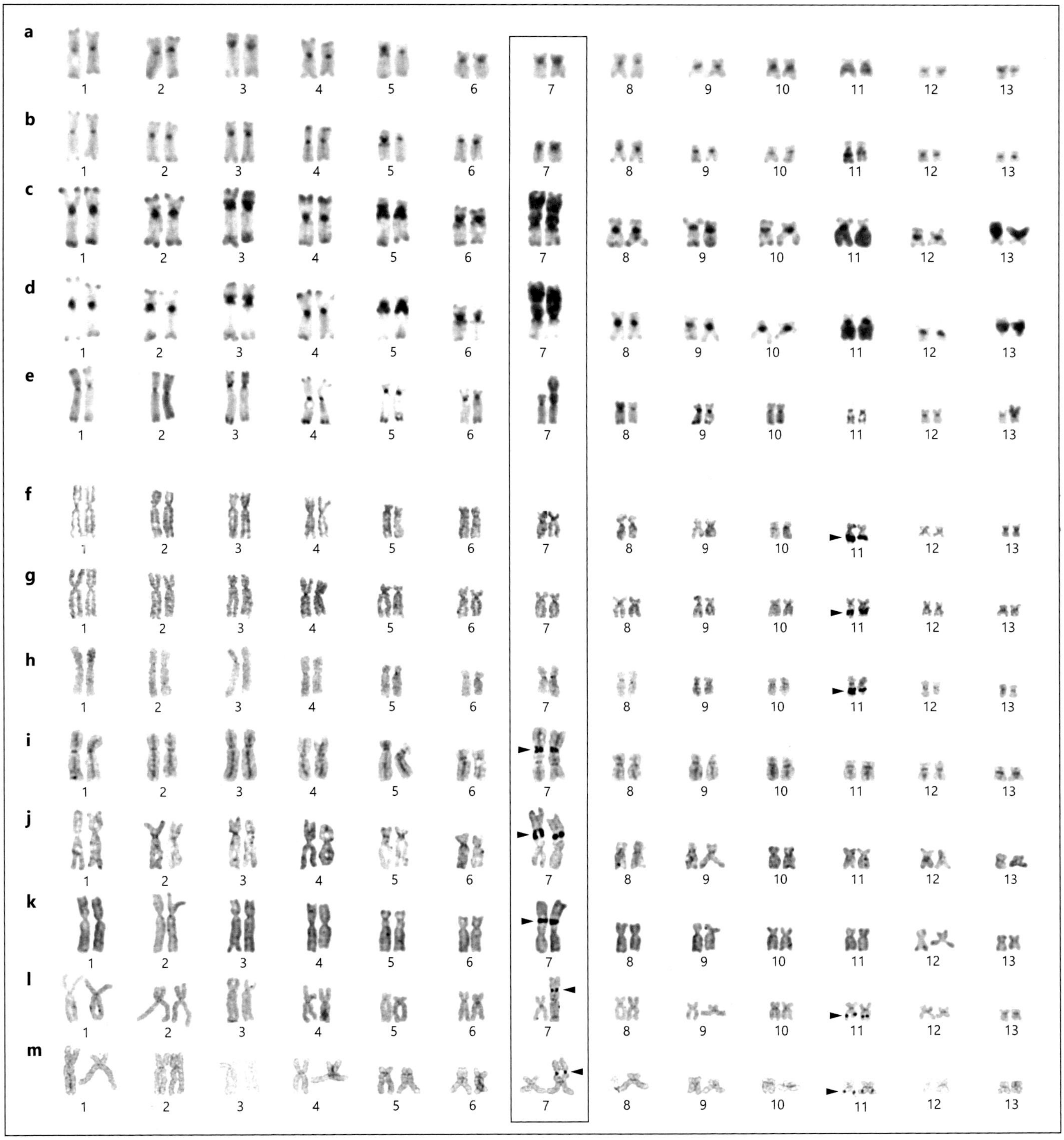

Fig. 42. Karyotypes of *Ranoidea wilcoxii* (**a**, **b**, **f–h**), *R. jungguy* (**c**, **d**, **i–k**), and their hybrid (**e**, **l**, **m**) showing C-banding (**a–e**) and Ag-staining of NORs (**f–m**). Note the heteromorphic homologues 7 in **e**, **l**, and **m**. For further explanations, see chapter 3.1.3.

karyotypes of the 2 southern groups, *R. lesueuri* and *R. wilcoxii*, differed only in minor details of their C-band patterns (Fig. 42a, b, 514–516). Locations with intergroup hybrids were identified in the Australian Capital Territory, mid-east Queensland, and 2 localities in Far North Queensland. The hybrids become immediately apparent by their karyotypes. Very impressive are the hybrids with *R. jungguy* as one of the parental species because of the chromosome 7 with the prominent C-band block in its short arm (Fig. 42e, 517, 518). Furthermore, the NORs occupy different positions in the karyotypes (Fig. 42f–m, 633–638).

3.1.4 Experimental in vitro Hybridization Experiments in Arboranae

Experimental in vitro hybridization experiments are easily accomplished with externally fertilizing vertebrates by combining gametes in a petri dish [Littlejohn, 1961]. The progressive development of artificially produced hybrids through embryological stages was information that was used by early workers to estimate genetic similarities and differences between the parental species [Moore, 1955; Blair, 1972] prior to any sequence-based phylogenetic analyses. Hybrids of closely related species of frogs may reach the adult stage, while more distantly related species may only progress to blastula, gastrula, or tadpole stages. Within the Arboranae, it is easy to obtain interspecies hybrids or recognize natural hybrids, and numerous reports have been published [e.g., Mecham, 1960; Ralin, 1970; Bogart and Bogart, 1971; Kuramoto, 1984; Anderson and Moler, 1986; Oliveira et al., 1991; Mable and Bogart, 1995]. This ability to hybridize suggests that few if any postzygotic isolating mechanisms exist for these frogs. It is important to examine the chromosomes of resulting hybrids because the possibility exists that a hybrid may be haploid, polyploid, gynogenetic diploid, or aneuploid. Such aberrant individuals can exist in hybridization experiments, and their occurrence must be considered when analyzing the results of those experiments [Bogart, 1972]. If the parental species have relatively similar karyotypes, it is difficult to determine the contribution of chromosomes from each parent. A gynogenetic diploid may easily be mistaken for a diploid hybrid, but the gynogen would not carry any male-derived chromosomes [Bogart et al., 1969]. Bi and Bogart [2006] used fluorescent genomic in situ hybridization (GISH) to visualize species-specific chromosomes in *Ambystoma* unisexual salamanders that possessed chromosomes from *A. jeffersonianum* and *A. laterale.* They applied a fluorescent genomic probe for one species (*A. laterale*) and blocking DNA from the other species (*A. jeffersonianum*) to chromosome spreads that contained both genomes in unisexual salamanders. By counterstaining with a fluorescent DNA stain, the chromosomes derived from each species had a distinctly different fluorescence.

Chromosome morphology is inherently variable with respect to the methods used to obtain the chromosomes and the inter-cell variation in chromosome condensation or coiling. Small, but evolutionarily meaningful, inversions and translocations that may be alternately fixed in related species could be unrecognized or not statistically significant based on standard deviations of arm length or the position of a constriction. GISH circumvents these problems because homoeologous chromosomes, from genetically dissimilar parents in the hybrid, can be distinguished and are derived from the same nucleus that has been subjected to the same methods. When such homoeologs are paired, subtle absolute differences in chromosome morphology are revealed. Hylid frogs are ideal subjects for GISH experimentation because many distinct species can be hybridized [Moore, 1955; Littlejohn, 1961; Pyburn and Kennedy, 1961; Ralin, 1970], and many hybrids reach the tadpole stage where chromosomes can be obtained from tail epithelial cells. Figure 43 shows the results of a preliminary feasibility study [Bogart JP, Bi K, Noble DWA, unpublished] that applied GISH and 5S rDNA FISH to chromosomes of *Pseudacris crucifer* (female) × *P. triseriata* (male) hybrid tadpoles. The methods for GISH and FISH are the same as those used by Bi and Bogart [2006] and Bi et al. [2008, 2009]. These species often breed in the same ponds in southern Ontario but have never been documented to hybridize in nature. In the artificial cross, only 5 tadpoles hatched from 115 eggs that were used for the cross. Chromosomes, which were obtained from all 5 tadpoles, showed that the hybrids were diploid (2n = 24) and each parent contributed 12 chromosomes (Fig. 43). The position of the 5S rRNA gene appears to be conserved as it is found at the same position on homoeologous chromosomes.

3.2 Meiosis, Heterochromatin, and Polymorphism of Heterochromatin in Anura and Arboranae

3.2.1 Male Meiosis in Anura and Arboranae

Morescalchi [1971, 1973] and Morescalchi and Galgano [1973] worked out the main characteristics of male meiosis in the Anura. According to these comparative studies, male meiosis in anuran species of the families Leiopelmatidae, Pipidae, and Discoglossidae is characterized by long diplotene and short diakinesis time periods, bivalents with a low degree of condensation, more than 2 chiasmata in the large bivalents which in most instances are located in an intercalary position, and incomplete terminalization of the chiasmata. These 3 families belong to the least advanced suborders of the Anura, the Leiopelmatidae are even considered to be the oldest family. These frogs with amphicoelous presacral vertebrae are *Ascaphus truei* from North America, the 6 known fossil and recent species of *Leiopelma* from New Zealand, and the Jurassic fossils of *Notobatrachus* and *Vieraella* from South America [Noble, 1931; Estes and Reig, 1973; Green et al., 1980; Cannatella, 1985; Duellman and Trueb, 1986; Worthy, 1987; Green et al., 1989; Green and Cannatella, 1993]. Considered from a phylogenetic point of view, Morescalchi [1973] proposed that the type of meiosis found in Leiopelmatidae, Pipidae, and Discoglossidae is the more ancient one in the Anura. A quantification of the frequency of chiasmata in male meiosis was performed by Chandler et al. [1993] for *Leiopelma hochstetteri, L. hamiltoni* (now *pakeka*, see Bell et al. [1998]) and *L. archeyi.* The mean chiasma frequency in *L. hochstetteri* individuals without B chromosomes was 24.81 ± 0.84, while in individuals with B chromosomes the mean was 23.40 ± 1.94. In *L. pakeka* and *L. archeyi*, the mean chiasma frequencies were 18.69 ± 1.16 and 19.71 ± 1.11, respectively. In all 3 species, chiasmata were found both at the ends of the bivalents as well as in interstitial sites. Up to 4 chiasmata were observed

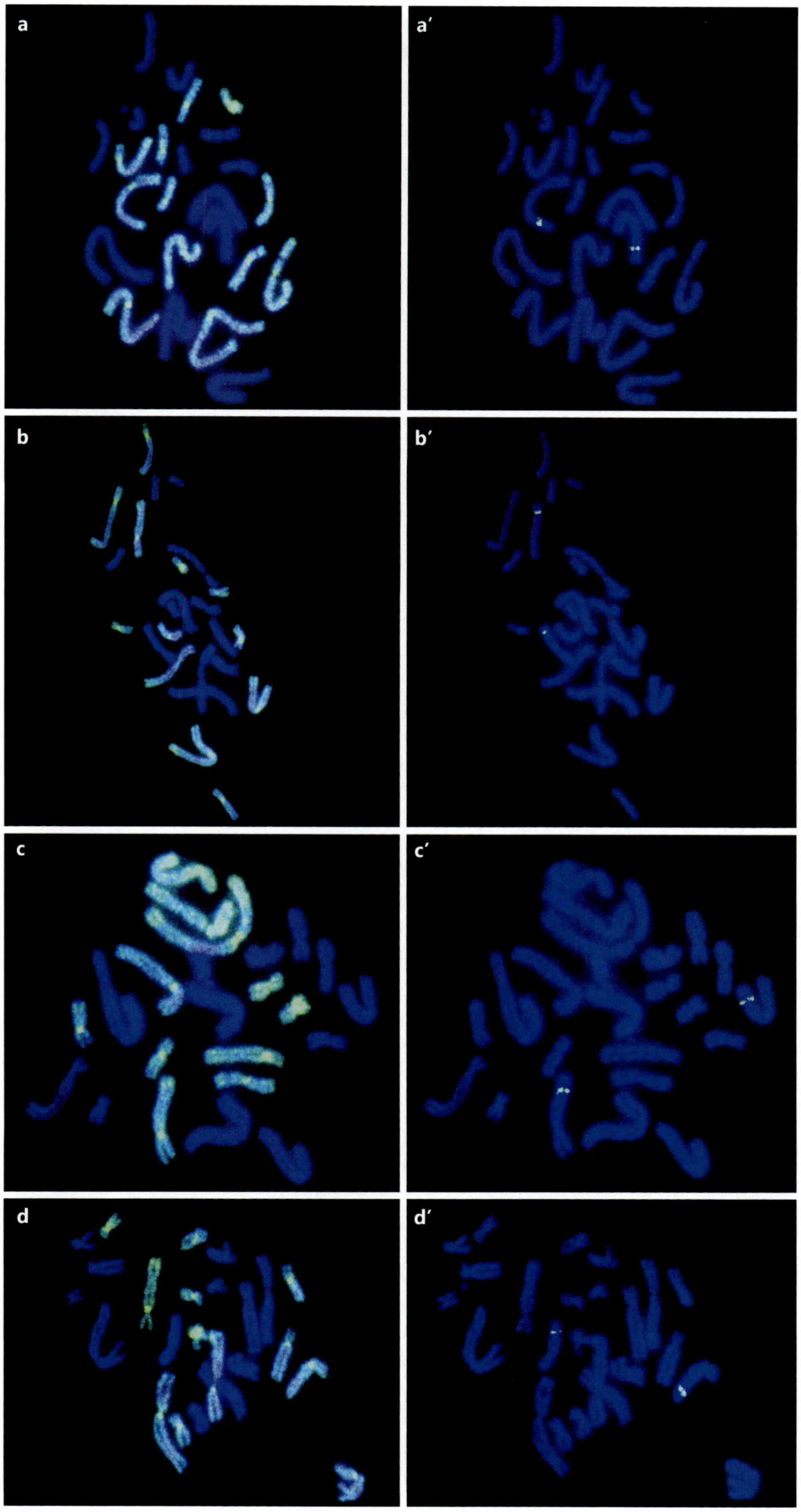

Fig. 43. GISH (**a–d**) and 5S rDNA FISH (**a′–d′**) applied to metaphase chromosome spreads of *Pseudacris crucifer* × *P. triseriata* hybrid tadpoles. For the GISH experiment, *P. crucifer* DNA probes were labeled with fluorescein isothiocyanate (FITC, green fluorescence). These probes and *P. triseriata* blocking DNA were applied to the metaphase spreads. The preparations were counterstained using DAPI (blue fluorescence). There is a clear distinction between the 12 *P. crucifer* chromosomes (green fluorescence) and the 12 *P. triseriata* chromosomes (blue fluorescence). A 5S rDNA FISH experiment was subsequently done on the same chromosome spreads to determine the position of the 5S rRNA gene on homoeologous chromosomes. The biotinylated 5S rDNA probes, developed from *P. crucifer* DNA, were also visualized by FITC, and the chromosomes were also counterstained with DAPI. The position of 5S rRNA genes in the karyotypes of the 2 species can be determined by comparing GISH-based homoeologous chromosomes. For further explanations, see chapter 3.1.4.

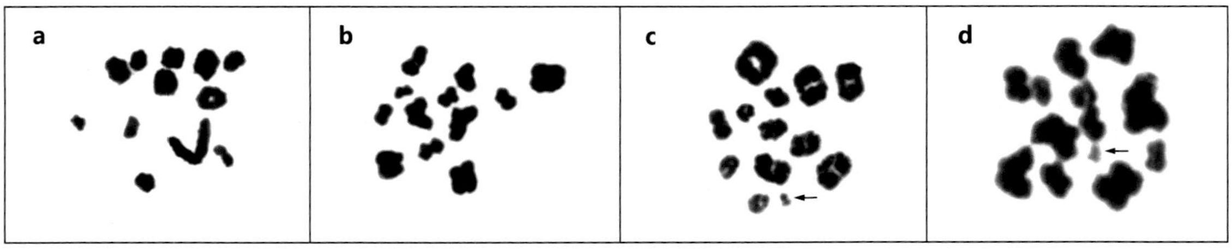

Fig. 44. Giemsa-stained male meiotic cells of *Boana prasina* (**a**, **b**) and *Bokermannohyla luctuosa* (**c**, **d**). **a–c** Diakineses. **d** Metaphase II. Note the B chromosome in **c** and **d** (arrows). From Gruber et al. [2012a, b]. For further explanations, see chapter 3.2.1.

in the larger bivalents. Thus, these 3 leiopelmatids have one of the highest recombination rates recorded for any studied frog species [Chandler et al., 1993].

In contrast, male meiosis of all other anuran families is distinguished by short diplotene and long diakinesis time periods, bivalents with a high degree of condensation, only 2 chiasmata in the bivalents that are always located at terminal positions, and complete terminalization of the chiasmata. But exceptions to these rules do exist, because in the male meiosis of *Pipa pipa* (family Pipidae), *Dendropsophus microps, D. rubicundulus* (family Hylidae), and *Flectonotus pygmaeus* (family Hemiphractidae), the bivalents also possess interstitial chiasmata [Beçak, 1968; Rabello, 1970; Morescalchi, 1973; Schmid et al., 1988]. Morescalchi [1973] and Schmid et al. [1988] realized that these species are peculiar in so far as they have telocentric chromosomes in their karyotypes. Actually, *P. pipa* and *F. pygmaeus* have only telocentric chromosomes, which is a relatively rare finding in the more highly evolved anuran families. Thus, both authors claimed that a correlation between intercalary chiasmata in male anuran meiosis and the presence of telocentric chromosomes in the karyotypes seems to exist. This hypothesis, although without any logical explanation, is supported by the results obtained in some terraranan species that have a majority of telocentric chromosomes in their karyotypes. A further exception is provided by some African pyxicephalids and hyperoliids (*Pyxicephalus adspersus, Kassina wealii, K. senegalensis, Hyperolius* sp.) which reveal diakinetic bivalents with no complete terminalization of proterminal or interstitial chiasmata [Morescalchi, 1973; Schmid, 1980b].

As emphasized by Morescalchi [1973], the chiasmata in female anuran meiosis generally have an unrestricted distribution in the lampbrush chromosomes. This has been shown in the frogs *Bombina variegata, Hyla arborea, Bufotes viridis, Pelophylus esculentus* complex [Morescalchi, 1973], *Bufo bufo* [Ullerich, 1970], and *Xenopus laevis* [Callan et al., 1987, 1988]. In male anuran meiosis, on the other hand, with the exception of the Leiopelmatidae and Discoglossidae, the chiasmata are restricted to the terminal regions of the bivalents. Therefore, Morescalchi [1973] argued that it is possible that different systems of genetic recombination are operating in most female and male anurans.

Male meiosis of 22 different arboranan species was examined in some detail in the present study. Some illustrations are shown in Figures 44 and 45. As a rule, the larger biarmed chromosomes form distinct ring-like bivalents with terminal chiasmata located in each of the 2 arms. In most of the C-banded bivalents in the diakinesis stage, the location of the centromeric regions can be recognized by the large blocks of centromeric heterochromatin. The NORs, which are inactive at diakinesis and therefore escape detection by Ag-staining, can be visualized by mithramycin staining.

With respect to the frequent telocentric chromosomes, 3 types of diakinetic configurations can be distinguished: (1) rod-like bivalents with terminalized chiasmata, (2) ring-like bivalents, and (3) cross-like bivalents. These 3 types of diakinetic configurations are not specific for each of the bivalents, but can vary between different diakineses. In the cross-like bivalents, the occurrence of interstitially located chiasmata can be assumed.

3.2.2 Constitutive Heterochromatin in Anura

The chromosomes of a large number of anuran species have been analyzed by the C-banding technique [for reviews, see Schmid, 1980c; King, 1990, 1991; Olmo, 1990; Schmid et al., 1990b; Green and Sessions, 1991a, 2007]. Although most anuran karyotypes have many constitutively heterochromatic regions, there are also species where positive C-banding is restricted to small pericentromeric regions. Generally, the number of heterochromatic regions is largely independent of the length of the chromosomes or their DNA content [Schmid, 1978a]. Small chromosomes can contain just as many, or even more, C-bands as the largest chromosomes in a karyotype. Centromeric, pericentromeric, and telomeric regions are the most common locations of C-bands. Interstitial heterochromatic bands of varying sizes may also be present. The NORs are almost always associated with constitutive heterochromatin. In most amphibian species, C-banding labels the heterochromatin of centromeric and pericentromeric chromosome regions as well as that associated with the NORs distinctly stronger than the heterochromatin in the telomeric regions, although there are many exceptions. Weakly C-banded telomeric regions are also observed in chromosomes of

Cytogenet Genome Res 2018;155:55–221
DOI: 10.1159/000489839

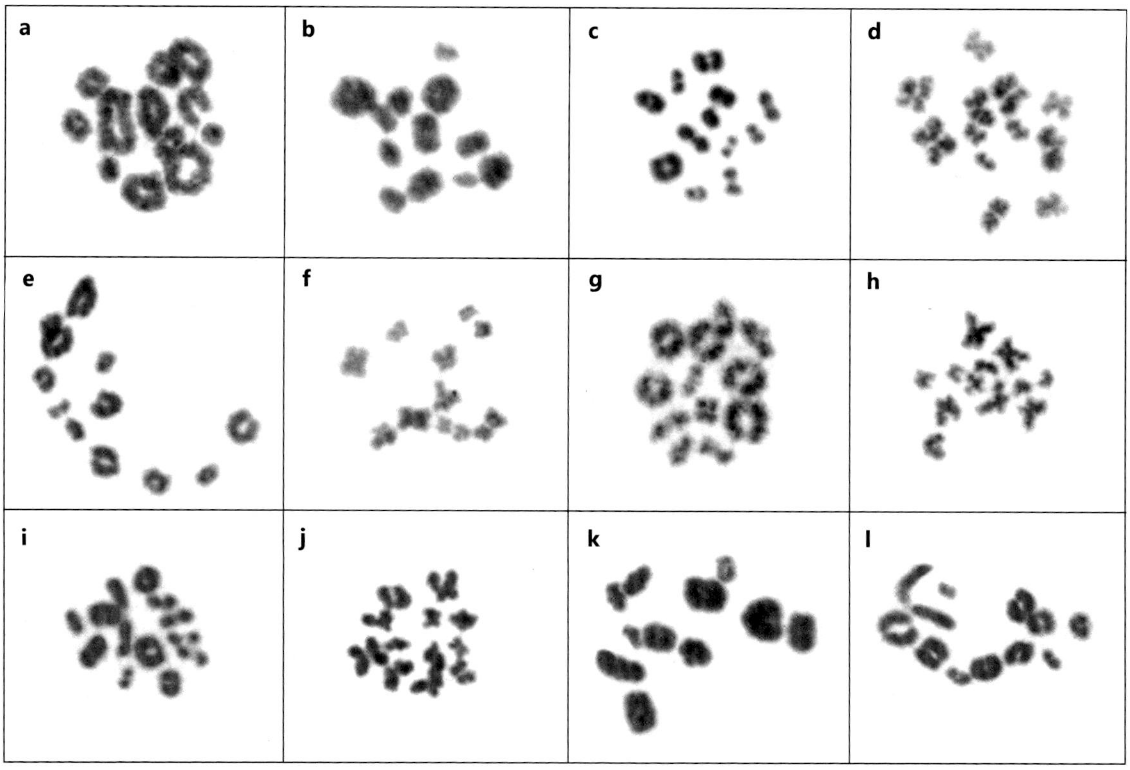

Fig. 45. Giemsa-stained male meiotic cells of *Aparasphenodon bokermanni* (**a**, **b**), *Itapotihyla langsdorffii* (**c**, **d**), *Trachycephalus* sp. (**e**, **f**), *T. mesophaeus* (**g**, **h**), *T. typhonius* (**i**, **j**), and *Phyllodytes luteolus* (**k**, **l**). **a–c**, **e**, **g**, **i**, **k**, **l** Diakineses. **d**, **f**, **h**, **j** Metaphases II. From Gruber et al. [2012b]. For further explanations, see chapter 3.2.1.

plants [Vosa, 1976], *Drosophila melanogaster* [Hsu, 1971], and mammals [Arrighi et al., 1974; Jalal et al., 1974]. As suggested by Hsu and Arrighi [1971], the difference in staining intensity between C-band-positive regions results from variable proportions of repetitive DNA sequences in these regions. According to the current explanation of the mechanism of C-banding, less DNA is removed from chromosome sites containing numerous highly repetitive DNA sequences during the C-banding procedure, which leads to a stronger binding of Giemsa stain. In contrast, more DNA is extracted from the telomeric regions that typically have lower amounts of repetitive DNA, resulting in greyish C-bands.

In some anuran karyotypes, such as ranids [Schmid, 1978b], the various heterochromatic regions have differing sensitivity to the alkaline denaturation. When preparations are treated according to the standard C-banding protocol of Sumner [1972], which is slightly modified for amphibian chromosomes [Schmid, 1978a], a maximum number of C-bands appear in the chromosomes. All telomeric C-bands, however, disappear when preparations are subjected to prolonged denaturation, and the centromeric, pericentromeric, and NOR-associated heterochromatin loses distinctiveness. Certain C-bands, however, always remain visible even after extreme denaturation. This differing sensitivity to alkaline denaturation and selective staining of certain C-bands is additional proof to support the heterogeneity of constitutive heterochromatin.

Interspecies comparisons of C-band patterns have been of considerable importance for cytotaxonomic studies. Although many species in various anuran genera share similarity in chromosome number and morphology, there can be extensive differences in the location and quantity of heterochromatin [Schmid, 1978a, b; King, 1980; Schmid et al., 1990b; Green and Sessions, 2007]. Except for the origin of heteromorphic sex chromosomes and some B chromosomes, an evolutionary relationship between the heterochromatin patterns in anuran chromosomes is not evident. Chromosomal positions and number of heterochromatic regions are variable factors that rapidly change during evolution [John, 1988]. As emphasized by King [1990], C-band patterns have significantly contributed to the differentiation of karyotypes in related species. Observed differences in the chromosomal positions, sizes, and fluorochrome binding affinities of C-bands among anuran karyotypes show that not even 2 species, even in the most related taxa, have identical C-band patterns. Thus, comparison of species in the anuran genera *Litoria* [King, 1980], *Bufo* [Schmid, 1978a; Matsui et al., 1985; Miura, 1995b], *Rana* [Schmid, 1978b], and

Neobatrachus demonstrate striking interspecific C-band variation between species with essentially the same karyotype.

C-banding has been helpful in the interpretation of anuran karyotypes of closely related species in which the number and morphology of the chromosomes are nearly identical, but the absolute length of the chromosomes and genome sizes show significant differences. Thus, chromosome banding and DNA flow cytometric analyses of the South American leptodactylid frog species *Pleurodema brachyops* and *P. thaul* have shown that the quantity of nuclear DNA in *P. brachyops* is 3.4 times larger than it is in *P. thaul* [Schmid et al., 1993b]. Nevertheless, the conventionally stained karyotypes of both species are nearly identical. C-banding revealed that almost 6 times more constitutive heterochromatin is present in the karyotype of *P. brachyops* than in the karyotype of *P. thaul*. The substantial differences in the genome sizes of these 2 anurans are explained by changes in the amount of highly repetitive (but also moderately repetitive and single copy) DNA sequences that occurred uniformly along all chromosomes. These changes in DNA content only altered the absolute quantity of DNA per chromosome, but did not affect the morphology (arm-length ratios) of the chromosomes [Schmid et al., 1993b]. Another most convincing example is provided by the 2 Australian myobatrachid frogs *Mixophyes fasciolatus* and *M. schevilli* [Schmid et al., 2002f]. In these species, karyotype differentiation occurred exclusively by considerable amplification of constitutive heterochromatin in *M. fasciolatus* (or, alternatively, its loss in *M. schevilli*). The amplification of the highly repetitive DNA sequences took place preferentially in most centromeric and pericentromeric regions and additionally in 12 telomeric regions in the *M. fasciolatus* karyotype (or, alternatively, deletions occurred in *M. schevilli* at the same sites). The major heterochromatic regions in the karyotypes of *M. fasciolatus* and *M. schevilli* amount to 30.2 and 20.7% of the karyotypes, respectively. The sizes and locations of intercalary heterochromatic regions, the sizes of all euchromatic segments, as well as the position of the centromeres remained nearly identical during their karyotype evolution. Addition or amplification of telomeric heterochromatin has also occurred in the karyotypes of the South American frogs *Caudiverbera caudiverbera* and *Telmatobius montanus*, where whole chromosome arms are heterochromatic [Veloso and Iturra, 1979]. Similar changes are also observed in the chromosome sets of *Leptodactylus ocellatus* [Bianchi et al., 1973], *Ceratophrys ornata*, *C. cranwelli* [Schmid et al., 1985], and the *Odontophrynus* species complex [Ruiz et al., 1981; Schmid et al., 1985]. Addition or amplification of repetitive DNA sequences (or, alternatively, their loss) in heterochromatic chromosome regions has also been documented for many other anuran species [Schmid, 1980c; King, 1990, 1991; Green and Sessions, 2007; Schmid et al., 2010].

There are amphibian species that exhibit considerable differences in their genome sizes, but show more or less the same amount of C-band-positive heterochromatin in their karyotypes. Thus, it has been shown that among the chromosomes of the European toads *Bufo bufo*, *Bufotes viridis*, and *Epidalea calamita* measurable differences in chromosome length exist [Ullerich, 1965, 1966]. Each of the chromosomes of *B. bufo* contains more DNA than the respective homoeologous chromosomes of *B. viridis* and *E. calamita*. *B. bufo* has 49% more nuclear DNA than *B. viridis* and 39% more than *E. calamita*. The DNA content of the homoeologous chromosomes of the 3 species approximately corresponds to their lengths, with the exception of the 5 small chromosomes [Schmid, 1978a]. It is unlikely that the large differences in nuclear DNA content in these species result from significant differences in the amount of genetically relevant DNA. A more plausible explanation is that different amounts of repetitive DNA sequences originated in their genomes during evolution. These should, however, be detectable by staining the constitutive heterochromatin in metaphase chromosomes. However, a comparison of the amounts of measurable C-bands in the karyotypes of these species showed no such differences. *B. bufo* does not have any more constitutive heterochromatin than *B. viridis* or *E. calamita*. In addition, the number of heterochromatic regions per chromosome is essentially the same in these species [Schmid, 1978a]. An explanation can be found in the study of Baldari and Amaldi [1976] who investigated DNA reassociation kinetics performed on amphibian genomes. They demonstrated that the large differences of DNA content in amphibian genomes are attributable not only to differing quantities of highly repetitive and foldback sequences but, to a far greater extent, to differing amounts of middle repetitive sequences. Additional proof that changes of genome sizes in amphibians mostly involve highly repetitive and middle repetitive DNA sequences has been compiled by Macgregor et al. [1976] and Olmo [1973, 1983]. It was further determined that the variations in the amount of middle repetitive DNA sequences between amphibian species result from differences in the copy numbers of the various sequence classes [Straus, 1971, 1976; Baldari and Amaldi, 1976]. Because C-banding only reveals those chromosome regions with highly repetitive DNA, the differences in the amounts of middle repetitive sequences remain undetected [Schmid, 1978a]. From DNA reassociation kinetics experiments with amphibian genomes, it can also be concluded that, as in all other vertebrate genomes, the various sequence classes have an interspersed organization [Davidson et al., 1972, 1973, 1975; Baldari and Amaldi, 1976]. The uniform increase in copy number of middle repetitive sequences (i.e., the repetition frequency) along the chromosomes changes their absolute sizes but not the length ratio of their arms. This explains why the morphology of the 10 or 11 chromosome pairs always found in the karyotypes of the numerous species in the family Bufonidae [Bogart, 1968, 1972; Cole et al., 1968; Bogart and Perret, 1977] has remained constant during evolution. If considerable increases in the amounts of repetitive DNA sequences (demonstrable by C-banding) had occurred in definite chromosome regions in some bufonid species, this would have changed the morphology of 1 or more chromosomes. Molecular studies on the various DNA fractions in amphibian genomes show that the highly repeated sequences range from 4 to 26% of the genome, moderately repeated sequences from 24 to 67%, and single-copy sequences from 20 to 63% [Straus, 1971; Davidson et al., 1972, 1973; Mizuno and Macgregor, 1974; Morescalchi and Serra, 1974; Baldari and Amaldi, 1976; Morescalchi and Olmo, 1982; Olmo et al., 1982].

Although repetitive DNA sequences are characterized by genetic inertness, there is evidence for the transcription of middle repetitive DNA sequences in anuran lampbrush chromosomes. Such read-through transcriptions of middle repetitive DNA sequences have been found in the lampbrush chromosomes of *Xenopus laevis* [Jamrich et al., 1983] and *Lithobates catesbeianus* [Wu et al., 1986]. These middle repetitive sequences are scattered throughout the length of all chromosomes. Transcription of middle repetitive DNA sequences has also been documented in sea urchins [Smith et al., 1974]. Other studies have detected transcription of highly repetitive DNA sequences in mice [Harel et al., 1968] and in cultured HeLa cells [Melli et al., 1975]. It has been proposed that read-through is the general explanation for such transcription [Wu et al., 1986].

The C-banding technique has a special value in both the identification and the analysis of amphibian sex chromosomes. As shown in Table 6, many anuran species have XY♂ or ZW♀ sex chromosomes in an early stage of morphological differentiation. In these, the only visible difference between the X and Y or the Z and W is a small heterochromatic C-band in the Y and W that is missing in the X and Z. Without C-banding, such tiny differences would escape detection, and the presence of sex chromosomes would never be known. In the more differentiated and highly heteromorphic amphibian XY♂ or ZW♀ sex chromosomes that can be recognized by conventional staining, C-banding provides information on the degree of heterochromatinization in the Y and W chromosomes.

Some heterochromatic regions in anuran chromosomes are not revealed by C-banding. Of the 21 heterochromatic blocks (recognized by the appearance of secondary constrictions) in the karyotype of the South American frog *Leptodactylus macrosternum* (formerly *L. ocellatus*), 4 have late DNA replication but are not differentially stained with the C-banding technique [Bianchi et al., 1973].

3.2.3 Constitutive Heterochromatin in Arboranae

Chromosomes of 90 treefrog species examined for the present monograph have been C-banded (Fig. 383–545). In general, the patterns of constitutive heterochromatin in arboranan chromosomes are like those in the karyotypes of other anurans. C-bands are preferentially located in centromeric, pericentromeric, and telomeric regions of the chromosomes. Additional C-bands of variable size and staining intensity can be located at intercalary positions and at the NORs. As in other anurans, in most arboranan karyotypes the constitutive heterochromatin of the centromeric and pericentromeric regions and the NOR-associated heterochromatin stain more intensely than the telomeric heterochromatin. A typical example for such heterochromatin distribution can be seen in the karyotype of *Agalychnis annae*, *A. callidryas*, and *A. dacnicolor* (Fig. 472–476). There are, however, exceptions to this rule. For example, most of the chromosome pairs of *Osteopilus ocellatus* lack large amounts of centromeric heterochromatin but have large interstitial and telomeric heterochromatic blocks that stain very intensely (Fig. 412–414). As has been demonstrated in other anuran species, the small chromosomes in the arboranan karyotypes usually contain just as much, or even more, constitutive heterochromatin as do the largest chromosomes. This is exemplified in the C-banded karyotypes of *Ololygon longilinea* (Fig. 422), *Bokermannohyla martinsi* (Fig. 447), and *Boana boans* (Fig. 460). Therefore, the number and the size of heterochromatic regions are not proportional to the lengths of the chromosomes.

The amount of centromeric and paracentromeric heterochromatin in arboranan karyotypes is variable. It is restricted to tiny C-bands in *Scinax ruber* (Fig. 436, 437), *Bokermannohyla ahenea* (Fig. 442), and *Boana albopunctata* (Fig. 456–458). In many species, its size is approximately the same as the width of 1 chromatid, as shown in the C-banded karyotypes of *Litoria pallida* (Fig. 492–494), *L. nasuta* (Fig. 498, 499), and *Ranoidea genimaculata* (Fig. 519–521). In some species, the size of centromeric and paracentromeric heterochromatin is 2 or 3 times the width of a chromatid, as in the chromosomes of *Boana boans* (Fig. 460). Finally, there are inflated amounts of centromeric heterochromatin in *Agalychnis* (Fig. 472–476).

As far as known, repetitive DNA sequences in the genomes of eukaryotes have no known genetic function. Therefore, sequence changes in reiterated DNA are considered to be neutral. Since they are not subjected to evolutionary selection pressure, sequence changes in repetitive DNAs can be accumulated and fixed very much faster than can the changes in functional (transcribed) DNA. As a consequence, constitutive heterochromatin is heterogeneous within and between species. It might be argued that closely related species sharing similar C-band patterns in their karyotypes are species that were phylogenetically separated from their common ancestor much more recently than other species that differ considerably in their C-banding patterns. Theoretically, such a hypothesis would require similar mutation rates to exist over long periods of time in different repetitive DNA families, in different chromosomes, in different species, under different environmental conditions, and in populations with different dynamics. This would lead to a synchronous evolution of constitutive heterochromatin in different species. However, such concerted evolution of highly repetitive DNA sequences is improbable. Therefore, any conclusions on phylogeny and speciation that are based solely on similarities and/or dissimilarities of constitutive heterochromatin in karyotypes are questionable.

In conclusion, constitutive heterochromatin in Arboranae behaves as it does in other amphibian and vertebrate taxa. By its nature, constitutive heterochromatin is extremely heterogeneous within and between species. It can increase spontaneously in specific chromosomes or in a whole karyotype, or it can decrease and almost disappear from a genome. As with other organisms, there seems to be no general rule that governs the different patterns of heterochromatin that have evolved in the autosomes of individual species or that could be used as synapomorphic characters to relate species at higher taxonomic levels in Arboranae. But, because of the species-specificity of the heterochromatin patterns, they are extremely useful as cytogenetic parameters for differentiating between closely related species that possess the same diploid number and chromosome morphology. Furthermore, the careful analysis of constitutive heterochromatin provides important information concerning the structure and the evolution of B and sex chromosomes.

Unfortunately, no molecular studies have yet been performed to elucidate the amounts of the various DNA fractions (highly repeated, moderately repeated, single-copy sequences) in the genome of any arboranan species. Such DNA denaturation-reassociation experiments [Straus, 1971; Davidson et al., 1972, 1973; Mizuno and Macgregor, 1974; Morescalchi and Serra, 1974; Baldari and Amaldi, 1976; Morescalchi and Olmo, 1982; Olmo et al., 1982] would be required to obtain the necessary information on DNA composition in treefrogs. As with other amphibians and other vertebrates [Schmid et al., 2010, 2012], not all categories of constitutively heterochromatic regions in the karyotypes of arboranans may be visualized by C-banding techniques and escape evaluation. Data obtained by C-banding and fluorescence staining of heterochromatin are sufficient to categorize arboranan karyotypes.

3.2.4 Polymorphisms of Constitutive Heterochromatin

The C-banding technique has revealed an extraordinary amount of variability in the size and/or staining intensities of heterochromatic regions in the chromosomes of virtually all plant and animal species that have been analyzed with this technique. This variability occurs intraindividually between homologous chromosomes, interindividually in the same population, and between different populations of the same species. The variability is best described by the term "polymorphism" which was defined by Ford [1960] as "the occurrence together in the same habitat of two or more discontinuous forms of a species in such proportions that the rarest of them cannot be maintained merely by recurrent mutation." The mode of inheritance of polymorphisms follows mendelian rules. The most spectacular polymorphism is that observed in the grasshopper *Atractomorpha similis* where all 10 chromosomes can carry numerous variations of heterochromatic regions, so that over 250 different cytomorphs are distinguished [John and King, 1983].

Apart from size variations, a further aspect of the variability of constitutive heterochromatin is its staining heterogeneity with AT- and GC-specific fluorochromes, presence or absence of 5-methylcytosine, variable C-banding density, and resistance to digestion by restriction endonucleases. Heterochromatin heterogeneity has been studied most extensively in human chromosomes. Not only do different C-bands in the human karyotype contain different fractions of repetitive DNA and show variable labeling with fluorochromes, but heterogeneity can be present within one and the same C-band [for references, see Babu, 1988; Verma, 1988; Sumner, 1990].

The most plausible explanation for the frequently observed size polymorphisms of constitutive heterochromatin is random unequal crossing-over. Evidence for this mechanism was compiled by Kurnit [1979]. The prerequisite for unequal crossing-over is the presence of highly repetitive DNA sequences in the constitutive heterochromatin. These promote recombination between homologous sequences that are separated by long distances. Crossing-over does not occur, however, in constitutive heterochromatin of autosomes during meiosis [for references, see John, 1976; Kurnit, 1979]. Therefore, if unequal crossing-over does indeed account for quantitative variation of heterochromatin, then it must occur in the mitotic cell lineages from which the meiotic germ cell lines derive. Some evidence supporting this assumption is provided by the studies of Craig-Holmes et al. [1975] and Simi and Tursi [1982] on human family pedigrees where they observed heteromorphic autosomal C-bands that were absent in the parents of the individuals, or by the observation of intraindividual mosaicism for heterochromatic variants. Since unequally sized C-bands have never been found in the sister chromatids of the same chromosome, the mosaics probably originated by somatic crossing-over between homologous chromosomes and not by replication errors during sister chromatid formation. Actually, by treating human lymphocyte cultures with distamycin A or 5-azacytidine, it is possible to maintain ectopic pairings between the constitutive heterochromatin in the chromosomes 1, 9, 15, 16, and Y existing in interphase nuclei until the metaphase stage [Schmid et al., 1981, 1983c; Haaf et al., 1986]. This analysis shows that ectopic pairings between the homologous pairs are more frequent than between nonhomologous chromosomes. Interindividual size polymorphisms of heterochromatic regions in Y and W sex chromosomes are explained by (1) duplication of repetitive DNA within the Y and W, (2) heterochromatinization of euchromatin as a result of its invasion by repetitive sequences from elsewhere, and (3) spontaneous deletions [John, 1988].

Extremely fast evolution and divergence of the different fractions of repetitive DNA in the genome are responsible for the large staining heterogeneity of constitutive heterochromatin with AT- and GC-specific fluorochromes. The apparent absence of selection pressure on these repetitive sequences permits a coincidental enrichment in either AT or GC base pairs. In the Anura, closely related species with differing fluorescence patterns of constitutive heterochromatin are not exceptional [Schmid et al., 2010].

3.2.4.1 Polymorphisms of Constitutive Heterochromatin in Arboranae

Like in other eukaryotes, polymorphisms of constitutive heterochromatin are not uncommon in the genomes of the arboranan frogs. They are the result of amplifications and deletions of reiterated DNA sequences located in the constitutive heterochromatin. Variable sizes of C-bands are present in the karyotypes of many arboranan species examined by C-banding. Size variations can be present in centromeric, telomeric, as well as in interstitial C-bands of the autosomes. A careful revision of the C-banded karyotypes in Figures 383–545 shows the variability of the sizes of C-bands between homologous chromosomes in some species. Polymorphisms of constitutive heterochromatin become also apparent between different populations of treefrogs.

3.3 Nucleolus Organizers, Telomeric DNA, and Hypermethylated DNA in Anura and Arboranae

3.3.1 Nucleolus Organizer Regions in Anura

Ag-staining has been applied to the chromosomes of a large number of anuran taxa [Schmid, 1980c; King, 1990; Olmo 1990; Green and Sessions, 1991a, 2007]. The cytochemical background of Ag-staining of NORs has been explained in detail in the 2 former monographs of this amphibian series. It must always be considered that, though Ag-staining is the most widely used technique for NOR detection in anuran metaphase chromosomes, inactive NORs can only be demonstrated by in situ hybridization with labeled 18S + 28S rDNA probes, or to a lesser extent, by mithramycin (MM)- or chromomycin A_3 (CMA_3)-positive staining. Thus, Wiley et al. [1989] found 2 additional "silent" NORs in *Dryophytes chrysoscelis* and 2 in *D. versicolor* that could only be detected by 18S + 28S rDNA in situ hybridization and not with Ag-staining.

The theoretical basis of fluorescence staining of NORs with MM or CMA_3 is much more complex. These fluorochromes stain the NORs, independent of their activity state, in fishes, amphibians, reptiles, and birds, but not those of mammals [Thiébaud, 1979a, b; Schmid, 1980a, c, 1982; Schmid and Guttenbach, 1988]. NORs in meiotic metaphases and mature sperm nuclei, as well as in the early stages of amphibian embryogenesis, prior to gastrulation, can be detected as brightly fluorescing blocks [Schmid, 1980a, c]. These fluorochromes stain the NORs because eukaryotic rDNA tends to have a higher GC content than the main band fraction DNA. MM and CMA_3 usually stain the NORs of anurans as extremely brightly fluorescing regions with a very sharp delineation of their fluorescence. In contrast, GC-rich constitutive heterochromatin, which also is highlighted by these fluorochromes, shows relatively diffuse boundaries. The fluorescence signal is always located within the nucleolar constriction and does not extend into adjacent NOR-associated heterochromatin. In metaphase chromosomes pre-stained with MM or CMA_3 and subsequently Ag-stained, the fluorescent and Ag labels are found in exactly the same chromosome location. A good agreement in the position of MM- or CMA_3-stained and Ag-stained NORs has also been established in somatic interphase nuclei. In bivalents of the diakinetic stage of male meiosis, NORs can be observed as paired, brightly fluorescing blocks. The MM/CMA_3-positive heterochromatin is no longer visible in these highly condensed meiotic chromosomes. The expected haploid number of MM/CMA_3-stained NORs is counted in the spermatids and mature sperm nuclei from anuran testes [Schmid, 1980a, c]. MM and CMA_3 staining of NORs has been successfully performed in a number of anurans [Schmid, 1980c; King, 1990; Green and Sessions, 1991a, 2007].

Counterstaining with the AT-specific distamycin A strongly enhances the contrast of differential MM or CMA_3 fluorescence of the NORs. The nonfluorescent distamycin A is a basic oligopeptide antibiotic which preferentially forms very stable bonds with AT-rich, double-stranded DNA. Beside distamycin A, other appropriate AT-specific counterstains, such as netropsin or methyl green, are also suitable for contrast enhancement.

Historically, anuran NORs have been detected by using the in situ hybridization technique, where either labeled rRNA or 18S + 28S rDNA obtained from anurans or other organisms was used as a probe to determine the chromosomal locations of NORs. In early studies, molecular probes were radioactively labeled (tritiated probes) and detected by autoradiography. Presently, the probes are labeled with biotin and detected by FISH [for references, see King, 1990; Green and Sessions, 2007; Schmid et al., 2010].

Hsu et al. [1975a] proposed that karyotypes with a single NOR pair are more "ancestral," while those in which the NORs are distributed over several chromosomes represent a "derived" condition. Such a presumption, however, considers only the NORs and not the other characters of the karyotypes. There are numerous examples where 2 related species considerably differ in their karyotypes, but still conserve a single NOR pair. Conversely, there are many species with identical karyotypes but differing numbers and chromosomal positions of NORs. Previous cytogenetic studies of NORs in a variety of anuran species and genera clearly demonstrate that a majority of species possess only a single "standard" pair of NORs [Schmid, 1980c; King, 1990; Green and Sessions, 2007].

Intraspecific polymorphisms in the number and location of the NORs have been discovered in several anurans. Foote et al. [1991] found that 11 out of 55 specimens of the North American toad *Anaxyrus terrestris* from various populations in eastern North Carolina had 1–3 ectopic NORs in addition to the standard NOR in chromosome 1. In all of these 11 specimens, the ectopic NORs occurred in

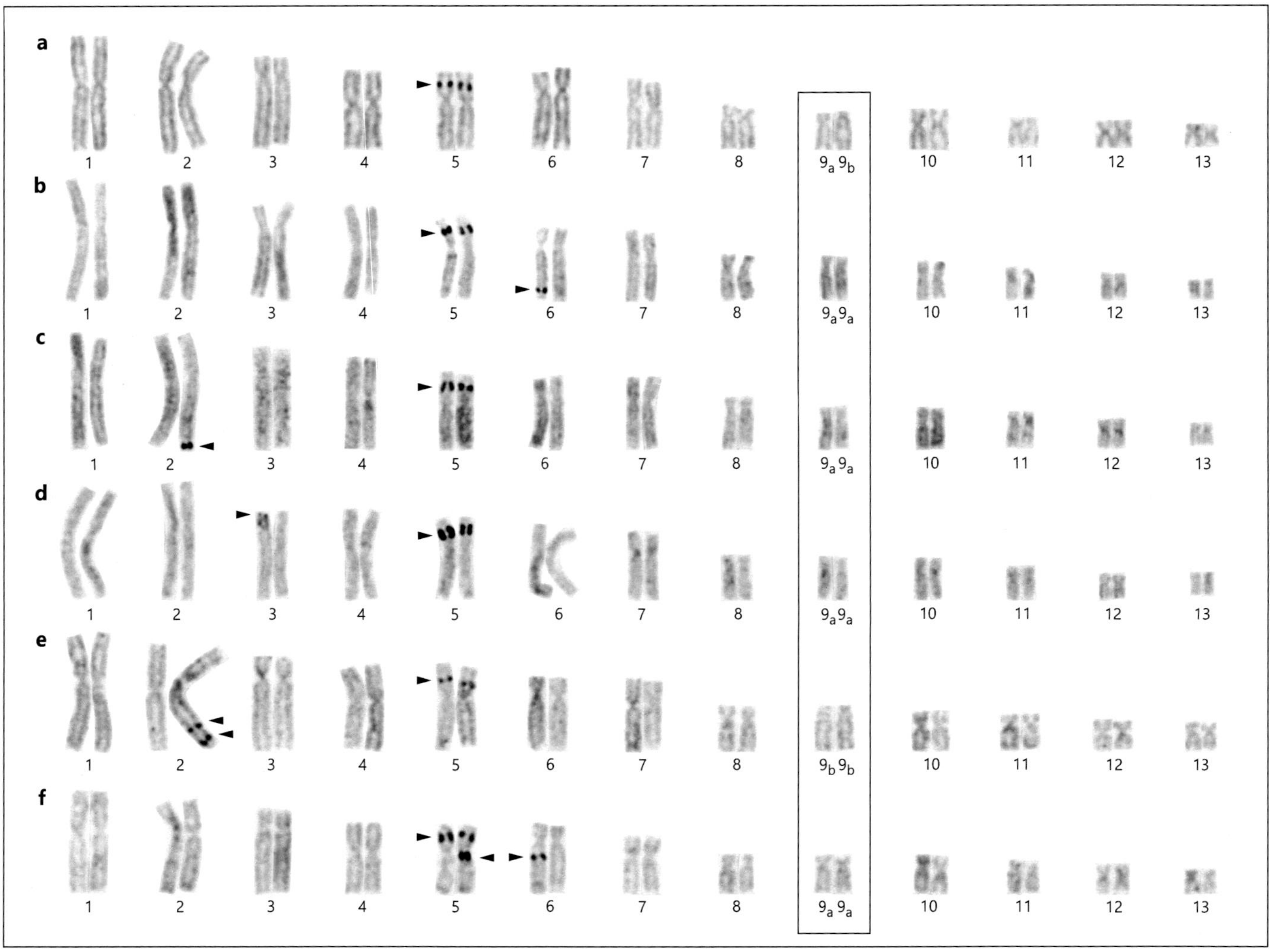

Fig. 46. Karyotypes of *Agalychnis callidryas* collected in the Caribbean lowlands of Costa Rica showing Ag-staining of NORs. The NORs are marked by arrowheads. **a** Standard pair of NORs located in the short arms of chromosome pair 5. These NORs are present in all specimens analyzed. **b–f** Karyotypes with either 1 (**b–d**) or 2 (**e**, **f**) additional NORs. Note that these extra NORs are located in only 1 of the homologues 2, 3, 5, or 6, each. The individuals in **b–d** and **f** are homozygous for the submetacentric chromosome 9_a, the specimen in **a** is heterozygous for the submetacentric chromosome 9_a and the subtelocentric chromosome 9_b, whereas the frog in **e** is homozygous for the telocentric chromosome 9_b. For further explanations, see chapter 3.3.1.

only 1 of the homologues, and had a paracentromeric, intercalary, or a telomeric location. In the North American treefrog *D. chrysoscelis* and its tetraploid sibling species, *D. versicolor*, the most common NOR is located in chromosome 6. After examination of a total of 236 *D. chrysoscelis* specimens from 34 populations in 15 US states and 100 *D. versicolor* from 15 populations in 12 US states, Wiley et al. [1989] discovered the existence of 13 additional NORs in *D. chrysoscelis* and 4 extra NORs in *D. versicolor*. In most of these *Dryophytes* karyomorphs, the NOR appeared to have been precisely excised from or inserted into the chromosomes without altering their morphology. In only 2 of the *D. chrysoscelis* karyomorphs, a paracentric or pericentric inversion had relocated the NOR within the same chromosome. This result was confirmed by the BrdU replication banding patterns of *D. chrysoscelis* and *D. versicolor* [Wiley and Little, 2000]. Homologous chromosome pairs of all studied *D. chrysoscelis* displayed no differences in replication banding patterns, nor did they differ from those found in the *D. versicolor* chromosomes. Although the NOR locations differed between the populations examined, no changes in the replication banding patterns were found, which indicates that structural rearrangements did not occur when the multiple NOR positions in *D. chrysoscelis* and *D. versicolor* originated.

In the Central American treefrog *Agalychnis callidryas* (Fig. 948a), the standard pair of NORs is located in chromosomes 5. In 5 of 19 specimens collected from 3 different populations of the Atlantic lowlands in Costa Rica, Schmid et al. [1995; present study] detected 6 additional ectopic NORs. These extra NORs occur in only one of the homologues in an intercalary position, at the telomeric region, or

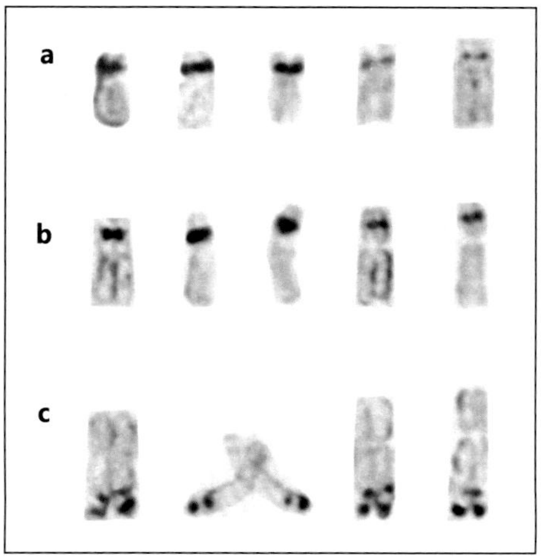

Fig. 47. Selected chromosomes 5 (**a**, **b**) and 2 (**c**) from mitotic metaphases of *Agalychnis callidryas* after in situ hybridization with a biotinylated 18S + 28S rDNA probe. The rDNA clusters in the short arm of chromosomes 5 are the standard NORs present in all specimens analyzed, whereas the duplicated signal in the distal long arm of chromosome 2 are additional NORs found in a single individual (compare with Fig. 46). For further explanations, see chapter 3.3.1.

in a paracentromeric position. They appear to have been inserted into the chromosomes without changing their morphologies. In 1 specimen, 2 additional NORs were both located in the same homologue (also confirmed by in situ hybridization with 18S +28S rDNA) (Fig. 46, 47).

A geographic variation in the location of NORs has been found in the Costa Rican frog *Dendropsophus ebraccatus* by Kaiser et al. [1996]. In 9 specimens sampled in the Caribbean lowlands, the NORs were located in the long arm telomeric region of the subtelocentric chromosome pair 10, whereas in 3 individuals from the Pacific lowlands, they were positioned in the paracentromeric region of the submetacentric chromosome pair 13. Because of a lack of heterozygous karyotypes, it is unlikely that this observation represents a polymorphism, but verifies that the heterogeneity of the NOR location has been fixed. The otherwise identical karyotypes in both populations as well as the external morphology and vocalization of the frogs showed that the analyzed specimens do indeed belong to the same species. The transposition of the NORs from chromosome 10 to 13, or vice versa, must have occurred via excision and insertion. Further cases of intraspecific polymorphisms in the number and location of Ag-stained NORs in anurans have been found in the genera *Odontophrynus* [Ruiz et al., 1981] and *Paratelmatobius* [Lourenço et al., 2000], in *Physalaemus petersi* [Lourenço et al., 1998], *P. cuvieri* [Silva et al., 1999], *Allobates caeruleodactylus* [Veiga-Menoncello et al., 2003], *Dendropsophus nanus* [Medeiros et al., 2003], *Leptodactylus petersii* [Amaro-Ghilardi et al., 2006], *L. mystacinus* [Silva et al., 2006], *Pithecopus rohdei* [Barth et al., 2009], *Eleutherodactylus atkinsis, E. erythroproctus, E. goini, E. johnstonei, E. pantoni, E. planirostris, E. varleyi, Craugastor bransfordii, C. mimus, C. talamancae, Pristimantis urichi*, and *P. fenestratus* [Siqueira et al., 2009; Schmid et al., 2010]. In an early cytogenetic study without the use of Ag-staining, Jotterand and Fischberg [1974] found an ectopic nucleolar constriction in one chromosome 7 of a male and a female of *Xenopus borealis* from the Lake Samburu area (northern Kenya) and explained this as the result of a reciprocal translocation between the chromosome 7 and the NOR-carrying chromosome 4 of this species. But, because the morphology of the chromosomes was unaltered in these specimens, an insertion of this ectopic NOR appears to be more likely.

A very unique and surprising mechanism of highly mobile NORs has been found in the terraranan frog *Craugastor fitzingeri* of Costa Rica and Panama [Schmid et al., 2010, 2017a]. In these studies, not a single pair of the many specimens examined by Ag-staining shared the same number and chromosomal locations of NORs. Each specimen possessed its own individual NOR pattern. It could be shown that (1) the NORs can be located in all 11 chromosome pairs of *C. fitzingeri*, (2) they always occupy a telomeric position in the chromosomes, (3) they can be located either in the short or long arm telomeric regions or even in both of them, (4) the number of Ag-stained NORs detected in the karyotypes of these specimens ranges from 1 to 6, and (5) there is no apparent common standard pair of NORs. Because these observations were made in metaphases obtained from 3 tissues (bone marrow cells, intestine epithelium cells, and spermatogonial cells), it is likely that all tissues of *C. fitzingeri* have the same hypervariable NOR patterns. FISH using biotinylated 18S + 28S rDNA from *Xenopus laevis* confirmed the results obtained by Ag-staining. Not only do the NOR patterns of *C. fitzingeri* differ at an interindividual level, they also show unexpected differences at the intraindividual level. In metaphases obtained from the same tissue from the same individual, different telomeric chromosome regions can be occupied with NORs. This phenomenon of extremely variable numbers and positions of the NORs in *C. fitzingeri* is best explained by highly mobile NORs that move freely between the 44 telomeres of the 22 chromosomes. Because of the high interindividual differences in the NOR patterns, this transposition must occur preferentially in premeiotic, meiotic, or postmeiotic stages, but also occurs at a lower frequency in the somatic tissues of these frogs. A question arises concerning the molecular mechanism that enables free movement of the NORs between the telomeres of *C. fitzingeri*. The 18S + 28S rDNA sequences per se are not endowed with molecular properties to break off from their original positions in the chromosomes and to reinsert somewhere in other sites of the genome. It can be hypothesized that transposable (mobile) elements are closely linked to the NORs or are inserted in the 18S + 28S rDNA spacers of *C. fitzingeri*. When such transposable elements spread by transposition they can carry with them complete or partial NORs. If this transposition occurs frequently, it enables further distribution of the 18S + 28S rDNA sequences throughout the genome. Therefore, the hypervariable NOR patterns of *C. fitzingeri* may merely represent markers for the existence of very active transposable elements in the genome of *C. fitzingeri*, rather than being the principal actors of the transposition process.

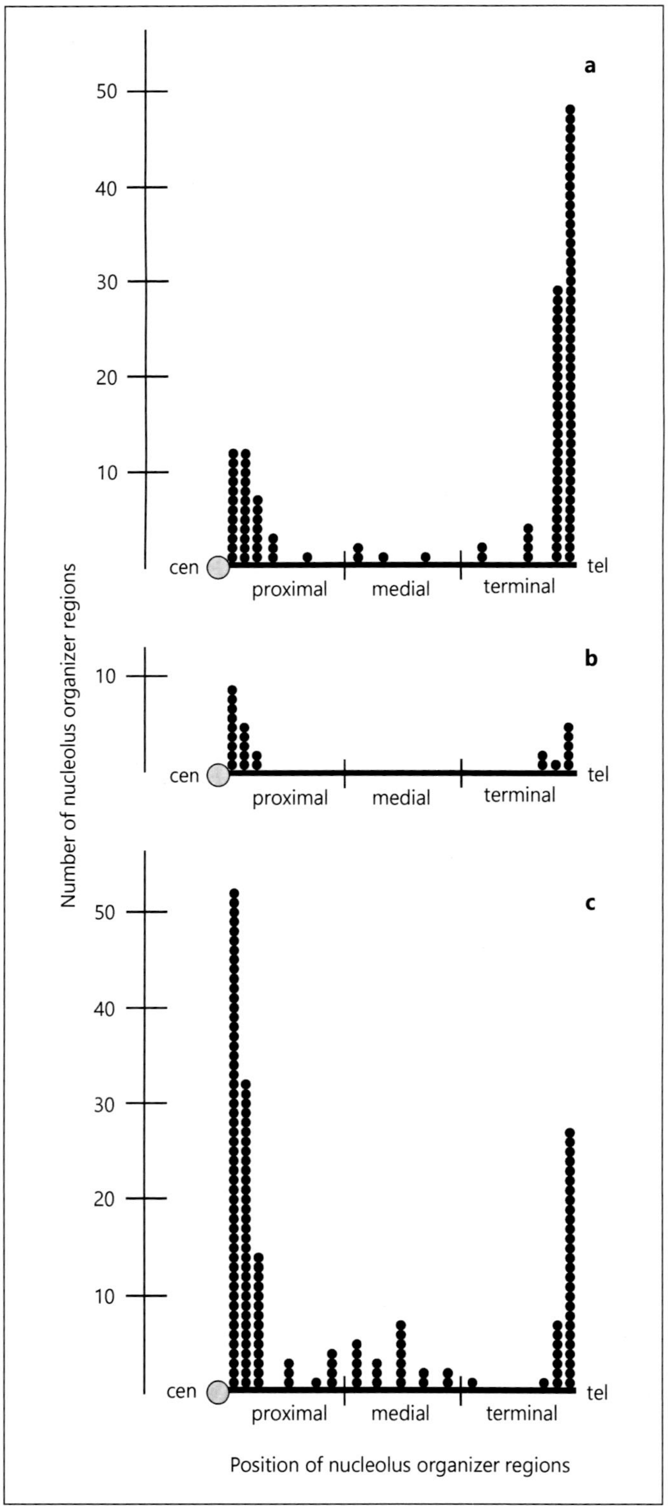

3.3.2 Nucleolus Organizer Regions in Arboranae

As is the rule in the majority of anuran species and genera [Schmid et al., 2010, 2012], most of the arboranan species analyzed with Ag-staining, mithramycin fluorescence or FISH with the 18S + 28S rDNA probe show the presence of 1 single standard pair of NORs in their karyotypes. However, exceptions to this rule are not rare and can be seen in *Smilisca baudinii* (Fig. 549–551) and *Osteopilus ocellatus* (Fig. 569, 570). These so-called ectopic NORs are believed to have originated by excision of 18S + 28S rDNA sequences from the standard NORs with subsequent random insertion to new chromosome sites. In agreement with that, the karyotype morphology is unaltered in these specimens. An alternative but less probable explanation is that in these individuals, 4 or more NORs were ancestrally in their karyotypes, and 1 or more were spontaneously deleted.

The question arises whether there are any regularities in the intrachromosomal locations of the NORs in the taxon Arboranae. Figure 48a shows the frequencies of the different intrachromosomal positions for all 122 detected NORs, including the ectopic ones, in arboranan chromosomes. Only the relative positions of the NORs to the centromeric and telomeric regions are evaluated. It is considered to be irrelevant whether the NORs are located in a short or long chromosome arm. It is immediately evident that the NORs do not occupy random positions along the chromosomes. They preferentially lie in the proximal (paracentromeric) and terminal (telomeric) regions of the chromosomes, whereas they are distinctly underrepresented in the medial segments. Similar conclusions have been drawn for terraranan and hemiphractid frogs (Fig. 48b, c) [Schmid et al., 2010, 2012] and other anurans [Schmid et al., in preparation]. On the other hand, the results obtained by Lima-de-Faria [1973a, b, 1976] indicate that the 18S + 28S rRNA genes in most eukaryotes studied in these investigations preferentially lie in the telomeric regions of the chromosomes. This conclusion, however, was based on the analysis of presumptive nucleolar constrictions in conventionally stained karyotypes, predominantly of plants, and not on in situ hybridization, Ag-staining, or labeling with fluorochromes that are much more precise techniques for the detection of 18S + 28S rDNA. In

Fig. 48. a Chromosomal locations of the standard pair of nucleolus organizer regions (NORs) and additional ectopic NORs in the arboranan species examined for the present monograph. Each dot on the graph plots the location of a NOR. Only Ag-positive NORs are considered. The positions of the NORs are given in relation to the position of the centromeric (cen) and the telomeric region (tel). The chromosome arms are divided in proximal, medial, and terminal segments. The locations of the NORs in the long or short arm of the chromosomes are irrelevant for this analysis. **b, c** For comparison, the NOR locations in the chromosomes of hemiphractid frogs (**b**) [Schmid et al., 2012] and a very large number of terraranan frogs from the families Eleutherodactylidae, Craugastoridae, Brachycephalidae, and Strabomantidae (**c**) [Schmid et al., 2010] are shown. In all 3 lineages, Arboranae, Hemiphractidae, and Terraranae (and possibly in all other anuran taxa), the distribution of the NORs along the centromere-telomere axis is distinctly nonrandom. For further explanations, see chapter 3.3.2.

any case, the data obtained for the Terraranae, Hemiphractidae, Arboranae, and other Anura show that there are factors which favor the pericentromeric and telomeric positions of the NORs in their chromosomes. The noncoincidental distribution of NORs along the chromosome arms in the karyotypes suggests an interaction between the individual chromosome regions. Although the functional significance of the linear arrangement of the various kinds of chromatin is still unknown, the observed patterns are reminiscent of the proposition of a "chromosome field" as hypothesized by Lima-de-Faria [1954, 1973a, b, 1976].

The amount of NOR-associated heterochromatin in arboranan karyotypes can be quite considerable, as found in *Smilisca sordida* (Fig. 386, 387), *Dryophytes japonicus* (Fig. 390), *Ranoidea dayi* (Fig. 533, 534), and *Ranoidea novaehollandiae* (Fig. 542–544). On the other hand, in most arboranans, the quantity of NOR-associated heterochromatin is rather small. King [1980, 1990] defined the NORs in anuran karyotypes by their mitotic behavior and their C-band patterns and distinguished 5 different types of nucleolar constrictions. This classification has not been adopted in the present monograph since any attempt to classify the NORs on the basis of their C-banding behavior at mitosis may be artificial because of the extreme variability of the NORs.

3.3.3 Telomeric DNA Sequences in Anura

In a FISH analysis of telomeric DNA repeats in 100 vertebrate species, Meyne et al. [1990] located the telomeric hexanucleotide $(TTAGGG)_n$ repeats in the karyotypes of *Xenopus laevis*, *Anaxyrus woodhousii fowleri*, *A. terrestris*, *Dryophytes chrysoscelis*, *D. versicolor*, and *D. squirellus*. In both species of *Anaxyrus*, the $(TTAGGG)_n$ repeats are located in the telomeric regions. In *X. laevis*, 1 chromosome pair has an additional heterochromatic intrachromosomal telomeric sequence (het-ITS), and all 3 *Dryophytes* species have several chromosome pairs with additional het-ITSs. Carvalho et al. [2009a, b] detected het-ITSs in the constitutive heterochromatin of all 11 chromosome pairs of the Brazilian hylid species *Aplastodiscus albofrenatus* and *A. arildae*.

The chromosomal distribution of the $(TTAGGG)_n$ repeats in the frogs *X. laevis*, *X. muelleri*, *X. borealis*, *X. clivii*, and *Silurana (Xenopus) tropicalis* were analyzed by Nanda et al. [2008]. The 4 species of *Xenopus* have a tetraploid origin and display karyotypes with 2n = 36 chromosomes that are remarkably similar and uniform, but *S. (X.) tropicalis* is a diploid (2n = 20) species [Tymowska, 1973, 1991]. In all species, FISH signals are present at the telomeric ends of all chromosomes. Furthermore, the in situ hybridization patterns demonstrate a varied organization of the $(TTAGGG)_n$ repeats in the different karyotypes. In *X. borealis* and *X. muelleri*, het-ITSs are located very close to the NORs or within the NOR-associated heterochromatin. The karyotype of *X. clivii* differs from other *Xenopus* karyotypes in displaying numerous het-ITSs. Out of 18 chromosome pairs, clear het-ITSs are visible in 12 pairs of homologues. Varying intensity of the het-ITSs indicates interchromosomal differences in the $(TTAGGG)_n$ repeat number. Unlike the het-ITSs in mammalian chromosomes, the het-ITSs in *X. clivii* chromosomes are located at non-centromeric regions. C-banding shows that the het-ITSs are part of the interstitial heterochromatin. Because the karyotypes of these species of *Xenopus* are so similar, the het-ITSs in *X. clivii* do not represent relics of ancestral telomeres resulting from chromosomal fusions or inversions, but are specific components of the highly reiterated DNA in heterochromatic regions. It is conceivable that the ITSs in the other 3 species of *Xenopus* and in *S. (X.) tropicalis* escape detection by FISH analysis because of a low number of $(TTAGGG)_n$ repeats. The het-ITSs are closely associated with the NORs in the karyotypes of *X. borealis* and *X. muelleri*. Similar colocalizations between $(TTAGGG)_n$ repeats and 18S + 28S rDNA have also been observed in mammals, such as Seba's fruit bat, wood lemming, and shrew [Meyne et al., 1990; Zhdanova et al., 2007], and even in some fishes [Salvadori et al., 1995; Abuín et al., 1996].

In the most extensive original comparative cytogenetic study on vertebrates ever reported, Schmid et al. [2010] examined the hybridization sites of the $(TTAGGG)_n$ repeats in the chromosomes of 43 anuran species belonging to the unranked taxon Terrarana (8 species of Craugastoridae, 21 species of Eleutherodactylidae, 14 species of Strabomantidae). The $(TTAGGG)_n$ repeats are located in the telomeric regions of all chromosomes. Twelve of these species carry homozygous centric fusion chromosomes, but het-ITSs are not observed at the presumed fusion sites. In 5 species, a certain proportion of the male individuals ($♂_a$ males) carry Y-autosome fusion chromosomes, and also in these het-ITSs are absent. Distinct het-ITSs with $(TTAGGG)_n$ repeats are present in only few of the terraranan species. They are located in the centromeric heterochromatin of chromosome pairs 1 and 2 in *Pristimantis riveroi*. In *Pristimantis fenestratus*, intrachromosomal $(TTAGGG)_n$ repeats are visible in the long arms of chromosome pairs 3, 7, 11, 14, and 16 and seem to correspond to very small intercalary heterochromatic regions. Very large het-ITSs are located in the centromeric and pericentromeric heterochromatic regions in all but 3 chromosome pairs of *Pristimantis cruentus* from a population at the Río Sucio River in the Braulio Carrillo National Park in Costa Rica. *Eleutherodactylus gundlachi* from the Pico Turquino National Park on the Sierra Maestra in Cuba has intraindividually variable numbers of mitotically unstable B chromosomes, ranging from 3 to 10 elements. Four different morphologies of B chromosomes can be differentiated. All these 4 types of B chromosomes are completely heterochromatic, and FISH experiments clearly demonstrate the presence of repetitive telomeric DNA sequences in these B chromosomes.

In a more recent study on intrachromosomal telomeric DNA sequences, Schmid and Steinlein [2016a] compiled the anuran species in which large ITSs were localized by FISH (Table 8). In that study, a detailed comparison between nu-

Table 8. Anuran species with large interstitial telomeric sequences and their proposed classification

Family Species[a]	2n[b]	Maximum numbers of large ITSs in diploid karyotypes, their category, and chromosome locations[c]					References
		n	het-ITSs		eu-ITSs		
			restricted	dispersed	restricted	dispersed	
Centrolenidae							
Vitreorana antisthenesi	20	3	2p 6(X)c				Schmid et al., 1989
Dicroglossidae							
Quasipaa boulengeri	26	2	3q[e]				Qing et al., 2013
Hylidae							
Aplastodiscus albofrenatus	22	22		**1c 2c 3c 4c 5c 6c 7c 8c 9c 10c 11c**			Carvalho et al., 2009b
Aplastodiscus arildae	22	22		**1c 2c 3c 4c 5c 6c 7c 8c 9c 10c 11c**			Carvalho et al., 2009b; Gruber et al., 2012a
Aplastodiscus eugenioi	22	22		**1c 2c 3c 4c 5c 6c 7c 8c 9c 10c 11c**			Gruber et al., 2012a
Aplastodiscus leucopygius	18	2	3c				Gruber et al., 2012a
Dryophytes cinereus	24	4			**4p 8p**		Wiley et al., 1992
Dryophytes femoralis	24	2			**4p 8pq**		Wiley et al., 1992; Schmid et al., unpubl.
Dryophytes squirellus	24	10			**4p 8pq** 9pq		Wiley et al., 1992
Dryophytes chrysoscelis	24	26				**1q[e] 2pq 3pq 4pq 5p 6q 7q 8pq 9q**	Wiley et al., 1992
Dryophytes versicolor (4n)	48	52				**1q[e] 2pq 3pq 4pq 5p 6q 7q 8pq 9q**	Wiley et al., 1992
Dryophytes chrysoscelis × *D. versicolor* (3n)	36	39				**1q[e] 2pq 3pq 4pq 5p 6q 7q 8pq 9q**	Wiley et al., 1992
Boana boans (Venezuela)	24	8	**5q** 7q		**2p** 9p		Schmid and Steinlein, 2016a; present study
Boana boans (Brazil)	24	4	**5q**		**2p**		Mattos et al., 2014
Boana wavrini	24	4	**5q**		**2p**		Mattos et al., 2014
Boana geographica	24	4	1c 5c				Mattos et al., 2014
Boana faber	24	24		1c 2c 3c 4c 5c 6c 7c 8c 9c 10c 11c 12c			Schmid and Steinlein, 2016a; present study
Itapotihyla langsdorffii	24	18		1c[f] 2c 3c 4c 5c 7c 8c 9c 11c			Gruber et al., 2012a; pers. comm.
Phyllodytes edelmoi	22	4	1pq 2pq				Gruber et al., 2012a
Pseudacris brimleyi	24	2			**4p 7p 11q**		Wiley et al., 1992
Pseudacris nigrita	24	2			**4p 7q 11q**		Wiley et al., 1992
Scarthyla goinorum	22	2	3c				Suárez et al., 2013
Sphaenorhynchus dorisae	24	2	B				Suárez et al., 2013
Sphaenorhynchus lacteus	24	4			2p		Suárez et al., 2013
Leptodactylidae							
Leptodactylus marmoratus	24	4	**1c 6c**				Gazoni et al., 2012
Leptodactylus podicipinus	22	10	**1c** 2c 3c 5c **6c**				Gazoni et al., 2012
Leptodactylus sp. (aff.) *podicipinus*	20	2	9c				Gazoni et al., 2012
Pelodryadidae							
Ranoidea australis	26	16	1p 3cq 5c 9c 10q 11q[g] 12c				Schmid and Steinlein, 2016a; present study
Ranoidea eucnemis	26	2	**4c**				present study
Ranoidea genimaculata	26	2	**4c**				Schmid and Steinlein, 2016a; present study
Litoria fallax	26		1c 3c **4c**				Schmid and Steinlein, 2016a; present study
Litoria verreauxii	26	2	**4c**				Schmid and Steinlein, 2016a; present study
Phyllomedusidae							
Agalychnis callidryas	26	5	1c[f] 3p[f] 5c[d] 7c[f]				Schmid and Steinlein, 2016a; present study
Phyllomedusa bahiana	26	6	**4c 6c 11c**				Barth et al., 2014; Bruschi et al., 2014a
Phyllomedusa vaillantii	26	6	**4c 6c**		13p		Bruschi et al., 2014a
Phyllomedusa distincta	26	4	8c **11c**				Bruschi et al., 2014a
Phyllomedusa distincta	26	6[h]	**6c 7c 11c**				Gruber et al., 2013
Phyllomedusa tetraploidea (4n)	52	6[h]	**6c 7c 11c**				Gruber et al., 2013
Phyllomedusa distincta × *P. tetraploidea* (3n)	39	3[h]	**6c 7c 11c**				Gruber et al., 2013
Phyllomedusa rohdei	26	2	**11c**				Barth et al., 2014
Pipidae							
Xenopus borealis	36	2	**4p[g]**				Nanda et al., 2008
Xenopus muelleri	36	2	**4p[g]**				Nanda et al., 2008
Xenopus clivii	36	30		1pq 3q 4pq 5q 6q 7q 8pq 10q 12q 13p 14p 17q			Nanda et al., 2008
Xenopus laevis	36	2	17q				Meyne et al., 1990; Nanda et al., 2008

DOI: 10.1159/000489839

Table 8 (continued)

Family Species[a]	2n[b]	Maximum numbers of large ITSs in diploid karyotypes, their category, and chromosome locations[c]					References
		n	het-ITSs		eu-ITSs		
			restricted	dispersed	restricted	dispersed	
Strabomantidae							
Pristimantis cruentus	34	28		1c 3c 4c 5c 6c 7c 8c 9c 10c 11c 12c 13c 14c 17c			Schmid et al., 2010
Pristimantis fenestratus	34	24		2c 3cp 4c 6c 7p 8c 10c 11p 14cp 16p			Schmid et al., 2010
Pristimantis riveroi		2	1c 2c				Schmid et al., 2003b, 2010

ITS, interstitial telomeric sequence.

[a] The systematics and nomenclature of species in the families Hylidae and Phyllomedusidae is according to Duellman et al. [2016], in the family Strabomantidae according to Hedges et al. [2008].

[b] Diploid chromosome numbers. *D. versicolor* and *P. tetraploidea* are tetraploid species. The natural hybrids of *D. versicolor* (4n) × *D. chrysoscelis* (2n) and of *P. tetraploidea* (4n) × *P. distincta* (2n) are triploid (3n).

[c] c, centromeric region; p, short arm; q, long arm. The ITSs in homoeologous chromosome regions of closely related species belonging to the same genera are in bold. In male specimens of *V. antisthenesi*, chromosomes 6 are the heteromorphic XY sex chromosomes, and an ITS is located exclusively in the X chromosome. In *S. dorisae*, 1–2 small heterochromatic supernumerary B chromosomes with ITSs are present in some individuals.

[d] In *A. callidryas*, an apparent tandem duplication of an ITS occurred in the centromeric region of chromosome 5.

[e] ITSs in heterozygous (+/–) and homozygous (+/+ and –/–) conditions in the diploid species *Q. boulengeri* and *D. chrysoscelis*. In the tetraploid *D. versicolor*, some specimens have the ITSs in 2 of the 4 homologues, but in individuals of other populations these are absent. In the single triploid *D. chrysoscelis* × *D. versicolor* hybrid examined, the ITS was found in 1 of the 3 homologues.

[f] ITS in heterozygous (+/–) condition in the single examined individuals of *I. langsdorffii* and *A. callidryas*.

[g] ITSs in NOR-associated constitutive heterochromatin in chromosome 11q of *D. australis* and chromosome 4p of *X. borealis* and *X. muelleri*.

[h] Due to diploidization, the tetraploid *P. tetraploidea* shows the same number of ITSs as the diploid *P. distincta*. In the triploid hybrid of *P. tetraploidea* (4n) × *P. distincta* (2n), only 1 each of the tercets 6, 7, and 11 exhibits an ITS.

merous published and unpublished data allowed a consistent classification of the various categories of large ITSs present in the genomes of anurans and other vertebrates. The classification takes into consideration the total numbers of large ITSs in the karyotypes, their chromosomal locations, and their specific distribution patterns. A new category of large ITSs was recognized to exist in anuran species. It consists of large clusters of ITSs located in euchromatic chromosome segments, which was in clear contrast to the large ITSs in heterochromatic chromosome regions known in vertebrates. The origin of the different categories of large ITSs in heterochromatic and euchromatic chromosome regions, their mode of distribution in the karyotypes and evolutionary fixation in the genomes, as well as their cytological detection were discussed by Schmid and Steinlein [2016a].

Up to date large ITSs were already detected in 43 anuran species belonging to the families Centrolenidae, Dicroglossidae, Hylidae, Leptodactylidae, Pelodryadidae, Phyllomedusidae, Pipidae, and Strabomantidae. These species and the corresponding studies are compiled in Table 8.

3.3.4 Telomeric DNA Sequences in Arboranae

The metaphase chromosomes of the 41 arboranan species which have been examined by in situ hybridization using $(GGGTTA)_7$ and $(TAACCC)_7$ oligonucleotide probes show the canonical telomeric DNA sequences (Fig. 872–937; Appendix 1). Wiley et al. [1992] studied the hybridization sites of $(TTAGGG)_n$ repeats in the karyotypes of several species of North American Hylidae (*Dryophytes chrysoscelis*, *D. versicolor* (4n), a triploid hybrid of *D. chrysoscelis* × *D. versicolor*, *D. femoralis*, *D. squirellus*, *D. cinereus*, *Pseudacris nigrita*, and *P. brimleyi*). In all of these species, telomeric $(TTAGGG)_n$ repeats and additional het-ITSs are present. With regard to the presence or absence of a het-ITS in the long arm paracentromeric region of chromosome 1, intra- and interspecific differences are observed in *D. chrysoscelis*, *D. versicolor*, and their triploid hybrid. Except for the variability of this het-ITS in chromosomes 1, *D. chrysoscelis*, *D. versicolor*, and their triploid hybrid share identical telomeric and het-ITSs. All the other *Dryophytes* species differ in the $(TTAGGG)_n$ patterns from these species and show their own, species-specific hybridization patterns. Both species of *Pseudacris* exhibit identical $(TTAGGG)_n$ patterns. All the species analyzed share a het-ITS in the subtelomeric region in the short arms of chromosomes 4. *D. chrysoscelis*, *D. versicolor*, and *D. femoralis* have identical $(TTAGGG)_n$ patterns in chromosome pairs (or quartets in *D. versicolor*) 5, 8, and 9. *D. cinereus* and *D. squirellus* display similar $(TTAGGG)_n$ patterns in the short arms of chromosomes 8 and 9, but those of *D. squirellus* show additional het-ITSs in the long arms of these chromosomes. All the diploid species of *Dryophytes* have 2n = 24 chromosomes with extremely similar morphology [Bogart, 1973; Wiley, 1982] (Appendix 1). Therefore, it is surprising to find so many differences in the het-ITS patterns. Because of the high evolutionary conservation of these karyotypes, the het-ITSs, with all probability, did not originate by structural chromosome rearrangements (reciprocal translocations, centric fusions, fissions, inversions). Such mutations would result in multiple differences in the morphology of the chromosomes. *D. chrysoscelis* and *D. versicolor* have the most het-ITSs of any amphibian that has previously been studied, confirming that the genomes of these species of

Dryophytes are evolutionarily "dynamic" [Wiley et al., 1989].

3.3.4.1 Large ITSs in Restricted Heterochromatic Regions (Restricted het-ITSs)

The majority of large ITSs found in the genomes of anurans and other vertebrates are located in only 1 or some few heterochromatic regions present in their karyotypes. Here, these are defined as "large ITSs in restricted heterochromatic regions (restricted het-ITSs)." They are always distributed in a random manner in a subset of the heterochromatic regions present in the karyotypes. All remaining heterochromatic regions are devoid of microscopically recognizable het-ITSs. As examples of such restricted het-ITSs, the karyotypes of 2 frog species belonging to the Australian family Pelodryadidae are presented here. In *Ranoidea genimaculata*, a single restricted het-ITS is found in the centromeric heterochromatin of chromosome 4 (Fig. 919–921). Fluorescence staining with the GC-specific mithramycin and the AT-specific quinacrine mustard indicate that this restricted het-ITS is embedded in GC-rich heterochromatin. In *Ranoidea australis*, a total of 8 restricted het-ITSs are detectable in centromeric and interstitial heterochromatic regions of chromosomes 1, 3, 5, and 9–12 (Fig. 935). Again, fluorescence staining with mithramycin and quinacrine mustard reveals that the restricted het-ITSs in the long arms of chromosomes 3 and 11 of this species are positioned in GC-rich heterochromatin. The restricted het-ITS in chromosome 11 is located in NOR-associated heterochromatin. The heterochromatic regions of the other restricted het-ITSs are not specifically labeled by mithramycin or quinacrine mustard. It is feasible that in these heterochromatic regions equal amounts of AT- and GC-pairs are present in the repetitive sequences, so that none of the DNA base pair-specific fluorochromes produces differential labeling [for review, see Schmid et al., 2010].

The further anuran species in whose genomes restricted het-ITSs were discovered up to date are listed in Table 8. Apparently, there is no obvious rule how these restricted het-ITSs are distributed in the karyotypes. They colocalize randomly with 1 or some few heterochromatic regions located in autosomes or sex chromosomes, in supernumerary B chromosomes, in centromeric or interstitial positions, in short and/or long chromosome arms, in some cases at NORs. Their numbers range from 2 to 16 per diploid karyotype.

In different populations of the same species, dissimilarities in the number and chromosomal position of restricted het-ITSs can occur. Hence, in the specimens of *Phyllomedusa distincta* examined by Bruschi et al. [2014a], restricted het-ITSs are located in the heterochromatic centromeric regions of chromosomes 8 and 11, whereas in the individuals studied by Gruber et al. [2013], these are found in the centromeric regions of chromosomes 6, 7, and 11 (Table 8). However, the possibility that these frogs actually belong to 2 different cryptic species should not be ignored.

Interestingly, in a single analyzed individual of *Agalychnis callidryas* (family Phyllomedusidae) collected in Costa Rica, the restricted het-ITSs in the centromeric regions of chromosomes 1 and 7 as well as in the short arms of chromosome 3 occur in heterozygous (+/–) condition (Fig. 888). This could indicate that these restricted het-ITSs are of more recent evolutionary origin. Whether such heterozygous restricted het-ITSs are finally fixed in, or eliminated from, a population by nondirectional and accidental fluctuations of the frequencies of the homologues with and without restricted het-ITSs is determined by the size of the population, the genetic flow between neighboring populations, and the selective neutrality of the restricted het-ITSs (genetic drift). It would be interesting to examine a representative number of *A. callidryas* specimens in that Costa Rican population with the $(TTAGGG)_n$ repeat in order to determine the degree of heterozygosity (+/–) and homozygosity (+/+ and –/–) of these restricted het-ITSs. In the same individual of *A. callidryas*, a possible tandem duplication of a restricted het-ITS was apparent in the centromeric heterochromatin of chromosome 5 (Fig. 888). The 2 resulting restricted het-ITSs are located in a mirror-like position on both sides of the centromeric constriction. Such duplicated ITSs were not observed in further anuran species until now.

It becomes apparent that in several related species belonging to the same genera nearly all restricted het-ITSs are located in homoeologous chromosome regions (Table 8). For example, in *Phyllomedusa bahiana*, *P. vaillantii*, *P. distincta*, and *P. tetraploidea*, the majority of restricted het-ITSs have homoeologous positions. Therefore, it seems likely that these restricted het-ITSs were already present in the common ancestor of these taxa and remained evolutionarily conserved during karyotype phylogeny.

3.3.4.2 Large ITSs in Dispersed Heterochromatic Regions (Dispersed het-ITSs)

A more exceptional pattern of het-ITSs, referred here to as "large ITSs in dispersed heterochromatic regions (dispersed het-ITSs)," is found in the genomes of some anurans and other vertebrates. An example of this category is the karyotype of the neotropical frog *Boana faber* (family Hylidae). In this Brazilian species, the heterochromatic centromeric regions of all 12 chromosome pairs are completely labeled after FISH with the telomeric DNA probe (Fig. 885). These dispersed het-ITSs perfectly colocalize with the centromeric heterochromatin in all chromosomes and do not extend into the adjacent pericentromeric heterochromatin present in chromosomes 3, 6, and 9. As shown by fluorescence staining with the GC-specific fluorochrome mithramycin and the AT-specific fluorochromes quinacrine mustard, DAPI, and Hoechst 33258, all dispersed het-ITSs of *B. faber* are located in AT-rich repetitive DNA sequences. There are no interstitial het-ITSs or NOR-associated het-ITSs in the karyotype of this species.

The other anurans in whose genomes dispersed het-ITSs could be recorded are compiled in Table 8. In these species, many dispersed het-ITSs colocalize with all or the majority of heterochromatic centromeric regions. Thus, like in *B. faber*, large dispersed het-ITSs are located in the constitutive heterochromatin of all centromeric regions in the karyotypes of the 3 hylid species *Aplastodiscus albofrenatus*, *A. arildae*, and *A. eugenioi*. With all probability, the dispersed het-ITSs in these closely related *Aplastodiscus* taxa can be traced back to a common ancestral karyotype. In the Brazilian hylid *Itapotihyla langsdorffii*, 18 of 24 heterochromatic centromeric regions are dispersed het-ITSs. The het-ITS in the chromosome 1 centromeric region was found in heterozygous (+/–) condition.

3.3.4.3 Large ITSs in Restricted Euchromatic Regions (Restricted eu-ITSs)

A new category of intrachromosomal telomeric sequences, which is in clear contrast to the known classes of ITSs in vertebrate genomes, was recognized in the present study. These are defined here as "large ITSs in restricted euchromatic regions (restricted eu-ITSs)." In the short arms of chromosome pairs 2 and 9 of the frog *Boana boans* collected in southern Venezuela, very large ITSs are detectable which are embedded in non-heterochromatic (euchromatic) chromosome segments (Fig. 49, 884). This is especially evident in preparations that were first hybridized in situ with the telomeric DNA probe and subsequently subjected to C-banding, which allows a direct comparison of FISH signals and C-band patterns in the same chromosomes (Fig. 49). More than half of the euchromatin in the short arm of chromosome 2 and the complete euchromatic short arm of chromosome 9 consist of ITSs. C-banding reveals that both chromosome segments are not heterochromatic (Fig. 49, 460). Furthermore, staining with AT- and GC-specific fluorochromes as well as distamycin A/DAPI counterstaining also failed in demonstrating constitutive heterochromatin in both chromosome segments.

In *B. boans*, 1 further very small ITS can be recognized in the middle of the long arms of chromosome pair 7, but this one is a genuine restricted het-ITS located in the NOR-associated heterochromatin (Fig. 49). Finally, in some preparations, a tiny ITS which may be associated with a small interstitial C-band can be recognized close to the genuine telomeric $(TTAGGG)_n$ repeats in the long arms of chromosome pair 5 (Fig. 49a).

Mattos et al. [2014] examined the metaphase chromosomes of 7 *Boana* species from the Amazonas State in Brazil using conventional staining, C-banding, silver staining, and FISH with telomeric DNA probes. Three of these species, *B. boans*, *B. wavrini*, and *B. geographica* are closely related and included in the *B. semilineatus* species group. The single male *B. boans* they analyzed was collected in São Sebastião do Uatumã. This locality is approximately 1,200 km (bee-line) apart from the site in Venezuela where the male *B. boans* studied here was sampled. The karyotypes in both *B. boans* populations are very similar concerning gross chromosome morphology, but differences exist in the C-banding patterns and location of the NORs. In the Brazilian *B. boans* population, the conspicuous large restricted eu-ITS in the short arm of chromosome 2 and the tiny ITS close to the telomeric $(TTAGGG)_n$ repeats in the long arms of chromosome pair 5 are also present. However, the large restricted eu-ITS in the short arm of chromosome 9 is clearly absent (Table 8). The same situation was encountered in *B. wavrini*. The similar patterns of ITSs and C-bands, and their chromosomal localization in the Brazilian *B. boans* and *B. wavrini* confirm the close phylogenetic relationship of both species [Faivovich et al., 2005]. Finally, in *B. geographica*, 2 restricted het-ITSs are located in the centromeric heterochromatin of chromosomes 1 and 5, but no eu-ITSs are present in this species (Table 8).

Interestingly, Mattos et al. [2014] did not realize that the ITSs present in the genomes of *B. boans* and *B. wavrini* differ from the well-known category of het-ITSs commonly found in other amphibians and vertebrates. This may be attributed to the lower resolution of the C-band patterns together with the higher degree of chromosome condensation obtained in their study, and because they did not perform FISH with the $(TTAGGG)_n$ repeat followed by C-banding on the same metaphases.

Table 8 lists the other Anura with restricted eu-ITSs detected up to date. Like in *B. boans* and *B. wavrini*, the Brazilian hylid *Sphaenorhynchus lacteus* possesses a very prominent restricted eu-ITS located in the same site of the chromosome 2 short arm. Unless no further species of the phylogenetically distant genera *Boana* and *Sphaenorhynchus* are cytogenetically analyzed with telomeric DNA repeats, it cannot be decided if this is a mere coincidence or a very ancient and highly conserved cluster of ITSs in some lineages of the family Hylidae. On the other hand, the smaller restricted eu-ITSs present in interstitial regions of chromosomes 4 and 8 of the phylogenetically closely related *Dryophytes cinereus*, *D. femoralis*, and *D. squirellus*, and in the chromosomes 4, 7, and 11 of *Pseudacris brimleyi* and *P. nigrita* are homoeologous and with high probability represent evolutionarily conserved markers that can be traced back to the karyotypes of common ancestors.

Concerning the origin of restricted het-ITSs, it is conceivable that they are relics of chromosomal repatternings that occurred during karyotype evolution and which have shifted telomeric $(TTAGGG)_n$ repeats into internal positions. From a theoretical point of view, those chromosome rearrangements capable of internalizing telomeric $(TTAGGG)_n$ repeats are mainly centric (robertsonian) fusions, followed by inversions, tandem translocations, and insertions [for review, see Schmid et al., 2010]. Definitely several of the many restricted het-ITSs found in vertebrates can be traced back to such chromosomal rearrangements which became fixed during karyotype evolution.

However, in the case of the dispersed het-ITSs of the frog *Boana faber* and certain other vertebrates (see above), it is

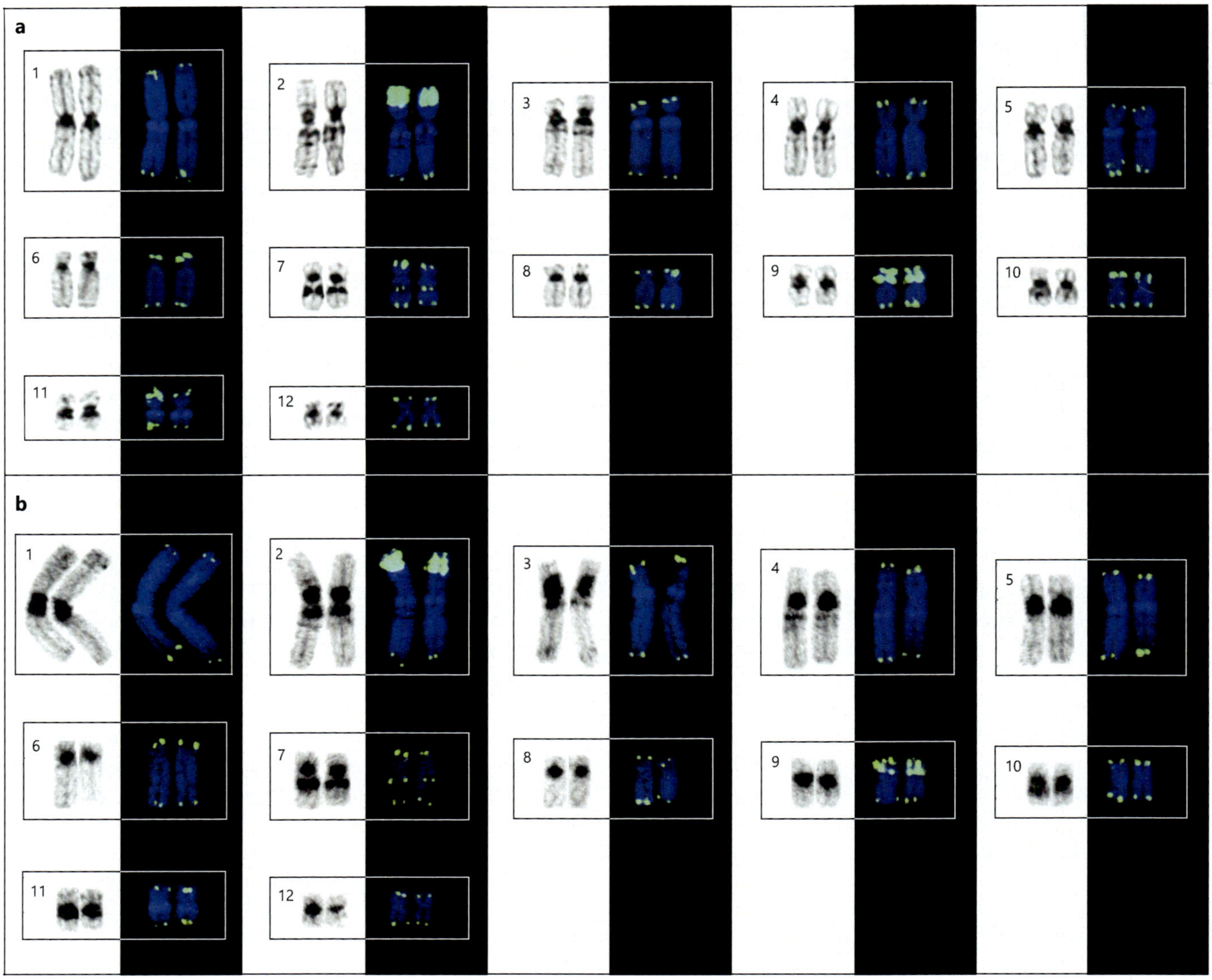

Fig. 49. Two karyotypes of a male *Boana boans* collected at km 99 on the road from El Dorado to Santa Elena de Uairén, Bolivar State, Venezuela (06°04′N, 61°24′W). The chromosomes to the left side show the C-banding patterns. On the right side, the same chromosomes exhibit the in situ hybridization signals obtained with a telomeric DNA probe. Note the 2 very large clusters of intrachromosomal telomeric sequences in the short arms of chromosomes 2 and 9. These are clearly located in C-band-negative (euchromatic) regions. For further explanations, see chapter 3.3.4.3.

impossible that so many $(TTAGGG)_n$ repeats, as estimated by the size and fluorescence intensity of the hybridization signals in the centromeric regions, have originated exclusively by centric fusions (compare size of telomeric and centromeric hybridization signals in Fig. 885). In the most favorable event, i.e., in case of a perfect telomeric fusion without terminal breaks or deletions, just a doubling of the telomeric hybridization signals, but not the observed remarkably large centromeric hybridization signals would be detected. Moreover, the karyotype of *B. faber* conforms to those of the vast majority of species in the family Hylidae, which all have diploid chromosome numbers of 2n = 24, fundamental numbers of FN = 48, and highly conserved chromosome morphologies. Therefore, the karyotype of *B. faber* was certainly not derived by a series of repeated centric fusions. As already proposed by Meyne et al. [1990], an alternative and more satisfactory explanation is that in such chromosomes, the $(TTAGGG)_n$ repeats are merely a major component of the repetitive DNA in the centromeric heterochromatin itself. The $(TTAGGG)_n$ repeat is known to be a main motif of the repetitive satellite DNAs of some vertebrates [Southern, 1970; Fry and Salser, 1977; Arnason et al., 1988; Arnason and Widegren, 1989; Adegoke et al., 1993; Garrido-Ramos et al., 1998; Rovatsos et al., 2011]. In other words, a partial or complete sequence identity of the telomeric DNA and the repetitive DNA in these heterochromatic regions is purely coincidental.

The same pattern of dispersed het-ITSs as occurring in *B. faber* was discovered in 3 exceptional lizards of the family Sphaerodactylidae. The karyotypes of the geckos *Gonatodes falconensis* and *G. taniae* from Venezuela are distinguished by the extraordinarily reduced chromosome number of 2n = 16, which is the lowest value found in reptiles up to date. Both species are characterized by considerable amounts of $(TTAGGG)_n$ sequences in the centromeric regions of all 8 chromosome pairs [Schmid et al., 1994, 2014]. In contrast, the Mediterranean *Euleptes europea* has 2n = 42 chromosomes, which is the highest value so far detected in the family Sphaerodactylidae, as well as one of the highest among all lizards. Also in this lizard, large amounts of $(TTAGGG)_n$ sequences are located in the centromeric regions of all telocentric/subtelocentric 21 chromosome pairs [Gornung et al., 2013]. Furthermore, in the lizards *Aspidoscelis sexlineata* and *A. gularis* of the family Teidae, $(TTAGGG)_n$ sequences colocalize with the centromeric heterochromatin of most chromosomes [Meyne et al., 1990]. The best-studied dispersed het-ITSs are those of the Chinese hamster (*Cricetulus griseus*). In this species, dispersed het-ITSs are a major constituent of the repetitive satellite DNA [Faravelli et al., 1998, 2002], which is located in the centromeric and pericentromeric regions of most chromosomes and which comprises about 5% of the hamster genome [Day et al., 1998]. These dispersed het-ITSs are arranged as extended tandem arrays that are not interrupted by any other sequences and do not contain restriction enzyme cleavage sites [Faravelli et al., 2002].

The question arises about the molecular structure and origin of eu-ITSs. It is possible that under certain conditions and promoted by whatsoever event(s), short ITSs in non-heterochromatic (euchromatic) chromosome regions can amplify excessively and extend into adjacent chromosome segments. This massive amplification could cause a size increase of the affected chromosome arm and change the original chromosome morphology. It is further tempting to hypothesize that transposable elements which are closely linked to telomeric $(TTAGGG)_n$ repeats spread by transposition carrying complete or partial $(TTAGGG)_n$ repeats and preferentially integrate into the chromosomal DNA of specific euchromatic regions containing appropriate target sites. In both cases, large numbers of reiterated sequences would accumulate and form a chromosome region with highly repetitive DNA, which can further degenerate and, in turn, form constitutive heterochromatin detectable by the classical C-banding technique.

In several stages of mitosis and meiosis, there is an alignment of the chromosomes with the centromeres grouped together in one part of the cell and their telomeres at the opposite side (Rabl-orientation). It has been proposed that this arrangement, together with unequal crossover, transposition, and amplification, leads to an equilocal spreading of constitutive heterochromatin throughout the genomes [Greilhuber and Loidl, 1983; Schweizer and Loidl, 1987]. This explains why in many species constitutive heterochromatin is located in the same or nearly the same relative regions in different chromosomes. By the same mechanism, both het-ITSs and eu-ITSs can multiply in a concerted manner throughout the genomes and reach equilocal sites in heterologous chromosomes. The dispersed het-ITSs found in centromeric and interstitial regions, as well as the dispersed eu-ITSs localized in interstitial regions of anuran chromosomes (Table 8) could have evolved in this way.

A formal classification of the different categories of $(TTAGGG)_n$ repeats in vertebrate genomes based on their locations in chromosomes and distribution patterns is presented in Table 9. Up to date, the anuran species analyzed by telomeric sequence mapping represent only a very small fraction of all recognized species. In order to substantiate this ITS classification, it is mandatory that many more species, representative of all genera, must be studied by FISH with the telomeric $(TTAGGG)_n$ repeat. These experiments must be extended to the chromosomes of species belonging to the amphibian orders Urodela and Apoda.

3.3.5 Hypermethylated DNA Sequences in Anura

An experimental approach using monoclonal anti-5-methylcytosine antibodies and indirect immunofluorescence was elaborated for detecting 5-methylcytosine-rich chromosome regions in vertebrate chromosomes. This technique was applied to mitotic metaphases of all 5 vertebrate classes, fishes [Schmid et al., 2015a], amphibians [Schmid and Steinlein, 2016b], reptiles [Schmid et al., in preparation], birds [Schmid et al., 2015b; Schmid and Steinlein, 2017], and mammals [Schmid et al., 2015c, 2017b]. In amphibians, 6 neotropical frog species belonging to 6 genera (*Craugastor, Eleutherodactylus, Gastrotheca, Stefania, Boana, Osteopilus*) and 4 families (Craugastoridae, Eleutherodactylidae, Hemiphractidae, Hylidae) were examined. The hypermethylation patterns were compared with a variety of banding patterns obtained by various techniques. The hypermethylated DNA sequences are species-specific and are located exclusively in constitutive heterochromatin. They are found in centromeric, pericentromeric, and interstitial positions of the chromosomes and adjacent to NORs. 5-methylcytosine-rich DNA sequences can be embedded both in AT- and GC-rich repetitive DNA. They are very useful as cytogenetic markers for easy and fast differentiation of related species with the same diploid chromosome number, chromosome morphology, and classical banding patterns.

The sizes and fluorescence intensities of the anti-5-methylcytosine antibody signals are specific for the respective heterochromatic regions in the karyotypes. This most probably reflects the amounts of hypermethylated DNA sequences present in them. Thus, in *Craugastor fitzingeri*, the centromeric or pericentromeric heterochromatin of chromosomes 2 and 3–8 fluoresces distinctly brighter than that in chromosomes 1 and 3. The chromosomal

Cytogenet Genome Res 2018;155:55–221
DOI: 10.1159/000489839

Table 9. Proposed classification of the known categories of $(TTAGGG)_n$ repeats in vertebrate genomes, their locations in chromosomes and distribution patterns according to Schmid and Steinlein [2016a]

Category	Chromosome location	Distribution patterns
Telomeric sequences	telomeric regions	in all chromosome arms
Intrachromosomal telomeric sequences (ITSs)	non-telomeric regions	
Short ITSs (s-ITSs)	scattered over all chromosomes	accumulate in some subterminal regions
Large ITSs in heterochromatic regions (het-ITSs)	heterochromatin	
- Restricted het-ITSs	restricted to few heterochromatic regions	random
- Dispersed het-ITSs	in many heterochromatic regions	non-random[a]
Large ITSs in euchomatic regions (eu-ITSs)	euchromatin	
- Restricted eu-ITSs	restricted to few euchromatic regions	random
- Dispersed eu-ITSs	in many euchromatic regions	non-random[b]

[a] Three distribution patterns: (1) in centromeric heterochromatin of all or most chromosomes, (2) in multiple interstitial heterochromatic regions, often at the same or nearly the same sites in heterologous chromosomes, and (3) mixed pattern with multiple sites in heterochromatin of centromeric and interstitial regions.

[b] In multiple euchromatic regions, often at equilocal or nearly equilocal sites in heterologous chromosomes.

distribution of 5-methylcytosine-rich heterochromatin in the anurans examined is highly variable. In *C. fitzingeri*, *Eleutherodactylus planirostris*, and *Stefania scalae*, it is located in centromeric and pericentromeric regions of 11, 3, and 8 chromosome pairs, respectively, whereas in *Gastrotheca walkeri*, it is restricted to subtelomeric or telomeric regions in the long arms of 2 telocentric chromosome pairs. In *Boana boans*, it is found in centromeric or pericentromeric regions of chromosomes 1 and 3, in an interstitial position in the long arm of chromosome 10, as well as at the NOR in the chromosome 7 long arm. Finally, in *Osteopilus septentrionalis*, 5-methylcytosine-rich DNA sequences are recognized in the telomeric and subtelomeric regions of 7 chromosome pairs and in the centromeric regions of the smallest chromosome pair.

3.3.6 Hypermethylated DNA Sequences in Arboranae

Heterochromatic regions showing bright fluorescence after staining with quinacrine mustard, DAPI, and Hoechst 33258 or after counterstaining with distamycin A/DAPI are believed to be rich in AT base pairs, whereas fluorescence staining with mithramycin or counterstaining with distamycin A/mithramycin specifically labels GC-rich heterochromatic regions [for review, see Schmid et al., 2010]. The comparison of the fluorescence patterns induced with these fluorochromes and the hypermethylation patterns in the karyotypes of the 6 anuran species shows that 5-methylcytosine-rich DNA sequences can be embedded both in AT- and GC-rich repetitive DNA [Schmid and Steinlein, 2016]. Thus, in *Boana boans*, the hypermethylated sequences in chromosomes 1, 3, and 10 are within AT-rich heterochromatin, whereas the NOR-associated hypermethylated sequences in chromosome 7 are within GC-rich heterochromatin.

3.4 Karyotype Evolution in Anura and Arboranae

In any group of eukaryotes an adequate knowledge of phylogenesis is a necessary prerequisite for interpreting chromosome evolution. As emphasized by Hillis [1991], most of the earlier studies dealing with amphibian chromosome evolution were restricted by an inadequate understanding of amphibian phylogenetic relationships. Exceptions include the studies of Hillis and Green [1990] and Schmid et al. [2010] who used much better amphibian phylogenies at the familial level [Hillis and Green, 1990] or generic and even subgeneric level [Schmid et al., 2010] to examine karyophylogeny.

The arboranan phylogenetic tree was constructed using sequences from 3 mitochondrial genes, 12S and 16S rRNA and the intervening tRNA-valine, and 2 nuclear genes (*RAG1* and tyrosinase precursor) in various combinations of species [Duellman et al., 2016]. For the convenience of the readers, the exact description of the Material and Methods section in the publication of Duellman et al. [2016] is reproduced in Appendix 2.

In all instances, this detailed phylogenetic tree resolved the relationships of species at generic and subgeneric levels. The karyotype characters of the examined arboranan species were mapped on this molecular phylogeny in the most parsimonious arrangements, i.e., with the fewest number of possible changes in a lineage.

The karyophylogeny in the Arboranae took place over a period of approximately 62 million years. It started in the Paleocene, after the split between the arboranan and the other nobleobatrachian taxa (Fig. 2). The major karyological changes that occurred after this separation were

(1) an overall reduction of the diploid chromosome number from 2n = 26 to 2n = 24 due to a tandem fusion, centric or nonreciprocal translocation in the common ancestor of the family Hylidae,

(2) further species-specific reduction in the chromosome number mainly in the genus *Aplastodiscus* (2n = 22 in *A. albofrenatus*, *A. arildae*, *A. erhardti*; 2n = 20 in *A. albosignatus*; and 2n = 18 in *A. cavicola*, *A. ibirapatinga*, and *A. leucopygius*) (Appendix 1). Furthermore, it can be observed in *Phyllodytes edelmoi* and *P. luteolus* with 2n = 22 chromosomes, in *Boana albopunctata* and *B. lanciformis* also with 2n = 22 (chapter 3.1.1.5; Appendix 1), and in *Scarthyla goinorum*.

(3) an increase of the diploid chromosome number in fewer species, like in *Osteopilus wilderi* (2n = 28), *O. ocellatus* (2n = 34), *Pseudis cardosoi* (2n = 28), and *Sphaenorhynchus carneus* (2n = 26) (Appendix 1). This was accomplished by the process of centric fission (Fig. 34, 35, 37; chapter 3.4.1).

The present study has a limitation that must be mentioned. Only a fraction of the total rearrangements that actually have occurred in the arboranan karyophylogeny could be detected with the staining techniques employed. These are centric fusions, tandem fusions, centric fissions, pericentric inversions, and formation of heteromorphic sex chromosomes. Reciprocal translocations, complex translocations, paracentric inversions, and insertions have, with certainty, escaped detection. An exact reconstruction of the karyophylogenesis in Arboranae will remain incomplete without the use of techniques that can identify homoeologous chromosomes and chromosome regions. Cross-species chromosome painting is a powerful and exact method for genome-wide comparisons of homoeologous chromosomes and chromosome regions. Chromosome-specific DNA probes are generated by flow-sorted chromosomes or microdissected chromosomes and are used as templates for DNA amplification by DOP-PCR [Telenius et al., 1992]. Employing this procedure, it is possible to identify ancient homoeologs shared by widely divergent species and to confirm the chromosomal repatternings that came about during karyotype evolution. Conserved homoeologous regions provide strong evidence for the likely constitution of ancestral chromosomes [Rens et al., 2006; Ferguson-Smith and Trifonov, 2007]. With this technique, chromosome rearrangements that cannot be resolved by multiple G- or R-bands may be accurately identified [Rens et al., 2001]. Cross-species chromosome painting has been extensively used in the analysis of mammalian [e.g., Yang et al., 1995, 1999, 2003; Rens et al., 2001, 2003, 2004; Grützner et al., 2004], avian [Griffin et al., 1999; Guttenbach et al., 2003; Nanda et al., 2006, 2007, 2008], and reptilian [Giovanotti et al., 2009] genomes. In arboranan amphibians, chromosome-specific DNA probes could be isolated, for example, in species with karyotypes that are considered to be more basic, such as the Phyllomedusidae or Pelodryadidae.

Cross-species chromosome painting of these DNA probes to species with derived karyotypes could definitively confirm or reject conclusions proposed in the present monograph. Furthermore, chromosome-specific DNA probes, isolated from basic species of Arboranae should be applied to related taxa in other nobleobatrachian lineages.

The karyotype formula of 2n = 26 and FN = 52 is commonly found in many species in other nobleobatrachian families (Pelobatidae, Leptodactylidae, Rhinodermatidae, Microhylidae, Hyperoliidae, Ranidae, and Rhacophoridae) (see compilations by King [1990] and Green and Sessions [2007]). Accordingly, the ancestral orthobatrachian karyotype with the supposed diploid chromosome number of 2n = 26 and the fundamental number of FN = 52 is the direct precursor.

The transspecific chromosome evolution of arboranans and other anuran clades becomes accessible to some degree when the sophisticated technique of chromosome replication banding with 5-BrdU is applied. Such late replication banding patterns on metaphase chromosomes can reveal homoeologous segments in the karyotypes of related and distant species. Thus, Miura [1995a] compared the BrdU replication banding pattern in the chromosomes of 9 species of Ranidae (*Rana nigromaculata*, *R. porosa*, *R. ridibunda*, *R. fukienensis*, *R. plancyi*, *R. lessonae*, *R. japonica*, *R. clamitans*, and *R. rugosa*), 1 species of Bufonidae (*Bufo japonicus*), and 1 species of Hylidae (*Dryophytes japonicus*). The replication banding patterns of all 13 chromosome pairs of the ranid species were homoeologous (Fig. 50). Minor differences were observed only in chromosomes 2, 7, and 13. In *B. japonicus* and *D. japonicus*, the 6 large chromosome pairs 1–6 showed striking banding homoeologies. Furthermore, among the species analyzed, the 4 large chromosome pairs 1–3 and 5 of *Rana* and *Dryophytes* and 1 and 3–5 of *Bufo* shared banding homoeologies (Fig. 50). These results convincingly show that the large chromosomes were highly conserved in the evolution of the genera *Rana*, *Bufo*, and *Dryophytes* [Miura, 1995a]. The homoeologies in these species can be extended to most of the other genera of Arboranae. It is even possible to extend the chromosomal homoeologies to some chromosomes of *Xenopus tropicalis* and *X. laevis*, although such comparisons must be treated with some caution, because the replication bands have their limits of resolution.

In the following, some categories of chromosome repatternings that may have played essential roles during the evolution of the Arboranae are explained. These have already been treated in much more detail, including schematic drawings, in the first volume of these 3 monographs on amphibian cytogenetics [Schmid et al., 2010], but are also very useful in the context of the present treatise.

3.4.1 Centric Fissions

By a centric fission, 2 telocentric chromosomes originate as a result of a single break through the centromeric region of a meta- or submetacentric chromosome. According to White [1973a], the centromere is "… undoubtedly a functionally compound body, in the sense that fragments of centromere still possess centromeric properties. Presumably it consists of a particular nucleotide sequence (or sequences)

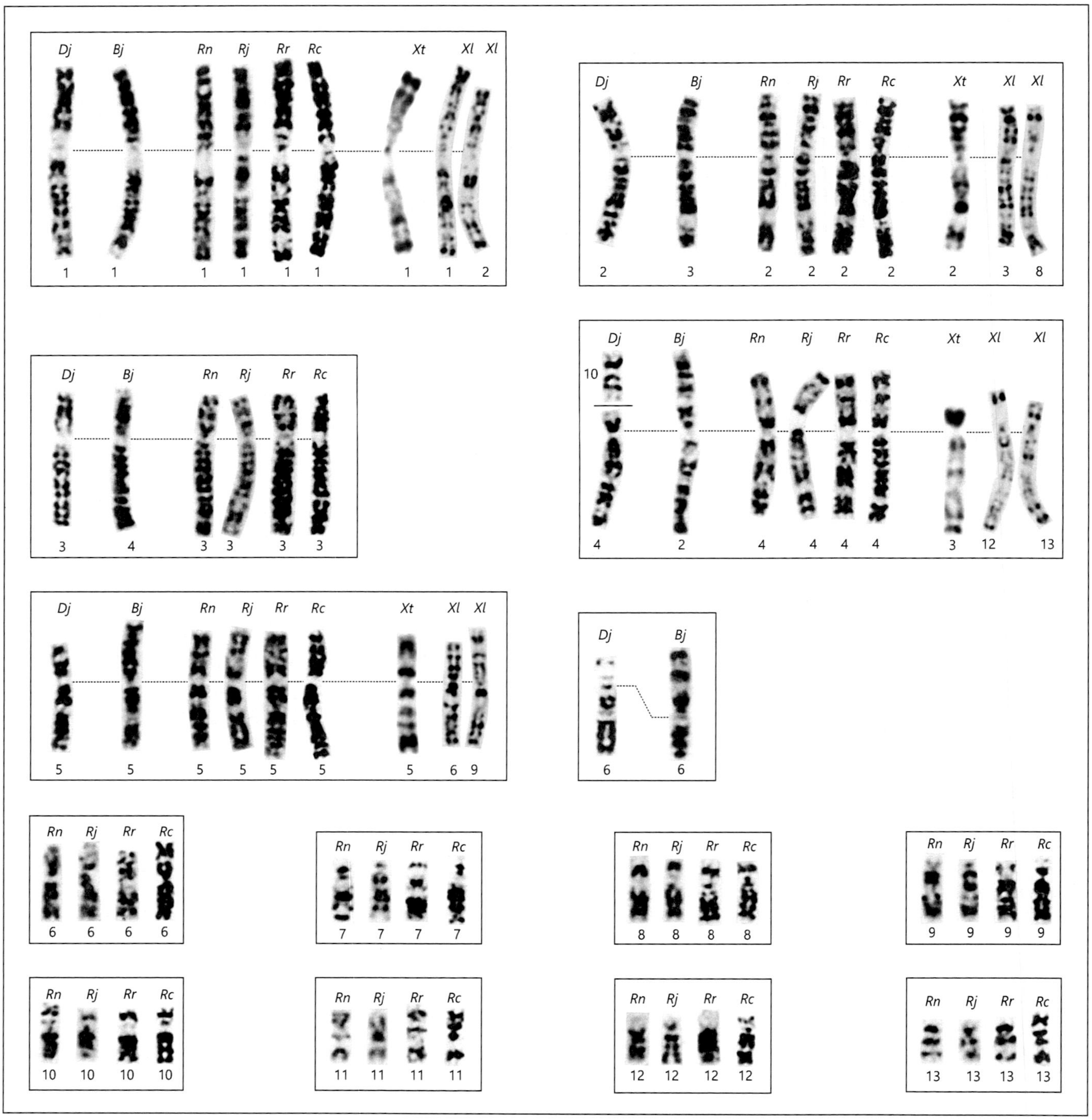

Fig. 50. Comparison of BrdU replication banding patterns in the chromosomes of *Dryophytes japonicus* (*Dj*), *Bufo japonicus* (*Bj*), *Rana nigromaculata* (*Rn*), *R. japonica* (*Rj*), *R. rugosa* (*Rr*), *R. clamitans* (*Rc*), *Xenopus tropicalis* (*Xt*), and *X. laevis* (*Xl*). A haploid karyotype of each of the 4 *Rana* species, one each of the chromosomes 1–6 and 10 of *D. japonicus*, 1–6 of *B. japonicus*, 1–3 and 5 of *X. tropicalis*, and the duetts 1 + 2, 3 + 8, 6 + 9, and 12 + 13 of *X. laevis* are arranged in order to demonstrate the probable evolutionarily conserved chromosomes. The dotted horizontal lines between the chromosomes connect the centromeric regions. Modified from Miura [1995a]; the chromosomes of *X. tropicalis* and *X. laevis* were added to the original figure. For further explanations, see chapter 3.4.

repeated many times …" White's hypothesis was later confirmed by electron microscopical and molecular studies. The molecular structure responsible for the centromeric function, the kinetochore, is fairly well-known [for reviews, see Pepper and Brinkley, 1977; Moroi et al., 1980; Brenner et al., 1981; Rieder, 1982; Earnshaw and Rothfield, 1985; Rattner, 1986; Pepper, 1988; Pluta et al., 1995; Choo, 1997a, b; De Wulf and Earnshaw, 2008; Schroeder-Reiter and Wanner, 2009].

Immediately after a centric fission event, 2 telocentric chromosomes devoid of the canonical $(TTAGGG)_n$ repeats at their centromeric ends are produced (Fig. 34a, 35a, 37a). Little is known about the stability of such telocentric elements or about the mechanism for recruitment of new telomeric DNA sequences that maintain their integrity. An analysis of spontaneously occurring centric fissions in somatic tissues of terraranan frogs showed that new telomeric repeats are relatively quickly reconstituted at the broken centromeric regions [Schmid et al., 2010]. Furthermore, extensive studies in mammalian cells have shown that healing of broken chromosome ends can actually take place, leading to the acquisition of new telomeric caps and to chromosome stabilization. Two different mechanisms have been proposed for telomere recruitment: (1) direct addition of telomeric $(TTAGGG)_n$ repeats onto the broken chromosome end by a telomerase-mediated de novo telomere addition [Wilkie et al., 1990; Flint et al., 1994; Sprung et al., 1999; Varley et al., 2000] or by a telomerase-independent recombination-based addition [Neumann and Reddel, 2002; Varley et al., 2002], and (2) telomere capture from other chromosomes, a process by which telomeres are translocated onto the broken ends [Meltzer et al., 1993; Bosco and Haber, 1998; Ning et al., 1998]. The molecular mechanism of telomere capture remains poorly understood [Ballif et al., 2004], but numerous chromosome aberrations with healing of broken chromosome ends have been observed in humans [Fortin et al., 2009; Yu and Graf, 2010].

Fission events that become fixed in natural populations lead to evolutionary chromosome number changes [Schubert et al., 1995]. Centric fissions have occurred in the bird genus *Bucephala* [Ray-Chaudhuri, 1973], the snake genera *Malpolon*, *Pseustes*, *Vipera*, and *Corallus*, the chamaeleon genera *Chamaeleo* and *Brooksiana* [Gorman, 1973], the amphisbaenid genera *Amphisbaena*, *Bipes*, and *Rhineura*, and the lizard genera *Anolis* and *Plica* [Gorman et al., 1967; Gorman, 1973, Olmo, 1986]. Centric fission of the original W sex chromosome into 2 smaller sex chromosomes (W_1, W_2) has been reported in a number of species of sea snakes [Singh, 1972a, b]. In eutherian mammals and in marsupials, variability of karyotypes can best be interpreted by assuming one or more centric fissions [Matthey, 1973; Hayman and Martin, 1974; Dementyeva et al., 2010]. Heterozygous centric fissions have been described for a single biarmed chromosome in the root vole *Microtus oeconomus* [Fredga and Bergström, 1970], the east African common zebra *Equus burchelli* [Whitehouse et al., 1984], and the primate *Presbytis entellus* [Egozcue, 1971]. Heterozygous centric fissions in multiple chromosomes are karyotype features of mole rats of the genus *Spalax* [Nevo and Bar-El, 1976] and the black rat *Rattus rattus* [Yosida et al., 1979]. A compilation of non-domestic mammalian species with centric fusion/fission polymorphisms has been published by Rubes et al. [2008]. Even in humans, there are reports describing the occurrence and inheritance of centric fissions [Sinha et al., 1972; Hansen, 1975; Dallapiccola et al., 1976; Imai et al., 1988].

A centric fission increases the diploid chromosome number by one (2n + 1) (Fig. 34a, 35a, 37a). During meiosis of individuals that are heterozygous for a centric fission, a chain of 3 chromosomes (trivalent) is formed, which consists of the 2 free fissioned telocentric chromosomes paired with their corresponding homologous arms of the normal meta- or submetacentric chromosome (Fig. 34c, 35c, 37c). Backcrossing of the founder individual carrying the centric fission with normal specimens produces 50% offspring with the 2 fission chromosomes, provided that segregation of the chromosomes forming the meiotic trivalent occurs preferentially in the alternate modus [for illustrations, see Schmid et al., 2010]. Crosses between these F_1 heterozygotes result in individuals that are homozygous for this centric fission (1 fission cycle). The fundamental number of the basic karyotype (FN_b) remains the same, but the heterozygotes for the centric fission have 1 more chromosome and an odd diploid chromosome number of $2n_b + 1$, and homozygotes for the centric fission have 2 more chromosomes and a diploid chromosome number of $2n_b + 2$ (Fig. 34a, 35a, 37a). If all centric fissions become homozygously fixed in a series of subsequent fission cycles, the number of meta- or submetacentric chromosomes will decrease in the following series

$$\left\{ m_b \rightarrow \; m_b^{fis_1} - 1 \rightarrow \; m_b^{fisc_1} - 2 \rightarrow \; m_b^{fis_2} - 3 \rightarrow \right.$$
$$\left. m_b^{fisc_2} - 4 \rightarrow \; \ldots \; m_b^{fis_x} - 2x + 1 \rightarrow \; m_b^{fisc_x} - 2x \right\}$$

where m_b = number of meta- or submetacentric chromosomes in the basic karyotype; $fis_1, fis_2, \ldots, fis_x$ = sequence of fissions; $fisc_1, fisc_2, \ldots, fisc_x$ = sequence of fission cycles; x = number of fissions or fission cycles.

With each new spontaneous centric fission, the number of meta- or submetacentric chromosomes decreases by one, and after each completed fission cycle it decreases by two (Fig. 34a, 35a, 37a). The fundamental number of the basic karyotype (FN_b) remains unaltered. Conversely, the number of telocentric chromosomes will increase in the following series

$$\left\{ t_b \rightarrow \; t_b^{fis_1} + 2 \rightarrow \; t_b^{fisc_1} + 4 \rightarrow \; t_b^{fis_2} + 6 \rightarrow \right.$$
$$\left. t_b^{fisc_2} + 8 \rightarrow \; \ldots \; t_b^{fis_x} + 4x - 2 \rightarrow \; t_b^{fisc_x} + 4x \right\}$$

where t_b = number of telocentric chromosomes in the basic karyotype; fis_1, fis_2, …, fis_x = sequence of fissions; $fisc_1, fisc_2, \ldots, fisc_x$ = sequence of fission cycles; and x = number of fissions and fission cycles.

Cytogenet Genome Res 2018;155:55–221
DOI: 10.1159/000489839

The above series when applied, for example, to 13 fission cycles will read for meta- or submetacentric and telocentric chromosomes, respectively

$$\{26 \rightarrow \; 26^{fis_1} - 1 \rightarrow \; 26^{fisc_1} - 2 \rightarrow \; 26^{fis_2} - 3 \rightarrow \; 26^{fisc_2} - 4 \rightarrow \ldots \; 26^{fisc_{13}} - 26 + 2 \rightarrow \; 26^{fisc_{13}} - 26\}$$

$$\{26 \rightarrow \; 25^{fis_1} \rightarrow \; 24^{fisc_1} \rightarrow \; 23^{fis_2} \rightarrow \; 22^{fisc_2} \rightarrow \ldots \; 1^{fis_{13}} \rightarrow \; 0^{fisc_{13}}\}$$

$$\{0 \rightarrow \; 0^{fis_1} + 2 \rightarrow \; 0^{fisc_1} + 4 \rightarrow \; 0^{fis_2} + 6 \rightarrow \; 0^{fisc_2} + 8 \rightarrow \ldots \; 0^{fis_{13}} + 26 - 2 \rightarrow \; 0^{fisc_{13}} + 26\}$$

$$\{0 \rightarrow \; 2^{fis_1} \rightarrow \; 4^{fisc_1} \rightarrow \; 6^{fis_2} \rightarrow \; 8^{fisc_2} \rightarrow \ldots \; 50^{fis_{13}} \rightarrow 52^{fisc_{13}}\}$$

In general terms, the total numbers of meta- or submetacentric (m_{tot}) and telocentric (t_{tot}) chromosomes after x fissions (*fis*) or fission cycles (*fisc*) in a basic karyotype originally containing m_b meta- or submetacentric and t_b telocentric chromosomes are calculated by the formulae

$$m_{tot}^{fis_x} = m_b - 2x + 1 \qquad m_{tot}^{fisc_x} = m_b - 2x$$

$$t_{tot}^{fis_x} = t_b + 4x - 2 \qquad t_{tot}^{fisc_x} = t_b + 4x$$

After the origin of a centric fission in the founder individual, 2 inbreeding generations are required before the first homozygous animals appear in a small population. Whether the centric fission is fixed in, or eliminated from, a population through nondirectional and stochastic fluctuations of the frequencies of the centric fission chromosome and the normal non-fissioned chromosomes (i.e., genetic drift) is determined by the size of the population, genetic flow between neighboring populations, and the selective value (neutrality, advantage, or disadvantage) of the centric fission. In species possessing many or only meta- or submetacentric chromosomes in their karyotypes, spontaneous centric fissions in 2 or more different chromosomes can originate simultaneously either in the same or in different individuals of the same population and generation. In such cases, the number of telocentric chromosomes can increase at a rapid pace. Converse to centric fusions, a series of spontaneous centric fissions and fission cycles can reduce a diversified biarmed karyotype consisting of meta- and submetacentric chromosomes with a variety of morphologies into a monotonous, completely uniarmed karyotype. During this series of repeated centric fissions, the original fundamental number (FN_b) remains the same, whereas the diploid chromosome number increases one by one until all biarmed chromosomes have been converted into telocentric elements.

Morescalchi [1973] suggested that the telocentric chromosomes found in the karyotypes of several neotropical Terraranae and Arboranae species were derived by centric fissions. More recently, in Asiatic ranids, diploid chromosome numbers of 2n = 26 (*Nanorana maculosa*), 2n = 40 (*N. unculuanus*), and 2n = 64 (*N. phrynoides*) were discovered [Liu and Zan, 1984; Wu and Zhao, 1984; Li and Hu, 1994]. All chromosomes are biarmed in the 2n = 26 karyotype of *N. maculosa*, which is typical of karyotypes found in the majority of ranid frogs. The 2n = 64 karyotype of *N. phrynoides* consists exclusively of telocentric chromosomes. The extremely derived karyotype of *N. phrynoides* cannot be explained without assuming the involvement of centric fissions in all biarmed chromosomes of an ancestral 2n = 26 karyotype, followed by the fragmentation of several of these "new" telocentric chromosomes and neocentromere activation. In the Japanese frog *Rana tagoi*, populations have been found with 2n = 26 and 2n = 28 chromosomes [Ryuzaki et al., 2006]. The 2n = 28 karyotype in the Chausu mountains has 2 pairs of telocentric chromosomes that were evidently derived by fission of a metacentric chromosome pair in the more usual 2n = 26 karyotype. In African frogs of the family Arthroleptidae, diploid chromosome numbers of 2n = 14, 16, 18, 28, and 54 exist [reviewed by King, 1990; Green and Sessions, 2007]. In the 2n = 14 karyotypes of *Arthroleptis poecilonotus*, *A. stenodactylus*, and *A.* sp. and in the 2n = 16 and 2n = 18 karyotypes of *Cardioglossa gracilis* and *C. leucomystax*, all or most chromosomes are meta- or submetacentric, whereas in the 2n = 28 and 2n = 54 karyotypes of *Nyctibates corrugatus* and *Astylosternus diadematus*, the majority of chromosomes have telocentric or subtelocentric morphology. Although Bogart and Tandy [1981] suggested that *A. diadematus* may be an allopolyploid that was derived from the interspecific mating of 2n = 28 and 2n = 26 individuals, it is also possible that multiple centric fissions have broken down an ancestral karyotype with biarmed chromosomes into a karyotype with many telocentric elements.

3.4.2 Tandem Fusions

There is no doubt that tandem fusions have occurred and were fixed in the karyotypes of eukaryotes. Tandem fusions are classical 2-break rearrangements. One break occurs within the long arm paracentromeric heterochromatin of a subtelocentric chromosome. The second break is in the long arm terminal (telomeric) heterochromatin of another subtelocentric chromosome, or in the short or long arm terminal heterochromatin of a biarmed chromosome. After reciprocal fusion, a large monocentric tandem fusion chromosome originates. There is evidence [Hsu et al., 1972] that the telomeric regions are rich in repetitive DNA sequences (among them the canonical $(TTAGGG)_n$ repeats) and might contain a high proportion of palindromes [Cavalier-Smith, 1974], so the loss of reiterated DNA sequences in this fusion chromosome would be expected to have no detrimental genetic effects. The other product of the tandem fusion would be a minute monocentric fusion chromosome consisting merely of a centromere, paracentromeric and terminal heterochromatin. The latter will be lost during a few mitotic divisions or during meiosis [for illustrations, see Schmid et al., 2010]. Following the origin of a tandem fusion in the founder individual, 2 inbreeding generations are required before the first homozygous individuals for this tandem fusion would appear in the population. Back-crossing of the founder individual with specimens having normal karyotypes produces 50% offspring carrying the tandem fusion. Matings between these F_1 heterozygous carriers of the tandem fusion (1 fusion cycle) will yield the first homozygous individuals for the tandem fusion. A tandem fusion does not necessarily require the loss of a centromere in the telocentric chromosome if one of the 2 centromeres in the fusion chromosome is functionally inactivated [Hsu et al., 1975b].

During meiosis of heterozygous carriers for a tandem fusion, a trivalent will be formed which consists of the 2 free telocentric chromosomes paired with their corresponding homologous segments in the tandem fusion chromosome. If in the large fusion chromosome a crossing-over regularly occurs between the centromere and the site of fusion, 50% of the resulting gametes will be aneuploid carrying a duplication or deficiency. Therefore, and provided there is no crossing-over suppression, selection against individuals heterozygous for tandem fusions is expected to be severe. Nevertheless, fixed tandem fusions (i.e., where all individuals are homozygous for the fusion) have been well documented in various species of *Drosophila* [Patterson and Stone, 1952], morabine grasshoppers [White, 1973a], cotton rats [Elder, 1980], cats [Wurster-Hill and Gray, 1973], and apes of the family Cercopithecidae [Stock and Hsu, 1973], and there are many other examples. The most impressive example for a series of homozygously fixed tandem fusions is found in barking deer of the genus *Muntiacus*. In the Indian deer *Muntiacus muntjak vaginalis*, the diploid chromosome number is 2n = 6♀, 7♂, which is the lowest diploid number found in mammals [Wurster and Benirschke, 1970]. In contrast, a closely related species, the Chinese deer *M. reevesi*, has a diploid number of 2n = 46, with exclusively telocentric chromosomes [Wurster and Benirschke, 1967]. The extreme karyotype of *M. m. vaginalis* is, with all probability, the derived one, but the reduction in chromosome number from 2n = 46 to 2n = 6♀, 7♂ cannot be explained by centric fusions between the telocentric chromosomes alone. Even the maximum number of centric fusions and fusion cycles will not reduce the diploid chromosome number to 2n = 6♀, 7♂. Therefore, it seems reasonable that each of the extremely large chromosomes of *M. m. vaginalis* is the result of multiple tandem fusions [Hsu et al., 1975b; Shi et al., 1980; Lin et al., 1991; Yang et al., 1995] in addition to centric fusions [Brinkley et al., 1984; Rattner, 1986; Latour et al., 1996]. As judged from chromosomal G-banding [Liming et al., 1980], 19 events of centric and tandem fusions must have occurred to match the present-day karyotype of *M. m. vaginalis* to a hypothetical ancestral karyotype of both species. A detailed molecular characterization of fusion sites in the chromosomes of the Indian muntjac has already been accomplished [Chi et al., 2005; Tsipouri et al., 2008]. Wurster and Atkin [1972] determined the nuclear DNA contents of *M. muntjak* and *M. reevesi* to be 72.4% and 89.4% of the human value, respectively. Liming et al. [1980] proposed that these differences in nuclear DNA content between the 2 species are the result of the loss of repetitive sequences in the course of genome evolution that led to *M. muntjak*. Using DNA reassociation kinetics, Schmidtke et al. [1981] have shown that the quantity of single-copy DNA sequences in *M. reevesi* and *M. muntjak* is almost identical, but there are pronounced differences in the quantity of repeated DNA, especially in the middle repetitive fraction. By comparing the rate constants of the middle repetitive DNA fractions of the 2 genomes, it becomes apparent that the mean repetition frequency of sequences constituting these components differs by one order of magnitude. Apparently, in the course of the *M. muntjak* evolution, a lowly repetitive subset of DNA sequences has been selectively lost from the chromosomes. Schmidtke et al. [1981] speculated that these DNA sequences were located in the centromeric and telomeric regions of the ancestral telocentric chromosomes, thus promoting repeated chromosome translocations by sequence-specific recognition. Other cases with multiple tandem and centric fusions (along with a variety of other rearrangements) are found in species of the South American rodent genus *Akodon* that demonstrates an impressive range of diploid numbers from 2n = 10 to 2n = 46 [for review, see Ventura et al., 2009] and in species of the American rodent genus *Sigmodon* [Elder, 1980; Swier et al., 2009].

Similar to centric fusions, individuals heterozygous for a tandem fusion would have 1 chromosome less and an odd diploid chromosome number of 2n – 1, and homozygous individuals would have 2 chromosomes less and a diploid chromosome number of 2n – 2. Unlike centric fusions, however, the fundamental number is also reduced by 1 in the heterozygous individuals (FN – 1) and by 2 in the homozygous animals (FN – 2). Provided all tandem fusions become homozygously fixed in a series of subsequent fusion cycles, the number of telocentric chromosomes will decrease in the following series

$$\{t_b \rightarrow t_b^{tfs_1} - 1 \rightarrow t_b^{tfsc_1} - 2 \rightarrow t_b^{tfs_2} - 3 \rightarrow t_b^{tfsc_2} - 4 \rightarrow \dots t_b^{tfs_x} - 2x + 1 \rightarrow t_b^{tfsc_x} - 2x\}$$

where t_b = number of telocentric chromosomes in the basic karyotype; $tfs_1, tfs_2, \dots, tfs_x$ = sequence of tandem fusions; $tfsc_1, tfsc_2, \dots, tfsc_x$ = sequence of tandem fusion cycles; and x = number of fusions or fusion cycles.

With each new tandem fusion, the number of telocentric chromosomes decreases by 1, and after each completed fusion cycle, it decreases by 2. The above series, when applied to the 8 tandem fusion cycles will read for telocentric chromosomes

$$\{52 \rightarrow 52^{tfs_1} - 1 \rightarrow 52^{tfsc_1} - 2 \rightarrow 52^{tfs_2} - 3 \rightarrow 52^{tfsc_2} - 4 \rightarrow \dots 52^{tfs_8} - 16 + 1 \rightarrow 52^{tfsc_8} - 16\}$$

$$\{52 \rightarrow 51^{tfs_1} \rightarrow 50^{tfsc_1} \rightarrow 49^{tfs_2} \rightarrow 48^{tfsc_2} \rightarrow \dots 37^{tfs_8} \rightarrow 36^{tfsc_8}\}$$

In general terms, the total number of telocentric chromosomes (t_{tot}) after x tandem fusions (*tfs*) or tandem fusion cycles (*tfsc*) in a karyotype with a basic karyotype originally containing t_b telocentric chromosomes is given by the simple formulae

$$t_{tot}^{tfs_x} = t_b - 2x + 1 \quad (1) \qquad t_{tot}^{tfsc_x} = t_b - 2x \quad (2)$$

After the origin of a tandem fusion in a single founder individual, 2 further inbreeding generations are required before the first individual that is homozygous for this tandem fusion appears in a restricted population. Whether this tandem fusion is finally fixed in, or eliminated from, a population through nondirectional and accidental fluctuations of the frequencies of the tandem fusion chromosome and the normal non-fused chromosomes (i.e., genetic drift), is determined by the size of the population, genetic flow between neighboring populations, and the selective value (neutrality, advantage, or disadvantage) of the tandem fusion. The majority of tandem fusion chromosomes have been efficiently fixed in the phylogeny of invertebrates and vertebrates.

Tandem fusions can also occur between the telomeric regions of 2 chromosomes. When 2 chromosomes fuse at their telomeres, the fusion chromosome will be either a dicentric element or a monocentric element if one of the kinetochores in the centromeric regions is functionally inactivated. There are well authenticated cases in human cytogenetics where tandemly fused dicentric chromosomes behave as monocentrics [Distèche et al., 1972; Pallister et al., 1974; Ruthner and Golob, 1974; Therman et al., 1974; Haaf and Schmid, 1990; Page and Shaffer, 1998]. One of the most impressive cases was published by Therman et al. [1974] and involved a fusion between 2 X chromosomes in a woman. The 2 X chromosomes had fused at the telomeres of their long arms, and there was no visible loss of euchromatic bands. In the majority of cells analyzed, only 1 centromeric constriction was present in the fusion chromosome, and in only a few cells was the fusion chromosome clearly dicentric. A further case of telomeric fusion between 2 human X chromosomes without chromatin loss and inactivation of one of the kinetochores is included in a review of 21 human X chromosome mutations [Rivera et al., 1984]. This review provides several hypotheses on the mechanism and the origin of terminal fusions. Terminal fusions between 2 human Y chromosomes with kinetochore inactivation have also been reported by Hermier et al. [1979], Butler et al. [1981], and Davis [1981].

There is no reason why inactivation of a centromere in a dicentric chromosome should only occur in mammalian cells. If centromeres can be inactivated into a latent state by some genetic mechanism, it is conceivable that they may be reactivated by the same, or a similar mechanism. Chromosomes that possess latent centromeres can break off into various fragments with functional centromeres on each of them, as long as the latent centromeres can be reactivated [Hsu et al., 1975b]. Activation of latent centromeres could explain karyotypes of species which increase their diploid chromosome number without engaging in centric fissions of biarmed chromosomes.

3.4.3 Centric Fusions

Centric fusions are reciprocal translocations and therefore 2-break rearrangements. Two subtelocentric chromosomes break, one of them close to the centromere in the short arm, the other one close to the centromere in the long arm. The subsequent reciprocal translocation yields a large meta- or submetacentric chromosome, as well as the reciprocal translocation product, a tiny chromosome consisting of a centromere with a kinetochore and minute arms. If the tiny reciprocal translocation product is not euchromatic but heterochromatic (i.e., if it does not contain essential genes), it will be lost during subsequent mitotic divisions or during meiosis. It can be argued that a centric fusion can also originate by 2 breaks directly in the centromeres of both telocentric chromosomes followed by a fusion of the broken centromeric regions. However, when the very rare event of a spontaneous de novo origin of centric fusions is observed, large biarmed centric fusion chromosomes plus the minute reciprocal translocation products are present [Schmid et al., 2010]. This strongly supports the assumption that centric fusions actually result from 2 breaks, 1 in the paracentromeric region of the short arm and the other in the paracentromeric region of the long arm of 2 subtelocentric chromosomes. Finally, the possibility exists that centric fusions arise by 2 breaks in the short arm paracentromeric regions of 2 subtelocentric chromosomes. By this nonreciprocal telomeric translocation, a dicentric centric fusion chromosome and 2 acentric fragments will result. Whereas the acentric fragments will be lost, the 2 centromeres in the dicentric fusion chromosome are so close to one another that they might function as 1 centromere. Although there is no experimental evidence that dicentric fusion chromosomes have been fixed in a homozygous condition during phylogeny, there are known cases of human dicentric fusion chromosomes [Niebuhr, 1972; Bross et al., 1973; Warburton et al., 1973; Dittes et al., 1975]. There is, however, the possibility that 1 of the 2 kinetochores in the dicentric translocation chromosome is functionally inactivated. In this case, the chromosome would behave like a monocentric element.

Through a centric fusion, 2 telocentric chromosomes become mutually attached at their centromeric regions forming a biarmed (meta- or submetacentric) chromosome. During meiosis of individuals that are heterozygous for a centric fusion, a chain of 3 chromosomes (trivalent) is formed, which consists of the 2 free telocentric chromosomes paired with their corresponding homologues of the meta- or submetacentric fusion chromosome. Backcrossing of the founder individual carrying this spontaneous centric fusion with normal specimens will produce 50% offspring carrying the centric fusion chromosome, provided that segregation of the chromosomes forming the meiotic trivalent occurs preferentially in the alternate modus [for illustrations, see Schmid et al., 2010]. Matings between these F_1 heterozygous carriers of the centric fusion (1 fusion cycle) will yield 25% homozygous individuals for that particular centric fusion. During this process, the fundamental number of the basic karyotype (FN_b) is conserved, but the individuals heterozygous for centric fusions have 1 chromosome less and an odd diploid chromosome number of $2n_b - 1$, whereas homozygous individuals have 2 chromosomes less and a diploid chromosome number of $2n_b - 2$. Provided all centric fusions become homozygously fixed in a series of subsequent fusion cycles, the number of telocentric chromosomes will decrease in the following series

$$\{t_b \rightarrow \; t_b^{fs_1} - 2 \rightarrow \; t_b^{fsc_1} - 4 \rightarrow \; t_b^{fs_2} - 6 \rightarrow \; t_b^{fsc_2} - 8 \rightarrow \; \dots \; t_b^{fs_x} - 4x + 2 \rightarrow \; t_b^{fsc_x} - 4x\}$$

where t_b = number of telocentric chromosomes in the basic karyotype; $fs_1, fs_2, \dots, fs_x$ = sequence of centric fusions; $fsc_1, fsc_2, \dots, fsc_x$ = sequence of centric fusion cycles; and x = number of fusions or fusion cycles.

With each new spontaneous centric fusion, the number of telocentric chromosomes decreases by 2, and after each completed fusion cycle it decreases by 4. The fundamental number of the basic karyotype (FN_b) remains unaltered. Conversely, the number of meta- or submetacentric chromosomes will increase in the following series

$$\{m_b \rightarrow \; m_b^{fs_1} + 1 \rightarrow \; m_b^{fsc_1} + 2 \rightarrow \; m_b^{fs_2} + 3 \rightarrow \; m_b^{fsc_2} + 4 \rightarrow \; \dots \; m_b^{fs_x} + 2x - 1 \rightarrow \; m_b^{fsc_x} + 2x\}$$

where m_b = number of meta- or submetacentric chromosomes in the basic karyotype; $fs_1, fs_2, \dots, fs_x$ = sequence of fusions; $fsc_1, fsc_2, \dots, fsc_x$ = sequence of fusion cycles; x = number of fusions and fusion cycles.

The above series, when applied, for example, to 5 fusion cycles, will read for telocentric and metacentric chromosomes, respectively

$$\{36 \rightarrow \; 36^{fs_1} - 2 \rightarrow \; 36^{fsc_1} - 4 \rightarrow \; 36^{fs_2} - 6 \rightarrow \; 36^{fsc_2} - 8 \rightarrow \; \dots \; 36^{fs_5} - 20 + 2 \rightarrow \; 36^{fsc_5} - 20\}$$

$$\{36 \rightarrow \; 34^{fs_1} \rightarrow \; 32^{fsc_1} \rightarrow \; 30^{fs_2} \rightarrow \; 28^{fsc_2} \rightarrow \; \dots \; 18^{fs_5} \rightarrow \; 16^{fsc_5}\}$$

$$\{0 \rightarrow \; 0^{fs_1} + 1 \rightarrow \; 0^{fsc_1} + 2 \rightarrow \; 0^{fs_2} + 3 \rightarrow \; 0^{fsc_2} + 4 \rightarrow \; \dots \; 0^{fs_5} + 10 - 1 \rightarrow \; 0^{fsc_5} + 10\}$$

$$\{0 \rightarrow \; 1^{fs_1} \rightarrow \; 2^{fsc_1} \rightarrow \; 3^{fs_2} \rightarrow \; 4^{fsc_2} \rightarrow \; \dots \; 9^{fs_5} \rightarrow \; 10^{fsc_5}\}$$

In general terms, the total number of telocentric (t_{tot}) and meta- or submetacentric (m_{tot}) chromosomes after x fusions (*fs*) or fusion cycles (*fsc*) in a karyotype with a basic karyotype originally containing t_b telocentric and m_b meta- or submetacentric chromosomes is given by the simple formulae

$$t_{tot}^{fs_x} = t_b - 4x + 2 \quad (1) \qquad t_{tot}^{fsc_x} = t_b - 4x \quad (2)$$

$$m_{tot}^{fs_x} = m_b + 2x - 1 \quad (3) \qquad m_{tot}^{fsc_x} = m_b + 2x \quad (4)$$

After the origin of a centric fusion in a single founder individual, 2 further inbreeding generations are required before the first individual that is homozygous for this centric fusion appears in a restricted population. Whether this centric fusion is finally fixed in, or eliminated from, a population through nondirectional and accidental fluctuations of the frequencies of the centric fusion chromosome and the normal non-fused chromosomes (i.e., genetic drift), is determined by the size of the population, genetic flow between neighboring populations, and the selective value (neutrality, advantage, or disadvantage) of the centric fusion. The majority of centric fusion chromosomes have been efficiently fixed in the phylogeny of invertebrates and vertebrates, provided that the meiotic trivalent segregates prevalently in the adjacent modus in both sexes.

In species possessing many or entirely telocentric chromosomes, spontaneous centric fusions between 2 or more different chromosomes can originate simultaneously either in the same or in different individuals of the same population and generation. Thus, the number of meta- or submetacentric chromosomes would increase, and the number of telocentric chromosomes would decrease in an accelerated manner. This means that after a few generations, centric fusions are present in a homozygous condition for each of the chromosome pairs involved in a frequency that is being regulated by the individual selective values of the centric fusions.

3.4.4 Alternating Centric Fusion-Inversion Cycles in Amphibia and Arboranae

Centric fusion cycles and pericentric inversion cycles can occur in a repeated and alternating manner. An initial centric fusion that involves 2 telocentric chromosomes is followed by inbreeding until the homozygous condition for the metacentric fusion chromosome has been attained by genetic drift. Then, a pericentric inversion takes place in 1 of the metacentric fusion chromosomes and, after further inbreeding, the large inverted chromosome also becomes homozygously fixed. Both processes, centric fusion and pericentric inversion, can occur in different chromosome pairs at the same time in the same population. This may lead to a considerable and rapid reduction of the diploid chromosome number in a karyotype with a concomitant dramatic increase in the chromosome lengths in a relatively short period of time.

Alternating centric fusion-inversion cycles are likely to have shaped the karyotypes of species in the anuran genera *Arthroleptis* and *Cardioglossa* (family Arthroleptidae, subfamily Arthroleptinae). *Arthroleptis poecilonotus*, *A. stenodactylus*, and *A.* sp. n. have the lowest diploid number of chromosomes (2n = 14, FN = 28) so far discovered in the Amphibia [Scheel, 1971; Morescalchi, 1973; Bogart and Tandy, 1981]. *Cardioglossa gracilis* and *C. leucomystax* have diploid chromosome numbers of 2n = 16 (FN = 32) and 2n = 18 (FN = 32), respectively [Bogart and Tandy, 1981]. All these 5 species have a very large metacentric chromosome pair 1, which is nearly twice the size of the other chromosomes. The smaller chromosomes gradually diminish in size. Scheel [1971] hypothesized that, in the genus *Arthroleptis*, the highly reduced karyotype of 2n = 14 could have been derived by a series of pericentric inversions producing telocentricity of the metacentric chromosomes, followed by centric fusions of the inverted telocentric chromosomes which reduced the chromosome number and reestablished metacentricity. Bogart and Tandy [1981] argued that karyotypes found in species of *Cardioglossa* are evidence of an intermediate stage that would be required according to Scheel's [1971] hypothesis.

In the anuran family Ranidae, karyotypes with diploid chromosome numbers of 2n = 20, 22, 24, 26, and 28 have been found [listed in King, 1990; Green and Sessions, 2007]. Many of the species possess 2n = 26 chromosomes. The single 2n = 28 karyotype found in *Rana tagoi* [Ryuzaki et al., 2006] is clearly derived by a fission event in a biarmed chromosome. The reduced 2n = 24, 22, and 20 karyotypes can be explained by 3 inversion-fusion cycles that gave rise to large biarmed elements. An alternative explanation given by Ullerich [1967] when discussing the chromosome number differences between *Rana temporaria* (2n = 26) and *R. arvalis* (2n = 24) are translocation events between different biarmed chromosomes in the 2n = 26 karyotypes. However, from a mechanistic point of view, it is impossible to reduce the chromosome number in a karyotype by conventional reciprocal translocations. If nonreciprocal translocations (e.g., tandem fusions) have been involved, a repeated inactivation of kinetochores would have been required (the probability of this being extremely low), and distinct submetacentric chromosomes with a short arm:long arm ratio of approximately 1:4 to 1:5 would have originated, which is not the case.

3.4.5 Inversions (General Principles)

Inversions are 2-break rearrangements. The chromosome segment in between the 2 breaks inverts by 180° and reinserts in the chromosome. The gene sequence in the inverted segment is reversed. If both breaks are located on the same side of the centromere, the inversion is paracentric, if the breaks are on the opposite sides of the centromere, the inversion is pericentric [for exact diagrammatic illustrations, see Schmid et al., 2010]. Paracentric inversions can be identified in chromosomes employing natural, inherent landmarks (polytene and lampbrush chromosomes) or in chromosomes with experimentally produced multiple Q-, G-, or R-bands. When multiple bands cannot be induced, however, as is the case in the mitotic chromosomes of cold-blooded vertebrates, paracentric inversions will escape detection in conventionally stained chromosomes. On the other hand, pericentric inversions can change the chromosome morphology provided the 2 breaks are not equidistant from the centromere. Therefore, most large pericentric inversions are detectable in conventionally stained preparations.

During meiosis of an individual heterozygous for an inversion, complete synapsis between the normal homologue and the homologue carrying the inversion leads to an "inversion loop." In paracentric inversions, the centromeres are located outside the loop, whereas in pericentric inversions they are inside the loop. If the inversion is sufficiently small, an inversion loop between the mutually inverted regions is suppressed because of limited chromosome flexibility. In meiotic bivalents with inversion loops, chiasmata (crossovers) can be formed either inside or outside the loop. Crossing-overs outside the inversion loops will not lead to structurally aberrant chromatids. However, a single crossing-over within the inversion loop leads to monocentric deletion-duplication chromatids in pericentric inversions and to acentric and dicentric deletion-duplication chromatids in paracentric inversions. More complicated situations occur when crossovers originate within the loop as well as between the loop region and the centromeres [for further details, see White, 1973a, b]. In any case, aneuploid gametes will originate in meiosis, which, if still capable of completing maturation, will produce aneuploid lethal zygotes or embryos, or offspring with malformations or reduced fitness. Therefore, if crossing-over is not suppressed between the mutually inverted chromosome regions, individuals heterozygous for inversions have a selective disadvantage. On theoretical grounds, one should expect natural selection to rapidly eliminate all newly arisen inversions. Nevertheless, experimental data leave no room for doubt that inversions have played a major role in phylogeny [for review, see White, 1973a, b].

In some cases, an inversion polymorphism is restricted to 1 autosome pair, where 3 cytogenetically different types of individuals will coexist (AA, AA^{inv}, $A^{inv}A^{inv}$). On the other end of the scale, there are instances such as the population of the grasshopper *Trimerotropis sparsa* found at Whitewater (Colorado, USA), which is polymorphic for inversions in at least 7 chromosome pairs [White, 1951], or the Californian *Pseudotrimerotropis thalassica*, which is polymorphic for inversions in as many as 10 chromosome pairs [King, 1923]. Theoretically, in these grasshopper populations, 3^7 (= 2,187) and 3^{10} (= 59,049) cytogenetically different types of individuals can originate. Inversion polymorphisms, either paracentric or pericentric, can be found in any group of eukaryotic organisms and are common among the well-examined mammals. Thus, in 8 specimens of the rodent *Mastomys natalensis*, Matthey [1966] discovered 2 individuals that were homozygous for a pericentric inversion, 2 individuals heterozygous for this inversion, and 4 animals with a normal karyotype. In the deer mice *Peromyscus maniculatus*, *P. sitkensis*, and *P. boylei* collected from wild populations, autosomal inversion polymorphisms have been found in a series of cytogenetic studies [Ohno et al., 1966; Greenbaum et al., 1978; Pengilly et al., 1983; Greenbaum and Reed, 1984; Davis et al., 1986; Gunn and Greenbaum, 1986; Hale, 1986]. In 3 geographical variants of the black rat *Rattus rattus*, inversion polymorphisms in 3 autosome pairs have been observed [Yosida et al., 1971; Yosida, 1977]. Further examples of inversion polymorphisms have been detected in the sand rat *Psammomys obesus* [Ashley et al., 1981] and in the bat *Mimon cretulatum* [Baker et al., 1972]. In birds, inversion polymorphisms have been unraveled in several species of the genus *Junco* [Shields, 1973, 1976], in the white-throated sparrow *Zonotrichia albicollis* [Thorneycroft, 1975], and in the cardinal *Cardinalis cardinalis* [Bass, 1979]. In reptiles and fishes, inversion polymorphisms were detected in lizards of the genus *Sceloporus* [Cole, 1970; Sites, 1983] and in the goodeid fish *Ilyodon furcidens* [Turner et al., 1985].

It is generally acknowledged that inversions have been of major importance during the transspecific evolution of amphibian genomes [for reviews, see King, 1990; Green and Sessions, 2007]. This has been reinforced by the study of Schmid et al. [2010] on terraranan frogs. Among the 227 species examined, various inversion polymorphisms were detected in 1 or several populations.

The diploid chromosome number of the basic karyotype ($2n_b$) remains unaltered by inversions. Individuals heterozygous for a pericentric inversion in a subtelocentric chromosome have 1 additional chromosome arm and an odd fundamental number of $FN_b + 1$, the homozygous individuals have 2 additional chromosome arms and a fundamental number of $FN_b + 2$. Provided all these inversions become homozygously fixed in a series of subsequent inversion cycles, the fundamental number will increase in the following series

$$\left\{ FN_b \rightarrow \quad FN_b^{i_1} + 1 \rightarrow \quad FN_b^{ic_1} + 2 \rightarrow \quad FN_b^{i_2} + 3 \rightarrow \right.$$
$$\left. FN_b^{ic_2} + 4 \rightarrow \ \ldots \ FN_b^{i_x} + 2x - 1 \rightarrow \quad FN_b^{ic_x} + 2x \right\}$$

where FN_b = fundamental number of the basic karyotype; $i_1, i_2, \ldots, i_x$ = sequence of inversions; $ic_1, ic_2, \ldots, ic_x$ = sequence of inversion cycles; x = number of inversions and inversion cycles.

With each new pericentric inversion, the FN increases by 1, and after each completed inversion cycle the FN increases by 2. The diploid chromosome number remains unaffected.

After the origin of a spontaneous inversion in a single individual in a restricted population, it takes at least 2 further generations until the first animals that are homozygous for this inversion appear. Whether this inversion will finally be fixed in, or eliminated from, a population through nondirectional and accidental fluctuations of the frequencies of the inverted and normal homologue (i.e., genetic drift) [Wright, 1932, 1942], is determined by various factors, such as the size of the population, genetic flow between neighboring populations, and the selective value (neutrality, advantage, or disadvantage) of the inversion. The number of generations required until, by chance, a selectively neutral inversion is fixed or eliminated in a population of finite and constant size can be calculated as if it were a mutated neutral allele by applying the statistical Monte-Carlo-simulations of Kimura and Ohta [1969]. In accordance with these calculations, the time required for fixation of an inversion t_{fix} is given by the equation

$$t_{fix} = 4 \times Ne \times g$$

where Ne = number of individuals in the population which actively participate in reproduction and g = generation time of the species.

3.4.6 Inversions in Anura

Inversions have been detected in a fixed and non-fixed (i.e., floating and polymorphic) condition in natural populations of the Anura. Formas [1978] identified 3 karyotypes in the Chilean leptodactylid frog *Eupsophus roseus*. All animals showed 15 chromosome pairs, but the number of telocentric pairs was either 7, 8 or 9. These differences can be interpreted to be the result of 3 homozygously fixed pericentric inversions. The North American frog *Rana muscosa* has a diploid chromosome number of 2n = 26 like all the other related *Rana* species, but the 3 smallest chromosomes are telocentric and derived by pericentric inversions from formerly metacentric chromosomes [Green, 1986]. In the Brazilian leptodactylid *Paratelmatobius* sp. aff. *cardosoi*, Lourenço et al. [2003a, b] described an apparent case of a pericentric inversion that resulted in a heteromorphic chromosome pair.

During meiosis of an individual that is heterozygous for an inversion, complete synapsis between the normal homologue and the homologue carrying the inversion leads to an "inversion loop." In paracentric inversions, the centromeres are located outside the loop, whereas in pericentric inversions they are inside the loop. Aneuploid gametes will originate in meiosis, which, if still capable of completing maturation, will produce aneuploid lethal zygotes or embryos, or offspring with congenital malformations or reduced fitness. Therefore, if crossing-over is not suppressed between the mutually inverted chromosome regions, individuals heterozygous for inversions have a strong selective disadvantage.

3.4.7 Inversions in Arboranae

It is pivotal to determine the naturally occurring inversion frequency in arboranan frogs. In the present monograph, clear inversions in heterozygous conditions were recognized in several of the karyotypes analyzed. Thus, pericentric inversions can be found in the chromosomes 6 of *Dryophytes avivocus* (Fig. 89), chromosomes 9 of *Agalychnis callidryas* (Fig. 51, 314–316), chromosomes 8 of *Agalychnis annae* (Fig. 472), chromosome 5 of *Litoria inermis* (Fig. 491), and chromosomes 12 of *Litoria coplandi* (Fig. 497).

As a rough estimate, 5% of the arboranan frogs examined and rexamined here are carriers of a heterozygous inversion. Smaller inversions, not detectable by microscopy may be present in the genomes as well. In conclusion, the inversion rate in the karyotypes of the Arboranae is estimated not to exceed the value of 10%. Up to date, in arboranan frog species no fixed differences of inversions have been found (Appendix 1). When compared to other anurans [for reviews, see Schmid et al., 2010, 2012], this value is moderately low. It is not known which intrinsic genetic, environmental, or behavioral factors promote or suppress the origin of inversions. Perhaps in the case of the Arboranae, the geographically broad distribution and the high pace with which new habitats are conquered and new populations are founded contribute to keep the inversion rate low and this, in turn, contributes to explain the evolutionarily conserved karyotypes of the Arboranae.

Agalychnis callidryas

The red-eyed tree frog *Agalychnis callidryas* (Fig. 948a) is distributed from the Atlantic lowlands of Veracruz and Oaxaca, Mexico, southeastward on the Caribbean lowlands to central Panama, Pacific lowlands of southern Costa Rica and eastern Panama to the Chocó of Colombia. It inhabits humid forests, primarily in lowlands and sometimes on premontane slopes up to 1,250 m [Amphibia Web, 2017; Frost, 2017].

The first and so far only report on the occurrence of a non-fixed inversion polymorphism at population level in Anura has been documented for *A. callidryas* [Schmid et al., 1995, 2012, present monograph]. Populations of *A. callidryas* in the Caribbean lowlands of southeastern Costa Rica are polymorphic for a pericentric inversion in chromosomes 9, which can have either submetacentric (9_a) or telocentric morphology (9_b) (Fig. 46, 51). Both males and females may be either homozygous ($9_a/9_a$ or $9_b/9_b$) or heterozygous ($9_a/9_b$). In the populations examined, the submetacentric chromosome 9_a has a significantly higher frequency (76%) than the telocentric 9_b variant (24%). Although the number of individuals examined was low (19), the cytogenetic data were sufficient to show that these populations are in Hardy-Weinberg equilibrium with respect to the 3 possible karyomorphs ($9_a/9_a$, $9_a/9_b$, $9_b/9_b$). Among the 8 *A. callidryas* specimens collected in the Pacific lowlands in Costa Rica, only 1 male showed the heteromorphic $9_a/9_b$ karyomorph, all others exhibited the homomorphic $9_a/9_a$ constitution. The frequency of the 9_b variant in the Pacific lowlands still remains to be determined. The ancestral stock of *Agalychnis* dispersed northward from South America through the Central American corridor during the Paleocene epoch [Savage, 1982; Duellman and Trueb, 1986]. The geographic range of *A. callidryas* is continuous from eastern and central Panama along the Atlantic and Pacific lowlands of Costa Rica, Nicaragua, Honduras, and Guatemala north to southern Mexico [Duellman, 1970, 1977]. In Costa Rica, the Atlantic and Pacific populations of *A. callidryas* are geographically separated from each other by the Central American Cordilleras. Nevertheless, the inverted 9_b chromosome occurs in both populations. It can be speculated that the inversions existed in ancestors of the extant *A. callidryas* prior to the geographic separation of the various populations, or else the inversion first occurred in one of the northern lowland populations and then spread southward along the Atlantic and Pacific slopes of the Cordilleras. This question can perhaps be resolved by a more extensive cytogenetic analysis of *A. callidryas* and its closely related species throughout the entire area of their distribution in Central America and Mexico. As in most other species of advanced Anura [Morescalchi, 1973; Schmid et al., 1990b, 2010], male diakinetic bivalents of *A. callidryas* are strongly condensed with terminal chiasmata in both arms producing a ring-like morphology. In most of the diakineses analyzed by Schmid et al. [1995], both the males with the homomorphic $9_a/9_a$ and $9_b/9_b$ karyomorphs, as well as the males with the heteromorphic $9_a/9_b$ constitution, all possessed bivalents with this ring-like configuration (Fig. 52). Intercalary or pericentromerically located chiasmata were not observed, neither in the larger nor in the smaller bivalents. Furthermore, no inversion loops were seen in the diplotene and diakinesis stages of these males, nor were acentric fragments or anaphase bridges detected at the anaphase stage. This leads to the conclusion that meiotic pairing and recombination between the heteromorphic submetacentric 9_a and telocentric 9_b chromosomes in males proceed just as smoothly as between the 2 homomorphic submetacentric 9_a chromosomes and between the 2 homomorphic telocentric 9_b chromosomes.

The cases of inversion polymorphisms in urodelan species [for review, see Schmid et al., 2010] and in *A. callidryas* (see above) show that the inverted chromosomes form normal bivalents in which chiasmata are restricted to regions outside the rearranged chromosome segments. Hence, these inversions would not result in meiotic disturbances in heterozygous carriers and so cannot act as reproductive barriers (producing, e.g., reduced fertility and/or chromosomally unbalanced, lethal offspring). The lack of reproductive barriers caused by chromosomal rearrangements would be a prerequisite for the spreading of a new chromosome type and for the formation of new populations that are not reproductively separated from the ancestral population [Hedrick, 1981]. The pericentric inversion in chromosome 9 of *A. callidryas* does not cause a selective disadvantage during meiosis of heterozygous males with the $9_a/9_b$ karyomorph because the chiasmata are restricted to the telomeric regions of the bivalents. The likelihood for recombination between the mutually inverted segments in the 9_a and 9_b homologues and of any aberrant meiotic products resulting from this event is very small. The absence of crossing-over between the inverted chromosome segments is a factor promoting the maintenance of fertility in heterozygous *A. callidryas* and the successful spreading of the inverted chromosome 9_b. However, it must always be considered that, unlike male meiosis, chiasmata have an unrestricted distribution in the lampbrush chromosomes of female meiosis. Therefore, it is possible that the pericentric inversion found in chromosome 9 of *A. callidryas* may actually be subjected to a selective disadvantage during female meiosis. This can only be resolved by examining the lampbrush chromosomes of mature oocytes of females that are heterozygous for the pericentric inversion.

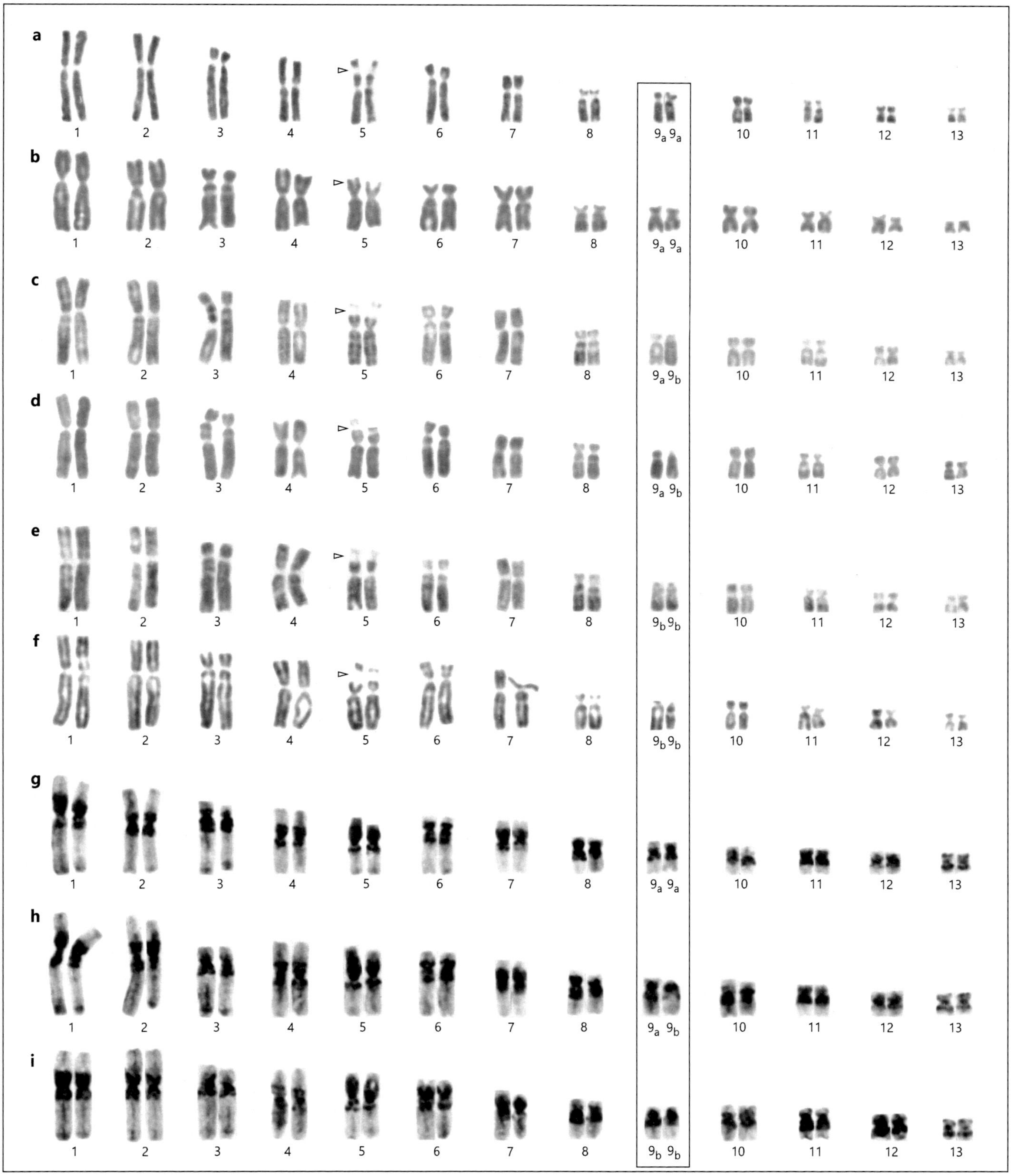

Fig. 51. Karyotypes of *Agalychnis callidryas* showing conventional Giemsa staining (**a–f**) and C-banding of constitutive heterochromatin (**g–i**). The chromosome pairs 9 are framed. **a**, **b**, **g** Individuals homozygous for the submetacentric chromosome 9_a. **c**, **d**, **h** Individuals heterozygous for the submetacentric chromosome 9_a and the subtelocentric chromosome 9_b. **e**, **f**, **i** Individuals homozygous for the subtelocentric chromosome 9_b. Note also the nucleolar constrictions in the short arms of chromosomes 5 (arrowheads) and the large amounts of centromeric and pericentromeric heterochromatin. For further explanations, see chapter 3.4.7.

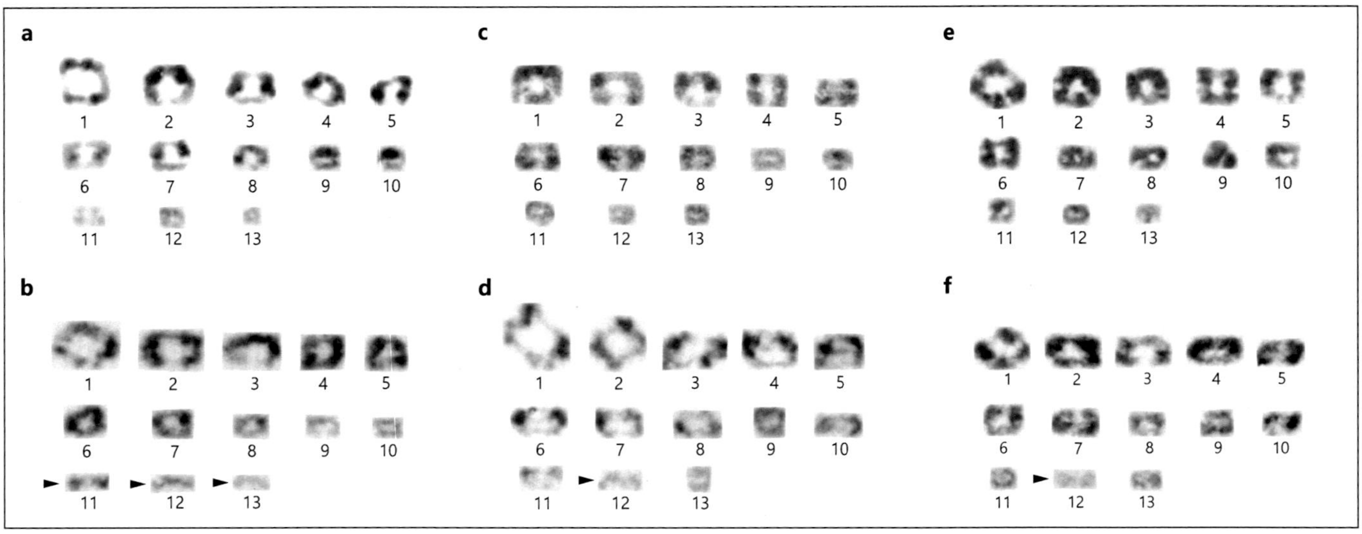

Fig. 52. Giemsa-stained diakineses from the male meiosis of *Agalychnis callidryas*. **a, b** Males homozygous for the submetacentric chromosome 9_a. **c, d** Males heterozygous for the submetacentric chromosome 9_a and the subtelocentric chromosome 9_b. **e, f** Males homozygous for the subtelocentric chromosome 9_b. In **a, c**, and **e**, all 13 bivalents exhibit a ring-shaped pairing configuration with telomeric chiasmata in both of their arms. In **b, d** and **f**, several of the medium-sized and/or small bivalents have attained a rod-like pairing arrangement due to the terminalization of the chiasma in one of their arms (arrowheads). For further explanations, see chapter 3.4.7.

3.4.8 Reciprocal Translocations

Reciprocal translocations belong to the category of 2-break rearrangements. They result from a break in each of 2 chromosomes followed by mutual interchange. If the translocation is symmetrical, the 2 new resulting chromosomes will be monocentric and can be transmitted unaltered through the mitotic cycles. In principle, genetic material is neither deleted from nor added to the genome by symmetrical reciprocal translocations. Theoretically, there is no selection against somatic cells or organisms carrying reciprocal translocations. But, problems can arise during meiosis. In reciprocal translocations, like in all other chromosomal rearrangements, the directional polarity of the DNA strands is conserved. If the reciprocally translocated chromosome segments are very different in size, the translocation becomes apparent even in conventionally stained preparations. If they are equally sized, they will escape detection if the chromosomes are not labeled with multiple Q-, G-, or R-bands.

The consequences of reciprocal translocations directly depend on the position of the 2 breaks in relation to the centromeres and telomeres. The lengths of the chromosome segments on both sides of the breaks and their crossing-over frequencies determine the pairing configuration and the course of the first meiotic division. After complete synapsis, the reciprocally translocated chromosomes have a cross-like configuration in the pachytene stage. The centromeric regions of the paired chromosomes are located in opposite arms of the cross. Crossovers can be formed in all 4 arms. In the arms where the centromeres are positioned, crossovers can be located either in the segment between the centromeres and the telomeres (external segment), or between the centromeres and the breakpoints (internal segment). When the crossovers are located in the external segments, the cross-like pairing configuration develops into a chromosome chain or ring after chiasmata terminalization. In anaphase, the chromosomes segregate either in alternate or adjacent modi. Alternate segregation yields euploid gametes, whereas adjacent segregation results in gametes with duplication-deficiencies which, in turn, produce non-viable or lethal zygotes or embryos, or offspring with congenital malformations and reduced fitness. Reciprocal translocations that have a relatively good chance of becoming evolutionarily fixed are those in which a single chiasma is formed in each of the 4 arms, and close to the end of the arms where it can be terminalized without difficulties. A chain or ring that is held together by terminal chiasmata may orient at the meiotic spindle in such a way that segregation of the chromosomes occurs preferentially in the alternate modus. In contrast, crossovers that are formed between the centromeres and the breakpoints (internal segment) confound the regularity of segregation and are unlikely to become phylogenetically fixed. There is also the possibility that, instead of the formation of a chain or ring of 4 chromosomes, 2 bivalents are formed during meiosis of the translocation heterozygote. These bivalents then will segregate in the non-favorable adjacent modus in 50% of the cells.

Because of the complexity of meiosis in heterozygotes with reciprocal translocations, aneuploid gametes carrying a chromosome duplication and deficiency will repeatedly be

Cytogenet Genome Res 2018;155:55–221
DOI: 10.1159/000489839

produced, unless for certain reason(s) the favorable alternate meiotic segregation is significantly preferred. Therefore, selective pressure against new reciprocal translocations originating de novo in wild populations is very strong. Three large heterozygous translocations that were initially observed in natural populations of *Drosophila ananassae*, *D. melanica*, and *D. prosaltans* were not found in later collections [Stone, 1955]. Despite these facts, translocation heterozygosity has been detected consistently in some natural animal populations [for review, see White, 1973a, b]. For example, in some Algerian populations of the rodent *Gerbillus pyramidum*, the meiotic XY pair is associated with an autosomal bivalent in a chain of 4 chromosomes which is probably the result of a reciprocal translocation [Matthey, 1954]. Although reciprocal translocations have likely played a minor role in chromosome evolution, not at all comparable with that of centric fusions and inversions, verified cases of homozygously fixed reciprocal translocations have been reported in primates [Dutrillaux, 1975; Yunis and Prakash, 1982; Marks, 1983; Stanyon et al., 1992].

With few exceptions (see below), there are no published reports of amphibians with a de novo reciprocal translocation heterozygosity in natural populations. This may be explained because only limited numbers of individuals per amphibian species have previously been examined, and also by the fact that multiple Q-, G-, or R-banding patterns cannot be induced in amphibian chromosomes. Surprisingly, the analysis of spontaneously arising chromosome aberrations in aging cultured kidney fibroblasts of the toads *Rhinella rubropunctata* and *Scaphiopus holbrookii* and of the marsupial frog *Gastrotheca riobambae* revealed a variety of chromosome rearrangements but not a single reciprocal translocation [Schmid et al., 2004]. But, in that study, the chromosomes were identified by the BrdU/dT replication banding technique where small reciprocal translocations between chromosome ends can escape microscopical detection.

In the most complete cytogenetic analysis of any one amphibian taxon performed prior to the present monograph, Schmid et al. [2010] determined the frequency of reciprocal translocations among 2,548 specimens of the taxon Terrarana. Three individuals (0.001%) showing a constitutional reciprocal translocation were found. An interesting non-fixed reciprocal translocation at the population level was discovered in the terraranan frog *Haddadus binotatus* from Brazil [Schmid et al., 2010]. In a male sampled from the Tijuca forest at Bom Retiro, Guanabara (Rio de Janeiro State), a reciprocal translocation between the long arm of an evolutionarily fixed inversion chromosome inv(8) and the short arm of a chromosome 2 was observed which was not present in the karyotypes of 7 other specimens collected in Tinguá, Nova Iguacu Municipality (Rio de Janeiro State). Siqueira et al. [2004] detected the same reciprocal translocation in 5 individuals (3 ♂, 2 ♀) out of 13. Two of these specimens were collected on the mainland, and 3 individuals came from the Ilha Belha Island located 1.76 km offshore the São Paulo State. Ilha Belha is a continental island that was formed approximately 11,000 years ago [Vanzolini, 1973]. If Ilha Belha was connected to the continent shortly after it emerged, this would allow an early and temporarily limited colonization of the island by amphibians. Therefore, the reciprocal translocation found in the *H. binotatus* of Ilha Belha has probably persisted for approximately 11,000 years. Although in the study of Schmid et al. [2010] only a few terraranan individuals (0.001%) have been found with a constitutional reciprocal translocation, it should be noted that reciprocal translocations in which the mutually exchanged chromosome segments are of equal size, would not be detected. Furthermore, reciprocal translocations that involve very small terminal segments would, in most instances, escape microscopical detection.

3.4.9 Amplification and Deletion of Constitutive Heterochromatin

Molecular studies on highly repetitive DNA sequences in constitutive heterochromatin showed that it is characterized by tandem repetition of DNA sequence units or sequences which differ in composition in different species and in which the repeat unit ranges in size from a few base pairs to several thousands [for review, see Singer, 1982]. In the highly repetitive DNA, the locations of restriction enzyme cleavage sites and the lengths of the DNA fragments cut by the restriction enzymes are preserved. This indicates that the repeat arrays of the tandem DNA sequences arose through a series of amplification events [Dennis et al., 1980; Arnason et al., 1984; John et al., 1986]. Amplification is a term that describes large-scale increases in the copy number of DNA sequences. It can be caused by multiple replication of a specific DNA sequence unit, or else by some form of exchange, most likely unequal sister chromatid exchange. It has been proposed that either multiple replication stimulates unequal exchange, or unequal exchange leads to multiple replication [Hamlin et al., 1984].

It is well-known that the amount of constitutive heterochromatin can vary significantly between species in different genera and even among closely related species in the same genus. Ground squirrels of the genera *Spermophilus* and *Ammospermophilus* have a diploid chromosome number of 2n = 32, but the metaphase chromosomes of the former are distinctly smaller than those of the latter. C-banding shows that in species of *Spermophilus*, constitutive heterochromatin is restricted to the centromeric regions, whereas in species of *Ammospermophilus* huge amounts of heterochromatin are present throughout their karyotypes [Deaven et al., 1977]. In 20 species of North American deer mice of the genus *Peromyscus*, Hsu and Arrighi [1966, 1968] found an invariable diploid chromosome number of 2n = 48, but the number of chromosome arms varied from FN = 56 in *P. crinitus* and *P. boylei* (mainly subtelocentric elements) to FN = 96 in *P. eremicus* and *P. collatus* (mainly submetacentric elements). The authors first explained these differences as being the result of pericentric inversions and unequal reciprocal translocations, but after reexamining the karyotypes using the C-banding technique, Pathak et al. [1973] found that, in the extreme karyotypes of *P. crinitus* and *P. boylei*, the constitutive heterochromatin is confined to the centromeric regions, whereas in *P. eremicus* and *P. collatus* all of the chromosome short arms are constitutively heterochromatic. Therefore, the variation in the number of chromosome arms in the genus *Peromyscus* can be attributed to amplification of constitutive heterochromatin in all the tiny short arms of the originally subtelocentric autosomes. From the genomic DNA of *P. eremicus*, a satellite DNA fraction with a buoyant density of 1.705 g/mL has been isolated in neutral CsCl density gradients. Radioactive in situ hybridization using ^{3}H-RNA complementary to this satellite DNA fraction showed that the heterochromatic short arms of most of the autosomes of *P. eremicus* contain this repetitive DNA sequence [Hazen et al., 1977]. Furthermore, G-banding of the chromosomes has shown that the euchromatic chromosome long arms are essentially the same in all *Peromyscus* species [Pathak et al., 1973]. But, as is usual in cytogenetics and evolution, there are always exceptions because in *P. leucopus*, *P. maniculatus*, and *P. sitkensis*, the short arms are not always heterochromatic, and G-banding has demonstrated that pericentric inversions were responsible for producing the euchromatic short arms in the autosomes [Arrighi et al., 1976; Pengilly et al., 1983]. In a most ambitious study on 188 specimens of 29 species of Australian murid rodents belonging to the subfamilies Pseudomyinae and Hydromyinae and the *Uromys/Melomys* group, Baverstock et al. [1977a] showed that all but 2 biarmed chromosomes of *Notomys cervinus* consisted of amplified heterochromatin. The amplification of heterochromatin in at least 13 autosomal pairs of *N. cervinus* likely represents one, or at most very few, evolutionary events and not 13 independent steps. In *Uromys caudimaculatus*, the specimens in the northern populations (Iron Range and Rocky River, Cape York) possess large blocks of terminal and intercalary heterochromatin in at least 12 pairs of autosomes. With high probability this heterochromatin originated by amplification as a single evolutionary step because the specimens in the southern populations (Cooktown, Atherton area, and Townsville), and species in the related genus *Melomys,* possess only a small amount of terminal or intercalary heterochromatin, and no intermediate stages were found [Baverstock et al., 1976, 1977a]. Further, Hsu and Arrighi [1971], Bostock et al. [1972], and Fredga and Mandahl [1973] have demonstrated completely heterochromatic short arms in many chromosomes of some mammalian genera (*Dipodomys*, *Mesocricetus*, and *Mustela*). Very large segments of amplified autosomal heterochromatin have been discovered in several chromosome pairs of the hedgehog genera *Erinaceus*, *Hemiechinus*, and *Aethechinus* [Gropp et al., 1969; Natarajan and Gropp, 1971; Mandahl, 1978]. In male meiosis, these heterochromatic segments pair as homologues until the end of pachytene, but separate during diplotene because they do not form crossovers. Lack of crossing-overs in heterochromatin has also been shown in *Drosophila* [Baker, 1958].

Amplification not only occurs at terminal and intercalary positions of the chromosomes, but also at centromeric and paracentromeric regions. This is best exemplified by the African bat *Glauconycteris beatrix* [Volleth and Heller, 2007; Schmid et al., 2017b]. Its reduced karyotype consists of only 2n = 22 chromosomes, and huge centromeric plus paracentromeric heterochromatic blocks are located in each chromosome pair extending at least over half of the length of each chromosome.

Amplification of constitutive heterochromatin has not only changed the morphology of autosomes but has also involved numerous sex chromosomes. In general, the highly differentiated mammalian Y and avian W sex chromosomes are completely heterochromatic and C-band-positive. The same is the case for well-differentiated Y and W chromosomes of the ectothermic vertebrates. One of the best-known examples of heterochromatin amplification is that found in the giant XY sex chromosomes in the European field vole *Microtus agrestis.* Three quarters of the X chromosome and the complete Y chromosome, except its tiny short arm, consist of constitutive heterochromatin that contains a large fraction of repetitive DNA [Arrighi et al., 1970; Yunis and Yasmineh, 1971; Pera, 1972]. These heterochromatic regions are identical to those recognized by Wolf et al. [1965] and Schmid et al. [1965] as late replicating in the S-phase. The Syrian hamster *Mesocricetus auratus* is another mammalian species with very large XY sex chromosomes, the X being the largest element in the karyotype. The entire Y, the whole X long arm, and the short arms of 16 autosomes consist of amplified heterochromatin [Hsu and Arrighi, 1971]. Ohno et al. [1964] first pointed out, that as a general rule, the euchromatic portion of the X chromosome of placental mammals accounts for about 5–6% of the total haploid female karyotype. X chromosomes that exceed this value, like those of *M. agrestis* and *M. auratus*, originate by amplification of constitutive heterochromatin. The study of Baverstock et al. [1977a] on 29 species of Australian murid rodents showed that, in most species, the amplified heterochromatin in the X chromosomes occupies a complete X arm (usually the shorter), and only in a few species does the amplification occur in intercalary positions. In at least 17 of their analyzed species, there was a clear correlation between the length of the X and Y or, to be more precise, between the amount of constitutive heterochromatin located in the X and Y chromosomes. This phenomenon was previously observed in other mammals by Schmid [1967], Fredga [1970], and Hayman and Martin [1974]. The amount of amplified heterochromatin in these X and Y chromosomes is highly correlated both within and between species. In 3 species of the Australian hopping mice of the genus *Notomys*, the variations in size and morphology of the X and Y chromosomes are entirely due to heterochromatin amplification [Baverstock et al., 1977b]. The Indian rodent *Bandicota bengalensis* also demonstrates a clear case of heterochromatin amplifi-

Table 10. Genome sizes in 45 arboranan species

Family Species	C-value	Method	Cell type
Hylidae			
Acris crepitans	4.08	FD	RBC
	4.15	FCM	RBC
Acris gryllus	3.72	FD	LV
	4.49	FD	RBC
Boana albomarginata	2.63	FD	RBC
Boana bischoffi	3.71	FD	RBC
Boana cinerascens	4.53	FCM	RBC
Boana crepitans	4.83	FD	LV
Boana faber	2.73	FD	RBC
Boana geographica	3.28	FD	LV
Boana heilprini	3.76	FD	LV
Boana lanciformis	6.61	FCM	RBC
Boana cf. *lanciformis* sp. 1	5.41	FCM	RBC
Boana cf. *lanciformis* sp. 2	7.45	FCM	RBC
Boana polytaenia	2.35	FD	RBC
Boana prasina	5.15	FD	RBC
Dendropsophus nanus	1.89	FD	RBC
Dryophytes chrysoscelis	4.67	FCM	RBC
Dryophytes cinereus	4.41	SCF	RBC
	4.69	FD	LV
Dryophytes femoralis	3.83	FD	LV
Dryophytes gratiosus	6.45	FD	LV
Dryophytes japonicus	4.00	FD	RBC
	4.64	FCM	RBC
	5.45	FD	LV
Dryophytes squirellus	3.95	FD	LV
	4.65	FD	LV
Dryophytes versicolor (4n)	9.20	FCM	RBC
	10.30	FD	LV
Hyla arborea	2.40	FD	RBC
	4.37	FD	RBC
	4.40	FD	RBC
	4.76	FCM	RBC
	4.99	FCM	RBC
	5.36	SCF	RBC
	7.00	NS	NS
Hyla savignyi	1.89	FCM	RBC
	4.06	FCM	RBC
	4.73	SCF	RBC
Osteocephalus taurinus	2.40	FCM	RBC
Osteopilus ocellatus	1.83	FD	LV
Osteopilus septentrionalis	1.95	FD	LV
Osteopilus vastus	2.53	FD	LV
Pseudacris brimleyi	4.14	FD	LV
Pseudacris crucifer	3.62	FD	LV
Pseudacris nigrita	4.05	FD	LV
Pseudacris ocularis	4.60	FD	LV
Pseudacris ornata	3.82	FD	LV
Pseudacris regilla	3.81	FD	LV
Pseudacris triseriata	4.37	FD	LV
Pseudis paradoxa	1.54	FD	LV
Scinax fuscomarginatus	2.28	FD	RBC
Scinax ruber	2.79	FCM	RBC
	3.00	FCM	RBC
	4.35	FCM	RBC
Scinax x-signatus	3.76	FCM	RBC
Trachycephalus typhonius	3.35	FCM	RBC
Phyllomedusidae			
Phyllomedusa cf. *bicolor*	7.79	FCM	RBC
Phyllomedusa burmeisteri	6.58	FD	RBC
Phyllomedusa tetraploidea (4n)	13.02	FD	RBC
Pelodryadidae			
Litoria bicolor	4.79	FD	RBC
Ranoidea alboguttata	2.85	FD	RBC

C values are given in pg DNA/diploid nucleus. The values have been calculated from the data compiled in the Animal Genome Size Database [Gregory, 2005b].

FCM, flow cytometry; FD, Feulgen densitometry; LV, liver cells; NS, not specified; RBC, red blood cells; SCF, static cell fluorometry.

cation in the X chromosome [Sharma and Raman, 1973]. Heterochromatin amplification in the Y chromosomes has been observed in the shrew *Suncus murinus* [Yong, 1974], the vole mouse *Akodon urichi* [Reig et al., 1971], the deer mouse *Peromyscus maniculatus* [Bradshaw and Hsu, 1972], and the spiny rat *Trichomys apereoides* [Souza and Yonenaga-Yassuda, 1982]. Heterochromatin amplification on an intrapopulation level has been found in the ground squirrel *Citellus parryi*. Three Y variants (small, medium, and large) have been observed in this species [Vorontsov et al., 1980]. The most frequent is a medium-sized telocentric Y^m chromosome, of the same size as the smallest autosome. This Y^m occurs in all populations in northeastern Siberia. But in some areas of northeastern Siberia and in Alaska, males with a large telocentric Y^l (twice the size of the smallest autosome) were collected. Finally, in the area of Verkhoyansk, a specimen having a dot-like Y^s was found. This Y^s is 3–4 times smaller than the smallest autosome. In the medium-sized Y^m, no constitutive heterochromatin can be demonstrated, but a very large heterochromatic segment is present in the Y^l chromosome. The minute Y^s was not analyzed by C-banding. Apparently, the large Y^l originated by de novo heterochromatin amplification in the Y^m, whereas the minute Y^s arose by a large deletion [Vorontsov et al., 1980].

The best documented example for variation and amplification of constitutive heterochromatin is the human Y chromosome. All the variation in length of the Y chromosome results from variable amounts of heterochromatin [Bobrow et al., 1971; Schnedl, 1971; Sumner, 1990]. The human Y chromosome can either contain very large amounts of heterochromatin or, as an extreme, completely lack any observable heterochromatin. Two specific repetitive DNA sequences (DYZ1, 3.4 kb; DYZ2, 2.1 kb) have been located in the Y heterochromatin [Cooke, 1976; Cooke and McKay, 1978; Schmidtke and Schmid, 1980; Cooke et al., 1983]. Both 3.4-kb and 2.1-kb fragments have been shown to increase in amount with increasing length (amplification) of the Y heterochromatin [McKay et al., 1978; Schmid et al., 1990c]. The question arises whether the length variability of the human Y heterochromatin is an evolving situation or is at equilibrium between amplifications and deletions. The length distribution of the human Y chromosome, within a defined population, seems to be Gaussian [Unnerus et al., 1967; Verma et al., 1978], which would be consistent with an equilibrium situation in which both products of unequal sister chromatid interchange are equally viable. The evolutionary lability of the Y heterochromatin is, however, indicated by interracial differences in the length of the human Y chromosome [for review, see Schmid, 1985]. A wide variety of constitutive heterochromatin variants have been observed in cultured human fibroblasts after mitomycin C treatment, including interstitially located heterochromatin and sister chromatid asymmetry of heterochromatic regions [Hoehn and Martin, 1972, 1973]. This sister chromatid asymmetry is cytological evidence for either unequal sister chromatid interchange or unequal replication of repetitive DNA sequences.

Amplifications (and deletions) of highly repetitive DNA sequences frequently cause polymorphisms of constitutive heterochromatin which are ubiquitous in many eukaryotes. Hence, they are one of the driving forces for karyotype evolution. The more heterochromatic regions are located in the genome of a given species, the higher is the probability that polymorphisms arise. Inbreeding in small natural populations contributes to the fixation of spontaneous amplifications (or deletions) of constitutive heterochromatin.

In the present study on arboranan chromosomes, spontaneous amplifications of constitutive heterochromatin are visible in heterozygous condition in individual frog specimens. These can be observed in one each of the homologues 12 of *Smilisca sordida* (Fig. 386, 387), in chromosomes 1 of *Tlalocohyla loquax* (Fig. 391), in homologues 5 of *Scinax fuscovarius* (Fig. 434), homologues 7 of *Bokermannohyla alvarengai* (Fig. 443), *Boana albopunctata* (Fig. 456–458), and *Agalychnis dacnicolor* (Fig. 476), homologues 11 of *Litoria meiriana* (Fig. 487), and homologues 5 of *Litoria inermis* (Fig. 491).

Amplifications of constitutive heterochromatin that evolutionarily became homozygous by inbreeding and genetic drift can be found in all individuals in a population. These are very impressive in both homologues 7 of *Ranoidea jungguy* (Fig. 511–513) and homologues 11 of *Ranoidea genimaculata* (Fig. 519–521), *R. eucnemis* (Fig. 522), and *R. novaehollandiae* (Fig. 542–544).

3.4.10 Genome Sizes in Arboranae

The genome sizes of 45 arboranan species that were analyzed range from 1.54 pg DNA/nucleus in *Pseudis paradoxa* to 7.79 pg DNA/nucleus in *Phyllomedusa* cf. *bicolor*, with a mean of 6.19 pg DNA/nucleus (Table 10). The tetraploid species *Dryophytes versicolor* and *Phyllomedusa tetraploidea* have values of 9.20 and 13.02, respectively.

Known genome sizes in the order Anura range from 1.90 pg DNA/nucleus to 19.71 pg DNA/nucleus. The frequencies of the genome sizes show a clear unimodal distribution, with a modal value between 8 and 10 pg DNA/nucleus. The genome sizes of arboranans seem to fit quite well in this frequency distribution (Table 10). It should be noted, however, that these 45 species represent only a small fraction of the extant species described and that in the majority of arboranan genera no genome size measurements were performed by now. Hence, not until the size of the nuclear genome of many more species of Arboranae has been measured, the possible variability of genome sizes in these frogs will remain incomplete. Quantification of nuclear genome sizes using DAPI- or ethidiumbromide-stained cell nuclei and DNA flow cytometric measurements is a noninvasive technique that circumvents major intervention or sacrification of animals [Lamatsch et al., 2000], so quantification of genome sizes in further treefrogs does not represent a major problem.

Figs. 53–937. The normal mitotic karyotypes of the examined arboranans showing conventional Giemsa staining, C-banding of constitutive heterochromatin, Ag-staining of NORs, quinacrine mustard staining, DAPI staining, distamycin A/DAPI counterstaining, distamycin A/mithramycin counterstaining, and FISH with telomeric oligonucleotide probes $(GGGTTA)_7/(TAACCC)_7$. The subfamilies and genera of all 3 families (Hylidae, Phyllomedusidae, Pelodryadidae) are arranged phyllogenetically according to the maximum likelihood phylogeny of arboranan frogs (Fig. 1) and the timetree of arboranan frogs (Fig. 2). Within the genera of the families Hylidae and Phyllomedusidae, the species are arranged alphabetically unless rearrangements in their karyotypes imply an alternative order. In the family Pelodryadidae (genera *Litoria* and *Ranoidea*), the species are grouped into several species groups which best differentiate their karyotype characteristics. The symbols and abbreviations used in the figures for these mitotic karyotypes are explained at the end of chapter 3.1.

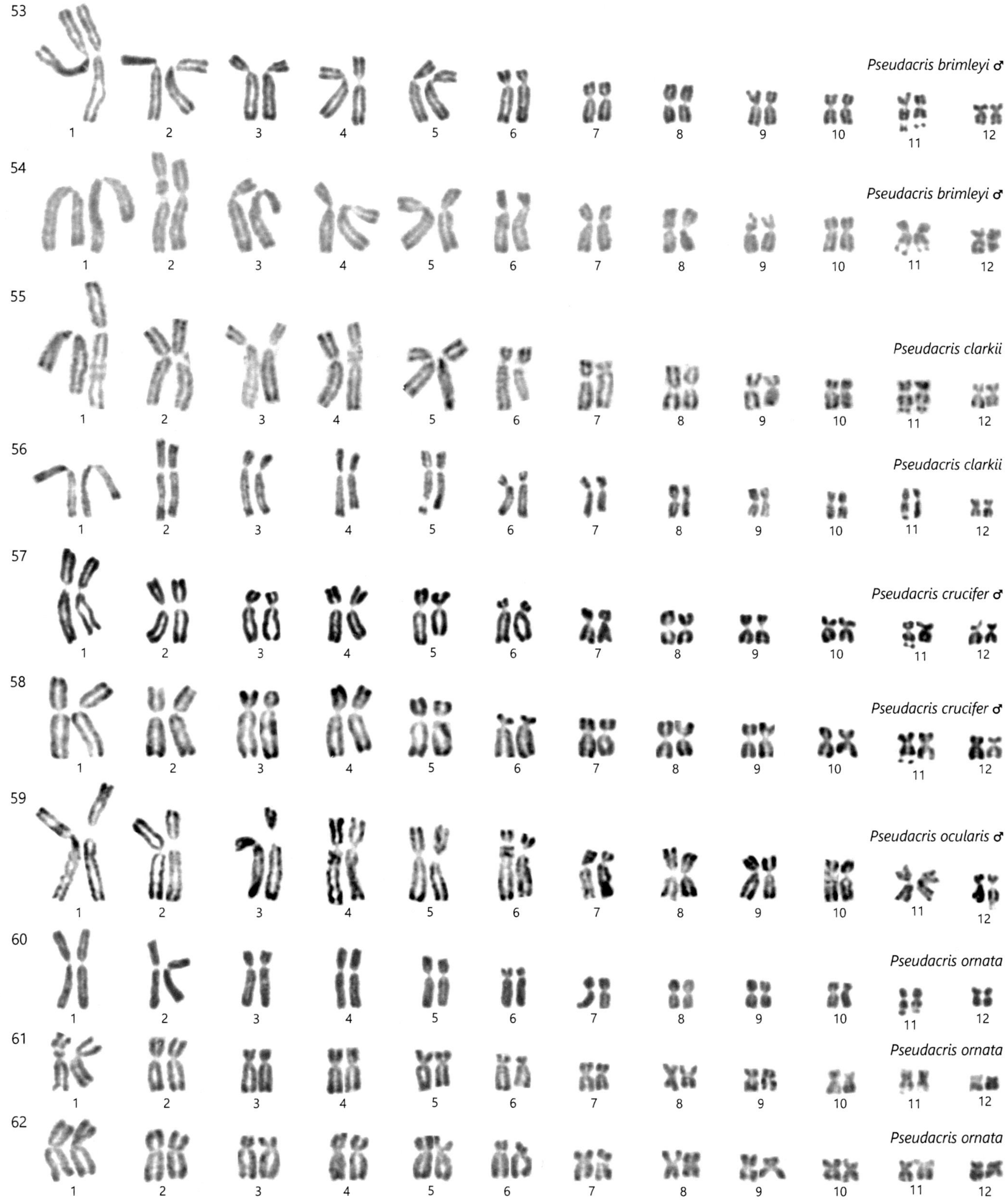

Cytogenet Genome Res 2018;155:55–221
DOI: 10.1159/000489839

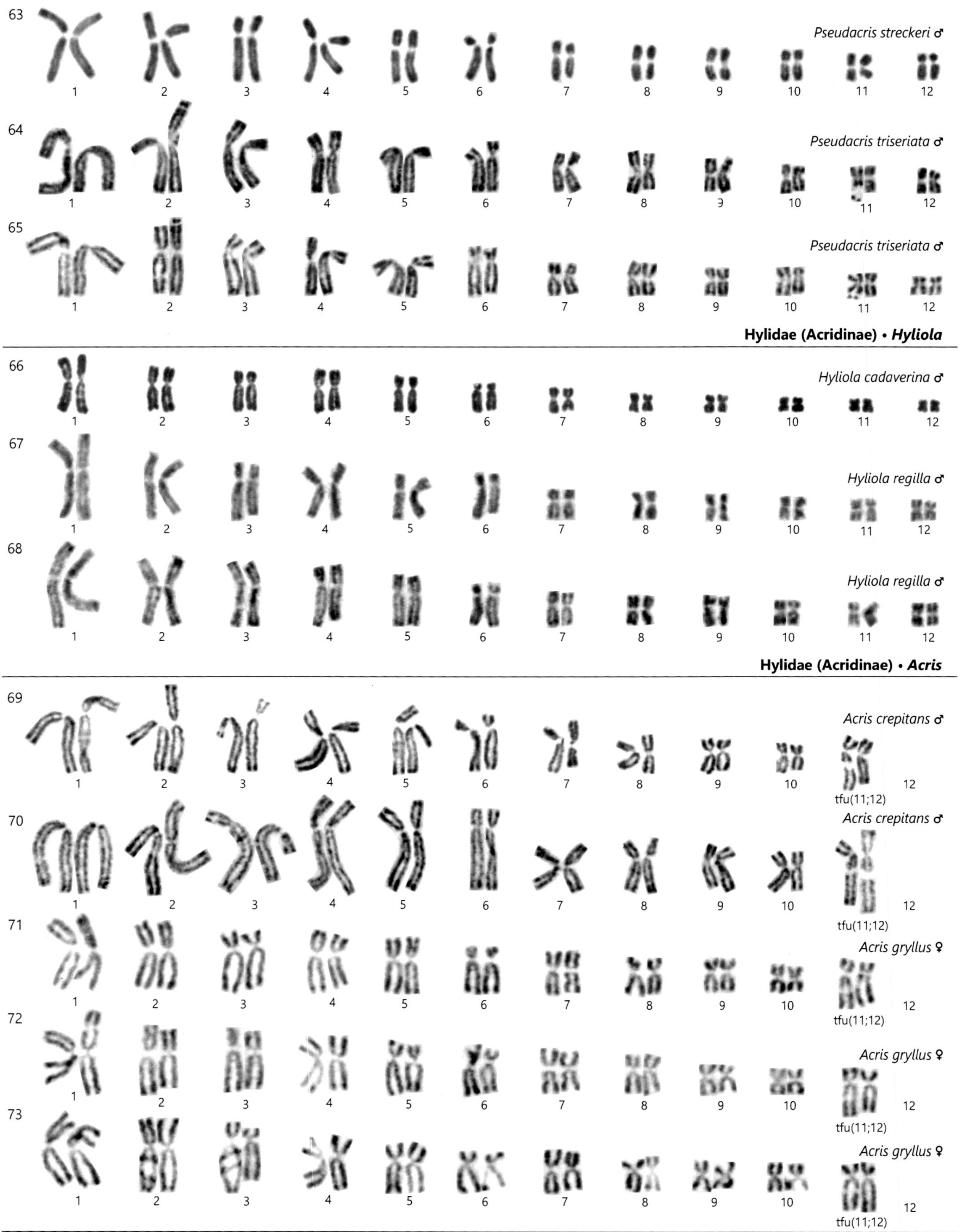
63 *Pseudacris streckeri* ♂
64 *Pseudacris triseriata* ♂
65 *Pseudacris triseriata* ♂
Hylidae (Acridinae) • *Hyliola*
66 *Hyliola cadaverina* ♂
67 *Hyliola regilla* ♂
68 *Hyliola regilla* ♂
Hylidae (Acridinae) • *Acris*
69 *Acris crepitans* ♂
tfu(11;12)
70 *Acris crepitans* ♂
tfu(11;12)
71 *Acris gryllus* ♀
tfu(11;12)
72 *Acris gryllus* ♀
tfu(11;12)
73 *Acris gryllus* ♀
tfu(11;12)

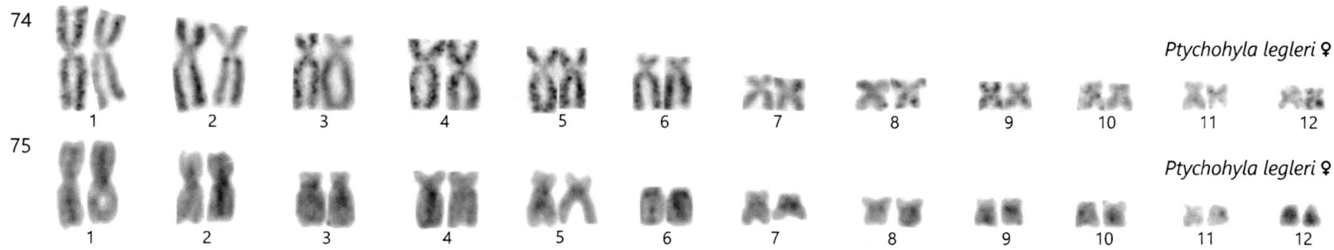

Hylidae (Hylinae) • *Smilisca*

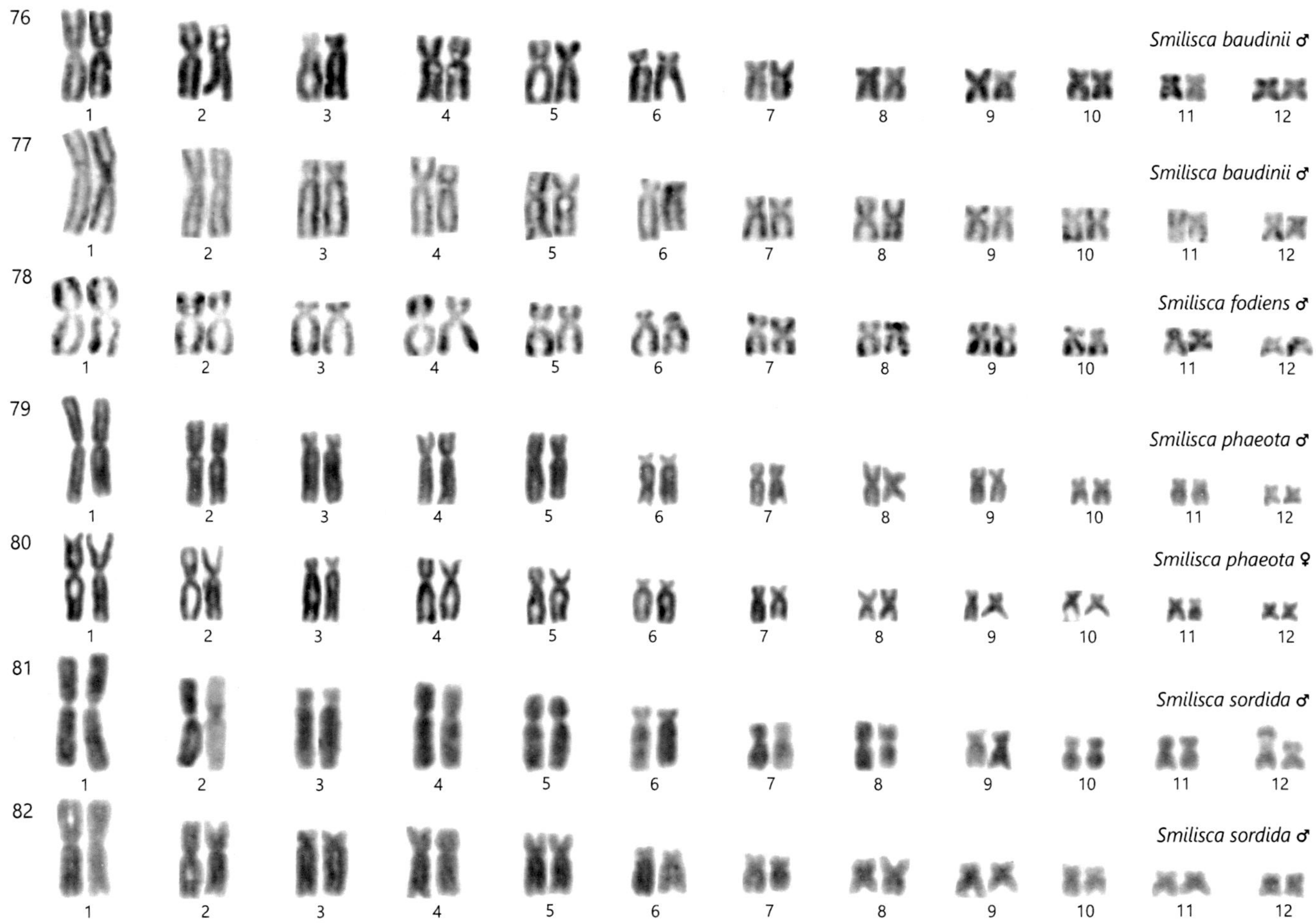

Hylidae (Hylinae) • *Anotheca*

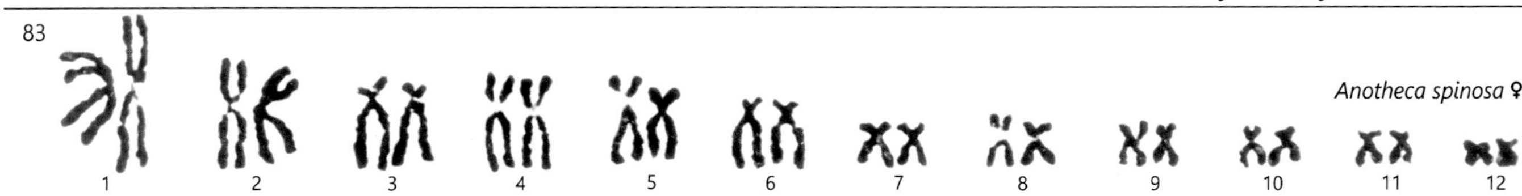

Hylidae (Hylinae) • *Isthmohyla*

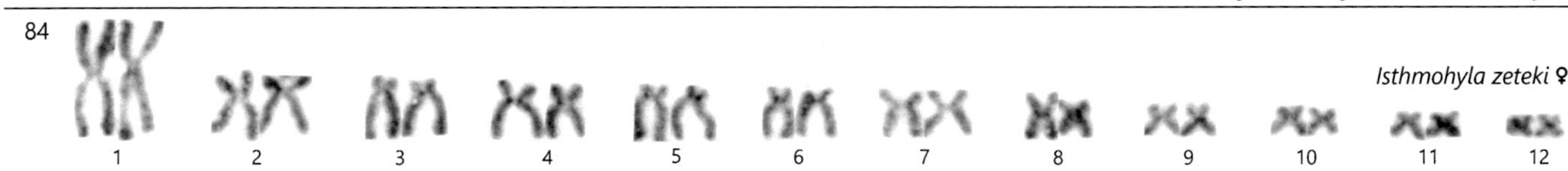

DOI: 10.1159/000489839

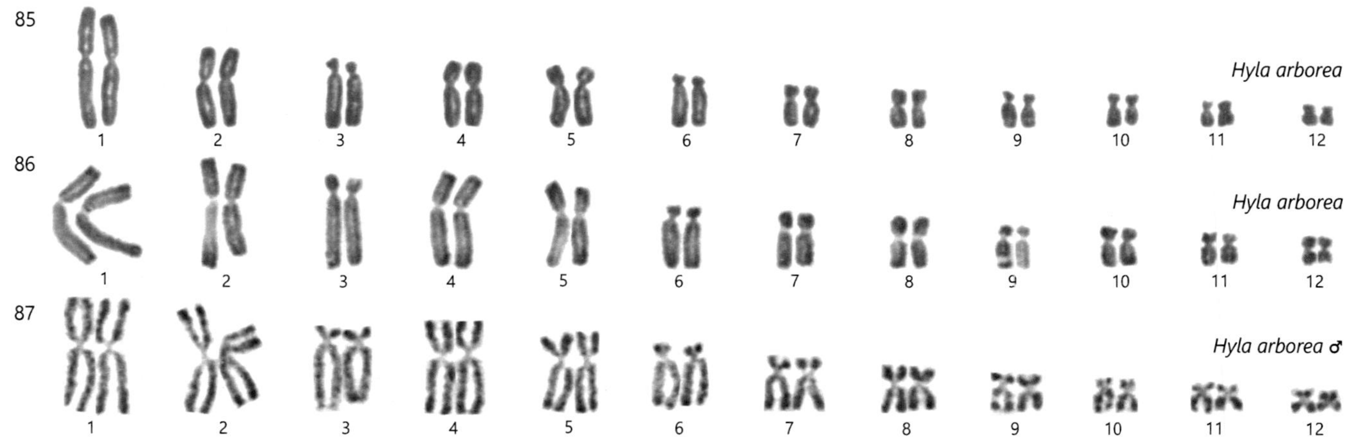

Hylidae (Hylinae) • *Dryophytes*

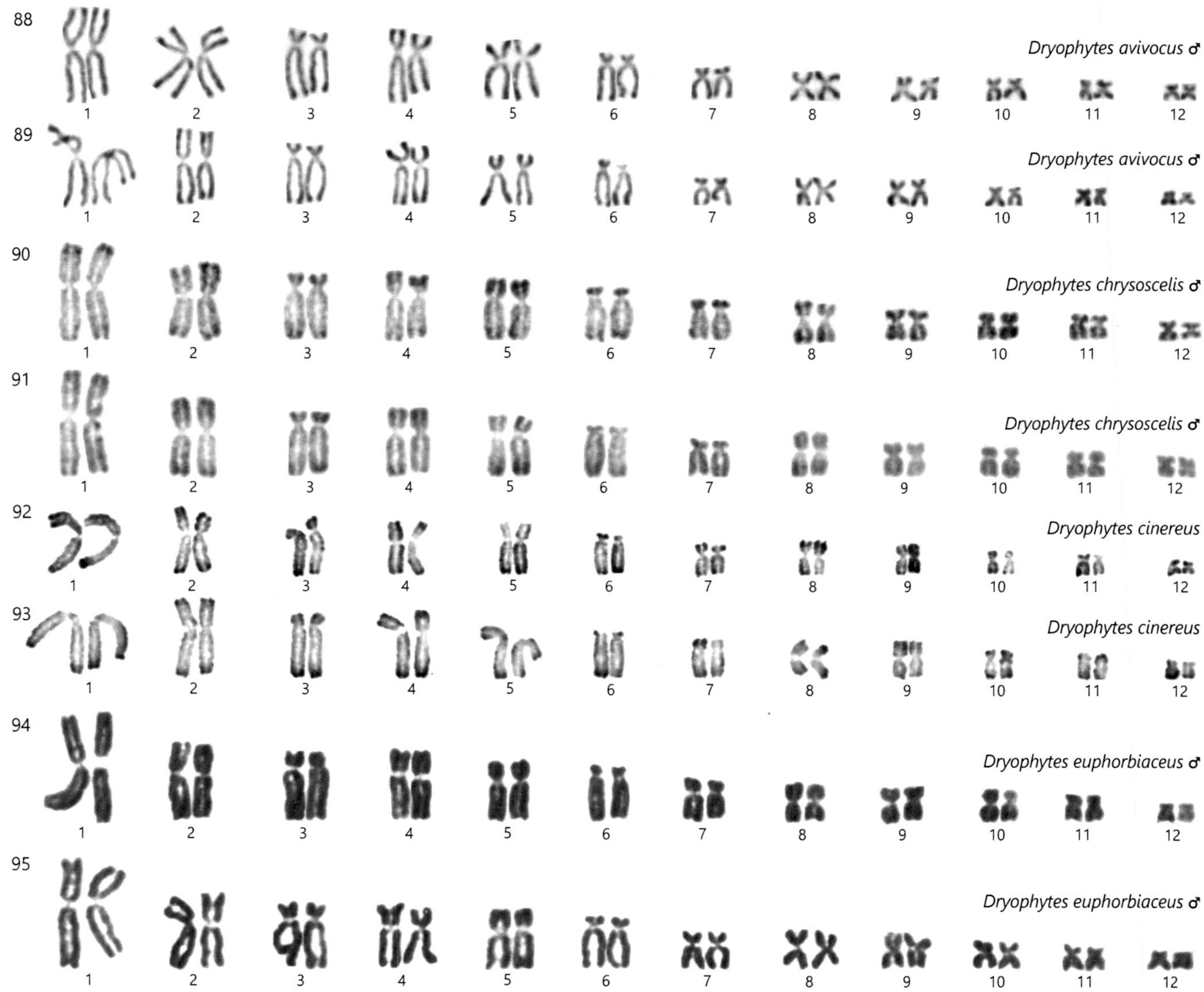

DOI: 10.1159/000489839

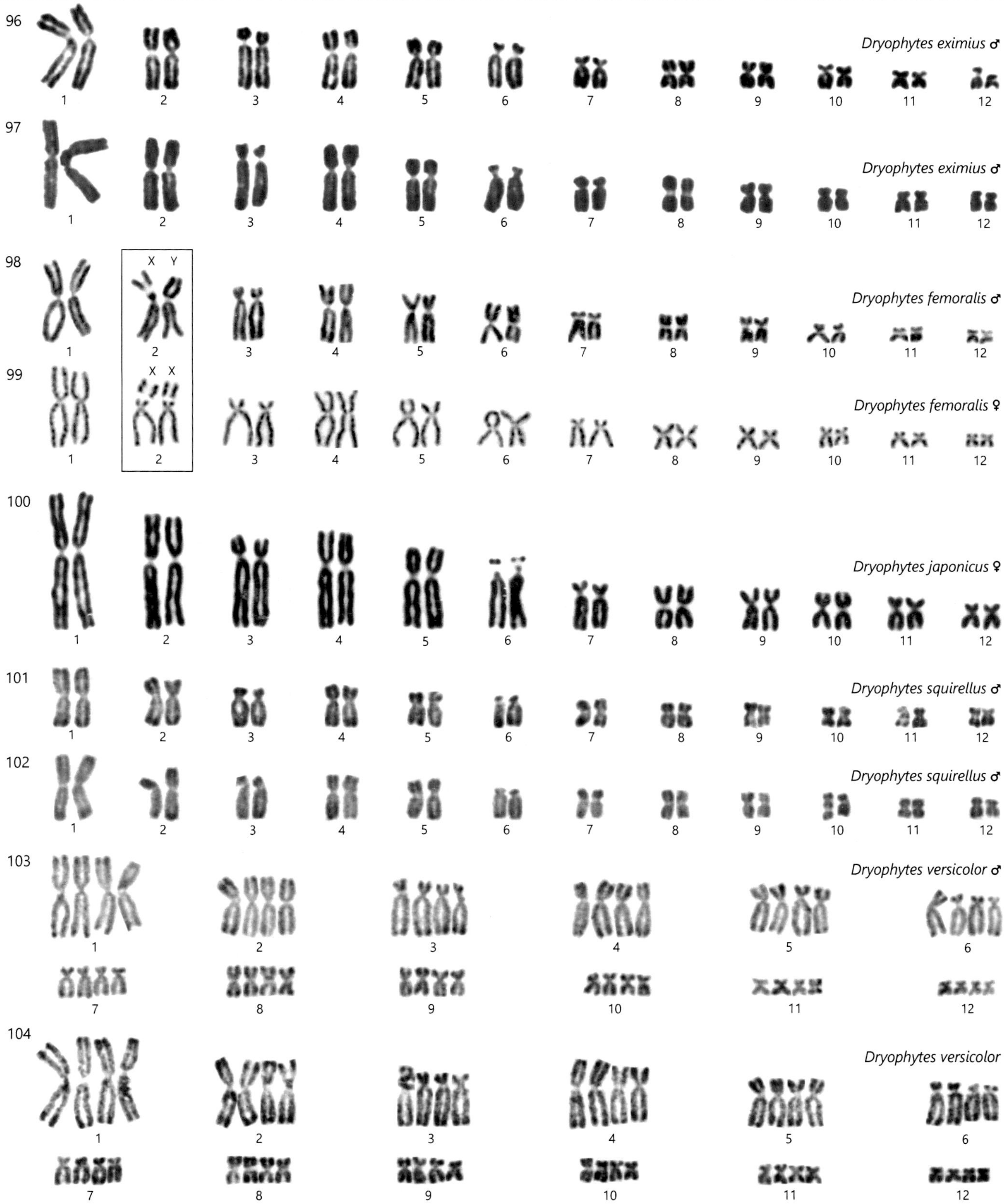
96
Dryophytes eximius ♂
97
Dryophytes eximius ♂
98
X Y
Dryophytes femoralis ♂
99
X X
Dryophytes femoralis ♀
100
Dryophytes japonicus ♀
101
Dryophytes squirellus ♂
102
Dryophytes squirellus ♂
103
Dryophytes versicolor ♂
104
Dryophytes versicolor

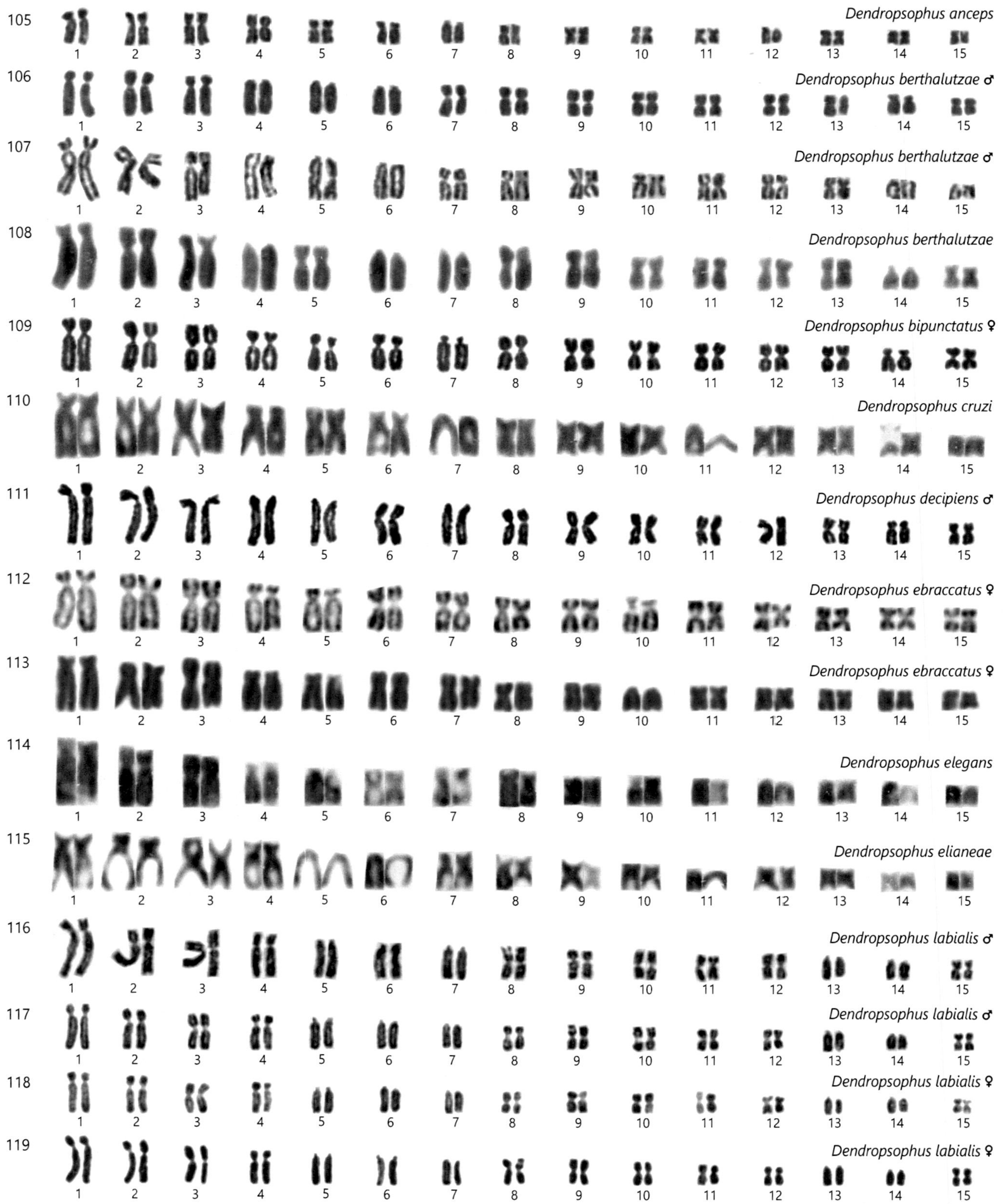

DOI: 10.1159/000489839

DOI: 10.1159/000489839

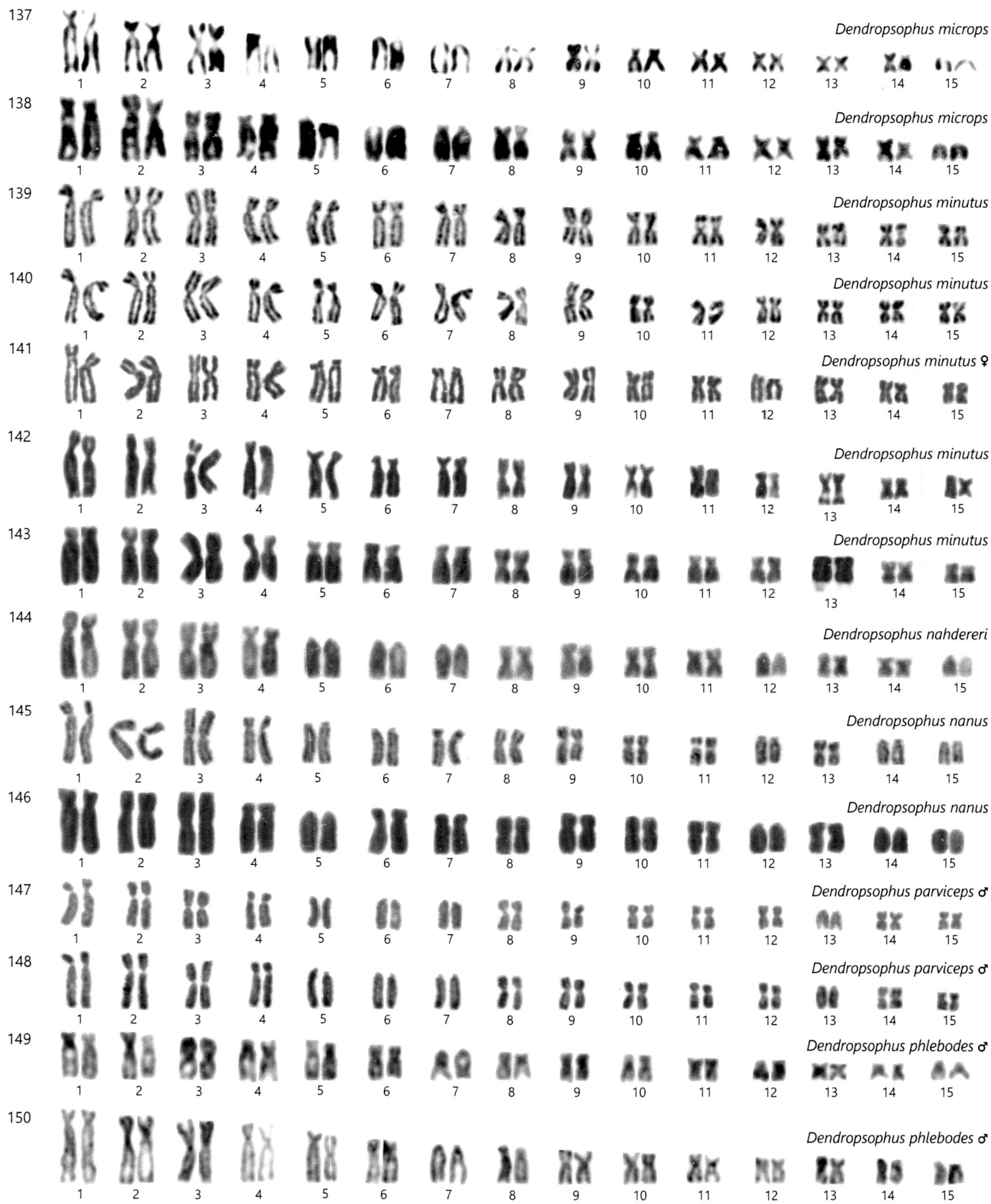
137 Dendropsophus microps
138 Dendropsophus microps
139 Dendropsophus minutus
140 Dendropsophus minutus
141 Dendropsophus minutus ♀
142 Dendropsophus minutus
143 Dendropsophus minutus
144 Dendropsophus nahdereri
145 Dendropsophus nanus
146 Dendropsophus nanus
147 Dendropsophus parviceps ♂
148 Dendropsophus parviceps ♂
149 Dendropsophus phlebodes ♂
150 Dendropsophus phlebodes ♂

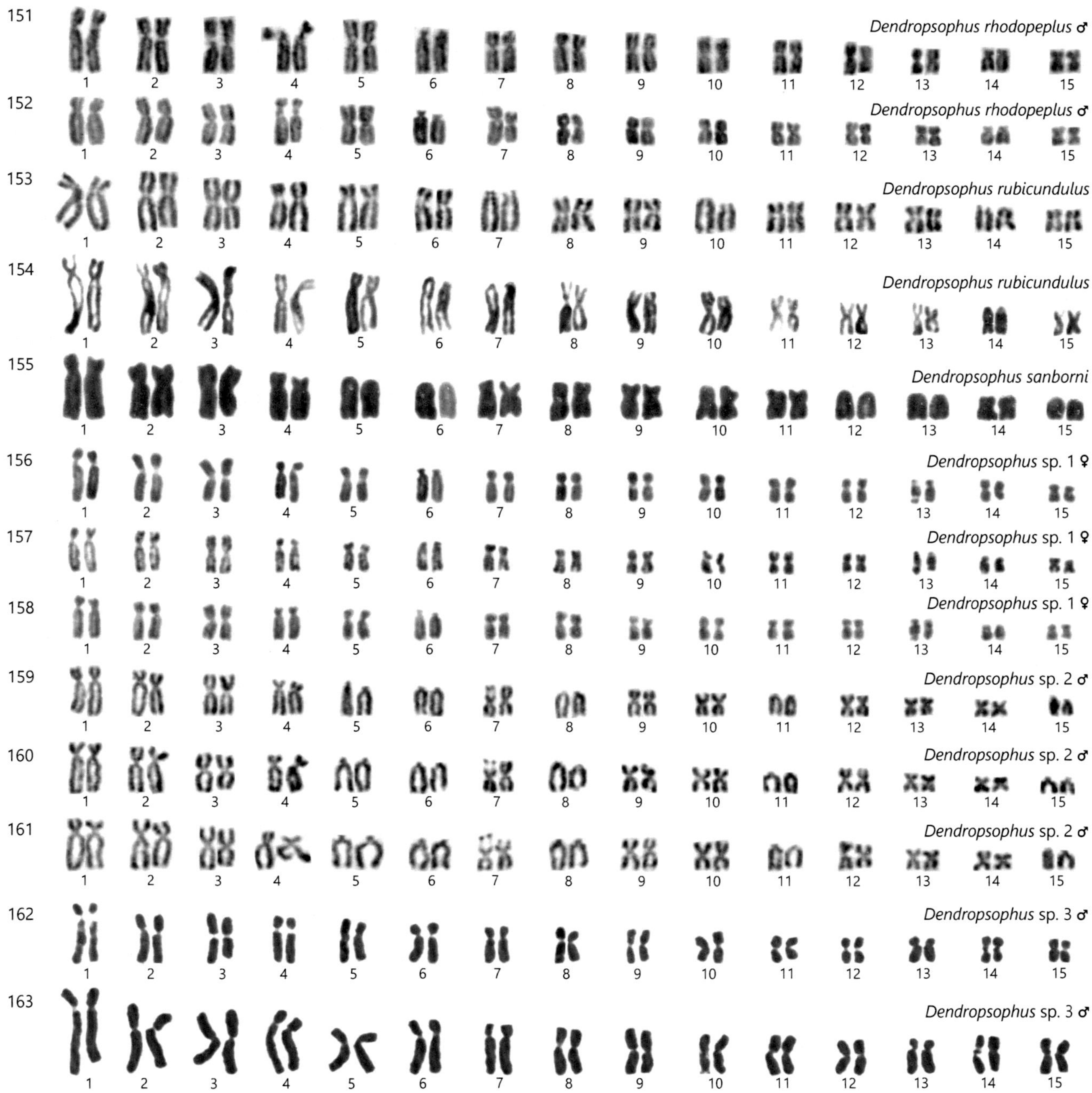

Hylidae (Lophyohylinae) • *Phyllodytes*

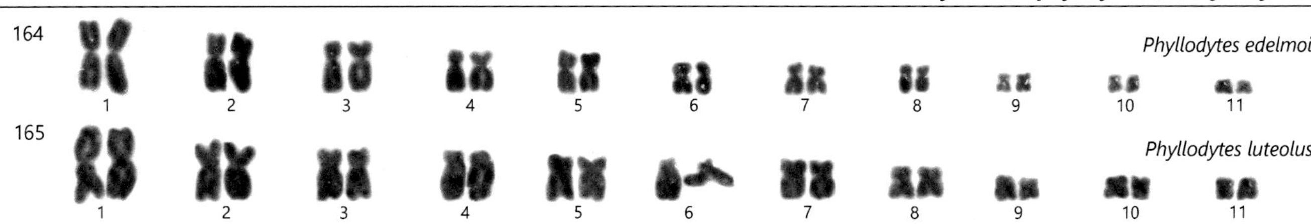

DOI: 10.1159/000489839

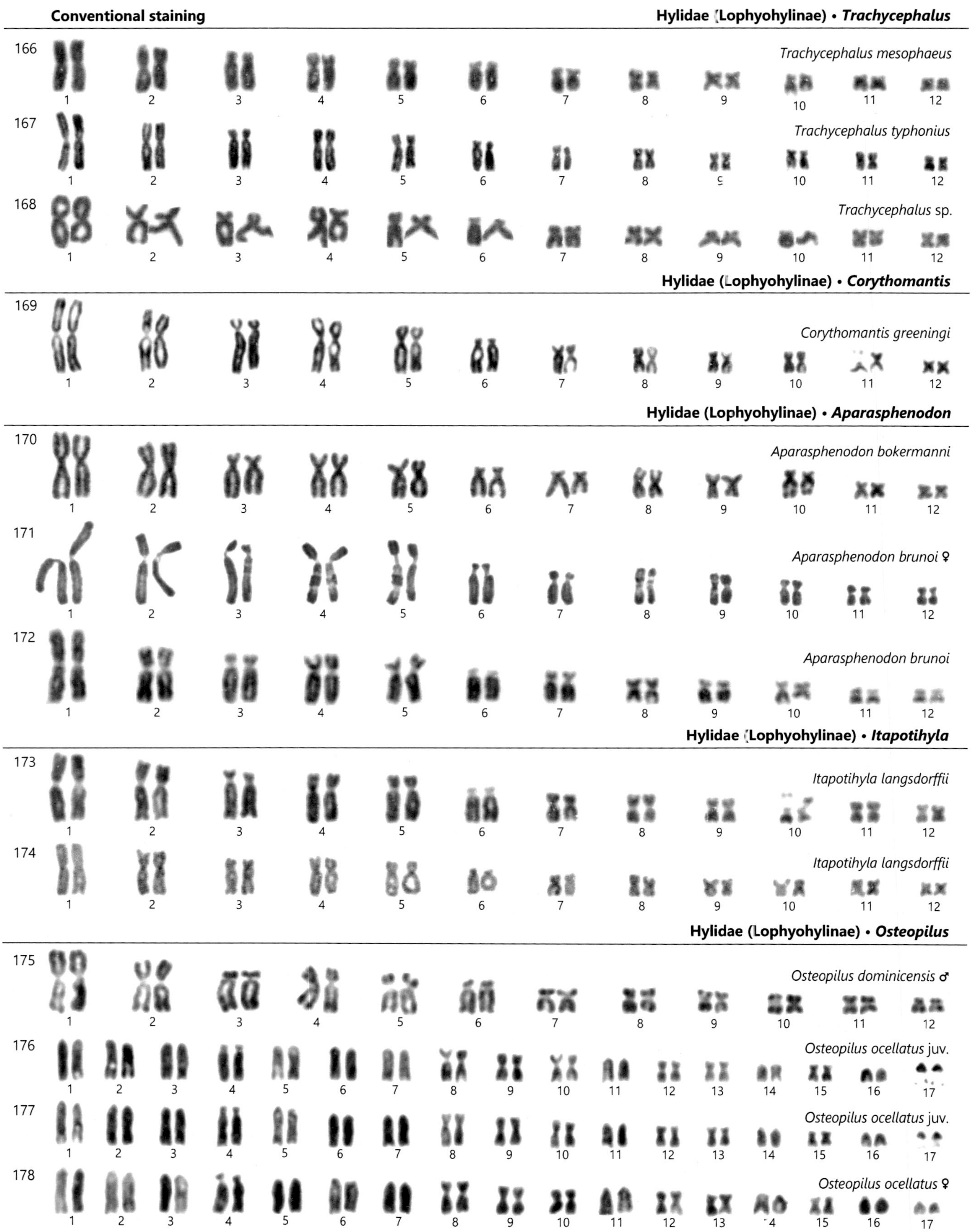

DOI: 10.1159/000489839

179
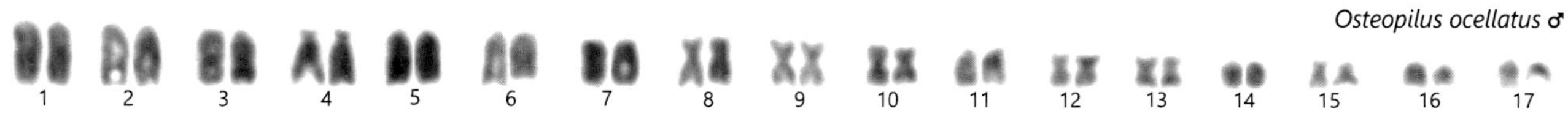

180
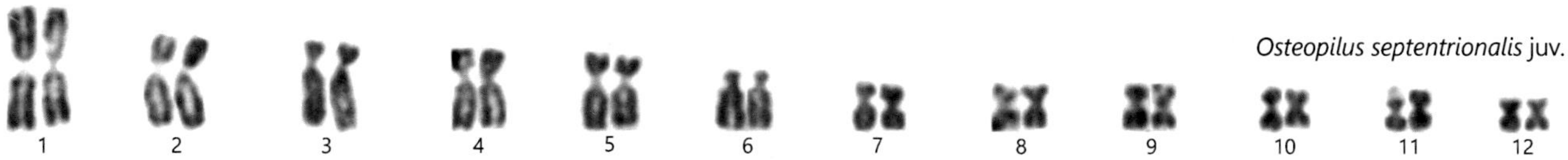

181
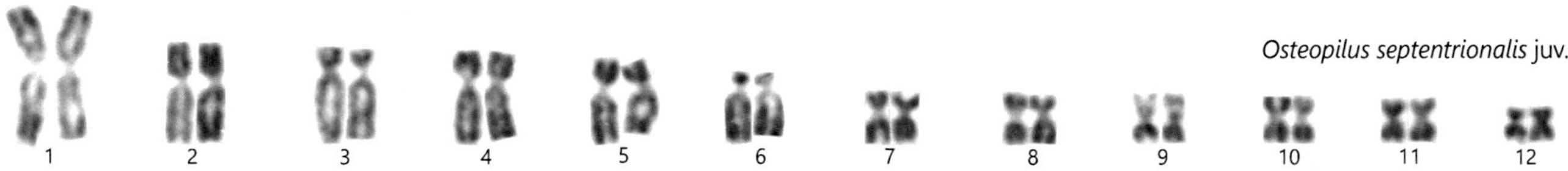

182
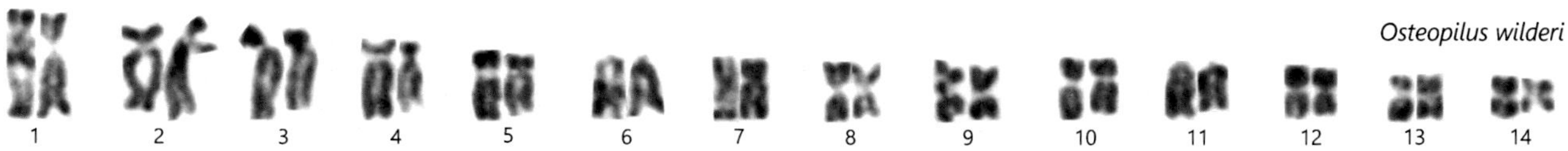

Hylidae (Lophyohylinae) • *Osteocephalus*

183
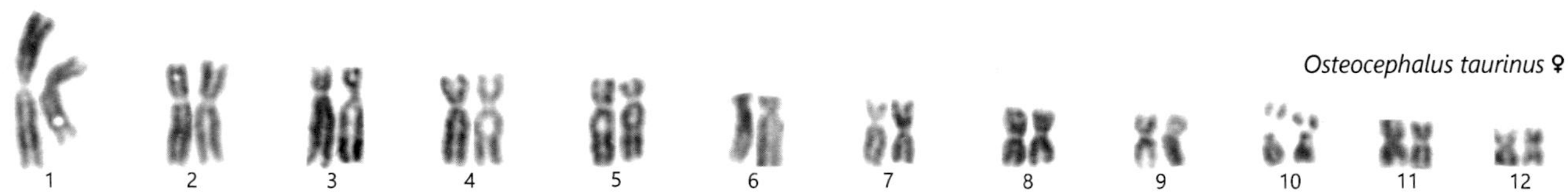

184
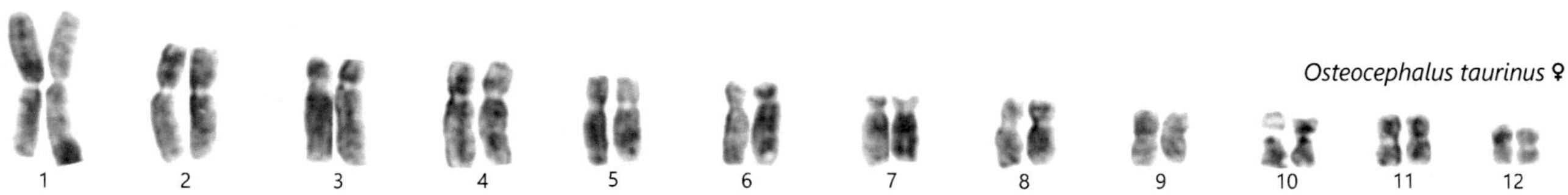

Hylidae (Scinaxinae) • *Sphaenorhynchus*

185
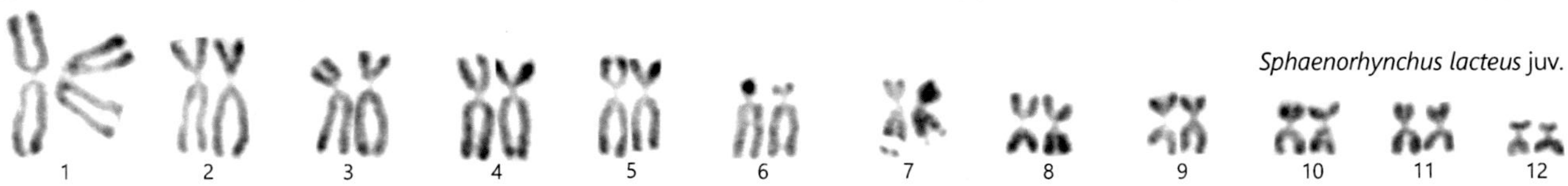

186
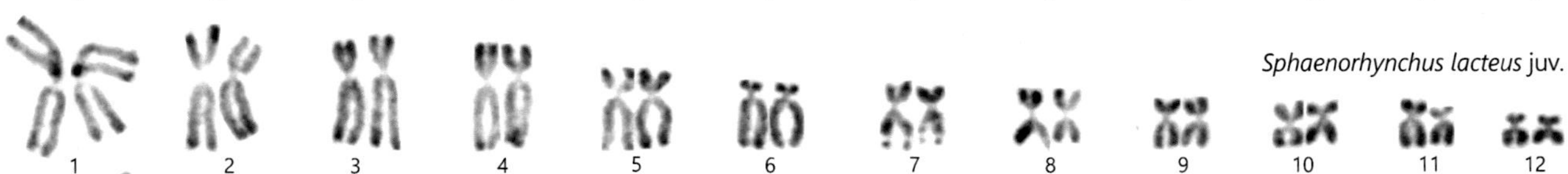

187
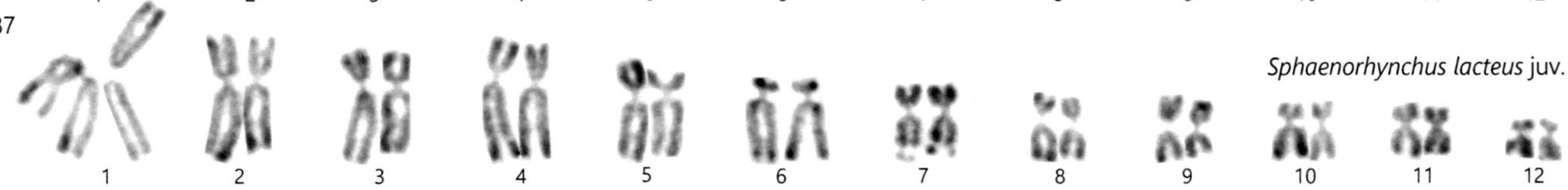

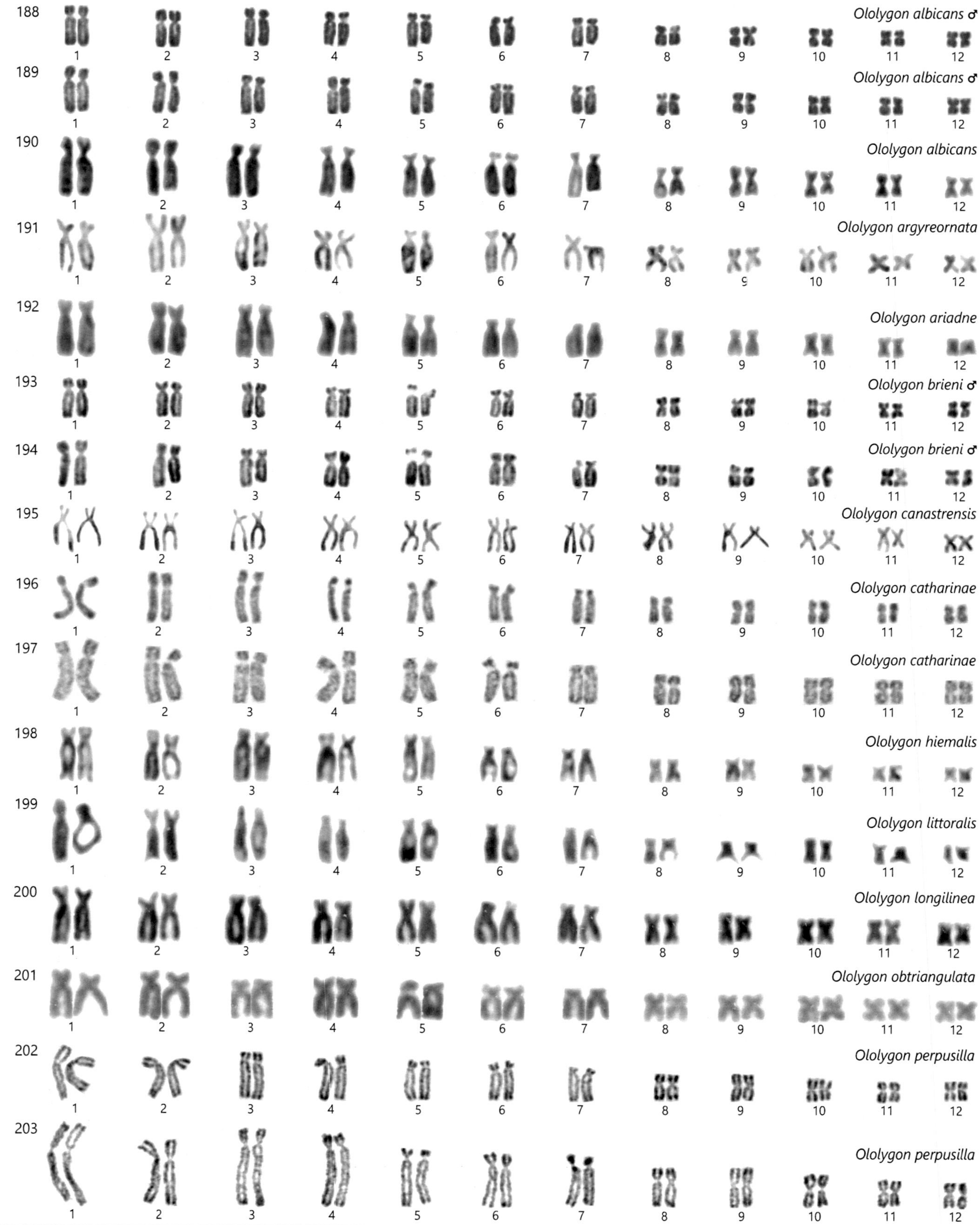
188 Ololygon albicans ♂
189 Ololygon albicans ♂
190 Ololygon albicans
191 Ololygon argyreornata
192 Ololygon ariadne
193 Ololygon brieni ♂
194 Ololygon brieni ♂
195 Ololygon canastrensis
196 Ololygon catharinae
197 Ololygon catharinae
198 Ololygon hiemalis
199 Ololygon littoralis
200 Ololygon longilinea
201 Ololygon obtriangulata
202 Ololygon perpusilla
203 Ololygon perpusilla

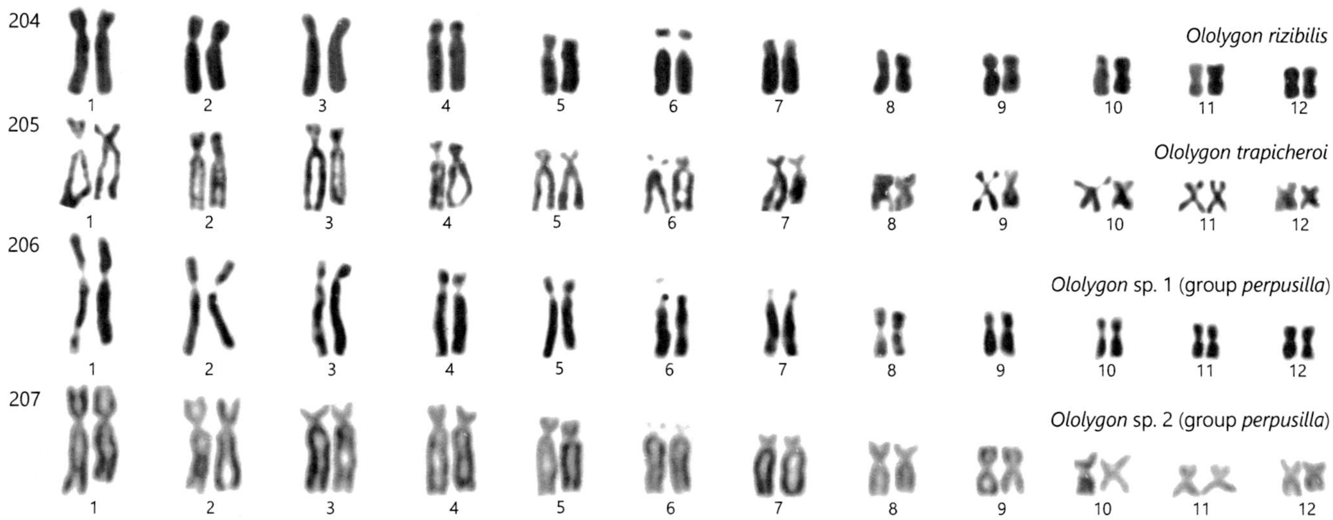

Hylidae (Scinaxinae) • *Scinax*

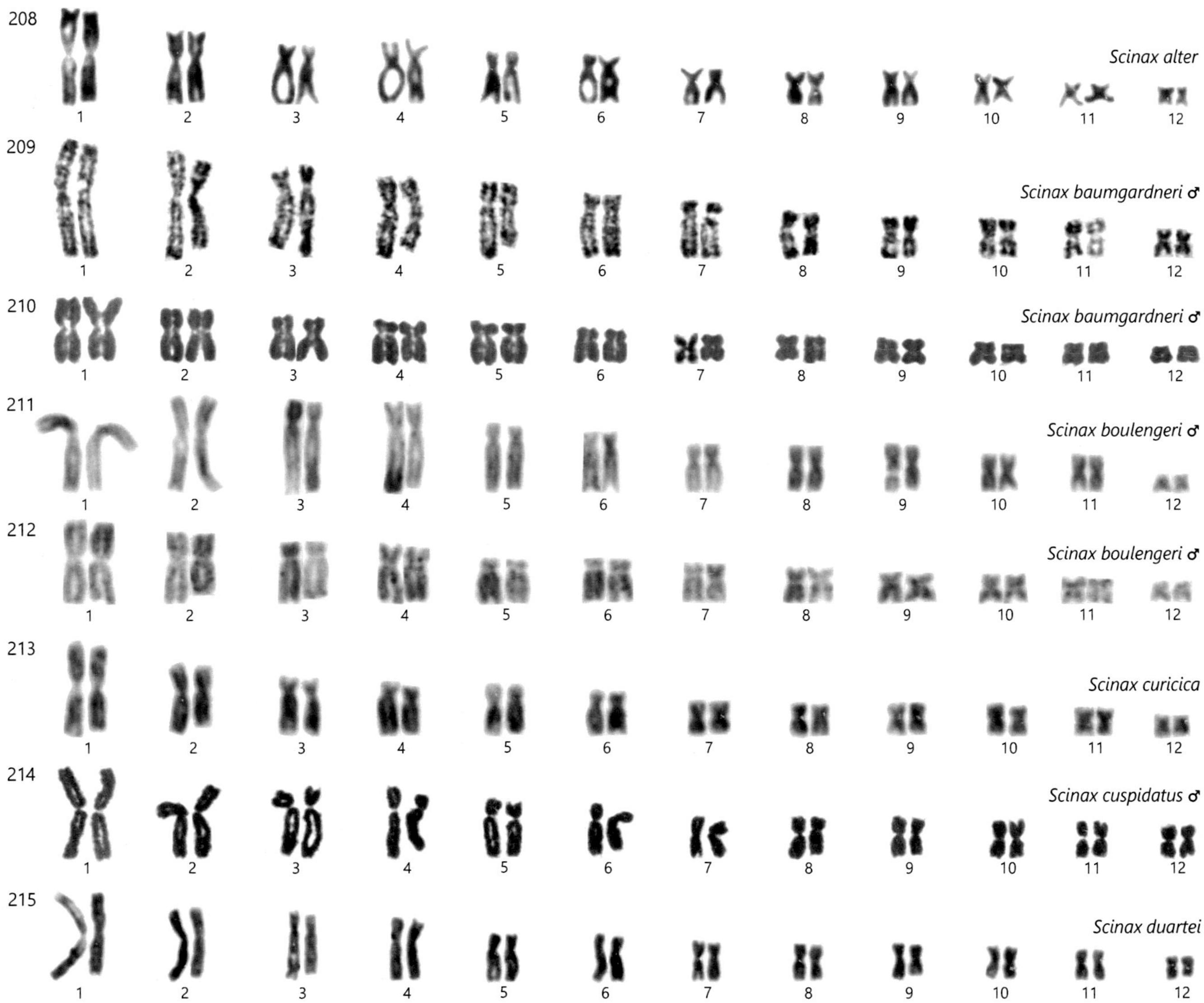

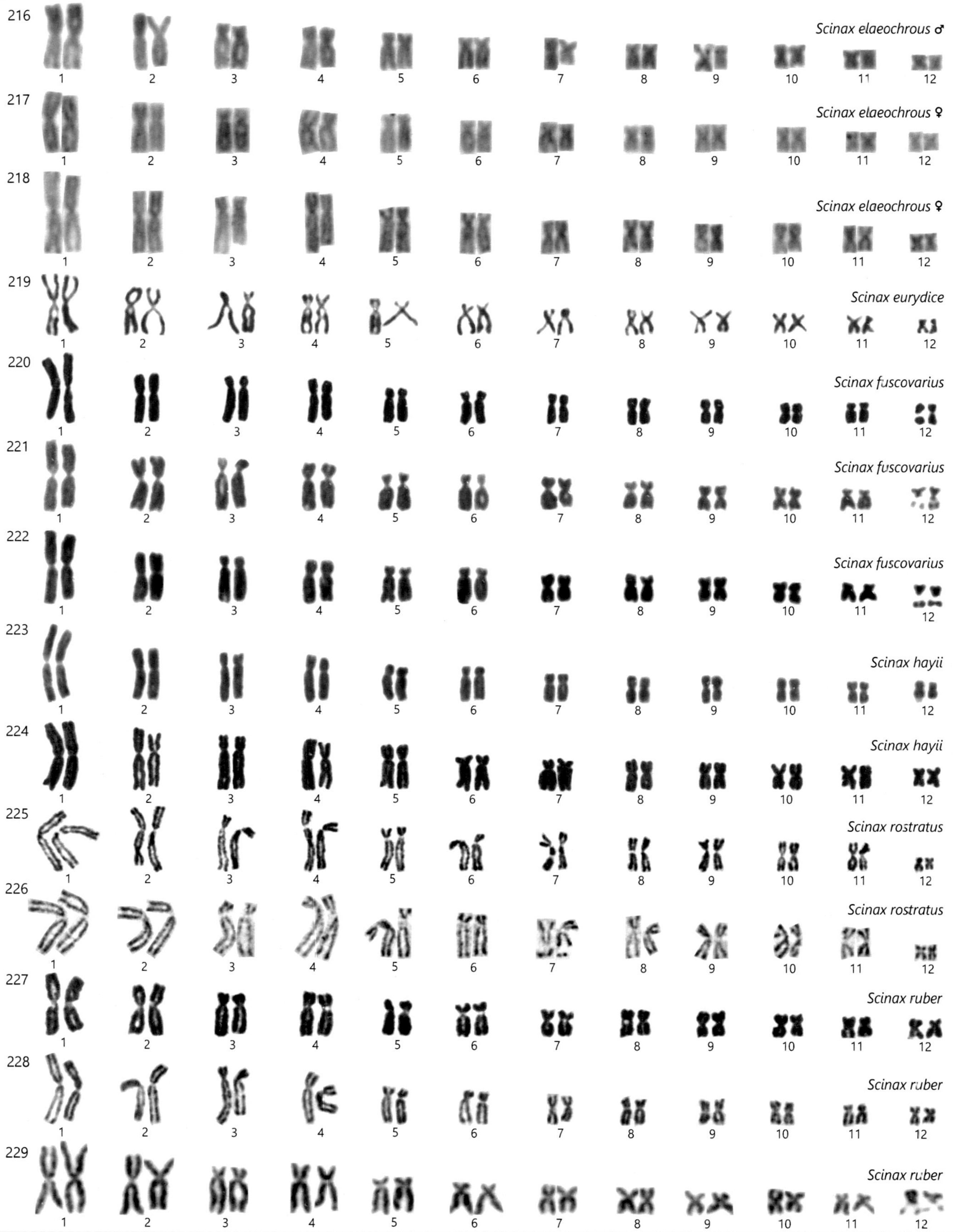
216
Scinax elaeochrous ♂
1 2 3 4 5 6 7 8 9 10 11 12
217
Scinax elaeochrous ♀
1 2 3 4 5 6 7 8 9 10 11 12
218
Scinax elaeochrous ♀
1 2 3 4 5 6 7 8 9 10 11 12
219
Scinax eurydice
1 2 3 4 5 6 7 8 9 10 11 12
220
Scinax fuscovarius
1 2 3 4 5 6 7 8 9 10 11 12
221
Scinax fuscovarius
1 2 3 4 5 6 7 8 9 10 11 12
222
Scinax fuscovarius
1 2 3 4 5 6 7 8 9 10 11 12
223
Scinax hayii
1 2 3 4 5 6 7 8 9 10 11 12
224
Scinax hayii
1 2 3 4 5 6 7 8 9 10 11 12
225
Scinax rostratus
1 2 3 4 5 6 7 8 9 10 11 12
226
Scinax rostratus
1 2 3 4 5 6 7 8 9 10 11 12
227
Scinax ruber
1 2 3 4 5 6 7 8 9 10 11 12
228
Scinax ruber
1 2 3 4 5 6 7 8 9 10 11 12
229
Scinax ruber
1 2 3 4 5 6 7 8 9 10 11 12

Hylidae (Scinaxinae) • *Scinax*

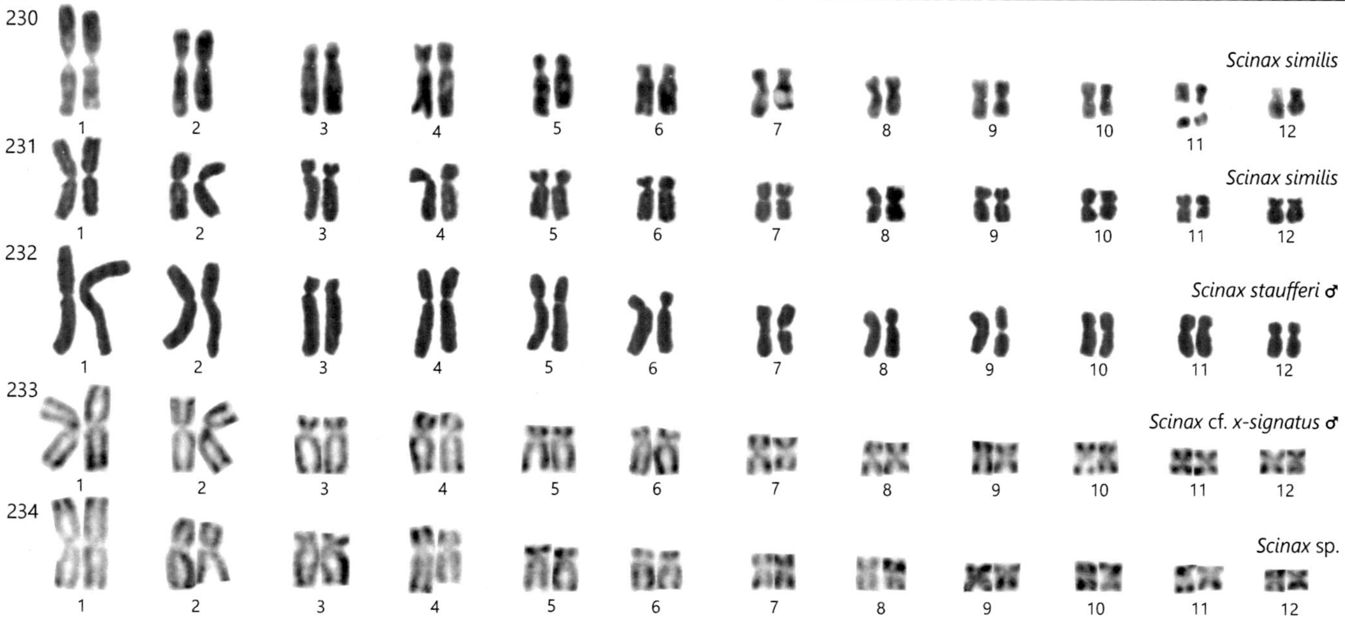

Hylidae (Cophomantinae) • *Hyloscirtus*

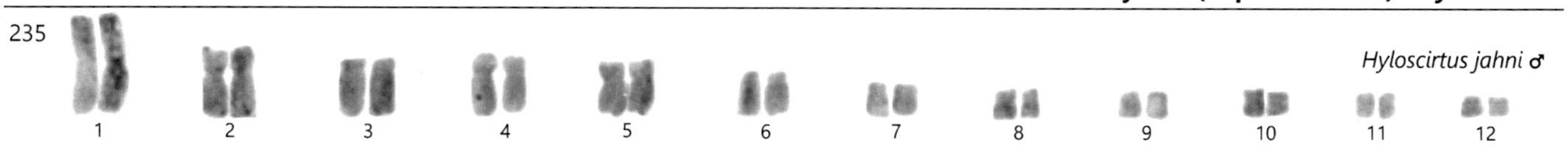

Hylidae (Cophomantinae) • *Bokermannohyla*

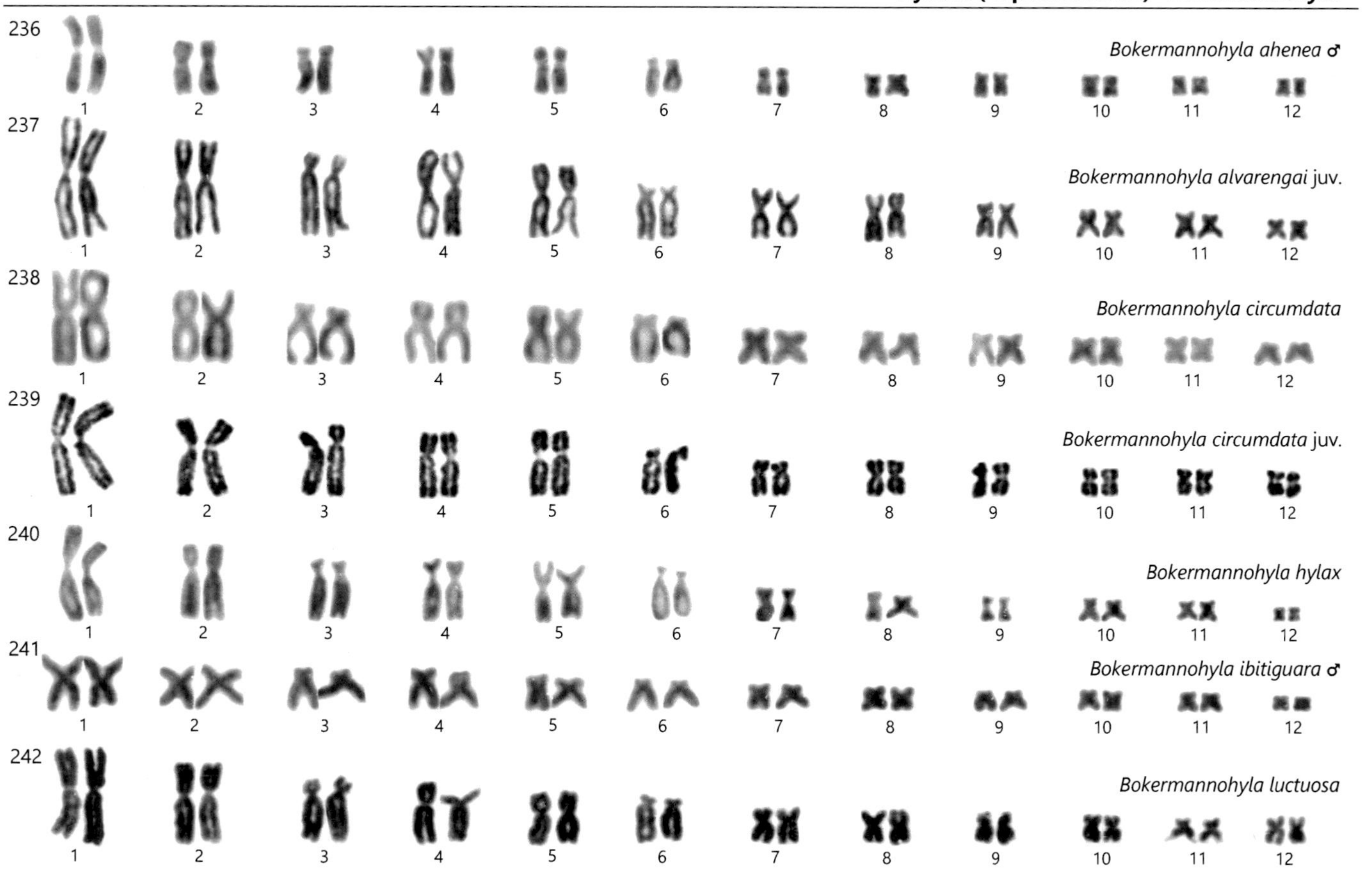

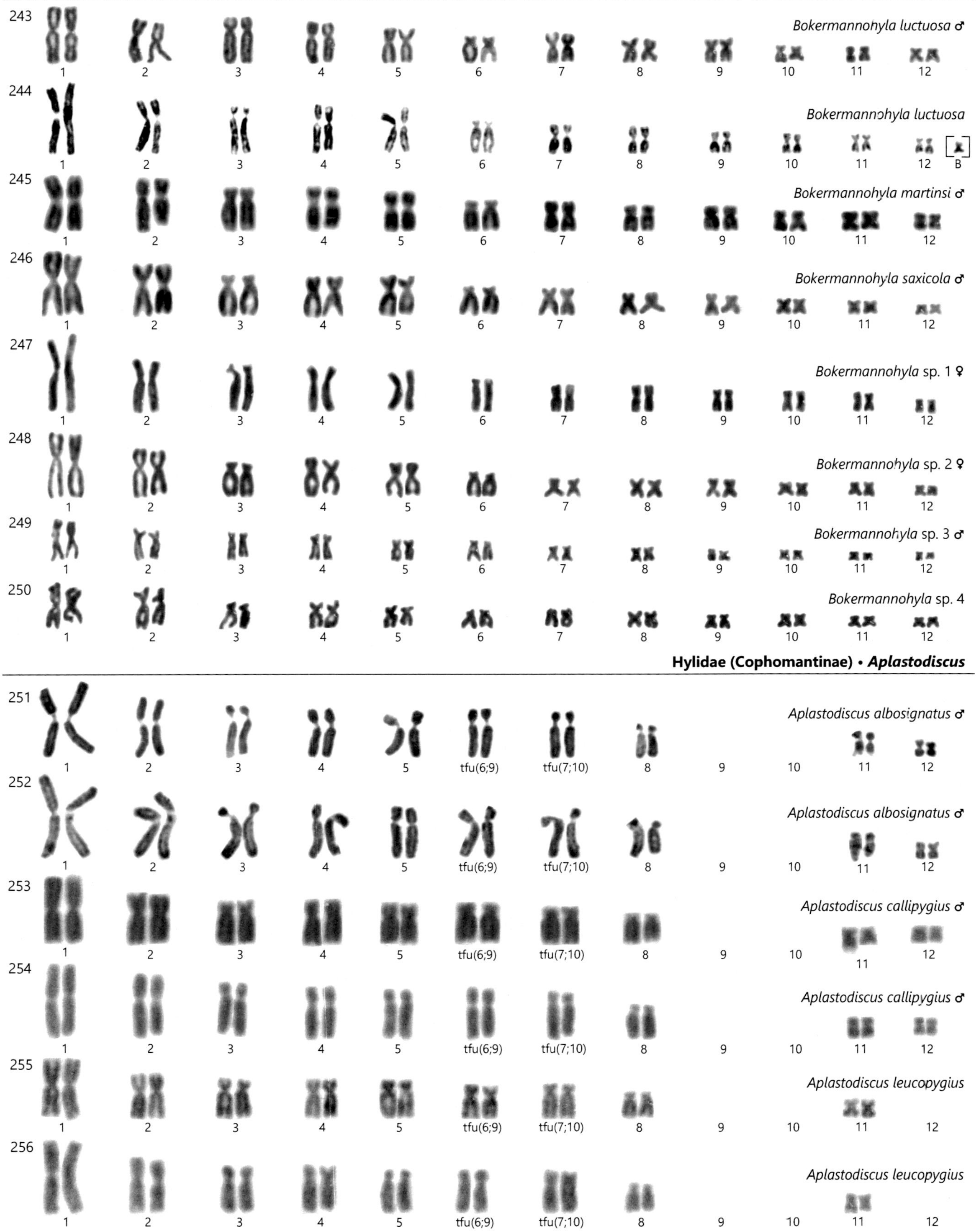
243
Bokermannohyla luctuosa ♂
1 2 3 4 5 6 7 8 9 10 11 12
244
Bokermannohyla luctuosa
1 2 3 4 5 6 7 8 9 10 11 12 B
245
Bokermannohyla martinsi ♂
1 2 3 4 5 6 7 8 9 10 11 12
246
Bokermannohyla saxicola ♂
1 2 3 4 5 6 7 8 9 10 11 12
247
Bokermannohyla sp. 1 ♀
1 2 3 4 5 6 7 8 9 10 11 12
248
Bokermannohyla sp. 2 ♀
1 2 3 4 5 6 7 8 9 10 11 12
249
Bokermannohyla sp. 3 ♂
1 2 3 4 5 6 7 8 9 10 11 12
250
Bokermannohyla sp. 4
1 2 3 4 5 6 7 8 9 10 11 12
Hylidae (Cophomantinae) • *Aplastodiscus*
251
Aplastodiscus albosignatus ♂
1 2 3 4 5 tfu(6;9) tfu(7;10) 8 9 10 11 12
252
Aplastodiscus albosignatus ♂
1 2 3 4 5 tfu(6;9) tfu(7;10) 8 9 10 11 12
253
Aplastodiscus callipygius ♂
1 2 3 4 5 tfu(6;9) tfu(7;10) 8 9 10 11 12
254
Aplastodiscus callipygius ♂
1 2 3 4 5 tfu(6;9) tfu(7;10) 8 9 10 11 12
255
Aplastodiscus leucopygius
1 2 3 4 5 tfu(6;9) tfu(7;10) 8 9 10 11 12
256
Aplastodiscus leucopygius
1 2 3 4 5 tfu(6;9) tfu(7;10) 8 9 10 11 12

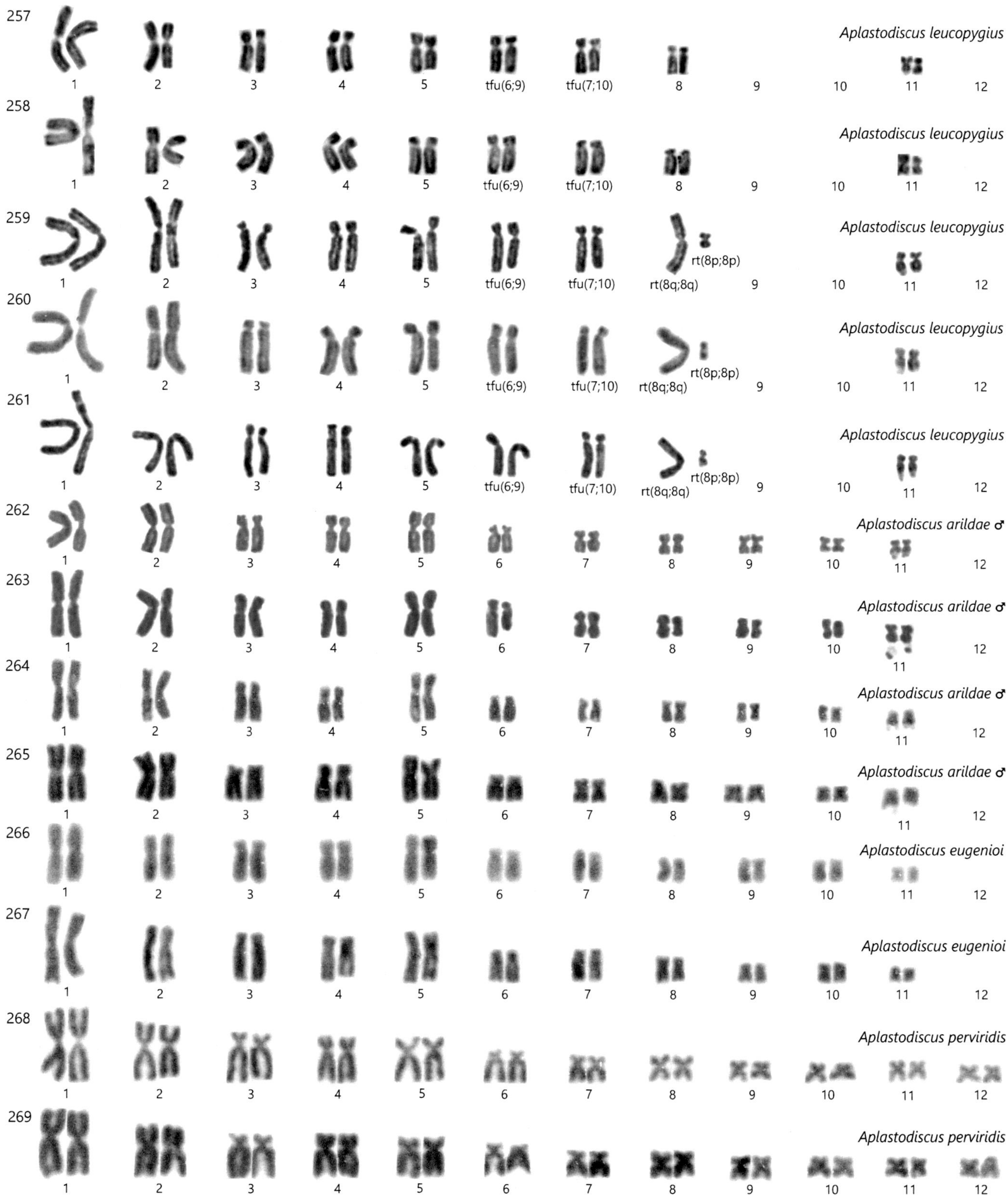

Cytogenet Genome Res 2018;155:55–221
DOI: 10.1159/000489839

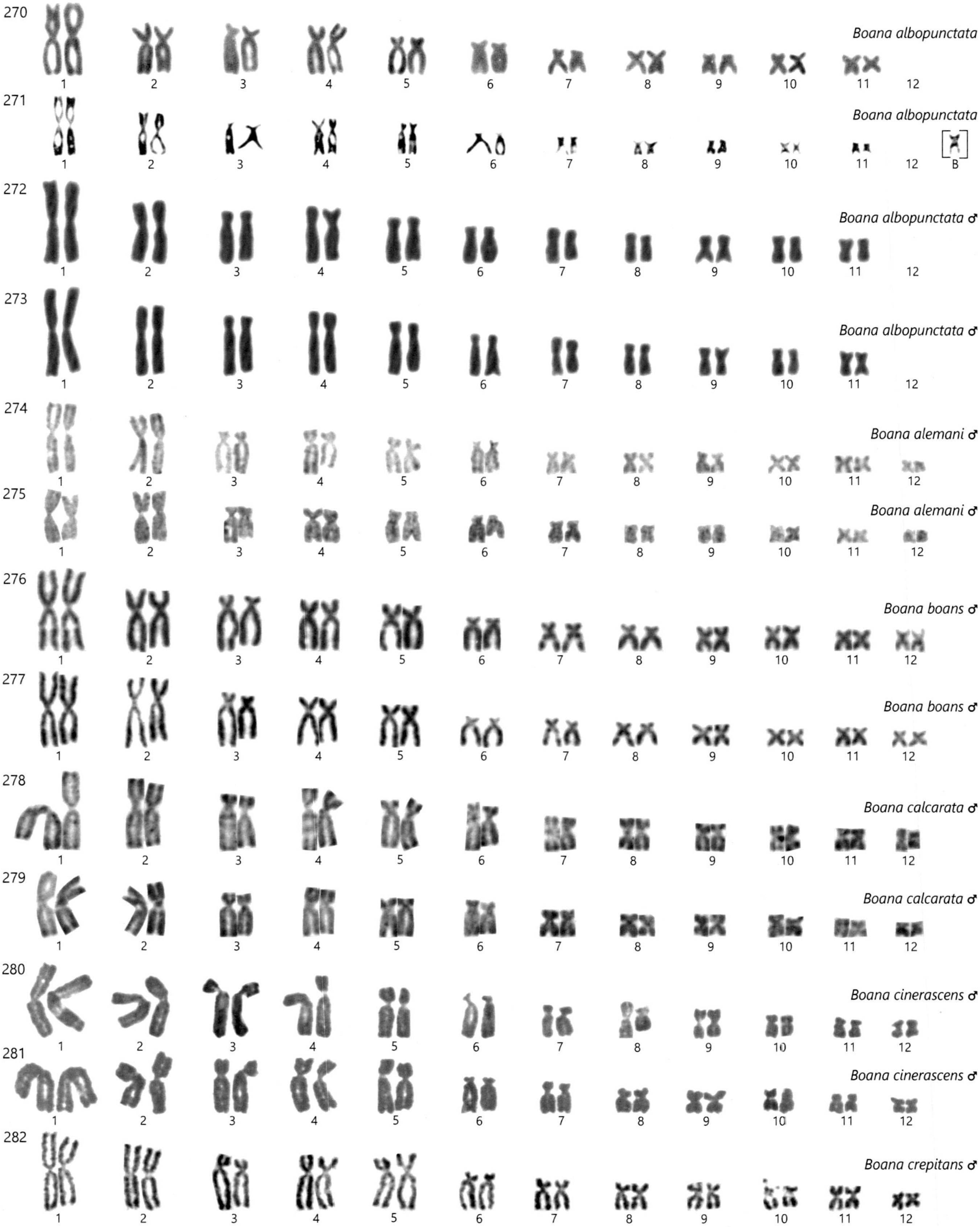
270 *Boana albopunctata*
271 *Boana albopunctata*
B
272 *Boana albopunctata* ♂
273 *Boana albopunctata* ♂
274 *Boana alemani* ♂
275 *Boana alemani* ♂
276 *Boana boans* ♂
277 *Boana boans* ♂
278 *Boana calcarata* ♂
279 *Boana calcarata* ♂
280 *Boana cinerascens* ♂
281 *Boana cinerascens* ♂
282 *Boana crepitans* ♂
1 2 3 4 5 6 7 8 9 10 11 12

283
1 2 3 4 5 6 7 8 9 10 11 12
Boana crepitans ♂
284
1 2 3 4 5 6 7 8 9 10 11 12
Boana crepitans
285
1 2 3 4 5 6 7 8 9 10 11 12
Boana crepitans ♂
286
1 2 3 4 5 6 7 8 9 10 11 12
Boana faber ♀
287
1 2 3 4 5 6 7 8 9 10 11 12
Boana faber ♀
288
1 2 3 4 5 6 7 8 9 10 11 12
Boana fasciata ♂
289
1 2 3 4 5 6 7 8 9 10 11 12
Boana fasciata ♂
290
1 2 3 4 5 6 7 8 9 10 11 12
Boana pardalis ♂
291
1 2 3 4 5 6 7 8 9 10 11 12
Boana pardalis ♂
292
1 2 3 4 5 6 7 8 9 10 11 12
Boana polytaenia ♂
293
1 2 3 4 5 6 7 8 9 10 11 12
Boana polytaenia ♂
294
1 2 3 4 5 6 7 8 9 10 11 12
Boana prasina ♂
295
1 2 3 4 5 6 7 8 9 10 11 12
Boana pulchella

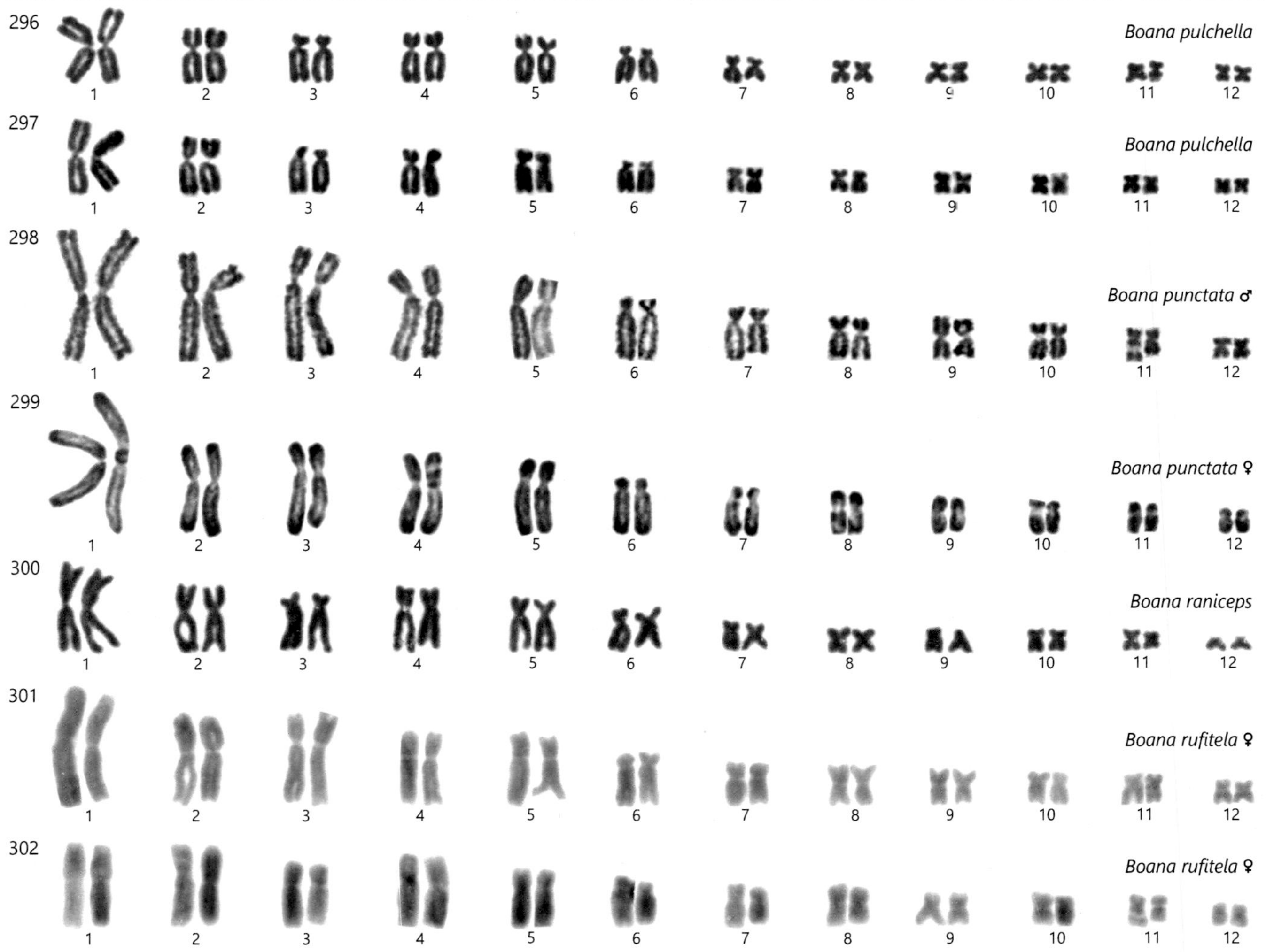
296
Boana pulchella
297
Boana pulchella
298
Boana punctata ♂
299
Boana punctata ♀
300
Boana raniceps
301
Boana rufitela ♀
302
Boana rufitela ♀

303
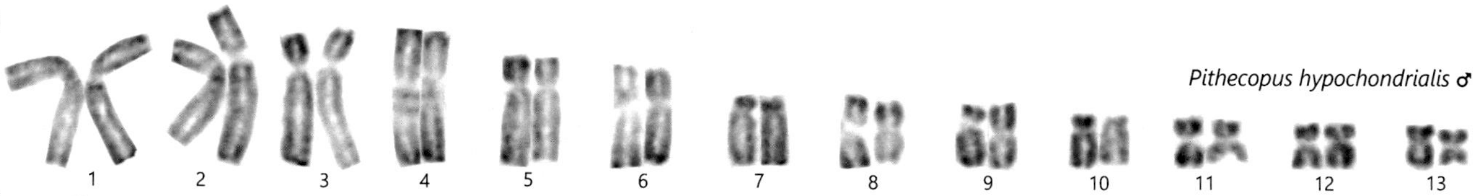

304
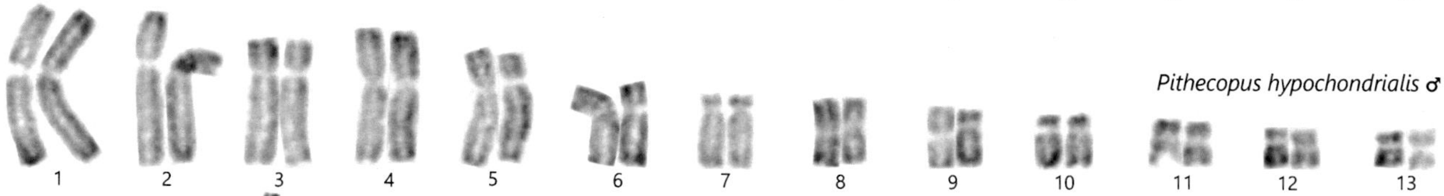

305
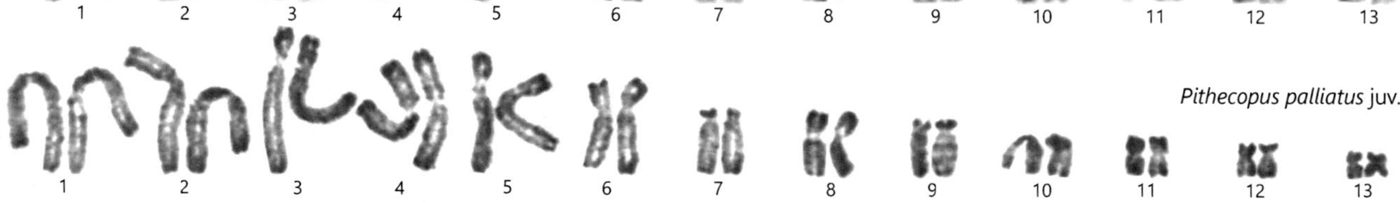

Phyllomedusidae • *Phyllomedusa*

306
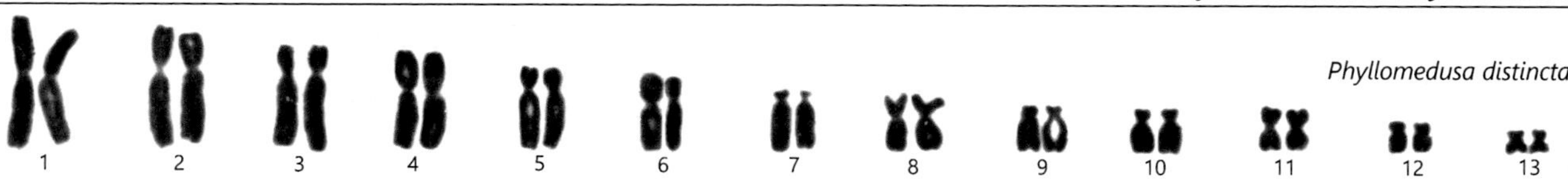

307
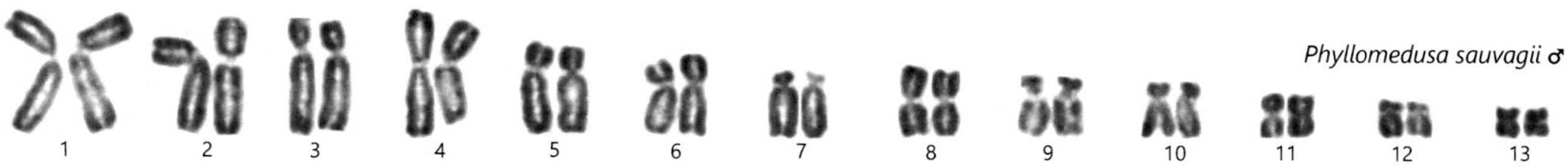

308
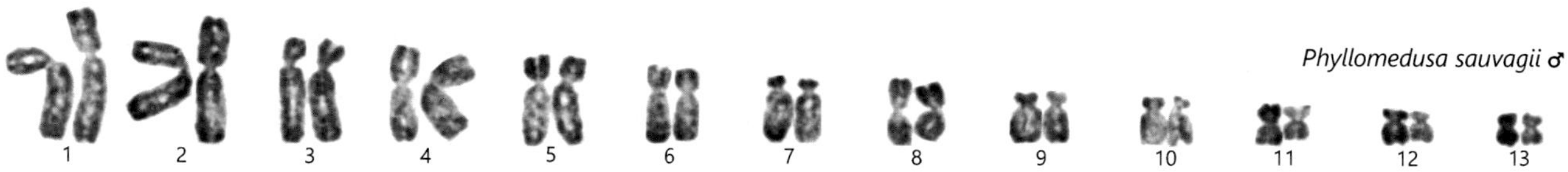

309
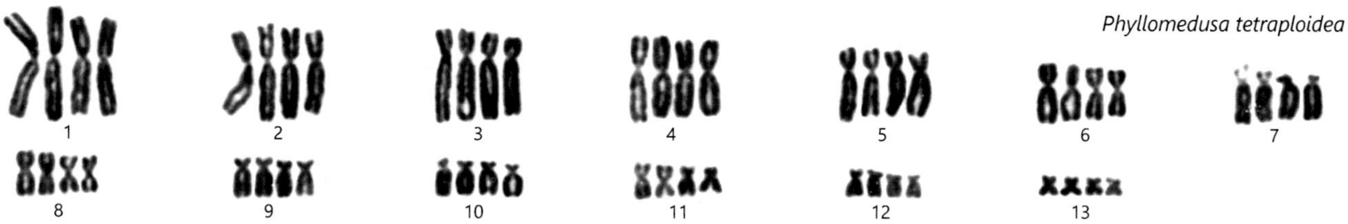

Phyllomedusidae • *Phasmahyla*

310
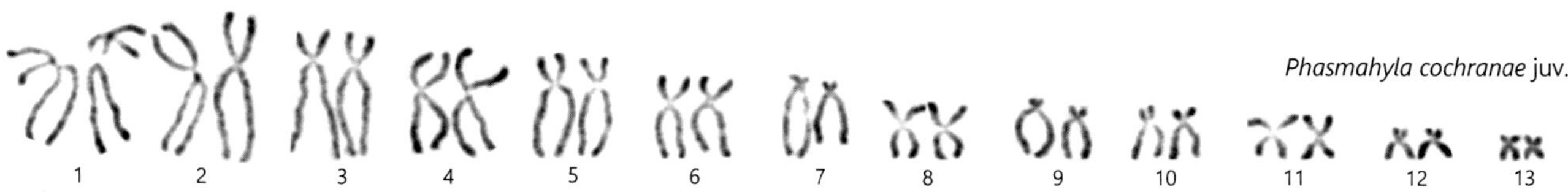

311
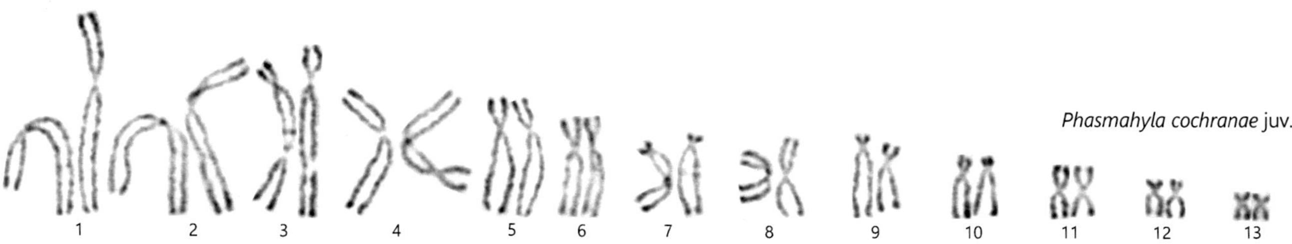

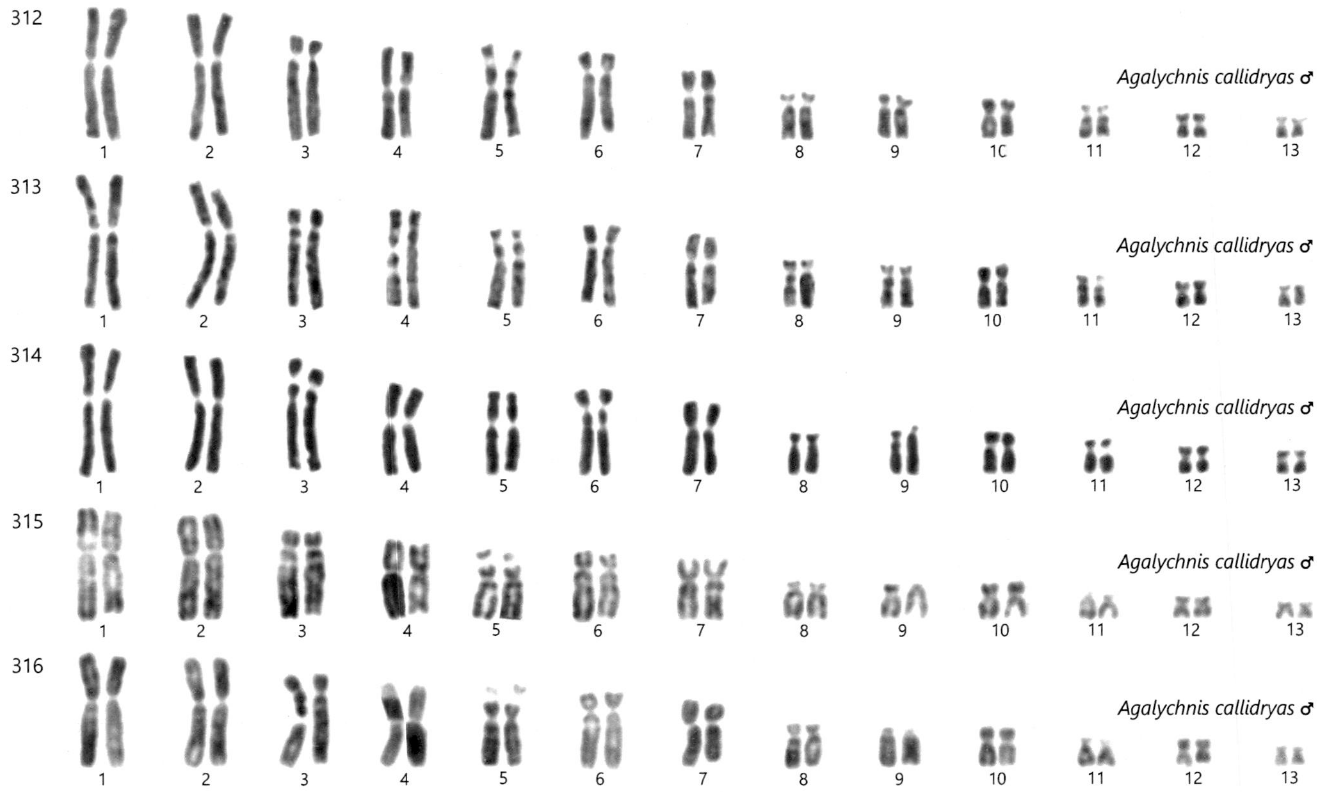
312
Agalychnis callidryas ♂
1 2 3 4 5 6 7 8 9 1C 11 12 13
313
Agalychnis callidryas ♂
1 2 3 4 5 6 7 8 9 10 11 12 13
314
Agalychnis callidryas ♂
1 2 3 4 5 6 7 8 9 10 11 12 13
315
Agalychnis callidryas ♂
1 2 3 4 5 6 7 8 9 10 11 12 13
316
Agalychnis callidryas ♂
1 2 3 4 5 6 7 8 9 10 11 12 13

Cytogenet Genome Res 2018;155:55–221
DOI: 10.1159/000489839

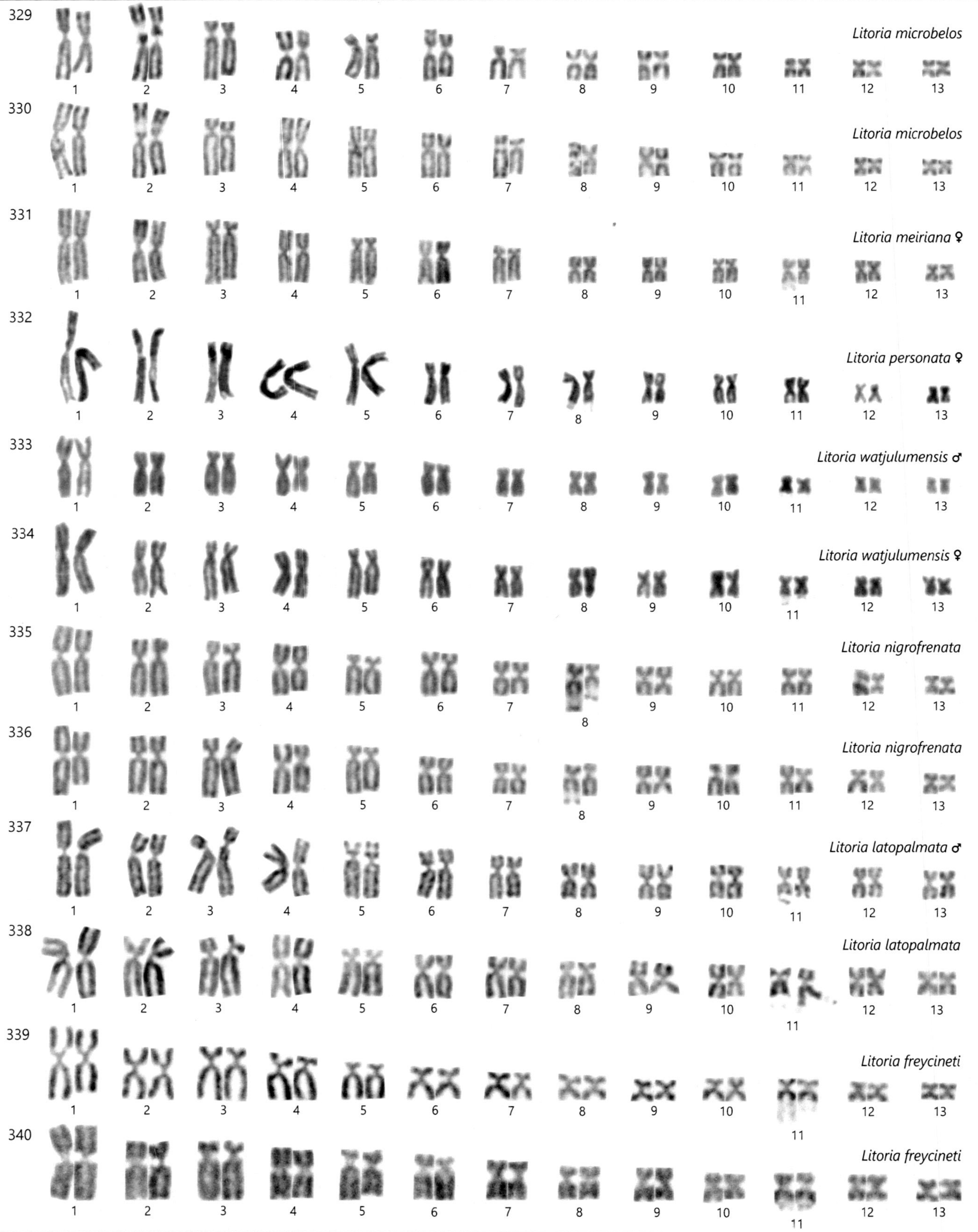
329
1 2 3 4 5 6 7 8 9 10 11 12 13
Litoria microbelos
330
1 2 3 4 5 6 7 8 9 10 11 12 13
Litoria microbelos
331
1 2 3 4 5 6 7 8 9 10 11 12 13
Litoria meiriana ♀
332
1 2 3 4 5 6 7 8 9 10 11 12 13
Litoria personata ♀
333
1 2 3 4 5 6 7 8 9 10 11 12 13
Litoria watjulumensis ♂
334
1 2 3 4 5 6 7 8 9 10 11 12 13
Litoria watjulumensis ♀
335
1 2 3 4 5 6 7 8 9 10 11 12 13
Litoria nigrofrenata
336
1 2 3 4 5 6 7 8 9 10 11 12 13
Litoria nigrofrenata
337
1 2 3 4 5 6 7 8 9 10 11 12 13
Litoria latopalmata ♂
338
1 2 3 4 5 6 7 8 9 10 11 12 13
Litoria latopalmata
339
1 2 3 4 5 6 7 8 9 10 11 12 13
Litoria freycineti
340
1 2 3 4 5 6 7 8 9 10 11 12 13
Litoria freycineti

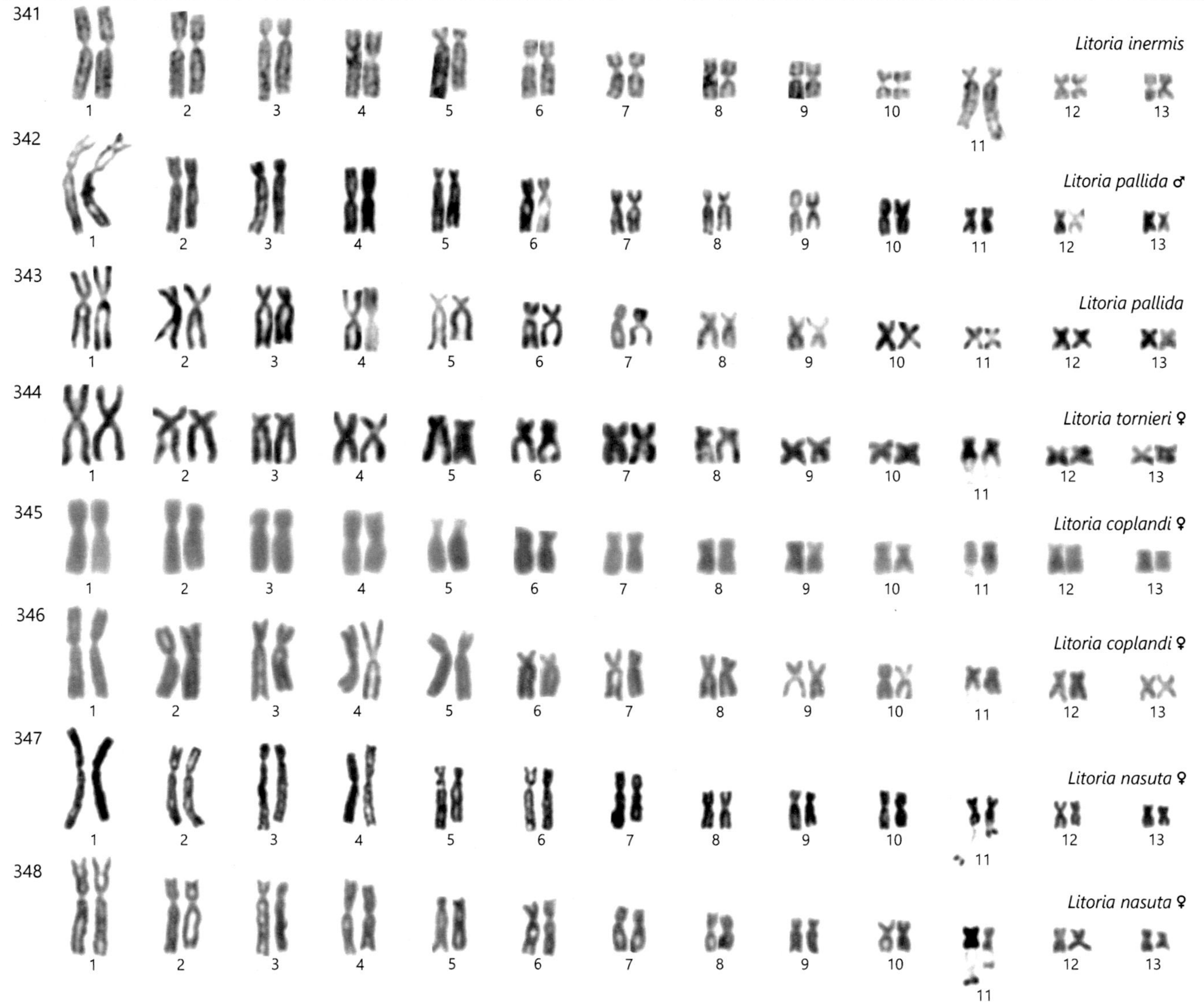

Pelodryadidae (Litoriinae) • *Litoria* • SGR 3

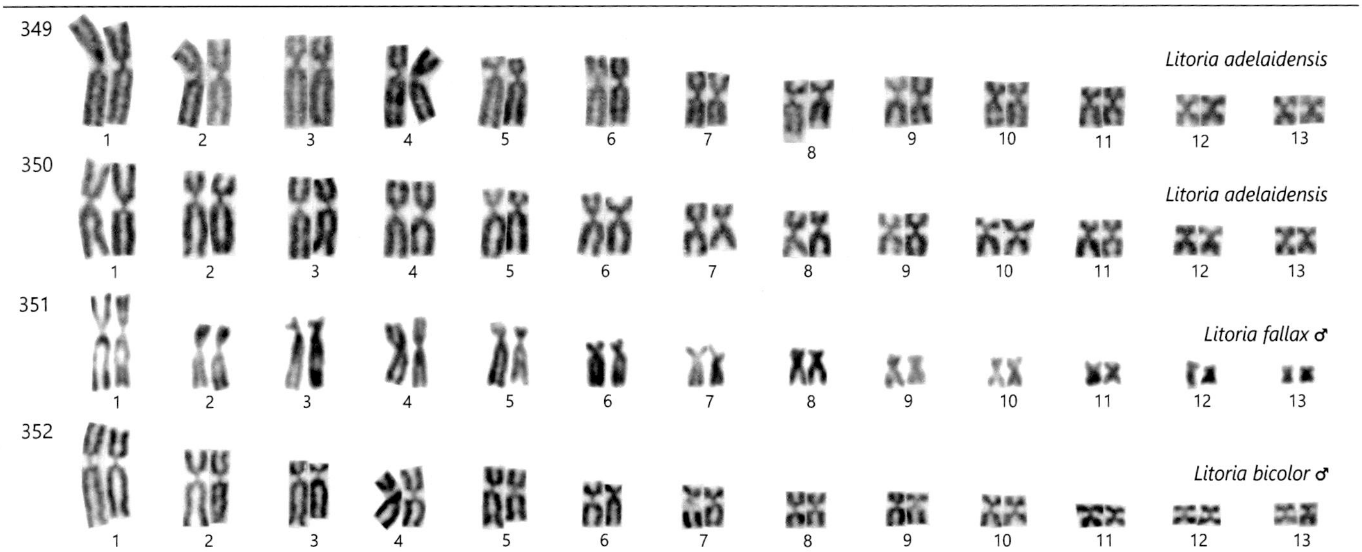

Pelodryadidae (Pelodryadinae) • *Nyctimystes*

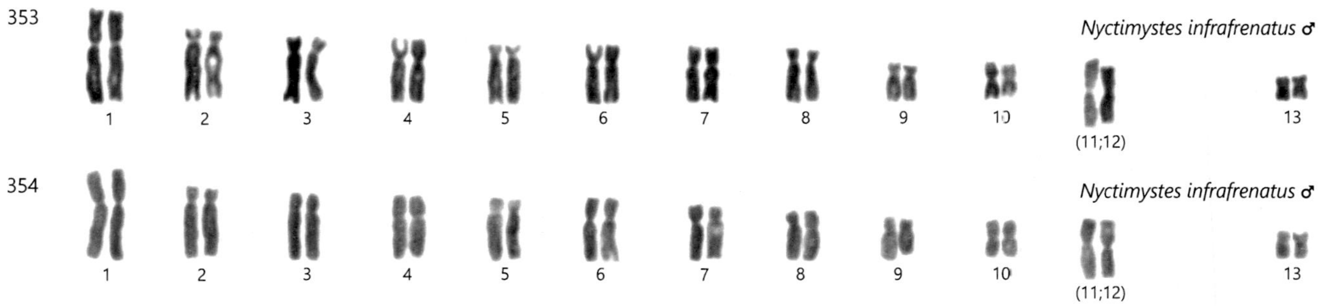

Pelodryadidae (Pelodryadinae) • *Ranoidea* • SGR 1

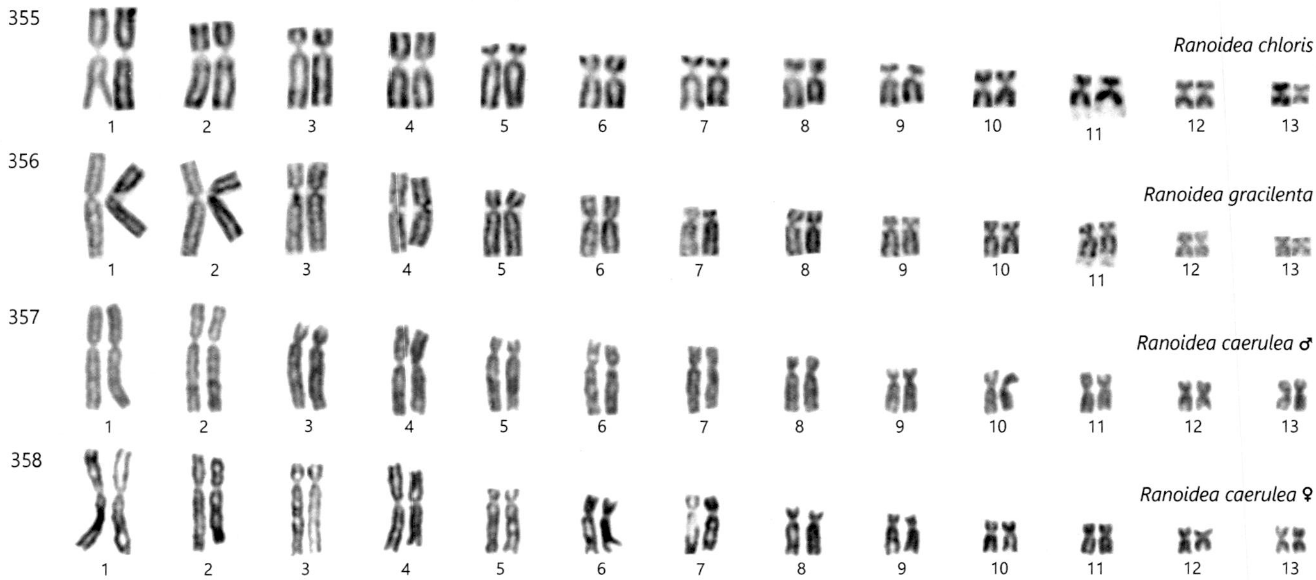

Pelodryadidae (Pelodryadinae) • *Ranoidea* • SGR 2

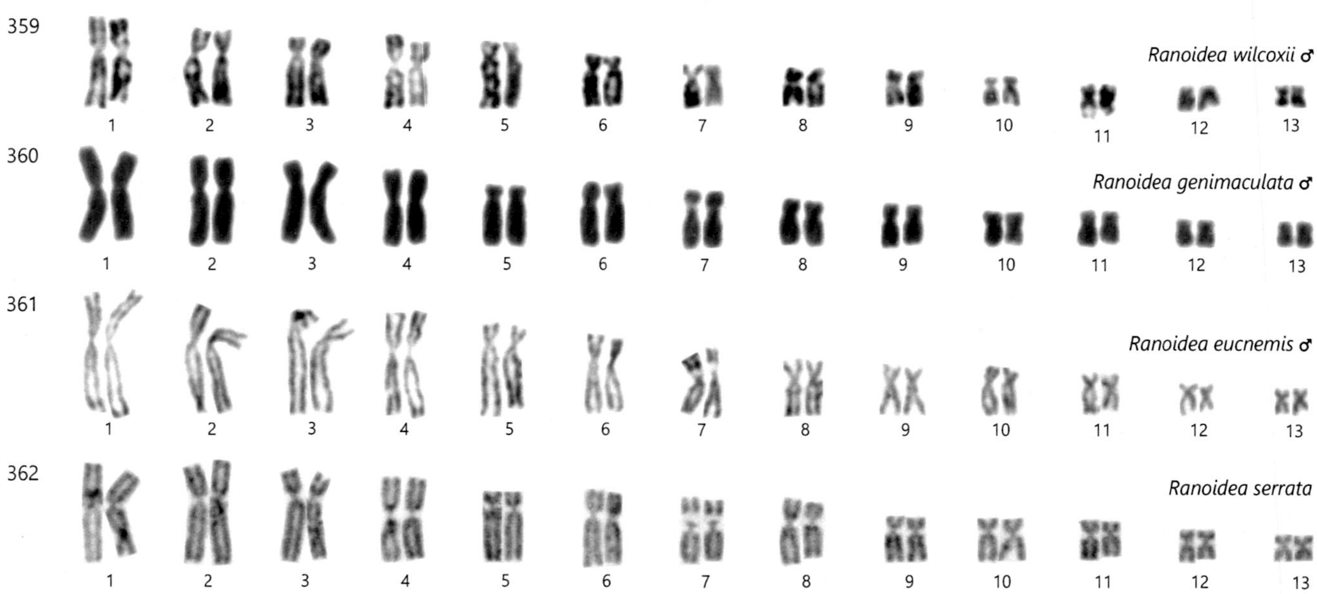

Pelodryadidae (Pelodryadinae) • *Ranoidea* • SGR 3

363 *Ranoidea barringtonensis* ♂

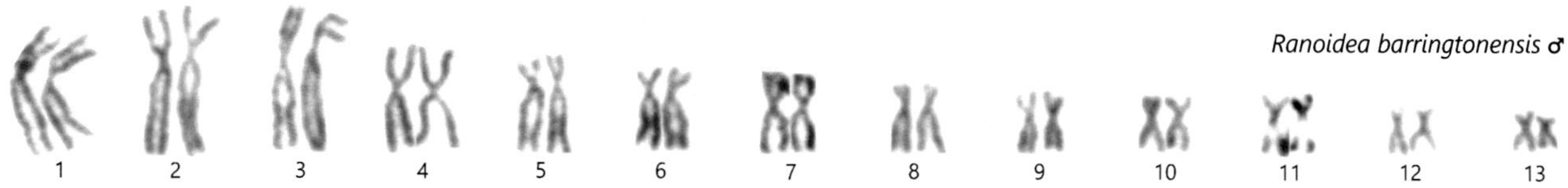

364 *Ranoidea citropa*

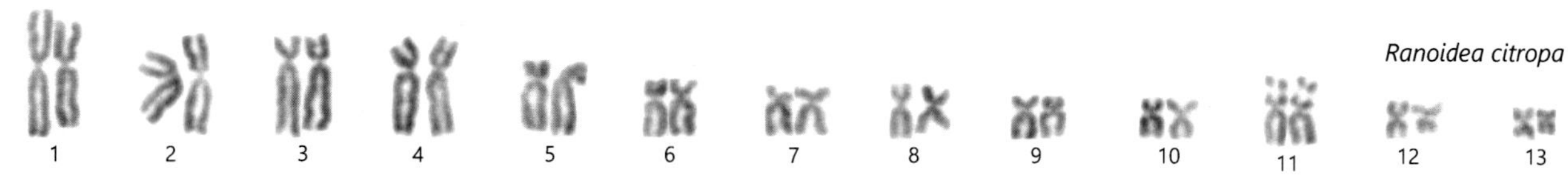

365 *Ranoidea citropa*

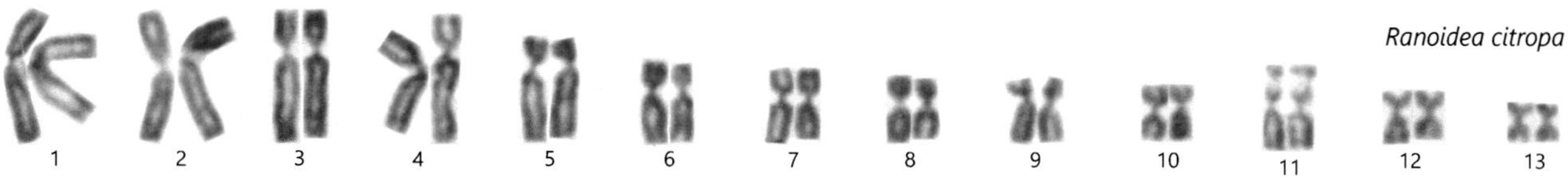

Pelodryadidae (Pelodryadinae) • *Ranoidea* • SGR 4

366 *Ranoidea rheocola* ♂

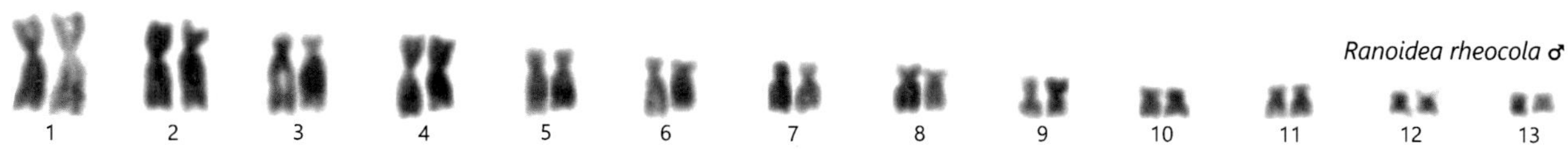

367 *Ranoidea nyakalensis* ♂

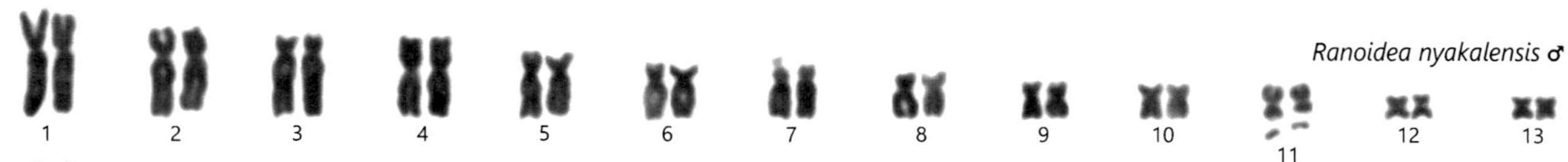

368 *Ranoidea nyakalensis* ♂

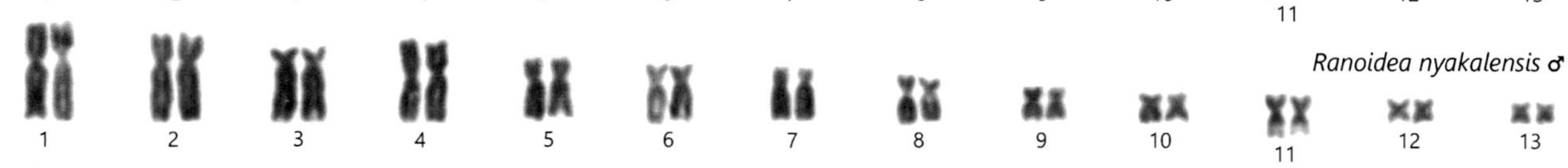

369 *Ranoidea nannotis* ♂

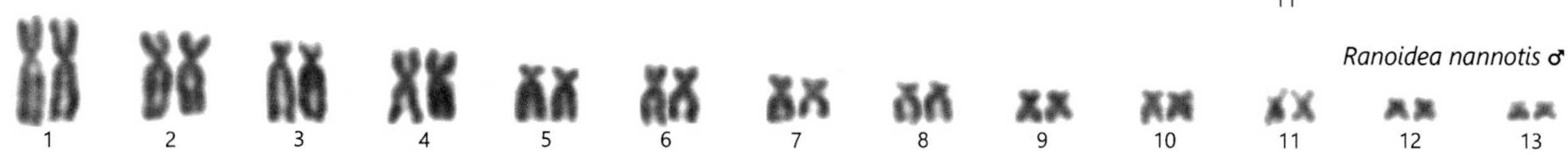

370 *Ranoidea nannotis* ♀

Pelodryadidae (Pelodryadinae) • *Ranoidea* • SGR 5

371 *Ranoidea cyclorhynchus* ♀

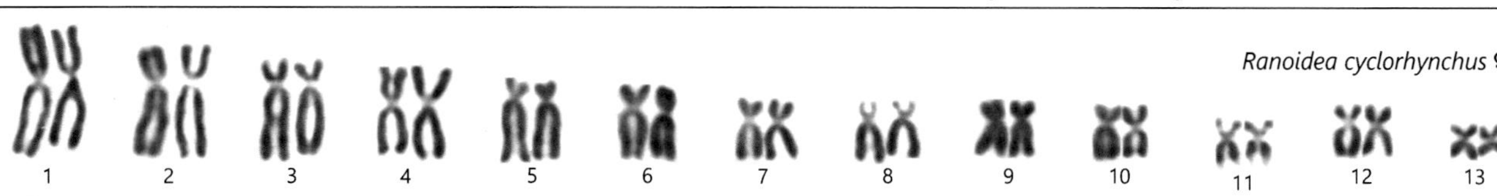

372 *Ranoidea cyclorhynchus* ♀

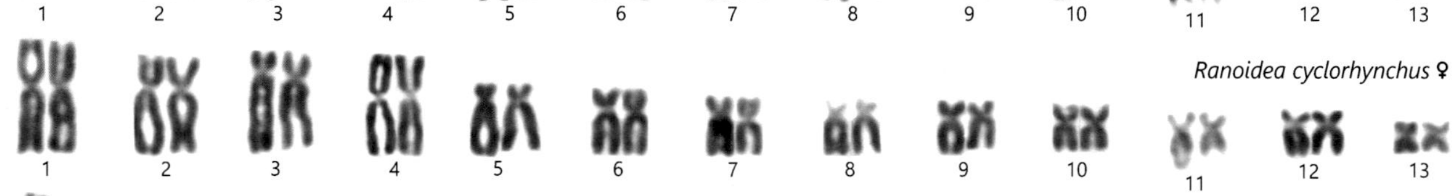

373 *Ranoidea raniformis*

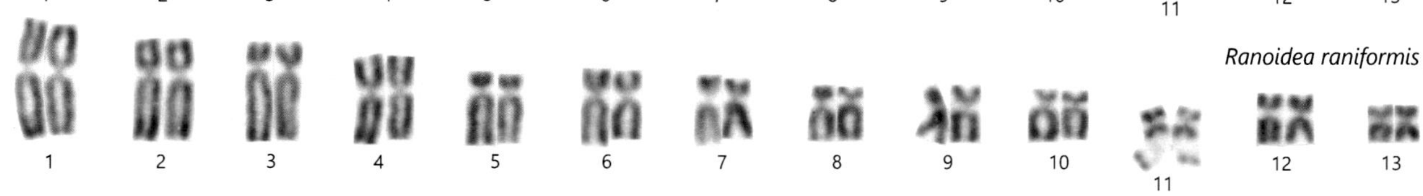

374
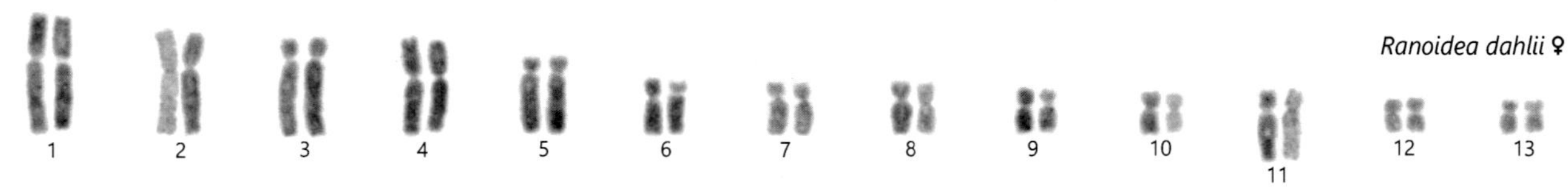

375
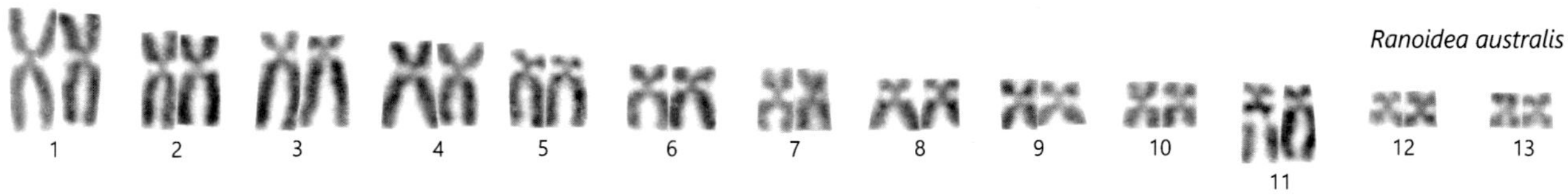

376
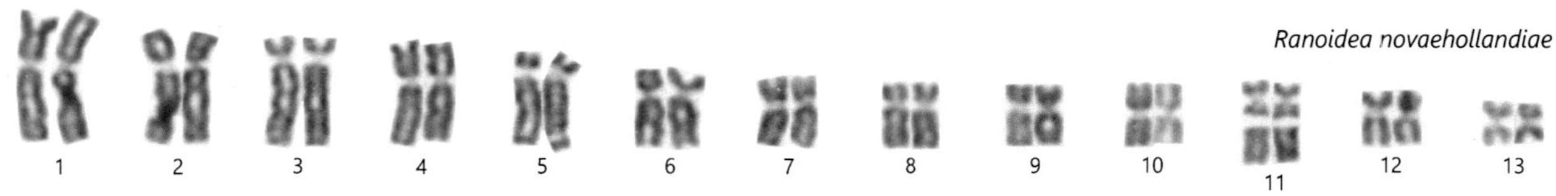

377
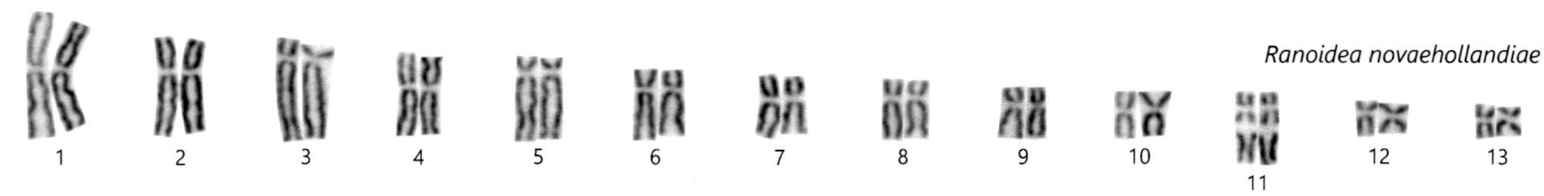

378
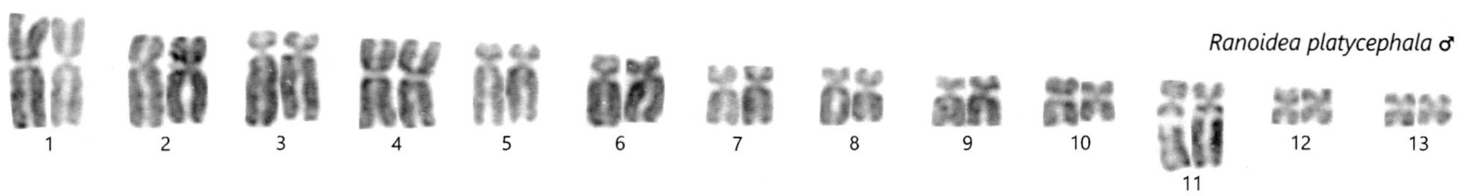

379
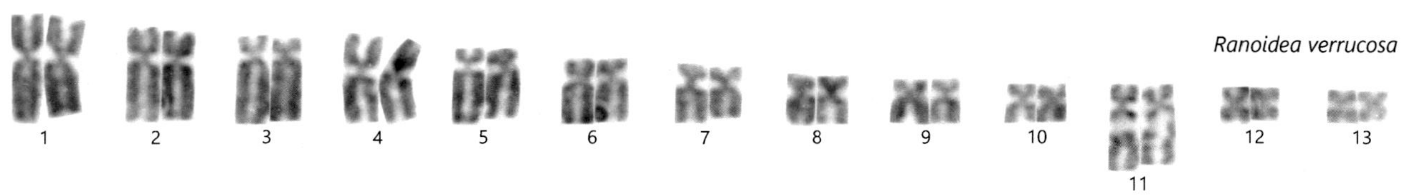

380
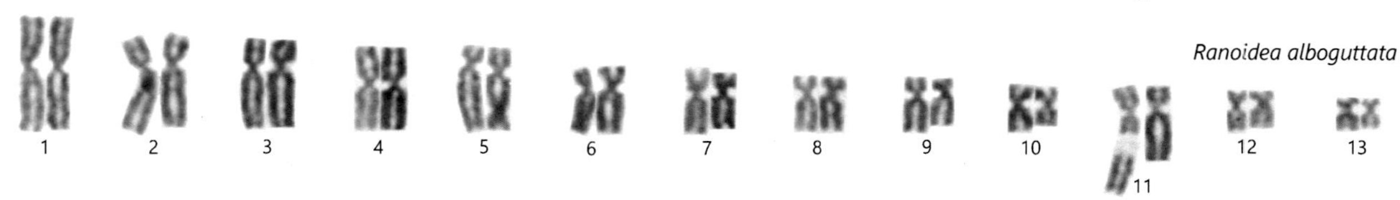

381
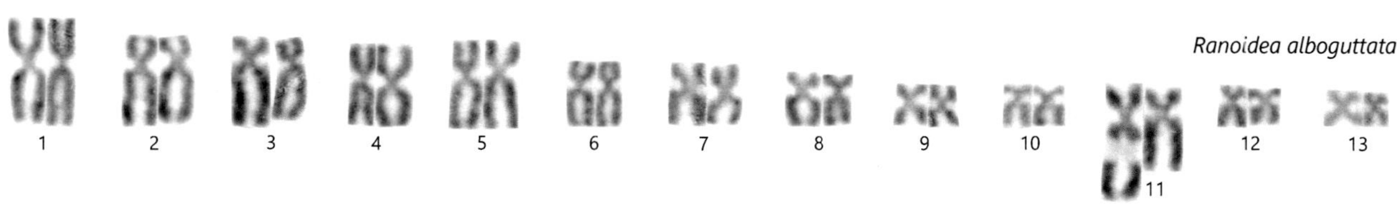

382
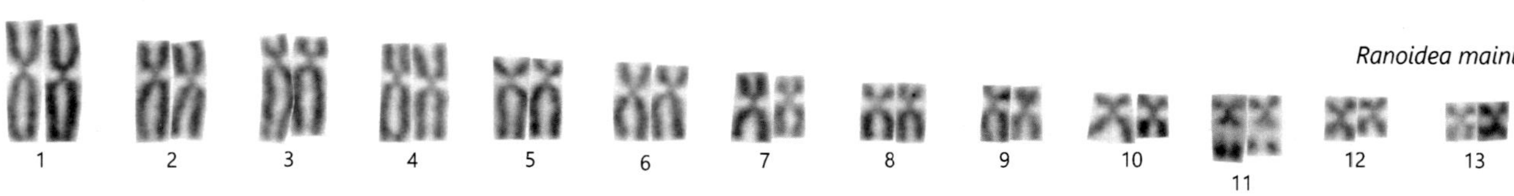

C-bands

Hylidae (Acridinae) • *Pseudacris*

383
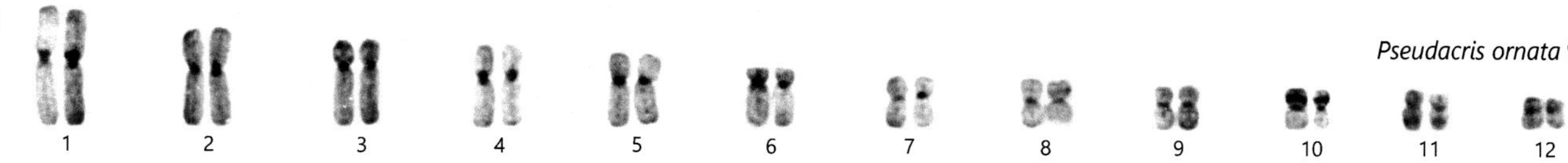

Pseudacris ornata ♀

Hylidae (Hylinae) • *Smilisca*

384
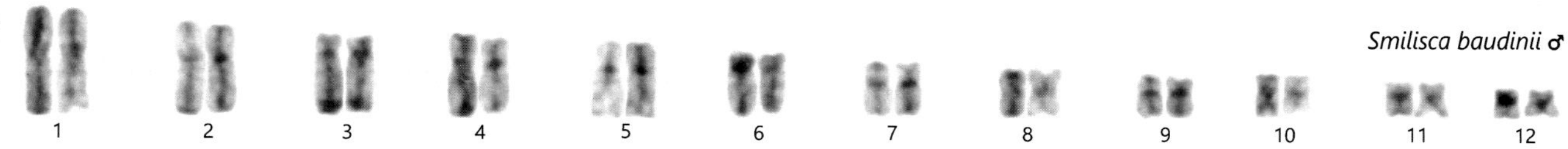

Smilisca baudinii ♂

385
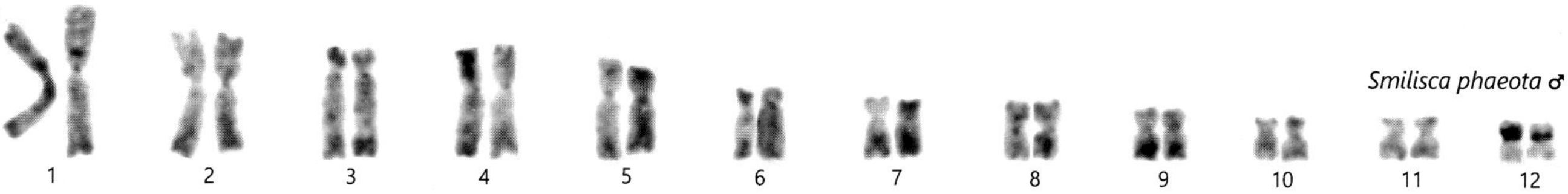

Smilisca phaeota ♂

386
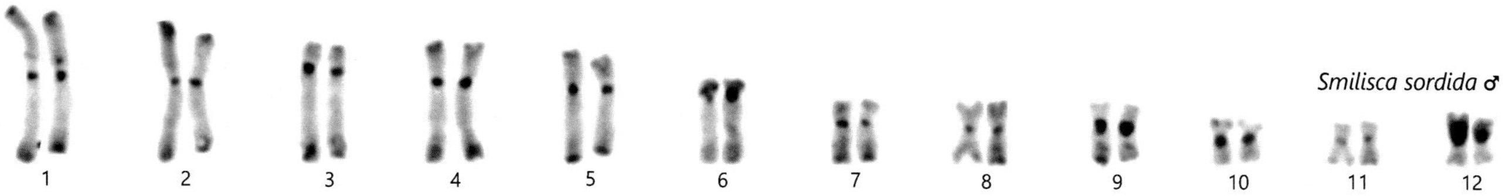

Smilisca sordida ♂

387
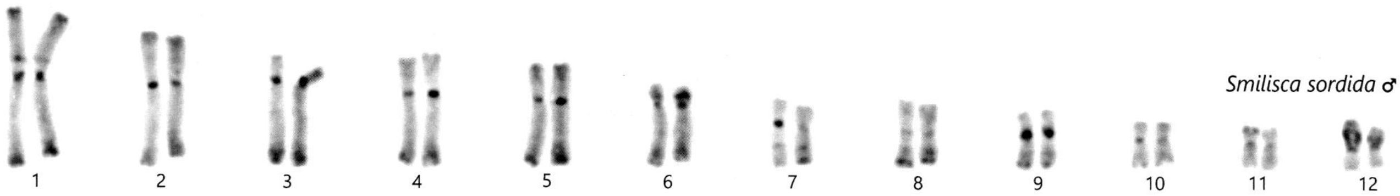

Smilisca sordida ♂

Hylidae (Hylinae) • *Isthmohyla*

388
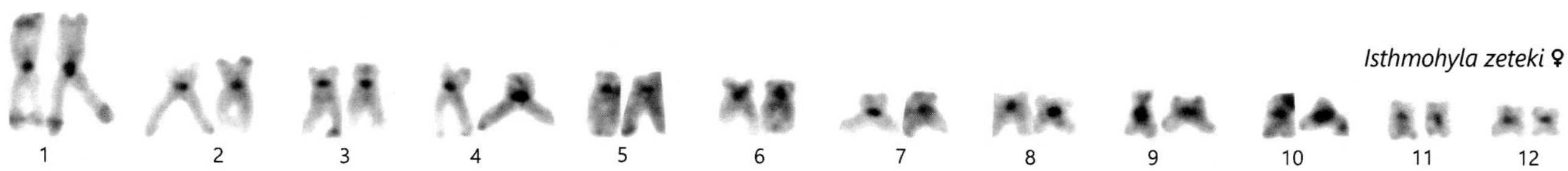

Isthmohyla zeteki ♀

Hylidae (Hylinae) • *Dryophytes*

389
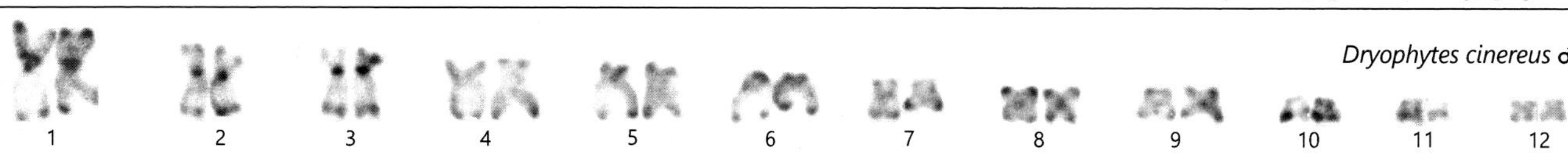

Dryophytes cinereus ♂

390
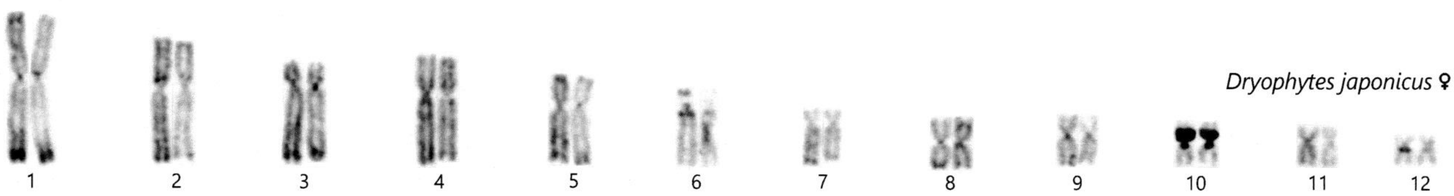

Dryophytes japonicus ♀

Hylidae (Hylinae) • *Tlalocohyla*

391
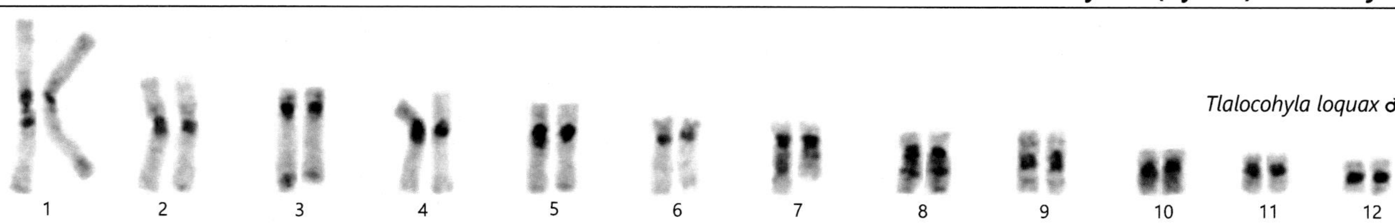

Tlalocohyla loquax ♂

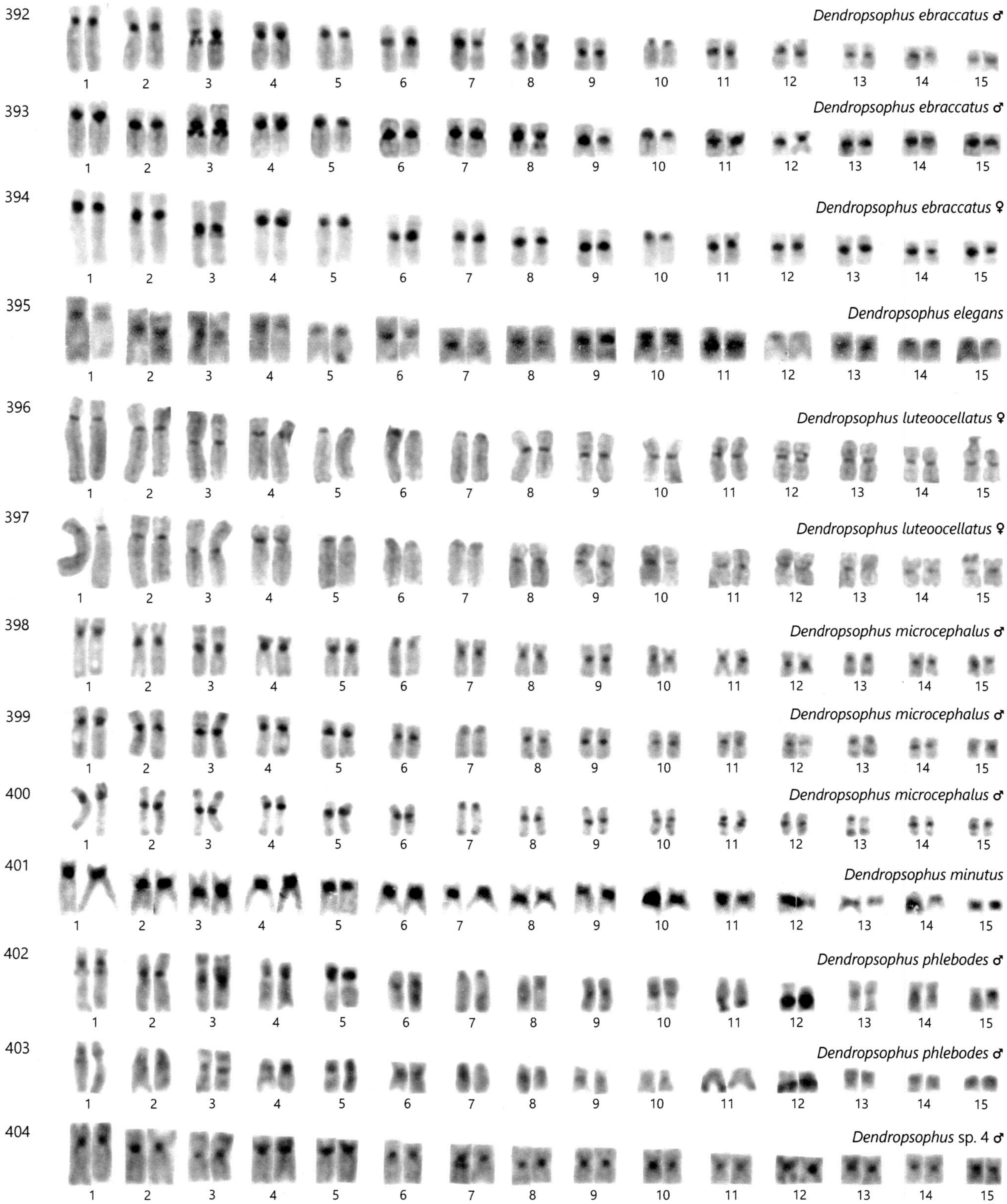
392 *Dendropsophus ebraccatus* ♂
393 *Dendropsophus ebraccatus* ♂
394 *Dendropsophus ebraccatus* ♀
395 *Dendropsophus elegans*
396 *Dendropsophus luteoocellatus* ♀
397 *Dendropsophus luteoocellatus* ♀
398 *Dendropsophus microcephalus* ♂
399 *Dendropsophus microcephalus* ♂
400 *Dendropsophus microcephalus* ♂
401 *Dendropsophus minutus*
402 *Dendropsophus phlebodes* ♂
403 *Dendropsophus phlebodes* ♂
404 *Dendropsophus* sp. 4 ♂

Hylidae (Lophyohylinae) • ***Trachycephalus***

405 *Trachycephalus mesophaeus*

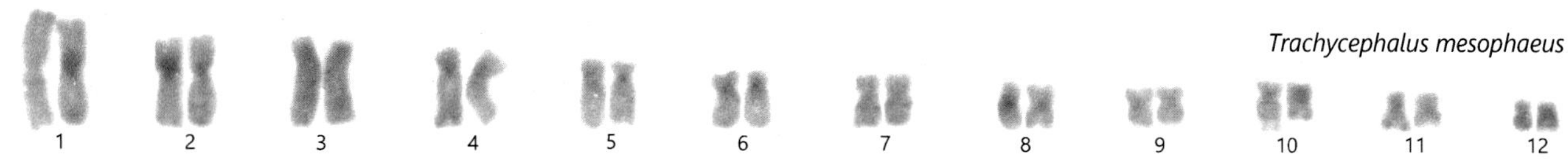

406 *Trachycephalus typhonius*

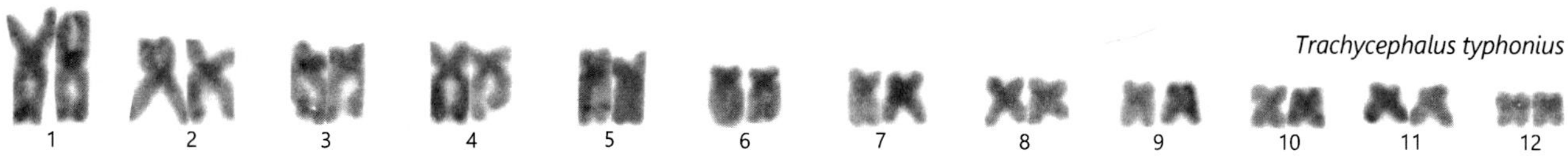

407 *Trachycephalus* sp.

Hylidae (Lophyohylinae) • ***Corythomantis***

408 *Corythomantis greeningi*

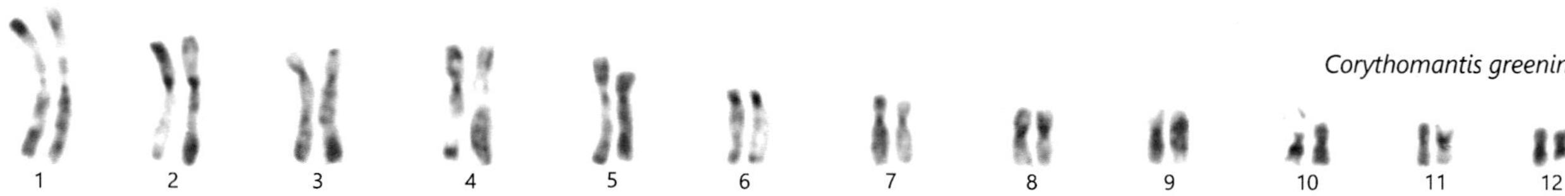

Hylidae (Lophyohylinae) • ***Aparasphenodon***

409 *Aparasphenodon bokermanni*

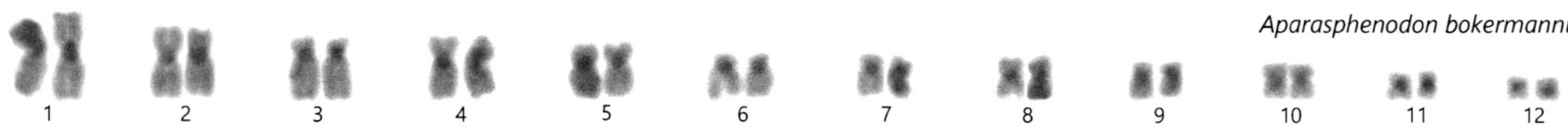

410 *Aparasphenodon brunoi* ♂

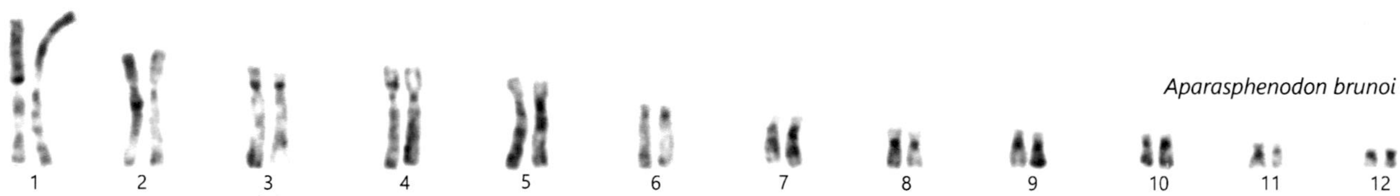

Hylidae (Lophyohylinae) • ***Itapotihyla***

411 *Itapotihyla langsdorffii*

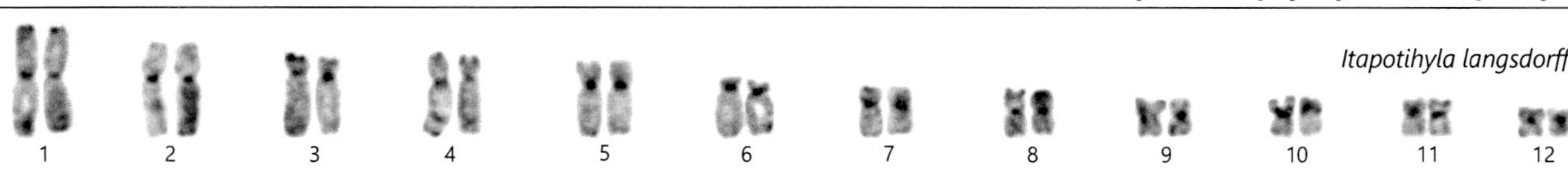

Hylidae (Lophyohylinae) • ***Osteopilus***

412 *Osteopilus ocellatus* ♀

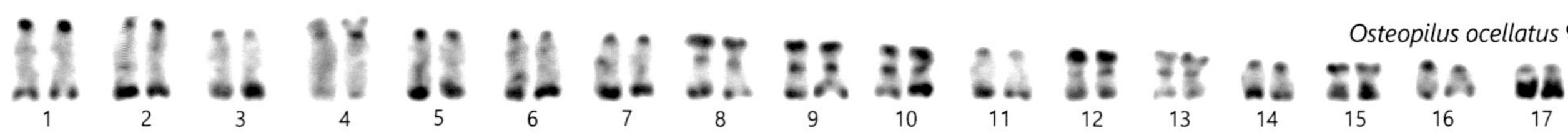

413 *Osteopilus ocellatus* ♀

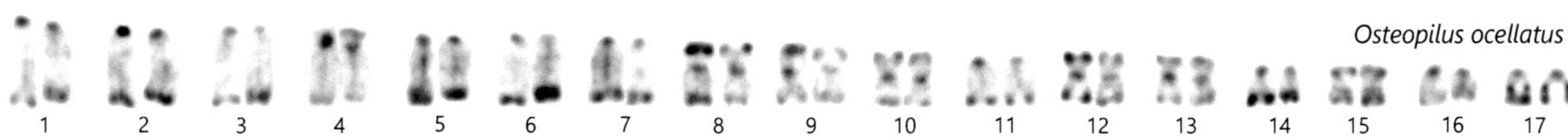

414 *Osteopilus ocellatus* ♀

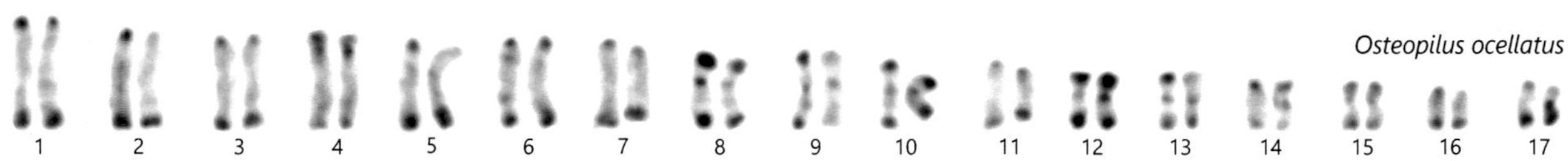

415 *Osteopilus septentrionalis*

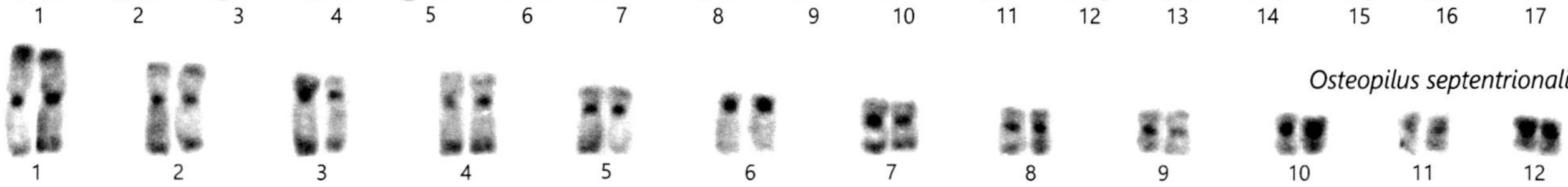

416
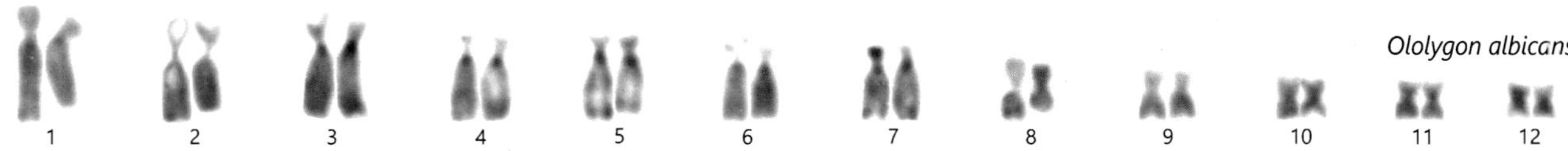

Ololygon albicans

417
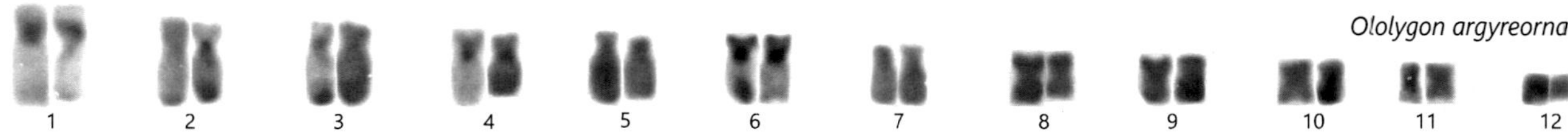

Ololygon argyreornata

418
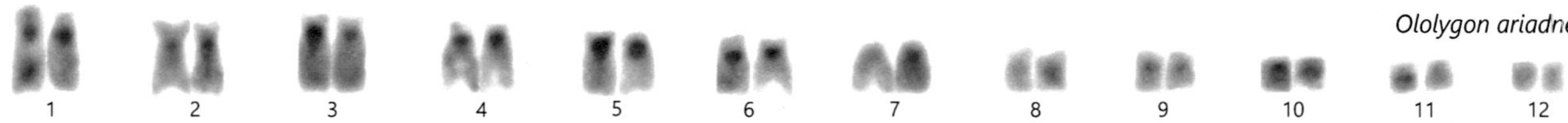

Ololygon ariadne

419
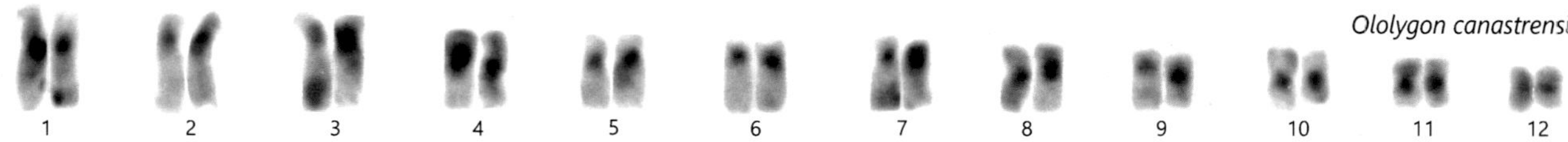

Ololygon canastrensis

420
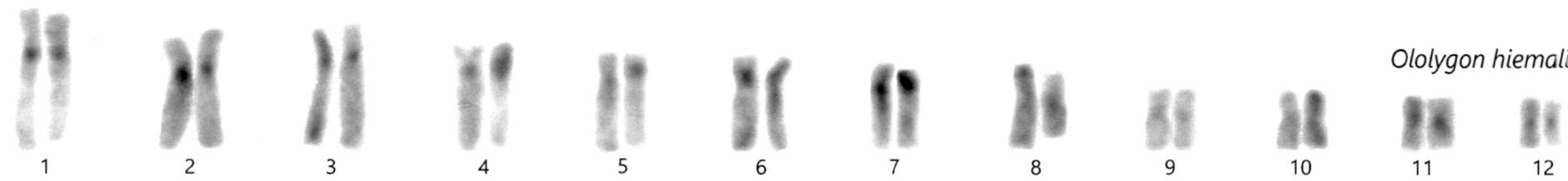

Ololygon hiemalis

421
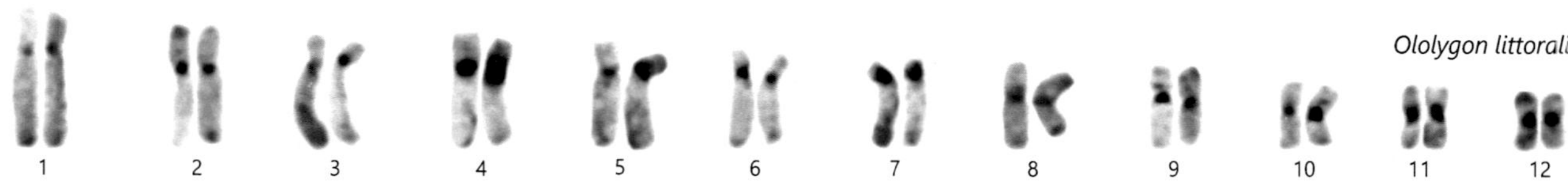

Ololygon littoralis

422
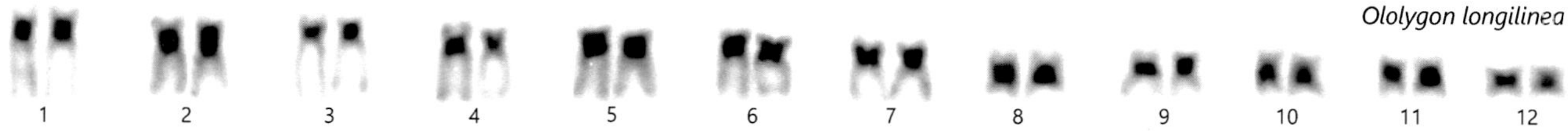

Ololygon longilinea

423
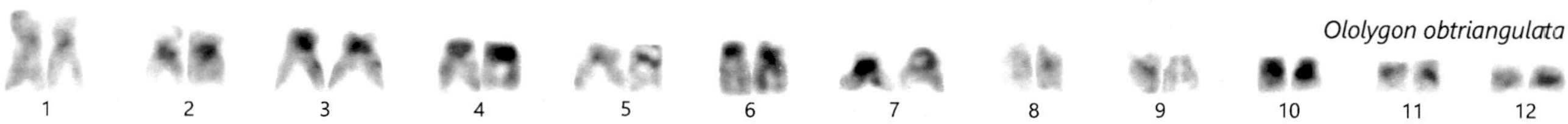

Ololygon obtriangulata

424
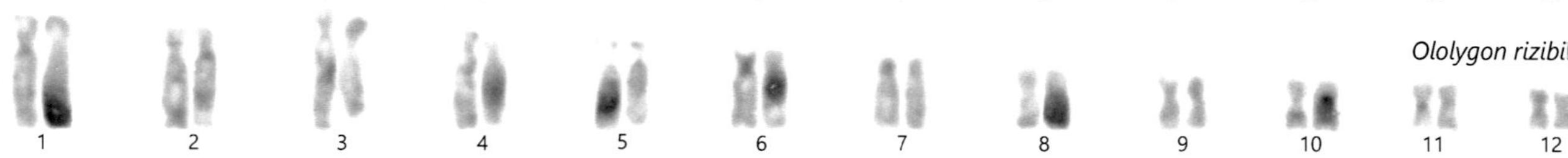

Ololygon rizibilis

425
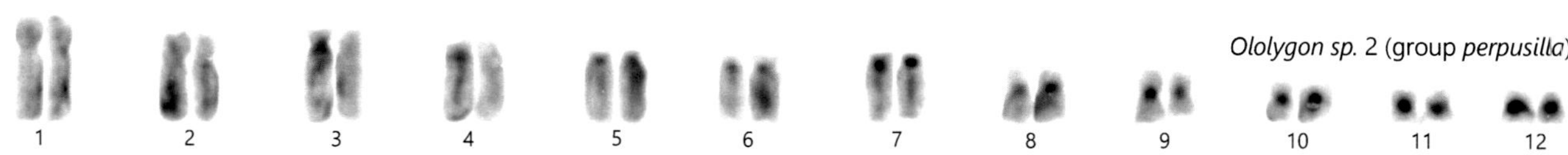

Ololygon sp. 2 (group *perpusilla*)

Hylidae (Scinaxinae) • ***Scinax***

426
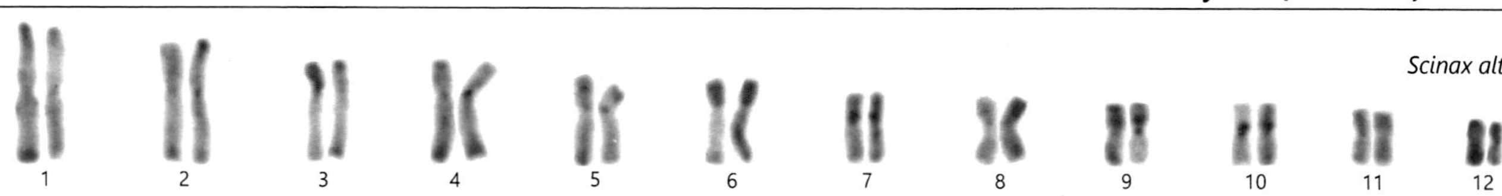

Scinax alter

427
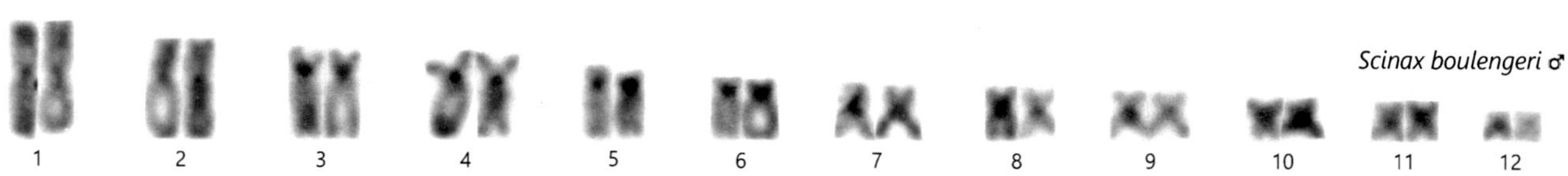

Scinax boulengeri ♂

DOI: 10.1159/000489839

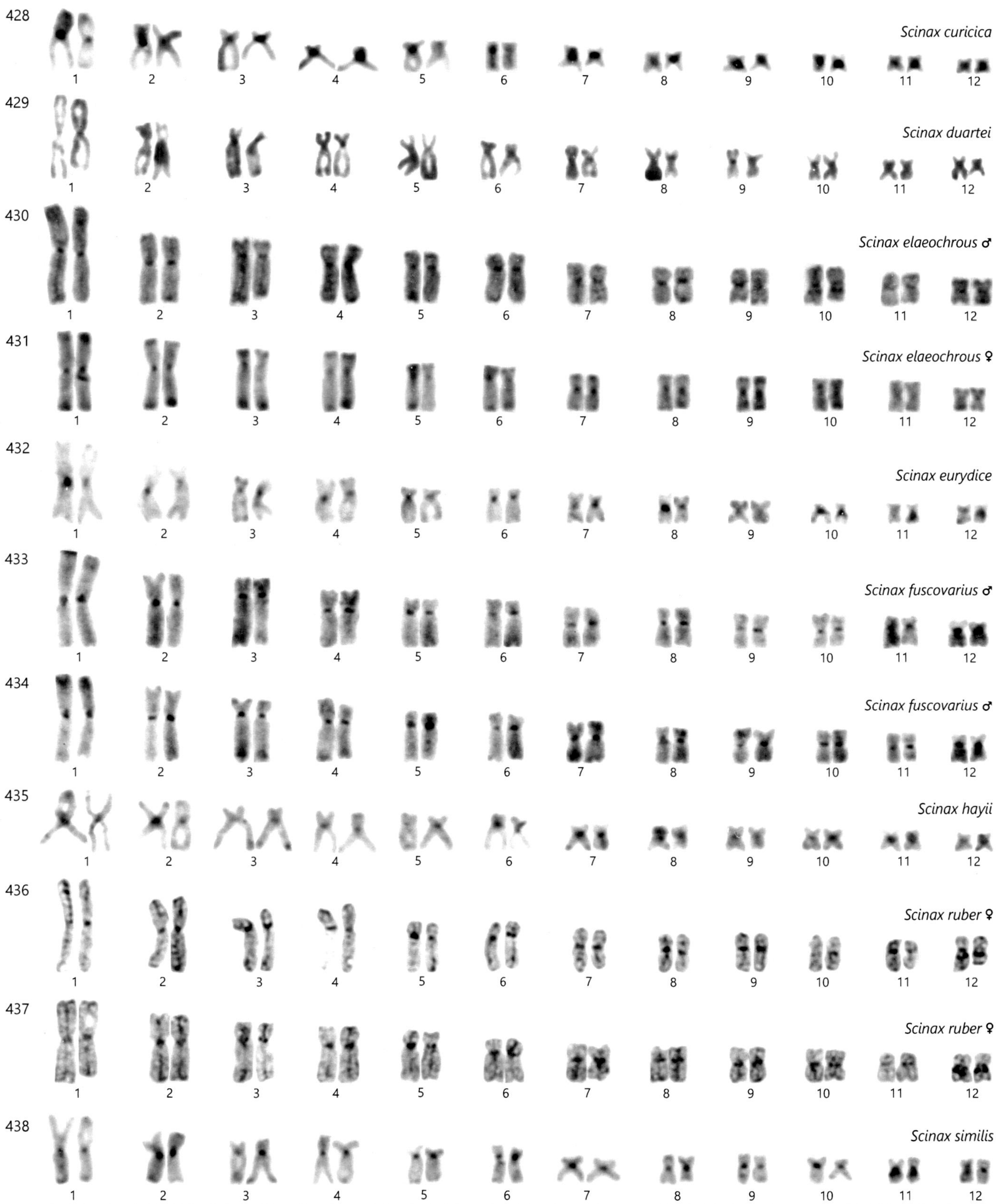
428 Scinax curicica
429 Scinax duartei
430 Scinax elaeochrous ♂
431 Scinax elaeochrous ♀
432 Scinax eurydice
433 Scinax fuscovarius ♂
434 Scinax fuscovarius ♂
435 Scinax hayii
436 Scinax ruber ♀
437 Scinax ruber ♀
438 Scinax similis

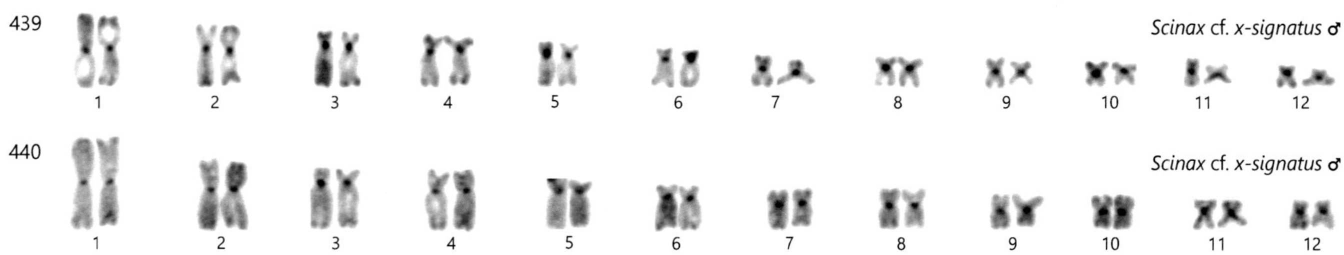

Hylidae (Cophomantinae) • *Hyloscirtus*

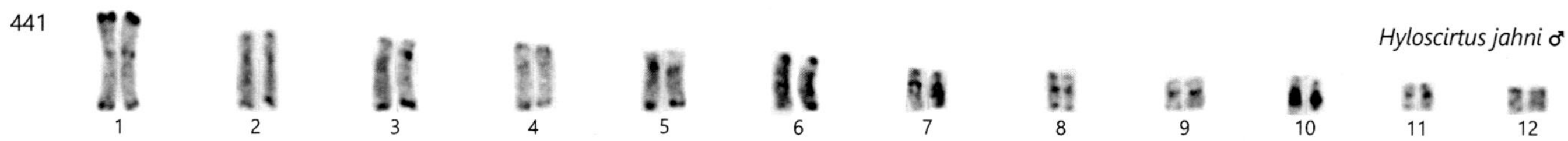

Hylidae (Cophomantinae) • *Bokermannohyla*

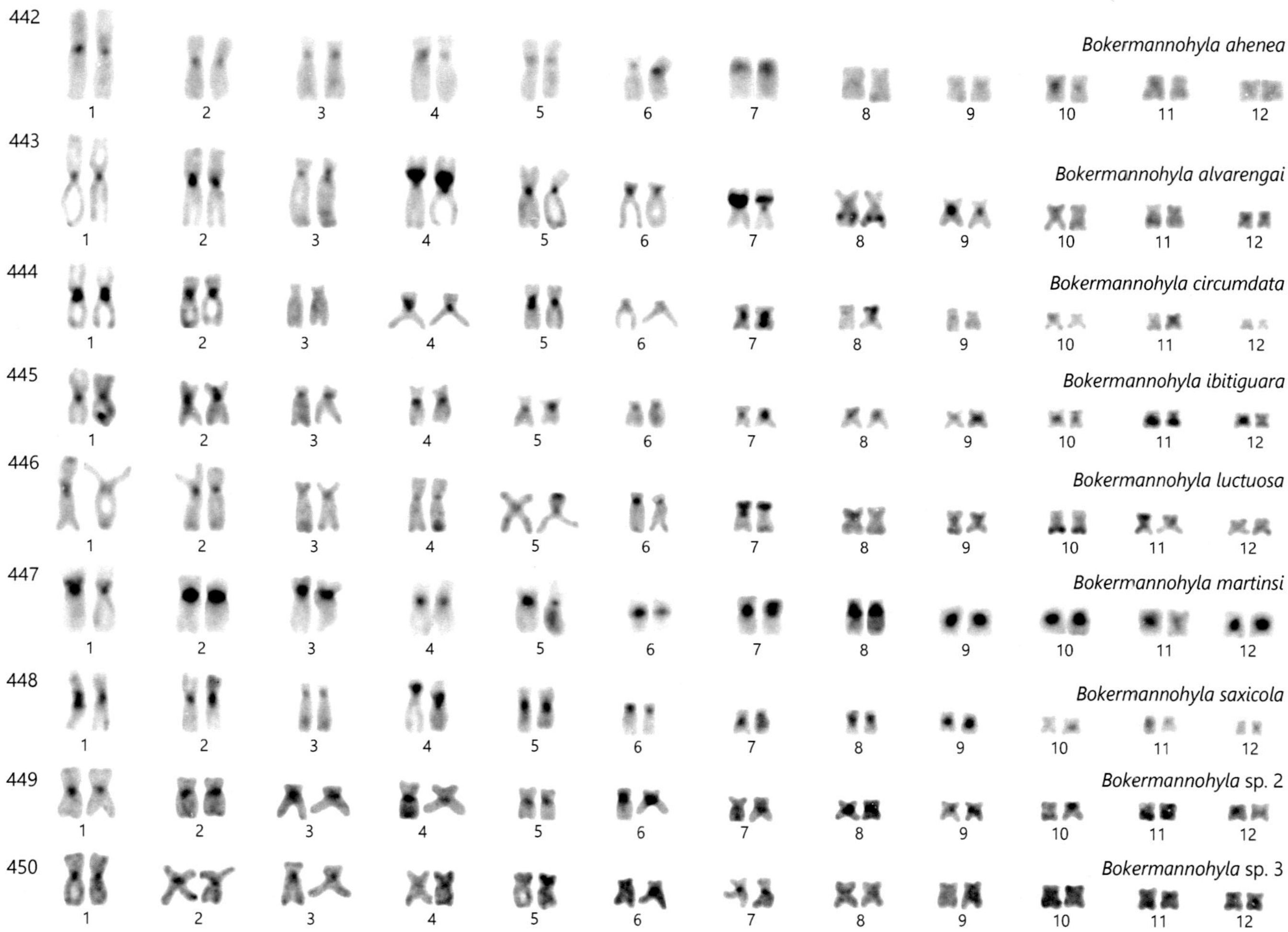

Cytogenet Genome Res 2018;155:55–221
DOI: 10.1159/000489839

451 452 453 454 455

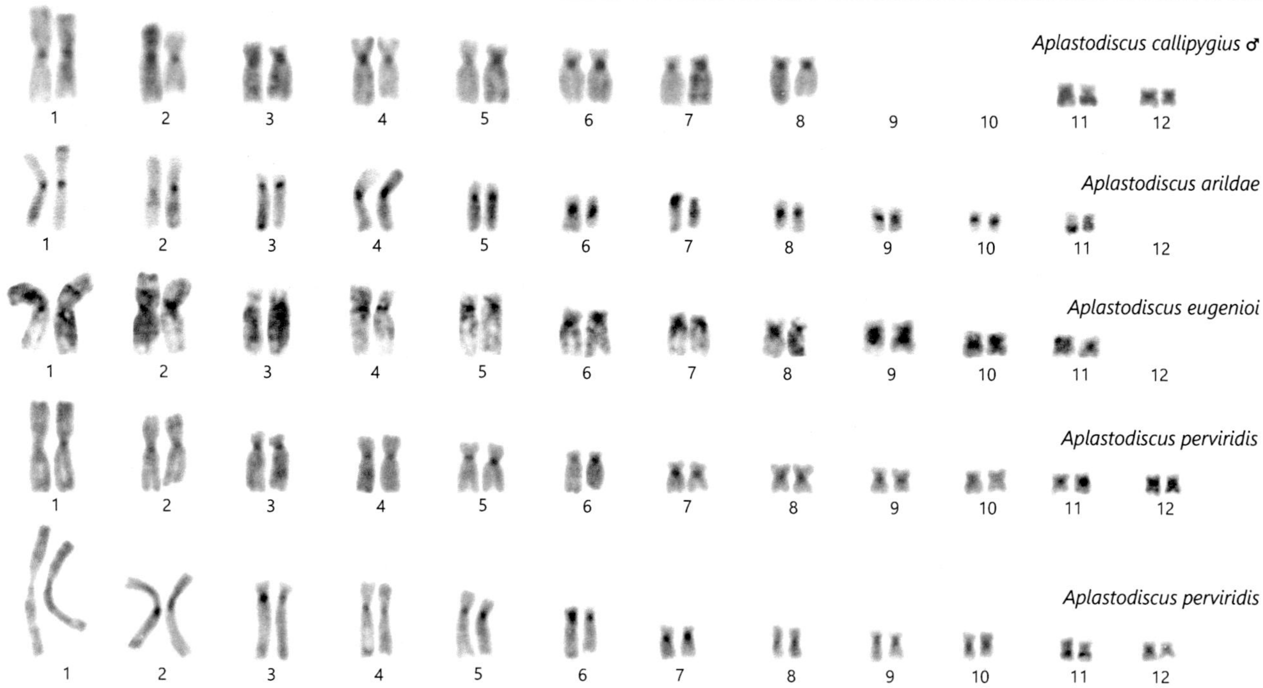

Hylidae (Cophomantinae) • *Boana*

456 457 458 459 460

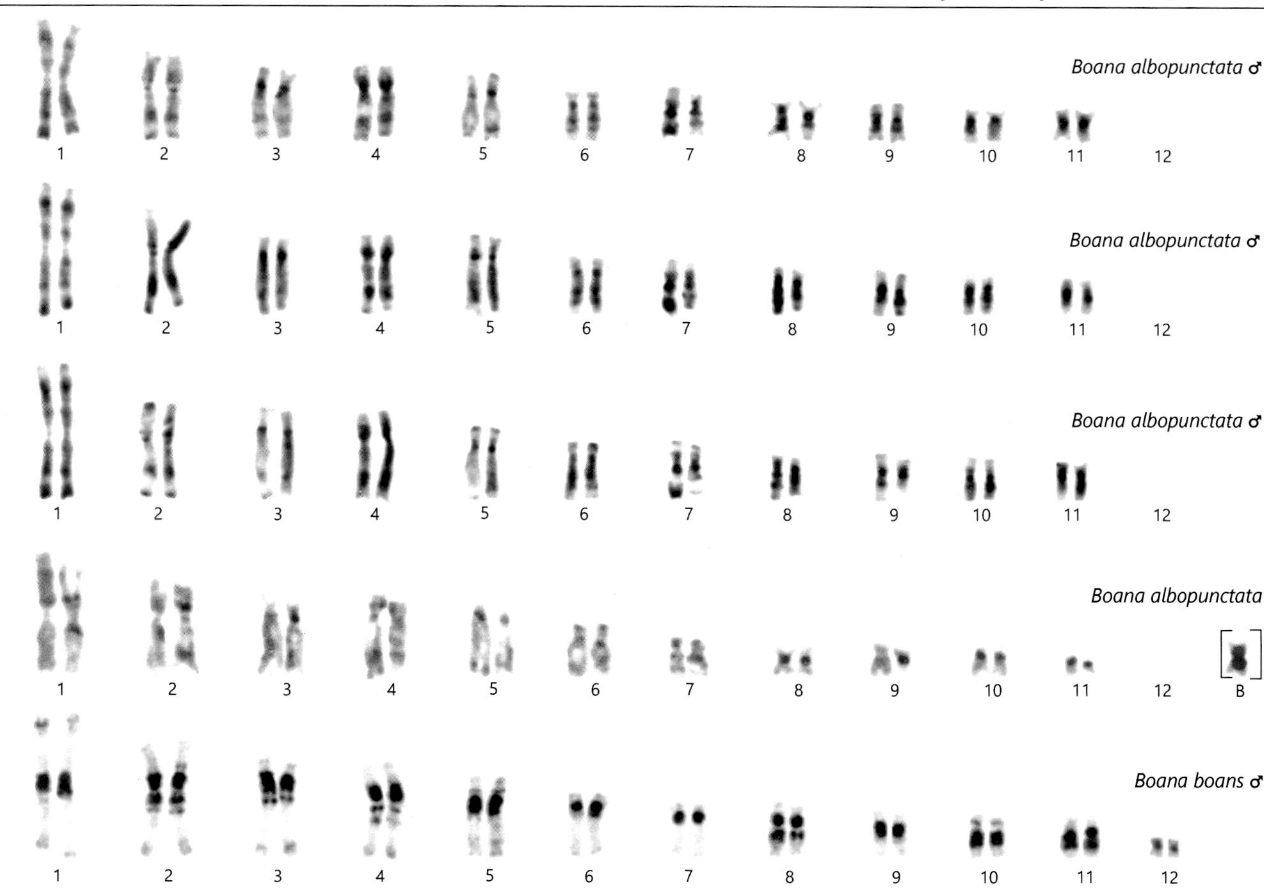

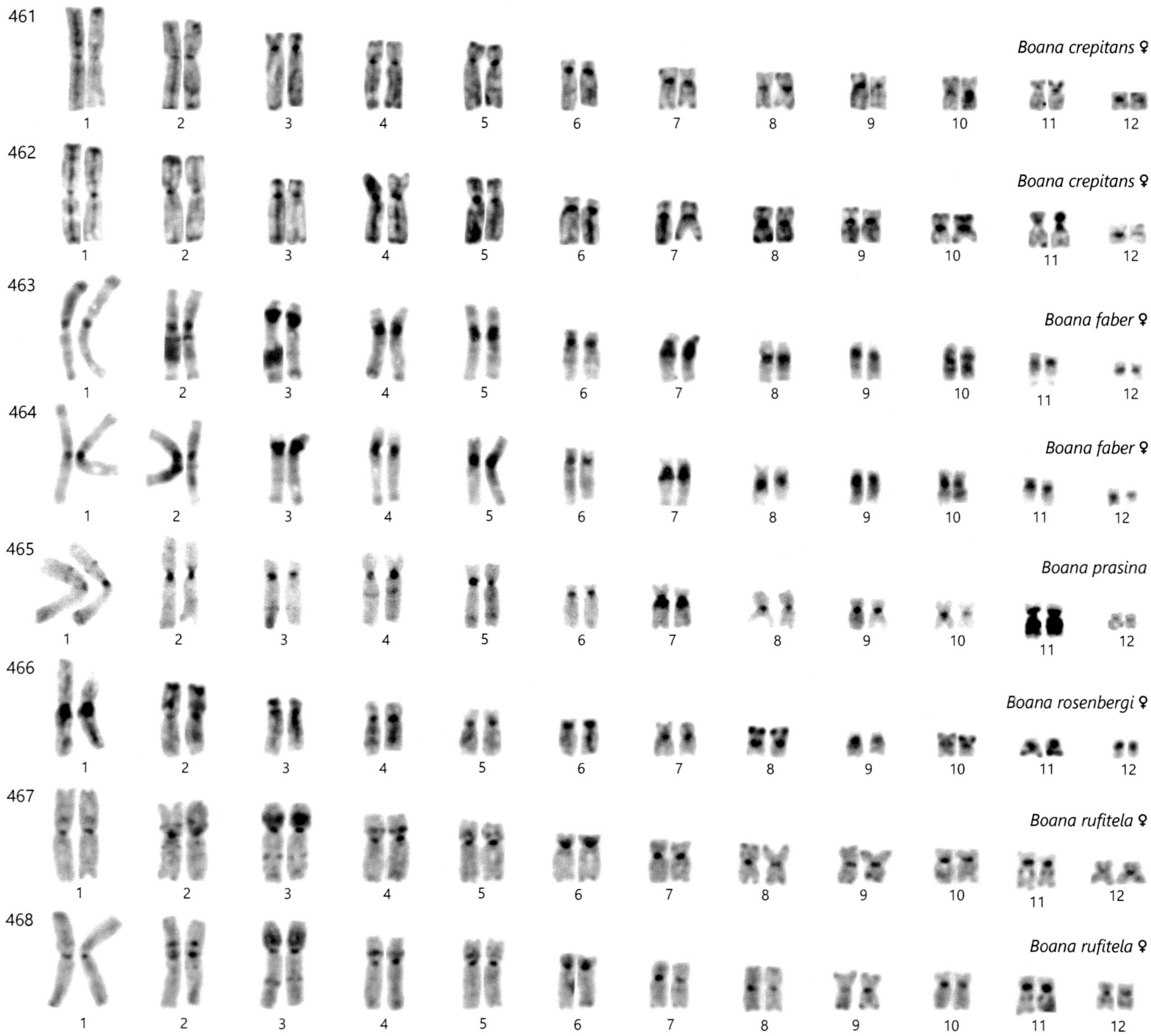

Phyllomedusidae • *Phyllomedusa*

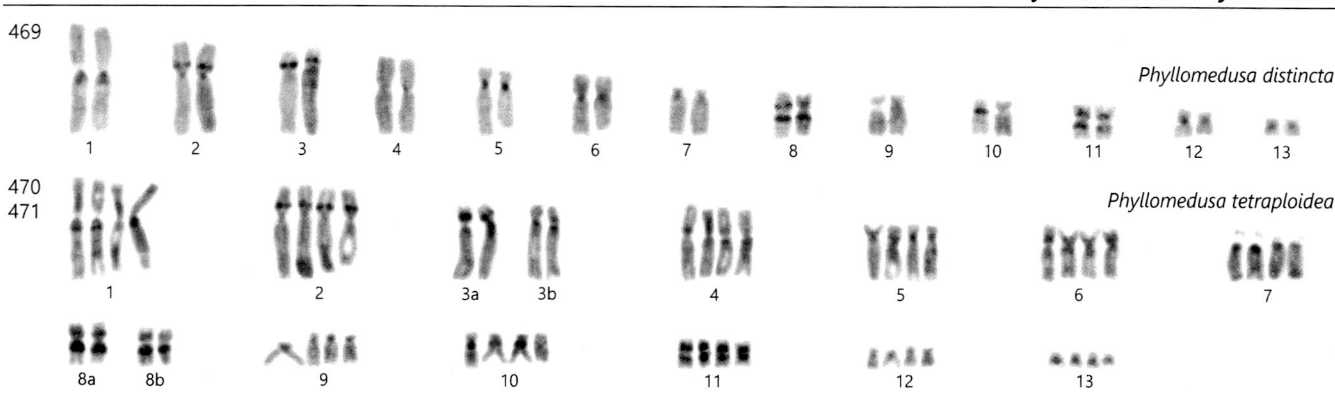

472

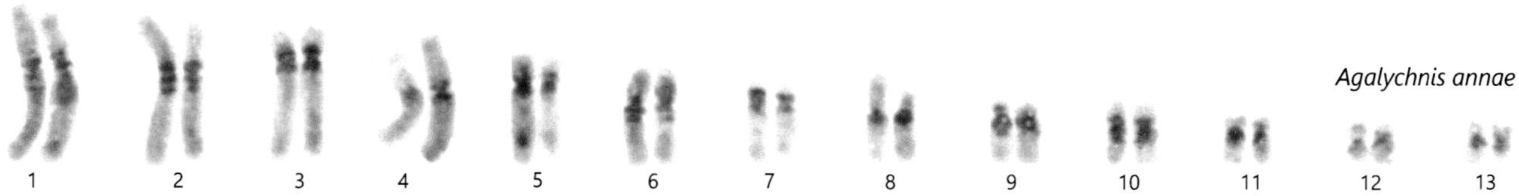

473

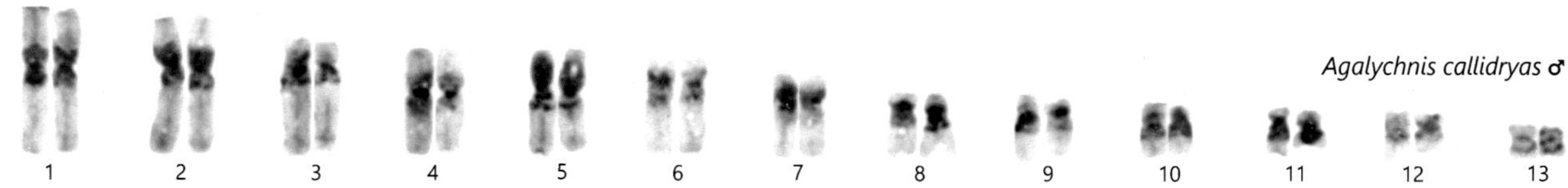

474

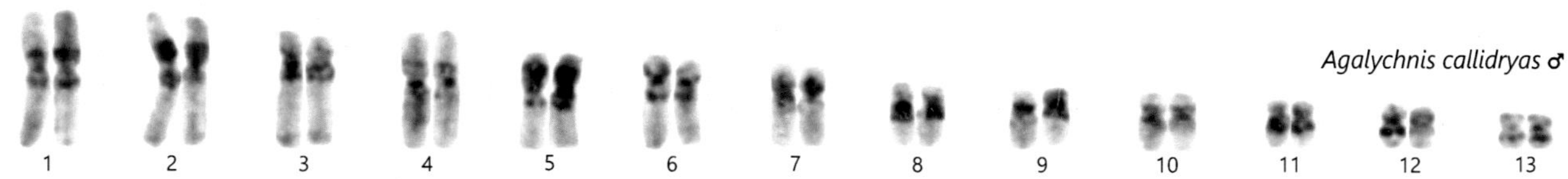

475

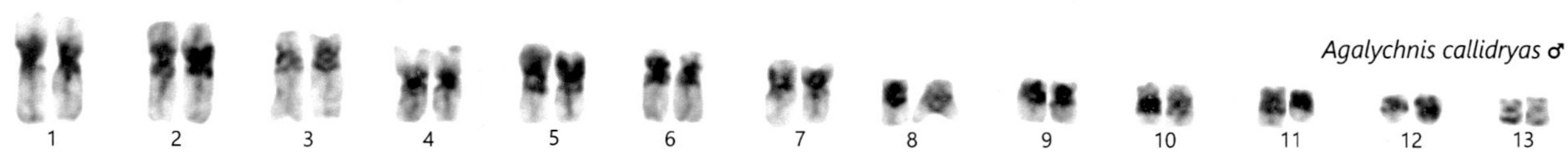

476

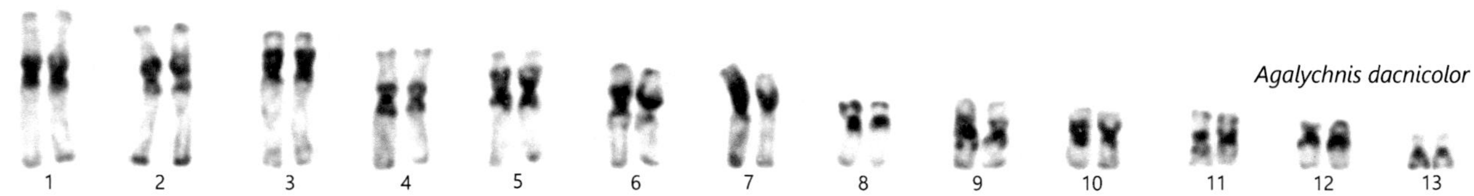

DOI: 10.1159/000489839

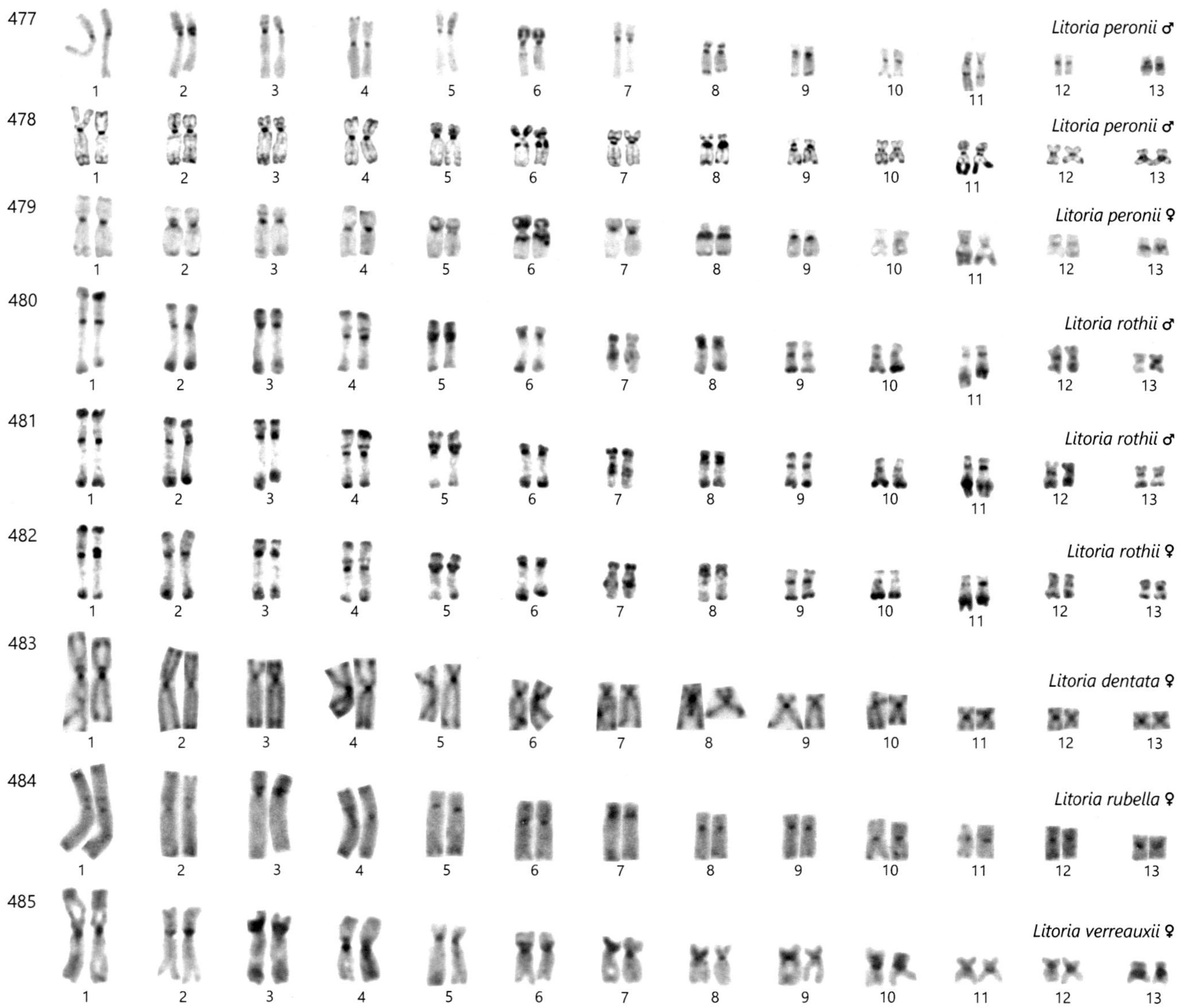

Pelodryadidae (Litoriinae) • *Litoria* • SGR 2

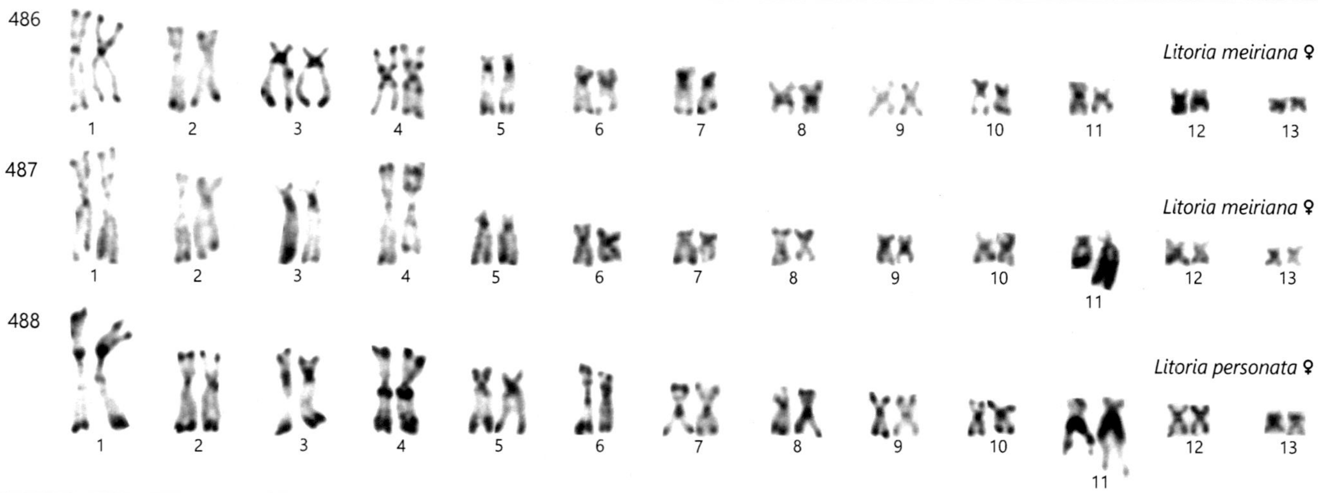

489
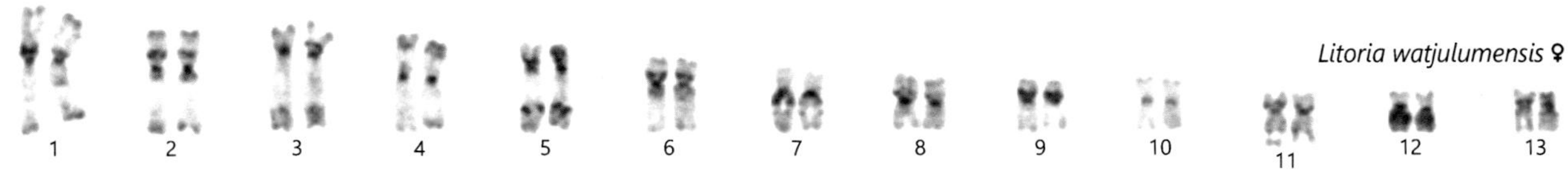

490
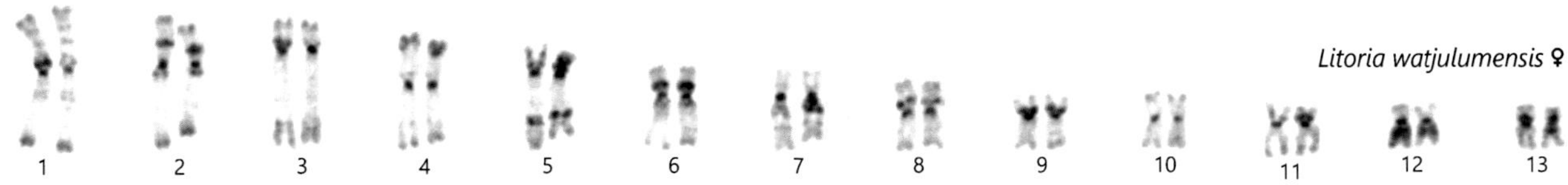

491
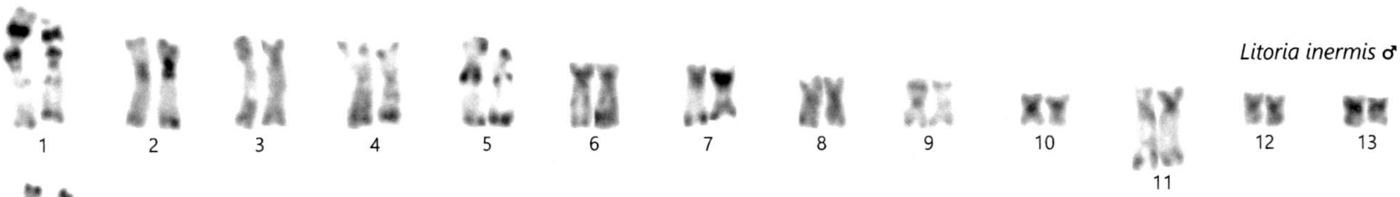

492
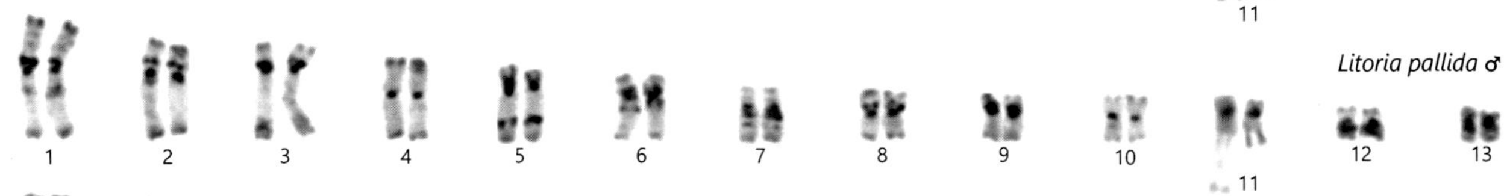

493
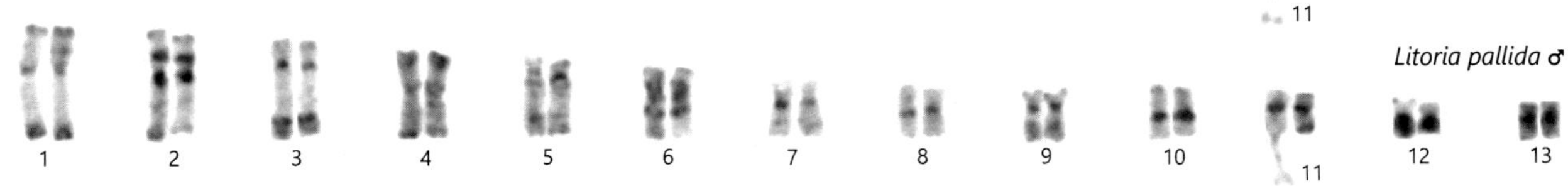

494
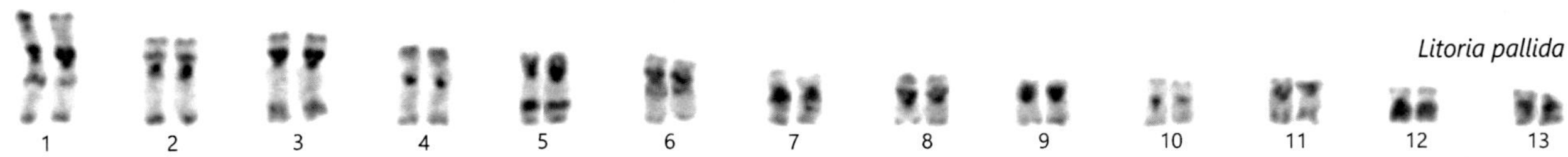

495
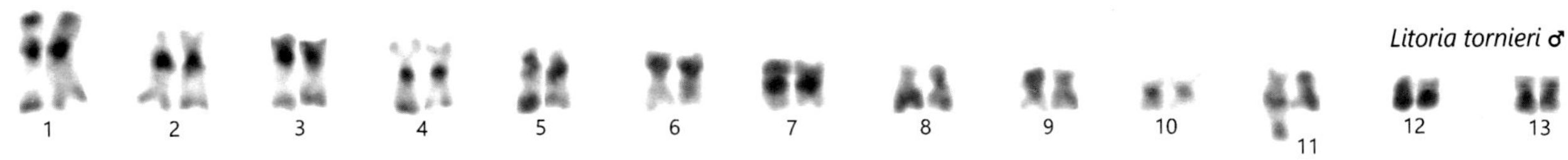

496
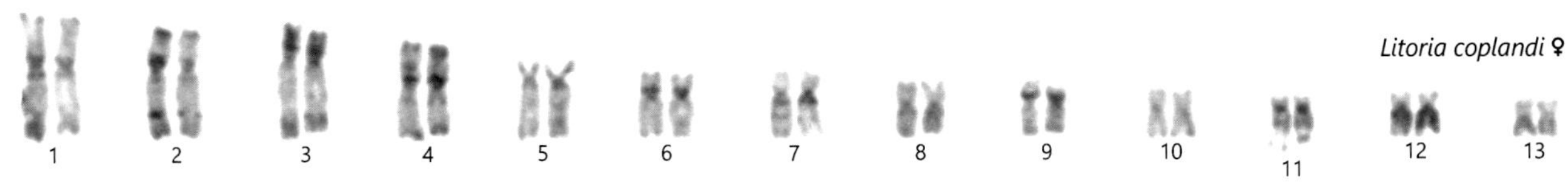

497
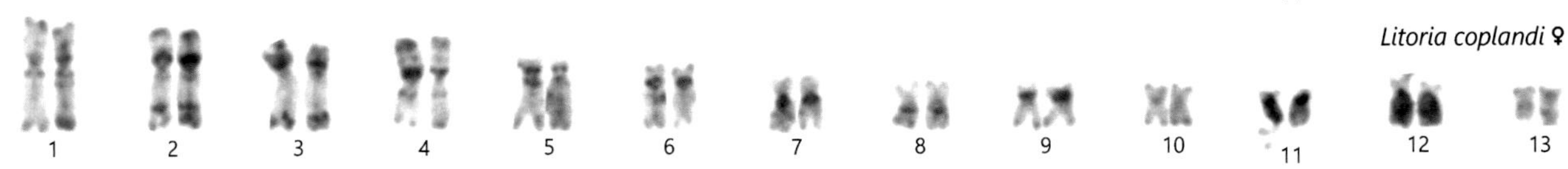

498
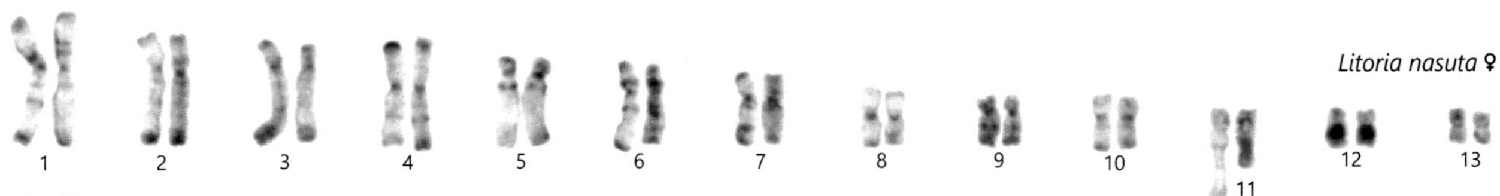

499
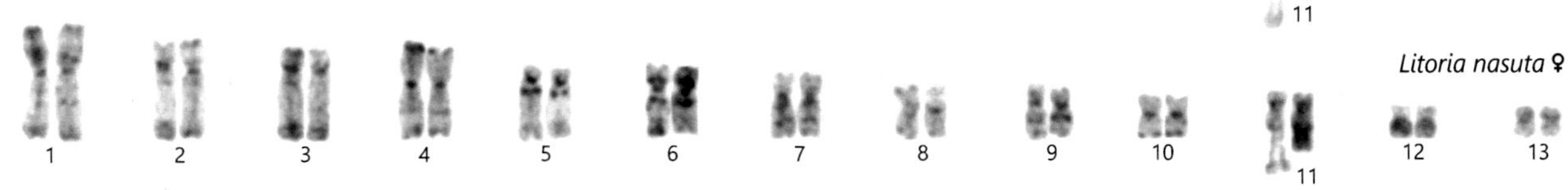

Cytogenet Genome Res 2018;155:55–221
DOI: 10.1159/000489839

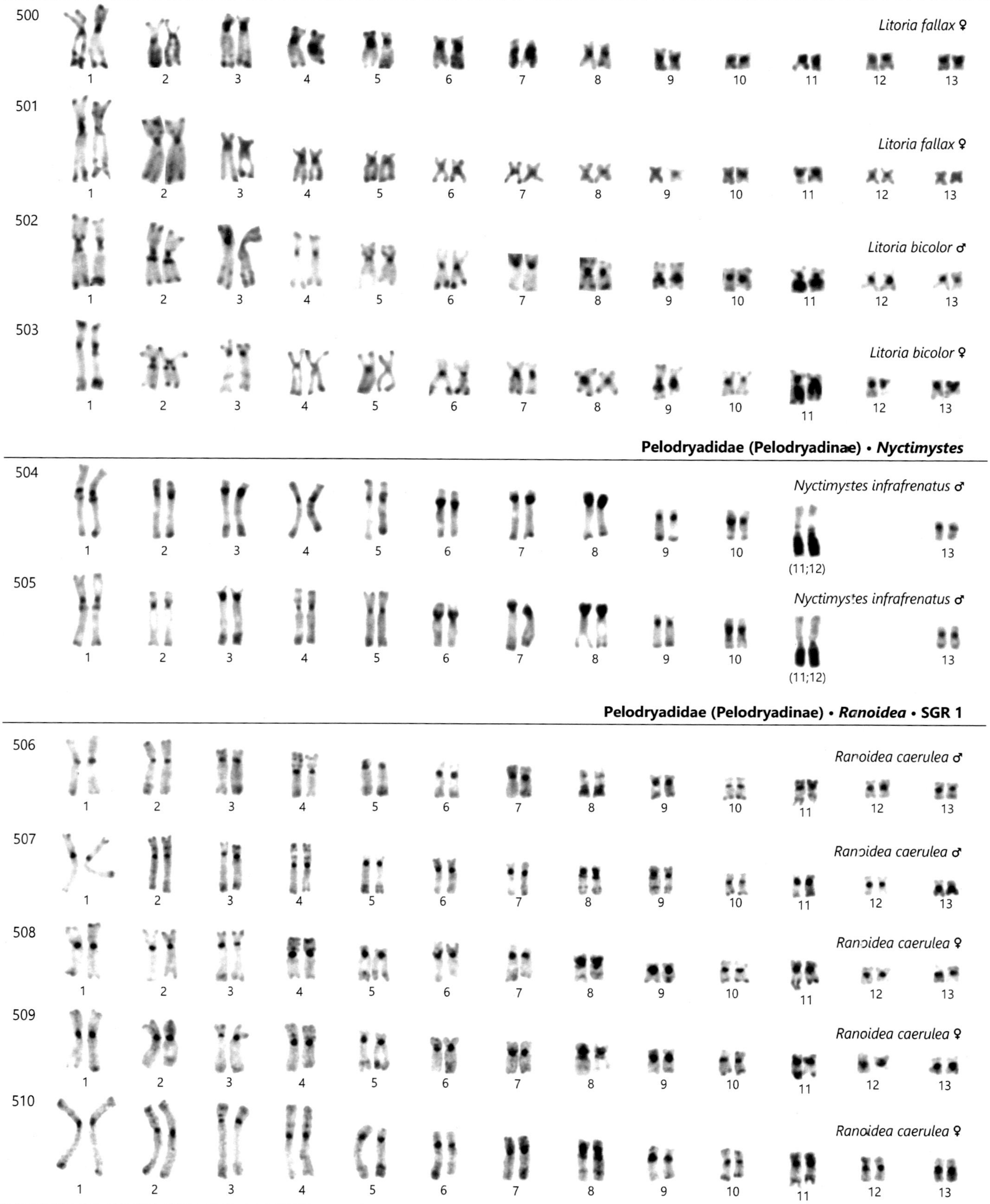
500
Litoria fallax ♀
1 2 3 4 5 6 7 8 9 10 11 12 13
501
Litoria fallax ♀
1 2 3 4 5 6 7 8 9 10 11 12 13
502
Litoria bicolor ♂
1 2 3 4 5 6 7 8 9 10 11 12 13
503
Litoria bicolor ♀
1 2 3 4 5 6 7 8 9 10 11 12 13
Pelodryadidae (Pelodryadinae) • *Nyctimystes*
504
Nyctimystes infrafrenatus ♂
1 2 3 4 5 6 7 8 9 10 (11;12) 13
505
Nyctimystes infrafrenatus ♂
1 2 3 4 5 6 7 8 9 10 (11;12) 13
Pelodryadidae (Pelodryadinae) • *Ranoidea* • SGR 1
506
Ranoidea caerulea ♂
1 2 3 4 5 6 7 8 9 10 11 12 13
507
Ranoidea caerulea ♂
1 2 3 4 5 6 7 8 9 10 11 12 13
508
Ranoidea caerulea ♀
1 2 3 4 5 6 7 8 9 10 11 12 13
509
Ranoidea caerulea ♀
1 2 3 4 5 6 7 8 9 10 11 12 13
510
Ranoidea caerulea ♀
1 2 3 4 5 6 7 8 9 10 11 12 13

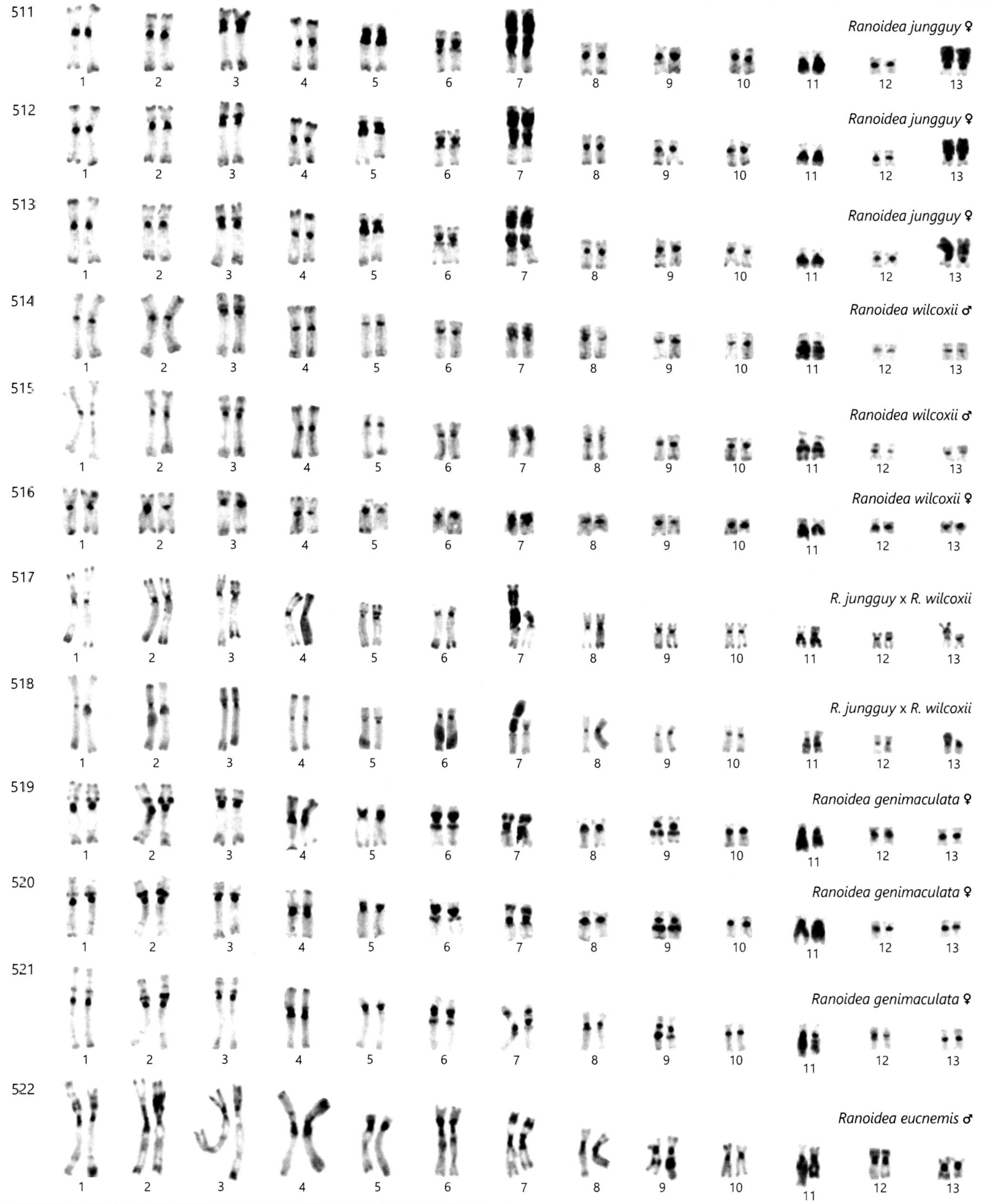
511
Ranoidea jungguy ♀
1 2 3 4 5 6 7 8 9 10 11 12 13
512
Ranoidea jungguy ♀
1 2 3 4 5 6 7 8 9 10 11 12 13
513
Ranoidea jungguy ♀
1 2 3 4 5 6 7 8 9 10 11 12 13
514
Ranoidea wilcoxii ♂
1 2 3 4 5 6 7 8 9 10 11 12 13
515
Ranoidea wilcoxii ♂
1 2 3 4 5 6 7 8 9 10 11 12 13
516
Ranoidea wilcoxii ♀
1 2 3 4 5 6 7 8 9 10 11 12 13
517
R. jungguy x R. wilcoxii
1 2 3 4 5 6 7 8 9 10 11 12 13
518
R. jungguy x R. wilcoxii
1 2 3 4 5 6 7 8 9 10 11 12 13
519
Ranoidea genimaculata ♀
1 2 3 4 5 6 7 8 9 10 11 12 13
520
Ranoidea genimaculata ♀
1 2 3 4 5 6 7 8 9 10 11 12 13
521
Ranoidea genimaculata ♀
1 2 3 4 5 6 7 8 9 10 11 12 13
522
Ranoidea eucnemis ♂
1 2 3 4 5 6 7 8 9 10 11 12 13

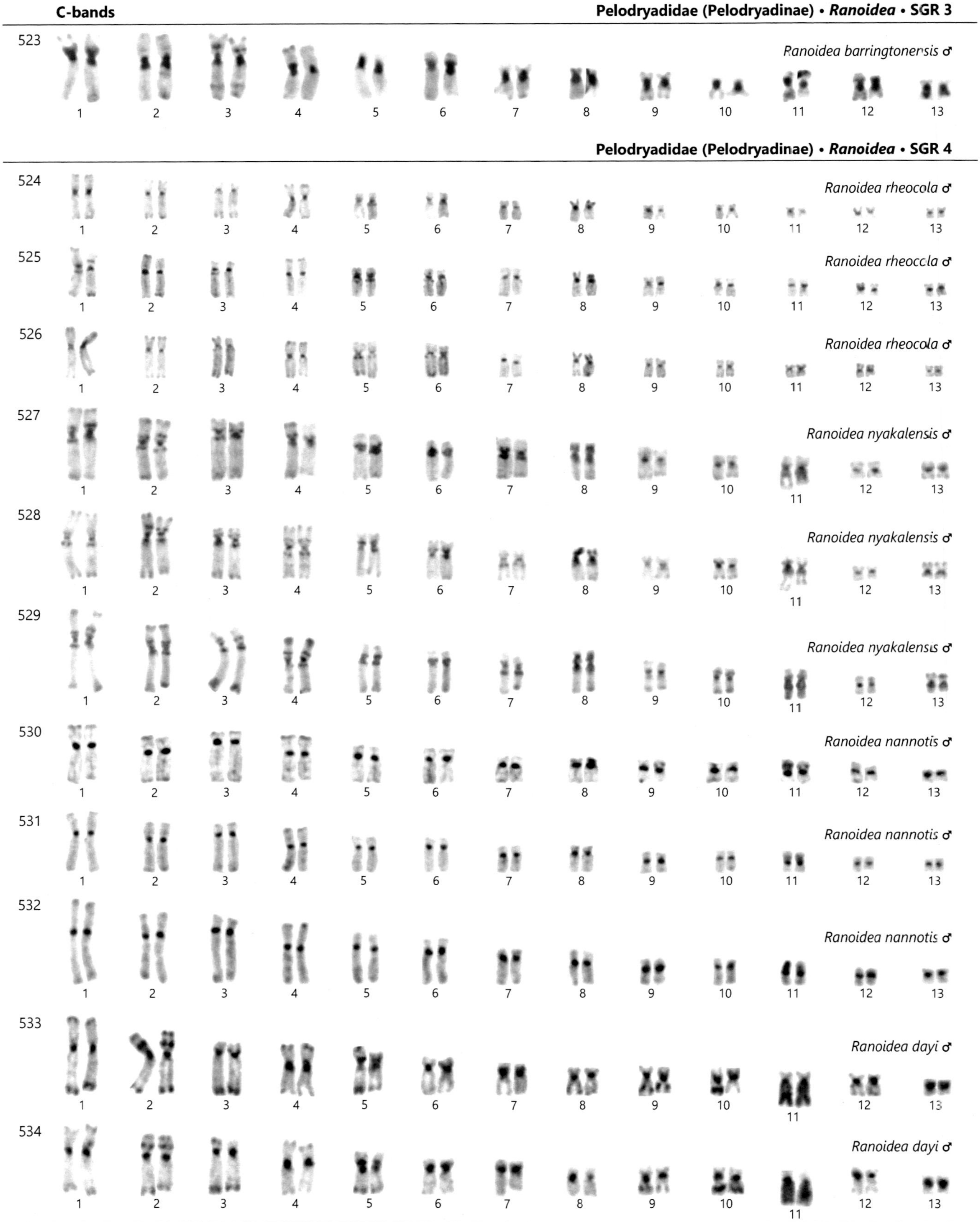
C-bands
Pelodryadidae (Pelodryadinae) • Ranoidea • SGR 3
523
Ranoidea barringtonensis ♂
1 2 3 4 5 6 7 8 9 10 11 12 13
Pelodryadidae (Pelodryadinae) • Ranoidea • SGR 4
524
Ranoidea rheocola ♂
525
Ranoidea rheocola ♂
526
Ranoidea rheocola ♂
527
Ranoidea nyakalensis ♂
528
Ranoidea nyakalensis ♂
529
Ranoidea nyakalensis ♂
530
Ranoidea nannotis ♂
531
Ranoidea nannotis ♂
532
Ranoidea nannotis ♂
533
Ranoidea dayi ♂
534
Ranoidea dayi ♂

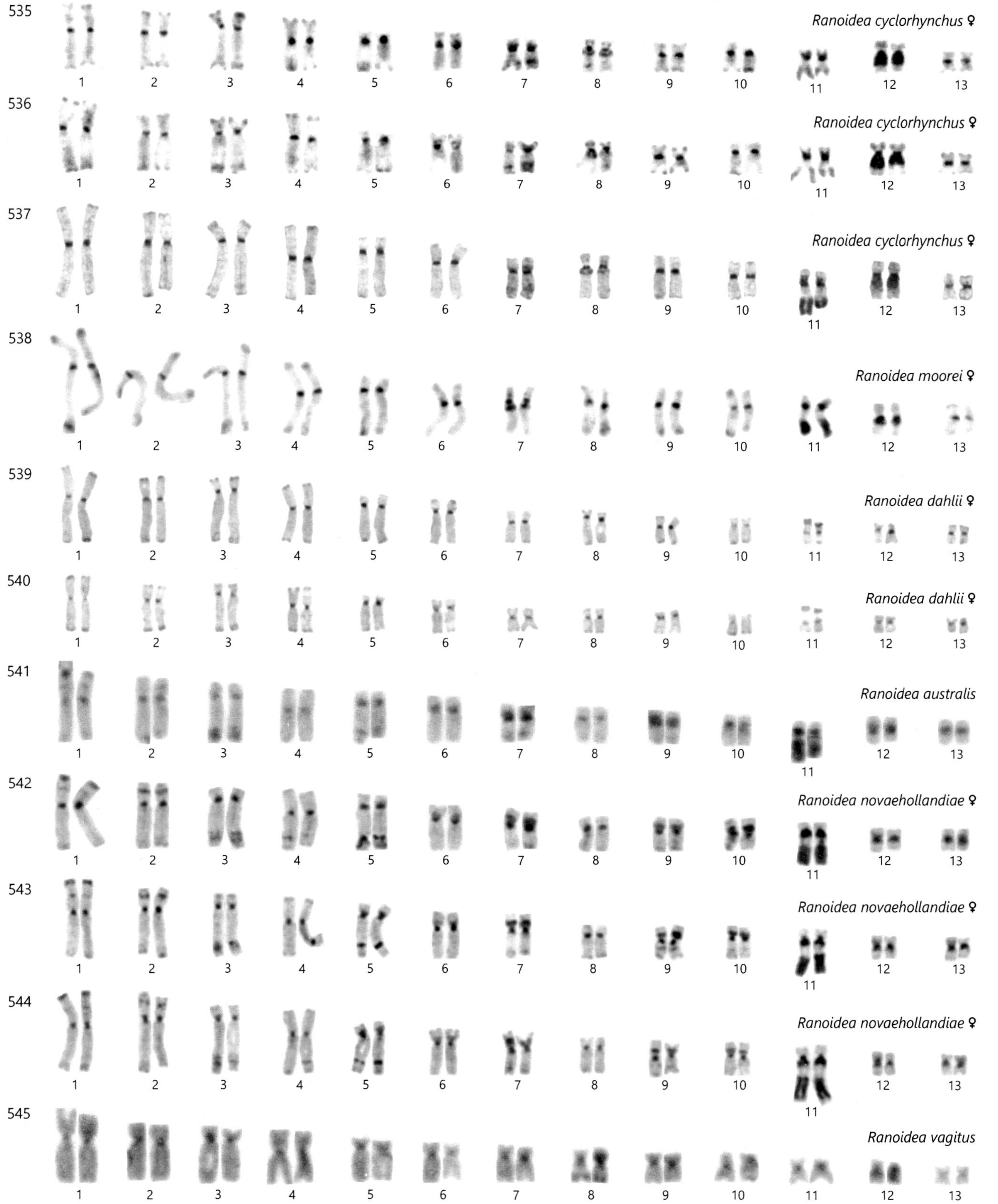
535 Ranoidea cyclorhynchus ♀
536 Ranoidea cyclorhynchus ♀
537 Ranoidea cyclorhynchus ♀
538 Ranoidea moorei ♀
539 Ranoidea dahlii ♀
540 Ranoidea dahlii ♀
541 Ranoidea australis
542 Ranoidea novaehollandiae ♀
543 Ranoidea novaehollandiae ♀
544 Ranoidea novaehollandiae ♀
545 Ranoidea vagitus

Hylidae (Acridinae) • *Pseudacris*

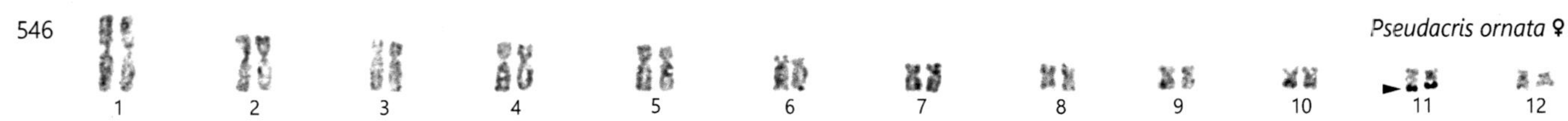

Hylidae (Hylinae) • *Ptychohyla*

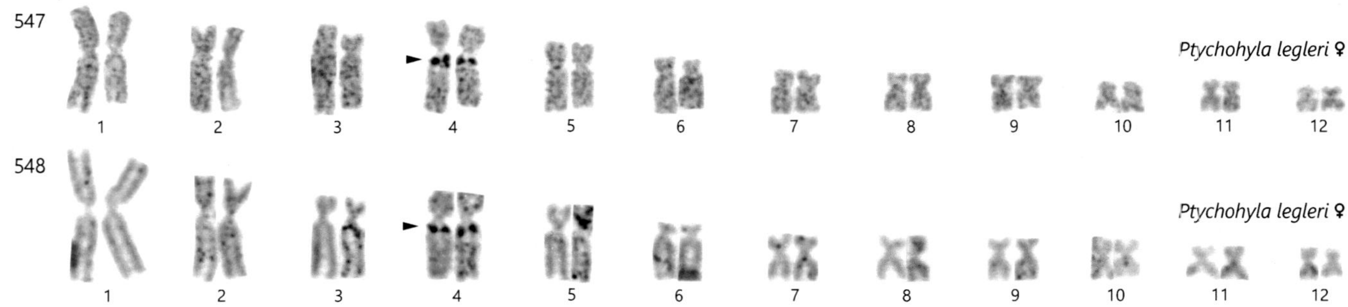

Hylidae (Hylinae) • *Smilisca*

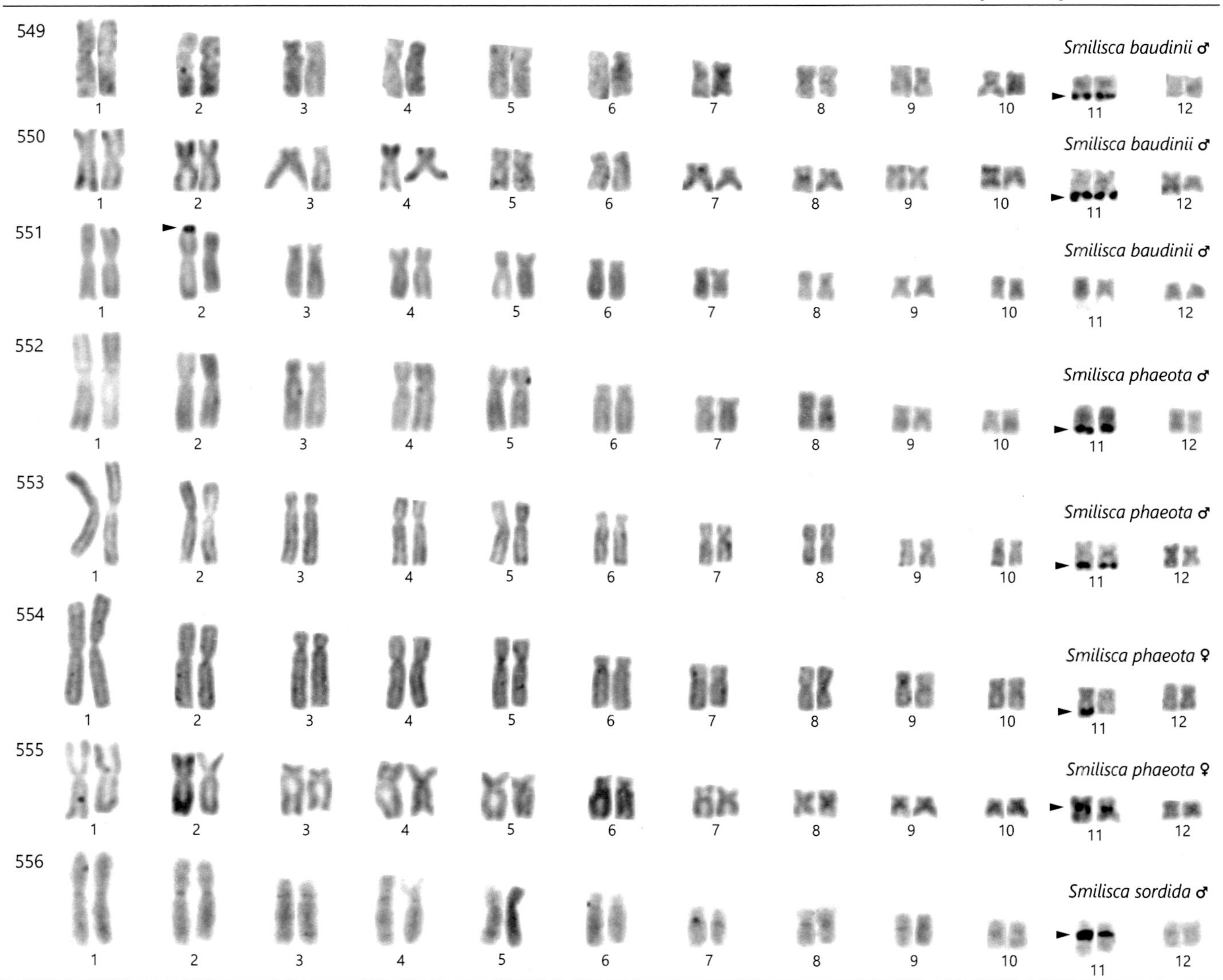

Hylidae (Hylinae) • *Smilisca*

557

558

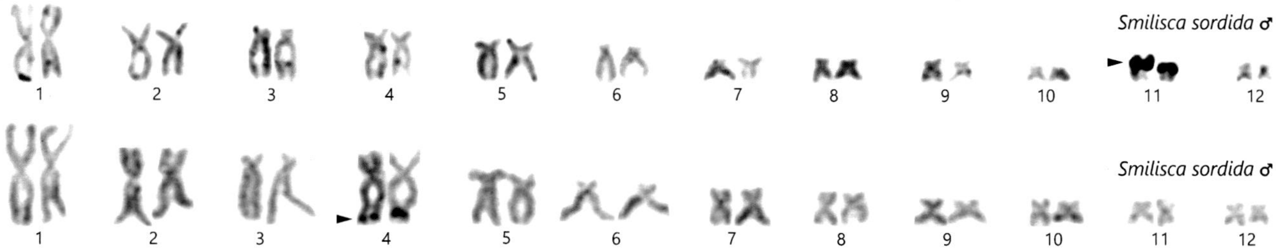

Hylidae (Hylinae) • *Isthmohyla*

559

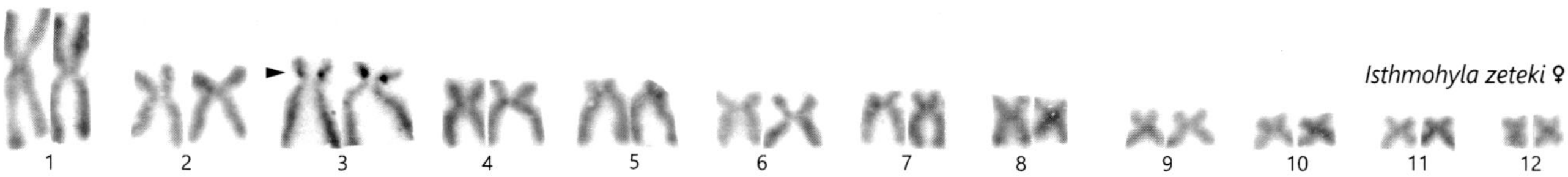

Hylidae (Hylinae) • *Dryophytes*

560

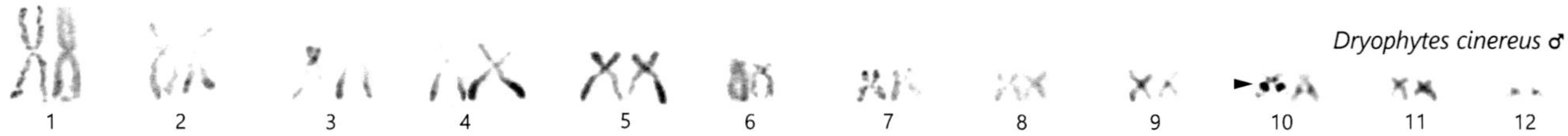

Hylidae• (Dendropsophinae) • *Dendropsophus*

561

562

563

564

565

566

567

568

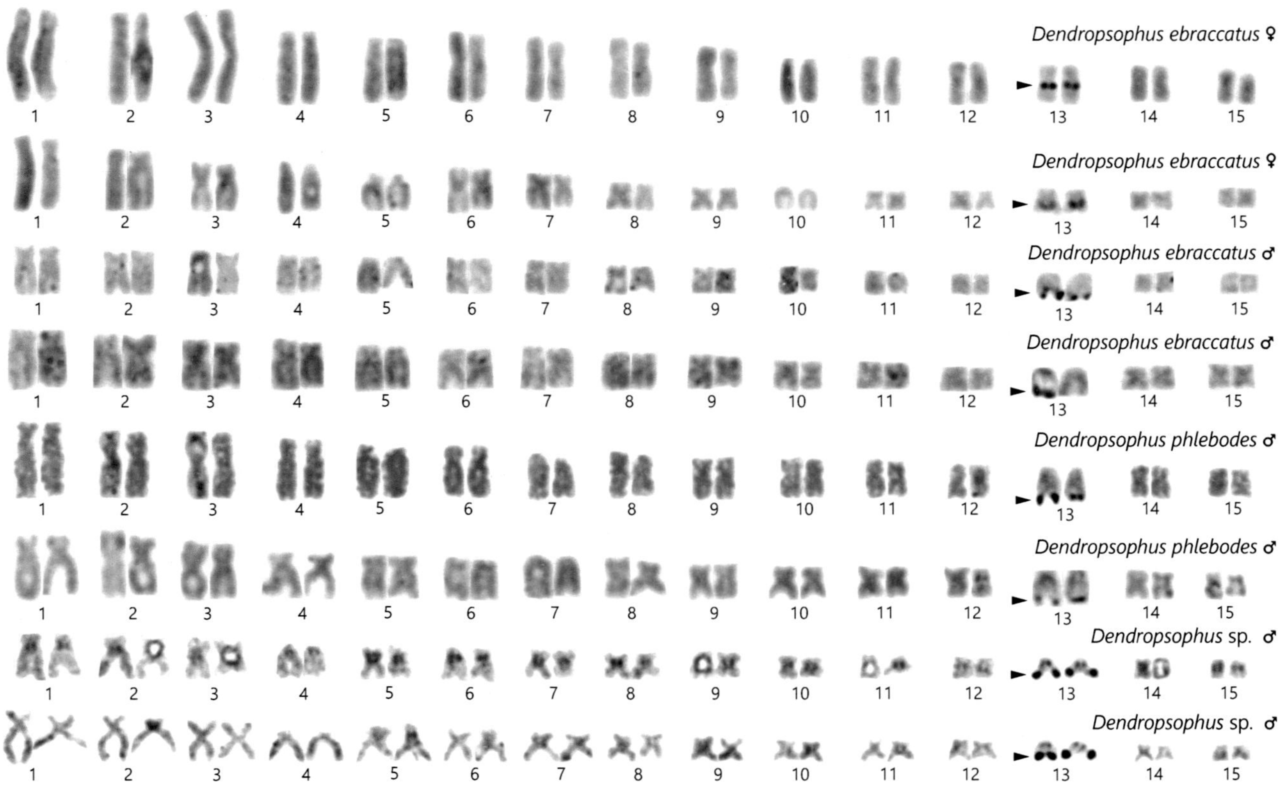

Hylidae (Lophyohylinae) • *Osteopilus*

569

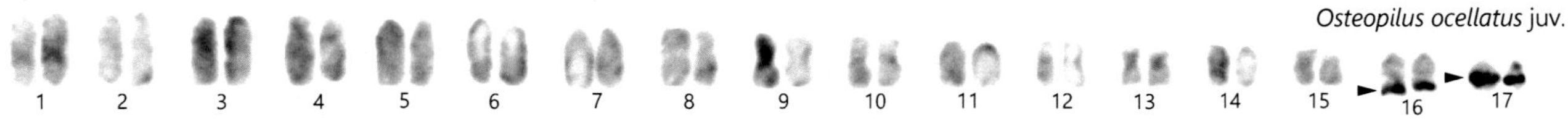

Cytogenet Genome Res 2018;155:55–221
DOI: 10.1159/000489839

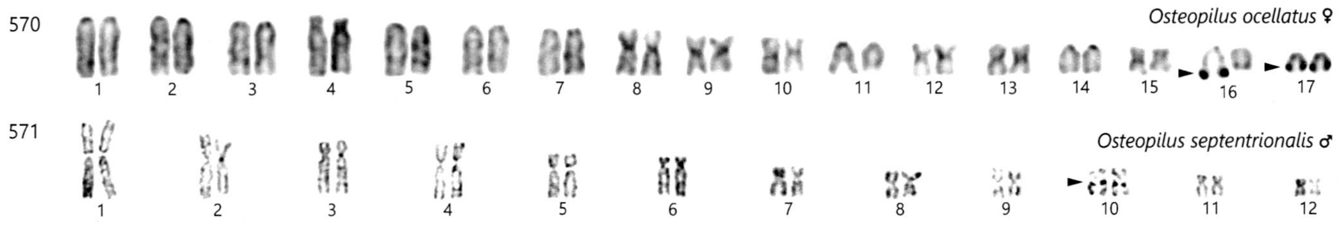

Hylidae (Scinaxinae) • *Scinax*

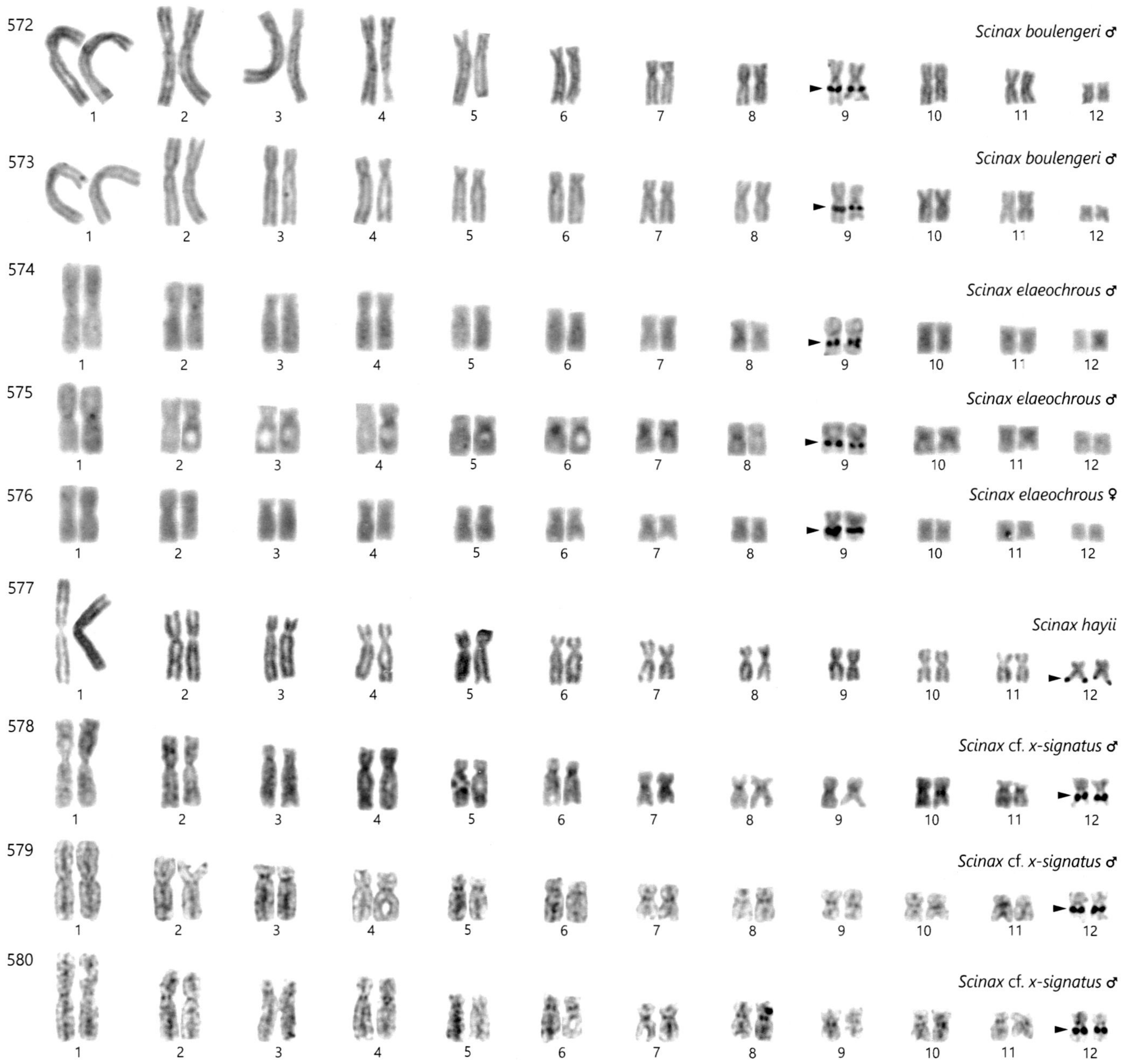

DOI: 10.1159/000489839

581
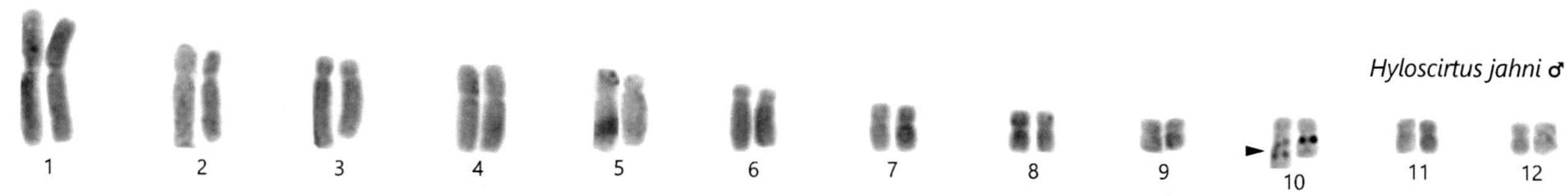

Hyloscirtus jahni ♂

Hylidae (Cophomantinae) • *Bokermannohyla*

582
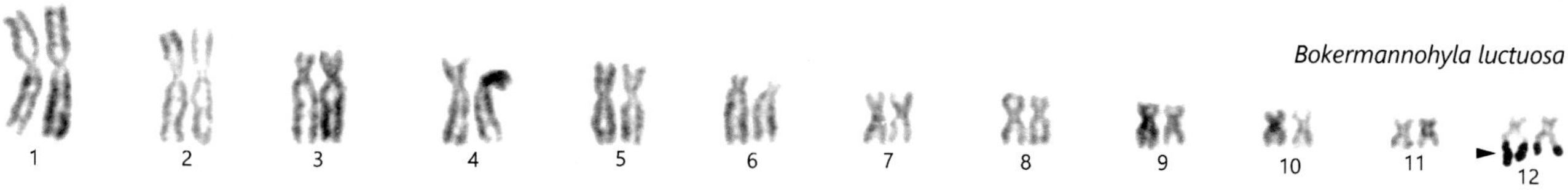

Bokermannohyla luctuosa

Hylidae (Cophomantinae) • *Boana*

583
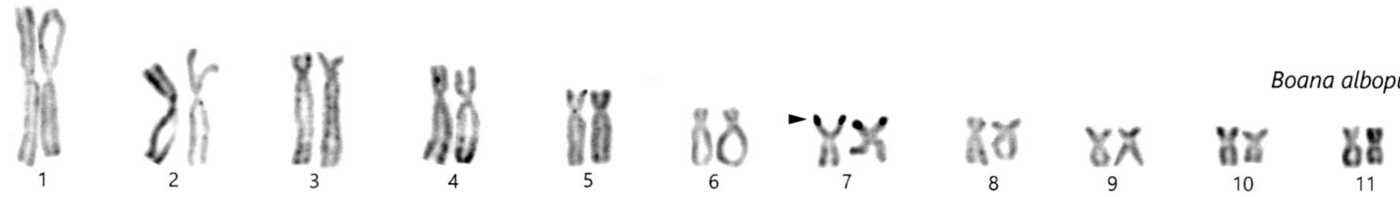

Boana albopunctata ♂

584
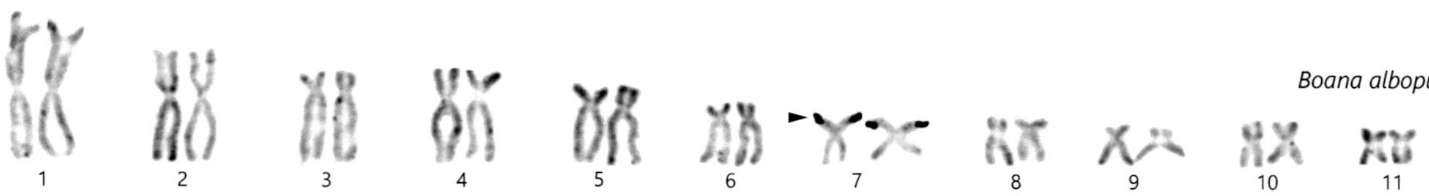

Boana albopunctata ♂

585
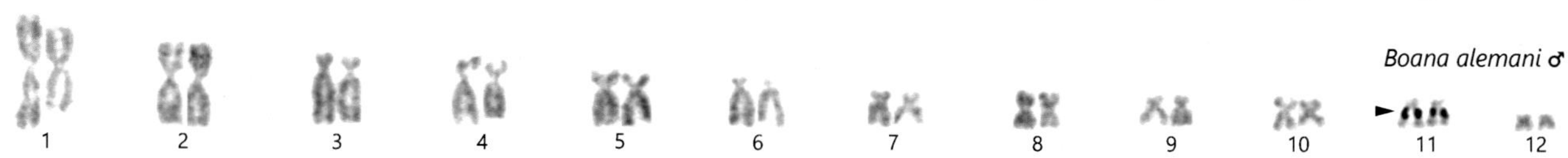

Boana alemani ♂

586
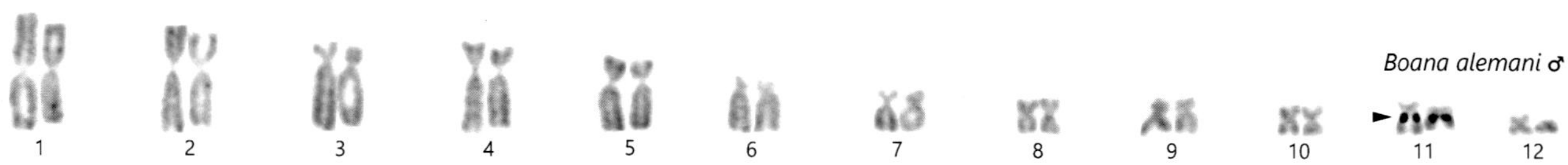

Boana alemani ♂

587
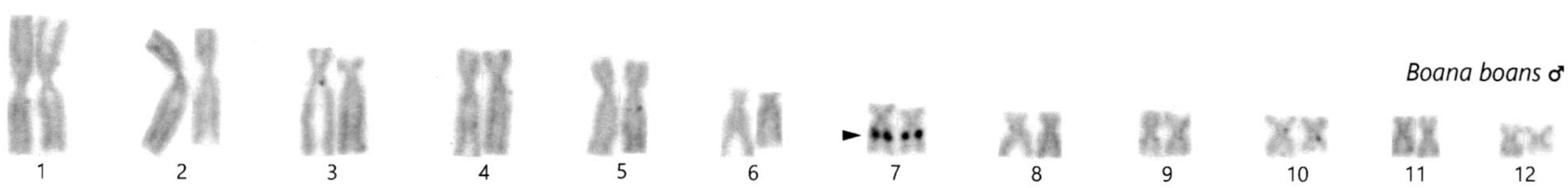

Boana boans ♂

588
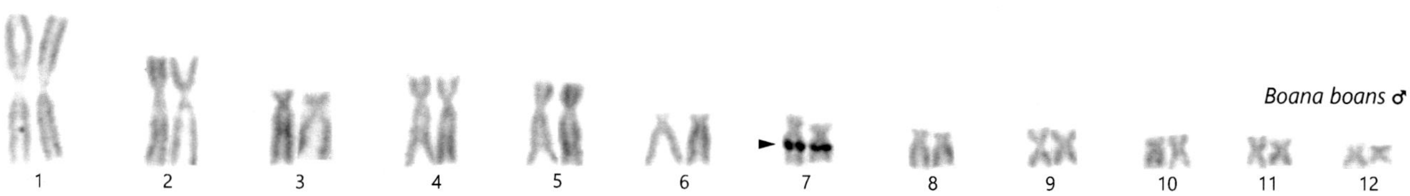

Boana boans ♂

589
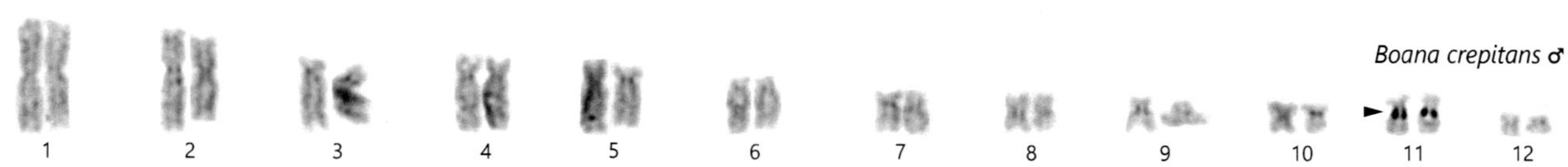

Boana crepitans ♂

590
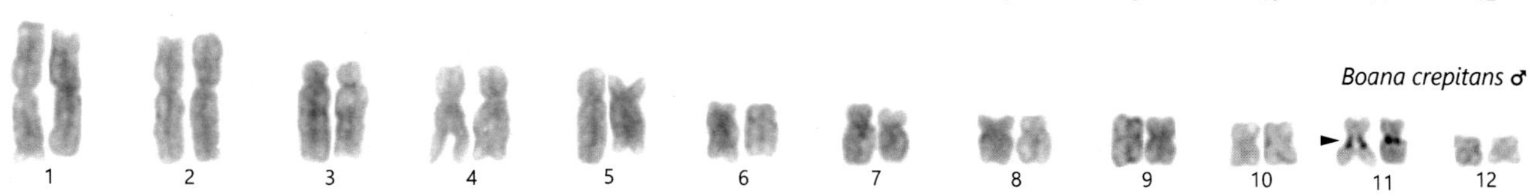

Boana crepitans ♂

Cytogenet Genome Res 2018;155:55–221
DOI: 10.1159/000489839

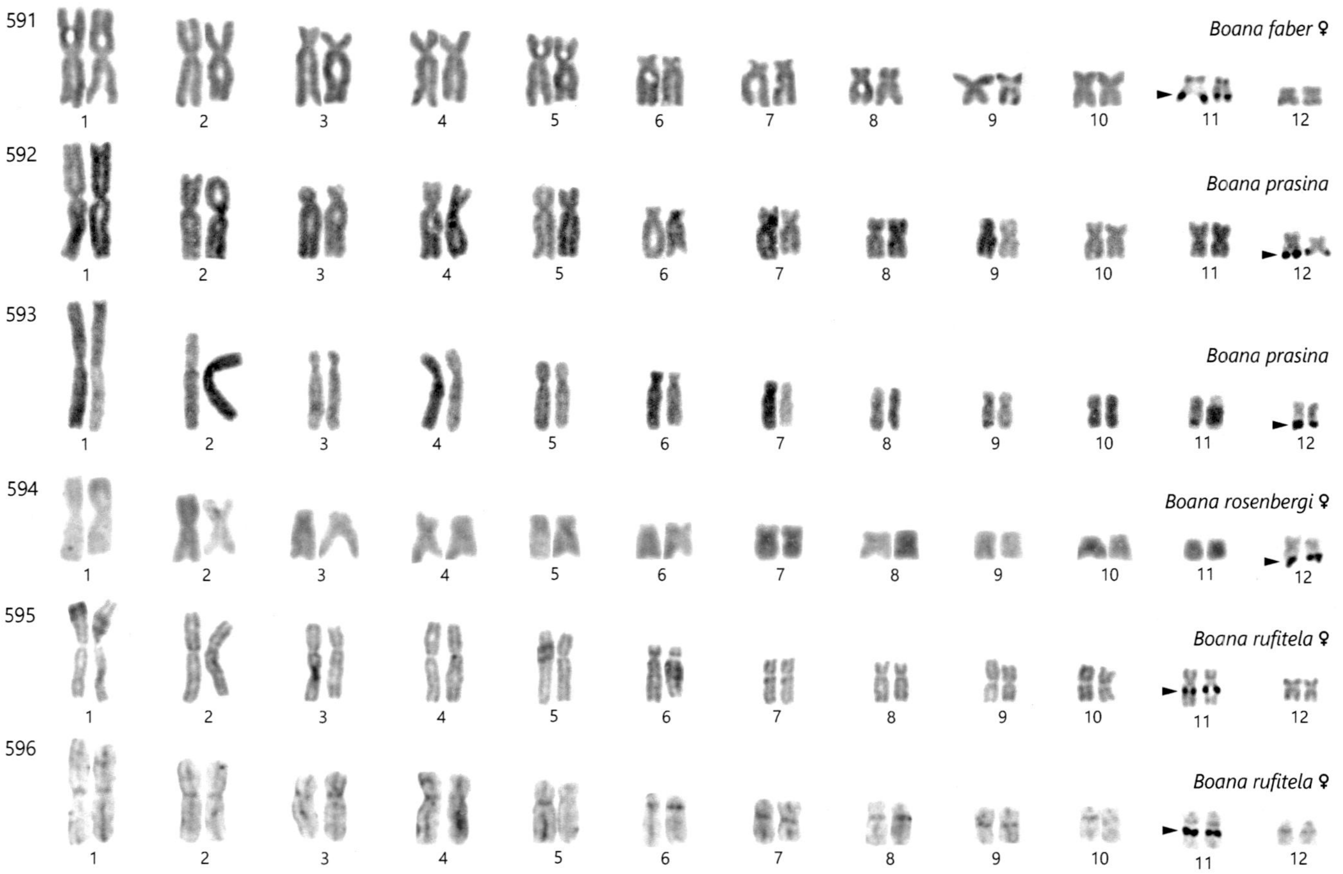
591
Boana faber ♀
1 2 3 4 5 6 7 8 9 10 11 12
592
Boana prasina
1 2 3 4 5 6 7 8 9 10 11 12
593
Boana prasina
1 2 3 4 5 6 7 8 9 10 11 12
594
Boana rosenbergi ♀
1 2 3 4 5 6 7 8 9 10 11 12
595
Boana rufitela ♀
1 2 3 4 5 6 7 8 9 10 11 12
596
Boana rufitela ♀
1 2 3 4 5 6 7 8 9 10 11 12

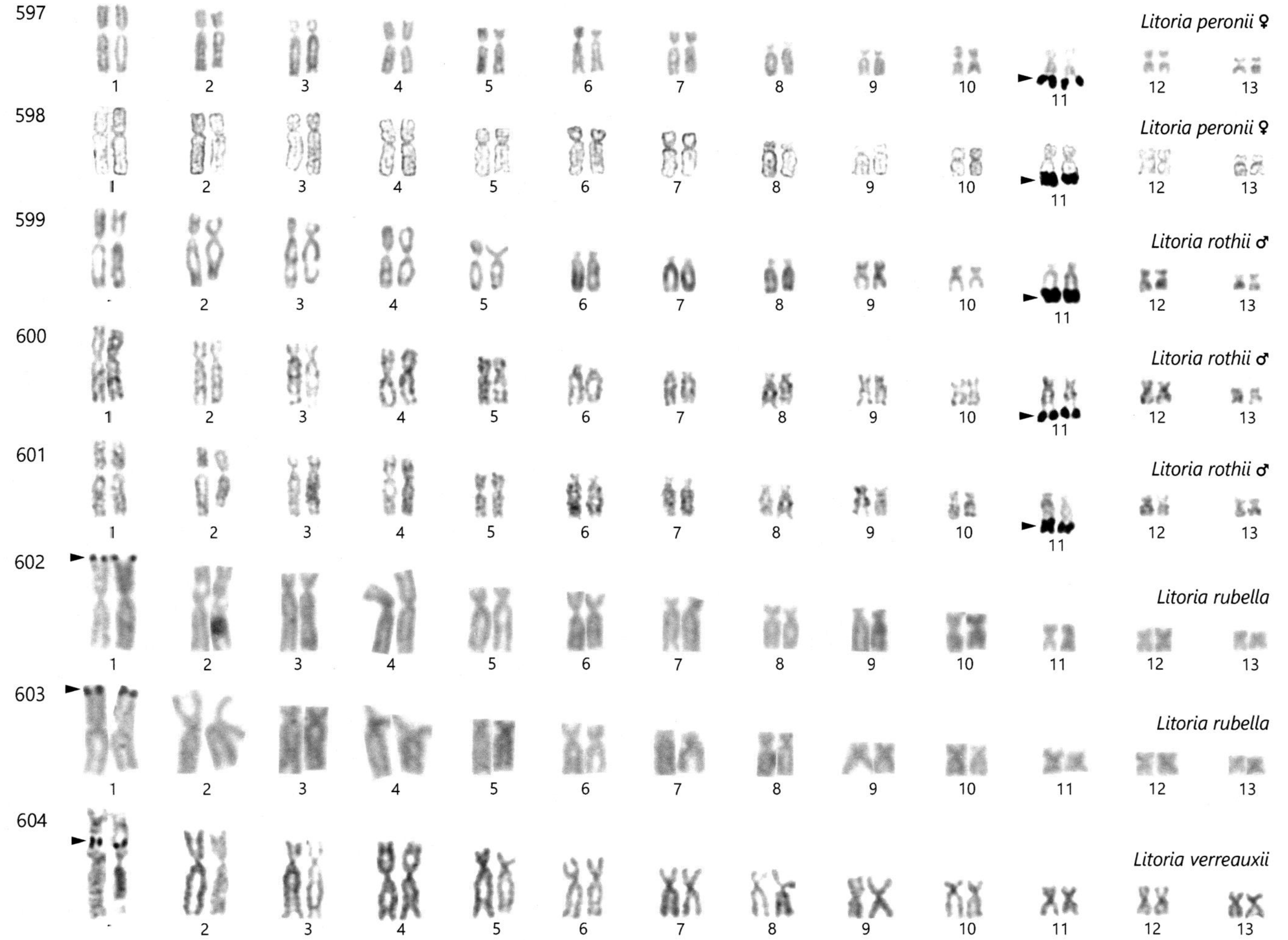

Pelodryadidae (Litoriinae) • *Litoria* • SGR 2

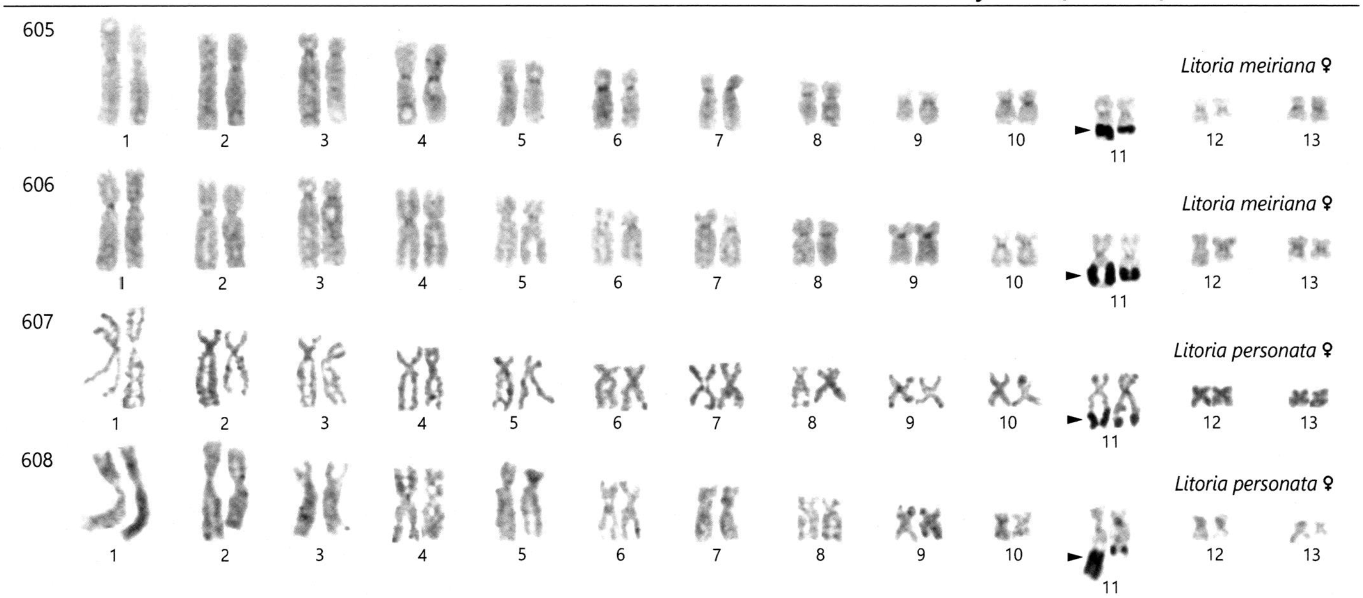

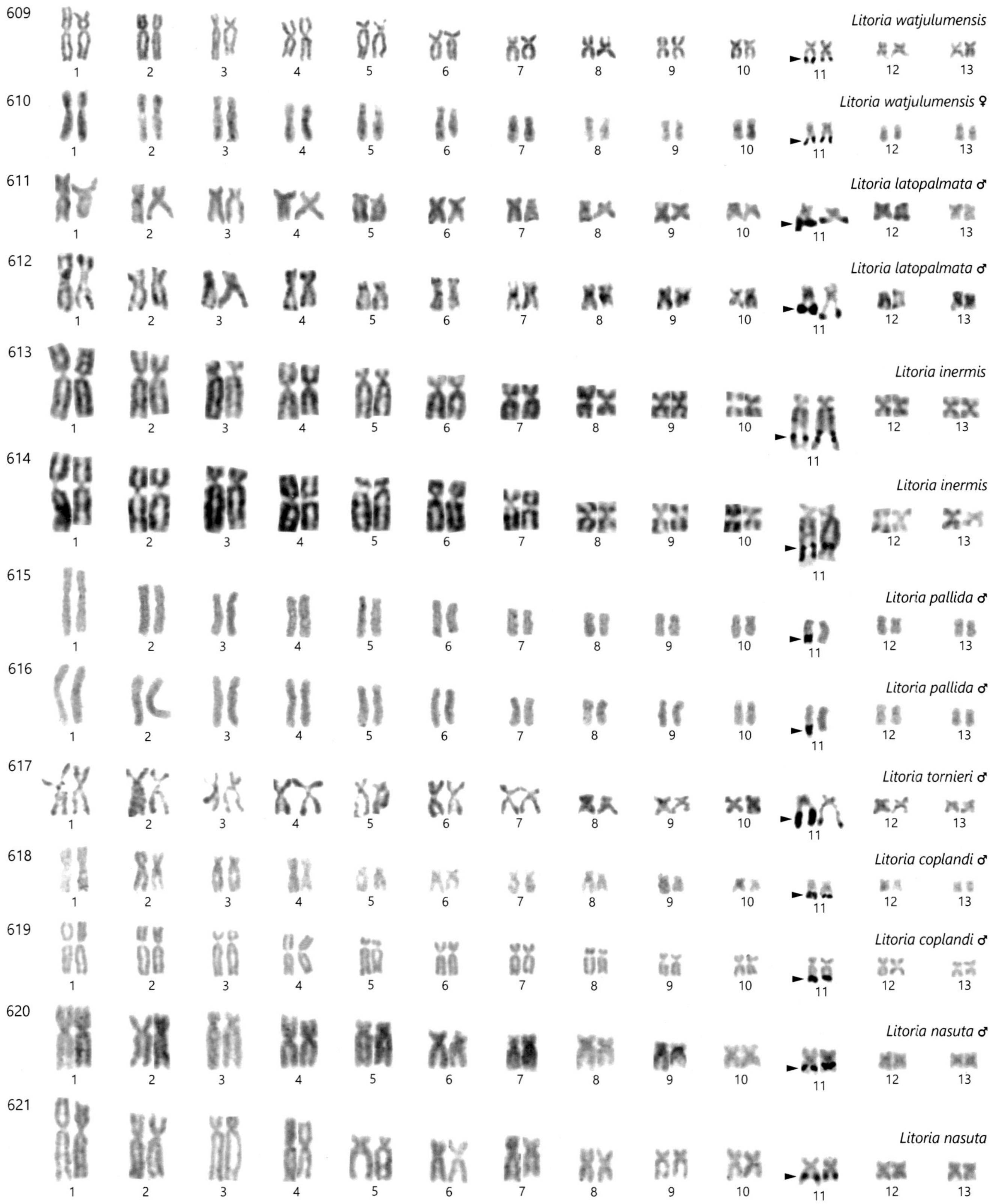
609 Litoria watjulumensis
610 Litoria watjulumensis ♀
611 Litoria latopalmata ♂
612 Litoria latopalmata ♂
613 Litoria inermis
614 Litoria inermis
615 Litoria pallida ♂
616 Litoria pallida ♂
617 Litoria tornieri ♂
618 Litoria coplandi ♂
619 Litoria coplandi ♂
620 Litoria nasuta ♂
621 Litoria nasuta
1 2 3 4 5 6 7 8 9 10 11 12 13

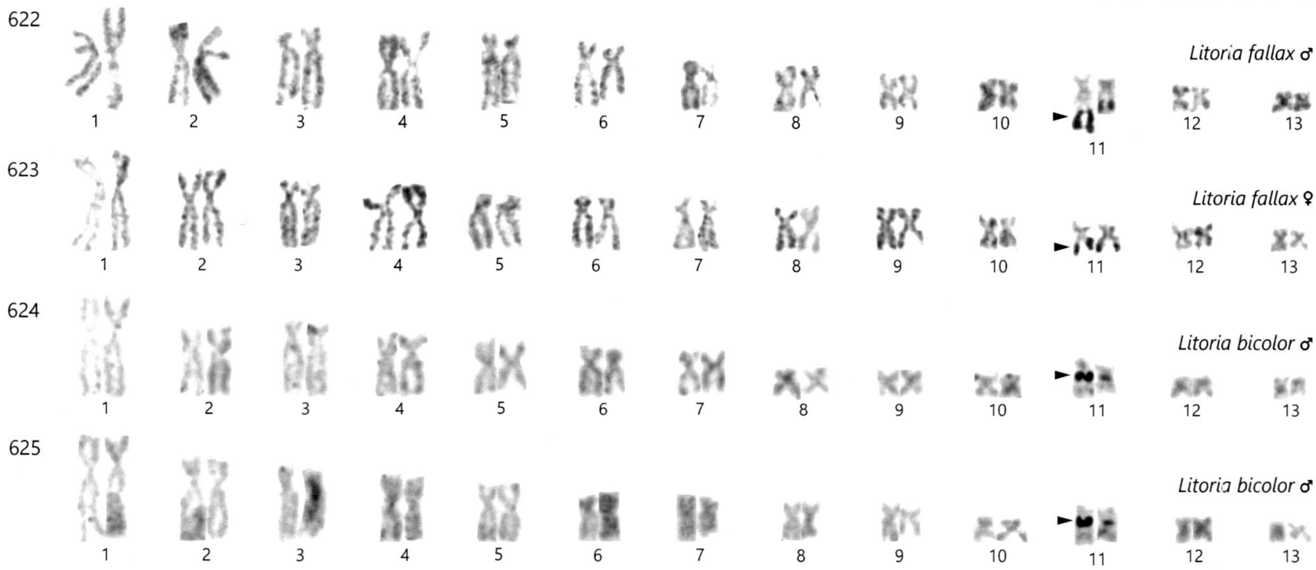

Pelodryadidae (Pelodryadinae) • *Nyctimystes*

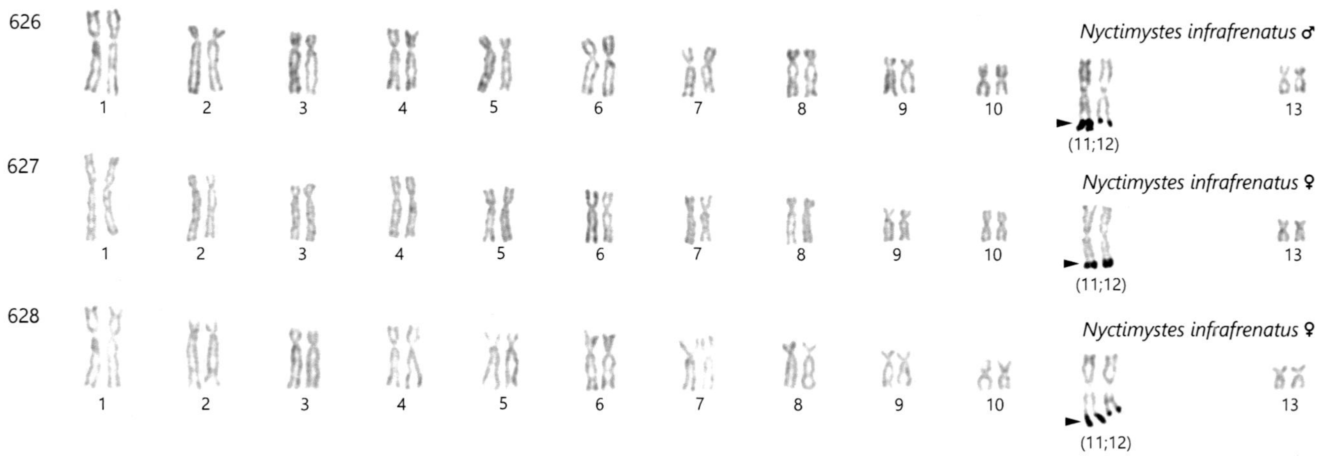

Pelodryadidae (Pelodryadinae) • *Ranoidea* • SGR 1

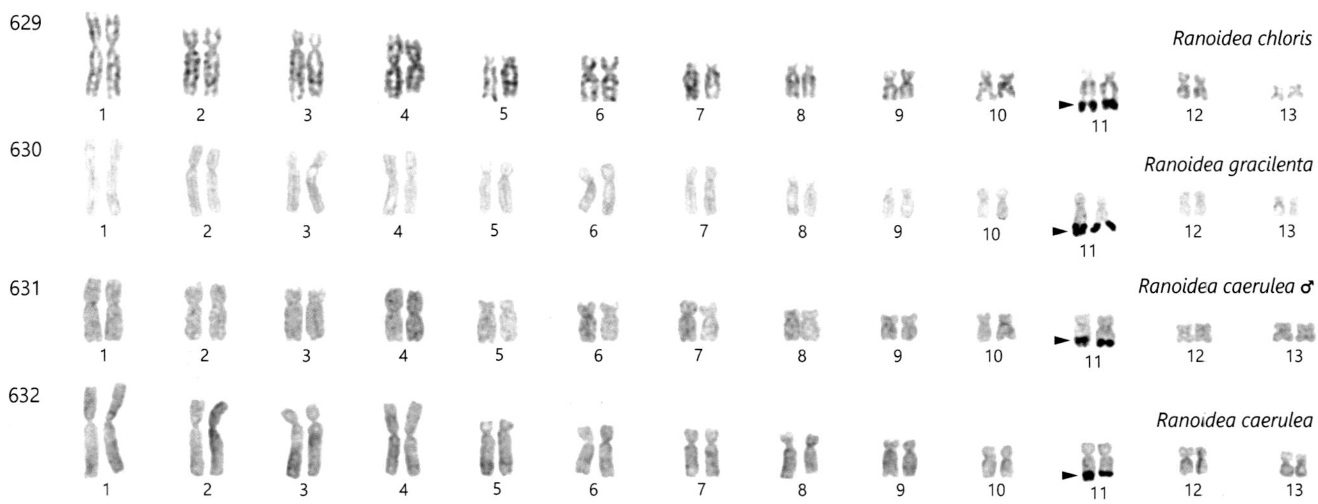

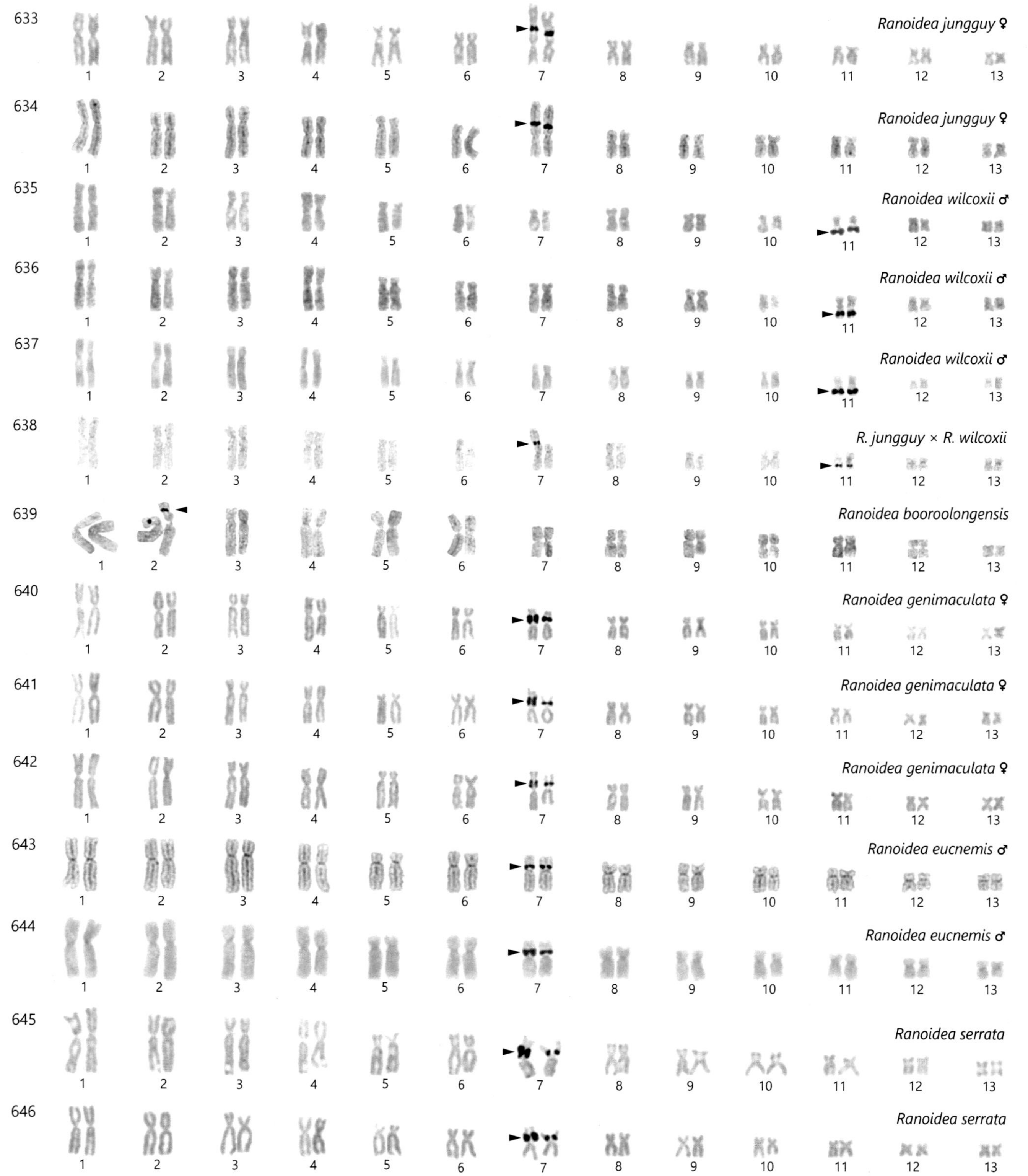
633 Ranoidea jungguy ♀
634 Ranoidea jungguy ♀
635 Ranoidea wilcoxii ♂
636 Ranoidea wilcoxii ♂
637 Ranoidea wilcoxii ♂
638 R. jungguy × R. wilcoxii
639 Ranoidea booroolongensis
640 Ranoidea genimaculata ♀
641 Ranoidea genimaculata ♀
642 Ranoidea genimaculata ♀
643 Ranoidea eucnemis ♂
644 Ranoidea eucnemis ♂
645 Ranoidea serrata
646 Ranoidea serrata
1 2 3 4 5 6 7 8 9 10 11 12 13

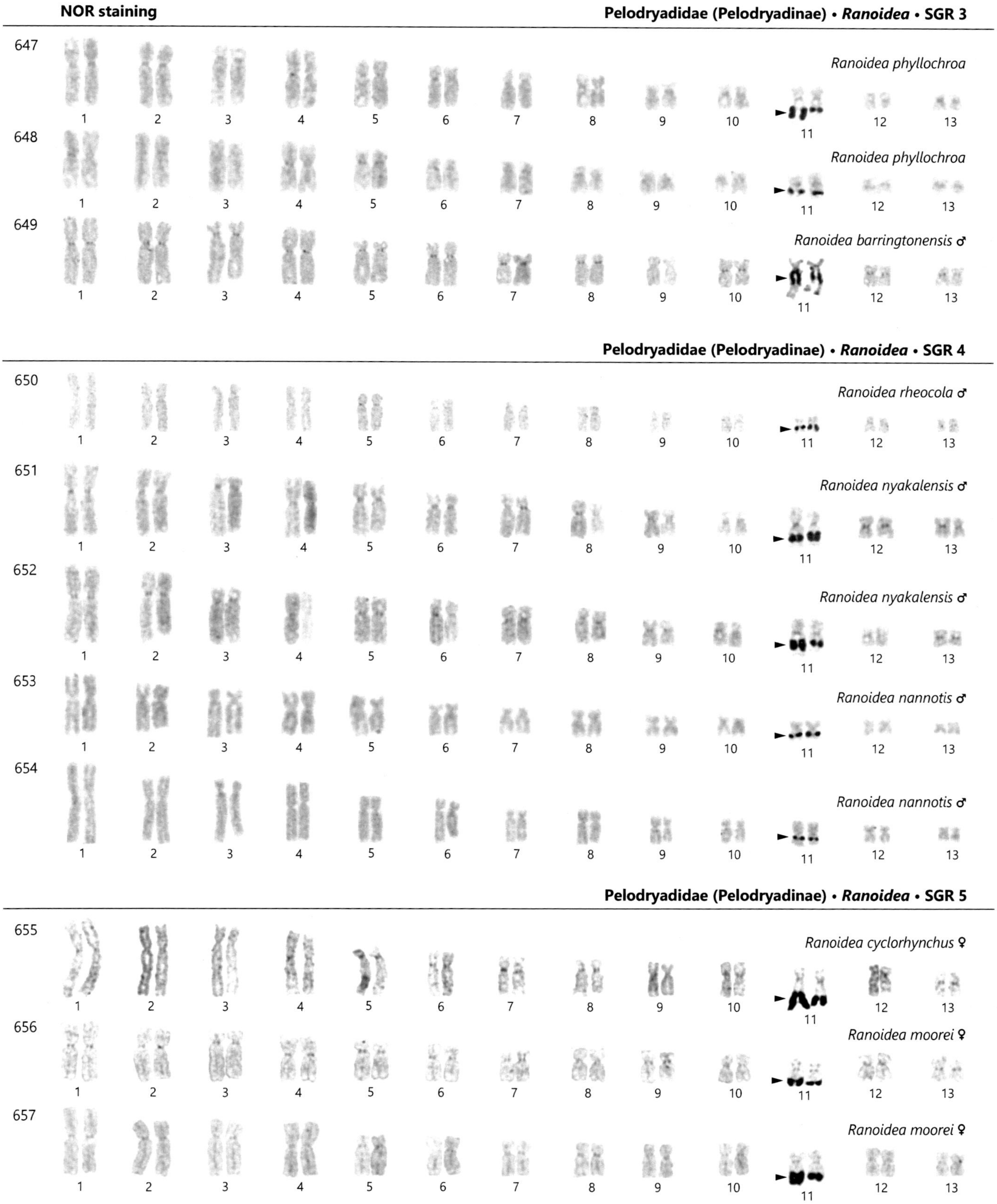
NOR staining
Pelodryadidae (Pelodryadinae) • *Ranoidea* • SGR 3
647 *Ranoidea phyllochroa*
648 *Ranoidea phyllochroa*
649 *Ranoidea barringtonensis* ♂
Pelodryadidae (Pelodryadinae) • *Ranoidea* • SGR 4
650 *Ranoidea rheocola* ♂
651 *Ranoidea nyakalensis* ♂
652 *Ranoidea nyakalensis* ♂
653 *Ranoidea nannotis* ♂
654 *Ranoidea nannotis* ♂
Pelodryadidae (Pelodryadinae) • *Ranoidea* • SGR 5
655 *Ranoidea cyclorhynchus* ♀
656 *Ranoidea moorei* ♀
657 *Ranoidea moorei* ♀
1 2 3 4 5 6 7 8 9 10 11 12 13

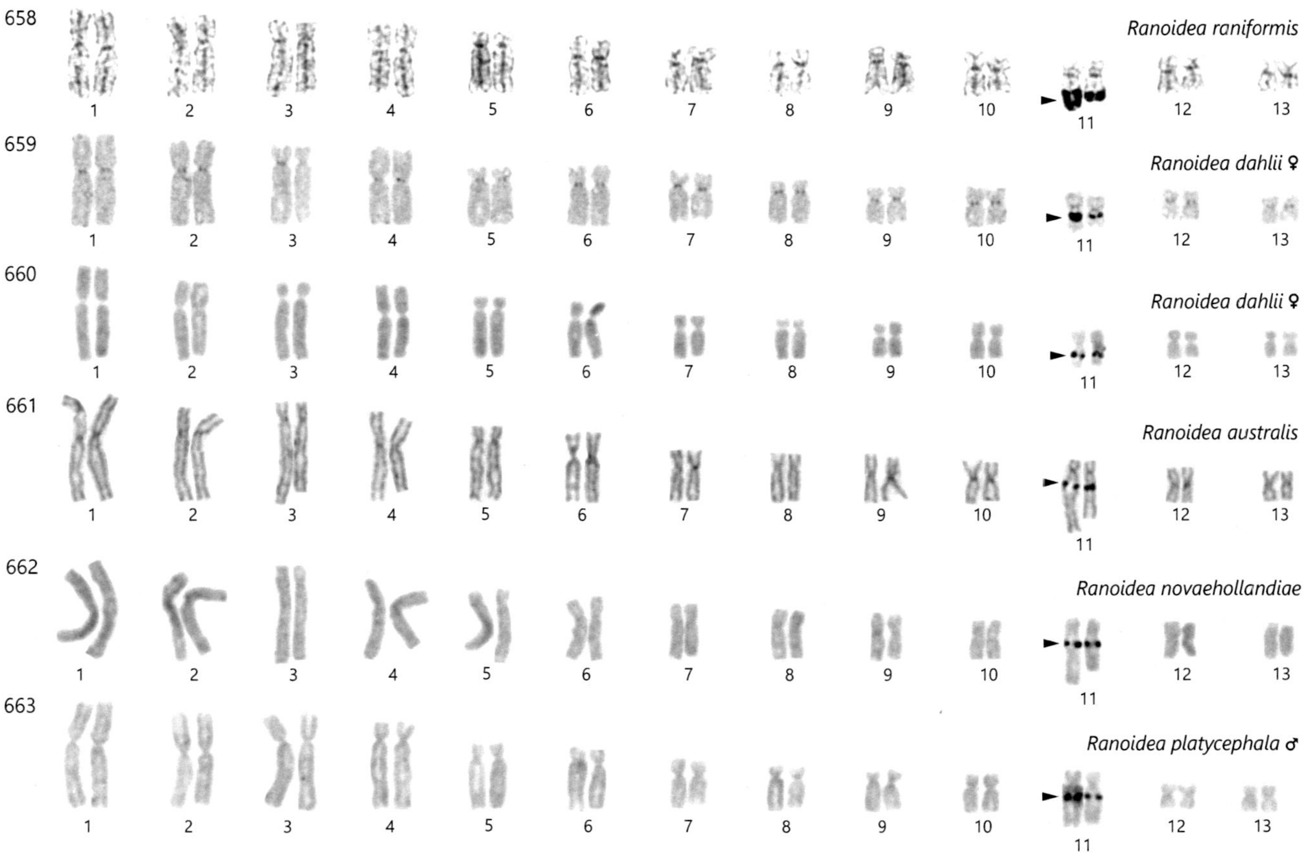
658
Ranoidea raniformis
659
Ranoidea dahlii ♀
660
Ranoidea dahlii ♀
661
Ranoidea australis
662
Ranoidea novaehollandiae
663
Ranoidea platycephala ♂
1 2 3 4 5 6 7 8 9 10 11 12 13

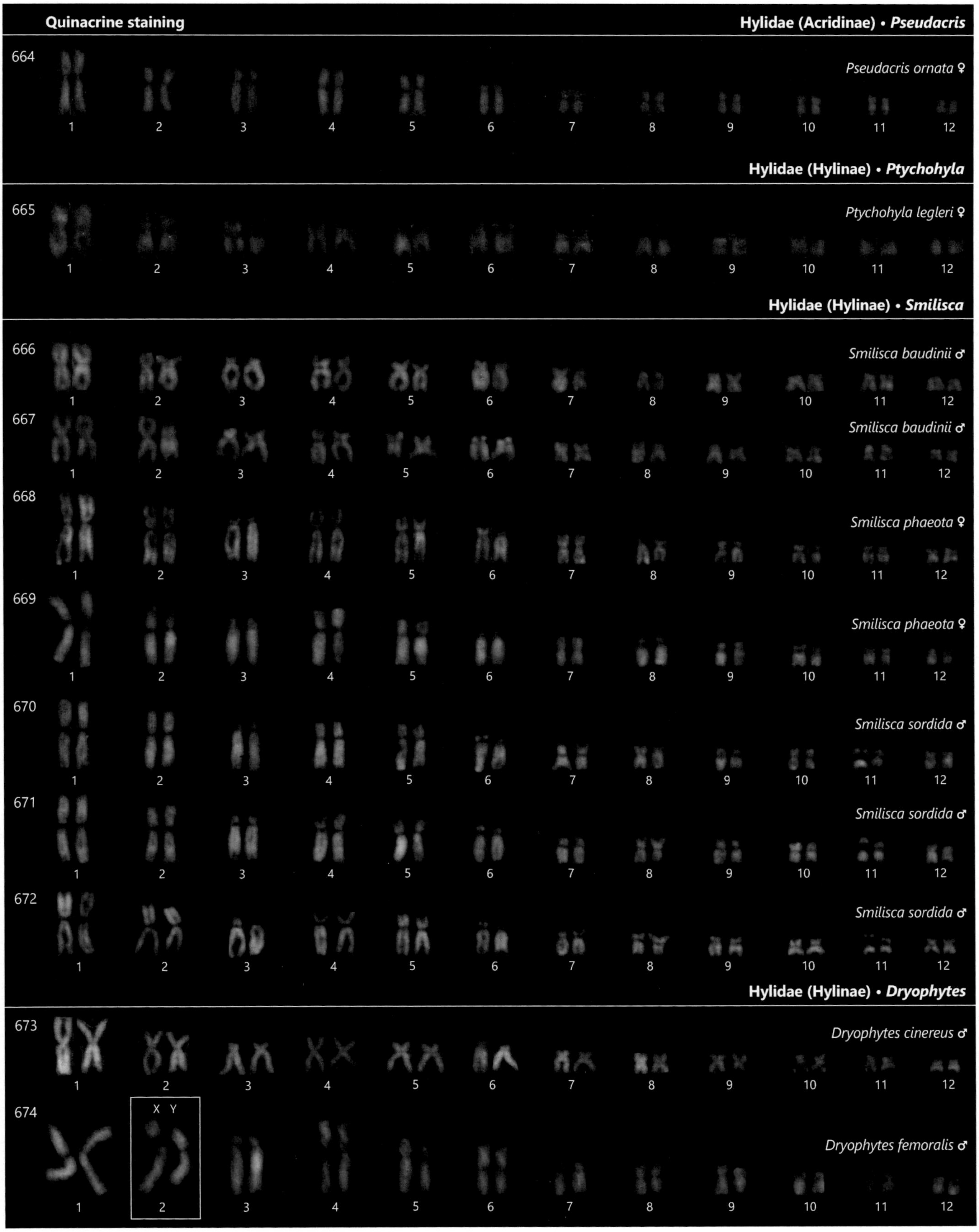

Cytogenet Genome Res 2018;155:55–221
DOI: 10.1159/000489839

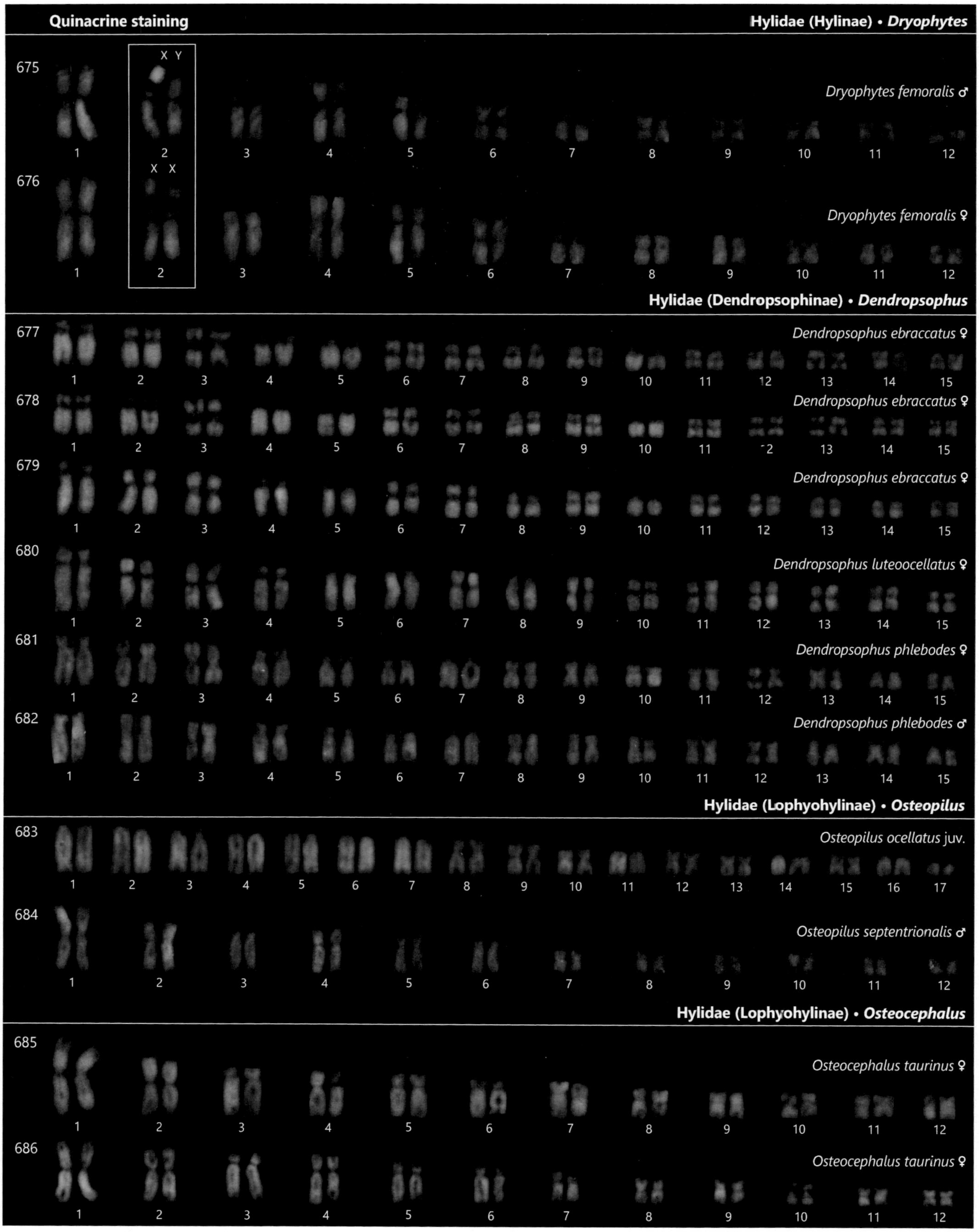
Quinacrine staining
Hylidae (Hylinae) • Dryophytes
675
X Y
Dryophytes femoralis ♂
1 2 3 4 5 6 7 8 9 10 11 12
676
X X
Dryophytes femoralis ♀
1 2 3 4 5 6 7 8 9 10 11 12
Hylidae (Dendropsophinae) • Dendropsophus
677
Dendropsophus ebraccatus ♀
1 2 3 4 5 6 7 8 9 10 11 12 13 14 15
678
Dendropsophus ebraccatus ♀
1 2 3 4 5 6 7 8 9 10 11 12 13 14 15
679
Dendropsophus ebraccatus ♀
1 2 3 4 5 6 7 8 9 10 11 12 13 14 15
680
Dendropsophus luteoocellatus ♀
1 2 3 4 5 6 7 8 9 10 11 12 13 14 15
681
Dendropsophus phlebodes ♀
1 2 3 4 5 6 7 8 9 10 11 12 13 14 15
682
Dendropsophus phlebodes ♂
1 2 3 4 5 6 7 8 9 10 11 12 13 14 15
Hylidae (Lophyohylinae) • Osteopilus
683
Osteopilus ocellatus juv.
1 2 3 4 5 6 7 8 9 10 11 12 13 14 15 16 17
684
Osteopilus septentrionalis ♂
1 2 3 4 5 6 7 8 9 10 11 12
Hylidae (Lophyohylinae) • Osteocephalus
685
Osteocephalus taurinus ♀
1 2 3 4 5 6 7 8 9 10 11 12
686
Osteocephalus taurinus ♀
1 2 3 4 5 6 7 8 9 10 11 12

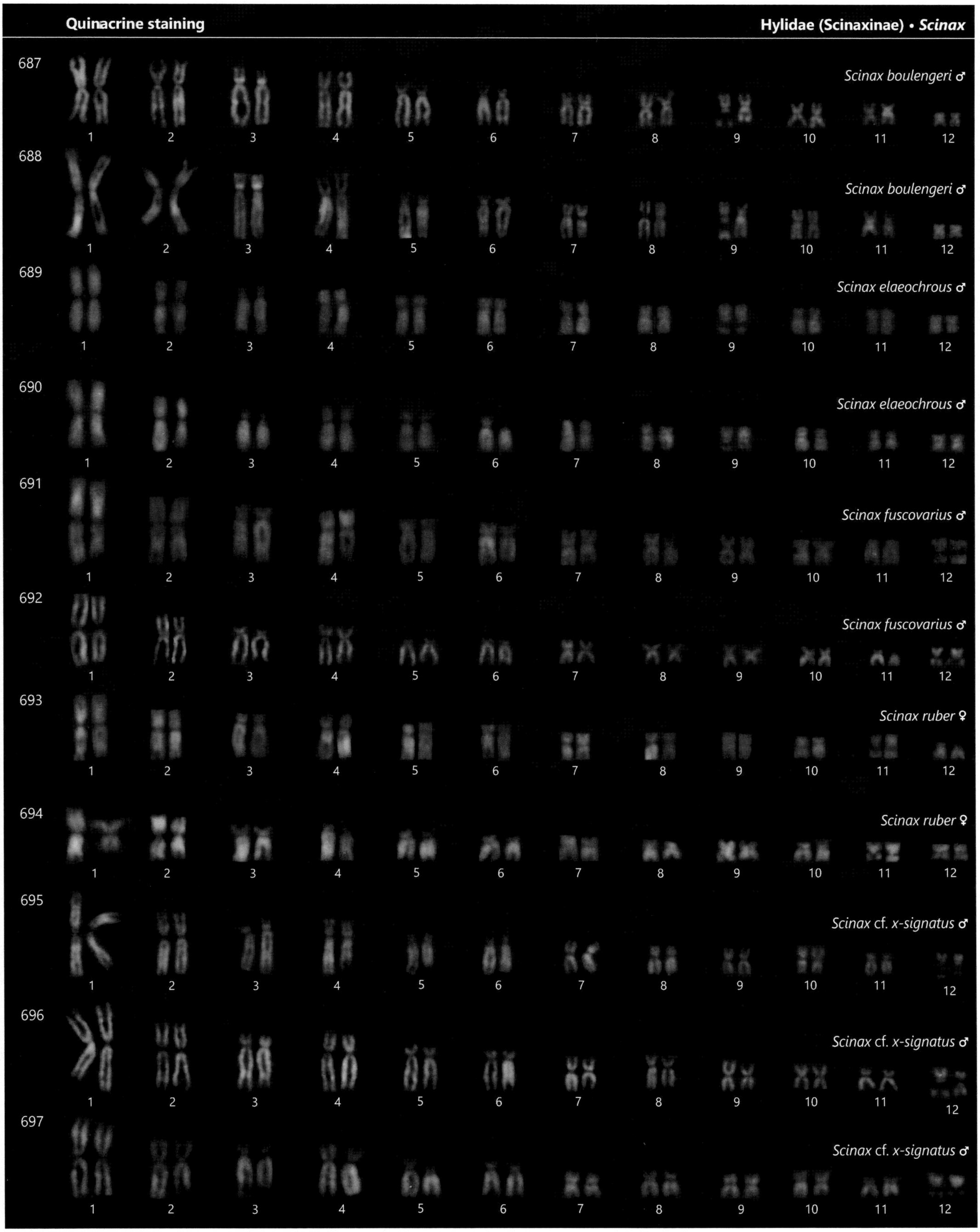
Quinacrine staining
Hylidae (Scinaxinae) • *Scinax*
687
Scinax boulengeri ♂
1 2 3 4 5 6 7 8 9 10 11 12
688
Scinax boulengeri ♂
1 2 3 4 5 6 7 8 9 10 11 12
689
Scinax elaeochrous ♂
1 2 3 4 5 6 7 8 9 10 11 12
690
Scinax elaeochrous ♂
1 2 3 4 5 6 7 8 9 10 11 12
691
Scinax fuscovarius ♂
1 2 3 4 5 6 7 8 9 10 11 12
692
Scinax fuscovarius ♂
1 2 3 4 5 6 7 8 9 10 11 12
693
Scinax ruber ♀
1 2 3 4 5 6 7 8 9 10 11 12
694
Scinax ruber ♀
1 2 3 4 5 6 7 8 9 10 11 12
695
Scinax cf. *x-signatus* ♂
1 2 3 4 5 6 7 8 9 10 11 12
696
Scinax cf. *x-signatus* ♂
1 2 3 4 5 6 7 8 9 10 11 12
697
Scinax cf. *x-signatus* ♂
1 2 3 4 5 6 7 8 9 10 11 12

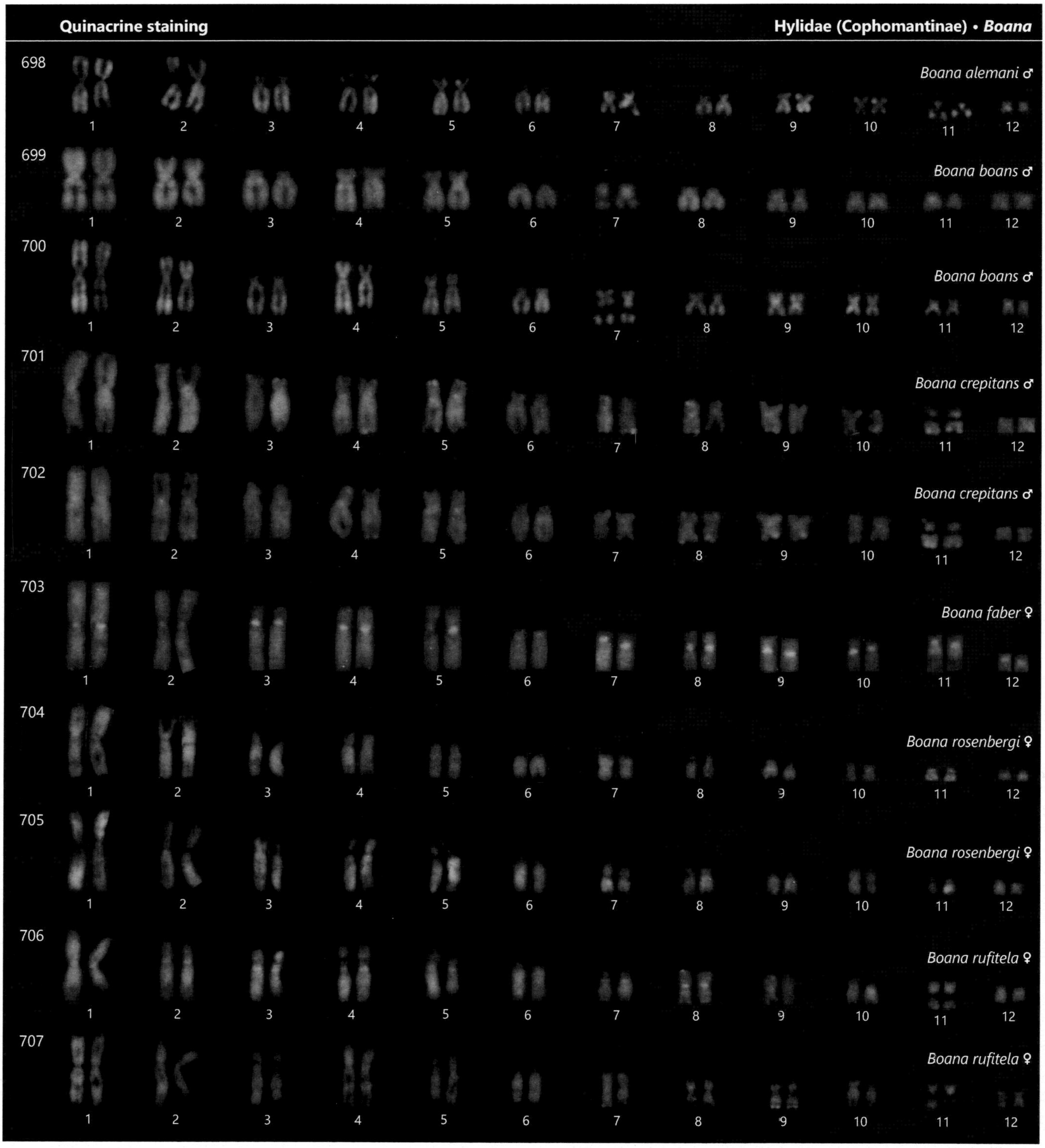
Quinacrine staining
Hylidae (Cophomantinae) • Boana
698
Boana alemani ♂
699
Boana boans ♂
700
Boana boans ♂
701
Boana crepitans ♂
702
Boana crepitans ♂
703
Boana faber ♀
704
Boana rosenbergi ♀
705
Boana rosenbergi ♀
706
Boana rufitela ♀
707
Boana rufitela ♀
1 2 3 4 5 6 7 8 9 10 11 12

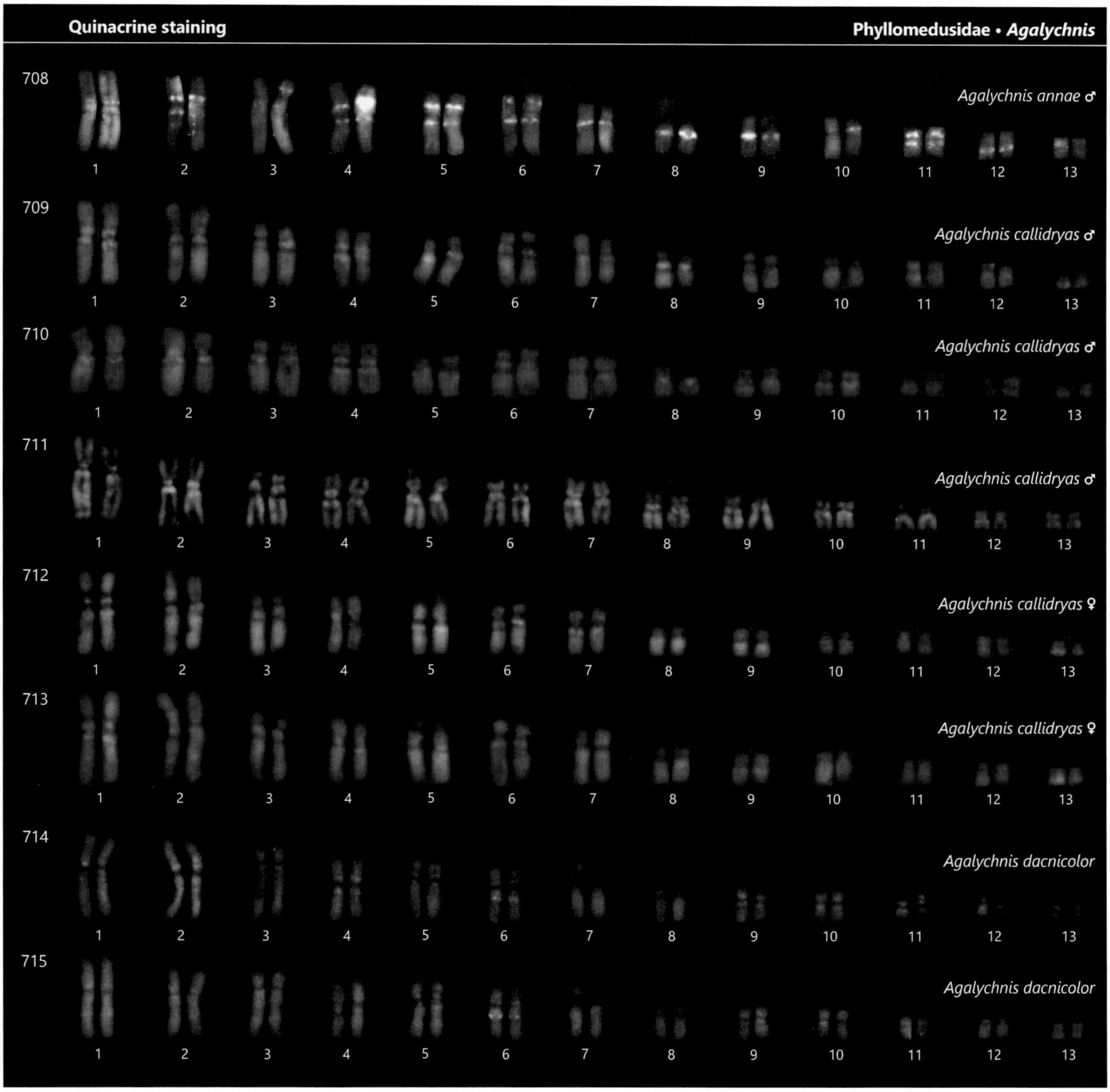
Quinacrine staining
Phyllomedusidae • Agalychnis
708
Agalychnis annae ♂
1
2
3
4
5
6
7
8
9
10
11
12
13
709
Agalychnis callidryas ♂
710
Agalychnis callidryas ♂
711
Agalychnis callidryas ♂
712
Agalychnis callidryas ♀
713
Agalychnis callidryas ♀
714
Agalychnis dacnicolor
715
Agalychnis dacnicolor

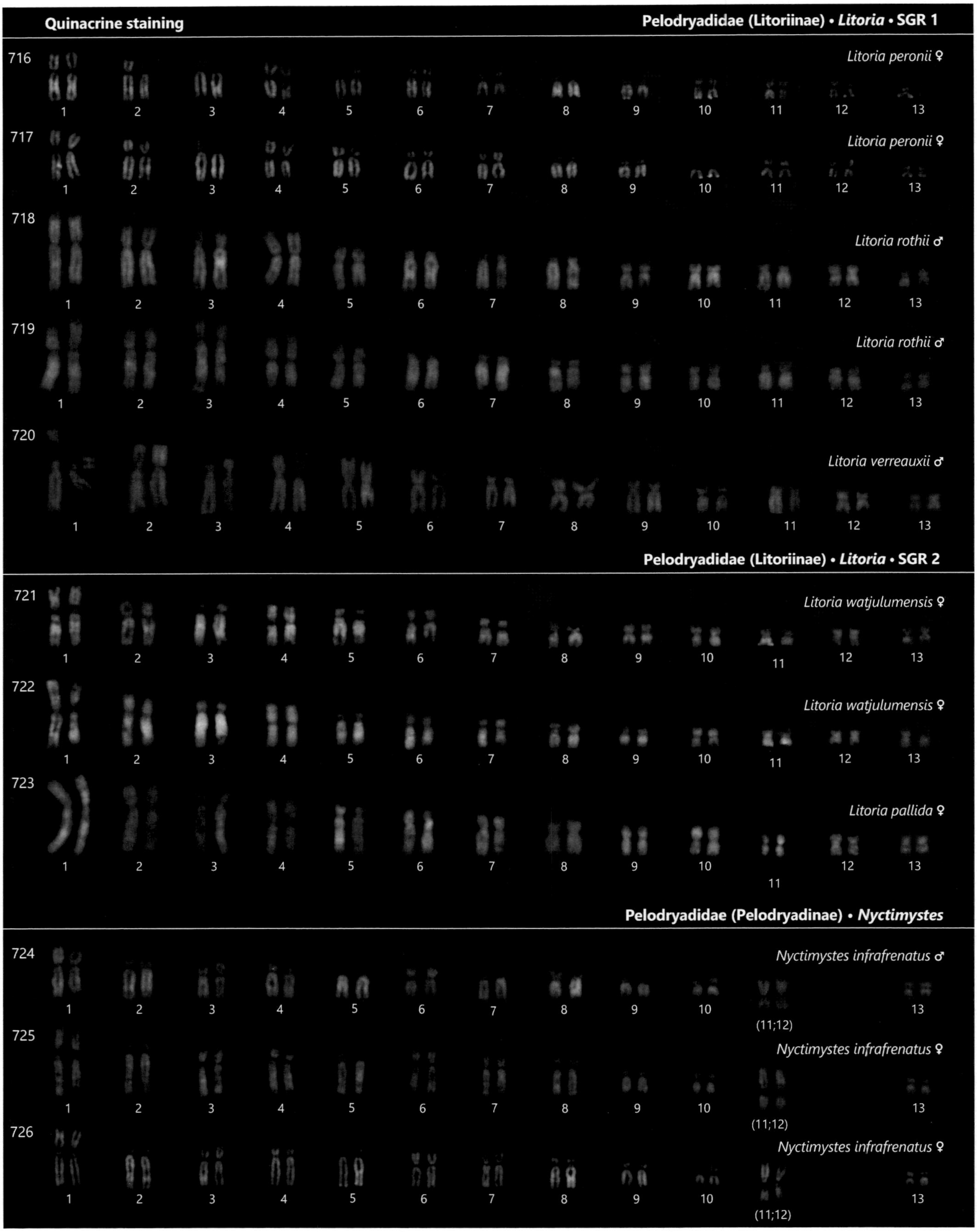
Quinacrine staining
Pelodryadidae (Litoriinae) • Litoria • SGR 1
716 Litoria peronii ♀
717 Litoria peronii ♀
718 Litoria rothii ♂
719 Litoria rothii ♂
720 Litoria verreauxii ♂
Pelodryadidae (Litoriinae) • Litoria • SGR 2
721 Litoria watjulumensis ♀
722 Litoria watjulumensis ♀
723 Litoria pallida ♀
Pelodryadidae (Pelodryadinae) • Nyctimystes
724 Nyctimystes infrafrenatus ♂
725 Nyctimystes infrafrenatus ♀
726 Nyctimystes infrafrenatus ♀
1 2 3 4 5 6 7 8 9 10 11 12 13
(11;12)

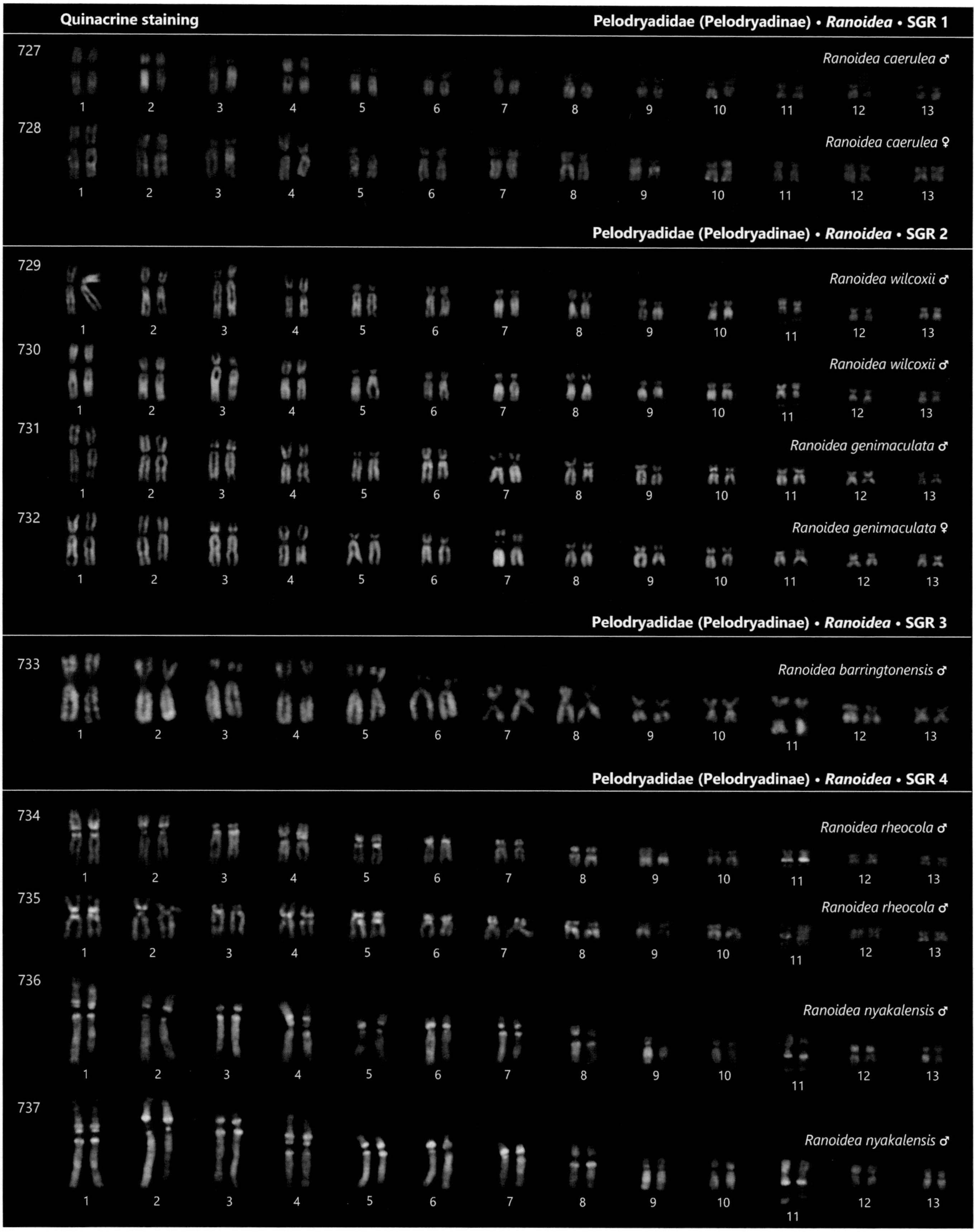
Quinacrine staining
Pelodryadidae (Pelodryadinae) • Ranoidea • SGR 1
727
Ranoidea caerulea ♂
728
Ranoidea caerulea ♀
Pelodryadidae (Pelodryadinae) • Ranoidea • SGR 2
729
Ranoidea wilcoxii ♂
730
Ranoidea wilcoxii ♂
731
Ranoidea genimaculata ♂
732
Ranoidea genimaculata ♀
Pelodryadidae (Pelodryadinae) • Ranoidea • SGR 3
733
Ranoidea barringtonensis ♂
Pelodryadidae (Pelodryadinae) • Ranoidea • SGR 4
734
Ranoidea rheocola ♂
735
Ranoidea rheocola ♂
736
Ranoidea nyakalensis ♂
737
Ranoidea nyakalensis ♂
1 2 3 4 5 6 7 8 9 10 11 12 13

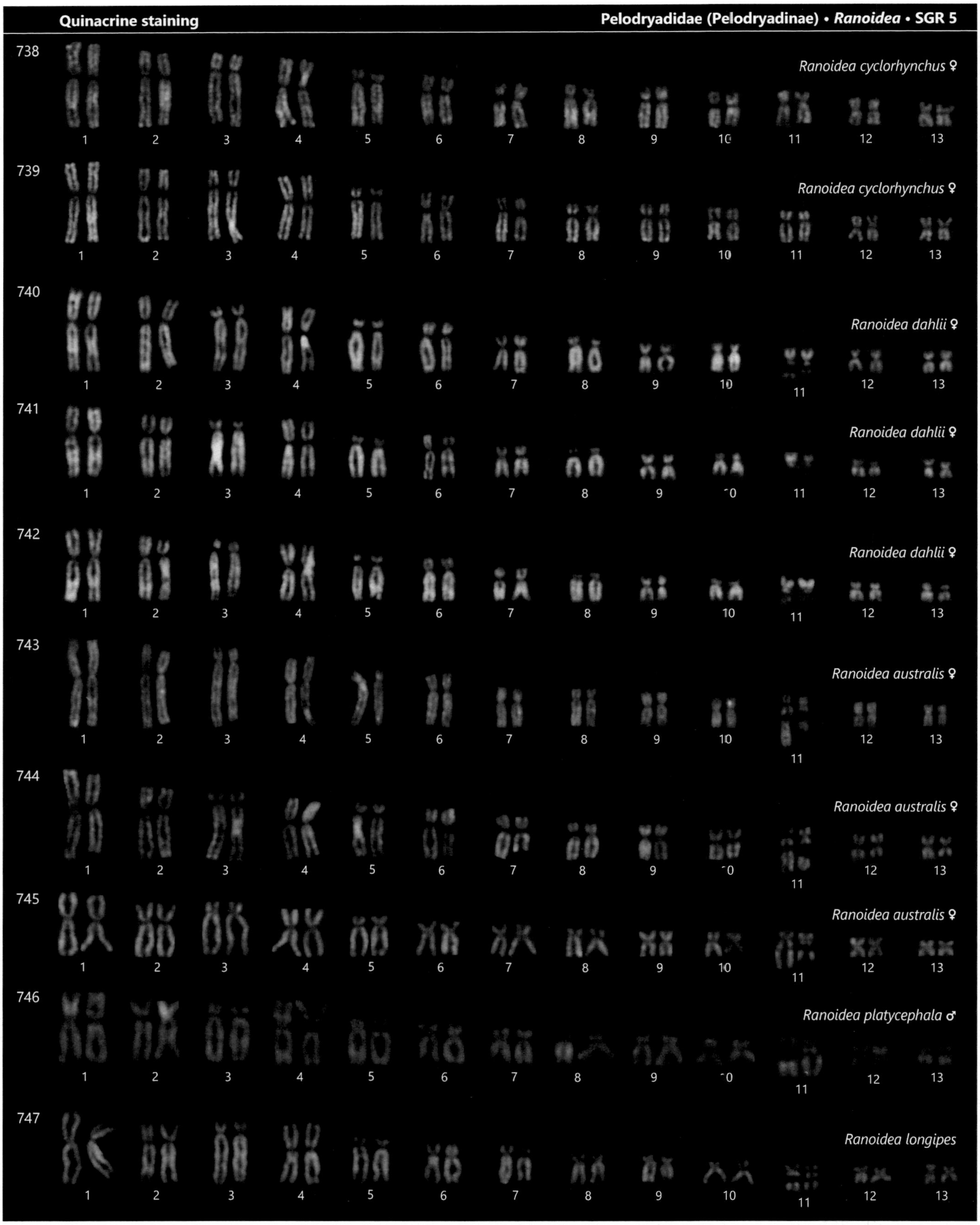
Quinacrine staining
Pelodryadidae (Pelodryadinae) • Ranoidea • SGR 5
738
Ranoidea cyclorhynchus ♀
739
Ranoidea cyclorhynchus ♀
740
Ranoidea dahlii ♀
741
Ranoidea dahlii ♀
742
Ranoidea dahlii ♀
743
Ranoidea australis ♀
744
Ranoidea australis ♀
745
Ranoidea australis ♀
746
Ranoidea platycephala ♂
747
Ranoidea longipes

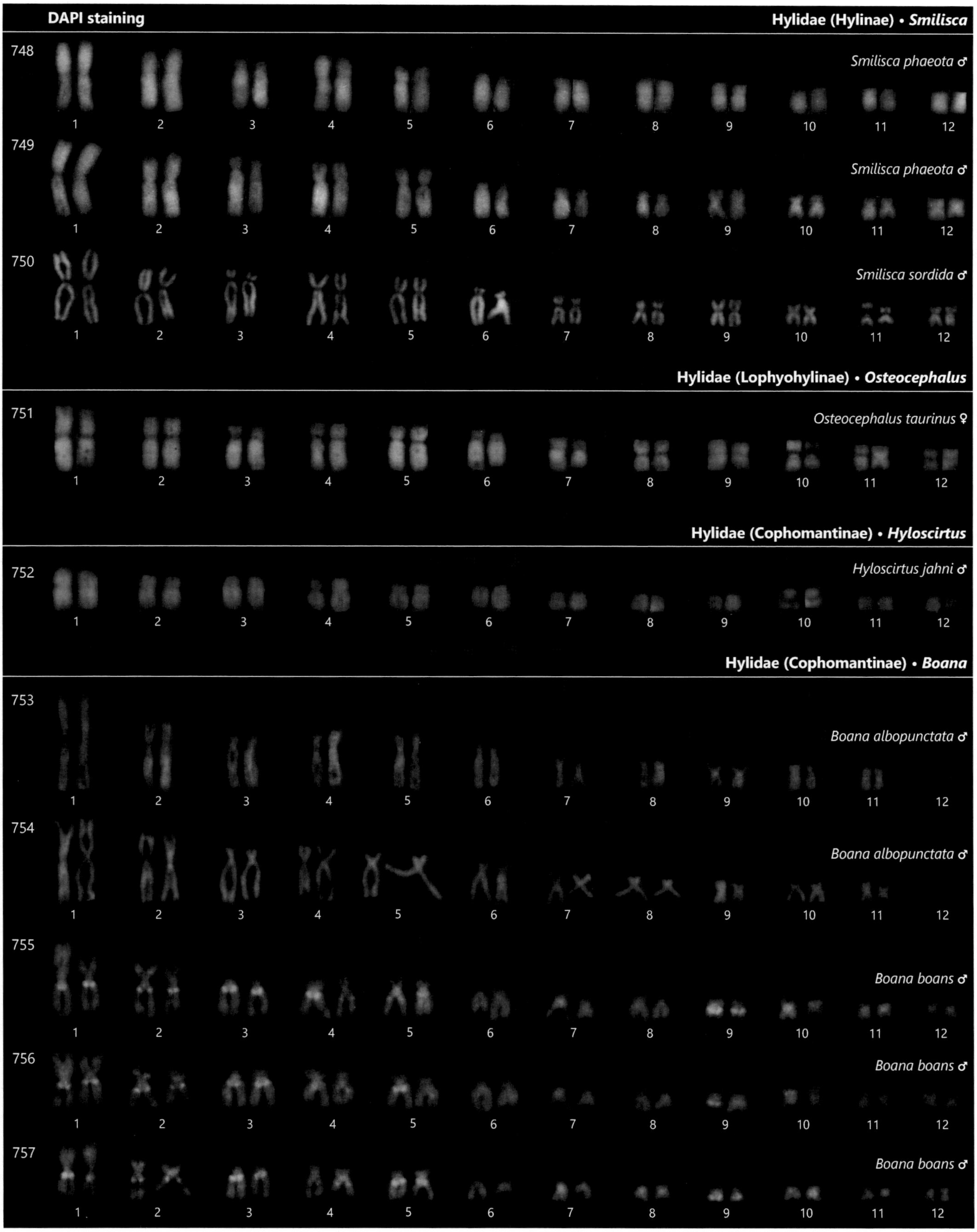

Cytogenet Genome Res 2018;155:55–221
DOI: 10.1159/000489839

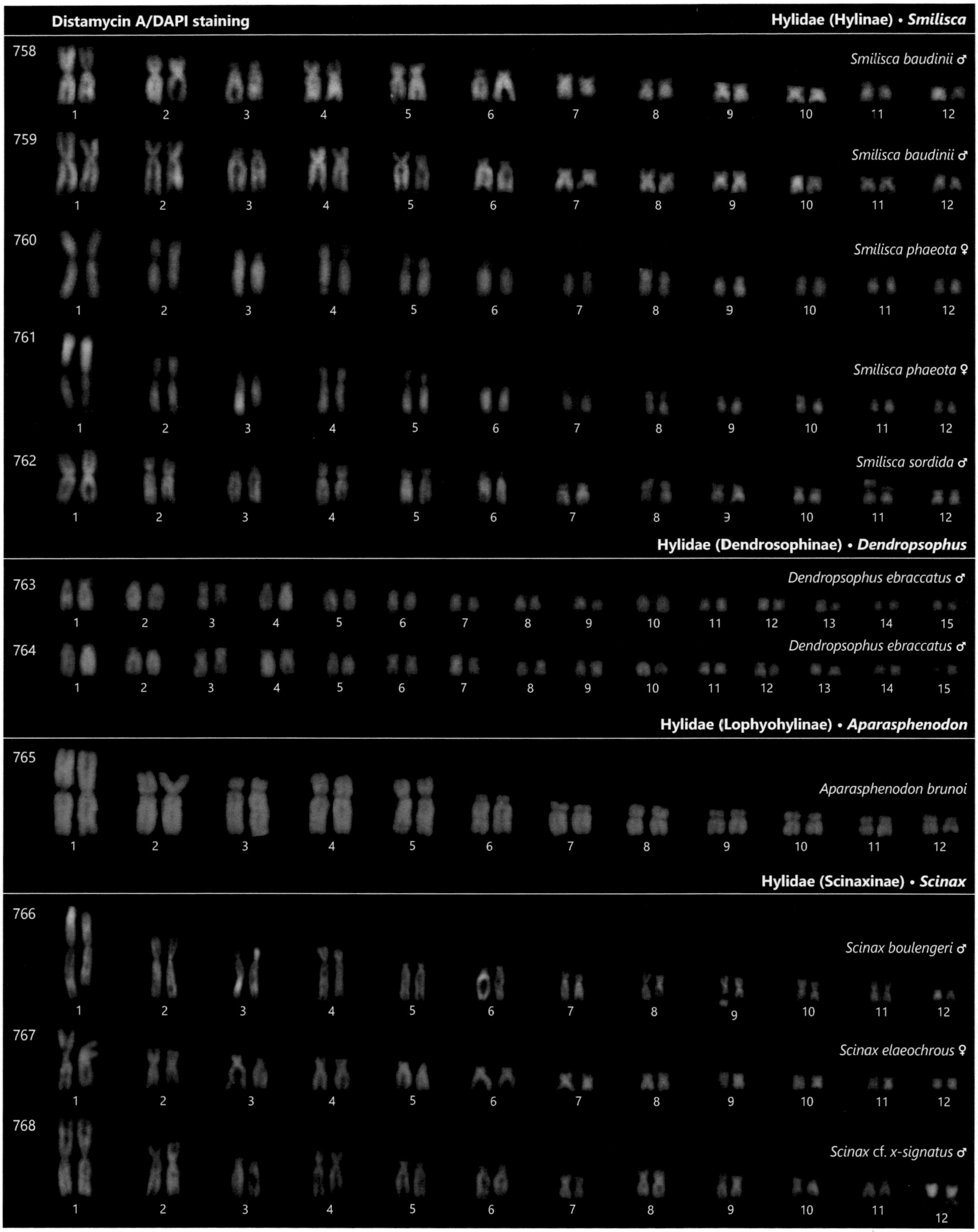
Distamycin A/DAPI staining
Hylidae (Hylinae) • Smilisca
758
Smilisca baudinii ♂
759
Smilisca baudinii ♂
760
Smilisca phaeota ♀
761
Smilisca phaeota ♀
762
Smilisca sordida ♂
Hylidae (Dendrosophinae) • Dendropsophus
763
Dendropsophus ebraccatus ♂
764
Dendropsophus ebraccatus ♂
Hylidae (Lophyohylinae) • Aparasphenodon
765
Aparasphenodon brunoi
Hylidae (Scinaxinae) • Scinax
766
Scinax boulengeri ♂
767
Scinax elaeochrous ♀
768
Scinax cf. x-signatus ♂

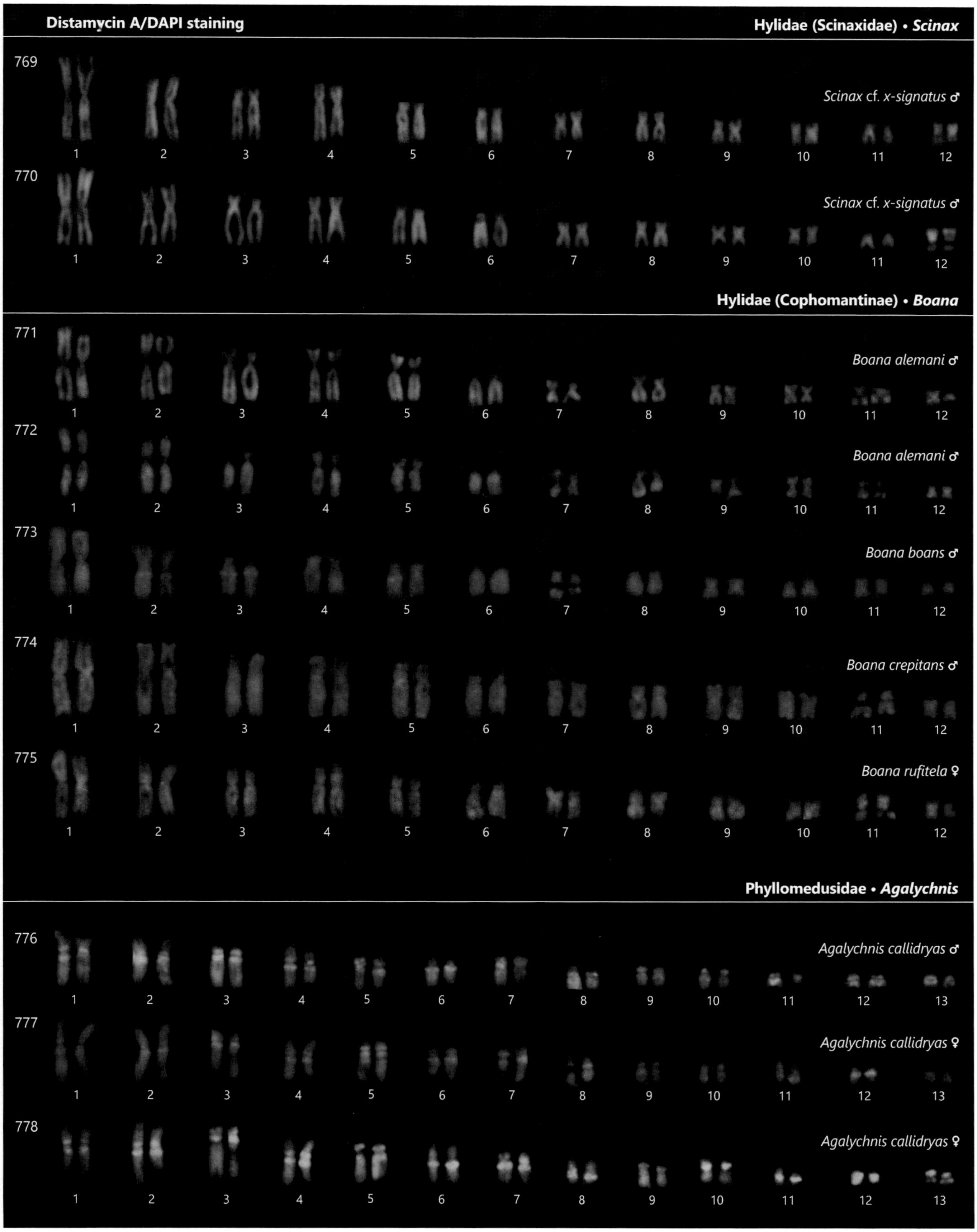
Distamycin A/DAPI staining
Hylidae (Scinaxidae) • Scinax
769
Scinax cf. x-signatus ♂
1 2 3 4 5 6 7 8 9 10 11 12
770
Scinax cf. x-signatus ♂
1 2 3 4 5 6 7 8 9 10 11 12
Hylidae (Cophomantinae) • Boana
771
Boana alemani ♂
1 2 3 4 5 6 7 8 9 10 11 12
772
Boana alemani ♂
1 2 3 4 5 6 7 8 9 10 11 12
773
Boana boans ♂
1 2 3 4 5 6 7 8 9 10 11 12
774
Boana crepitans ♂
1 2 3 4 5 6 7 8 9 10 11 12
775
Boana rufitela ♀
1 2 3 4 5 6 7 8 9 10 11 12
Phyllomedusidae • Agalychnis
776
Agalychnis callidryas ♂
1 2 3 4 5 6 7 8 9 10 11 12 13
777
Agalychnis callidryas ♀
1 2 3 4 5 6 7 8 9 10 11 12 13
778
Agalychnis callidryas ♀
1 2 3 4 5 6 7 8 9 10 11 12 13

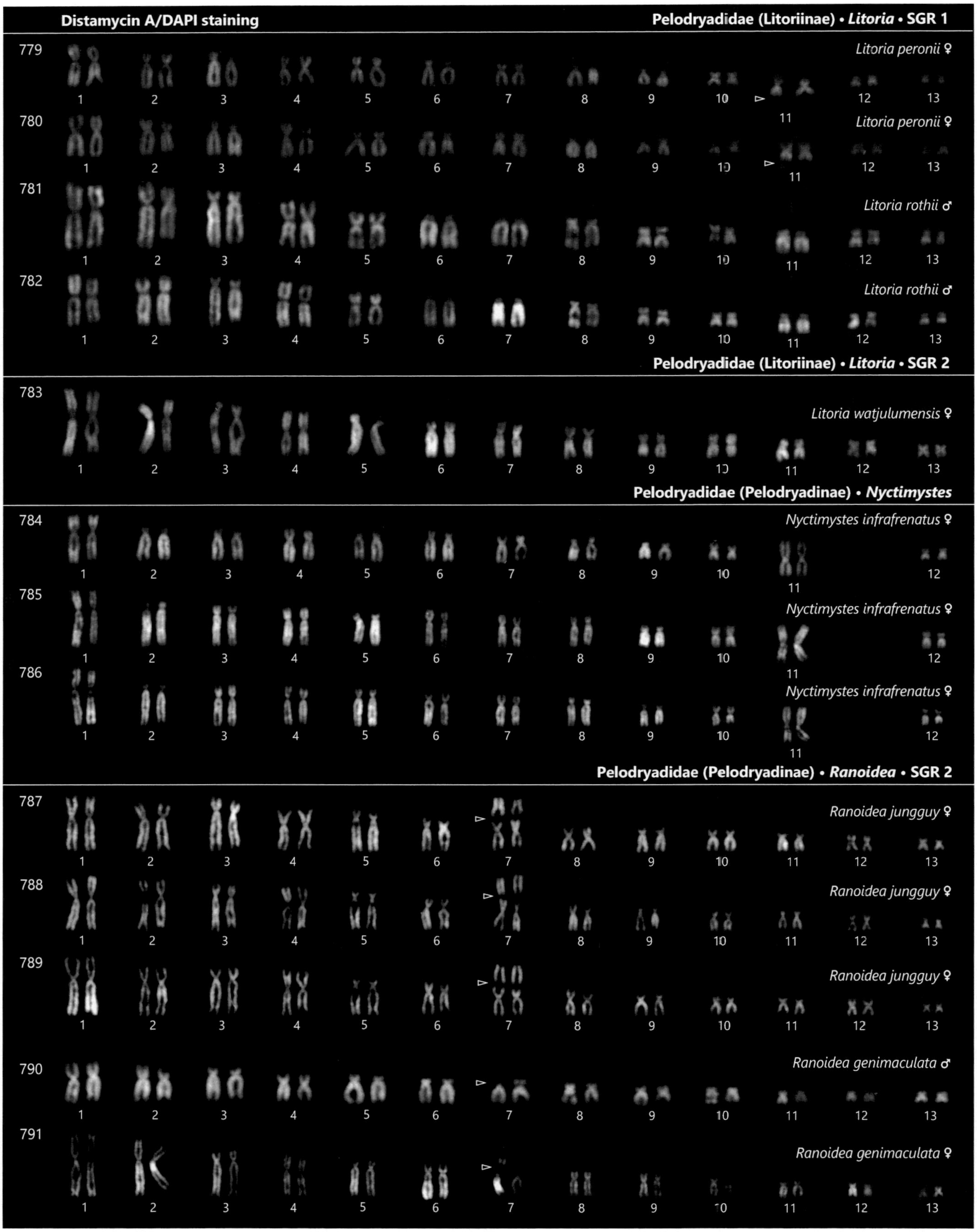

DOI: 10.1159/000489839

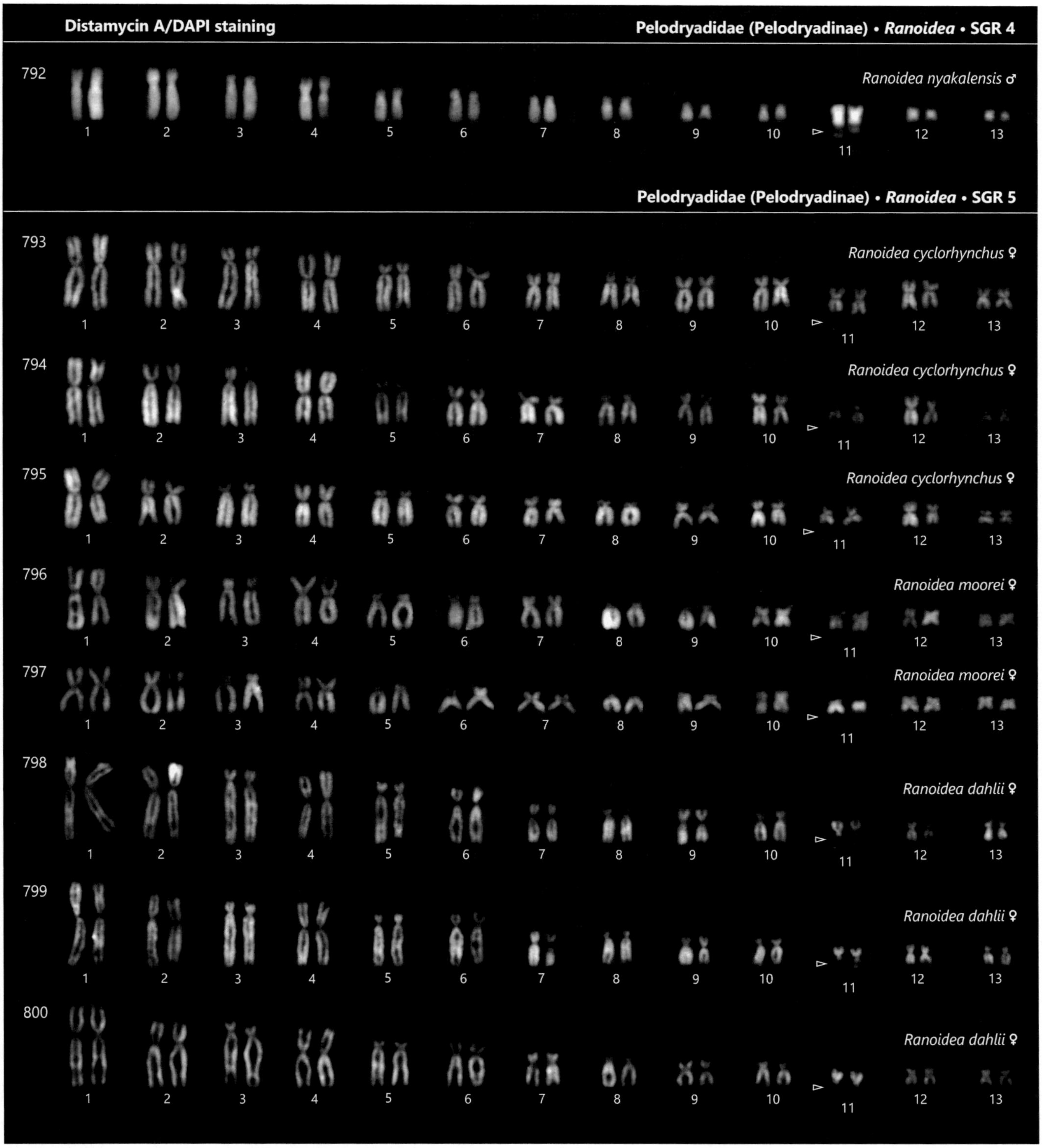
Distamycin A/DAPI staining
Pelodryadidae (Pelodryadinae) • *Ranoidea* • SGR 4
792
Ranoidea nyakalensis ♂
1 2 3 4 5 6 7 8 9 10 11 12 13
Pelodryadidae (Pelodryadinae) • *Ranoidea* • SGR 5
793
Ranoidea cyclorhynchus ♀
1 2 3 4 5 6 7 8 9 10 11 12 13
794
Ranoidea cyclorhynchus ♀
1 2 3 4 5 6 7 8 9 10 11 12 13
795
Ranoidea cyclorhynchus ♀
1 2 3 4 5 6 7 8 9 10 11 12 13
796
Ranoidea moorei ♀
1 2 3 4 5 6 7 8 9 10 11 12 13
797
Ranoidea moorei ♀
1 2 3 4 5 6 7 8 9 10 11 12 13
798
Ranoidea dahlii ♀
1 2 3 4 5 6 7 8 9 10 11 12 13
799
Ranoidea dahlii ♀
1 2 3 4 5 6 7 8 9 10 11 12 13
800
Ranoidea dahlii ♀
1 2 3 4 5 6 7 8 9 10 11 12 13

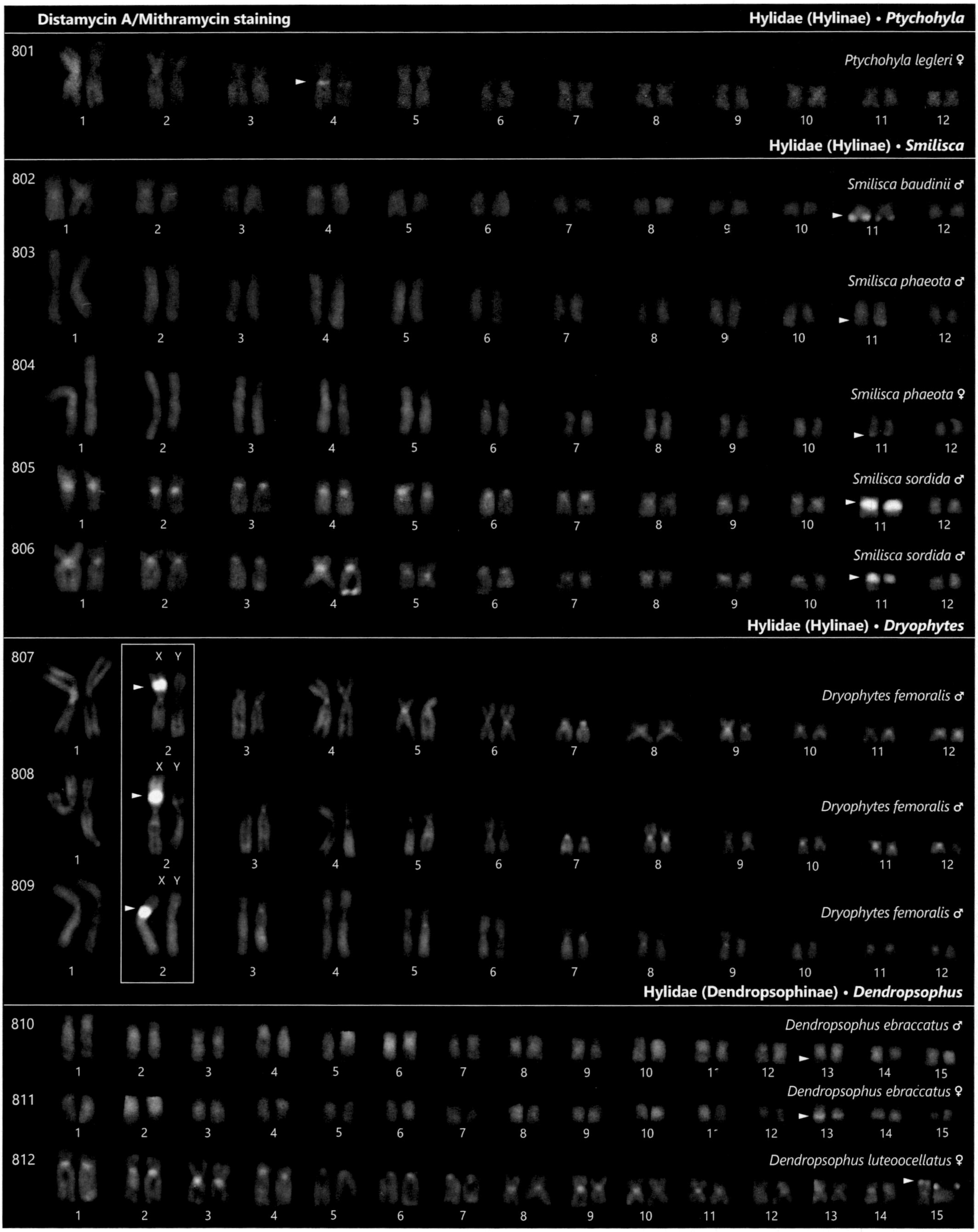
Distamycin A/Mithramycin staining
Hylidae (Hylinae) • Ptychohyla
801
Ptychohyla legleri ♀
Hylidae (Hylinae) • Smilisca
802
Smilisca baudinii ♂
803
Smilisca phaeota ♂
804
Smilisca phaeota ♀
805
Smilisca sordida ♂
806
Smilisca sordida ♂
Hylidae (Hylinae) • Dryophytes
807
X Y
Dryophytes femoralis ♂
808
X Y
Dryophytes femoralis ♂
809
X Y
Dryophytes femoralis ♂
Hylidae (Dendropsophinae) • Dendropsophus
810
Dendropsophus ebraccatus ♂
811
Dendropsophus ebraccatus ♀
812
Dendropsophus luteoocellatus ♀

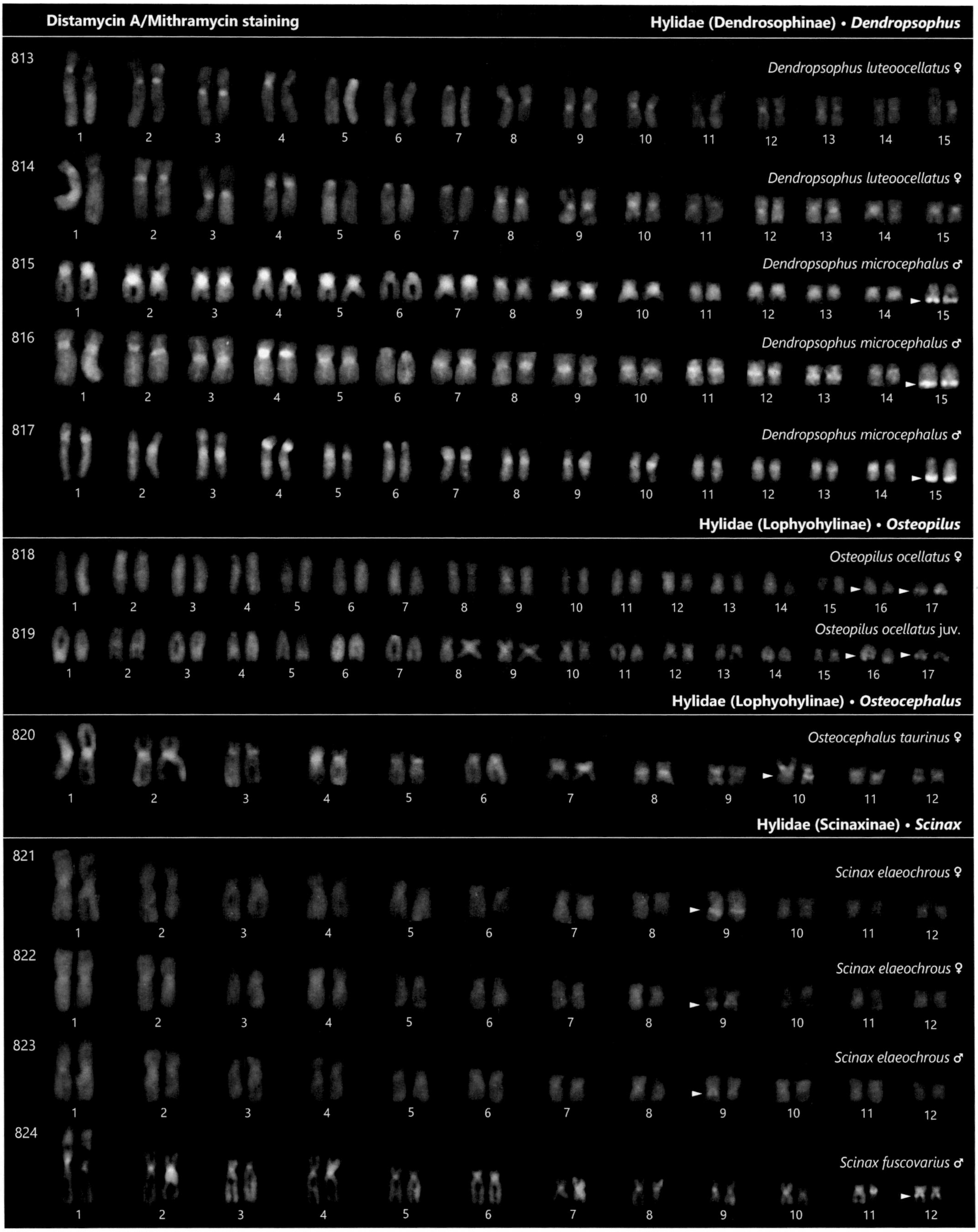
Distamycin A/Mithramycin staining
Hylidae (Dendrosophinae) • Dendropsophus
813
Dendropsophus luteoocellatus ♀
1 2 3 4 5 6 7 8 9 10 11 12 13 14 15
814
Dendropsophus luteoocellatus ♀
1 2 3 4 5 6 7 8 9 10 11 12 13 14 15
815
Dendropsophus microcephalus ♂
1 2 3 4 5 6 7 8 9 10 11 12 13 14 15
816
Dendropsophus microcephalus ♂
1 2 3 4 5 6 7 8 9 10 11 12 13 14 15
817
Dendropsophus microcephalus ♂
1 2 3 4 5 6 7 8 9 10 11 12 13 14 15
Hylidae (Lophyohylinae) • Osteopilus
818
Osteopilus ocellatus ♀
1 2 3 4 5 6 7 8 9 10 11 12 13 14 15 16 17
819
Osteopilus ocellatus juv.
1 2 3 4 5 6 7 8 9 10 11 12 13 14 15 16 17
Hylidae (Lophyohylinae) • Osteocephalus
820
Osteocephalus taurinus ♀
1 2 3 4 5 6 7 8 9 10 11 12
Hylidae (Scinaxinae) • Scinax
821
Scinax elaeochrous ♀
1 2 3 4 5 6 7 8 9 10 11 12
822
Scinax elaeochrous ♀
1 2 3 4 5 6 7 8 9 10 11 12
823
Scinax elaeochrous ♂
1 2 3 4 5 6 7 8 9 10 11 12
824
Scinax fuscovarius ♂
1 2 3 4 5 6 7 8 9 10 11 12

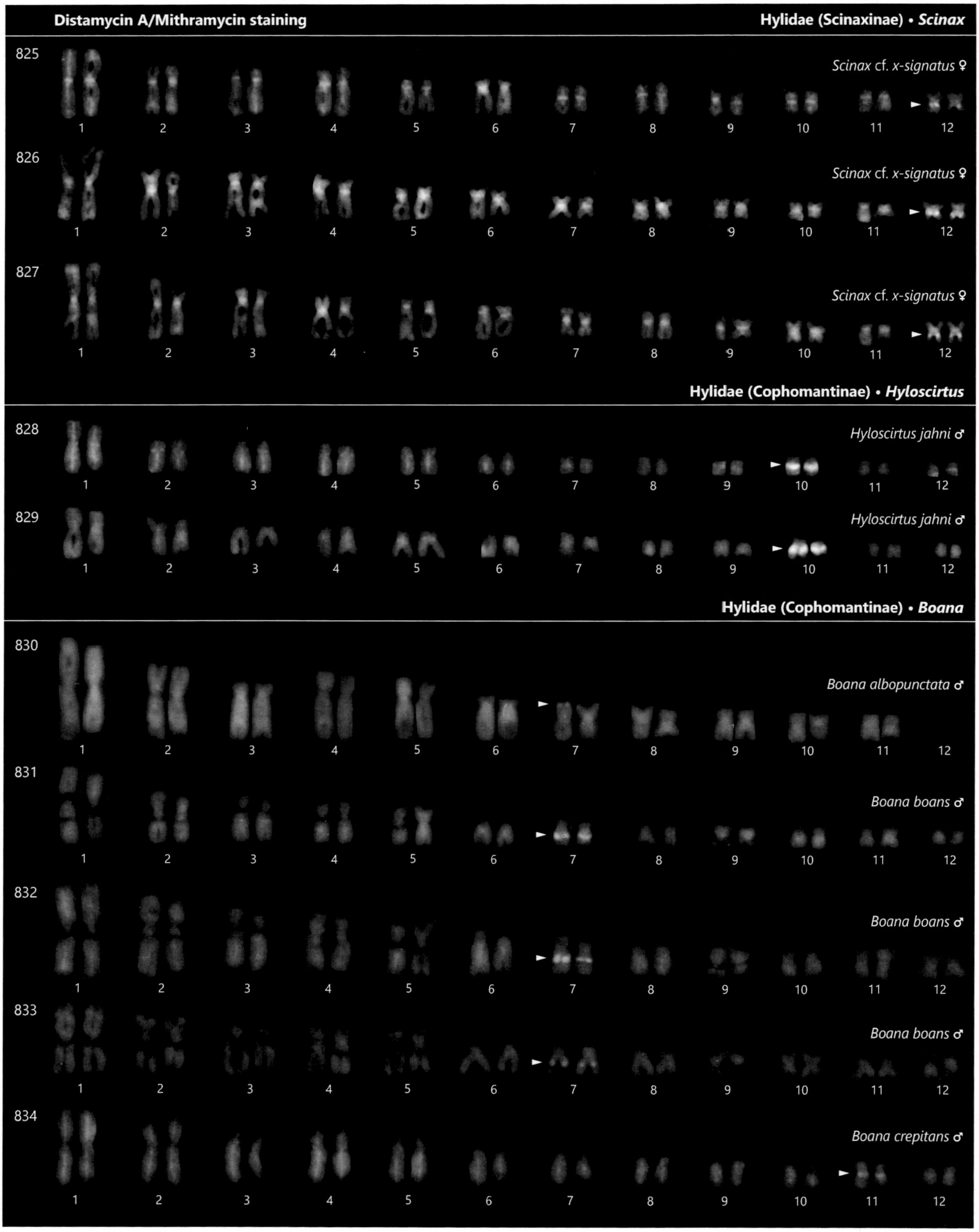
Distamycin A/Mithramycin staining
Hylidae (Scinaxinae) • Scinax
825
Scinax cf. x-signatus ♀
1 2 3 4 5 6 7 8 9 10 11 12
826
Scinax cf. x-signatus ♀
1 2 3 4 5 6 7 8 9 10 11 12
827
Scinax cf. x-signatus ♀
1 2 3 4 5 6 7 8 9 10 11 12
Hylidae (Cophomantinae) • Hyloscirtus
828
Hyloscirtus jahni ♂
1 2 3 4 5 6 7 8 9 10 11 12
829
Hyloscirtus jahni ♂
1 2 3 4 5 6 7 8 9 10 11 12
Hylidae (Cophomantinae) • Boana
830
Boana albopunctata ♂
1 2 3 4 5 6 7 8 9 10 11 12
831
Boana boans ♂
1 2 3 4 5 6 7 8 9 10 11 12
832
Boana boans ♂
1 2 3 4 5 6 7 8 9 10 11 12
833
Boana boans ♂
1 2 3 4 5 6 7 8 9 10 11 12
834
Boana crepitans ♂
1 2 3 4 5 6 7 8 9 10 11 12

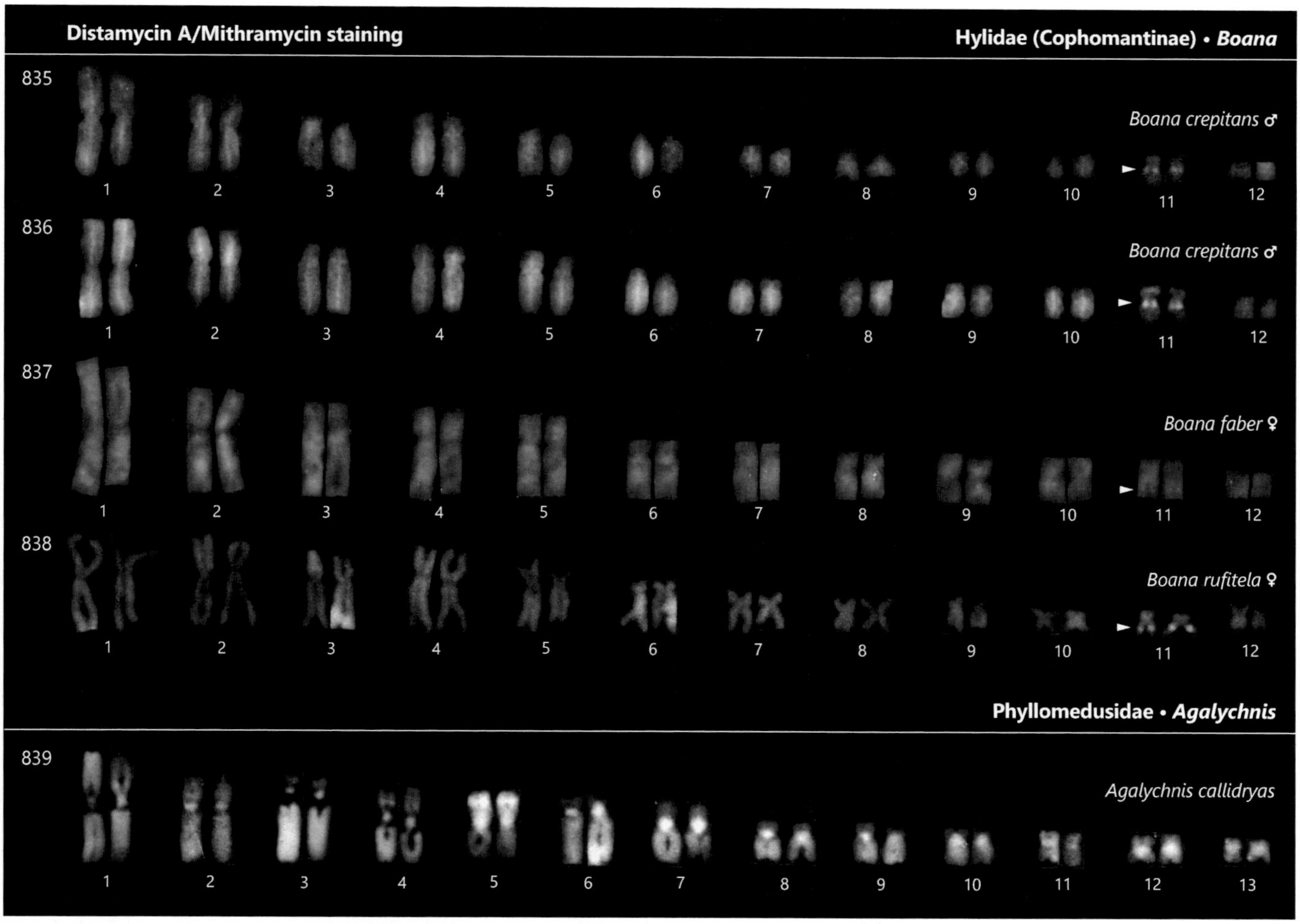

Distamycin A/Mithramycin staining
Hylidae (Cophomantinae) • Boana
835
Boana crepitans ♂
1 2 3 4 5 6 7 8 9 10 11 12
836
Boana crepitans ♂
1 2 3 4 5 6 7 8 9 10 11 12
837
Boana faber ♀
1 2 3 4 5 6 7 8 9 10 11 12
838
Boana rufitela ♀
1 2 3 4 5 6 7 8 9 10 11 12
Phyllomedusidae • Agalychnis
839
Agalychnis callidryas
1 2 3 4 5 6 7 8 9 10 11 12 13

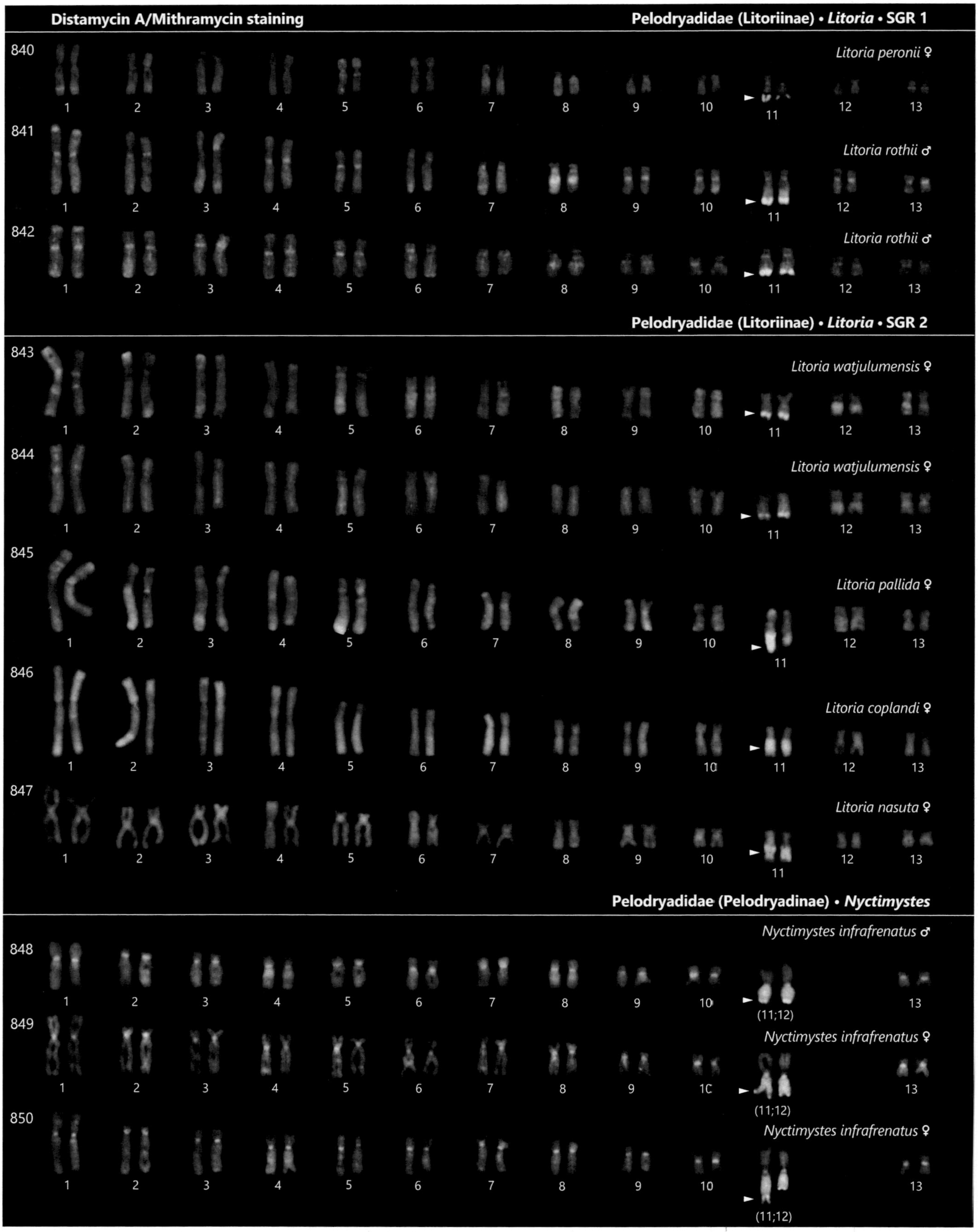

DOI: 10.1159/000489839

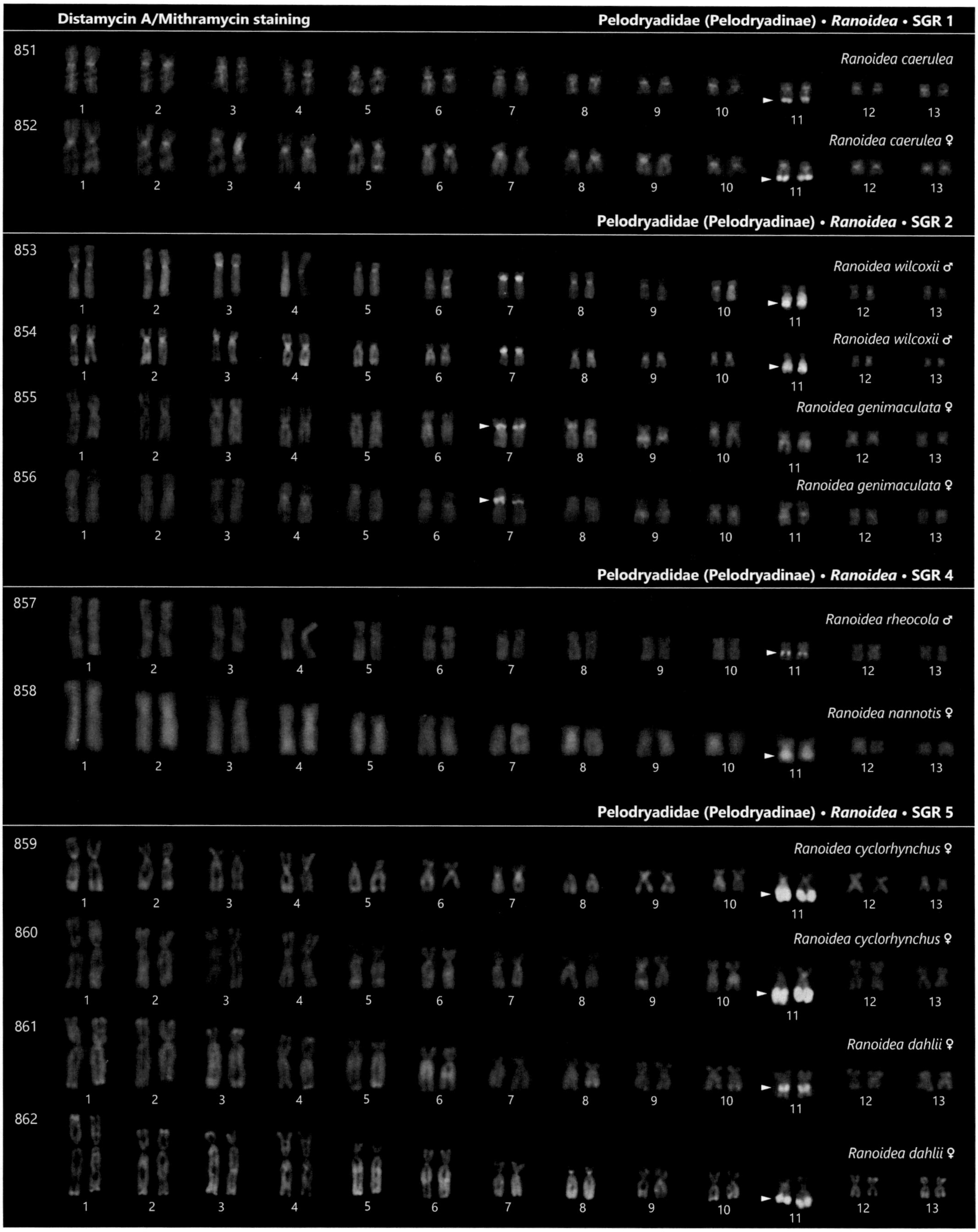
Distamycin A/Mithramycin staining
Pelodryadidae (Pelodryadinae) • Ranoidea • SGR 1
851
Ranoidea caerulea
852
Ranoidea caerulea ♀
Pelodryadidae (Pelodryadinae) • Ranoidea • SGR 2
853
Ranoidea wilcoxii ♂
854
Ranoidea wilcoxii ♂
855
Ranoidea genimaculata ♀
856
Ranoidea genimaculata ♀
Pelodryadidae (Pelodryadinae) • Ranoidea • SGR 4
857
Ranoidea rheocola ♂
858
Ranoidea nannotis ♀
Pelodryadidae (Pelodryadinae) • Ranoidea • SGR 5
859
Ranoidea cyclorhynchus ♀
860
Ranoidea cyclorhynchus ♀
861
Ranoidea dahlii ♀
862
Ranoidea dahlii ♀
1 2 3 4 5 6 7 8 9 10 11 12 13

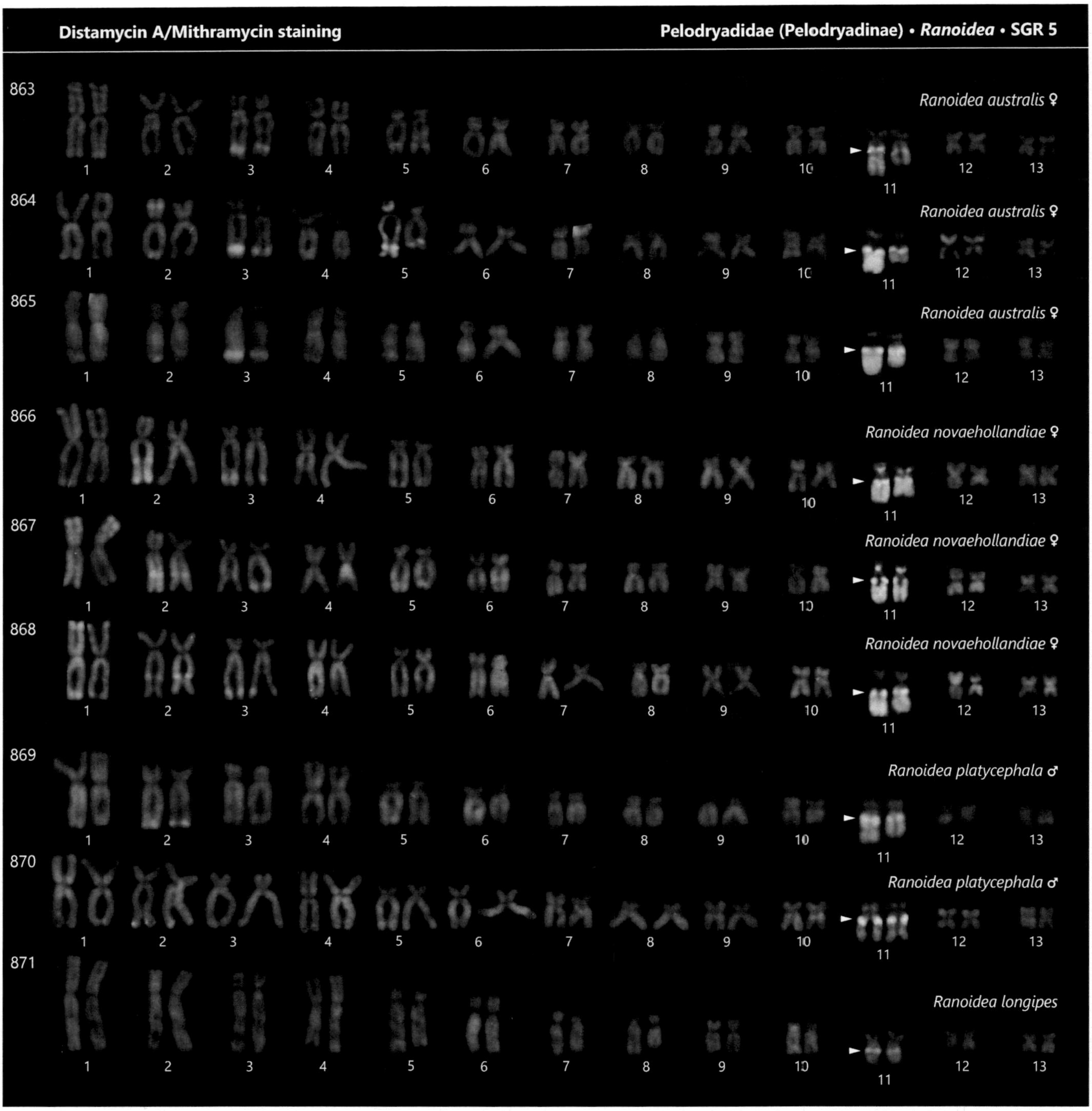
Distamycin A/Mithramycin staining
Pelodryadidae (Pelodryadinae) • Ranoidea • SGR 5
863 Ranoidea australis ♀
864 Ranoidea australis ♀
865 Ranoidea australis ♀
866 Ranoidea novaehollandiae ♀
867 Ranoidea novaehollandiae ♀
868 Ranoidea novaehollandiae ♀
869 Ranoidea platycephala ♂
870 Ranoidea platycephala ♂
871 Ranoidea longipes

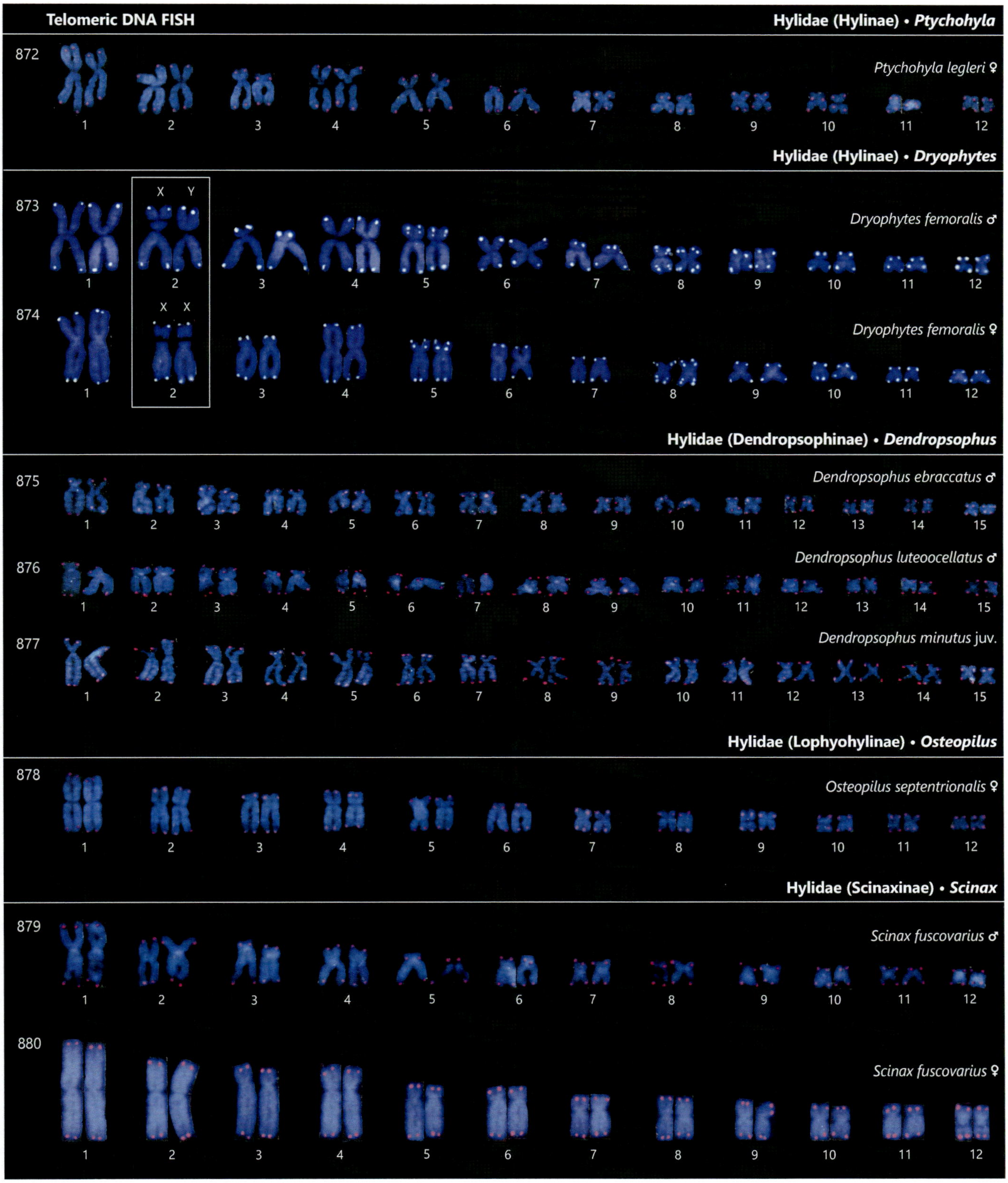

Telomeric DNA FISH
Hylidae (Hylinae) • Ptychohyla
872
Ptychohyla legleri ♀
Hylidae (Hylinae) • Dryophytes
873
Dryophytes femoralis ♂
X Y
874
Dryophytes femoralis ♀
X X
Hylidae (Dendropsophinae) • Dendropsophus
875
Dendropsophus ebraccatus ♂
876
Dendropsophus luteoocellatus ♂
877
Dendropsophus minutus juv.
Hylidae (Lophyohylinae) • Osteopilus
878
Osteopilus septentrionalis ♀
Hylidae (Scinaxinae) • Scinax
879
Scinax fuscovarius ♂
880
Scinax fuscovarius ♀

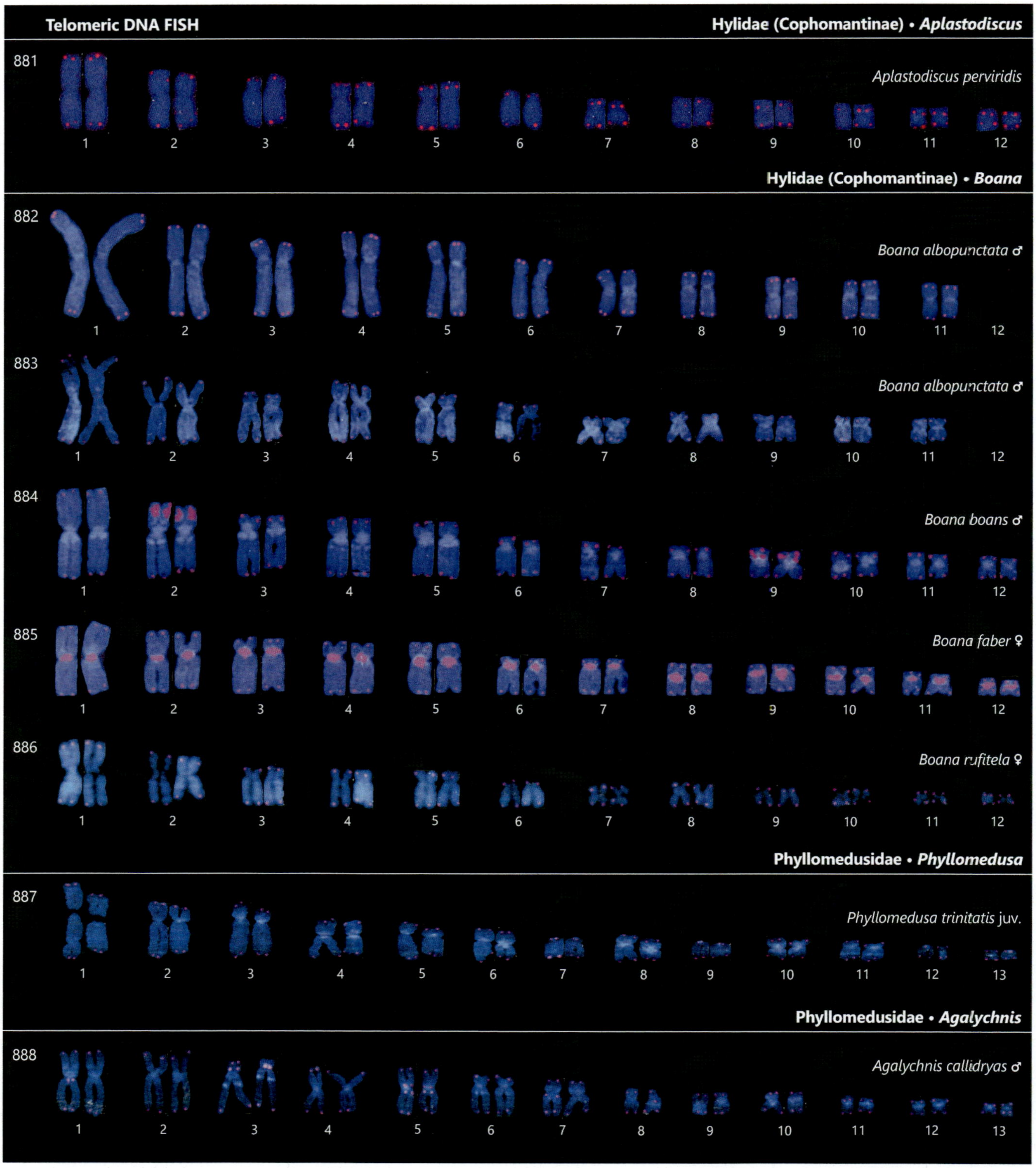

DOI: 10.1159/000489839

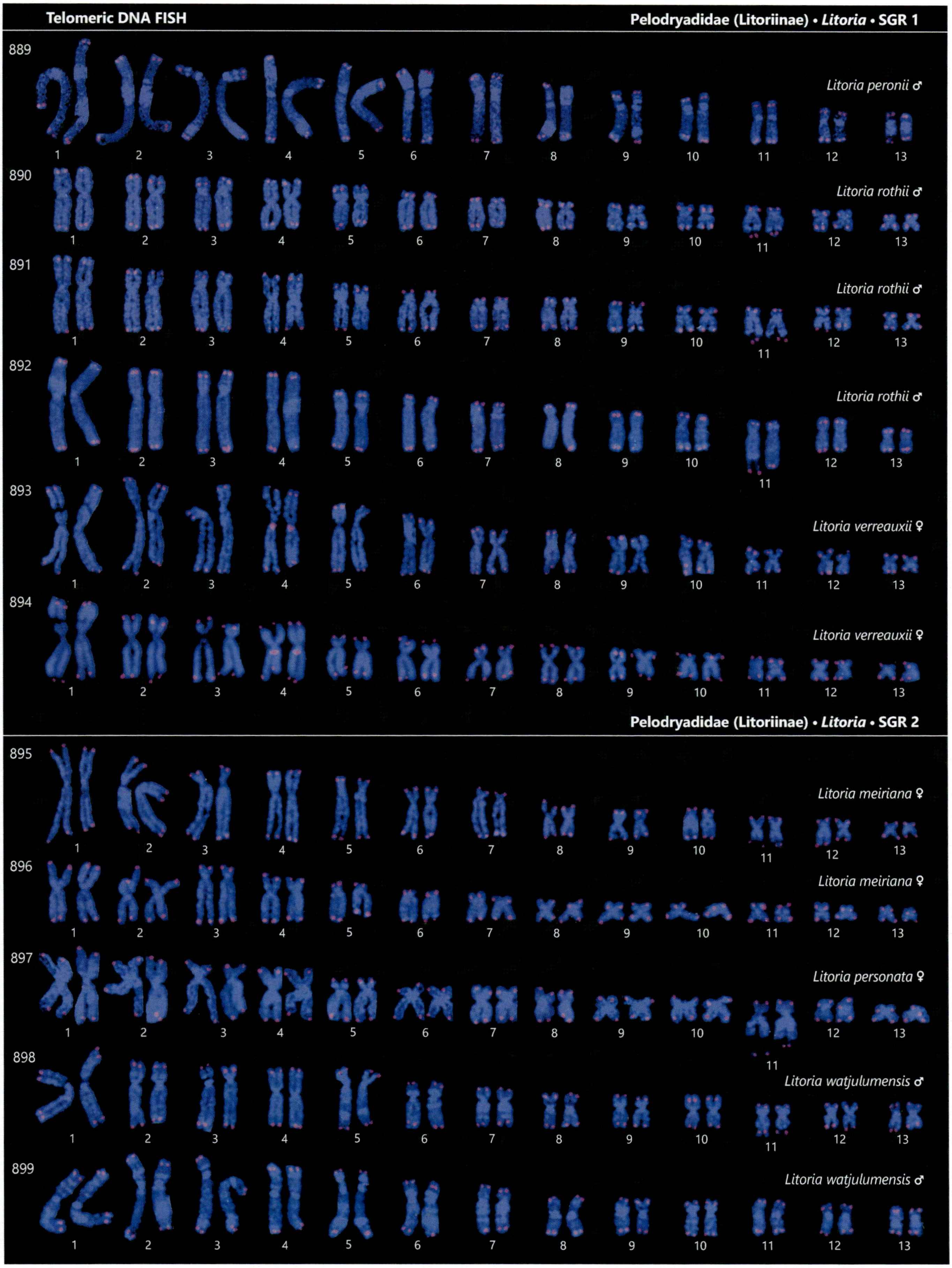
Telomeric DNA FISH
Pelodryadidae (Litoriinae) • *Litoria* • SGR 1
889
Litoria peronii ♂
890
Litoria rothii ♂
891
Litoria rothii ♂
892
Litoria rothii ♂
893
Litoria verreauxii ♀
894
Litoria verreauxii ♀
Pelodryadidae (Litoriinae) • *Litoria* • SGR 2
895
Litoria meiriana ♀
896
Litoria meiriana ♀
897
Litoria personata ♀
898
Litoria watjulumensis ♂
899
Litoria watjulumensis ♂
1
2
3
4
5
6
7
8
9
10
11
12
13

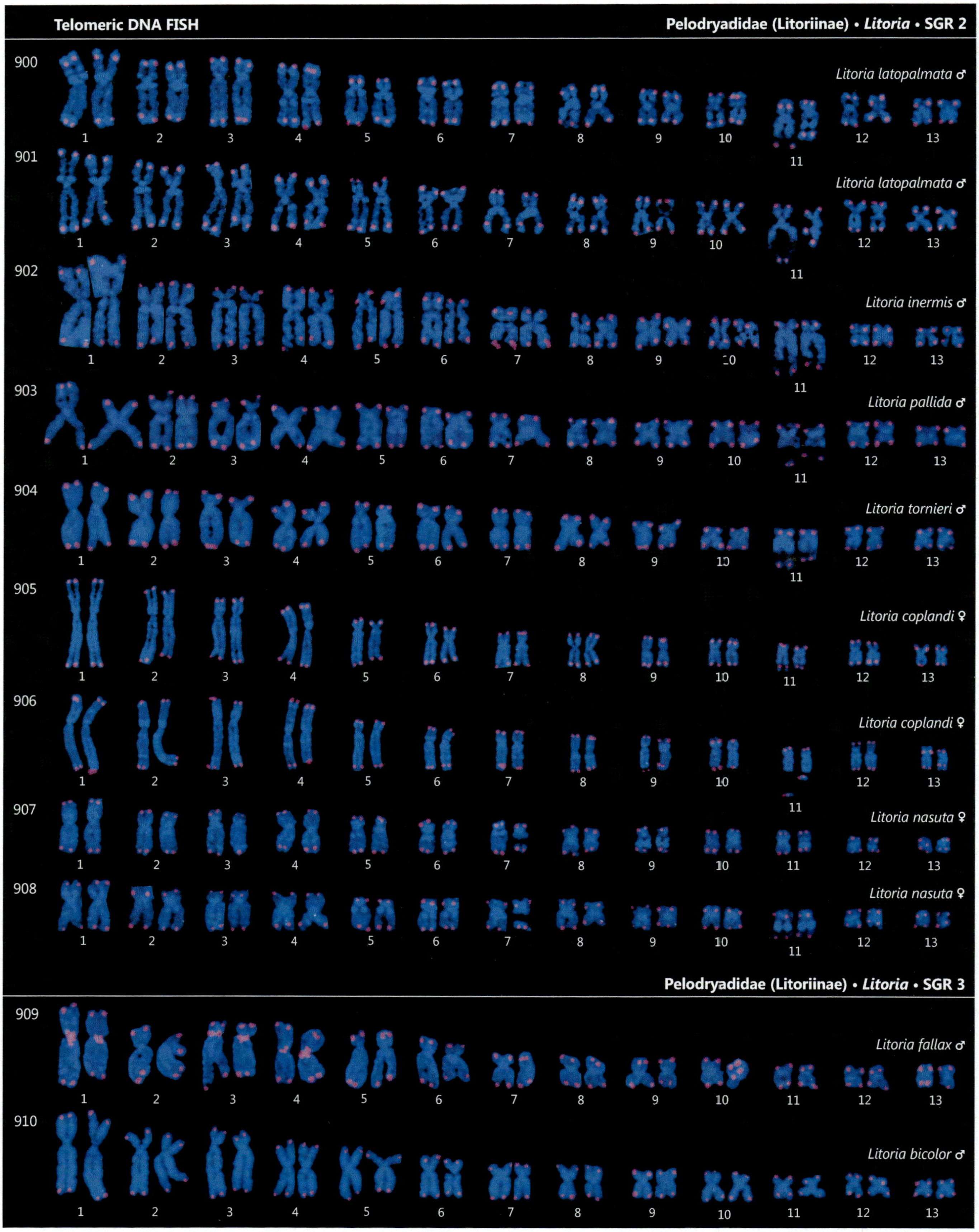
Telomeric DNA FISH
Pelodryadidae (Litoriinae) • Litoria • SGR 2
900 Litoria latopalmata ♂
901 Litoria latopalmata ♂
902 Litoria inermis ♂
903 Litoria pallida ♂
904 Litoria tornieri ♂
905 Litoria coplandi ♀
906 Litoria coplandi ♀
907 Litoria nasuta ♀
908 Litoria nasuta ♀
Pelodryadidae (Litoriinae) • Litoria • SGR 3
909 Litoria fallax ♂
910 Litoria bicolor ♂

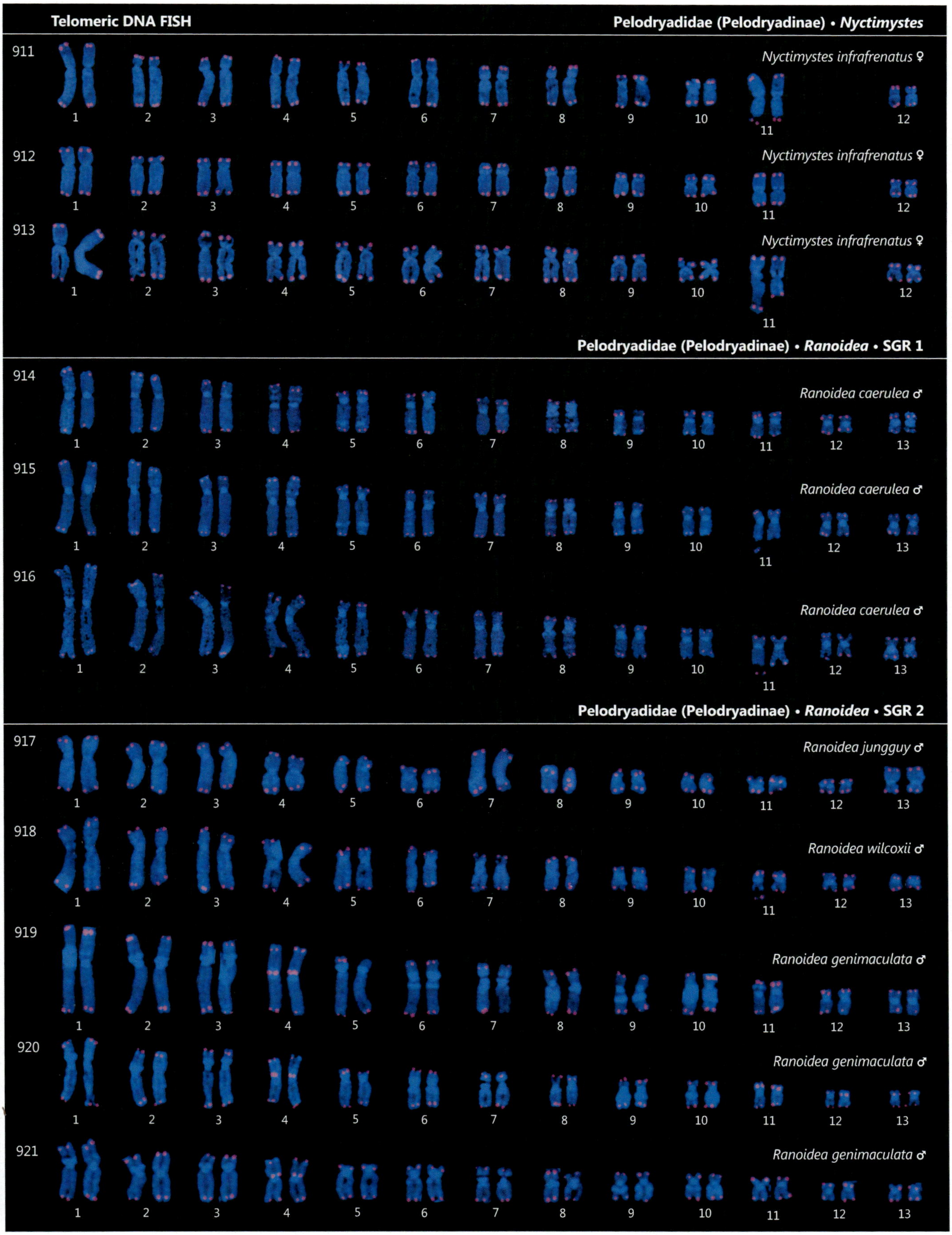
Telomeric DNA FISH
Pelodryadidae (Pelodryadinae) • Nyctimystes
911
Nyctimystes infrafrenatus ♀
912
Nyctimystes infrafrenatus ♀
913
Nyctimystes infrafrenatus ♀
Pelodryadidae (Pelodryadinae) • Ranoidea • SGR 1
914
Ranoidea caerulea ♂
915
Ranoidea caerulea ♂
916
Ranoidea caerulea ♂
Pelodryadidae (Pelodryadinae) • Ranoidea • SGR 2
917
Ranoidea jungguy ♂
918
Ranoidea wilcoxii ♂
919
Ranoidea genimaculata ♂
920
Ranoidea genimaculata ♂
921
Ranoidea genimaculata ♂

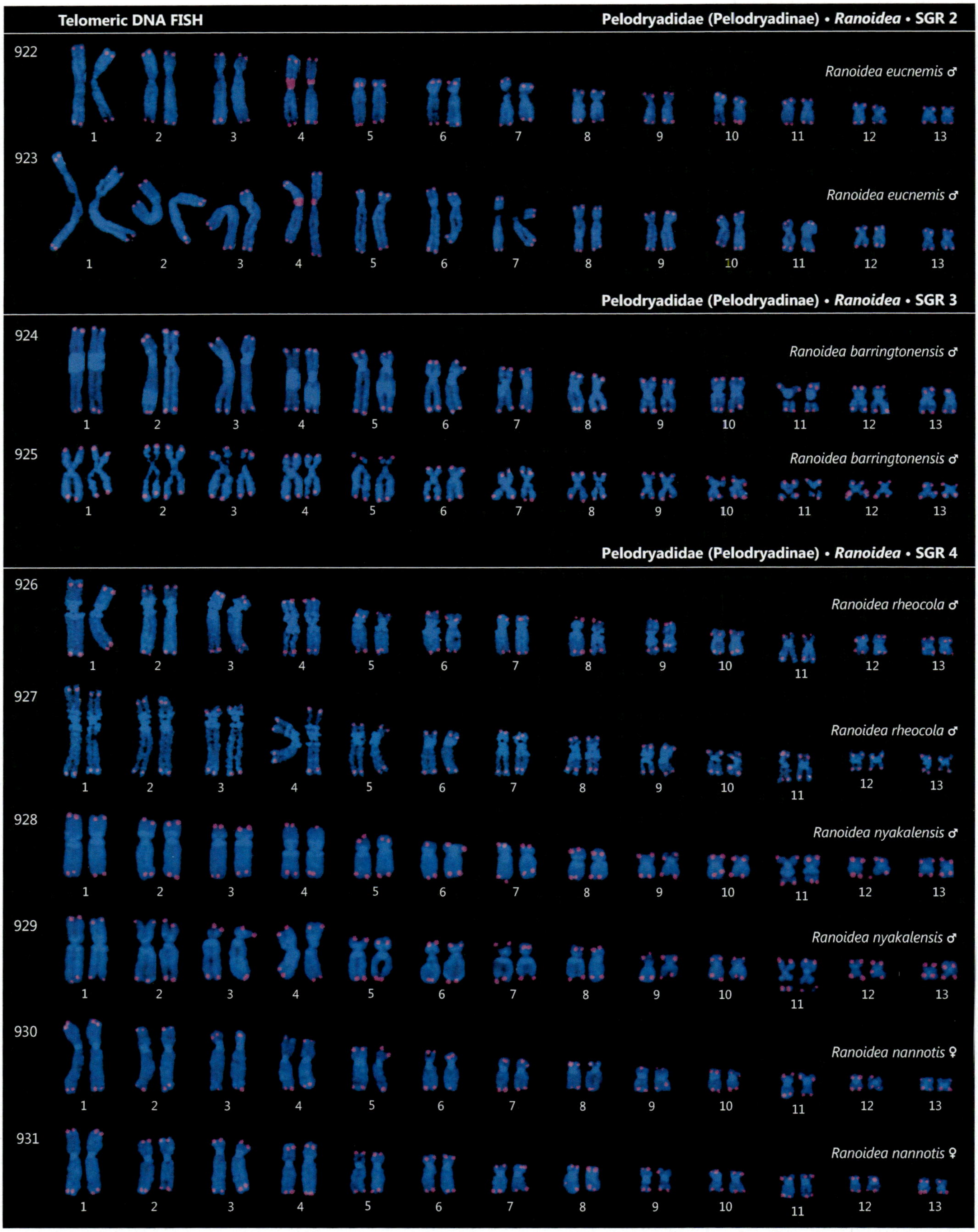
Telomeric DNA FISH
Pelodryadidae (Pelodryadinae) • Ranoidea • SGR 2
922
Ranoidea eucnemis ♂
1 2 3 4 5 6 7 8 9 10 11 12 13
923
Ranoidea eucnemis ♂
1 2 3 4 5 6 7 8 9 10 11 12 13
Pelodryadidae (Pelodryadinae) • Ranoidea • SGR 3
924
Ranoidea barringtonensis ♂
1 2 3 4 5 6 7 8 9 10 11 12 13
925
Ranoidea barringtonensis ♂
1 2 3 4 5 6 7 8 9 10 11 12 13
Pelodryadidae (Pelodryadinae) • Ranoidea • SGR 4
926
Ranoidea rheocola ♂
1 2 3 4 5 6 7 8 9 10 11 12 13
927
Ranoidea rheocola ♂
1 2 3 4 5 6 7 8 9 10 11 12 13
928
Ranoidea nyakalensis ♂
1 2 3 4 5 6 7 8 9 10 11 12 13
929
Ranoidea nyakalensis ♂
1 2 3 4 5 6 7 8 9 10 11 12 13
930
Ranoidea nannotis ♀
1 2 3 4 5 6 7 8 9 10 11 12 13
931
Ranoidea nannotis ♀
1 2 3 4 5 6 7 8 9 10 11 12 13

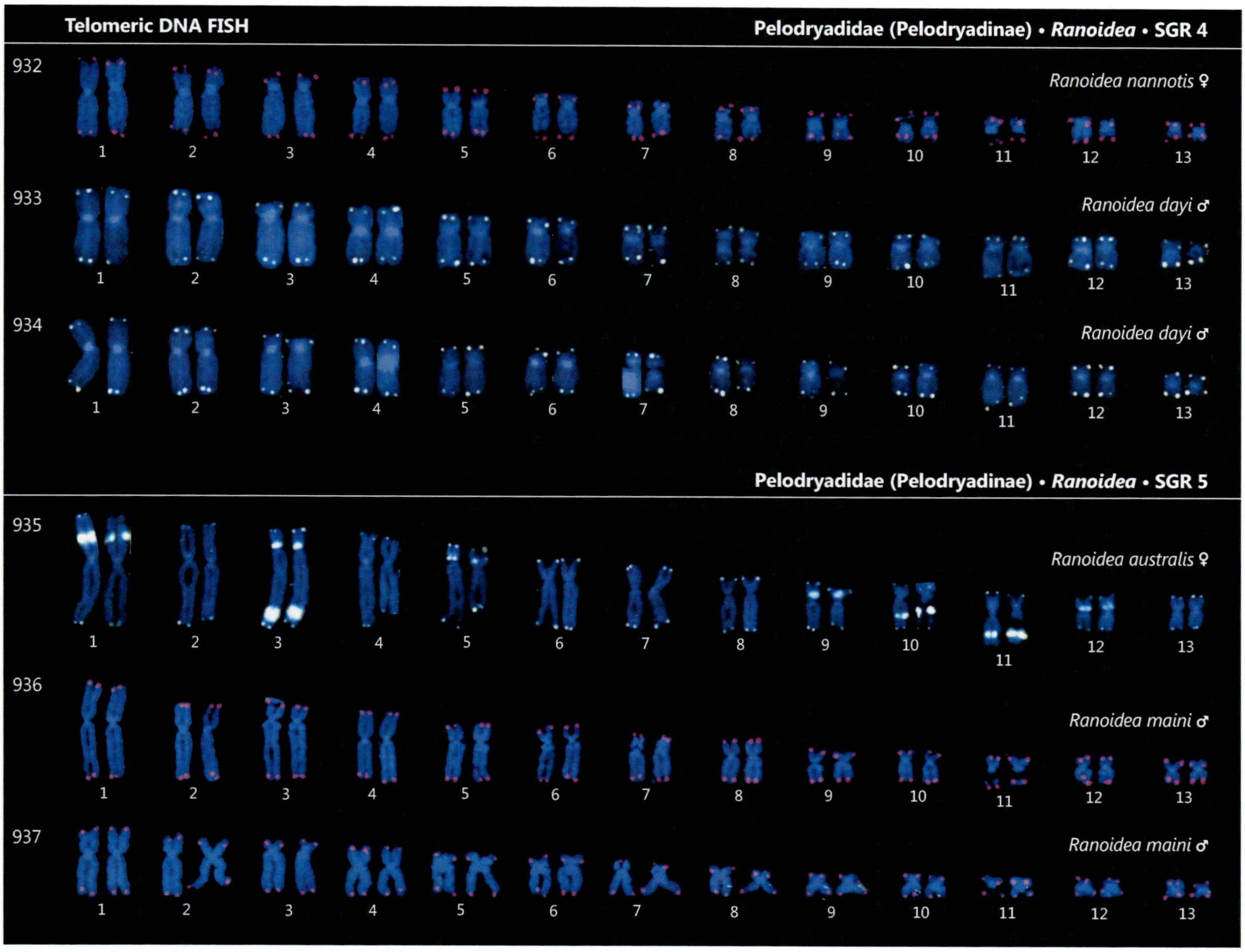

Fig. 938–950. Photographs of 78 different species of Arboranae belonging to all 59 arboranan genera (see Fig. 1). At least 1 species per genus is depicted. The collection sites, the museum catalogue numbers of the preserved specimens, and the names of the colleagues who kindly contributed the illustrations are included in the legends below the figures. The order of the 59 genera in these 13 figures follows the order of the genera listed in Figures 1 and 2.

Cytogenet Genome Res 2018;155:55–221
DOI: 10.1159/000489839

Fig. 938. Examples of Nearctic Acrisinae and Hylinae. **a** *Acris blanchardi* (KU 117930) from Lawrence, Douglas County, KS, USA. **b** *Hyliola regilla* (KU 207435) from Santa Ynez River, Santa Barbara County, CA, USA. **c** *Pseudacris clarkii* (KU 207413) from Sec. 22, T29S, R17W, Kiowa County, KS, USA. **d** *Hyla savignyi* (KU 207337) from near Kawkaban, ca. 50 km NW Sanàa, Yemen. **e** *Dryophytes cinereus* (KU 207357) from Bluff Lake, Noxubee County, MS, USA. **f** *Dryophytes walkeri* (KU 57832) from 19 km NW Comitán, Chiapas State, Mexico. Photographs by W.E. Duellman.

Fig. 939. Examples of some Middle American genera of Hylinae. **a** *Smilisca baudinii* (UMMZ 115179) from 1 km W Xicotencatl, Tamaulipas State, Mexico. **b** *Ecnomiohyla miliaria* (KU 101610) from Finca Santa Clara, Chiriquí Province, Panama. **c** *Ecnomiohyla valancifer* (KU 95416) from south slope Volcán San Martín, Veracruz State, Mexico. **d** *Tlalocohyla loquax* (KU 103686) from Puerto Viejo, Heredia Province, Costa Rica. **e** *Diaglena spatulata* (UMMZ 119157) from Salina Cruz, Oaxaca State, Mexico. **f** *Triprion petasatus* from Chichén Itza, Yucatán State, Mexico. Photograph by J.C. Lee (**f**), all others by W.E. Duellman.

Fig. 940. Some genera of Hylinae endemic to Middle America. **a** *Charadrahyla altipotens* (KU 101105) from 37 km N San Gabriel Mixtepec, Oaxaca State, Mexico. **b** *Megastomatohyla mixomaculata* (UMMZ 118158) from 4.7 mi SW Huatusco, Veracruz State, Mexico. **c** *Rheohyla miotympanum* (KU100965) from Salto Cola de Caballo, Nuevo León State, Mexico. **d** *Exerodonta bivocata* (KU 58432) from 5.6 km S Rayon Mescalapa, Chiapas State, Mexico. **e** *Duellmanohyla uranochroa* (KU 101727) from north slope of Cerro Pando, Bocas del Toro Province, Panama. **f** *Ptychohyla hypomykter* (KU 58532) from Finca Los Alpes, Alta Verapaz Department, Guatemala. Photographs by W.E. Duellman.

Fig. 941. More genera of Hylinae endemic to Middle America. **a** *Sarcohyla bistincta* (UMMZ 119193) from Dos Aguas, Michoacán State, Mexico. **b** *Plectrohyla acanthodes* (KU 58822) from 5.6 km S Rayon Mescalapa, Chiapas State, Mexico. **c** *Isthmohyla tica* (KU 91691) from 14.2 km N San Isidro el General, San José Department, Costa Rica. **d** *Isthmohyla calypsa* (KU 101736) from north slope Cerro Pando, Bocas del Toro Province, Panama. **e** *Bromeliohyla dendroscarta* (UMMZ 118167) from Mirador, Veracruz State, Mexico. **f** *Anotheca spinosa* (UMMZ 118173) from south slope Volcán San Martín, Veracruz State, Mexico. Photographs by W.E. Duellman.

Fig. 942. Genera of South American stream-breeding frogs in the subfamily Cophomantinae. **a** *Aplastodiscus leucopygius* from Domingo Martins, Espiríto Santo State, Brazil. **b** *Bokermannohyla circumdata* from Estação Biológica de Boracéia, São Paulo State, Brazil. **c** *Colomascirtus armatus* (KU 173219) from Buenos Aires, Cusco Región, Peru. **d** *Colomascirtus lindae* (KU 164202) from 11 km E Papallacta, Napo Province, Ecuador. **e** *Hyloscirtus alytolylax* (KU 164251) from 3 km NE Mindo, Pichincha Province, Ecuador. **f** *Myersiohyla chamaeleo* (USNM 562949) from Cerro de la Neblina, Amazonas State, Venezuela. Photographs by I. Sazima (**a**), M. Teixeira Jr. (**b**), and R.W. McDiarmid (**f**), all others by W.E. Duellman.

Fig. 943. Examples of the hylid frog genera *Boana* and *Dendropsophus*. **a** *Boana gladiator* (KU 139212) from 4 km WSW Santa Isabel, Cusco Región, Peru. **b** *Boana boans* (KU 220287) from Explorama, Loreto Región, Peru. **c** *Boana cinerascens* (KU 205486) from Cusco Amazónico, Río Madre de Dios, Madre de Dios Región, Peru. **d** *Boana heilprini* from 10.5 km W Jayaco, La Vega Province, Dominican Republic. **e** *Dendropsophus carnifex* (KU 164283) from 4 km NE Dos Rios, Pichincha Province, Ecuador. **f** *Dendropsophus ebraccatus* (KU 95981) from Río Tuira at Río Mono, Darién Province, Panama. Photograph by S.B. Hedges (**d**), all others by W.E. Duellman.

Fig. 944. Frogs of the subfamilies Dendropsophinae and Pseudinae. **a** *Dendropsophus marmoratus* (KU 215280) from Cusco Amazónico, Río Madre de Dios, Madre de Dios Región, Peru. **b** *Dendropsophus sarayacuensis* (KU 221814) from 1.5 km N Teniente López, Loreto Región, Peru. **c** *Xenohyla truncata* from Restinga da Mariacá, Rio de Janeiro State, Brazil. **d** *Pseudis paradoxa* (KU 183283) from Okinawa 1, Santa Cruz Department, Bolivia. **e** *Lysapsus laevis* (KU 130015) from Obidos, Pará State, Brazil. **f** *Scarthyla goinorum* (KU 215412) from Cusco Amazónico, Río Madre de Dios, Madre de Dios Región, Peru. Photographs by J.M. Pombal Jr. (**c**), D.C. Cannatella (**d**), and M.L. Crump (**e**), all others by W.E. Duellman.

Fig. 945. Members of the subfamily Scinaxinae. **a** *Scinax staufferi* (KU 117861) from 6 km SSW Penonome, Coclé Province, Panama. **b** *Scinax garbei* (KU 221053) from San Jacinto, Loreto Región, Peru. **c** *Ololygon argyreornata* from Aracruz, Espiríto Santo State, Brazil. **d** *Ololygon humilis* from Aracrackensauz, Espiríto Santo State, Brazil. **e** *Julianus uruguayus* from Paso Centurión, Cerro Largo Department, Uruguay. **f** *Sphaenorhynchus dorisae* (KU 220317) from Explorama, Loreto Región, Peru. Photographs by J.L. Gasparini (**c**, **d**) and by R. Maneyro (**e**), all others by W.E. Duellman.

Fig. 946. Examples of species from genera in the subfamily Lophyohylinae. **a** *Aparasphenodon brunoi* from Conceição da Barra, Espiríto Santo State, Brazil. **b** *Argenteohyla siemersi* from Corrientes Province, Argentina. **c** *Corythomantis greeningi* from Exu, Pernambuco State, Brazil. **d** *Itapotihyla langsdorffii* from Vita Dois Rios, Ilha Grande, Rio de Janeiro State, Brazil. **e** *Phyllodytes luteolus* from Mata Escura Reserve, Minas Gerais State, Brazil. **f** *Phytotriades auratus* from Cerro Humo, Serranía de Paría, Sucre State, Venezuela. Photographs by J.L. Gasparini (**a**), by V.H. Zaracho (**b**), by L.J. Vitt (**c**), by L. Boquimpani-Freita (**d**), by M. Teixeira Jr. (**e**), and by M. de Feitas (**f**).

Fig. 947. Examples of species from genera in the subfamily Lophyohylinae. **a** *Nyctimantis rugiceps* (KU 125930) from Santa Cecilia, Sucumbíos Province, Ecuador. **b** *Tepuihyla rodriguezi* (KU 167007) from km 144 El Dorado-Santa Elena de Uairén road, Bolívar State, Venezuela. **c** *Dryaderces pearsoni* from Río Maije, Iturralde, La Paz Department, Bolivia. **d** *Osteocephalus taurinus* (KU 205403) from Cusco Amazónico, Río Madre Dios, Madre de Dios Región, Peru. **e** *Osteopilus crucialis* from Mandeville, Manchester Parish, Jamaica. **f** *Trachycephalus typhonius* (KU 220328) from Explorama, Loreto Región, Peru. Photographs by J.M. Padial (**c**) and S.B. Hedges (**e**), all others by W.E. Duellman.

Fig. 948. Frogs from the family Phyllomedusidae. **a** *Agalychnis callidryas* (KU 63934) from Laguna Monte Alegre, Alajuela Province, Costa Rica. **b** *Agalychnis medinae* (KU 167186) from Estación Biológica Rancho Grande, Aragua State, Venezuela. **c** *Callimedusa duellmani* (KU 181813) from 8 km NNE Balzapata, Amazonas Región, Peru. **d** *Callimedusa tomopterna* (KU 220332) from Río Sucusari, Loreto Región, Peru. **e** *Cruziohyla calcarifer* (KU 78415) from Laguna, Darién Province, Panama. **f** *Hylomantis aspera* from Uruçuca, Bahia State, Brazil. Photographs by W.E. Duellman (**a–e**) and by M. Teixeira Jr. (**f**).

Fig. 949. Frogs from the family Phyllomedusidae. **a** *Phasmahyla cochranae* from Pozos de Caldas, Minas Gerais State, Brazil. **b** *Phrynomedusa marginata* from Santa Teresa, Espiríto Santo State, Brazil. **c** *Phyllomedusa bicolor* (KU 124907) from Leticia, Amazonas Department, Colombia. **d** *Phyllomedusa sauvagii* (KU 160426) from Tucumán, Tucumán Department, Argentina. **e** *Pithecopus azureus* (KU 183441) from Okinawa 1, Santa Cruz Department, Bolivia. **f** *Pithecopus palliatus* (KU 143240) from Santa Cecilia, Sucumbíos Province, Ecuador. Photographs by W.C.A. Bokermann (**a**), by E. Izecksohn (**b**), by D.C. Cannatella (**e**), and all others by W.E. Duellman.

Fig. 950. Frogs from the family Pelodryadidae. **a** *Litoria olongburensis* (KU 179858) from Stradbroke Island, Queensland, Australia. **b** *Litoria peronii* (KU 179888) from Bellthorpe State Forest, QLD, Australia. **c** *Nyctimystes trachydermis* from Mt. Obree, Central Province, Papua New Guinea. **d** *Nyctimystes infrafrenatus* from Port Moresby, National Capital District, Papua New Guinea. **e** *Ranoidea caerulea* (KU 180664) from Mitchell Plateau, WA, Australia. **f** *Ranoidea dahlii* from Darwin, NT, Australia. Photographs by F. Kraus (**c**, **d**), all others by W.E. Duellman.

Fig. 951. Egg clutches of hylid frogs. **a** Globular mass of *Isthmohyla pseudopuma* from Tapantí, Cartago Province, Costa Rica. **b** Surface mass of *Trachycephalus coriaceus* from Cusco Amazónico, Río Madre de Dios, Madre de Dios Región, Peru. **c** Mass on leaf by *Dendropsophus leucophyllatus* from Cusco Amazónico, Río Madre de Dios, Madre de Dios Región, Peru. **d** On leaf of bromeliad by *Bromeliohyla bromeliacia* from Finca Chicoyou, Alta Verapaz Department, Guatemala. **e** Nest of *Boana boans* from Río Cuyuní, Bolívar State, Venezuela. **f** Eggs of *Anotheca spinosa* on inner wall of container; descendants from individuals from San Vito de Java, Puntarenas Province, Costa Rica. Photograph by K.-H. Jungfer (**f**), all others by W.E. Duellman.

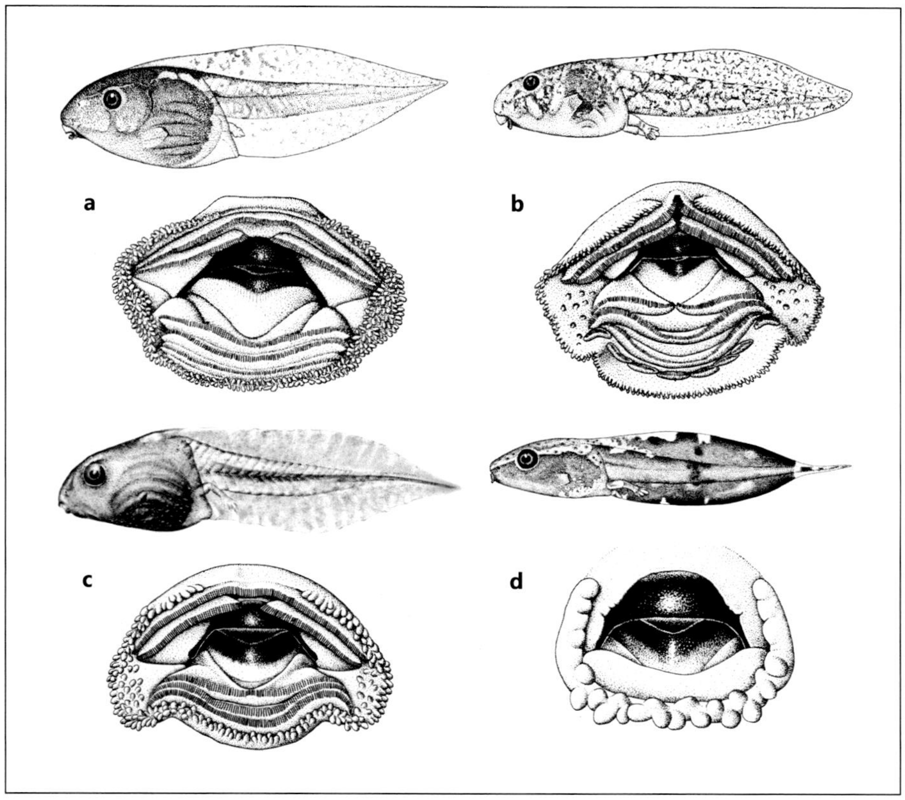

Fig. 952. Pond-dwelling tadpoles of hylid frogs. **a** *Smilisca baudinii* from Chinajá, Alta Verapaz Department, Guatemala. **b** *Boana boans* from Camp Sasardi, San Blas Province, Panama. **c** *Dendropsophus columbianus* from 8 km E Popoyan, Cauca Department, Colombia. **d** *Dendropsophus subocularis* from Río Ucurganti, Darién Province, Panama. Drawings in **c** from Duellman and Trueb [1983], all others from Duellman [2001].

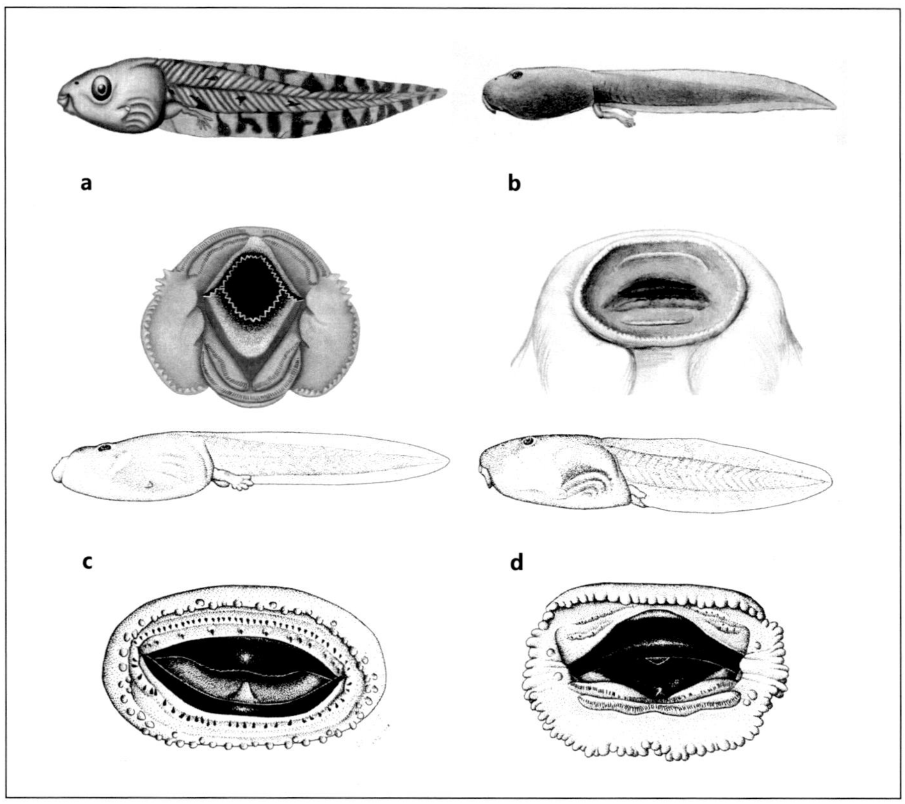

Fig. 953. a *Scarthyla goinorum* from Cusco Amazónico, Río Madre de Dios, Madre de Dios Región, Peru. **b** *Osteopilus marianae* from Spaldings, Clarendon Parish, Jamaica. **c** *Isthmohyla zeteki* from La Palma, San José Province, Costa Rica. **d** *Anotheca spinosa* from Vista Hermosa, Oaxaca State, Mexico. Not drawn to scale. Drawings in **a** from Duellman and de Sá [1988], **b** from Dunn [1926], and **c**, **d** from Duellman [2001].

Cytogenet Genome Res 2018;155:55–221
DOI: 10.1159/000489839

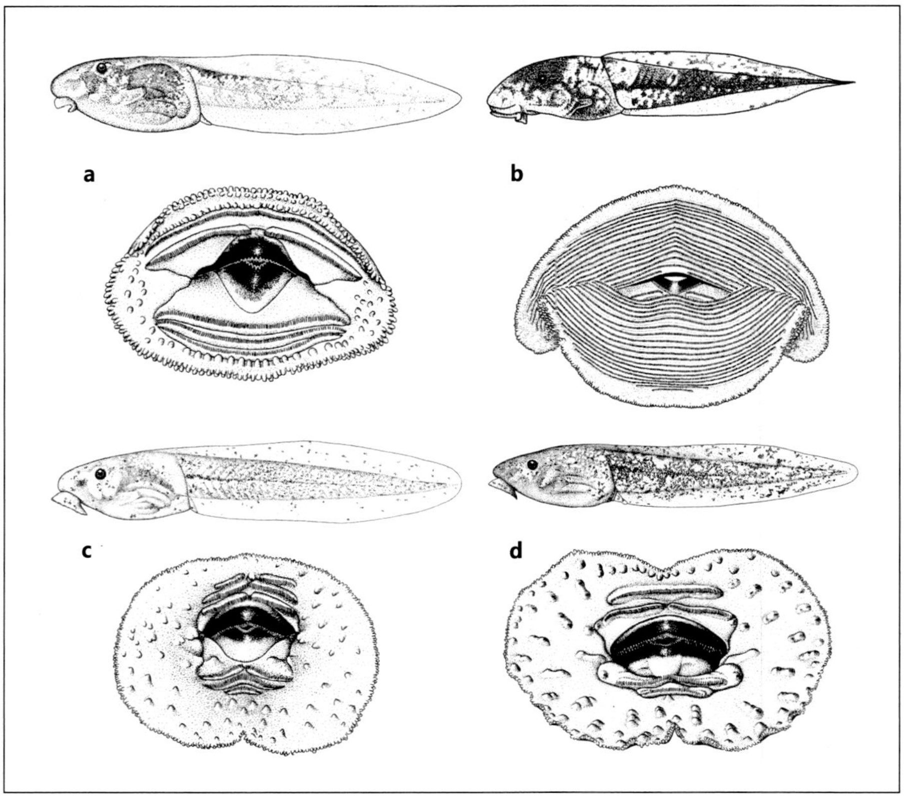

Fig. 954. Stream-dwelling tadpoles of hylid frogs. **a** *Sarcohyla pentheter* from 37 km N San Gabriel Mixtepec, Oaxaca State, Mexico. **b** *Myersiohyla neblinaria* from Cerro de la Neblina, Amazonas Department, Venezuela. **c** *Ptychohyla ignicolor* from Vista Hermosa, Oaxaca State, Mexico. **d** *Duellmanohyla uranochroa* from north slope Cerro Pando, Bocas de Toro Province, Panama. Drawings in **b** from Faivovich et al. [2013], all others from Duellman [2001].

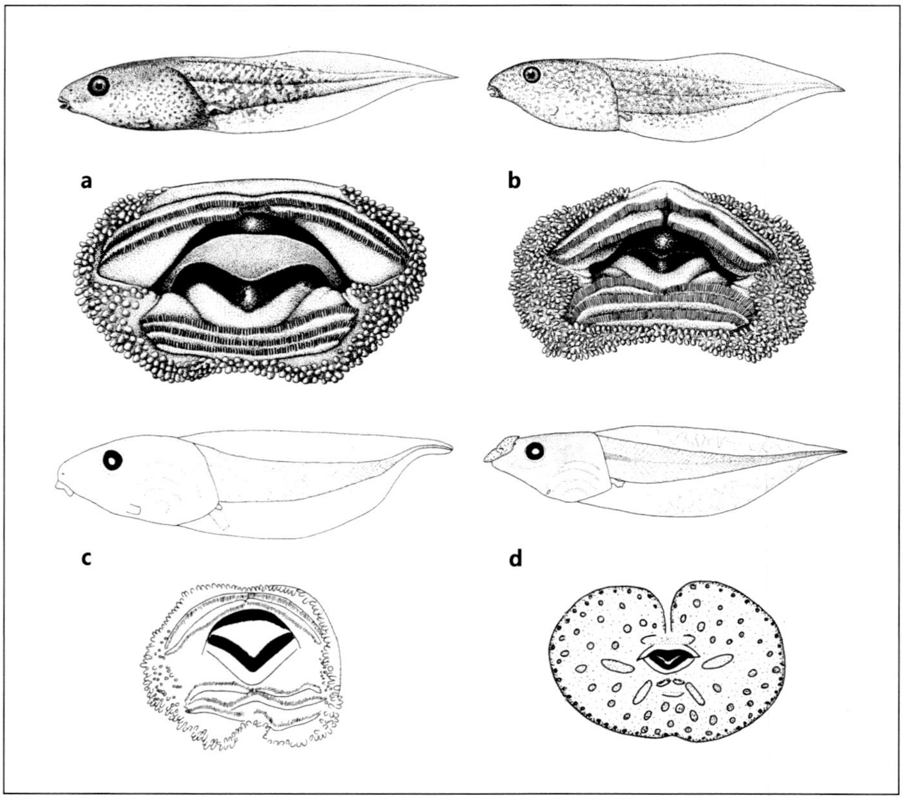

Fig. 955. Tadpoles of phyllomedusid frogs. **a** *Agalychnis spurrelli* from 4.5 km W Rincón de Osa, Puntarenas Province, Costa Rica. **b** *Agalychnis lemur* from El Tigre, Limón Province, Costa Rica. **c** *Phrynomedusa vanzolinii* from Boracéia, São Paulo State, Brazil. **d** *Phasmahyla cochranae* from Boracéia, São Paulo State, Brazil. Drawings in **a** and **b** from Duellman [2001], in **c** and **d** from Heyer et al. [1990].

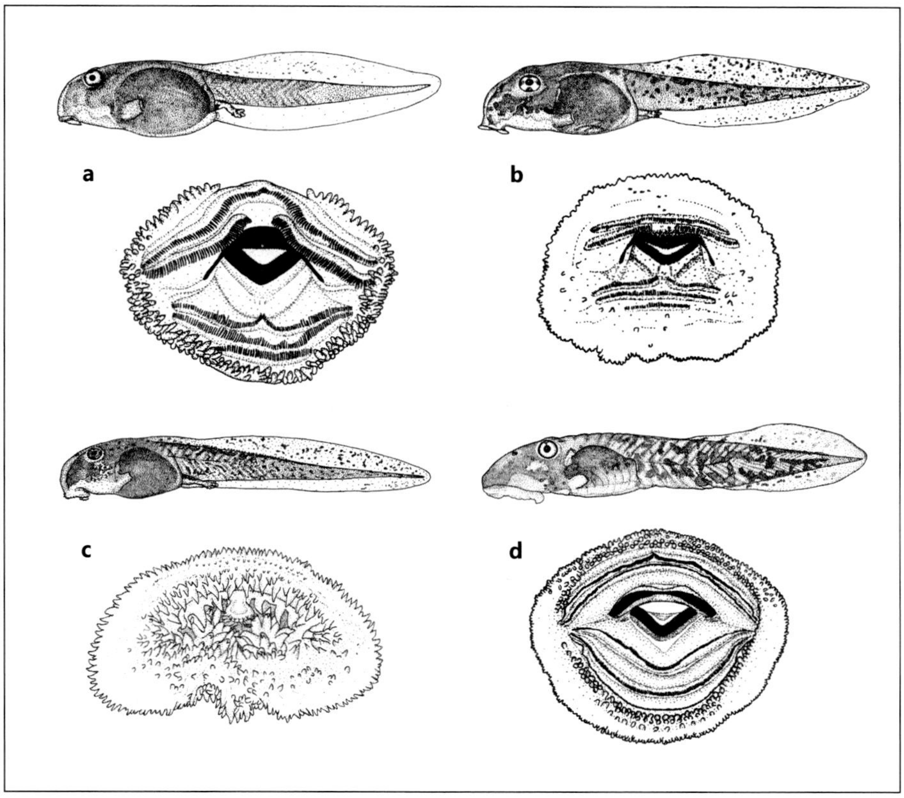

Fig. 956. Tadpoles of Australian pelodryadid frogs. **a** *Litoria bicolor* from 15 km SE Darwin, NT. **b** *Litoria citropa* from Darkes Forest, NSW. **c** *Ranoidea daviesae* from Tirrill Creak, Bulga State Forest, NSW. **d** *Ranoidea nannotis* from Tully Valley, QLD. Not drawn to scale. From Antsis [2013], New Holland Publishers.

Cytogenet Genome Res 2018;155:223–247
DOI:10.1159/000490328

The Arboranan Frogs:

References

KARGER

E-Mail karger@karger.com
www.karger.com/cgr

Abramyan J, Ezaz T, Marshall Graves JA, Koopman P: Z and W sex chromosomes in the cane toad. Chromosome Res 17:1015–1024 (2009).
Abuín M, Martínez P, Sánchez L: Localization of the repetitive telomeric sequence $(TTAGGG)_n$ in four salmonid species. Genome 39:1035–1038 (1996).
Acevedo M, Smith E: *Exerodonta perkinsi*. http://amphibiaweb.org/ (2004).
Adegoke JA, Arnason U, Widegren B: Sequence organization and evolution, in all extant whalebone whales, of a satellite DNA with terminal chromosome localization. Chromosoma 102:382–388 (1993).
Alexander MA, Eischeid JK: Climate variability in regions of amphibian declines. Conserv Biol 15:930–942 (2001).
Alfonso IP, Gambale PG, Oda FH, Bastos RP: Anurans from a degraded area in a seasonally semi-deciduous forest in southern Brazil. Herpetropicos 9:69–78 (2013).
Al-Shehri AH, Al-Saleh AA: Karyotype of amphibians in Saudi Arabia 2: the karyotype of *Hyla savignyi*. J Biol Sci 5:768–770 (2005).
Altig R, McDiarmid RW: Diversity: familial and generic characterizations, in McDiarmid RW, Altig R (eds): Tadpoles: The Biology of Anuran Larvae, pp 295–337 (University of Chicago Press, Chicago 1999).
Altig R, McDiarmid RW: Morphological diversity and evolution of egg and clutch structure in amphibians. Herpetol Monogr 21:1–32 (2007).
Alves-Silva R, da Silva HR: Life in bromeliads: reproductive behaviour and the monophyly of the *Scinax perpusillus* species group (Anura: Hylidae). J Nat Hist 43:205–217 (2009).
Amaro-Ghilardi RC, Skuk G, De Sá RO, Rodrigues MT, Yonenaga-Yassuda Y: Karyotypes of eight species of *Leptodactylus* (Anura, Leptodactylidae) with a description of a new karyotype for the genus. Phyllomedusa 5:119–133 (2006).
AmphibiaWeb 2017 (University of California, Berkeley, CA, USA). https://amphibiaweb.org. Accessed December 2017.
Ananias F: Caracterizção cromossômica de espécies e subespécies do grupo pulchella (Amphibia, Anura, Hylidae). Master thesis, Universidade Estadual de Campinas, Campinas (1996).
Ananias F, Garcia PCA, Recco-Pimentel SM: Conserved karyotypes in the *Hyla pulchella* species group (Anura, Hylidae). Hereditas 140:42–48 (2004).
Ananias F, Modesto ADS, Mendes SC, Napoli MF: Unusual primitive heteromorphic ZZ/ZW sex chromosomes in *Proceratophrys boiei* (Anura, Cycloramphidae, Alsodinae), with description of C-band interpopulational polymorphism. Hereditas 144:206–212 (2007).
Anderson K: A cytotaxonomic analysis of the holarctic treefrogs in the genus *Hyla*. PhD thesis, University of New York (1986).
Anderson K: Chromosome evolution in Holarctic treefrogs, in Green DM, Sessions SK (eds): Amphibian Cytogenetics and Evolution, pp 299–331 (Academic Press, San Diego 1991).
Anderson K: A karyological perspective on the monophyly of the hylid genus *Osteopilus*, in Powell R, Henderson RW (eds): Contributions to West Indian Herpetology: A Tribute to Albert Schwartz, pp 157–168 (Society for the Study of Amphibians and Reptiles, New York 1996).
Anderson K, Moler PE: Natural hybrids of the pine barren treefrog, *Hyla andersoni* with *Hyla cinerea* and *H. femoralis* (Anura; Hylidae): morphological and chromosomal evidence. Copeia 1986:70–76 (1986).
Anstis M: Tadpoles and Frogs of Australia (New Holland Publishers, London 2013).
Arendt D, Nübler-Jung K: Rearranging gastrulation in the name of yolk: evolution of gastrulation in yolk-rich amniote eggs. Mech Dev 81:3–22 (1999).
Arnason U, Widegren B: Composition and chromosomal localization of cetacean highly repetitive DNA with special reference to the blue whale, *Balaenoptera musculus*. Chromosoma 98:323–329 (1989).
Arnason U, Höglund M, Widegren B: Conservation of highly repetitious DNA in cetaceans. Chromosoma 89:238–242 (1984).
Arnason U, Allerdice PW, Lien J, Widegren B: Highly repetitive DNA in the baleen whale genera *Balaenoptera* and *Megaptera*. J Mol Evol 27:217–221 (1988).
Arrighi FE, Hsu TC, Saunders P, Saunders GF: Localization of repetitive DNA in the chromosomes of *Microtus agrestis* by means of in situ hybridization. Chromosoma 32:224–236 (1970).
Arrighi FE, Hsu TC, Pathak S, Sawada H: The sex chromosomes of the Chinese hamster: constitutive heterochromatin deficient in repetitive DNA sequences. Cytogenet Cell Genet 13:268–274 (1974).
Arrighi FE, Stock AD, Pathak S: Chromosomes of *Peromyscus* (Rodentia, Cricetidae). V. Evidence of pericentric inversions, in Pearson PL, Lewis KR (eds): Chromosomes Today, vol 5, pp 323–329 (John Wiley & Sons, Hoboken 1976).
Ashley T, Moses MJ, Solari AJ: Fine structure and behaviour of a pericentric inversion in the sand rat, *Psammomys obesus*. J Cell Sci 50:105–119 (1981).
Austin CR, Bishop MWH: Some features of the acrosome and perforatorium in mammalian spermatozoa. Proc R Soc London B Biol Sci 149:234–240 (1958).
Babu A: Heterogeneity of heterochromatin of human chromosomes as demonstrated by restriction endonuclease treatment, in Verma RS (ed): Heterochromatin. Molecular and Structural Aspects, pp 250–275 (Cambridge University Press, Cambridge 1988).
Baccetti B: Evolution of the sperm cells, in Metz C, Monroy A (eds): Biology of Fertilization, Vol 2, pp 3–47 (Academic Press, Orlando 1985).
Bachmann K, Bogart JP: Comparative cytochemical measurements in the diploid-tetraploid species pair of hylid frogs *Hyla chrysoscelis* and *H. versicolor*. Cytogenet Cell Genet 15:186–194 (1975).
Baker RJ, Gardner AL, Patton JL: Chromosomal polymorphism in the phyllostomatid bat, *Mimon crenulatum* (Geoffroy). Experientia 28:969–970 (1972).
Baker WK: Crossing-over in heterochromatin. Am Nat 92:59–60 (1958).
Baldari CT, Amaldi F: DNA reassociation kinetics in relation to genome size in four amphibian species. Chromosoma 59:13–22 (1976).
Baldissera FA, Oliveira PSL, Kasahara S: Estudos cromossômicos com anfíbios da fauna brasileira (Hylidae e Leptodactylidae). XVII Congreso Brasileiro de Zoologia. Abstracts, p 305 (1991).
Baldissera FA, de Oliveira PSL, Kasahara S: Cytogenetics of four Brazilian *Hyla* species (Amphibia-Anura) and description of a case with a supernumerary chromosome. Rev Brasil Gen 16:335–345 (1993).
Ballif BC, Wakui K, Gajecka M, Shaffer LG: Translocation breakpoint mapping and sequence analysis in three monosomy 1p36 subjects with der(1)t(1;1)(p36;q44) suggest mechanisms for telomere capture in stabilizing de novo terminal rearrangements. Hum Genet 114:198–206 (2004).
Baraquet M, Salas NE, Martino AL: C-banding patterns and meiotic behavior in *Hypsiboas pulchellus* and *H. cordobae* (Anura, Hylidae). J Basic Appl Genet 24:32–39 (2013).
Barr HJ, Esper H: Nucleolar size in cells of *Xenopus laevis* in relation to nucleolar competition. Exp Cell Res 31:211–214 (1963).
Barrio A: Estudio cariotípico y análisis audioespectrográfico de los cantos de las especies de *Phyllomedusa* (Anura, Hylidae) que habitan en la Argentina. Physis 35:65–74 (1976).
Barrio A, Pistol de Rubel D: Caracteristicas del cariotipo de los pseudidos (Amphibia, Anura). Physis 29:505–510 (1970).
Barth A, Solé M, Costa MA: Chromosome polymorphism in *Phyllomedusa rohdei* (Anura: Hylidae). J Herpetol 43:676–679 (2009).
Barth A, Souza VA, Solé M, Costa MA: Molecular cytogenetics of nucleolar organizer regions in *Phyllomedusa* and *Phasmahyla* species (Hylidae, Phyllomedusinae): a cytotaxonomic contribution. Genet Mol Res 12: 2400–2408 (2013).
Barth A, Vences M, Solé M, Costa MA: Molecular cytogenetics and phylogenetic analysis of Brazilian leaf frog species of the genera *Phyllomedusa* and *Phasmahyla* (Hylidae: Phyllomedusinae). Can J Zool 92:795–802 (2014).
Barth LG: Embryology (Holt Rinehart and Winston, New York 1953).
Bass RA: Chromosomal polymorphism in cardinals *Cardinalis cardinalis*. Can J Genet Cytol 21:549–553 (1979).
Bastos RP, Haddad CFB, Pombal JP: Foam nest in *Scinax rizibilis* (Amphibia: Anura: Hylidae). Zoologia 27:881–886 (2010).
Batistic RF: Aspectos citogenéticos da evulução em *Phyllomedusa* (Anura-Amphibia). PhD thesis, Faculdade de Medicina de Ribeirão Preto, Universidade de São Paulo (1989).
Batistic RF, Soma M, Beçak ML, Beçak W: Further studies on polyploid amphibians. A diploid population of *Phyllomedusa burmeisteri*. J Hered 66:160–162 (1975).

Baverstock PR, Watts CHS, Hogarth JT: Heterochromatin variation in the Australian rodent *Uromys caudimaculatus*. Chromosoma 57:397–403 (1976).

Baverstock PR, Watts CHS, Hogarth JT: Chromosome evolution in Australian rodents. I. The Pseudomyinae, the Hydromyinae and the *Uromys/Melomys* group. Chromosoma 61:95–125 (1977a).

Baverstock PR, Watts CHS, Hogarth JT: Polymorphism of the X-chromosome, Y-chromosome and autosomes in the Australian hopping mice, *Notomys alexis*, *N. cervinus* and *N. fuscus* (Rodentia, Muridae). Chromosoma 61:243–256 (1977b).

Beçak ML: Cariótipos e evolução cromossômica em Amphibia – Anura. DSc thesis, Faculdade de Medicina de Ribeirão Preto, Universidade São Paulo, São Paulo (1967).

Beçak ML: Chromosomal analysis of eighteen species of Anura. Caryologia 21:191–208 (1968).

Beçak ML, Beçak W, Rabello MN: Cytological evidence of constant tetraploidy in the bisexual South American frog *Odontophrynus americanus*. Chromosoma 19:188–193 (1966).

Beçak ML, Beçak W, Rabello MN: Further studies on polyploid amphibians (Ceratophrydidae). I. Mitotic and meiotic aspects. Chromosoma 22:192–201 (1967).

Beçak ML, Denaro L, Beçak W: Polyploidy and mechanisms of karyotypic diversification in Amphibia. Cytogenetics 9:225–238 (1970a).

Beçak ML, Beçak W, Vizotto LD: A diploid population of the polyploid amphibian *Odontophrynus americanus* and an artificial intraspecific triploid hybrid. Experientia 26:545–546 (1970b).

Beebee TJC: Amphibian breeding and climate change. Nature 374:219–220 (1995).

Belcheva RG, Sofianidou TS: Karyological investigation of the brown frog species (Anura – Ranidae) from Bulgaria and Greece, in Olmo E (ed): Cytogenetics of Amphibians and Reptiles, pp 141–146 (Birkhäuser, Basel 1990).

Bell BD, Daugherty CH, Hay JM: *Leiopelma pakeka* n. sp. (Anura, Leiopelmatidae), a cryptic species of frog from Maud Island, New Zealand, and a reassessment of the conservation status of *L. hamiltoni* from Stephens Island. J R Soc New Zeal 28:39–54 (1998).

Bell G, Burt A: B-chromosomes: germ-line parasites which induce changes in host recombination. Parasitology 100:S19–S26 (1990).

Berend SA, Hale DW, Enstrom MD, Greenbaum IF: Cytogenetics of collared lemmings (*Dicrostonyx groenlandicus*). II. Meiotic behaviour of B chromosomes suggests a Y-chromosome origin of supernumerary chromosomes. Cytogenet Cell Genet 95:85–91 (2001).

Berger L, Speare R, Daszak P, Green DE, Cunningham AA, et al: Chytridiomycosis causes amphibian mortality associated with population declines in the rain forests of Australia and Central America. Proc Natl Acad Sci USA 95:9031–9036 (1998).

Berger L, Speare R, Hyatt AD: Chytrid fungi and amphibian declines: overview, implications and future directions, in Campbell A (ed): Declines and Disappearances of Australian Frogs, pp 21–31 (Environment Australia, Canberra 2000).

Berset-Brändli L, Jaquiéry J, Dubey S, Perrin N: A sex-specific marker reveals male heterogamety in European tree frogs. Mol Biol Evol 23:1104–1106 (2006).

Beukeboom LW: Bewildering B's: an impression of the 1st B chromosome conference. Heredity 73:328–336 (1994).

Bi K, Bogart JP: Identification of intergenomic recombinations in unisexual salamanders of the genus *Ambystoma* by genomic in situ hybridization (GISH). Cytogenet Genome Res 112:307–312 (2006).

Bi K, Bogart JP, Fu J: Genealogical relationships of unisexual salamanders of the genus *Ambystoma* inferred from intergenomic exchanges and 45S rDNA cytotypes. Chromosome Res 16:275–289 (2008).

Bi K, Bogart JP, Fu J: A populational survey of 45S rDNA polymorphism in the Jefferson salamander *Ambystoma jeffersonianum* as revealed by fluorescence in situ hybridization (FISH). Curr Zool 55:145–149 (2009).

Bianchi NO, Bianchi MS, Vidal-Rioja L: Heterochromatin late replication and secondary constrictions in the chromosome complement of *Leptodactylus ocellatus*. Caryologia 26:397–403 (1973).

Bickham JW, Baker RJ: Chromosome homology and evolution of emydid turtles. Chromosoma 54:201–219 (1976).

Bickmore W, Craig J: Chromosome Bands. Patterns in the Genome (Springer, Heidelberg 1997).

Birnstiel ML, Wallace H, Sirlin JL, Fischberg M: Localization of the ribosomal DNA complements in the nucleolus organizer region of *Xenopus laevis*, in Vincent WS, Miller OL (eds): International Symposium on the Nucleolus: Its Structure and Function, pp 431–448 (Nat Cancer Inst Monograph 23, Bethesda 1966).

Blair WF: Evidence from hybridization, in Blair WF (ed): Evolution in the Genus *Bufo*, pp 196–232 (Univ Texas Press, Austin 1972).

Blaustein AR, Kiesecker JM: Complexity in conservation: lessons from the global decline of amphibian populations. Ecol Lett 5:597–608 (2002).

Blaustein AR, Wake DB: Declining amphibian populations: a global phenomenon? Trends Ecol Evol 5:203 (1990).

Blaustein AR, Hokit DG, O'Hara RK, Holt RA: Pathogenic fungus contributes to amphibian losses in the Pacific Northwest. Biol Conserv 67:251–254 (1994a).

Blaustein AR, Hoffman PD, Hokit DG, Kiesecker JM, Walls SC, Bays JB: UV repair and resistance to UV-B in amphibian eggs: a link to population declines. Proc Natl Acad Sci USA 91:1791–1795 (1994b).

Blaxter M: Molecular systematics – counting angels with DNA. Nature 421:122–124 (2003).

Blum M, Beyer T, Weber T, Vick P, Andre P, et al: *Xenopus*, an ideal model system to study vertebrate left-right asymmetry. Dev Dyn 238:1215–1225 (2009).

Bobrow M, Pearson PL, Pike MC, El Alfi OS: Length variation in the quinacrine-binding segment of human Y chromosomes of different sizes. Cytogenetics 10:190–198 (1971).

Bobrow M, Collacott HEAC, Madan K: Chromosome banding with acridine orange. Lancet 2:1311 (1972).

Bogart JP: Chromosomes of the South American amphibian family Ceratophridae with a reconsideration of the taxonomic status of *Odontophrynus americanus*. Can J Genet Cytol 9:531–542 (1967).

Bogart JP: Chromosome number differences in the amphibian genus *Bufo*: the *Bufo regularis* species group. Evolution 22:42–45 (1968).

Bogart JP: Cromosomas en algunos géneros de anuros. Act IV Congr Latin Zool 1:79–86 (1970).

Bogart JP: Karyotypes, in Blair WF (ed): Evolution in the Genus *Bufo*, pp 171–195 (University of Texas Press, Austin 1972).

Bogart JP: Evolution of anuran karyotypes, in Vial JL (ed): Evolutionary Biology of the Anurans, pp 337–349 (University of Missouri Press, Columbia 1973).

Bogart JP: Evolutionary implications of polyploidy in amphibians and reptiles, in Lewis WH (ed): Polyploidy: Biological Relevance, pp 341–378 (Plenum Press, New York 1980).

Bogart JP: The influence of life history on karyotypic evolution in frogs, in Green DM, Sessions SK (eds): Amphibian Cytogenetics and Evolution, pp 233–258 (Academic Press, San Diego 1991).

Bogart JP, Bi K: Genetic and genomic interactions of animals with different ploidy levels. Cytogenet Genome Res 140:117–136 (2013).

Bogart JP, Bogart JE: Genetic compatibility experiments between some South American anuran amphibians. Herpetologica 27:229–235 (1971).

Bogart JP, Perret JL: The karyotype of *Bufo danielae* Perret. Rev Suisse Zool 84:501–504 (1977).

Bogart JP, Tandy M: Polyploid amphibians: three more diploid-tetraploid cryptic species of frogs. Science 193:334–335 (1976).

Bogart JP, Tandy M: Chromosome lineages in African ranoid frogs. Monit Zool Ital Suppl 15:55–91 (1981).

Bogart JP, Wasserman AO: Diploid-tetraploid cryptic species pairs: a possible clue to evolution by polyploidization in anuran amphibians. Cytogenetics 11:7–24 (1972).

Bogart JP, Blair WF, Hubbs C: Comments on the "cytologically verified" hybrids between *Bufo* and *Rana*, two genera of anuran amphibians. Cytogenetics 8:30–34 (1969).

Bonaccorso E, Guayasamin JM, Méndez D, Speare R: Chytridiomycosis in a Venezuelan amphibian (Bufonidae: *Atelopus cruciger*). Herpetol Rev 34:331–334 (2003).

Borkin LY: Distribution of amphibians in North Africa, Europe, western Asia, and the former Soviet Union, in Duellman WE (ed): Patterns of Distribution of Amphibians, pp 329–420 (Johns Hopkins University Press, Baltimore 1999).

Borkin LY, Kuzmin SL: Amphibians of Mongolia, species account, in Vorob'eva EI, Darevski IS (eds): Amphibian and Reptiles of the Mongolian Peoples Republic. General Problems, pp 30–197 (Nauka, Moskow 1988).
Borkin LY, Terbish C, Caune IA: Tetraploidnaya i diploidnye populyacii zhab gruppy *Bufo viridis* iz Mongolii. Doklady Akedemii Nauk SSSR 287:760–764 (1986a).
Borkin LY, Caune IA, Pisanets EM, Rozhanov YM: Karyotype and genome size in the *Bufo viridis*-group, in Rocek Z (ed): Proceed 6th Ord Gen Meet Soc Eur Herp, pp 137–141 (Prague, 1986b).
Borkin LY, Eremchenko VK, Helfenberger N, Panfilov AM, Rozanov JM: On the distribution of diploid, triploid and tetraploid green toads in south-eastern Kazakhstan. Russ J Herpetol 8:45–53 (2001).
Bosco G, Haber JE: Chromosome break-induced DNA replication leads to nonreciprocal translocations and telomere capture. Genetics 150:1037–1047 (1998).
Bostock CJ, Prescott DM, Hatch FT: Timing and replication of the satellite and main band DNAs in cells of the kangaroo rat (*Dipodomys ordii*). Exp Cell Res 74:487–495 (1972).
Bradshaw WN, Hsu TC: Chromosomes of *Peromyscus* (Rodentia, Cricetidae). III. Polymorphism in *Peromyscus maniculatus*. Cytogenetics 11:436–451 (1972).
Brannelly LA, Webb R, Skerratt LF, Berger L: Amphibians with infectious disease increase their reproductive effort: evidence for the terminal investment hypothesis. Open Biol 6:1–24 (2016).
Brassaloti RA, Rossa-Feres DC, Bertoluci J: Anurofauna da Floresta Estacional Semidecidual da Estação Ecológica dos Cactetus, sudeste do Brasil. Biota Neotrop 10:275–292 (2010).
Brelsford A, Stöck M, Betto-Collard C, Dubey S, Dufresnes C, et al: Homologous sex chromosomes in three deeply divergent anuran species. Evolution 67–68:2434–2440 (2013).
Brenner S, Pepper D, Berns MW, Tan E, Brinkley BR: Kinetochore structure, duplication and distribution in mammalian cells: analysis by human autoantibodies from scleroderma patients. J Cell Biol 91:95–102 (1981).
Brinkley BR, Valdivia MM, Tousson A, Brenner SL: Compound kinetochores of the Indian muntjac: evolution by linear fusion of unit kinetochores. Chromosoma 91:1–11 (1984).
Brinkman JN, Sessions SK, Houben A, Green DM: Structure and evolution of supernumerary chromosomes in the Pacific giant salamander, *Dicamptodon tenebrosus*. Chromosome Res 8:477–485 (2000).
Brooks TM, Mittermeier RA, Mittermeier CG, da Fonesca GAB, Rylands AB, et al: Habitat loss and extinction in the hotspots of biodiversity. Conserv Biol 16:909–923 (2002).
Bross K, Dittes H, Krone W, Schmid M, Vogel W: Biochemical and cytogenetic studies on the nucleolus organizing regions (NOR) of man. I. Comparison of trisomy 21 with balanced translocations t(DqGq). Humangenetik 20: 223–229 (1973).
Brown DD, Gurdon JB: Absence of ribosomal RNA synthesis in the anucleolate mutant of *Xenopus laevis*. Proc Natl Acad Sci USA 51: 139–147 (1964).
Brown DD, Wensink PC, Jordan E: A comparison of the ribosomal DNA's of *Xenopus laevis* and *Xenopus mulleri*: the evolution of tandem genes. J Mol Biol 63:57–73 (1972).
Brunes TO, Sequeira F, Haddad CFB, Alexandrino J: Gene and species trees of a neotropical goup of treefrogs: genetic diversification in the Atlantic Forest and the origin of a polyploid species. Mol Phylogenet Evol 57: 1120–1133 (2010).
Bruschi DP, Busin CS, Siqueira S, Recco-Pimentel SM: Cytogenetic analysis of two species in the *Phyllomedusa hypochondrialis* group (Anura, Hylidae). Hereditas 149:34–40 (2012).
Bruschi DP, Busin CS, Toledo LF, Vasconcellos GA, Strussman C, et al: Evaluation of the taxonomic status of populations assigned to *Phyllomedusa hypochondrialis* (Anura, Hylidae, Phyllomedusinae) based on molecular, chromosomal, and morphological approach. BMC Genetics 14:70 (2013).
Bruschi DP, Rivera M, Lima AP, Zuñiga AB, Recco-Pimentel SM: Interstitial telomeric sequences (ITS) and major rDNA mapping reveal insights into the karyotypical evolution of neotropical leaf frogs species (*Phyllomedusa*, Hylidae, Anura). Mol Cytogenet 7:22 (2014a).
Bruschi DP, Lucas EM, Garcia PCA, Recco-Pimentel SM: Molecular and morphological evidence reveals a new species in the *Phyllomedusa hypochondrialis* group (Hylidae, Phyllomedusinae) from the Atlantic Forest of the highlands of southern Brazil. PloS One 9:e105608 (2014b).
Busch H, Lischwe MA, Michalik J, Chan PK, Busch RK: Nucleolar proteins of special interest: silver-staining proteins B23 and C23 and antigens of human tumour nucleoli, in Jordan EG, Cullis CA (eds): The Nucleolus, pp 43–71 (Cambridge University Press, Cambridge 1982).
Bushnell RJ, Bushnell EP, Parker MV: A chromosome study of five members of the family Hylidae. J Tennessee Acad Sci 14:209–215 (1939).
Busin CS, Vinciprova G, Recco-Pimentel SM: Chromosomal rearrangements as the source of variation in the number of chromosomes in *Pseudis* (Amphibia, Anura). Genetica 110: 131–141 (2001).
Busin CS, Lima AP, Prado CPA, Strüssmann C, Siqueira S, Recco-Pimentel SM: Chromosomal differentiation of populations of *Lysapsus limellus limellus*, *L. l. bolivianus*, and of *Lysapsus caraya* (Hylinae, Hylidae). Micron 37:355–362 (2006).
Busin CS, Andrade GV, Bertoldo J, Del Grande ML, Uetanabaro M, Recco-Pimentel SM: Cytogenetic analysis of four species of *Pseudis* (Anura, Hylidae), with the description of ZZ/ZW sex chromosomes in *P. tocantins*. Genetica 133:119–127 (2008).
Butler MG, Sanger WG, Walzak MP: A unique Y/Y translocation in an infertile man. Cytogenet Cell Genet 31:175–177 (1981).
Buys CHCM, Osinga J: Selective staining of the same set of nucleolar phosphoproteins by silver and Giemsa. A combined biochemical and cytochemical study on staining of NORs. Chromosoma 89:387–396 (1984).
Cabrero J, Alché JD, Camacho JPM: Effects of B chromosomes of the grasshopper *Eyprepocnemis plorans* on nucleolus organiser activity. Activation of a latent NOR on a B chromosome fused to an autosome. Genome 29: 116–121 (1987).
Cadle JE, Altig R: Two lotic tadpoles from the Andes of southern Peru: *Hyla armata* and *Bufo veraguensis*, with notes on the call of *Hyla armata* (Amphibia: Anura: Hylidae and Bufonidae). Stud Neotrop Fauna Environ 26: 45–53 (1991).
Caldwell J: Diversity of reproductive modes in anurans: facultative nest construction in gladiator frogs, in Hamlett WC (ed): Reproductive Biology of South American Vertebrates, pp 85–97 (Springer, New York 1992).
Callan HG: Lampbrush Chromosomes (Springer, Heidelberg 1986).
Callan HG, Gall JG, Berg CA: The lampbrush chromosomes of *Xenopus laevis*: preparation, identification, and distribution of 5S DNA sequences. Chromosoma 95:236–250 (1987).
Callan HG, Gall JG, Murphy C: The distribution of oocyte 5S, somatic 5S and 18S + 28S rDNA sequences in the lampbrush chromosomes of *Xenopus laevis*. Chromosoma 97:43–54 (1988).
Camacho JPM (ed): B Chromosomes in the Eukaryote Genome. Cytogenet Genome Res 106:147–411 (2004).
Camacho JPM, Sharbel TF, Beukeboom LW: B-chromosome evolution. Phil Trans R Soc Lond B 355:163–178 (2000).
Camacho JPM, Schmid M, Cabrero J: B chromosomes and sex in animals. Sex Dev 5:155–166 (2011).
Caminer MA, Mila B, Jansen M, Fouquet A, Venegas PJ, et al: Systematics of the *Dendropsophus leucophyllatus* species complex (Anura: Hylidae): cryptic diversity and the description of two new species. PLoS One 12: e0171785 (2017).
Campos JRC, Martins IA, Haddad CFB, Kasahara S: The karyotype of *Holoaden luederwaldti* (Anura, Strabomantidae), with report of natural triploidy. Folia Biol 58:144–150 (2012).
Cannatella DC: A Phylogeny of Primitive Frogs (Archaeobatrachians). PhD thesis, University of Kansas, Kansas (1985).
Cano JM, Laurila A, Vilkki J, Merilä J: First-generation linkage map for the common frog *Rana temporaria* reveals a sex-linkage group. Heredity 107:530–536 (2011).
Cardoso AJ: Utilização de recursos para reprodução em comunidade de anuros no sudeste do Brasil. PhD thesis, Universidade Estadual de Campinas, Campinas (1986).
Cardozo DE, Leme DM, Bortoleto JF, Catroli GF, Baldo D, et al: Karyotypic data on 28 species of *Scinax* (Amphibia: Anura: Hylidae): diversity and informative variation. Copeia 2011:252–263 (2011).

Carezzano FJ, Dorflinger K, Alonso M, Bardone SP: Testicular morphohistology of *Hypsiboas pulchellus* (Amphibia, Hylidae) during the breeding season. Acta Biológica Colomb 20: 225–228 (2014).
Carvalho KA, Garcia PCA, Recco-Pimentel SM: Cytogenetic comparison of tree frogs of the genus *Aplastodiscus* and *Hypsiboas faber* group (Anura, Hylidae). Genet Mol Res 8: 1498–1508 (2009a).
Carvalho KA, Garcia PCA, Recco-Pimentel SM: NOR dispersion, telomeric sequence detection in centromeric regions and meiotic multivalent configurations in species of the *Aplastodiscus albofrenatus* group. Cytogenet Genome Res 126:359–367 (2009b).
Carvalho MA, Rodrigues MT, Siqueira S, Garcia C: Dynamics of chromosomal evolution in the genus *Hypsiboas* (Anura: Hylidae). Genet Mol Res 13:7826–7838 (2014).
Cash MN, Bogart JP: Cytological differentiation of the diploid-tetraploid species pair of North American treefrogs (Amphibia, Anura, Hylidae). J Herpetol 12:555–558 (1978).
Castro AB, Moura LHD, Berneck B, Yonenaga-Yasuda Y, Rodrigues MT, et al: Chromosomal studies in six species of *Aplastodiscus* treefrogs (Amphibia, Anura, Hylidae). Resumos do 58° Congresso Brasileiro de Genetica, Foz de Iguaçu, Paraná (2012).
Catroli GF: Cariótipo de seis espécies de *Bokermannohyla* dos grupos de *B. circumdata* e *B. pseudopseudis* (Anura, Hylidae). Master thesis, Universidade Estadual Paulista, Rio Claro (2008).
Catroli GF, Kasahara S: Cytogenetic data on species of the family Hylidae (Amphibia, Anura): results and perspectives. Publ UEPG Biol Health Sci 15:67–86 (2009).
Catroli GF, Faivovich J, Haddad CFB, Kasahara S: Conserved karyotypes in Cophomantini: cytogenetic analysis of 12 species from 3 species groups of *Bokermannohyla* (Amphibia: Anura: Hylidae). J Herpetol 45:120–128 (2011).
Cavalier-Smith T: Palindromic base sequences and replication of eukaryote chromosome ends. Nature 250:467–470 (1974).
Cavallo D, de Vita R, Eleuteri P, Borkin L, Eremchenko V, et al: Karyological and flow cytometric evidence of triploid specimens in *Bufo viridis* (Amphibia, Anura). Eur J Histochem 46:159–164 (2002).
Cervini M, Carvalho MA, Siqueira S, Rodrigues MT, Garcia C: Comparative cytogenetic studies in *Hypsiboas* (Anura, Hylidae). Resumos do 58° Congresso Brasileiro de Genetica, Foz de Iguaçu, Paraná, p 59 (2012).
Chakrabarti S, Banerjee SN, Neogi LN, Ray-Chaudhuri S: C-band positive W chromosome in the female Indian frog. Experientia 39:321–322 (1983).
Chandler M, Green DM, Taylor A, Zeyl CW, Sharbel TF: Chiasma frequency in frogs of the genus *Leiopelma* (Amphibia; Anura) and a test of the hypothesis of inducible recombination. Hereditas 118:205–210 (1993).
Chang CY, Witschi E: Breeding of sex-reversed males of *Xenopus laevis* Daudin. Proc Soc Exp Biol Med 89:150–152 (1955).
Chang CY, Witschi E: Gene control and hormonal reversal of sex differentiation in *Xenopus*. Proc Soc Exp Biol Med 93:140–144 (1956).
Channing A, Bogart JP: Description of a tetraploid *Tomopterna* (Anura: Ranidae) from South Africa. S Afr J Zool 31:80–85 (1996).
Chi JX, Huang L, Nie W, Wang J, Su B, Yang F: Defining the orientation of the tandem fusions that occurred during the evolution of the Indian muntjac chromosomes by BAC mapping. Chromosoma 114:167–172 (2005).
Chieffi G, Rastogi RK, Milone M, Iela L: Amphibian reproduction: reproductive physiology in the male *Rana esculenta* L. Boll Zool 47:63–70 (1980).
Chipman AD, Haas A, Khaner O: Variations in anuran embryogenesis: yolk-rich embryos of *Hyperolius puncticulatus* (Hyperoliidae). Evol Dev 1:49–61 (1999).
Choo KHA: The Centromere (Oxford University Press, Oxford 1997a).
Choo KHA: Centromere DNA dynamics: latent centromeres and neocentromere formation. Am J Hum Genet 61:1225–1233 (1997b).
Coggins LW: An ultrastructural and radioautographic study of early oogenesis in the toad *Xenopus laevis*. J Cell Sci 12:71–93 (1973).
Cole CJ: The chromosomes of *Acris crepitans blanchardi* Harper (Anura: Hylidae). Copeia 1966:578–580 (1966).
Cole CJ: Karyotypes and evolution of the *spinosus* group of lizards in the genus *Sceloporus*. Am Mus Novit 2431:1–47 (1970).
Cole CJ: Chromosomes of the hylid frog, *Phyllomedusa dacnicolor* Cope. Herpetol Rev 3: 55–56 (1971).
Cole CJ: Chromosome evolution in selected treefrogs, including casque-headed species (*Pternohyla*, *Triprion*, *Hyla*, and *Smilisca*). Amer Mus Nat Hist Nov 2541:1–10 (1974).
Cole CJ, Lowe CH, Wright JW: Karyotypes of eight species of toads (genus *Bufo*) in North America. Copeia 1:96–100 (1968).
Collazo A, Keller R: Early development of *Ensatina eschscholtzii*: an amphibian with a large, yolky egg. Evo Devo 1:1–13 (2010).
Collins JP, Crump ML: Extinction in Our Time: Global Amphibian Decline (Oxford University Press, New York 2009).
Coloma LA: Ecuadorian frogs of the genus *Colostethus* (Anura: Dendrobatidae). Misc Publ Univ Kansas Mus Nat Hist 87:1–72 (1995).
Coloma LA, Carvajal-Endara S, Dueñas JF, Paredes-Recalde A, Morales-Mite M, et al: Molecular phylogenetics of stream treefrogs of the *Hyloscirtus larinopygion* group (Anura: Hylidae) and description of two new species from Ecuador. Zootaxa 3364:1–78 (2012).
Comings DE: Mechanisms of chromosome banding. VIII. Hoechst 33258-DNA interaction. Chromosoma 52:229–243 (1975).
Comings DE: Mechanisms of chromosome banding and implications for chromosome structure. Annu Rev Genet 12:25–46 (1978).
Cooke HJ: Repeated sequences specific of human males. Nature 262:182–186 (1976).
Cooke HJ, McKay RDG: Evolution of a human Y chromosome-specific repeated sequence. Cell 13:453–460 (1978).
Cooke HJ, Fantes J, Green D: Structure and evolution of human Y chromosome DNA. Differentiation 23 (Suppl):48–55 (1983).
Costa GC, Vieira GHC, Teixeira RD, Garda AA, Colli GR, Báo SN: An ultrastructural comparative study of the sperm of *Hyla pseudopseudis*, *Scinax rostratus*, and *S. squalirostris* (Amphibia: Anura: Hylidae). Zoomorphology 123:191–197 (2004).
Covert SF: Supernumerary chromosomes in filamentous fungi. Curr Genet 33:311–319 (1998).
Craig-Holmes AP, Moore FB, Shaw MW: Polymorphism of human C-band heterochromatin. II. Family studies with suggestive evidence for somatic crossing-over. Am J Hum Genet 27:178–189 (1975).
Crump ML: Quantitative analysis of the ecological distribution of a tropical herpetofauna. Occ Pap Nat Hist Mus Univ Kansas 3:1–62 (1971).
Crump ML: Reproductive strategies in a tropical anuran community. Misc Publ Kansas Univ 61:1–68 (1974).
Crump ML: Anuran reproductive modes: evolving perspectives. J Herpetol 49:1–16 (2015).
Cuevas CC, Formas JR: Heteromorphic sex chromosomes in *Eupsophus insularis* (Amphibia, Anura, Leptodactylidae). Chromosome Res 4:467–470 (1996).
Cuny G, Soriano P, Macaya G, Bernardi G: The major components of the mouse and human genomes: preparation, basic properties, and compositional heterogeneity. Eur J Biochem 111:227–233 (1981).
Curi LM, Olea GB, Álvarez BB, Céspedez JA, Lombardo DM: Reproductive aspects of *Dendropsophus sanborni* (Anura, Hylidae) in northeastern Argentina. Iheringia Série Zool 104:432–438 (2014).
Dallapiccola B, Mastroiacovo P, Gandini E: Centric fission of chromosome No. 4 in the mother of two patients with trisomy 4p. Hum Genet 31:121–125 (1976).
da Silva FR, Almeida-Neto M, do Prado VHM, Haddad CFB, Rossa-Feres D: Humidity levels drive reproductive modes and phylogenetic diversity of amphibians in the Brazilian Atlantic Forest. J Biogeogr 39:1720–1732 (2012).
Daszak P, Scott DE, Kilpatrick AM, Faggioni C, Gibbons JW, Porter D: Amphibian population declines at Savannah River site are linked to climate, not chytridiomycosis. Ecology 86:3232–3237 (2005).
Davidson EH: Gene Activity in Early Development, ed 2 (Academic Press, New York 1976).
Davidson EH, Amenson CS, Hough BR, Britten RJ: Interspersion of repetitive sequences in *Xenopus* DNA. Carnegie Inst Wash Year Book 71:273–276 (1972).
Davidson EH, Hough BR, Amenson CS, Britten RJ: General interspersion of repetitive and non-repetitive sequence elements in the DNA of *Xenopus*. J Mol Biol 77:1–23 (1973).
Davidson EH, Galau GA, Angerer RC, Britten RJ: Comparative aspects of DNA organization in Metazoa. Chromosoma 51:253–259 (1975).

Davis KM, Smith SA, Greenbaum IF: Evolutionary implications of chromosomal polymorphisms in *Peromyscus boylii* from southwestern Mexico. Evolution 40:645–649 (1986).

Davis RM: Localisation of male determining factors in man: a thorough review of structural anomalies of the Y chromosome. J Med Genet 18:161–195 (1981).

Day JP, Limoli CL, Morgan WF: Recombination involving interstitial telomere repeat-like sequences promotes chromosomal instability in Chinese hamster cells. Carcinogenesis 19: 259–265 (1998).

Deaven LL, Vidal-Rioja L, Jett JH, Hsu TC: Chromosomes of *Peromyscus* (Rodentia, Cricetidae). VI. The genome size. Cytogenet Cell Genet 19:241–249 (1977).

de la Riva I: Ecología de una comunidad neotropical de anfibios durante la estación lluviosa. Doc Disser Univ Compl Madrid (1993).

de la Riva I: A new *Phyllomedusa* from southwestern Amazonia (Amphibia: Anura: Hylidae). Rev Española Herpetol 13:123–131 (1999).

del Pino E, Humphries AA: Multiple nuclei during early oogenesis in *Flectonotus pygmaeus* and other marsupial frogs. Biol Bull 154: 198–212 (1978).

del Pino EM, Loor-Vela S: The pattern of early cleavage of the marsupial frog *Gastrotheca riobambae*. Development 110:781–789 (1990).

del Pino EM, Ávila ME, Pérez OD, Benítez MS, Alarcón I, et al: Development of the dendrobatid frog *Colostethus machalilla*. Int J Dev Biol 48:663–670 (2004).

del Pino EM, Venegas-Ferrín M, Romero-Carvajal A, Montenegro-Larrea P, Sáenz-Ponce N, et al: Comparative analysis of frog early development. Proc Natl Acad Sci USA 104: 11882–11888 (2007).

de Mattos TL, Coelho AC, Menin M, Gross MC: Cytogenetic analysis of testicle cells from *Hypsiboas cinerascens* (Hylidae, Anura) collected in Manaus, Amazonas. Resumos do 58° Congresso Brasileiro de Genetica, Foz de Iguaçu, Paraná, p 241 (2012).

de Mattos TL, Coelho AC, Schneider CH, Telles DOC, Menin M, Gross MC: Karyotypic diversity in seven Amazonian anurans in the genus *Hypsiboas* (family Hylidae). BMC Genetics 15:43 (2014).

Dementyeva PV, Trifonov VA, Kulemzina AI, Graphodatsky AS: Reconstruction of the putative Cervidae ancestral karyotype by chromosome painting of Siberian roe deer (*Capreolus pygargus*) with dromedary probes. Cytogenet Genome Res 128:228–235 (2010).

Deng C, Shang K: A cytogenetics demonstration of XY sex determination in *Bufo raddei*. Acta Gen Sin 11:395–399 (1984).

Dennis ES, Dunsmuir P, Peacock WJ: Segmental amplification in a satellite DNA: restriction enzyme analysis of the major satellite of *Macropus rufogriseus*. Chromosoma 79:179–198 (1980).

de Oliveira C, de Souza Santos LR: Histological characterization of cellular types during *Scinax fuscovarius* oogenesis (Lutz) (Anura, Hylidae). Rev Bras Zool 21:919–923 (2004).

de Oliveira HHP, Souza CCN, Ribeiro CL, da Cruz AD, Bastos RP, de Melo e Silva D: Citogenética comparativa das familias Leptodactylidae e Hylidae do cerrado goiano. Estudos (Pontificia Universidade Católica de Goiás) 39:123–131 (2012).

De Wulf P, Earnshaw WC: The Kinetochore: From Molecular Discoveries to Cancer Therapy (Springer, New York 2008).

Dias I, Medeiros T, Vila Nova M, Solé M: Amphibians of Serra Bonita, southern Bahia: a new hotpoint within Brazil's Atlantic Forest hotspot. ZooKeys 449:105–130 (2014).

Distèche C, Hagemeijer A, Frederic J, Progneaux D: An abnormally large human chromosome identified as an end-to-end fusion of two X's by combined results of the new banding techniques and microdensitometry. Clin Genet 3:388–395 (1972).

Dittes H, Krone W, Bross K, Schmid M, Vogel W: Biochemical and cytogenetic studies on the nucleolus organizing regions (NOR) of man. II. A family with the 15/21 translocation. Humangenetik 26:47–59 (1975).

Dodd CK, Smith LL: Habitat destruction and alteration, in Semlitsch RD (ed): Amphibian Conservation, pp 94–112 (Smithsonian Institution Press, Washington 2003).

Donnellan SC, Mahony MJ: Allozyme, chromosomal and morphological variability in the *Litoria lesueuri* species group (Anura: Hylidae), including a description of a new species. Austr J Zool 52:1–28 (2004).

Downie JR, Sams K, Walsh PT: The paradoxical frog *Pseudis paradoxa*: larval anatomical characteristics, including gonadal maturation. Herpetol J 19:1–10 (2009).

Downie JR, Nokhbatolfoghahai M, Bruce D, Smith JM, Orthmann-Brask N, MacDonald-Allan I: Nest structure, incubation and hatching in the Trinidadian leaf-frog *Phyllomedusa trinitatis* (Anura: Hylidae). Phyllomedusa 12:13–32 (2013).

Drummond AJ, Rambaut A: BEAST: Bayesian evolutionary analysis by sampling trees. BMC Evol Biol 7:214 (2007).

Duarte TC, Veiga-Menoncello ACP, Lima JFR, Strüssmann C, Del-Grande ML, et al: Chromosome analysis in *Pseudopaludicola* (Anura, Leiuperidae) with description of sex chromosomes XX/XY in *P. saltica*. Hereditas 147:43–52 (2010).

Dubois A, Frétey T: A new nomen for a subfamily of frogs (Amphibia, Anura). Dumerilia 6: 17–23 (2016).

Dubois A, Duellman WE, Ohler A: Taxonomic and nomenclatural notes on the subfamily Acridinae of the family Hylidae (Amphibia, Anura). Bionomina 11:49–61 (2017).

Duellman WE: A distributional study of the amphibians of the Isthmus of Tehuantepec, México. Univ Kansas Pub Mus Nat Hist 13: 19–72 (1960).

Duellman WE: The amphibians and reptiles of Michoacán, México. Univ Kansas Pub Mus Nat Hist 15:1–148 (1961).

Duellman WE: Amphibians and reptiles of the rainforests of southern El Petén, Guatemala. Univ Kansas Pub Mus Nat Hist 14:205–240 (1963).

Duellman WE: Additional studies of chromosomes of anuran amphibians. Syst Zool 16: 38–43 (1967).

Duellman WE: The hylid frogs of Middle America. Monogr Mus Nat Hist Univ Kansas 1:1–753 (1970).

Duellman WE: Liste der rezenten Amphibien und Reptilien: Hylidae, Centrolenidae, Pseudidae, in Mertens R, Hennig W (eds): Das Tierreich 95 (de Gruyter, Berlin 1977).

Duellman WE: The biology of an equatorial herpetofauna in Amazonian Ecuador. Misc Pub Nat Hist Mus Univ Kansas 65:1–352 (1978).

Duellman WE: A new species of small yellow *Hyla* from Perú (Anura: Hylidae). Amphibia-Reptilia 3:153–160 (1982).

Duellman WE: Reproductive strategies of frogs. Sci Am 267:80–87 (1992).

Duellman WE: Amphibians of La Escalera region, southeastern Venezuela: taxonomy, ecology, and biogeography. Sci Pap Nat Hist Mus Univ Kansas 2:1–52 (1997).

Duellman WE: Distribution patterns of amphibians in South America, in Duellman WE (ed): Patterns of Distribution of Amphibians, pp 255–328 (Johns Hopkins University Press, Baltimore 1999).

Duellman WE: Hylid Frogs of Middle America (Society for the Study of Amphibians and Reptiles, Ithaca 2001).

Duellman WE: Cusco Amazónico. The Lives of Amphibians and Reptiles in an Amazonian Rainforest (Cornell University Press, Ithaca 2005).

Duellman WE, Cole CJ: Studies of chromosomes of some anuran amphibians (Hylidae and Centrolenidae). Syst Zool 14:139–143 (1965).

Duellman WE, de Sá RO: A new genus and species of South American hylid frog with a highly modified tadpole. Trop Zool 1:117–136 (1988).

Duellman WE, Trueb L: Neotropical hylid frogs, genus *Smilisca*. Univ Kansas Nat Hist Mus Publ 17:281–375 (1966).

Duellman WE, Trueb L: The systematic status and relationships of the hylid frog *Nyctimantis rugiceps* Boulenger. Occ Pap Mus Nat Hist Univ Kansas 58:1–14 (1976).

Duellman WE, Trueb L: Frogs of the *Hyla columbiana* group: taxonomy and phylogenetic relationships, in Rhodin AGJ, Miyata K (eds): Advances in Herpetology and Evolutionary Biology, pp 33–51 (Museum of Comparative Zoology, Harvard University, Cambridge 1983).

Duellman WE, Trueb L: Biology of Amphibians (The Johns Hopkins University Press, Baltimore 1986).

Duellman WE, Cadle JE, Cannatella DC: A new species of terrestrial *Phyllomedusa* (Anura: Hylidae) from southern Peru. Herpetologica 44:91–95 (1988).

Duellman WE, Marion AB, Hedges SB: Phylogenetics, classification, and biogeography of the treefrogs (Amphibia, Anura, Arboranae). Zootaxa 4104:1–109 (2016).

Dumont JN: Oogenesis in *Xenopus laevis* (Daudin). I. Stages of oocyte development in laboratory maintained animals. J Morphol 136:153–179 (1972).

Duryea MC, Brasileiro CA, Zamudio KR: Characterization of microsatellite markers for snouted treefrogs in the *Scinax perpusillus* species group (Anura, Hylidae). Conserv Genet 10:1053–1056 (2009).
Duryee WR: Chromosomal physiology in regard to nuclear structure. Ann NY Acad Sci 50: 920–953 (1950).
Dutrillaux B: Sur le nature et l'origine des chromosomes humains (L'Expansion Scientifique Française, Paris 1975).
Dutrillaux B, de Grouchy J, Finaz C, Lejeune J: Mise en évidence de la structure fine des chromosomes humains par digestion enzymatique (pronase en particulier). C R Acad Sci Ser D 273:587–588 (1971).
Dutrillaux B, Rethoré MO, Prieur M, Raoul O, Berger R, Lejeune J: Reconnaissance des chromosomes du groupe G par la méthode de dénaturation ménagée. Exp Cell Res 70: 453–455 (1972).
Eakin RM: Determination and regulation of polarity in the retina of *Hyla regilla*. Univ Calif Publ Zool 51:245–288 (1946).
Earnshaw WC, Rothfield N: Identification of a family of human centromere proteins using autoimmune sera from patients with scleroderma. Chromosoma 91:313–321 (1985).
Eaton TH, Imagawa RM: Early development of *Pseudacris clarkii*. Copeia 1948:263–266 (1948).
Ecker A: The Anatomy of the Frog (Clarendon Press, Oxford 1889).
Edgar C: MUSCLE: multiple sequence alignment with high accuracy and high throughput. Nucleic Acids Res 32:1792–1797 (2004).
Egozcue J: A possible case of centric fission in a primate. Experientia 27:969–970 (1971).
Elder FFB: Tandem fusion, centric fusion, and chromosomal evolution in the cotton rats, genus *Sigmodon*. Cytogenet Cell Genet 26: 199–210 (1980).
Elinson RP: Genetic analysis of developmental arrest in an amphibian hybrid (*Rana catesbeiana*, *Rana clamitans*). Dev Biol 81:167–176 (1981).
Elinson RP: Inheritance and expression of a sex-linked enzyme in the frog, *Rana clamitans*. Biochem Genet 21:435–442 (1983).
Elinson RP, Beckham Y: Development in frogs with large eggs and the origin of amniotes. Zoology 105:105–117 (2002).
Elinson RP, del Pino EM: Cleavage and gastrulation in the egg-brooding, marsupial frog, *Gastrotheca riobambae*. Embryol Exp Morph 90:223–232 (1985).
Elinson RP, del Pino EM: Developmental diversity of amphibians. Rev Dev Biol 1:345–369 (2012).
Elsdale TR, Fischberg M, Smith SA: A mutation that reduces nucleolar number in *Xenopus laevis*. Exp Cell Res 14:642–643 (1958).
Engel W, Kobel HR: The Z-chromosome is involved in the regulation of H-W (H-Y) antigen gene expression in *Xenopus*. Cytogenet Cell Genet 35:28–33 (1983).
Engel W, Schmid M: H-Y antigen as a tool for the determination of the heterogametic sex in Amphibia. Cytogenet Cell Genet 30:130–136 (1981).
Engel W, Zenzes MT, Schmid M: Activation of mouse ribosomal RNA genes at the 2-cell stage. Hum Genet 38:57–63 (1977).
Estes R, Reig OA: The early fossil record of frogs. A review of the evidence, in Vial JL (ed): Evolutionary Biology of the Anurans, pp 11–63 (University of Missouri Press, Columbia 1973).
Evans BJ, Carter TF, Tobias ML, Kelley DB, Hanner R, Tinsley RC: A new species of clawed frog (genus *Xenopus*) from the Itombwe Massif, Democratic Republic of the Congo: implications for DNA barcodes and biodiversity conservation. Zootaxa 1780:55–68 (2008).
Evans BJ, Carter TF, Greenbaum E, Gvoždik V, Kelley DB, et al: Genetics, morphology, advertisement calls, and historical records distinguish six new polyploid species of African clawed frog (*Xenopus*, Pipidae) from West and Central Africa. PLoS One 10: e0142823 (2015).
Ewald AJ, Peyrot SM, Tyszka JM, Fraser SE, Wallingford JB: Regional requirements for dishevelled signaling during *Xenopus* gastrulation: separable effects on blastopore closure, mesendoderm internalization and archenteron formation. Development 131: 6195–6209 (2004).
Fabrezi M, Quinzio SI, Goldberg J: The ontogeny of *Pseudis platensis* (Anura, Hylidae): heterochrony and the effects of larval development on postmetamorphic life. J Morphol 271:496–510 (2010).
Faivovich J, Haddad CFB, García PCA, Frost DR, Campbell JA: Systematic review of the frog family Hylidae, with special reference to Hylinae: phylogenetic analysis and taxonomic revision. Bull Am Mus Nat Hist 294:1–240 (2005).
Faivovich J, Haddad CFB, Baêta D, Jungfer K-H, Álvares GFR, et al: The phylogenetic relationships of the charismatic poster frogs, Phyllomedusinae (Anura, Hylidae). Cladistics 26:227–261 (2010).
Faivovich J, McDiarmid RW, Myers CW: Two new species of *Myersiohyla* (Anura: Hylidae) from Cerro de la Neblina, Venezuela, with comments on other species of the genus. Am Mus Nov 3792:1–63 (2013).
Fan H, Penman S: Regulation of synthesis and processing of nucleolar components in metaphase-arrested cells. J Mol Biol 59:27–42 (1971).
Faravelli M, Moralli D, Bertoni L, Attolini C, Chernova O, et al: Two extended arrays of a satellite DNA sequence at the centromere and the short-arm telomere of Chinese hamster chromosome 5. Cytogenet Cell Genet 83: 281–286 (1998).
Faravelli M, Azzalin CM, Bertoni L, Chernova O, Attolini C, et al: Molecular organization of internal telomeric sequences in Chinese hamster chromosomes. Gene 283:11–16 (2002).
Feitosa VLC, Recco-Pimentel SM, Cardoso AJ: Structural chromosomal alterations in *Hyla albopunctata* and *Aplastodiscus perviridis* (Anura, Hylidae) from the Morro do Ferro region of Minas Gerais. Rev Brasil Genet 18: 191–197 (1995).
Ferguson-Smith MA, Trifonov V: Mammalian karyotype evolution. Nat Rev Genet 8:950–962 (2007).
Ferreira A, Mehanna M: Seasonal testicular changes in *Dendropsophus minutus* Peters, 1872 (Anura, Hylidae). Biocell 36:57–62 (2012).
Ferreira A, Mehanna M, Prado CPA: Morphologic and morphometric analysis of testis of *Pseudis limellum* (Cope, 1862) (Anura, Hylidae) during the reproductive cycle in the Pantanal, Brazil. Biocell 32:185–194 (2008).
Ferreira A, dos Santos AB, Mehanna M: Organizaçao celular dos testiculos em Hylidae e Leptodactylidae, no Pantanal (Estado do Mato Grosso do Sul, Brasil). Acta Sci Biol Sci 31:447–452 (2009).
Ferro JM, Marti DA, Bidau CJ, Suárez P, Nagamachi CY, et al: B chromosomes in the tree frog *Hypsiboas albopunctatus* (Anura: Hylidae). Herpetologica 68:482–490 (2012).
Ferro JM, Taffarel A, Cardozo D, Grosso J, Puig MP, et al: Cytogenetic characterization and B chromosome diversity in direct-developing frogs of the genus *Oreobates* (Brachycephaloidea, Craugastoridae). Comp Cytogenet 10:141–156 (2016).
Filipski J, Thiery JP, Bernardi G: An analysis of the bovine genome by Cs_2SO_4/Ag^+ density gradient centrifugation. J Mol Biol 80:177–197 (1973).
Fischberg M, Wallace H: A mutation which reduces nucleolar number in *Xenopus laevis*, in Mitchell JS (ed): The Cell Nucleus, pp 30–34 (Butterworths, London 1960).
Fisher RA: The Genetical Theory of Natural Selection (Clarendon Press, Oxford 1930).
Fitch WM: Toward defining the course of evolution: minimum change for a specific tree topology. Syst Zool 20:406–416 (1971).
Fleming RI, Mackenzie CD, Cooper A, Kennedy MW: Foam nest components of the Túngara frog: a cocktail of proteins conferring physical and biological resilience. Proc R Soc London B Biol Sci 276:1787–1795 (2009).
Flint J, Craddock CF, Villlegas A, Bentley DP, Williams HJ, et al: Healing of broken human chromosome ends by the addition of telomeric repeats. Am J Hum Genet 55:505–512 (1994).
Foote DL, Wiley JE, Little ML, Meyne J: Ribosomal RNA gene site polymorphism in *Bufo terrestris*. Cytogenet Cell Genet 57:196–199 (1991).
Ford EB: Mendelism and Evolution (Methuen, London 1960).
Foresti F: Aspectos cromossômicos da família Hylidae (Amphibia-Anura). Master thesis, Universidade de São Paulo, São Paulo (1972).
Formas JR: Systematic problems in the frog species *Eupsophus roseus* (Anura: Leptodactylidae) detected by karyological analysis. Experientia 34:446 (1978).
Formas JR: A triploid individual of the Chilean leptodactylid frog *Eupsophus vertebralis*. J Herpetol 28:394–395 (1994).
Fortin F, Beaulieu Bergeron M, Fetni R, Lemieux N: Frequency of chromosome healing and interstitial telomeres in 40 cases of constitutional abnormalities. Cytogenet Genome Res 125:176–185 (2009).

Fouquet A, Vences M, Salducci MD, Meyer A, Marty C, et al: Revealing cryptic diversity using molecular phylogenetics and phylogeography in frogs of the *Scinax ruber* and *Rhinella margaritifera* species groups. Mol Phylogenet Evol 43:567–582 (2007).

Fouquette MJ, Delahoussaye AT: Sperm morphology in the *Hyla rubra* group (Amphibia, Anura, Hylidae), and its bearing on generic status. J Herpetol 11:387–396 (1977).

Fox H: Amphibian Morphogenesis (Humana Press, Clifton 1984).

Fredga K: Unusual sex chromosome inheritance in mammals. Phil Trans R Soc Lond B 259: 15–36 (1970).

Fredga K, Bergström U: Chromosome polymorphism in the root vole (*Microtus oeconomus*). Hereditas 66:145–152 (1970).

Fredga K, Mandahl N: Autosomal heterochromatin in some carnivores, in Caspersson T, Zech L (eds): Chromosome Identification – Technique and Applications in Biology and Medicine. Nobel Symp 23, pp 104–117 (Academic Press, London 1973).

Frost DR: Amphibian Species of the World: an Online Reference. Version 6.0 (accessed December 7, 2017). Electronic Database accessible at http://research.amnh.org/herpetology/amphibia/index.html. American Museum of Natural History, New York, USA (2017).

Frost DR, Grant T, Faivovich J, Bain RH, Haas A, et al: The amphibian tree of life. Bull Amer Mus Nat Hist 297:1–371 (2006).

Fry K, Salser W: Nucleotide sequences of HS-α satellite DNA from kangaroo rat *Dipodomys ordi* and characterization of similar sequences in other rodents. Cell 12:1069–1084 (1977).

Galgano M: Evoluzione degli spermatociti di I ordine e cromosomi pseudosessuali in alcune specie di anfibi. Arch Ital Anat Embriol 32:171–200 (1933).

Gallien L: Inversion totale du sexe chez *Xenopus laevis* Daud. à la suite d'un traitement gynogène par le benzoate d'oestradiol administré pendant la vie larvaire. C R Acad Sci Hebd Seances Acad Sci D 237:1565–1566 (1953).

Garda A, Colli GR, Aguiar O, Recco-Pimentel SM, Báo SN: The ultrastructure of the spermatozoa of *Epipedobates flavopictus* (Amphibia, Anura, Dendrobatidae), with comments on its evolutionary significance. Tissue Cell 34:356–364 (2002).

Garda AA, Costa GC, Colli GR, Báo SN: Spermatozoa of Pseudinae (Amphibia, Anura, Hylidae), with a test of the hypothesis that sperm ultrastructure correlates with reproductive modes in anurans. J Morphol 261:196–205 (2004).

Garrido-Ramos MA, Herrán R, Rejón R, Rejón MR: A satellite DNA of the Sparidae family (Pisces, Perciformes) associated with telomeric sequences. Cytogenet Cell Genet 83: 3–9 (1998).

Gatto KP, Busin CS, Lourenço LB: Unravelling the sex chromosome heteromorphism of the paradoxical frog *Pseudis tocantins*. PLoS One 11:e0156176 (2016).

Gazoni T, Gruber SL, Silva APZ, Araújo OGS, Narimatsu H, et al: Cytogenetic analyses of eight species in the genus *Leptodactylus* Fitzinger, 1843 (Amphibia, Anura, Leptodactylidae), including a new diploid number and a karyotype with dispersed translocations. BMC Genetics 13:109 (2012).

Gerchen JF, Reichert SJ, Röhr JT, Dieterich C, Kloas W, Stöck M: A single transcriptome of a green toad (*Bufo viridis*) yields candidate genes for sex determination and -differentiation and non-anonymous population genetic markers. PLoS One 11:e0156419 (2016).

Giovanotti M, Caputo V, O'Brien PCM, Lovell FL, Trifonov V, et al: Skinks (Reptilia: Scincidae) have highly conserved karyotypes as revealed by chromosome painting. Cytogenet Genome Res 127:224–231 (2009).

Gitlin D: The development of *Eleutherodactylus portoricensis*. Copeia 1944:91–98 (1944).

Goldberg J: Gonadal differentiation and development in the snouted treefrog, *Scinax fuscovarius* (Amphibia, Anura, Hylidae). J Herpetol 49:468–478 (2015).

Goldberg J, Barrasso DA, Agostini MG, Quinzio S: Vocal sac development and accelerated sexual maturity in the lesser swimming frog, *Pseudis minuta* (Anura, Hylidae). Zoology 119:489–499 (2016).

Gomez-Mestre I, Warkentin KM: To hatch and hatch not: similar selective trade-offs but different responses to egg predators in two closely related syntopic treefrogs. Oecologia 153:197–206 (2007).

Gomez-Mestre I, Saccoccio VL, Iijima T, Collins EM, Rosenthal GG, Warkentin KM: The shape of things to come: linking developmental plasticity to post-metamorphic morphology in anurans. J Evol Biol 23:1364–1373 (2010).

Goodpasture C, Bloom SE: Visualization of nucleolar organizer regions in mammalian chromosomes using silver staining. Chromosoma 53:37–50 (1975).

Gordo M, Toledo LF, Suárez P, Kawashita-Ribeiro RA, Avila RW, et al: A new species of milk frog *Trachycephalus tschudi* (Anura, Hylidae) from the Amazonian rainforest. Herpetologica 69:466–479 (2013).

Gorman GC: The chromosomes of the Reptilia, a cytotaxonomic interpretation, in Chiarelli AB, Capanna E (eds): Cytotaxonomy and Vertebrate Evolution, pp 349–424 (Academic Press, London 1973).

Gorman GC, Atkins L, Holzinger T: New karyotypic data on 15 genera of lizards in the family Iguanidae, with a discussion of taxonomic and cytological implications. Cytogenetics 6:286–299 (1967).

Gornung E, Mosconi F, Annesi F, Castiglia R: The first cytogenetic description of *Euleptes europaea* (Gené, 1839) from Northern Sardinia reveals the highest diploid chromosome number among sphaerodactylid geckos (Sphaerodactylidae, Squamata). Comp Cytogenet 7:153–161 (2013).

Gosner KA: Simplified table for staging anuran embryos and larvae with notes on identification. Herpetologica 16:183–190 (1960).

Green DM: Systematics and evolution of Western North American frogs allied to *Rana aurora* and *Rana boyli*: karyological evidence. Syst Zool 35:273–282 (1986).

Green DM: Heteromorphic sex chromosomes in the rare and primitive frog *Leiopelma hamiltoni* from New Zealand. J Hered 79:165–169 (1988a).

Green DM: Cytogenetics of the endemic New Zealand frog, *Leiopelma hochstetteri*: extraordinary supernumerary chromosome variation and a unique sex-chromosome system. Chromosoma 97:55–70 (1988b).

Green DM: Muller's ratchet and the evolution of supernumerary chromosomes. Genome 33: 818–824 (1990).

Green DM: Supernumerary chromosomes in amphibians, in Green DM, Sessions SK (eds): Amphibian Cytogenetics and Evolution, pp 333–358 (Academic Press, San Diego 1991).

Green DM: Genetic and cytogenetic diversity in Hochstetter's frog, *Leiopelma hochstetteri*, and its importance for conservation management. New Zealand J Zool 21:417–424 (1994).

Green DM: Chromosome polymorphism in Archey's frog (*Leiopelma archeyi*) from New Zealand. Copeia 2002:204–207 (2002).

Green DM: Structure and evolution of B chromosomes in amphibians. Cytogenet Genome Res 106:235–242 (2004).

Green DM, Cannatella DC: Phylogenetic significance of the amphicoelous frogs, Ascaphidae and Leiopelmatidae. Ethol Ecol Evol 5: 233–245 (1993).

Green DM, Sessions SK (eds): Amphibian Cytogenetics and Evolution (Academic Press, San Diego 1991a).

Green DM, Sessions SK: Nomenclature for chromosomes, in Green DM, Sessions SK (eds): Amphibian Cytogenetics and Evolution, pp 431–432 (Academic Press, San Diego 1991b).

Green DM, Sessions SK: Karyology and cytogenetics, in Heatwole H, Tyler M (eds): Amphibian Biology, vol 7 (Surrey Beatty & Sons, Chipping Norton 2007).

Green DM, Sharbel TF: Comparative cytogenetics of the primitive frog, *Leiopelma archeyi* (Anura, Leiopelmatidae). Cytogenet Cell Genet 47:212–216 (1988).

Green DM, Daugherty CH, Bogart JP: Karyology and systematic relationships of the tailed frog *Ascaphus truei*. Herpetologica 36:346–352 (1980).

Green DM, Kezer J, Nussbaum RA: Supernumerary chromosome variation and heterochromatin distribution in the endemic New Zealand frog, *Leiopelma hochstetteri*. Chromosoma 95:339–344 (1987).

Green DM, Sharbel TF, Hitchmough RA, Daugherty CH: Genetic variation in the genus *Leiopelma* and relationships to other primitive frogs. J Zool Syst Evol Res 27:65–79 (1989).

Green DM, Zeyl CW, Sharbel TF: The evolution of hypervariable sex and supernumerary chromosomes in the relict New Zealand frog, *Leiopelma hochstetteri*. J Evol Biol 6: 417–441 (1993).

Greenbaum IF, Reed MJ: Evidence for heterosynaptic pairing of the inverted segment in pericentric inversion heterozygotes of the deer mouse (*Peromyscus maniculatus*). Cytogenet Cell Genet 38:106–111 (1984).

Greenbaum IF, Baker RJ, Bowers JH: Chromosomal homology and divergence between sibling species of deer mice: *Peromyscus maniculatus* and *P. melanotis* (Rodentia, Cricetidae). Evolution 32:334–341 (1978).

Gregory TR: Genome size evolution in animals, in Gregory TR (ed): The Evolution of the Genome, pp 3–87 (Elsevier Academic Press, Burlington 2005a).

Gregory TR: Animal Genome Size Database. http://www.genomesize.com (2005b).

Greilhuber J, Loidl J: On regularities of C-banding patterns and their possible cause, in Brandham PE, Bennett MD (eds): Kew Chromosome Conference II (Allen and Unwin, London 1983).

Griffin DK, Haberman F, Masabanda J, O'Brien P, Bagga M, et al: Micro- and macrochromosome paints generated by flow cytometry and microdissection: tools for mapping the chicken genome. Cytogenet Cell Genet 87: 278–281 (1999).

Gropp A, Citoler P, Geisler M: Karyotypvariationen und Heterochromatinmuster bei Igeln (*Erinaceus* und *Hemiechinus*). Chromosoma 27:288–307 (1969).

Gruber SL: Estudos citogenéticos em espécies do gênero *Hyla* (Anura, Hylidae) dos grupos com 2n = 24 e 2n = 30, com técnicas de coloração diferencial. Master thesis, Universidade Estadual de São Paulo, São Paulo (2002).

Gruber SL, Haddad CFB, Kasahara S: Evaluating the karyotypic diversity in species of *Hyla* (Anura: Hylidae) with 2n = 30 chromosomes based on the analysis of ten species. Folia Biologica 51:68–75 (2005).

Gruber SL, Haddad CFB, Kasahara S: Chromosome banding in three species of *Hypsiboas* (Hylidae, Hylinae), with special reference to a new case of a B-chromosome in anuran frogs and to the reduction of the diploid number of 2n = 24 to 2n = 22 in the genus. Genetica 130:281–291 (2007).

Gruber SL, Zina J, Narimatsu H, Haddad CFB, Kasahara S: Comparative karyotype analysis and chromosome evolution in the genus *Aplastodiscus* (Cophomantini, Hylinae, Hylidae). BMC Genetics 13:28 (2012a).

Gruber SL, Haddad CFB, Kasahara S: Karyotype analysis of seven species of the tribe Lophiohylini (Hylinae, Hylidae, Anura), with conventional and molecular cytogenetic techniques. Comp Cytogenet 6:409–423 (2012b).

Gruber SL, Silva APZ, Haddad CFB, Kasahara S: Cytogenetic analysis of *Phyllomedusa distincta* Lutz, 1950 (2n = 2x = 26), *P. tetraploidea* Pombal and Haddad, 1992 (2n = 4x = 52), and their natural triploid hybrids (2n = 3x = 39) (Anura, Hylidae, Phyllomedusinae). BMC Genet 14:75 (2013).

Gruber SL, Diniz D, Sobrinho-Scudeler PE, Foresti F, Haddad CFB, Kasahara S: Possible interspecific origin of the B chromosome of *Hypsiboas albopunctatus* (Spix, 1824) (Anura, Hylidae), revealed by microdissection, chromosome painting, and reverse hybridisation. Comp Cytogenet 8:185–197 (2014).

Gruber SL, de Oliveira GIG, Silva APZ, Narimatsu H, Haddad CFB, Kasahara S: Comparative analysis based on replication banding reveals the mechanism responsible for the difference in the karyotype constitution of treefrogs *Ololygon* and *Scinax* (Arboranae, Hylidae, Scinaxinae). Comp Cytogenet 11:267–283 (2017).

Grützner F, Rens W, Tsend-Ayush E, E-Mogharbell N, O'Brien PCM, et al: In the platypus a meiotic chain of ten sex chromosomes shares genes with the bird Z and mammal X chromosomes. Nature 432:913–917 (2004).

Guerrero RF, Kirkpatrick M, Perrin N: Cryptic recombination in the ever-young sex chromosomes of hylid frogs. J Evol Biol 25:1947–1954 (2012).

Guimarães C, Peixoto MA, Lacerda J, Feio R: The tadpole of *Scinax cosenzai* (Anura: Hylidae). Salamandra 50:99–104 (2014).

Gunn SJ, Greenbaum IF: Systematic implications of karyotypic and morphologic variation in mainland *Peromyscus* from the Pacific Northwest. J Mammal 67:294–304 (1986).

Gurdon JB: Egg cytoplasm and gene control in development. Proc R Soc Lond B 198:211–247 (1977).

Gutknecht L, Sperlich D, Bachmann L: A species specific satellite DNA family of *Drososphila subsilvestris* appearing predominantly in B chromosomes. Chromosoma 103:539–544 (1995).

Guttenbach M, Nanda I, Feichtinger W, Masabanda JS, Griffin DK, Schmid M: Comparative chromosome banding of chicken autosomal paints 1–9 in nine different bird species. Cytogenet Genome Res 103:173–184 (2003).

Guyer C: The herpetofauna of La Selva, Costa Rica, in Gentry AH (ed): Four Neotropical Rainforests, pp 371–385 (Yale University Press, New Haven 1990).

Haaf T, Schmid M: Y isochromosome associated with a mosaic karyotype and inactivation of the centromere. Hum Genet 85:486–490 (1990).

Haaf T, Steinlein K, Schmid M: Preferential somatic pairing between homologous heterochromatic regions of human chromosomes. Am J Hum Genet 38:319–329 (1986).

Haaf T, Hayman DL, Schmid M: Quantitative determination of rDNA units in vertebrate cells. Exp Cell Res 193:78–86 (1991).

Haczkiewicz K, Ogielska M: Gonadal sex differentiation in frogs: how testes become shorter than ovaries. Zool Sci 30:125–134 (2013).

Haddad CFB, Prado CP: Reproductive modes in frogs and their unexpected diversity in the Atlantic Forest of Brazil. Bioscience 55:207–217 (2005).

Haddad CFB, Sawaya RJ: Reproductive modes of Atlantic Forest hylid frogs : a general overview and the description of a new mode. Biotropica 32:862–871 (2000).

Haddad CFB, Pombal JP, Batistic RF: Natural hybridization between diploid and tetraploid species of leaf-frogs, genus *Phyllomedusa* (Amphibia). J Herpetol 28:425–430 (1994).

Haddad CFB, Faivovich J, Garcia PCA: The specialized reproductive mode of the treefrog *Aplastodiscus perviridis* (Anura: Hylidae). Amphibia-Reptilia 26:87–92 (2005).

Hale DW: Heterosynapsis and suppression of chiasmata within heterozygous pericentric inversions of the Sitka deer mouse. Chromosoma 94:425–432 (1986).

Hamlin JL, Milbrandt JD, Heintz NH, Azizkhan JC: DNA sequence amplification in mammalian cells. Int Rev Cytol 90:31–82 (1984).

Hanada H: G and C banding show structural differences between the Z and W chromosomes in the frog *Buergeria buergeri*. Hereditas 136: 151–154 (2002).

Hansen S: A case of centric fission in man. Humangenetik 26:257–259 (1975).

Hardy LM, McDiarmid RW: The amphibians and reptiles of Sinaloa, México. Univ Kansas Pub Mus Nat Hist 18:39–252 (1969).

Harel J, Hanania N, Tapiero H, Harel L: RNA replication by nuclear satellite DNA in different mouse cells. Biochem Biophys Res Commun 33:696–701 (1968).

Harland RM, Grainger RM: *Xenopus* research: metamorphosed by genetics and genomics. Trends Genet 27:507–515 (2011).

Harms JW: Beobachtungen über Geschlechtsumwandlung reifer Tiere und deren F_1 Generation. Zool Anz 67:67–79 (1926).

Hass CA, Maxson LR, Hedges SB: Relationships and divergence times of West Indian amphibians and reptiles: insights from albumin immunology, in Woods CA, Sergile FE (eds): Biogeography of the West Indies: Patterns and Perspectives, pp 157–174 (CRC Press, Boca Raton 2001).

Hayman DL, Martin PG: Mammalia I: Monotremata and Marsupialia, in John B (ed): Animal Cytogenetics, Vol 4/4 (Gebrüder Borntraeger, Berlin 1974).

Hayman DL, Rofe R: Marsupial sex chromosomes, in Calaby JH, Tyndale-Biscoe CH (eds): Reproduction and Evolution, pp 69–79 (Australian Academy of Science, Canberra 1977).

Hayman D, Sharp P: Verification of the structure of the complex sex chromosome system in *Lagorchestes conspicillatus* Gould (Marsupialia: Mammalia). Chromosoma 83:263–274 (1981).

Hayman DL, Ashworth LK, Carrano AV: The relative DNA contents of the eutherian and marsupial X chromosomes. Cytogenet Cell Genet 34:265–270 (1982).

Hazen MW, Kuo MT, Arrighi FE: Genome analysis of *Peromyscus* (Rodentia, Cricetidae). VII. Localization of satellite DNA sequences and cytoplasmic poly(A) RNA sequences of *P. eremicus* on metaphase chromosomes. Chromosoma 64:133–142 (1977).

Hedges SB: An electrophoretic analysis of Holarctic hylid frog evolution. Syst Zool 35:1–21 (1986).

Hedges SB: The number of replications needed for accurate estimation of the bootstrap p-value in phylogenetic studies. Mol Biol Evol 9:366–369 (1992).
Hedges SB: Global amphibian declines: a perspective from the Caribbean. Biodivers Conserv 2:290–303 (1993).
Hedges SB: The origin of West Indian amphibians and reptiles, in Powell R, Henderson RW (eds): Contributions to West Indian Herpetology: A Tribute to Albert Schwartz, pp 95–128 (Society for the Study of Amphibians and Reptiles, Ithaca 1996).
Hedges SB, Hass CA, Maxson LR: Caribbean biogeography: molecular evidence for dispersal in West Indian terrestrial vertebrates. Proc Natl Acad Sci USA 89:1909–1913 (1992).
Hedges SB, Duellman WE, Heinicke MP: New World direct-developing frogs (Anura: Terrarana): molecular phylogeny, classification, biogeography, and conservation. Zootaxa 1737:1–182 (2008).
Hedrick PW: The establishment of chromosomal variants. Evolution 35:322–332 (1981).
Henriques-Gil N, Santos JL, Giraldez R: Genotype-dependent effect of B-chromosomes on chiasma frequency in *Eyprepocnemis plorans* (Acrididae: Orthoptera). Genetica 59:223–227 (1982).
Henriques-Gil N, Arana P, Santos JL: Spontaneous translocations between B chromosomes and the normal complement in the grasshopper *Eyprepocnemis plorans*. Chromosoma 88:145–148 (1983).
Heppich S: Hybridogenesis in *Rana esculenta*: C-band karyotypes of *Rana ridibunda*, *Rana lessonae* and *Rana esculenta*. J Zool Syst Evol Res 16:27–39 (1978).
Heppich S, Tunner HG: Chromosomal constitution and C-banding in homotypic *Rana esculenta* crosses. Mitt Zool Mus Berl 55:111–114 (1979).
Hermier M, Philippe N, François R: Sujet masculin pubère et sans amnbiguïté génitale, 45 X/46 X dic(Yq). Archs Fr Pédiat 36:162–172 (1979).
Hernández-Guzmán J, Arias-Rodriguez L, Indy JR: Los cromosomas meióticos de la rana arborícola *Smilisca baudinii* (Anura: Hylidae). Rev Biol Trop 59:355–362 (2011).
Hernandez-Verdun D, Derenzini M, Bouteille M: The morphological relationship in electron microscopy between NOR-silver proteins and intranucleolar chromatin. Chromosoma 85:461–473 (1982).
Hervas F, Torres KP, Montenegro-Larrea P, del Pino EM: Development and gastrulation in *Hyloxalus vertebralis* and *Dendrobates auratus* (Anura: Dendrobatidae). Amphib Reptile Conserv 8:121–135 (2015).
Hewitt GM: The integration of supernumerary chromosomes into the orthopteran genome. Cold Spring Harbor Symp Quant Biol 38: 183–194 (1974).
Hewitt GM, East TM, Shaw MV: Sperm dysfunction produced by B-chromosomes in the grasshopper *Myrmeleotettix maculatus*. Heredity 58:59–68 (1987).
Heyer WR, Rand AS, da Cruz CAG, Peixoto OL: Decimations, extinctions, and colonizations of frog populations in southeast Brazil and their evolutionary implications. Biotropica 20:230–235 (1988).
Heyer WR, Rand AS, Cruz CAG, Peixoto OL, Nelson CE: Frogs of Boracéia. Arq Zool Mus Zool Univ São Paulo 31:231–419 (1990).
Hillis DM: The phylogeny of amphibians: current knowledge and the role of cytogenetics, in Green DM, Sessions SK (eds): Amphibian Cytogenetics and Evolution, pp 7–31 (Academic Press, San Diego 1991).
Hillis DM, Green DM: Evolutionary changes of heterogametic sex in the phylogenetic history of amphibians. J Evol Biol 3:49–64 (1990).
Höbel G: Plasticity and geographic variation in the reproductive ecology of gladiator frogs, particularly *Hypsiboas rosenbergi*. Stapfia 88:329–334 (2008).
Hödl W: An analysis of foam nest construction in the neotropical frog *Physalaemus ephippifer* (Leptodactylidae). Copeia 1990:547 (1990).
Hoehn H, Martin GM: Heritable alteration of human constitutive heterochromatin induced by mitomycin C. Exp Cell Res 75:275–278 (1972).
Hoehn H, Martin GM: Clonal variants of constitutive heterochromatin of human fibroblasts after recovery from mitomycin C treatment. Chromosoma 43:203–210 (1973).
Hoffman H, Kubicki B: The tadpole of *Ecnomiohyla sukia* Savage & Kubicki, 2010 (Amphibia: Hylidae). Zootaxa 2793:63–66 (2011).
Hofgärtner FJ, Schmid M, Krone W, Zenzes MT, Engel W: Pattern of activity of nucleolus organizers during spermatogenesis in mammals as analyzed by silver staining. Chromosoma 71:197–216 (1979a).
Hofgärtner FJ, Krone W, Jain K: Correlated inhibition of ribosomal RNA synthesis and silver staining by actinomycin D. Hum Genet 47:329–333 (1979b).
Holloway AK, Cannatella DC, Gerhardt HC, Hillis DM: Polyploids with different origins and ancestors form a single sexual polyploid species. Am Nat 167:e88–e110 (2006).
Holmquist GP: DNA sequences in G-bands and R-bands, in Adolph KW (ed): Chromosomes and Chromatin, vol II, pp 76–121 (CRC Press, Boca Raton 1988).
Houlahan JE, Findlay CS, Schmidt BR, Meyer AH, Kuzmin SL: Quantitative evidence for global amphibian population declines. Nature 404:752–755 (2000).
Howell WM: Visualization of ribosomal gene activity: silver stains proteins associated with rRNA transcribed from oocyte chromosomes. Chromosoma 62:361–367 (1977).
Hsu TC: Heterochromatin pattern in metaphase chromosomes of *Drosophila melanogaster*. J Hered 62:285–287 (1971).
Hsu TC, Arrighi FE: Chromosomal evolution in the genus *Peromyscus* (Cricetidae, Rodentia). Cytogenetics 5:355–359 (1966).
Hsu TC, Arrighi FE: Chromosomes of *Peromyscus* (Rodentia, Cricetidae). I. Evolutionary trends in 20 species. Cytogenetics 7:417–446 (1968).
Hsu TC, Arrighi FE: Distribution of constitutive heterochromatin in mammalian chromosomes. Chromosoma 34:243–253 (1971).
Hsu TC, Brinkley BR, Arrighi FE: The structure and behavior of the nucleolus organizer in mammalian cells. Chromosoma 23:137–153 (1967).
Hsu TC, Arrighi FE, Saunders G: Compositional heterogeneity of human heterochromatin. Proc Natl Acad Sci USA 69:1464–1466 (1972).
Hsu TC, Spirito SE, Pardue ML: Distribution of 18S + 28S ribosomal genes in mammalian genomes. Chromosoma 53:25–36 (1975a).
Hsu TC, Pathak S, Chen TR: The possibility of latent centromeres and a proposed nomenclature system for total chromosome and whole arm translocations. Cytogenet Cell Genet 15:41–49 (1975b).
Huckembeck S, Claudino M, Correa F, Bastos R, Loebmann D, et al: The activity patterns and microhabitat use of *Pseudis minuta* Günther, 1858 (Anura, Hylidae) in the Lagoa do Peixe National Park, a biosphere reserve of the Brazilian subtropics. Braz J Biol 72:331–336 (2012).
Hudson AP, Cuny G, Cortadas J, Haschemeyer AEV, Bernardi G: An analysis of fish genomes by density gradient centrifugation. Eur J Biochem 112:203–210 (1980).
Hunter AS, Murcia de Valdivieso B: La reproducción de la rana, *Hyla labialis*. Caldasia 8: 573–583 (1962).
Ikebe C, Kohno S, Nambu H: Cytogenetic studies of Hynobiidae (Urodela). XVII. Banded karyotypes and sex chromosomes in *Hynobius tenuis* Nambu. Chromosome Sci 4:111–115 (2000).
Ikebe C, Kuro-o M, Ohtani H, Kawase Y, Matsui T, Kohno S: Cytogenetic studies of Hynobiidae. XIX. Morphological variation of sex chromosomes pairing behavior of sex lampbrush chromosomes in *Hynobius quelpaertensis* (Mori) from Cheju Island, South Korea. Chromosome Res 13:157–167 (2005).
Imai HT, Takahata N, Maruyama T, Daniel A, Honda T, et al: Theoretical bases for karyotype evolution. II. The fusion burst in man and mouse. Jpn J Genet 63:313–342 (1988).
Iriki S: Studies on amphibian chromosomes. I. On the chromosomes of *Hyla arborea japonica* Guenther. Mem Coll Sci Kyoto Imp Univ, Ser B, 5:1–17 (1930).
Iriki S: Studies on amphibian chromosomes. V. Explanation of the sex-chromosome type of *Hyla arborea japonica* from the standpoint of spiral structure. Sci Rep Tokyo Bunrika Daigaku Sec B 1:73–79 (1932a).
Iriki S: Studies on amphibian chromosomes. VIII. On the fixation of the chromosomes of *Hyla arborea japonica* and some other animals. Sci Rep Tokyo Bunrika Daigaku Sec B 1:105 (1932b).
Iturra P, Veloso A: Evidence for heteromorphic sex chromosomes in male amphibians (Anura: Leptodactylidae). Cytogenet Cell Genet 31:108–110 (1981).
Iturra P, Veloso A: Further evidence for early sex chromosome differentiation of anuran species. Genetica 78:25–31 (1989).

Iturralde-Vinent MA, MacPhee RDE: Paleogeography of the Caribbean region: implications for Cenozoic biogeography. Bull Am Mus Nat Hist 238:1–95 (1999).

IUCN: The IUCN Red List of Threatened Species. International Union for the Conservation of Nature, Cambridge, UK. http://www.iucnredlist.org (accessed January 1, 2014).

Iwasawa H, Futagami J: Normal stages of development of a tree frog, *Hyla japonica* Günther. Jpn J Herpetol 14:129–142 (1992).

Jalal SM, Clark RW, Hsu TC, Pathak S: Cytological differentiation of constitutive heterochromatin. Chromosoma 48:391–403 (1974).

Jamilena M, Ruíz-Rejón C, Ruíz-Rejón M: A molecular analysis of the origin of the *Crepis capillaris* B chromosome. J Cell Sci 107:703–708 (1994).

Jamilena M, Garrido-Ramos M, Ruíz-Rejón M, Ruíz-Rejón C, Parker JS: Characterisation of repeated sequences from microdissected B chromosomes of *Crepis capillaris*. Chromosoma 104:113–120 (1995).

Jamrich M, Warrior R, Steele R, Gall JG: Transcription of repetitive sequences on *Xenopus* lampbrush chromosomes. Proc Natl Acad Sci USA 80:3364–3367 (1983).

Jared C, Mailho-Fontana PL, Antoniazzi MM, Mendes VA, Barbaro KC, et al: Venomous frogs use heads as weapons. Curr Biol 25:1–5 (2015).

John B: Myths and mechanisms of meiosis. Chromosoma 54:295–325 (1976).

John B: The biology of heterochromatin, in Verma RS (ed): Heterochromatin. Molecular and Structural Aspects, pp 1–147 (Cambridge University Press, Cambridge 1988).

John B, King M: Population cytogenetics of *Atractomorpha similis*. I. C band variation. Chromosoma 88:57–68 (1983).

John B, Appels R, Contreras N: Population cytogenetics of *Atractomorpha similis*. II. Molecular characterisation of the distal C-band polymorphisms. Chromosoma 94:45–58 (1986).

Johnson C: Species recognition in the *Hyla versicolor* complex. Texas J Sci 18:361–364 (1966).

Jones KW: Evolutionary conservation of sex-specific DNA sequences. Differentiation 23 (Suppl):56–59 (1983a).

Jones KW: Evolution of sex chromosomes, in Johnson MH (ed): Development in Mammals, vol 5, pp 297–319 (Elsevier Science Publishers, Amsterdam 1983b).

Jones KW, Singh L: Conserved repeated DNA sequences in vertebrate sex chromosomes. Hum Genet 58:46–53 (1981).

Jones RN: Are B chromosomes selfish?, in Cavalier-Smith T (ed): The Evolution of Genome Size, pp 397–425 (John Wiley & Sons, London 1985).

Jones RN: B-chromosome drive. Am Nat 137: 430–442 (1991).

Jones RN: B chromosomes in plants. Tansley review No. 85. New Phytol 131:411–434 (1995).

Jones RN, Rees H: B Chromosomes (Academic Press, London 1982).

Jotterand M, Fischberg M: A chromosome mutation affecting the number of nucleoli in *Xenopus borealis* Parker. Experientia 30:1003–1005 (1974).

Jungfer K-H: Reproduction and parental care of the coronated treefrog, *Anotheca spinosa* (Steindachner, 1864) (Anura: Hylidae). Herpetologica 52:25–32 (1996).

Kahn J: The nucleolar organizer in the mitotic chromosome complement of *Xenopus laevis*. Quart J Micr Sci 103:407–409 (1962).

Kaiser H, Mais C, Bolaños F, Steinlein C, Feichtinger W, Schmid M: Chromosomal investigation of three Costa Rican frogs from the 30-chromosome radiation of *Hyla* with the description of a unique geographic variation in nucleolus organizer regions. Genetica 98: 95–102 (1996).

Kakampuy W, Supraprom T, Steinlein C, Schmid M: Chromosome banding of two *Litoria* species (Anura, Hylidae). Songklanakarin J Sci Technol 35:383–387 (2013).

Kasahara S, Silva APZ, Gruber SL: Use of lymphocyte cultures for BrdU replication banding patterns in anuran species (Amphibia). Genet Mol Biol 21:471–476 (1998).

Kasahara S, Silva APZ, Gruber SL, Haddad CFB: Comparative cytogenetic analysis on four tree frog species (Anura, Hylidae, Hylinae) from Brazil. Cytogenet Genome Res 103: 155–162 (2003).

Kasahara S, Campos JRC, Catroli GF, Haddad CFB: Cytogenetics of *Phyllomedusa distincta* (2n = 2x = 26), *Phyllomedusa tetraploidea* (2n = 4x = 52) and their triploid hybrids (2n = 3x = 39). 16th International Chromosome Conference, Amsterdam (2007).

Kawamura T: Artificial parthenogenesis in the frog. II. The sex of parthenogenetic frogs. J Sci Hiroshima Univ Ser B 7:39–86 (1939).

Kawamura T: The offspring of triploid males of the frog, *Rana nigromaculata*. J Sci Hiroshima Univ Ser B 12:11–20 (1951).

Kawamura T, Nishioka M: Aspects of the reproductive biology of Japanese anurans, in Taylor DH, Guttman SI (eds): The Reproductive Biology of Amphibians, pp 103–139 (Plenum Press, New York 1977).

Kawamura T, Yokota R: The offspring of sex-reversed females of *Rana japonica* Guenther. J Sci Hiroshima Univ Ser B 18:31–38 (1959).

Kawamura T, Nishioka M, Ueda H: Inter- and intraspecific hybrids among Japanese, European and American toads. Sci Rep Lab Amphibian Biol Hiroshima Univ 4:1–125 (1980).

Keller RE: The cellular basis of amphibian gastrulation, in LW Bowder (ed): The Cellular Basis of Morphogenesis, pp 241–327 (Springer, Boston 1986).

Keller R, Shook D: Gastrulation in amphibians, in CD Stern (ed): Gastrulation: from Cells to Embryo, pp 171–204 (CSHL Press, Cold Spring Harbor 2004).

Keyl HG, Hägele K: B-chromosomes in *Chironomus*. Chromosoma 35:403–417 (1971).

Kiesecker JM, Blaustein AR, Belden LK: Complex causes of amphibian declines. Nature 410:681–683 (2001).

Kimura M, Ohta T: The average number of generations until fixation of a mutant gene in a finite population. Genetics 61:763–771 (1969).

King M: C-banding studies on Australian hylid frogs: secondary constriction structure and the concept of euchromatin transformation. Chromosoma 80:191–217 (1980).

King M: A cytotaxonomic analysis of Australian hylid frogs of the genus *Litoria*, in Banks CB, Martin AA (eds): Proceedings of the Melbourne Herpetological Symposium May 19–21, 1980, pp 169–175 (Zoological Board of Victoria & Royal Melbourne Zoological Gardens, Melbourne 1981).

King M: Amphibia, in John B (ed): Animal Cytogenetics, vol 4/2 (Gebrüder Borntraeger, Berlin 1990).

King M: The evolution of heterochromatin in the amphibian genome, in Green DM, Sessions SK (eds): Amphibian Cytogenetics and Evolution, pp 359–391 (Academic Press, San Diego 1991).

King M, Rofe M: Karyotypic variation in the Australian gekko *Phyllodactylus marmoratus* (Gray) (Gekkonidae: Reptilia). Chromosoma 54:75–87 (1976).

King M, Tyler MJ, Davies M, King D: Karyotypic studies on *Cyclorana* and associated genera of Australian frogs. Aust J Zool 27:699–708 (1979).

King M, Contreras N, Honeycutt RL: Variation within and between nucleolus organizer regions in Australian hylid frogs (Anura) shown by 18S + 28S in-situ hybridization. Genetica 80:17–29 (1990).

King RL: Heteromorphic homologous chromosomes in three species of *Pseudotrimerotropis* (Orthoptera: Acrididae). J Morphol 38: 19–63 (1923).

Kluge AG: The life history, social organization and parental behavior of *Hyla rosenbergi* Boulenger, a nest building gladiator frog. Misc Pub Mus Zool Univ Mich 160:1–170 (1981).

Kobel HR, Du Pasquier L, Fischberg M, Gloor H: *Xenopus amieti* sp. nov. (Anura: Pipidae) from the Cameroons, another case of tetraploidy. Rev Suisse Zool 87:919–926 (1980).

Kochan KJ, Wright DA, Schroeder LJ, Shen J, Morizot DC: Genetic linkage maps of the West African clawed frog *Xenopus tropicalis*. Dev Dyn 226:99–102 (2003).

Kohno S, Ikebe C, Kuro-o M: Banding karyotypes and phylogenetic relationships of Hynobiidae from Japan and Korea, in Olmo E (ed): Cytogenetics of Amphibians and Reptiles, pp 52–56 (Birkhäuser, Basel 1990).

Kohno S, Kuro-o M, Ikebe C: Cytogenetics and evolution of hynobiid salamanders, in Green DM, Sessions SK (eds): Amphibian Cytogenetics and Evolution, pp 67–88 (Academic Press, San Diego 1991).

Kriger KM, Hero JM: The chytrid fungus *Batrachochytrium dendrobatidis* is non-randomly distributed across amphibian breeding habitats. Divers Distrib 13:781–788 (2007a).

Kriger KM, Hero JM: Large-scale seasonal variation in the prevalence and severity of chytridiomycosis. J Zool 27:352–359 (2007b).

Kuramoto M: Karyotypes of three leptodactylid frogs from Chile with a discussion on the chromosome numbers of the family Leptodactylidae. Bull Fukuoka Univ Educ, Part III, 21:133–141 (1972).

Kuramoto M: Systematic implications of hybridization experiments with some Eurasian treefrogs (genus *Hyla*). Copeia 1984:609–616 (1984).

Kuramoto M: Karyological studies on some Philippine frogs, in Matsui M, Kikida T, Goris RC (eds): Current Herpetology in East Asia, pp 115–121 (Herpetological Society of Japan, Kyoto 1989).

Kuramoto M: A list of chromosome numbers of anuran amphibians. Bull Fukuoka Univ Educ, Part III, 39:83–127 (1990).

Kuramoto M, Allison A: Karyotypes of five hylid frogs from Papua New Guinea, with a discussion on their systematic implications. Jpn J Herpetol 14:6–11 (1991).

Kuramoto M, Furuya E, Takegami M, Yano K: Karyotypes of several species of frogs from Japan and Taiwan. Bull Fukuoka Univ Educ 23:67–78 (1974).

Kurnit DM: Satellite DNA and heterochromatin variants: the case for unequal mitotic crossing over. Hum Genet 47:169–186 (1979).

Kuro-o M: Cytogenetic studies of the genus *Hynobius* by means of DNA replication patterns (R-banding). MSc thesis, Toho University, Funabashi (1986).

Kuro-o M, Ikebe C, Kohno S: Cytogenetic studies of Hynobiidae (Urodela). IV. DNA replication bands (R-banding) in the genus *Hynobius* and the banding karyotype of *Hynobius nigrescens* Stejneger. Cytogenet Cell Genet 43:14–18 (1986).

Kuro-o M, Ikebe C, Kohno S: Cytogenetic studies of Hynobiidae (Urodela). VI. R-banding patterns in five pond-type *Hynobius* from Korea and Japan. Cytogenet Cell Genet 44: 69–75 (1987).

Kuro-o M, Ikebe C, Izumisawa Y, Fujinuki Y, Sasaki K, et al: Cytogenetic studies of Hynobiidae (Urodela). XVIII. A ZZ/ZW sex-determining mechanism in a hynobiid salamander species, *Hynobius tokyoensis* Tago. Cytogenet Genome Res 99:194–199 (2002).

La Marca E, Lötters S: Monitoring of declines in Venezuelan *Atelopus* (Amphibia: Anura: Bufonidae), in Böhme W, Bischoff W, Ziegler T (eds): Herpetologia Bonnensis, pp 207–231 (SEH, Bonn 1997).

La Marca E, Reinthaler HP: Population changes in *Atelopus* species of the Cordillera de Mérida, Venezuela. Herpetol Rev 22:125–128 (1991).

La Marca E, Lips KR, Lotters S, Puschendorf R, Ibañez R, et al: Catastrophic population declines and extinctions in neotropical Harlequin frogs (Bufonidae: *Atelopus*). Biotropica 37:190–201 (2005).

Lamatsch DK, Steinlein C, Schmid M, Schartl M: Noninvasive determination of genome size and ploidy level in fishes by flow cytometry: detection of triploid *Poecilia formosa*. Cytometry 39:91–95 (2000).

Lanfear R, Calcott B, Ho SYW, Guindon S: PartitionFinder: combined selection of partitioning schemes and substitution models for phylogenetic analyses. Mol Biol Evol 29: 1695–1701 (2012).

Lanfear R, Calcott B, Kainer D, Mayer C, Stamatakis A: Selecting optimal partitioning schemes for phylogenomic datasets. BMC Evol Biol 14:82 (2014).

Lanzone C, Baldo D, Rosset SD: Meiotic differentiation in two allopatric population groups of the tetraploid frog *Odontophrynus americanus* from Argentina. Herpetol J 18:213–222 (2008).

Latour DR, Vig BK, Finze EM, Paweletz N: The centromeres of the Indian muntjac: evidence for the existence of multiple centromeres. Mutat Res 356:187–195 (1996).

Latt SA, Munroe SH, Disteche C, Rogers WE, Cassel DM: Uses of fluorescent dyes to study chromosome structure and replication, in De la Chapelle A, Sorsa M (eds): Chromosomes Today, vol 6, pp 27–36 (Elsevier, Amsterdam 1977).

Laurance WF: A crisis in the making: responses of Amazonian forests to land use and climate change. Trends Ecol Evol 13:411–415 (1998).

Laurance WF: Tropical logging and human invasions. Conserv Biol 15:4–5 (2001).

Ledbetter DH: NOR-bearing Y chromosome in a primate, *Hylobates (Symphalangus) syndactylus*. Cytogenet Cell Genet 29:250–252 (1981).

Lee JC: The Amphibians and Reptiles of the Yucatán Peninsula (Cornell University Press, Ithaca 1996).

Lee M, Jamieson B: The ultrastructure of the spermatozoa of bufonid and hylid frogs (Anura, Amphibia): implications for phylogeny and fertilization biology. Zool Scr 22:309–323 (1993).

Lee YH, Kwon AS: Ultrastructure of spermiogenesis in *Hyla japonica* (Anura, Amphibia). Acta Zool 73:49–55 (1992).

Leite GB, Franco-Belussi L, Provete DB, Oliveira C: Comparative testis morphology of Neotropical anurans. Zool Anzeiger 257:29–38 (2015).

León PE: Report of the chromosome numbers of some Costa Rican anurans. Rev Biol Trop 17: 119–124 (1970).

Li SS, Hu JS: On the karyotypes and Ag-NORs of three sympatrically [sic] *Paa* frogs in Yunnan Province. Acta Zool Sinica 40:317–323 (1994).

Li SS, Wang CY, Li CY, Wang RF, Liu GZ: A comparative investigation of the karyotypes from four amphibian species. Zool Res 2:17–24 (1981).

Li ZX, Powell CA: An outline of the palaeogeographic evolution of the Australasian region since the beginning of the Neoproterozoic. Earth-Sci Rev 53:237–277 (2001).

Lima AP, Magnusson WE, Menin M, Erdtmann LK, Rodrigues DJ, et al: Guia de Sapos da Reserva Adolpho Ducke (INPA, Manaus 2006).

Lima-de-Faria A: Chromosome gradient and chromosome field in *Agapanthus*. Chromosoma 6:330–370 (1954).

Lima-de-Faria A: Equations defining the position of ribosomal cistrons in the eukaryotic chromosome. Nat New Biol 241:136–139 (1973a).

Lima-de-Faria A: The specific location of ribosomal genes in the eukaryotic chromosome, in Pfeiffer RA (ed): Modern Aspects of Cytogenetics: Constitutive Heterochromatin in Man. Symp Med Hoechst 6, pp 39–44 (Schattauer, Stuttgart 1973b).

Lima-de-Faria A: The chromosome field. I. Prediction of the location of ribosomal cistrons. Hereditas 83:1–22 (1976).

Liming S, Yingying Y, Xingsheng D: Comparative cytogenetic studies on the red muntjak, Chinese muntjak, and their F_1 hybrids. Cytogenet Cell Genet 26:22–27 (1980).

Lin CC, Sasi R, Fan YS, Chen ZQ: New evidence for tandem chromosome fusions in the karyotypic evolution of Asian muntjacs. Chromosoma 101:9–24 (1991).

Lips KR: Decline of a tropical montane amphibian fauna. Conserv Biol 12:106–117 (1998).

Lips KR: Mass mortality of the anuran fauna at an upland site in Panama. Conserv Biol 13: 117–125 (1999).

Lips KR, Green DE, Papendick R: Chytridiomycosis in wilds frogs from southern Costa Rica. J Herpetol 37:215–218 (2003).

Lips KR, Brem F, Brenes R, Reeve JD, Alford RA, et al: Emerging infectious disease and the loss of biodiversity in a neotropical amphibian community. Proc Natl Acad Sci USA 103: 3165–3170 (2006).

Littlejohn MJ: Artificial hybridization between some hylid frogs of the United States. Texas J Sci 13:176–184 (1961).

Litvinchuk SN, Rosanov JM: The first case of natural spontaneous triploidy in the family Bombinatoridae. Amphibia-Reptilia 37: 243–245 (2016).

Litvinchuk SN, Mazepa GO, Pasynkova RA, Saidov A, Satorov T, et al: Influence of environmental conditions on the distribution of Central Asian green toads with three ploidy levels. J Zool Syst Evol Res 49:233–239 (2011).

Litvinchuk SN, Rosanov JM, Schepina, NA, Kazakov VI, Skorinov DV, Borkin LJ: The first case of natural triplody in the mongolian toad, *Bufo raddei*. Russ J Herpetol 4:333–336 (2012).

Litvinchuk SN, Skorinov DV, Rosanov JM: Natural spontaneous autotriploidy in the genus *Pelophylax* (Anura: Ranidae). Russ J Herpetol 22:318–320 (2015).

Liu T, Song M, Xia Y, Zeng X: A tandemly arranged pattern of two 5S rDNA arrays in *Amolops mantzorum* (Anura, Ranidae). Cytogenet Genome Res 151:161–170 (2017).

Liu WG, Zan RG: A special karyotype in the genus *Rana*. An investigation of the karyotype, C-banding and Ag-stained NORs of *Rana phrynoides* Boulenger. Acta Genetica Sinica 11:61–67 (1984).

Lofts B: Seasonal changes in the functional activity of the interstitial and spermatogenetic tissues of green frog, *Rana esculenta*. Gen Comp Endocrinol 4:550–562 (1964).

Lofts B: Reproduction, in Lofts B (ed): Physiology of the Amphibia. Volume II, B, pp 115–157 (Academic Press, San Diego 1974).

Lombardi J: Comparative Vertebrate Reproduction (Springer, Boston 1998).

Longcore JE, Pessier AP, Nichols DK: *Batrachochytrium dendrobatidis*, gen. et sp. nov., a chytrid pathogenic to amphibians. Mycologia 91:219–227 (1999).

Lopez-León MD, Neves N, Schwarzacher T, Heslop-Harrison TS, Hewitt CM, Camacho JPM: Possible origin of a B chromosome deduced from its DNA composition using double FISH technique. Chromosome Res 2:87–92 (1994).

Loumont C: Deux espèces nouvelles de *Xenopus* du Cameroun (Amphibia, Pipidae). Rev Suisse Zool 90:169–177 (1983).

Loumont C: *Xenopus pygmaeus*, a new diploid pipid frog from rain forest of equatorial Africa. Rev Suisse Zool 93:755–764 (1986).

Lourenço ACC, Zina J, Catroli GF, Kasahara S, Faivovich J, Haddad CFB: A new species of the *Scinax catharinae* group (Anura: Hylidae) from southeastern Brazil. Zootaxa 4154: 415–435 (2016).

Lourenço LB, Recco-Pimentel SM, Cardoso AJ: Polymorphism of the nucleolus organizer regions (NORs) in *Physalaemus petersi* (Amphibia, Anura, Leptodactylidae) detected by silver staining and fluorescence in situ hybridization. Chromosome Res 6:621–628 (1998).

Lourenço LB, Recco-Pimentel SM, Cardoso AJ: Two karyotypes and heteromorphic sex chromosomes in *Physalaemus petersi* (Anura, Leptodactylidae). Can J Zool 77:624–631 (1999).

Lourenço LB, Garcia PCA, Recco-Pimentel SM: Cytogenetics of two species of *Paratelmatobius* (Anura: Leptodactylidae), with phylogenetic comments. Hereditas 133:201–210 (2000).

Lourenço LB, Garcia PCA, Recco-Pimentel SM: Cytogenetics of a new species of *Paratelmatobius cardosoi* group (Anura: Leptodactylidae), with the description of an apparent case of pericentric inversion. Amphibia-Reptilia 24:47–55 (2003a).

Lourenço LB, Garcia PCA, Recco-Pimentel SM: Restriction fragment analysis of the ribosomal DNA of *Paratelmatobius* and *Scythrophrys* species (Anura, Leptodactylidae). Genet Mol Biol 26:139–143 (2003b).

Lynch JD: Discovery of the richest frog fauna in the world – an exploration of the forests to the north of Leticia. Rev Acad Colomb Cienc 29:581–588 (2005).

Lynch JD: The amphibian fauna in the Villavicencio Region of eastern Colombia. Caldasia 28:135–155 (2006).

Lyttle TW: Segregation disorders. Annu Rev Genet 25:511–557 (1991).

Mable BK, Bogart JP: Hybridization between tetraploid and diploid species of treefrogs (genus *Hyla*). J Hered 86:432–440 (1995).

Macaya G, Thiery JP, Bernardi G: An approach to the organization of eukaryotypic genomes at a macromolecular level. J Mol Biol 108: 237–254 (1976).

Macgregor HC, del Pino EM: Ribosomal gene amplification in multinucleate oocytes of the egg brooding hylid frog *Flectonotus pygmaeus*. Chromosoma 85:475–488 (1982).

Macgregor HC, Kezer J: Gene amplification in oocytes with 8 germinal vesicles from the tailed frog *Ascaphus truei* Stejneger. Chromosoma 29:189–206 (1970).

Macgregor HC, Mizuno S, Vlad M: Chromosomes and DNA sequences in salamanders, in Pearson PL, Lewis KR (eds): Chromosomes Today, vol 5, pp 531–539 (John Wiley & Sons, Hoboken 1976)

Machado RA, Bernarde PS, Abrahão Mirado SA, dos Anjos L: Análise comparada da riqueza de anuros entre duas áreas com diferentes estados de conservação no municipio de Londrina, Paraná, Brazil (Amphibia, Anura). Rev Bras Zool 16:997–1004 (1999).

Magnusson WE, Hero JM: Predation and the evolution of complex oviposition behaviour in Amazon rainforest frogs. Oecologia 86: 310–318 (1991).

Mahony ML: Heteromorphic sex chromosomes in the Australian frog *Crinia bilingua* (Anura: Myobatrachidae). Genome 34:334–337 (1991).

Mahony MJ, Roberts JD: Two new species of desert burrowing frogs of the genus *Neobatrachus* (Anura: Myobatrachidae). Rec West Aust Mus 13:155–170 (1986).

Mahony MJ, Robinson ES: Polyploidy in the Australian leptodactylid frog genus *Neobatrachus*. Chromosoma 81:199–212 (1980).

Makinen A, Fredga K: Banding analysis of the somatic chromosomes of racoon dogs, *Nyctereutes procyonoides*, from Finland. Proc 4th Eur Collog Cytogenet Domestic Anim, pp 420–430 (1980).

Mandahl N: Variation in C-stained chromosome regions in European hedgehogs (Insectivora: Mammalia). Hereditas 89:107–128 (1978).

Marinetti GV, Bagnara JT: Characterization of blue and yellow pigments in eggs of the Mexican leaf frog. Biochemistry 22:5651–5655 (1983).

Marks J: Hominoid cytogenetics and evolution. Yearbook Phys Anthropol 26:131–159 (1983).

Marshall E: Taxonomy – will DNA bar codes breathe life into classification. Science 307: 1037 (2005).

Martins EP: A comparative study of the evolution of *Sceloporus* push-up displays. Am Nat 142:994–1018 (1993).

Martins M: Observations on nest dynamics and embryonic and larval development in the nest building gladiator frog, *Hyla faber*. Amphibia-Reptilia 14:411–421 (1993).

Martins M, Haddad CF: Vocalizations and reproductive behaviour in the smith frog, *Hyla faber* Wied (Amphibia: Hylidae). Amphibia-Reptilia 9:409–460 (1988).

Martirosyan A, Stepanyan I: The karyotype of *Hyla savignyi* Audouin, 1827 (Amphibia, Anura) from Southern Armenia. Comp Cytogenet 1:107–112 (2007).

Matsui M, Seto T, Kohsaka Y, Borkin LJ: Bearing of chromosome C-banding patterns on the classification of Eurasian toads of the *Bufo bufo* complex. Amphibia-Reptilia 6:23–33 (1985).

Matthey R: Un cas nouveau des chromosomes sexuels multiples dans le genre *Gerbillus* (Rodentia-Muridae-Gerbillinae). Experientia 10:464–465 (1954).

Matthey R: Une inversion péricentrique à l'origine d'un polymorphisme chromosomique non-robertsonien dans une population de *Mastomys* (Rodentia-Murinae). Chromosoma 18:188–200 (1966).

Matthey R: The chromosome formulae of eutherian mammals, in Chiarelli AB, Capanna E (eds): Cytotaxonomy and Vertebrate Evolution, pp 531–616 (Academic Press, London 1973).

Maxson LR, Jameson DL: A comparison of the chromosomes of *Hyla regilla* and *Hyla californiae*. Copeia 1968:704–707 (1968).

Maxson LR, Wilson AC: Albumin evolution and organismal evolution in tree frogs (Hylidae). Syst Zool 24:1–15 (1975).

Mazik EY, Kadyrova BK, Tokotosunov AT: Osobennosti kariotipa zelenoi zhaby (*Bufo viridis*) v Kirgizii. Zool Zhurn 55:1740–1742 (1976).

McAllister BF, Werren JH: Hybrid origin of a B chromosome (PSR) in the parasitic wasp *Nasonia vitripennis*. Chromosoma 106:243–253 (1997).

McKay RDG, Bobrow M, Cooke HJ: The identification of a repeated DNA sequence involved in the karyotype polymorphism of the human Y chromosome. Cytogenet Cell Genet 21:19–32 (1978).

Mecham J: Introgressive hybridization between two southeastern treefrogs. Evolution 14: 445–457 (1960).

Medeiros LR, Rossa-Feres DC, Recco-Pimentel SM: Chromosomal differentiation of *Hyla nana* and *Hyla sanborni* (Anura, Hylidae) with a description of NOR polymorphism in *H. nana*. J Hered 94:149–154 (2003).

Medeiros LR, Rossa-Feres DC, Jim J, Recco-Pimentel SM: B-chromosomes in two Brazilian populations of *Dendropsophus nanus* (Anura, Hylidae). Genet Mol Biol 29:257–262 (2006).

Medeiros LR, Lourenço LB, Rossa-Feres DC, Lima AP, Andrade GV, et al: Comparative cytogenetic analysis of some species of the *Dendropsophus microcephalus* group (Anura, Hylidae) in the light of phylogenetic inferences. BMC Genetics 14:59 (2013).

Medrano L, Bernardi G, Couturier J, Dutrillaux B, Bernardi G: Chromosome banding and genome compartmentalization in fishes. Chromosoma 96:178–183 (1988).

Melli M, Ginelli E, Corneo G, Di Lernia R: Clustering of the DNA sequences complementary to repetitive nuclear RNA of Hela cells. J Mol Biol 93:23–38 (1975).

Meltzer PS, Guan XY, Trent JM: Telomere capture stabilizes chromosome breakage. Nat Genet 4:252–255 (1993).

Mengden GA: Linear differentiation of the C-band pattern of the W chromosome in snakes and birds. Chromosoma 83:275–287 (1981).

Mengden GA, Stock AD: Chromosomal evolution in Serpentes: a comparison of G and C chromosome banding patterns of some colubrid and boid genera. Chromosoma 79:53–64 (1980).

Menzies J: The frogs of New Guinea and the Solomon Islands (Pensoft, Sofia 2006).
Menzies JI, Tippett J: Chromosome numbers of Papuan hylid frogs and the karyotype of *Litoria infrafrenata* (Amphibia, Anura, Hylidae). J Herpetol 10:167–173 (1976).
Mercadal de Barrio IT: *Ceratophrys joazeirensis* sp.n. (Ceratophryidae, Anura) del noreste de Brasil. Amphibia-Reptilia 7:313–334 (1986).
Merry DE, Pathak S, VandeBerg JL: Differential NOR activities in somatic and germ cells of *Monodelphis domestica* (Marsupialia, Mammalia). Cytogenet Cell Genet 35:244–251 (1983).
Mestriner CA, Galetti PM, Valentini SR, Ruiz IRG, Abel LDS: Structural and functional evidence that a B chromosome in the Characidae fish *Astyanax scabripinnis* is an isochromosome. Heredity 85:1–9 (2000).
Meyer E, Jamieson BGM, Scheltinga DM: Sperm ultrastructure of six Australian hylid frogs from two genera (*Litoria* and *Cyclorana*): phylogenetic implications. J Submicrosc Cytol Pathol 29:443–451 (1997).
Meyne J, Baker RJ, Hobart HH, Hsu TC, Ryder OA, et al: Distribution of non-telomeric sites of the $(TTAGGG)_n$ telomeric sequence in vertebrate chromosomes. Chromosoma 99: 3–10 (1990).
Middleton EM, Herman JR, Celarier EA, Wilkinson JW, Carey C, Ruskin RJ: Evaluating ultraviolet radiation exposure with satellite data at sites of amphibian declines in Central and South America. Conserv Biol 15: 914–929 (2001).
Milani M, Cassini CS, Recco-Pimentel SM, Lourenço LB: Karyotypic data detect interpopulational variation in *Physalaemus olfersii* and the first case of supernumerary chromosomes in the genus. Animal Biol J 2:21–28 (2011).
Miller DA, Dev VG, Tantravahi R, Miller OJ: Suppression of human nucleolus organizer activity in mouse-human somatic hybrid cells. Exp Cell Res 101:235–243 (1976).
Miller MA, Pfeiffer W, Schwartz T: Creating the CIPRES Science Gateway for inference of large phylogenetic trees, in: Proceedings of the Gateway Computing Environments Workshop (GCE) (IEEE, New Orleans, 2010). http://dx.doi.org/10.1109/GCE.2010.5676129.
Miller OJ, Miller DA, Dev VG, Tantravahi R, Croce CM: Expression of human and suppression of mouse nucleolus organizer activity in mouse-human somatic cell hybrids. Proc Natl Acad Sci USA 73:4531–4535 (1976).
Mittermeier RA, Myers N, Thompsen JG, da Fonseca GAB, Olivieri S: Biodiversity hotspots and major wilderness areas: approaches to setting conservation priorities. Conserv Biol 12:56–520 (1998).
Miura I: Sex chromosome differentiation in the Japanese brown frog, *Rana japonica*. I. Sex-related heteromorphism of the distribution pattern of constitutive heterochromatin in chromosome No. 4 of the Wakuya population. Zool Sci 11:797–806 (1994).
Miura I: The late replication banding patterns of chromosomes are highly conserved in the genera *Rana*, *Hyla* and *Bufo* (Amphibia, Anura). Chromosoma 103:567–574 (1995a).
Miura I: Two differentiated groups of the Japanese toad *Bufo bufo japonicus*, demonstrated by C-banding analysis of chromosomes. Caryologia 48:123–136 (1995b).
Miura I, Ohtani H, Hanada H, Ichikawa Y, Kashiwagi A, Nakamura M: Evidence for two successive pericentric inversions in sex lampbrush chromosomes of *Rana rugosa* (Anura: Ranidae). Chromosoma 106:178–182 (1997).
Mizuno S, Macgregor HC: Chromosomes, DNA sequences, and evolution in salamanders of the genus *Plethodon*. Chromosoma 48:239–296 (1974).
Molinari de Bastiani VI, Lucas EM: Anuran diversity (Amphibia, Anura) in a seasonal forest fragment in southern Brazil. Biota Neotrop 13:255–264 (2013).
Montiel EE, Cabrero J, Camacho JPM, López-León MD: *Gypsy*, *RTE* and *Mariner* transposable elements populate *Eyprepocnemis plorans* genome. Genetica 140:365–374 (2012).
Moore JA: Abnormal combinations of nuclear and cytoplasmic systems in frogs and toads, in Demerec M (ed): Advances in Genetics, vol VII, pp 139–182 (Academic Press, New York 1955).
Morand M, Hernando A: Cariotipo y región organizadora del nucléolo en *Argenteohyla siemersi pederseni* (Anura: Hylidae). Facultad de Ciencias Exactas y Naturales y Agrimensura 12:141–144 (1996).
Morand M, Hernando A: Localización cromosómica de genes ribosomales activos en *Phyllomedusa hypochondrialis* y *P. sauvagii* (Anura: Hylidae). Cuadernos Herpetología 11:31–36 (1997).
Morescalchi A: Conferma della presenza di eterocromosomi in *Xenopus laevis* Daudin. Rend Accad Sci Fis Mat Soc Naz Sci Lett Arti 30:310–314 (1963).
Morescalchi A: Il corredo cromosomico di *Discoglossus pictus* Otths: cromosomi sessuali, spiralizzazione cromosomica e zone eteroctomatiche. Caryologia 17:317–345 (1964).
Morescalchi A: Il corredo cromosomico di *Hyla arborea arborea* (L.) e di *Hyla regilla* (Baird and Girard): il problema dei cromosomi sessuali e l'evoluzione del cariotipo di *Hyla*. Caryologia 18:193–206 (1965).
Morescalchi A: Initial cytotaxonomic data on certain families of amphibious Anura (Diplasiocoela, after Noble). Experientia 24: 280–283 (1968a).
Morescalchi A: The karyotype of two specimens of *Leiopelma hochstetteri* Fitz. (Amphibia, Salientia). Caryologia 22:97–102 (1968b).
Morescalchi A: Some trends of karyological evolution in the Anura. Act IV Congr Latin Zool 1:159–166 (1970).
Morescalchi A: Comparative karyology of the Amphibia. Boll Zool 38:317–320 (1971).
Morescalchi A: Amphibia, in Chiarelli AB, Capanna E (eds): Cytotaxonomy and Vertebrate Evolution, pp 233–348 (Academic Press, London 1973).
Morescalchi A: New developments in vertebrate cytotaxonomy. I. Cytotaxonomy of the amphibians. Genetica 50:179–193 (1979).
Morescalchi A, Galgano M: Meiotic chromosomes and their taxonomic value in Amphibia Anura. Caldasia 11:41–50 (1973).
Morescalchi A, Ingram GJ: New chromosome numbers in Australian Leptodactylidae (Amphibia, Salientia). Experientia 30:1134–1135 (1974).
Morescalchi A, Olmo E: Single-copy DNA and vertebrate phylogeny. Cytogenet Cell Genet 34:93–101 (1982).
Morescalchi A, Serra V: DNA renaturation kinetics in some paedogenetic urodeles. Experientia 30:487–489 (1974).
Morgan GT, Macgregor HC, Colman A: Multiple ribosomal gene sites revealed by in situ hybridization of *Xenopus* rDNA to *Triturus* lampbrush chromosomes. Chromosoma 80: 309–330 (1980).
Moritz C: The evolution of a highly variable sex chromosome in *Gehyra purpurascens* (Gekkonidae). Chromosoma 90:111–119 (1984).
Moriwaki T: Studies on matured parthogenetic frogs. I. The development and reproductive ability. J Sci Hiroshima Univ Ser B 17:13–32 (1959).
Moroi Y, Peebles C, Fritzler MJ, Steigerwald J, Tan EM: Autoantibody to centromere (kinetochore) in scleroderma sera. Proc Natl Acad Sci USA 77:1627–1631 (1980).
Moura MR, Motta AP, Feio RN: An unusual reproductive mode in *Hypsiboas* (Anura: Hylidae). Zool 28:142–144 (2011).
Moya IM, Alarcón I, del Pino EM: Gastrulation of *Gastrotheca riobambae* in comparison with other frogs. Dev Biol 304:467–478 (2007).
Nali RC, Prado CPA: Habitat use, reproductive traits and social interactions in a stream-dweller treefrog endemic to the Brazilian Cerrado. Amphibia-Reptilia 33:337–347 (2012).
Nali RC, Faivovich J, Prado CPA: The occurrence of unpigmented mature oocytes in *Hypsiboas* (Anura: Hylidae). Salamandra 50: 53–56 (2014).
Nanda I, Karl E, Volobouev V, Griffin DK, Schartl M, Schmid M: Extensive gross genomic rearrangements between chicken and Old World vultures (Falconiformes: Accipitridae). Cytogenet Genome Res 112:286–295 (2006).
Nanda I, Karl E, Griffin DK, Schartl M, Schmid M: Chromosome repatterning in three representative parrots (Psittaciformes) inferred from comparative chromosome painting. Cytogenet Genome Res 117:43–53 (2007).
Nanda I, Fugate M, Steinlein C, Schmid M: Distribution of $(TTAGGG)_n$ telomeric sequences in karyotypes of the *Xenopus* species complex. Cytogenet Genome Res 122:396–400 (2008).
Nascimento J, Quinderé YRSD, Recco-Pimentel SM, Lima JRF, Lourenço LB: Heteromorphic Z and W sex chromosomes in *Physalaemus ephippifer* (Steindachner, 1864) (Anura, Leiuperidae). Genetica 138:1127–1132 (2010).
Natarajan AT, Gropp A: The meiotic behaviour of autosomal heterochromatic segments in hedgehogs. Chromosoma 35:143–152 (1971).

Nelsen OE: Comparative Embryology of the Vertebrates (McGraw Hill, London 1953).

Neri TS, Medrado AS, Oliveira RV, Freire IV, Costa MA, Siqueira S: Estudo citogenético de três espécies de *Scinax* (Anura, Hylidae) no municipio de Jequié-BA. Resumos do 55° Congresso Brasileiro de Genetica, Águas de Lindóia, São Paulo, p 112 (2009).

Neumann AA, Reddel RR: Telomere maintenance and cancer – look, no telomerase. Nat Rev Cancer 2:879–884 (2002).

Nevo E, Bar-El H: Hybridization and speciation in fossorial mole rats. Evolution 30:831–840 (1976).

Niebuhr E: Dicentric and monocentric Robertsonian translocations in man. Humangenetik 16:217–226 (1972).

Nikolov EN, Dabeva MD, Nikolov TK: Turnover of ribosomes in regenerating rat liver. Int J Biochem 15:1255–1260 (1983).

Ning Y, Liang JC, Nagarajan J, Schröck E, Ried T: Characterization of 5q deletions by subtelomeric probes and spectral karyotyping. Cancer Genet Cytogenet 103:170–172 (1998).

Nishioka M, Hanada H: Sex of reciprocal hybrids between the Murakami (ZZ-ZW type) population and Hamakita (XX-XY type) population in *Rana rugosa*. Sci Rep Lab Amphibian Biol Hiroshima Univ 13:35–50 (1994).

Nishioka M, Miura I, Saitoh K: Sex chromosomes of *Rana rugosa* with special reference to local differences in sex determining mechanism. Sci Rep Lab Amphibian Biol Hiroshima Univ 12:55–81 (1993a).

Nishioka M, Kodama Y, Sumida M, Ryuzaki M: Systematic evolution of 40 populations of *Rana rugosa* distributed in Japan elucidated by electrophoresis. Sci Rep Lab Amphibian Biol Hiroshima Univ 12:83–131 (1993b).

Nishioka M, Hanada H, Miura I, Ryuzaki M: Four kinds of sex chromosomes in *Rana rugosa*. Sci Rep Lab Amphibian Biol Hiroshima Univ 13:1–34 (1994).

Noble GK: The Biology of the Amphibia (Dover Publishing, New York 1931).

Nogueira L, Paim F, Diniz D, Solé M, Affonso P, et al: Cytogenetic analysis of *Scinax auratus* and *Scinax eurydice* (Anura, Hylidae) with emphasis on cytotaxonomy. Comp Cytogenet 9:227–236 (2015a).

Nogueira L, Zanoni JB, Solé M, de Mello Affonso PRA, et al: Cytogenetic studies in six species of *Scinax* (Anura, Hylidae) clade *Scinax ruber* from northern and northeastern Brazil. Genet Mol Biol 38:156–161 (2015b).

Nordgren H, Stenram U: Decreased half-life of the RNA of free and membrane-bound ribosomes in the liver of protein-deprived rats. Hoppe-Seyler's Z Physiol Chem 353:1832–1836 (1972).

Nunes RRA: Citogenética de anfíbios da familia Hylidae do Espírito Santo. Master thesis, Universidade Federal do Espírito Santo, Vitória (2006).

Nunes RRA, Fagundes V: Patterns of ribosomal DNA distribution in hylid frogs from the *Hypsiboas faber* and *H. semilineatus* groups. Genet Mol Biol 31:982–987 (2008a).

Nunes RRA, Fagundes V: Cariótipos de oito espécies de anfíbios das subfamílias Hylinae e Phyllomedusinae (Anura, Hylidae) do Espírito Santo, Brasil. Bol Mus Biol Mello Leitão 23:21–36 (2008b).

Nunes RRA, Muhl EL, Fagundes V: Comparative cytogenetics of three species of *Dendrosophus* (Anura, Hylidae, Dendrosophini) showed species-specific karyotypes. 2ª Reunião Brasileira de Citogenética, Águas de Lindóia, São Paulo, CA08 (2011).

Nur U, Brett BLH: Genotypes suppressing meiotic drive of a B chromosome in the mealy bug *Pseudococcus obscurus*. Genetics 110: 73–92 (1985).

Nur U, Brett BLH: Control of meiotic drive of B chromosomes in the mealybug *Pseudococcus affinis (obscurus)*. Genetics 115:499–510 (1987).

Nur U, Brett BLH: Genotypes affecting the condensation and transmission of heterochromatic B-chromosomes in the mealybug *Pseudococcus affinis*. Chromosoma 96:205–212 (1988).

Nur U, Nevo E: Supernumerary chromosomes in the cricket frog *Acris crepitans*. Caryologia 22:97–102 (1969).

Ochs RL, Busch H: Further evidence that phosphoprotein C23 (110 kD/pI 5.1) is the nucleolar silver staining protein. Exp Cell Res 152: 260–265 (1984).

Ochs RL, Lischwe M, O'Leary P, Busch H: Localization of nucleolar phosphoproteins B23 and C23 during mitosis. Exp Cell Res 146: 139–149 (1983).

Oda FH, Batista VG, Gambale PG, Mise FT, de Souza F, et al: Anuran species richness, composition, and breeding habitat preferences: a comparison between forest remnants and agricultural landscapes in Southern Brazil. Zool Stud 55:1–14 (2016).

Odierna G, Aprea G, Capriglione T, Parisi P, Arribas O, Morescalchi MA: Chromosomal and molecular analysis of some repeated DNA families in *Discoglossus* Otth, 1837 (Anura, Discoglossidae): taxonomic and phylogenetic implications. Ital J Zool 66: 273–283 (1999).

Odierna G, Aprea G, Capriglione T, Castellano S, Balleto E: Cytological evidence for population-specific sex chromosome heteromorphism in Palaearctic green toads (Amphibia. Anura). Indian Acad Sci 32:763–768 (2007).

Ogielska M, Kotusz A: Pattern and rate of ovary differentiation with reference to somatic development in anuran amphibians. J Morphol 259:41–54 (2004).

Ogielska M, Kotusz A, Augustyńska R, Ihnatowicz J, Paśko L: A stockpile of ova in the grass frog *Rana temporaria* is established once for the life span. Do ovaries in amphibians and in mammals follow the same evolutionary strategy? Anat Rec 296:638–653 (2013).

Oguma K, Makino S: A new list of the chromosome numbers in Vertebrata. J Facul Sci Hokkaido Imperial Univ Ser VI Zool 5:297–356 (1937).

Ohno S, Beçak W, Beçak ML: X-autosome ratio and the behaviour pattern of individual X-chromosomes in placental mammals. Chromosoma 15:14–30 (1964).

Ohno S, Weiler C, Poole J, Christian L, Stenius C: Autosomal polymorphism due to pericentric inversions in the deer mouse (*Peromyscus maniculatus*) and some evidence of somatic segregation. Chromosoma 18:177–187 (1966).

Ohta S: Sex determining mechanism in *Buergeria buergeri* (Schlegel). I. Heterozygosity of chromosome pair No. 7 in the female. Sci Rep Lab Amphibian Biol Hiroshima Univ 8: 29–43 (1986).

Ohta S: Sex determining mechanism in *Buergeria buergeri* (Schlegel). II. The effects of sex hormones on the differentiation of gonads and the offspring of sex-reversed females. Sci Rep Lab Amphibian Biol Hiroshima Univ 9: 213–238 (1987).

Ohta S, Sumida M, Nishioka M: Sex-determining mechanism in *Buergeria buergeri* (Anura, Rhacophoridae). III. Does the ZZW triploid frog become female or male? J Exp Zool 283:295–306 (1999).

Oko R, Sutovsky P: Biogenesis of sperm perinuclear theca and its role in sperm functional competence and fertilization. J Reprod Immunol 83:2–7 (2009).

Oliveira C, Vicentini CA, Taboga S: Structural characterization of nuclear phenotypes during *Scinax fuscovarius* spermatogenesis (Anura, Hylidae). Caryologia 56:75–83 (2003).

Oliveira ME, Paillete M, Rosa HD, Crespo EG: A natural hybrid between *Hyla arborea* and *Hyla meridionalis* detected by mating calls. Amphibia-Reptilia 12:15–20 (1991).

Olmo E: Quantitative variations in the nuclear DNA and phylogenesis of the Amphibia. Caryologia 26:43–68 (1973).

Olmo E: Nucleotype and cell size in vertebrates. A review. Basic Appl Histochem 27:227–256 (1983).

Olmo E: Reptilia, in John B (ed): Animal Cytogenetics, vol 4/3 (Gebrüder Borntraeger, Berlin 1986).

Olmo E (ed): Cytogenetics of Amphibians and Reptiles (Birkhäuser, Basel 1990).

Olmo E, Morescalchi A, Stingo V, Odierna G: Genome characteristics and the systematics of the Discoglossidae (Amphibia, Salientia). Monit Zool Ital 16:283–299 (1982).

Olofsson B, Bernardi G: Organization of nucleotide sequences in the chicken genome. Eur J Biochem 130:241–245 (1983).

Orlova VF, Uteshev VK: The tetraploid toad of the *Bufo viridis* group from Dzungarian Gobi, Mongolia, in Rocek Z (ed): Proceed 6th Ord Gen Meet Soc Eur Herp, pp 143–146 (Prague, 1986).

Padial JM, Grant T, Frost DR: Molecular systematics of terraranans (Anura: Brachycephaloidea) with an assessment of the effects of alignment and optimality criteria. Zootaxa 3825:1–132 (2014).

Page SL, Shaffer LG: Chromosome stability is maintained by short intercentromeric distance in functionally dicentric human Robertsonian translocations. Chromosome Res 6:115–122 (1998).

Pahkala M, Rasanen K, Laurila A, Johanson U, Bjorn LO, Merila J: Lethal and sublethal effects of UV-B/pH synergism on common frogs. Conserv Biol 16:1063–1073 (2002).

Paiva CR, Nascimento J, Silva APZ, Bernarde PS, Ananias F: Karyotypes and Ag-NORs in *Phyllomedusa camba* De la Riva, 1999 and *P. rohdei* Mertens 1926 (Anura, Hylidae, Phyllomedusinae): cytotaxonomic considerations. Ital J Zool 77:116–121 (2010).

Pallister PD, Patau K, Inhorn SL, Opitz JM: A woman with multiple congenital anomalies, mental retardation and mosaicism for an unusual translocation chromosome t(6;19). Int J Clin Genet 5:188–195 (1974).

Pansanota A, Mott T, Strüssmann C: Anuran amphibians' diversity in a northwestern area of the Brazilian Pantanal. Biota Neotrop 11: 77–86 (2011).

Pardue ML: Localization of repeated DNA sequences in *Xenopus* chromosomes. Cold Spring Harbor Symp Quant Biol 38:475–482 (1974).

Park EH, Grimm H: Distribution of C-band heterochromatin in the ZW sex chromosomes of European and American eels (Anguillidae, Teleostomi). Cytogenet Cell Genet 31: 167–174 (1981).

Parker JS, Ainsworth CC, Taylor S: The B-chromosome system of *Hypochoeris maculata*. II. B-effects on meiotic A-behaviour. Chromosoma 67:123–143 (1981).

Parmelee JR: Trophic ecology of a tropical anuran assemblage. Sci Pap Nat Hist Mus Univ Kansas 1:1–59 (1999).

Patawang I, Tanomtong A, Phimphan S, Chuaynkern V, Chuaynkern C, et al: The identification of sex chromosomes and karyological analysis of rice frog, *Fejervarya limnocharis* from northeast Thailand. Cytologia 79:141–150 (2014).

Pathak S, Hsu TC, Arrighi FE: Chromosomes of *Peromyscus* (Rodentia, Cricetidae). IV. The role of heterochromatin in karyotypic evolution. Cytogenet Cell Genet 12:315–326 (1973).

Pathak S, Van Tuinen P, Merry DE: Heterochromatin, synaptonemal complex, and NOR activity in the somatic and germ cells of a male domestic dog, *Canis familiaris* (Mammalia, Canidae). Cytogenet Cell Genet 34:112–118 (1982).

Patterson JT, Stone WS: Evolution in the Genus *Drosophila* (Macmillan, New York 1952).

Patton JL, Gardner AL: Parallel evolution of multiple sex-chromosome systems in the phyllostomatid bats, *Carollia* and *Choeroniscus*. Experientia 27:106–106 (1971).

Pauls E, Bertollo LA: Evidence for a system of supernumerary chromosomes in *Prochilotus scrofa* Steindachner, 1981 (Pisces, Prochilodontidae). Caryologia 36:307–314 (1983).

Pederassi J, Lima MSCS, Peixoto O, Souza CAS: The choice of bromeliads as a microhabitat by *Scinax argyreornatus* (Anura, Hylidae). Braz J Biol 72:229–233 (2012).

Peixoto MAA, Lacerda JVA, Coelho-Augusto C, Feio RN, Dergam JA: The karyotypes of five species of the *Scinax perpusillus* group (Amphibia, Anura, Hylidae) of southeastern Brazil show high levels of chromosomal stabilization in this taxon. Genetica 143:729–739 (2015).

Peixoto MAA, Oliveira MPC, Feio RN, Dergam JA: Karyological study of *Ololygon tripui* (Lourenço, Nascimento and Pires, 2009), (Anura, Hylidae) with comments on chromosomal traits among populations. Comp Cytogenet 10:505–516 (2016).

Peixoto OL: Caracterização do grupo perpusilla e revalidação da posição taxonômica de *Ololygon perpusilla perpusilla* e *Ololygon perpusilla v-signata* (Amphibia, Anura, Hylidae). Arq da Univ Federal Rural do Rio Janeiro 10:37–49 (1987).

Pengilly D, Jarrell GH, MacDonald SO: Banded karyotypes of *Peromyscus sitkensis* from Baranof Island, Alaska. J Mammal 64:682–685 (1983).

Pepper DA: The mammalian kinetochore, in Verma RS (ed): Heterochromatin. Molecular and Structural Aspects, pp 187–202 (Cambridge University Press, Cambridge 1988).

Pepper DA, Brinkley BR: Localization of tubulin in the mitotic apparatus of mammalian cells by immunofluorescence and immunoelectron microscopy. Chromosoma 60:223–235 (1977).

Pera F: Pattern of repetitve DNA of the chromosomes of *Microtus agrestis*. Chromosoma 36: 263–271 (1972).

Peres CA: Paving the way to the future of Amazonia. Trends Ecol Evol 16:217–219 (2001).

Perrin N: Sex reversal: a fountain of youth for sex chromosomes. Evolution 63:3043–3049 (2009).

Pigozzi MI, Solari AJ: Germ cell restriction and regular transmission of an accessory chromosome that mimics a sex body in the zebra finch *Taenopygia guttata*. Chromosome Res 6:105–113 (1998).

Pimm SL, Raven P: Extinction by numbers. Nature 403:843–845 (2000).

Pinto-Erazo MA, Goldberg J, Jerez A: Gonadal development in the neotropical high Andean frog *Dendropsophus labialis* (Amphibia : Hylidae). Cuad Herpetol 30:57–68 (2016).

Pinto-Viveros MA, Garcia-Durán DM, Mano-Cuellar K, Pinto-Ledezma JN: Listado de los anfibios y reptiles de la propiedad Juan Deriba, Santa Cruz, Bolivia. Kempffiana 11:70–75 (2015).

Piotrowski JS, Annis SL, Longcore JE: Physiology of *Batrachochytrium dendrobatidis*. A chytrid pathogen of amphibians. Mycologia 96:9–15 (2004).

Pisanets EM: O novom poliploidnom vide zhab Bufo *danatensis* Pisanets sp. n. iz. Turkmenii. Doklady Akad Nauk Ukrainskoi SSR Ser B 3:280–284 (1978).

Pitnick S, Hosken DJ, Birkhead T: Sperm morphological diversity, in Birkhead T, Hosken DJ, Pitnick S (eds): Sperm Biology : An Evolutionary Perspective, pp 69–119 (Academic Press, New York 2009).

Pizon V, Cuny G, Bernardi G: Nucleotide sequence organization in the very small genome of a tetraodontid fish *Arothron diadematus*. Eur J Biochem 140:25–30 (1984).

Pluta AF, Mackay AM, Ainsztein AM, Goldberg IG, Earnshaw WC: The centromere: hub of chromosomal activities. Science 270:1591–1594 (1995).

Pombal JP, Haddad CFB: Espécies de *Phyllomedusa* do grupo *burmeisteri* do Brasil oriental, com descrição de uma espécie nova (Amphibia, Hylidae). Rev Brasil Biol 52:217–229 (1992).

Pombal JP, Haddad CFB, Kasahara S: A new species of *Scinax* (Anura: Hylidae) from southeastern Brazil, with comments on the genus. J Herpetol 29:1–6 (1995).

Ponse K: La proportion sexuelle dans la descendance tissue des œufs produits par l'organe de Bidder de craupauds femelles (Note preliminaire). Rev Suisse Zool 48:541–544 (1941).

Ponse K: Sur la digametie du crapaud hermaphrodite. Rev Suisse Zool 49:185–189 (1942).

Ponse K: La génétique du chez les batraciens avec un aperçu des travaux de RR Humphrey. Arch d'Anat Micr Morph Exp 39:183–214 (1950).

Portela-Castro ALB, Júlio HF, Nishiyama PB: New occurrence of microchromosomes B in *Moenkhausia sanctaefilomenae* (Pisces, Characidae) from the Paran river of Brazil: analysis of the synaptonemal complex. Genetica 110:277–283 (2001).

Pounds JA: Climate and amphibian declines. Nature 410:639–640 (2001).

Pounds JA, Crump ML: Amphibian declines and climate disturbance: the case of the golden toad and the harlequin frog. Conserv Biol 8: 72–85 (1994).

Pounds JA, Fogden MP, Savage JM, Gorman GC: Test of null models for amphibian declines on a tropical mountain. Conserv Biol 11: 1307–1322 (1997).

Pounds JA, Fogden MP, Campbell JH: Biological response to climate change on a tropical mountain. Nature 398:611–615 (1999).

Pounds JA, Bustamante MR, Coloma LA, Consuegra JA, Fogden MPL, et al: Widespread amphibian extinctions from epidemic disease driven by global warming. Nature 439: 161–167 (2006).

Prado CP, Uetanabaro M, Haddad CFB: Breeding activity patterns, reproductive modes, and habitat use by anurans (Amphibia) in a seasonal environment in the Pantanal, Brazil. Amphibia-Reptilia 26:211–221 (2005).

Pritts GR, Milller BT: Identification of Gray treefrog populations in middle Tennessee: chromosome counts from squashed tadpole tail tips. J Tenn Acad Sci 70:65–67 (1976).

Prokofieva AA: Morphologische Struktur der Chromosomen von *Drosophila melanogaster*. Z Zellforsch 22:225–262 (1935a).

Prokofieva AA: The structure of the chromocenter. Cytologia 6:438–443 (1935b).

Prokofieva-Belgovskaya AA: The inert region in the subterminal part of the X-chromosome of *Drosophila melanogaster*. Bull Acad Sci URSS Ser Biol 4:97–103 (1938).

Ptacek MB, Gerhardt HC, Sage RD: Speciation by polyploidy in tree frogs: multiple origins of the tetraploid, *Hyla versicolor*. Evolution 48:898–908 (1994).

Pyburn WF: Breeding behavior of the leaf frogs *Phyllomedusa callidryas* and *Phyllomedusa dacnicolor* in Mexico. Copeia 2:209–218 (1970).

Pyburn WF, Kennedy JP: Hybridization of U.S. treefrogs of the genus *Hyla*. Proc Biol Soc Washington 74:157–160 (1961).

Pyron A, Wiens JJ: A large-scale phylogeny of Amphibia including over 2800 species, and a revised classification of extant frogs, salamanders, and caecilians. Mol Phylogenet Evol 61:543–583 (2011).

Pyron RA, Burbrink FT, Colli GR, Montes de Oca AN, Vitt J, et al: The phylogeny of advanced snakes (Colubroidea), with discovery of a new subfamily and comparison of support methods for likelihood trees. Mol Phyl Evol 58:329-342 (2011).

Pyron RA, Burbrink FT, Wiens JJ: A phylogeny and revised classification of Squamata, including 4161 species of lizards and snakes. BMC Evol Biol 13:93 (2013).

Qing LY, Xia Y, Zheng YC, Zeng XM: A de novo case of floating chromosomal polymorphism by translocation in *Quasipaa boulengeri* (Anura, Ranidae). PLoS One 7:e46163 (2012).

Qing LY, Xia Y, Zheng YC, Zeng XM: FISH of 5S and telomeric (TTAGnsGG) repeats in normal and translocated populations of the frog *Quasipaa boulengeri* (Anura, Ranidae). Chinese Sci Bull 58:2168–2173 (2013).

Rabello MN: Chromosomal studies in Brazilian anurans. Caryologia 23:45–59 (1970).

Rabello MN, Beçak ML, Beçak W: Contribuição à citotaxonomia da familia Hylidae. Arquivos do Museu Nacional 54:285–286 (1971).

Raber SC, Carvalho KA, Garcia PCA, Vinciprova G, Recco-Pimentel SM: Chromosomal characterization of *Hyla bischoffi* and *Hyla guentheri* (Anura, Hylidae). Phyllomedusa 3: 43–49 (2004).

Rafferty KA, Sherwin RW: The length of secondary chromosomal constrictions in normal individuals and in a nucleolar mutant of *Xenopus laevis*. Cytogenetics 8:427–438 (1969).

Ralin DB: Genetic compatibility and a phylogeny of the temperate North American hylid fauna. PhD Dissertation, University of Texas, Austin (1970).

Ralin DB, Selander RK: Evolutionary genetics of diploid-tetraploid species of tree frogs of the genus *Hyla*. Evolution 33:595–608 (1979).

Rand AS, Myers CW: The herpetofauna of Barro Colorado Island, Panama: an ecological summary, in Gentry AH (ed): Four Neotropical Rainforests, pp 386–409 (Yale University Press, New Haven 1990).

Rastogi RK, Bagnara JT, Iela L, Krasovich M: Reproduction in the Mexican leaf frog, *Pachymedusa dacnicolor*. J Morphol 197:277–302 (1988).

Rattner JB: Organization within the mammalian kinetochore. Chromosoma 93:515–520 (1986).

Ray-Chaudhuri R: Cytotaxonomy and chromosome evolution in birds, in Chiarelli AB, Capanna E (eds): Cytotaxonomy and Vertebrate Evolution, pp 424–483 (Academic Press, London 1973).

Ray-Chaudhuri SP, Singh L, Sharma T: Evolution of sex-chromosomes and formation of W-chromatin in snakes. Chromosoma 33: 239–251 (1971).

Reading CJ: Linking global warming to amphibian declines through its effects on female body condition and survivorship. Oecologia 151:125–131 (2007).

Reeder AL, Foley GL, Nichols DK, Hansen LG, Wikoff B, et al: Forms and prevalence of intersexuality and effects of environmental contaminants on sexuality in cricket frogs (*Acris crepitans*). Environ Health Perspect 106:261–266 (1998).

Reig OA, Olivo N, Kiblisky P: The idiogram of the Venezuelan vole mouse, *Akodon urichi venezuelensis* Allen (Rodentia, Cricetidae). Cytogenetics 10:99–114 (1971).

Rengel D, Pisanó A, Lavilla E: Diferenciacion sexual de *Phyllomedusa boliviana* (Anura. Hylidae). Cuad Herpetol 9:15–20 (1995).

Rens W, O'Brien PCM, Yang F, Solanky N, Perelman P, et al: Karyotype relationships between distantly related marsupials from South America and Australia. Chromosome Res 9:301–308 (2001).

Rens W, O'Brien PCM, Fairclough H, Harman L, Graves JAM, Ferguson-Smith MA: Reversal and convergence in marsupial chromosome evolution. Cytogenet Genome Res 102:282–290 (2003).

Rens W, Grützner F, O'Brien PCM, Fairclough H, Graves JAM, Ferguson-Smith MA: Resolution and evolution of the duck-billed platypus karyotype with an X1Y1X2Y2X3Y3X4Y4X5Y5 male sex chromosome constitution. Proc Natl Acad Sci USA 101:16257–16261 (2004).

Rens W, Fu B, O'Brien PCM, Ferguson-Smith MA: Cross-species chromosome painting. Nat Protoc 1:783–790 (2006).

Richards CM, Nace GW: Gynogenetic and hormonal sex reversal used in tests of the XX-XY hypothesis of sex determination in *Rana pipiens*. Growth 42:319–331 (1978).

Rieder CL: The formation, structure and composition of the mammalian kinetochore and kinetochore fiber. Int Rev Cytol 79:1–58 (1982).

Rivera H, Solé MT, García-Cruz D, Martínez-Wilson M, Cantú JM: On telomere replication and fusion in eukaryotes: apropos of a case of 45,X/46,X,ter rea(X;X)(p22.3;p22.3). Cytogenet Cell Genet 38:23–28 (1984).

Rodrigues DJ, Uetanabaro M, Lopes FS: Breeding biology of *Phyllomedusa azurea* Cope, 1862 and *P. sauvagii* Boulenger, 1882 (Anura) from the Cerrado, Central Brazil. J Nat Hist 41:1841–1851 (2007).

Rodríguez LO, Cadle JE: A preliminary overview of the herpetofauna of Cocha Cashu, Manu National Park, Peru, in Gentry AH (ed): Four Neotropical Rainforests, pp 410–425 (Yale University Press, New Haven 1990).

Rodríguez LO, Duellman WE: Guide to the frogs of the Iquitos Region, Amazonian Ecuador. Univ Kans Nat Hist Mus Spec Pub 22:1–80 (1994).

Rodríguez-Gómez Y, Rodríguez-Ortiz R, Sanz-Ochotorena A, Casas-Guillama L, Fernández-Jimenez I, Estupiñán-Díaz B: Desarrollo folicular en la rana platanera *Osteopilus septentrionalis* (Anura: Hylidae). Morfovirtual 2014 1:1–9 (2014).

Romera F, Jimenez MM, Puertas MJ: Genetic control of the rate of transmission of rye B-chromosomes. I. Effects in 2B × 0B crosses. Heredity 66:61–65 (1991).

Ron SR, Merino A: Declinación de anfibios del Ecuador: información general y primer reporte de chytridiomycosis para Sudamérica. Froglog 42:2–3 (2000).

Ron SR, Duellman WE, Coloma LA, Bustamante MR: Population decline of the Jambato toad *Atelopus ignescens* (Anura, Bufonidae) in the Andes of Ecuador. J Herpetol 37:117–126 (2003).

Rosa C, Aguiar O, Giaretta AA, Recco-Pimentel SM: Karyotypic variation in *Megaelosia* (Anura, Hylodinae) with the first description of a B-chromosome in a leptodactylid frog. Copeia 2003:166–174 (2003).

Rosset SD, Baldo D, Lanzone C, Basso NG: Review of the geographic distribution of diploid and tetraploid populations of the *Odontophrynus americanus* species complex (Anura, Leptodactylidae). J Herpetol 40: 464–477 (2006).

Rot-Nikcevic I: Arrested development in *Xenopus laevis* tadpoles: how size constrains metamorphosis. J Exp Biol 207:2133–2145 (2004).

Roth P, Ráb P: Karyotype analysis of the *Bufo viridis*-group: systematic implications, in Rocek Z (ed): Studies in Herpetology: Proceedings of the European Herpetological Meeting (3rd Ordinary General Meeting of the Societas Europaea Herpetologica) Prague 1985, pp 131–136 (Prague, 1986).

Roth P, Ráb P: Sequential chromosome banding studies in the natterjack toad, *Bufo calamita*, in van Gelder JJ, Strijbosch H, Bergers PJM (eds): Proceedings of the 4th Ordinary General Meeting of the Societas Europaea Herpetologica, pp 335–338 (Nijmegen, 1987).

Rovatsos MT, Marchal JA, Romero-Fernández I, Fernández FJ, Giagia-Athanosopoulou EB, Sánchez A: Rapid, independent, and extensive amplification of telomeric repeats in pericentromeric regions in karyotypes of arvicoline rodents. Chromosome Res 19:869–882 (2011).

Rubes J, Musilova P, Mastromonaco GF: Cytogenetics of wild and captive bred non-domestic animals. Cytogenet Genome Res 120:61–68 (2008).

Rugh R: Experimental Embryology. Techniques and Procedures (Burgess Publishers, Minneapolis 1962).

Ruiz IRG, Soma M, Beçak W: Nucleolar organizer regions and constitutive heterochromatin in polyploid species of the genus *Odontophrynus* (Amphibia, Anura). Cytogenet Cell Genet 29:84–98 (1981).

Ruthner U, Golob E: Fusion of the short arms of one X chromosome in a patient with gonadal dysgenesis. Humangenetik 24: 159–160 (1974).

Ryan MJ, Lips KR, Eichholz MW: Decline and extirpation of an endangered Panamanian stream frog population (*Craugastor punctariolus*) due to an outbreak of chytridiomycosis. Biol Conserv 141:1636–1647 (2008).

Ryuzaki M, Hanada H, Okumoto H, Takizawa N, Nishioka M: Evidence for heteromorphic sex chromosomes in males of *Rana tagoi* and *Rana sakuraii* in Nishitama district of Tokyo (Anura: Ranidae). Chromosome Res 7:31–42 (1999).

Ryuzaki M, Nishioka M, Kawamura T: Karyotypes of *Rana tagoi* Okada with diploid number of 28 in the Chausu mountains of the Minamishinshu district of Nagano prefecture, Japan (Anura: Ranidae). Cytogenet Genome Res 114:56–65 (2006).

Saenz-Ponce N, Santillana-Ortiz JD, del Pino EM: The gastrocoel roof plate in embryos of different frogs. Differentiation 83:S62–S66 (2012).

Saez FA, Brum N: Chromosomes of South American amphibians. Nature 185:945 (1960).

Salica MJ, Haad MB, Candioti FV, Faivovich J: Early development of two species of *Phyllomedusa* (Anura: Phyllomedusinae). Salamandra 47:144–154 (2011).

Salvadori S, Deiana A, Elisabetta C, Floridia G, Rossi E, Zuffardi O: Colocalization of $(TTAGGG)_n$ telomeric sequences and ribosomal genes in Atlantic eels. Chromosome Res 3:54–58 (1995).

Sanchiz B: Vertebrates from the Early Miocene lignite deposits of the opencast mine Oberdorf (western Styrian Basin, Austria). 2. Amphibia. Ann Naturhist Mus Wien 99A:13–29 (1998).

Sanmartín I, Ronquist F: Southern hemisphere biogeography inferred by event-based models: plant versus animal patterns. Syst Biol 53: 216–243 (2004).

Santos LR, de Oliveira C: Histological aspects and structural characteristics of the testes of *Dendropsophus minutus* (Anura, Hylidae). Micron 39:1266–1270 (2008).

Santos LR, Franco-Belussi L, de Oliveira C: Germ cell dynamics during the annual reproductive cycle of *Dendropsophus minutus* (Anura: Hylidae). Zool Sci 28:840–844 (2011).

Sarria F: Tabla de desarrollo embrionario de la rana sabanera *Dendropsophus labialis*: Hylidae en cautiverio. PhD thesis, Pontificia Universidad Javeriana, Bogotá (2010).

Savage JM: The enigma of the Central American herpetofauna: dispersals or vicariance? Ann Missouri Bot Gard 69:464–547 (1982).

Sayim F, Kaya U: Embryonic development of the tree frog, *Hyla arborea*. Biologia (Bratisl) 63: 588–593 (2008).

Schartl M, Nanda I, Schlupp I, Wilde B, Epplen JT, et al: Incorporation of subgenomic amounts of DNA as compensation for mutational load in a gynogenetic fish. Nature 373: 68–71 (1995).

Schechtman AM: The Mechanism of Amphibian Gastrulation (University of California Press, Los Angeles 1942).

Scheel JJ: The seven-chromosome karyotype of the African frog *Arthroleptis*, a probable derivative of the thirteen-chromosome karyotype of *Rana*. Hereditas 67:287–290 (1971).

Scheltinga DM, Jamieson BGM: Spermatogenesis and the mature spermatozoon: form, function and phylogenetic implications, in Jamieson BGM (ed): Reproductive Biology and Phylogeny of Anura, pp 119–251 (Science Publishers, New Hampshire 2003).

Scheltinga DM, Jamieson BGM, McDonald K: Ultrastructure of the spermatozoa of *Litoria longirostris* (Hylidae, Anura, Amphibia): modifications for penetration of a gelatinous layer surrounding the arboreal egg clutch. Mem Queensl Mus 48:215–220 (2002).

Schempp W, Schmid M: Chromosome banding in Amphibia. VI. BrdU-replication patterns in Anura and demonstration of XX/XY sex chromosomes in *Rana esculenta*. Chromosoma 83:697–710 (1981).

Schlüter A, Icochea J, Peres JM: Amphibians and reptiles of the lower Río Llullapichis, Amazonian Peru: update species list with ecological and biogeographical notes. Salamandra 40:141–160 (2004).

Schmid M: Chromosome banding in Amphibia. I. Constitutive heterochromatin and nucleolus organizer regions in *Bufo* and *Hyla*. Chromosoma 66:361–388 (1978a).

Schmid M: Chromosome banding in Amphibia. II. Constitutive heterochromatin and nucleolus organizer regions in Ranidae, Microhylidae and Rhacophoridae. Chromosoma 68: 131–148 (1978b).

Schmid M: Chromosome banding in Amphibia. IV. Differentiation of GC- and AT-rich chromosome regions in Anura. Chromosoma 77: 83–103 (1980a).

Schmid M: Chromosome banding in Amphibia. V. Highly differentiated ZW/ZZ sex chromosomes and exceptional genome size in *Pyxicephalus adspersus* (Anura, Ranidae). Chromosoma 80:69–96 (1980b).

Schmid M: Chromosome evolution in Amphibia, in Müller H (ed): Cytogenetics of Vertebrates, pp 4–27 (Birkhäuser, Basel 1980c).

Schmid M: Chromosome banding in Amphibia. VII. Analysis of the structure and variability of NORs in Anura. Chromosoma 87:327–344 (1982).

Schmid M: Variations of content and structure of the mammalian Y chromosome, in Sandberg AA (ed): The Y Chromosome, Part A: Basic Characteristics of the Y Chromosome, pp 373–401 (Alan R Liss, New York 1985).

Schmid M, Bachmann K: A frog with highly evolved sex chromosomes. Experientia 37: 242–244 (1981).

Schmid M, de Almeida CG: Chromosome banding in Amphibia. XII. Restriction endonuclease banding. Chromosoma 96:283–290 (1988).

Schmid M, Guttenbach M: Evolutionary diversity of reverse (R) fluorescent chromosome bands in vertebrates. Chromosoma 97:101–114 (1988).

Schmid M, Klett R: Chromosome banding in Amphibia. XX. DNA replication patterns in *Gastrotheca riobambae* (Anura, Hylidae). Cytogenet Cell Genet 65:122–126 (1994).

Schmid M, Steinlein C: Chromosome banding in Amphibia. XVI. High-resolution replication banding patterns in *Xenopus laevis*. Chromosoma 101:123–132 (1991).

Schmid M, Steinlein C: Chromosome banding in Amphibia. XXIX. The primitive XY/XX sex chromosomes of *Hyla femoralis*. Cytogenet Genome Res 101:74–79 (2003).

Schmid M, Steinlein C: Chromosome banding in Amphibia. XXXII. The genus *Xenopus* (Anura, Pipidae). Cytogenet Genome Res 145: 201–217 (2015).

Schmid M, Steinlein C: Chromosome banding in Amphibia. XXXIV. Intrachromosomal telomeric DNA sequences in Anura. Cytogenet Genome Res 148:211–226 (2016a).

Schmid M, Steinlein C: Chromosome banding in Amphibia. XXXIII. Demonstration of 5-methylcytosine-rich heterochromatin in Anura. Cytogenet Genome Res 148:35–43 (2016b).

Schmid M, Steinlein C: The hypermethylated regions in avian chromosomes. Cytogenet Genome Res 151:216–227 (2017).

Schmid M, Hofgärtner FJ, Zenzes MT, Engel W: Evidence for postmeiotic expression of ribosomal RNA genes during male gametogenesis. Hum Genet 38:279–284 (1977).

Schmid M, Poppen A, Schmid W, Engel W: Somatic pairings of the Y heterochromatin in human XYY and XYqi cells. Cytogenet Cell Genet 29:203–214 (1981).

Schmid M, Löser C, Schmidtke J, Engel W: Evolutionary conservation of a common pattern of activity of nucleolus organizers during spermatogenesis in vertebrates. Chromosoma 86:149–179 (1982).

Schmid M, Haaf T, Geile B, Sims S: Chromosome banding in Amphibia. VIII. An unusual XY/XX-sex chromosome system in *Gastrotheca riobambae* (Anura, Hylidae). Chromosoma 88:69–82 (1983a).

Schmid M, Haaf T, Geile B, Sims S: Unusual heteromorphic sex chromosomes in a marsupial frog. Experientia 39:1153–1155 (1983b).

Schmid M, Grunert D, Haaf T, Engel W: A direct demonstration of somatically paired heterochromatin of human chromosomes. Cytogenet Cell Genet 36:554–561 (1983c).

Schmid M, Haaf T, Schempp W: Chromosome banding in Amphibia. IX. The polyploid karyotypes of *Odontophrynus americanus* and *Ceratophrys ornata* (Anura, Leptodactylidae). Chromosoma 91:172–184 (1985).

Schmid M, Sims SH, Haaf T, Macgregor HC: Chromosome banding in Amphibia. X. 18S and 28S ribosomal genes, nucleolus organizers and nucleoli in *Gastrotheca riobambae*. Chromosoma 94:139–145 (1986).

Schmid M, Vitelli L, Batistoni R: Chromosome banding in Amphibia. XI. Constitutive heterochromatin, nucleolus organizers, 18S+28S and 5S ribosomal RNA genes in Ascaphidae, Pipidae, Discoglossidae and Pelobatidae. Chromosoma 95:271–284 (1987).

Schmid M, Steinlein C, Feichtinger W, de Almeida CG, Duellman WE: Chromosome banding in Amphibia. XIII. Sex chromosomes, heterochromatin and meiosis in marsupial frogs (Anura, Hylidae). Chromosoma 97:33–42 (1988).

Schmid M, Steinlein C, Feichtinger W: Chromosome banding in Amphibia. XIV. The karyotype of *Centrolenella antisthenesi* (Anura, Centrolenidae). Chromosoma 97:434–438 (1989).
Schmid M, Steinlein C, Friedl R, de Almeida CG, Haaf T, et al: Chromosome banding in Amphibia. XV. Two types of Y chromosomes and heterochromatin hypervariability in *Gastrotheca pseustes* (Anura, Hylidae). Chromosoma 99:413–423 (1990a).
Schmid M, Steinlein C, Nanda I, Epplen JT: Chromosome banding in Amphibia, in Olmo E (ed): Cytogenetics of Amphibians and Reptiles, pp 21–45 (Birkhäuser, Basel 1990b).
Schmid M, Guttenbach M, Nanda I, Studer R, Epplen JT: Organization of DYZ2 repetitive DNA on the human Y chromosome. Genomics 6:212–218 (1990c).
Schmid M, Steinlein C, Feichtinger W: Chromosome banding in Amphibia. XVII. First demonstration of multiple sex chromosomes in amphibians: *Eleutherodactylus maussi* (Anura, Leptodactylidae). Chromosoma 101: 284–292 (1992).
Schmid M, Ohta S, Steinlein C, Guttenbach M: Chromosome banding in Amphibia. XIX. Primitive ZW/ZZ sex chromosomes in *Buergeria buergeri* (Anura, Rhacophoridae). Cytogenet Cell Genet 62:238–246 (1993a).
Schmid M, Steinlein C, Feichtinger W, Poot M: Chromosome banding in Amphibia. XVIII. Karyotype evolution and genomic size variation in *Pleurodema* (Anura, Leptodactylidae). Cytogenet Cell Genet 62:42–48 (1993b).
Schmid M, Feichtinger W, Nanda I, Schakowski R, Visbal García R, et al: An extraordinarily low diploid chromosome number in the reptile *Gonatodes taniae* (Squamata, Gekkonidae). J Hered 85:255–260 (1994).
Schmid M, Feichtinger W, Weimer R, Mais C, Bolaños F, León P: Chromosome banding in Amphibia. XXI. Inversion polymorphism and multiple nuleolus organizer regions in *Agalychnis callidryas* (Anura, Hylidae). Cytogenet Cell Genet 69:18–26 (1995).
Schmid M, Feichtinger W, Steinlein C, Nanda I, Mais C, et al: Chromosome banding in Amphibia. XXII Atypical Y chromosomes in *Gastrotheca walkeri* and *Gastrotheca ovifera* (Anura, Hylidae). Cytogenet Genome Res 96: 228–238 (2002a).
Schmid M, Ziegler CG, Steinlein C, Nanda I, Haaf T: Chromosome banding in Amphibia. XXIV. The B chromosomes of *Gastrotheca espeletia* (Anura, Hylidae). Cytogenet Genome Res 97:205–218 (2002b).
Schmid M, Feichtinger M, Steinlein C, Haaf T, Visbal García R, Fernández Badillo A: X chromosomes of American marsupials contain minimal amounts of euchromatin. Cytogenet Genome Res 99:315–322 (2002c).
Schmid M, Feichtinger W, Steinlein C, Rupprecht A, Haaf T, Kaiser H: Chromosome banding in Amphibia. XXIII. Giant W sex chromosomes and extremely small genome sizes in *Eleutherodactylus euphronides* and *Eleutherodactylus shrevei* (Anura, Leptodactylidae). Cytogenet Genome Res 97:81–94 (2002d).
Schmid M, Feichtinger W, Steinlein C, Haaf T, Schartl M, et al: Chromosome banding in Amphibia. XXVI. Coexistence of homomorphic XY sex chromosomes and a derived Y-autosome translocation in *Eleutherodactylus maussi* (Anura, Leptodactylidae). Cytogenet Genome Res 99:330–343 (2002e).
Schmid M, Haaf T, Steinlein C, Nanda I, Mahony M: Chromosome banding in Amphibia. Karyotype evolution and heterochromatin characterization in Australian *Mixophyes* (Anura, Myobatrachidae). Cytogenet Genome Res 97:239–253 (2002f).
Schmid M, Steinlein C, Haaf T: Chromosome banding in Amphibia. XXVII. DNA replication banding patterns in three anuran species with greatly differing genome sizes. Cytogenet Genome Res 101:54–61 (2003a).
Schmid M, Feichtinger W, Steinlein C, Visbal García R, Fernández Badillo A: Chromosome banding in Amphibia. XXVIII. Homomorphic XY sex chromosomes and a derived Y-autosome translocation in *Eleutherodactylus riveroi* (Anura, Leptodactylidae). Cytogenet Genome Res 101:62–73 (2003b).
Schmid M, Steinlein C, Haaf T: Chromosome banding in Amphibia. XXX. Karyotype aberrations in cultured fibroblast cells. Cytogenet Genome Res 104:277–282 (2004).
Schmid M, Ziegler CG, Steinlein C, Nanda I, Schartl M: Cytogenetics of the bleak (*Alburnus alburnus*), with special emphasis on the B chromosomes. Chromosome Res 14:231–242 (2006).
Schmid M, Steinlein C, Bogart JP, Feichtinger W, León P, et al: The chromosomes of terraranan frogs – insights into vertebrate cytogenetics. Cytogenet Genome Res 130/131:1–568 (2010).
Schmid M, Steinlein C, Bogart JP, Feichtinger W, Haaf T, et al: The hemiphractid frogs – phylogeny, embryology, life history, and cytogenetics. Cytogenet Genome Res 138:69–384 (2012).
Schmid M, Steinlein C, Feichtinger W, Haaf T, Mijares Urrutia A, et al: Cytogenetic studies in *Gonatodes* (Reptilia, Squamata, Sphaerodactylidae). Cytogenet Genome Res 144:47–61 (2014).
Schmid M, Steinlein C, Yano CF, Cioffi MB: Hypermethylated chromosome regions in nine fish species with heteromorphic sex chromosomes. Cytogenet Genome Res 147:169–178 (2015a).
Schmid M, Smith J, Burt DW, Aken BL, Antin PB, et al: Third report on chicken genes and chromosomes 2015. Cytogenet Genome Res 145:75–180 (2015b).
Schmid M, Steinlein C, Lomb C, Neitzel H, Sperling K: 5-Methylcytosine-rich heterochromatin in the Indian muntjac. Cytogenet Genome Res 147:240–246 (2015c).
Schmid M, Steinlein C, Feichtinger W, Nanda I: Chromosome banding in Amphibia. XXXV. Highly mobile nucleolus organizing regions in *Craugastor fitzingeri* (Anura, Craugastoridae). Cytogenet Genome Res 152:180–193 (2017a).
Schmid M, Steinlein C, Lomb C, Volleth M: Demonstration of 5-methylcytosine-rich DNA sequences in Chiroptera. Cytogenet Genome Res 152:38–45 (2017b).
Schmid W: Heterochromatin in mammals. Arch J Klaus Stift Vererbf 42:1–60 (1967).
Schmid W, Smith DW, Theiler K: Chromatinmuster in verschiedenen Zelltypen und Lokalisation von Heterochromatin auf Metaphase-Chromosomen bei *Microtus agrestis*, *Mesocricetus auratus*, *Cavia cobaya* und beim Menschen. Arch J Klaus Stift Vererbf 40:35–49 (1965).
Schmidtke J, Schmid M: Regional assignment of a 2.1-kb repetitive sequence to the distal part of the human Y heterochromatin. Hum Genet 55:255–257 (1980).
Schmidtke J, Brennecke H, Schmid M, Neitzel H, Sperling K: Evolution of muntjac DNA. Chromosoma 84:187–193 (1981).
Schnedl W: Fluoreszenzuntersuchungen über die Längenvariabilität des Y-Chromosoms beim Menschen. Humangenetik 12:188–194 (1971).
Schroeder-Reiter E, Wanner G: Chromosome centromeres: structural and analytical investigations with high resolution scanning electron microscopy in combination with focussed ion beam milling. Cytogenet Genome Res 124:239–250 (2009).
Schubert I, Rieger R, Fuchs J: Alteration of basic chromosome number by fusion-fission cycles. Genome 38:1289–1292 (1995).
Schultz A, Jansen M, Köhler G: Diversity and ecology of anuran communities in San Sebastián (Chiquitano region, Bolivia). Salamandra 45:75–90 (2009).
Schwarzacher HG, Mikelsaar AV, Schnedl W: The nature of the Ag-staining of nucleolus organizer regions. Electron- and light microscopic studies on human cells in interphase, mitosis, and meiosis. Cytogenet Cell Genet 20:24–39 (1978).
Schweizer D: Reverse fluorescent chromosome banding with chromomycin and DAPI. Chromosoma 58:307–324 (1976).
Schweizer D, Loidl J: A model for heterochromatin dispersion and the evolution of C-band patterns, in Stahl A, Luciani JM, Vagner-Capodano AM (eds): Chromosomes Today, vol 9, pp 61–79 (Allen and Unwin, London 1987).
Seabright M: The use of proteolytic enzymes for the mapping of structural rearrangements in the chromosomes of man. Chromosoma 36: 204–210 (1972).
Sekiya K, Nakagawa H: Cytogenetics of *Xenopus laevis*. I. G-banding pattern of *Xenopus laevis* chromosomes. Experientia 39:786–787 (1983).
Sessions S: The chromosomes of *Anotheca spinosa* (Stejneger), family Hylidae. Herpetologica 34:70–73 (1978).
Seto T: The karyotype of *Hyla arborea japonica* with some remarks on heteromorphism of the sex chromosomes. J Fac Sci Hokkaido Univ Ser VI Zool 15:366–373 (1964).
Seto T, Makino S: On the heteromorphism in an autosomal pair in the Japanese tree frog. Proc Jap Acad 40:862–865 (1964).

Shalan AG, Bradshaw SD, Withers PC, Thompson G, Bayomy MFF, et al: Spermatogenesis and plasma testosterone levels in Western Australian burrowing desert frogs, *Cyclorana platycephala*, *Cyclorana maini*, and *Neobatrachus sutor*, during aestivation. Gen Comp Endocrinol 136:90–100 (2004).
Shang K, Deng C: A cytogenetic demonstration of ZW sex determination in *Bufo bufo gargarizans*. Acta Genet Sinica 10:298–305 (1983).
Sharbel TF, Green DM, Houben A: B chromosome origin in the endemic New Zealand frog *Leiopelma hochstetteri* through sex chromosome devolution. Genome 41:14–22 (1998).
Sharma T, Raman R: Variation of constitutive heterochromatin in the sex chromosomes of the rodent *Bandicota bengalensis bengalensis* (Gray). Chromosoma 41:75–84 (1973).
Shi L, Ye Y, Duan X: Comparative cytogenetic studies on the red muntjac, Chinese muntjac and their F1 hybrids. Cytogenet Cell Genet 26:22–27 (1980).
Shi L, Tang L, Ma K, Ma C: Synaptonemal complex formation among supernumerary B chromosomes: an electron microscopic study on spermatocytes of Chinese racoon dogs. Chromosoma 97:178–183 (1988).
Shields GF: Chromosomal polymorphism common to several species of *Junco* (Aves). Can J Genet Cytol 15:461–471 (1973).
Shields GF: Meiotic evidence for pericentric inversion polymorphism in *Junco* (Aves). Can J Genet Cytol 18:747–751 (1976).
Shoemaker VH, McClanahan LL Jr: Evaporative water loss, nitrogen excretion, and osmoregulation in phyllomedusine frogs. J Comp Physiol 100:331–345 (1975).
Shook DR, Majer C, Keller R: Urodeles remove mesoderm from the superficial layer by subduction through a bilateral primitive streak. Dev Biol 248:220–239 (2002).
Shook DR, Majer C, Keller R: Pattern and morphogenesis of presumptive superficial mesoderm in two closely related species, *Xenopus laevis* and *Xenopus tropicalis*. Dev Biol 270: 163–185 (2004).
Silva APZ, Haddad CFB, Kasahara S: Nucleolus organizer regions in *Physalaemus cuvieri* (Anura, Leptodactylidae), with evidence of a unique case of Ag-NOR variability. Hereditas 131:135–141 (1999).
Silva APZ, Haddad CFB, Galassi GG, Kasahara S: Multiple nucleolus organizer regions in *Leptodactylus mystacinus* (Amphibia, Anura) and comments on its systematic position in the *L. fuscus* group based on cytogenetic and molecular analyses. Genetica 127: 35–44 (2006).
Simi S, Tursi F: Polymorphism of human chromosomes 1, 9, 16, Y: variations, segregation and mosaicism. Hum Genet 62:217–220 (1982).
Singer MF: Highly repeated sequences in mammalian genomes. Int Rev Cytol 76:67–112 (1982).
Singh L: Evolution of karyotypes in snakes. Chromosoma 38:185–236 (1972a).
Singh L: Multiple W chromosome in a sea snake, *Enhydrina schistosa* Daudin. Experientia 28: 95–97 (1972b).
Singh L, Purdom IF, Jones KW: Satellite DNA and evolution of sex chromosomes. Chromosoma 59:43–62 (1976).
Sinha AK, Pathak S, Norra JJ: A human family suggesting evidence for centric fission and stability of a telocentric chromosome. Hum Hered 22:423–429 (1972).
Siqueira S, Ananias F, Recco-Pimentel SM: Cytogenetics of three Brazilian species of *Eleutherodactylus* (Anura, Leptodactylidae) with 22 chromosomes and re-analysis of multiple translocations in *E. binotatus*. Genet Mol Biol 27:363–372 (2004).
Siqueira S, Aguiar O, Souza MB, Lima AP, Recco-Pimentel SM: Unusual intra-individual karyotypical variation and evidence of cryptic species in Amazonian populations of *Pristimantis*. Hereditas 146:141–151 (2009).
Siripiyasing P, Chulalaksaananukul W, Pariyanonth P, Kaewsri S, Sittigut S, et al: The identification of the sex chromosome and karyotype of four toad species (genus *Bufo* in Thailand by T-lymphocyte culture. Cytologia 73:229–241 (2008).
Sites JW: Chromosome evolution in the iguanid lizard *Sceloporus grammicus*. I. Chromosome polymorphisms. Evolution 37:38–53 (1983).
Sitjà-Bobadilla A: Can myxosporean parasites compromise fish and amphibian reproduction? Proc Biol Sci 276:2861–2870 (2009).
Skuk G, Langone JA: Comunicación preliminar sobre la citogenética de *Hyla microcephala meridiana* B. Lutz, 1954 (Anura: Hylidae). Res XIII Congr Brasil Zool, Anfibios, p 165 (1986).
Skuk G, Langone JA: Los cromosomas de cuatro especies del género *Hyla* (Anura: Hylidae) con número diploide de 2N = 30. Acta Zool Lilloana 41:165–171 (1992).
Smith MJ, Hough BR, Chamberlin ME, Davidson EH: Repetitive and non-repetitive sequences in sea urchin heterogeneous nuclear RNA. J Mol Biol 85:103–126 (1974).
Sodhi NS, Bickford D, Diesmos AC, Lee TM, Koh LP, et al: Measuring the meltdown: drivers of global amphibian extinction and decline. PLoS One 3:e1636 (2008).
Sousa IMC: Caracterizão Citogenética de Anfibios Anuros do bioma Cerrado em Municipios do Estado de Goiás. Monografía pra conclusão do Curso de Biologia. Dept de Biologia, Universidade Catolica de Goiás (2007).
Southern EM: Base sequence and evolution of Guinea pig α-satellite DNA. Nature 227:794–798 (1970).
Souza MJ, Yonenaga-Yassuda Y: Chromosomal variability of sex chromosomes and NORs in *Trichomys apereoides* (Rodentia, Echimyidae). Cytogenet Cell Genet 33:197–203 (1982).
Souza VT, Necker M, De Bastiani V, Lucas EM, Busin CS: Cytogenetic study of two species of the genus *Scinax* (Anura, Hylidae). Resumos do 57º Congresso Brasileiro de Genetica, Águas de Lindóia, São Paulo, p 54 (2011).
Springer MS, Westerman M, Kavanagh JR, Burk A, Woodburne, et al: The origin of the Australasian marsupial fauna and the phylogenetic affinities of the enigmatic monito del monte and marsupial mole. Proc Biol Sci 265: 2381–2386 (1998).
Sprung CN, Reynolds GE, Jasin M, Murnane JP: Chromosome healing in mouse embryonic stem cells. Proc Natl Acad Sci USA 96:6781–6786 (1999).
Stamatakis A: RAxML version 8: a tool for phylogenetic analysis and post-analysis of large phylogenies. Bioinformatics 30:1312–1313 (2014).
Stamatakis A, Hoover P, Rougemont J: A rapid bootstrap algorithm for the RAxML Web Servers. Syst Biol 57:758–771 (2008).
Stanyon R, Wienberg J, Romagno D, Bigoni F, Jauch A, Cremer T: Molecular and classical cytogenetic analyses demonstrate an apomorphic reciprocal chromosomal translocation in *Gorilla gorilla*. Am J Phys Anthropol 88:245–250 (1992).
Stephenson EM, Stephenson NG: Karyotypes of two Australian hylids. Chromosoma 30:38–50 (1970).
Stephenson EM, Robinson ES, Stephenson NG: Karyotype variation within the genus *Leiopelma* (Amphibia, Anura). Can J Genet Cytol 14:691–702 (1972).
Stern R: Satellite DNAs of *Xenopus mulleri*. Carnegie Inst Wash Yearbook 71:22 (1972).
Stock AD: Chromosome banding pattern homology and its phylogenetic implications in the bat genera *Carollia* and *Choeroniscus*. Cytogenet Cell Genet 14:34–41 (1975).
Stock AD, Hsu TC: Evolutionary conservatism in arrangement of genetic material: a comparative analysis of chromosome banding between the rhesus macaque (2n = 42, 84 arms) and the African green monkey (2n = 60, 120 arms). Chromosoma 42:211–224 (1973).
Stock AD, Mengden GA: Chromosome banding pattern conservatism in birds and nonhomology of chromosome banding patterns between birds, turtles, snakes and amphibians. Chromosoma 50:69–77 (1975).
Stock AD, Mengden GA: The origin of mitotic sex-chromosome association in the brush-tailed possum, *Trichosurus vulpecula* (Marsupialia: Phalangeridae). Cytogenet Cell Genet 34:305–309 (1982).
Stöck M: Tetraploid toads (*Bufo viridis* complex) from north-western China and preliminary taxonomic conclusions for *Bufo nouettei* Mocquard, 1910. Z Feldherp 5:139–166 (1998).
Stöck M, Schmid M, Steinlein C, Grosse WR: Mosaicism in somatic triploid specimens of the *Bufo viridis* complex in the Karakoram with examination of calls, morphology and taxonomic conclusions. Ital J Zool 66:215–232 (1999).
Stöck M, Frynta D, Grosse WR, Steinlein C, Schmid M: A review of the distribution of diploid, triploid and tetraploid green toads (*Bufo viridis* complex) in Asia including new data from Iran and Pakistan. Asiatic Herpetol Res 9:77–100 (2001).

Stöck M, Lamatsch DK, Steinlein C, Epplen JT, Grosse WR et al: A bisexually reproducing all-triploid vertebrate. Nat Genet 30:325–328 (2002).
Stöck M, Steinlein C, Lamatsch DK, Schartl M, Schmid M: Multiple origins of tetraploid taxa in the Eurasian *Bufo viridis* subgroup. Genetica 124:255–272 (2005).
Stöck M, Horn A, Grossen C, Lindtke D, Sermier R, et al: Ever-young sex chromosomes in European tree frogs. PLoS Biol 9:e1001062 (2011).
Stöck M, Savary R, Zaborowska A, Górecki G, Brelsford A, et al: Maintenance of ancestral sex chromosomes in Palearctic tree frogs: direct evidence from *Hyla orientalis*. Sex Dev 7:261–266 (2013a).
Stöck M, Savary R, Betto-Colliard C: Low rates of X-Y recombination, not turnovers, account for homomorphic sex chromosomes in several diploid species of Palearctic green toads (*Bufo viridis* subgroup). J Evol Biol 26: 674–682 (2013b).
Stone WS: Genetic and chromosomal variability in *Drosophila*. Cold Spring Harb Symp Quant Biol 20:256–269 (1955).
Straus NA: Comparative DNA renaturation kinetics in amphibians. Proc Natl Acad Sci USA 68:799–802 (1971).
Straus NA: Repeated DNA in eukaryotes, in King RC (ed): Handbook of Genetics. V. Molecular Genetics, pp 3–29 (Plenum Press, New York 1976).
Stuart SN, Chanson JS, Cox NA, Young BE, Rodrigues ASL, et al: Status and trends of amphibian declines and extinctions worldwide. Science 306:1783–1786 (2004).
Suárez P, Cardozo D, Baldo D, Pereyra MO, Faivovich J, et al: Chromosome evolution in Dendropsophini (Amphibia: Anura: Hylinae). Cytogenet Genome Res 141:295–308 (2013).
Sugai JLMM, Souza Terra J, Ferreira VL: Anurans of a threatened savanna area in western Brazil. Biota Neotrop 14:1 (2014).
Sumner AT: A simple technique for demonstrating centromeric heterochromatin. Exp Cell Res 75:304–306 (1972).
Sumner AT: Chromosome Banding (Unwin Hyman, London 1990).
Sunnucks P: Efficient genetic markers for population biology. Trends Ecol Evol 15:199–203 (2000).
Swier VJ, Bradley RD, Rens W, Elder FFB, Baker RJ: Patterns of chromosomal evolution in *Sigmodon*, evidence from whole chromosome paints. Cytogenet Genome Res 125: 54–66 (2009).
Switonski M, Gustavsson I, Hojer K, Ploen L: Synaptonemal complex analysis of the B-chromosomes in spermatocytes of the silver fox (*Vulpes vulpes* Desm.). Cytogenet Cell Genet 45:84–92 (1987).
Syromyatnikova EV: Anurans of the Tagay locality (Baikal Lake, Russia: Miocene): Bombitoridae, Hylidae, and Ranidae. Russian J Herp 23:145–157 (2016).
Tamschick S, Rozenblut-Kościsty B, Ogielska M, Kekenj D, Gajewski F, et al: The plasticizer bisphenol A affects somatic and sexual development, but differently in pipid, hylid and bufonid anurans. Environ Pollut 216:282–291 (2016a).
Tamschick S, Rozenblut-Kościsty B, Ogielska M, Lehmann A, Lymberakis P, et al: Sex reversal assessments reveal different vulnerability to endocrine disruption between deeply diverged anuran lineages. Sci Rep 6:23825 (2016b).
Tamschick S, Rozenblut-Kościsty B, Ogielska M, Lehmann A, Lymberakis P, et al: Impaired gonadal and somatic development corroborate vulnerability differences to the synthetic estrogen ethinylestradiol among deeply diverged anuran lineages. Aquat Toxicol 177: 503–514 (2016c).
Tamura K, Peterson D, Peterson N, Stecher G, Nei M, Kumar S: MEGA5: Molecular evolutionary genetics and analysis using maximum likelihood, evolutionary distance, and maximum parsimony methods. Mol Biol Evol 28:2731–2739 (2011).
Tamura K, Stecher G, Peterson D, Filipski A, Kumar S: MEGA6: Molecular Evolutionary Genetics Analysis version 6.0. Mol Biol Evol 30: 2725–2729 (2013).
Tandy M, Bogart JP, Largen MJ, Feener DJ: A tetraploid species of *Bufo* (Anura Bufonidae) from Ethiopia. Monit Zool Ital N.S. Suppl. 17, No. 1:1–79 (1980).
Tang XR, Wang XB, Fang JJ: Preliminary observations on the karyotype of *Hyla arborea immaculata*. Acta Herpetol Sinica 3:37–40 (1984).
Teixeira LSR, Tacioli A, Lourenço LB: Cytogenetic analysis of the species *Dendropsophus seniculus* (Anura, Hylidae): a contribution for the 30-chromosome genus *Dendropsophus*. 2ª Reunião Brasileira de Citogenética, Águas de Lindóia, São Paulo, CA012 (2011).
Teixeira LSR, Seger KR, Targueta CP, Orrico VGD, Lourenço LB: Comparative cytogenetics of tree frogs of the *Dendropsophus marmoratus* (Laurenti, 1768) group: conserved karyotypes and interstitial telomeric sequences. Comp Cytogenet 10:753–767 (2016).
Teixeira RD, Colli GR, Báo SN: The ultrastructure of the spermatozoa of the lizard *Micrablepharus maximiliani* (Squamata, Gymnophthalmidae), with considerations on the use of sperm ultrastructure characters in phylogenetic reconstruction. Acta Zool 80: 47–59 (1999).
Telenius H, Pelmear AH, Tunnacliffe A, Carter NP, Behmel A, et al: Cytogenetic analysis by chromosome painting using DOP-PCR amplified flow-sorted chromosomes. Genes Chromosomes Cancer 4:257–263 (1992).
Teruel M, Cabrero J, Montiel EE, Acosta MJ, Sánchez A, Camacho JPM: Microdissection and chromosome painting of X and B chromosomes in *Locusta migratoria*. Chromosome Res 17:11–18 (2009).
Therman E, Sarto GE, Patau K: Apparently isodicentric but functionally monocentric X chromosome in man. Am J Hum Genet 26: 83–92 (1974).
Thiébaud CH: The intra-nucleolar localization of amplified rDNA in *Xenopus laevis* oocytes. Chromosoma 73:29–36 (1979a).
Thiébaud CH: Quantitative determination of amplified rDNA and its distribution during oogenesis in *Xenopus laevis*. Chromosoma 73:37–44 (1979b).
Thiery JP, Macaya G, Bernardi G: An analysis of eukaryotic genomes by density gradient centrifugation. J Mol Biol 108:219–235 (1976).
Thorneycroft HB: A cytogenetic study of the white-throated sparrow, *Zonotrichia albicollis* (Gmelin). Evolution 29:611–621 (1975).
Toktosunov AT: Ecological Basis of Altitude Adaptation of the Vertebrates of Tian-Shan (Nauka, Leningrad 1984).
Touchon JC, Warkentin KM: Reproductive mode plasticity: aquatic and terrestrial oviposition in a treefrog. Proc Natl Acad Sci USA 105:7495–7499 (2008).
Tracy CR, Reynolds SJ, McArthur L, Tracy CR, Christian KA: Ecology of aestivation in a cocoon-forming frog, *Cyclorana australis* (Hylidae). Copeia 2007:901–912 (2007).
Tsipouri V, Schueler MG, Hu S, Dutra A: Comparative sequence analyses reveal sites of ancestral chromosome fusions in the Indian muntjac genome. Genome Biol 9:R155 (2008).
Tsurugi K, Morita T, Ogata K: Mode of degradation of ribosomes in regenerating rat liver in vivo. Eur J Biochem 45:119–126 (1974).
Turner BJ, Grudzien TA, Adkisson KP, Worrell RA: Extensive chromosomal divergence within a single river basin in the goodeid fish, *Ilyodon furcidens*. Evolution 39:122–134 (1985).
Tyler M: Australian Frogs (Penguin Books Australia Ltd., Ringwood 1989).
Tyler MJ: Australian Frogs: A Natural History (Cornell University Press, Ithaca 1998).
Tyler MJ, Davies MM, King M: The Australian frog *Chiroleptes dahlii* Boulenger: its systematic position, morphology, chromosomes and distribution. Trans R Soc S Aust 102:17–24 (1978).
Tymowska J: Karyotypic analysis of *Xenopus tropicalis* Gray (Pipidae). Cytogenet Cell Genet 12:297–304 (1973).
Tymowska J: A karyotype analysis of the genus *Xenopus* (Anura, Pipidae). PhD thesis, University of Geneva (1976).
Tymowska J: A comparative study of the karyotypes of eight *Xenopus* species and subspecies possessing a 36-chromosome complement. Cytogenet Cell Genet 18:165–181 (1977).
Tymowska J: Polyploidy and cytogenetic variation in frogs of the genus *Xenopus*, in Green DM, Sessions SK (eds): Amphibian Cytogenetics and Evolution, pp 259–297 (Academic Press, New York 1991).
Tymowska J, Fischberg M: Chromosome complements of the genus *Xenopus*. Chromosoma 44:335–342 (1973).
Tymowska J, Fischberg M: The karyotype of *Xenopus wittei* Tinsley, Kobel and Fischberg, another tetraploid anuran species (Pipidae). Cytogenet Cell Genet 28:208–212 (1980).

Tymowska J, Fischberg M: A comparison of the karyotype, constitutive heterochromatin, and nucleolar organizer regions of a new tetraploid species *Xenopus epitropicalis* Fischberg and Picard with those of *Xenopus tropicalis* Gray (Anura, Pipidae). Cytogenet Cell Genet 34:149–157 (1982).

Tymowska J, Kobel HR: Karyotype analyis of *Xenopus muelleri* (Peters) and *Xenopus laevis* (Daudin), Pipidae. Cytogenetics 11:270–278 (1972).

Tymowska J, Fischberg M, Tinsley RC: The karyotype of the tetraploid species *Xenopus vestitus* Laurent (Anura: Pipidae). Cytogenet Cell Genet 19:344–354 (1977).

Ueda H: The sex of triploids and gynogenetic diploids in *Bombina orientalis*. Sci Rep Lab Amphibian Biol Hiroshima Univ 4:185–200 (1980).

Ueda H: Sex ratio of the offspring of sex-reversed males in *Bufo viridis* from Turkey. Jap J Genet 60:638 (1985).

Ueda H: Offspring of sex-reversed males in *Bufo viridis* Laur. Sci Rep Lab Amphibian Biol Hiroshima Univ 10:155–164 (1990).

Ueda Y, Yoshizaki N, Iwao Y: Acrosome reaction in sperm of the frog, *Xenopus laevis*: its detection and induction by oviductal pars recta secretion. Dev Biol 243:55–64 (2002).

Ullerich FH: Unterschiede im DNS-Gehalt der Genome von *Bufo bufo* und *Bufo viridis*. Z Naturforsch 20:720–722 (1965).

Ullerich FH: Karyotyp und DNS-Gehalt von *Bufo bufo*, *B. viridis*, *B. bufo* × *B. viridis* und *B. calamita* (Amphibia, Anura). Chromosoma 18:316–342 (1966).

Ullerich FH: Weitere Untersuchungen über Chromosomenverhältnisse und DNS-Gehalt bei Anuren (Amphibia). Chromosoma 21:345–368 (1967).

Ullerich FH: DNS-Gehalt und Chromosomenstruktur bei Amphibien. Chromosoma 30: 1–37 (1970).

Unnerus V, Fellman J, de la Chapelle A: The length of the human Y chromosome. Cytogenetics 6:213–227 (1967).

Vanzolini PE: Distribution and differentiation of animals along the coast and on continental islands of the state of São Paulo, Brazil. 1. Introduction to the area and problems. Pap Avul Zool 26:281–294 (1973).

Vargas-Salinas F, Bolaños ME, Berrío H: Notas sobre la ecología reproductiva de *Agalychnis spurrelli* (Anura: Hylidae) en una población de Anchicayá, Pacífico Colombiano. Rev La Acad Colomb Ciencias Exactas Físicas Nat 24:85–99 (2000).

Varley H, Di S, Scherer SW, Royle NJ: Characterization of terminal deletions at 7q32 and 22q13.3 healed by de novo telomere addition. Am J Hum Genet 67:610–622 (2000).

Varley H, Pickett HA, Foxon JL, Reddel RR, Royle NJ: Molecular characterization of inter-telomere and intra-telomere mutations in human ALT cells. Nat Genet 30:301–305 (2002).

Varley JM, Morgan GT: Silver staining of the lampbrush chromosomes of *Triturus cristatus carnifex*. Chromosoma 67:233–244 (1978).

Veiga-Menoncello ACP, Lima AP, Recco-Pimentel SM: Cytogenetic analysis of four Central Amazonian species of *Colostethus* (Anura, Dendrobatidae) with a diploid complement of 22 chromosomes. Hereditas 139:189–198 (2003).

Velo-Antón G, Burrowes PA, Joglar RL, Martínez-Solano I, Beard KH, Parra-Olea G: Phylogenetic study of *Eleutherodactylus coqui* (Anura: Leptodactylidae) reveals deep genetic fragmentation in Puerto Rico and pinpoints origins of Hawaiian populations. Mol Phylogenet Evol 45:716–728 (2007).

Veloso A, Iturra P: Posibilidades del análisis citogenético en un estudio de bandeo cromosómico en dos especies de anfibios (Anura – Leptodactylidae). Arch Biol Med Exper 2:91–96 (1979).

Veloso A, Galleguillos R, Diaz N: Karyotypic analysis of allopatric populations of *Pleurodema thaul* (Lesson) Amphibia, Leptodactylidae. Caryologia 26:69–76 (1973).

Ventura K, O'Brien PCM, Yonenaga-Yassuda Y, Ferguson-Smith MA: Chromosome homologies of the highly rearranged karyotypes of four *Akodon* species (Rodentia, Cricetidae) resolved by reciprocal chromosome painting: the evolution of the lowest diploid number in rodents. Chromosome Res 17:1063–1078 (2009).

Verma RS: Heteromorphism of heterochromatin, in Verma RS (ed): Heterochromatin. Molecular and Structural Aspects, pp 276–292 (Cambridge University Press, Cambridge 1988).

Verma RS, Dosik H, Scharf T, Lubs HA: Length heteromorphisms of fluorescent (f) and nonfluorescent (nf) segments of human Y chromosome: classification, frequencies, and incidence in normal Caucasians. J Med Genet 15:277–281 (1978).

Vicente VE, Moreira-Filho O, Camacho JPM: Sex-ratio distortion associated with the presence of a B chromosome in *Astyanax scabripinnis* (Teleostei, Characidae). Cytogenet Cell Genet 74:70–75 (1996).

Vilela VMFN, Brassaloti, RA, Bertoluci J: Anurofauna da floresta de restinga do Parque Estadual da Ilha do Cardoso. Sueste do Brazil: composição de espécies e uso de sítios reprodutivos. Biota Neotrop 11:1 (2011).

Vitousek PM, Mooney HA, Lubchenco J, Melillo JM: Human domination on Earth's ecosystems. Science 277:494–499 (1997).

Voitovich A, Yeliseeva KG, Afonin VY, Novitsky RV, Kazhur YI: B-chromosomes in populations of *Rana temporaria* (Amphibia: Anura) from Belarus. Acta Zool Bulgarica 58:109–116 (2006).

Volleth M, Heller KG: Chromosome number reduction accompanied by extensive heterochromatin addition in the bat *Glauconycteris beatrix* (Mammalia; Chiroptera, Vespertilionidae). Cytogenet Genome Res 119:245–247 (2007).

Volpe P, Wilkens M, Dobie J: Embryonic and larval development of *Hyla avivoca*. Copeia 66: 21–26 (1961).

Vorontsov NN, Lyapunova EA, Borissov YM, Dovgal VE: Variability of sex chromosomes in mammals. Genetica 52/53:361–272 (1980).

Vos P, Hogers R, Bleeker M, Reijans M, van de Lee T, et al: AFLP, a new technique for DNA fingerprinting. Nucleic Acids Res 23:4407–4414 (1995).

Vosa CG: Heterochromatic banding patterns in *Allium*. II. Heterochromatin variation in species of the *paniculatum* group. Chromosoma 57:119–133 (1976).

Wachtel SS, Koo GC, Boyse EA: Evolutionary conservation of H-Y (male) antigen. Nature 254:270–272 (1975).

Wake DB: Action on amphibians. Trends Ecol Evol 13:379–380 (1998).

Wallace H: The development of anucleolate embryos of *Xenopus laevis*. J Embryol Exp Morphol 8:405–413 (1960).

Warburton D, Henderson AS, Shapiro LR, Hsu LJF: A stable human dicentric chromosome, t dic (12,14)(p13;p13) including an intercalary satellite region between centromeres. Am J Hum Genet 25:439–445 (1973).

Warkentin KM: Adaptive plasticity in hatching age: a response to predation risk trade-offs. Proc Natl Acad Sci USA 92:3507–3510 (1995).

Warkentin KM: Effects of hatching age on development and hatchling morphology in the red-eyed tree frog, *Agalychnis callidryas*. Biol J Linn Soc 68:443–470 (1999).

Warkentin KM: Environmental and developmental effects on external gill loss in the red-eyed tree frog, *Agalychnis callidryas*. Physiol Biochem Zool 73:557–565 (2000).

Warkentin KM: Hatching timing, oxygen availability, and external gill regression in the tree frog, *Agalychnis callidryas*. Physiol Biochem Zool 75:155–164 (2002).

Warkentin KM: Oxygen, gills, and embryo behavior: mechanisms of adaptive plasticity in hatching. Comp Biochem Physiol A Mol Integr Physiol 148:720–731 (2007).

Warkentin KM, Gomez-Mestre I, McDaniel JG: Development, surface exposure, and embryo behavior affect oxygen levels in eggs of the red-eyed treefrog, *Agalychnis callidryas*. Comp Physiol Biochem 78:956–966 (2005).

Warkentin KM, Caldwell MS, McDaniel JG: Temporal pattern cues in vibrational risk assessment by embryos of the red-eyed treefrog, *Agalychnis callidryas*. J Exp Biol 209: 1376–1384 (2006).

Wasserman AO: Polyploidy in the common tree toad *Hyla versicolor* Le Conte. Science 167: 385–386 (1970).

Weiler C, Ohno S: Cytological confirmation of female heterogamety in the African water frog (*Xenopus laevis*). Cytogenetics 1:217–223 (1962).

Wells KD: The Ecology and Behavior of Amphibians (University of Chicago Press, Chicago 2007).

Wen C, Lu Q, Xiang W: Studies of chromosome banding and sister chromatid exchange in *Bufo bufo gargarizans*. Acta Genet Sinica 10: 291–297 (1983).

Weygoldt P: Changes in the composition of mountain stream frog communities in the Atlantic mountains of Brazil: frogs as indicators of environmental deteriorations? Stud Neotrop Fauna Env 243:249–255 (1989).

White MJD: Structural heterozygosity in natural populations of the grasshopper *Trimerotropis sparsa*. Evolution 5:376–394 (1951).
White MJD: Some general problems of chromosomal evolution and speciation in animals. Surv Biol Progr 3:109–147 (1957).
White MJD: Animal Cytology and Evolution (Cambridge University Press, Cambridge 1973a).
White MJD: Chromosomal rearrangements in mammalian population polymorphism and speciation, in Chiarelli AB, Capanna E (eds): Cytotaxonomy and Vertebrate Evolution, pp 95–128 (Academic Press, London 1973b).
White MJD, Blackith RE, Blackith RM, Cheney J: Cytogenetics of the *viatica* group of morabine grasshoppers. I. The "coastal" species. Aust J Zool 15:263–302 (1967).
Whitehouse DB, Evans EP, Putt W, George AM: Karyotypes of the East African common zebra, *Equus burchelli*: centric fission in a pedigree. Cytogenet Cell Genet 38:171–175 (1984).
Whitfield SM, Bell KE, Philippi T, Sasa M, Bolaños F, et al: Amphibian and reptile declines over 35 years at La Selva, Costa Rica. Proc Natl Acad Sci USA 104:8352–8356 (2007).
Whitmore TC: Tropical forest disturbance, disappearance, and species loss, in Laurance WF, Bierregaard RO (eds): Tropical Forest Remnants: Ecology, Management, and Conservation of Fragmented Communities, pp 3–12 (University of Chicago Press, Chicago 1997).
Wiberg UH: Sex determination in the European eel (*Anguilla anguilla* L.). Cytogenet Cell Genet 36:589–598 (1983).
Wickbom T: Cytological studies on Dipnoi, Urodela, Anura and Emys. Hereditas 31:241–346 (1945).
Wiens JJ, Kuczynski CA, Hua X, Moen DS: An expanded phylogeny of treefrogs (Hylidae) based on nuclear and mitochondrial sequence data. Mol Phylogenet Evol 55:871–882 (2010).
Wiley JE: Chromosome banding patterns of treefrogs (Hylidae) of the eastern United States. Herpetologica 38:507–520 (1982).
Wiley JE: Chromosome polymorphism in *Hyla chrysoscelis*. Copeia 1:273–274 (1983).
Wiley JE: Replication banding and FISH analysis reveal the origin of the *Hyla femoralis* karyotype and XY/XX sex chromosomes. Cytogenet Cell Genet 101:80–83 (2003).
Wiley JE, Little ML: Replication banding patterns of the diploid-tetraploid treefrogs *Hyla chrysoscelis* and *H. versicolor*. Cytogenet Cell Genet 88:11–14 (2000).
Wiley JE, Little ML, Romano MA, Blount DA, Cline GR: Polymorphism in the location of the 18S and 28S rRNA genes on the chromosomes of the diploid-tetraploid treefrogs *Hyla chrysoscelis* and *H. versicolor*. Chromosoma 97:481–487 (1989).
Wiley JE, Meyne J, Little M, Stout JC: Interstitial hybridization sites of the $(TTTGGG)_n$ telomeric sequence on the chromosomes of some North American hylid frogs. Cytogenet Cell Genet 61:55–57 (1992).
Wilkie AO, Lamb J, Harris PC, Higgs DR: A truncated chromosome 16 associated with alpha thalassaemia is stabilized by addition of telomeric repeat $(TTAGGG)_n$. Nature 346: 868–871 (1990).
Wilson EB: Studies on chromosomes. V. The chromosomes of *Metapodius*. A contribution to the hypothesis of genetic continuity of chromosomes. J Exp Zool 6:147–205 (1906).
Wilson LD, McCranie JR: Amphibian population decline in a Honduran national park. Froglog 25:1–2 (1998).
Wilson LD, McCranie JR, Williams KL: Two new species of fringe-limbed hylid frogs from Nuclear Middle America. Herpetologica 41: 141–150 (1985).
Winkelmann L, Noleto RB: Caracterização cariotípica de *Phyllomedusa tetraploidea* (Anura, Hylidae), proveniente da mata Atlântica Paranaense. Luminária 17:146–157 (2015).
Witschi E: Ergebnisse der neueren Arbeiten über die Geschlechtschromosomen bei Amphibien. Z Induk Abstamm Vererbungsl 31:287–312 (1923).
Witschi E: Studies on sex differentiation and sex determination in amphibians: III. Rudimentary hermaphroditism and Y chromosome in *Rana temporaria*. J Exp Zool 54:157–223 (1929).
Wolf U, Flinspach G, Böhm R, Ohno S: DNS-Reduplikationsmuster bei den Riesen-Geschlechtschromosomen von *Microtus agrestis*. Chromosoma 16:609–617 (1965).
Wong D: *Phyllomedusa camba*. http://amphibiaweb.org/ (2013).
Woodruff DS: Australian anuran chromosome numbers. Herpetol Rev 4:208 (1972).
Worthy TH: Osteology of *Leiopelma* (Amphibia, Leiopelmatidae) and descriptions of three new subfossil *Leiopelma* species. J Roy Soc New Zealand 17:201–251 (1987).
Wright DA, Richards CM: Two sex-linked loci in the leopard frog *Rana pipiens*. Genetics 103: 249–261 (1983).
Wright DA, Richards CM, Frost JS, Camozzi AM, Kunz BJ: Genetic mapping in amphibians, in Rattazzi MC, Scandalios JG, Whitt GS (eds): Isozymes. Current Topics in Biological and Medical Research, Vol 10, pp 287–311 (Alan R Liss, New York 1983).
Wright SE: The roles of mutation, inbreeding, crossbreeding, and selection in evolution. Proc 6th Int Congr Genet Ithaca 1:356–366 (1932).
Wright SE: Statistical genetics and evolution. Bull Am Math Soc 48:223–246 (1942).
Wu GF, Zhao EM: A rare karyotype of anurans, the karyotype of *Rana phrynoides*. Acta Herpetol Sinica 3:29–32 (1984).
Wu GF, Zhao EM: Preliminary studies on karyotypes of the genus *Amolops* of the Hengduan Mountains. Acta Herpetol Sinica 4:276–282 (1985).
Wu H, Zhang R: A study of sex chromosome in *Rana nigromaculata* by BrdU-Hoechst 33258-Giemsa technique. Acta Genetica Sinica 12:462–469 (1985).
Wu M, Zhao Y: A preliminary study of the karyotype of *Bufo viridis* Laurenti in Xinjiang. Zool Res (Kunming) 8:339–342 (1987).
Wu Z, Murphy C, Gall JG: A transcribed satellite DNA from the bullfrog *Rana catesbeiana*. Chromosoma 93:291–297 (1986).
Wurster DH, Atkin NB: Muntjac chromosomes: a new karyotype for *Muntiacus muntjak*. Experientia 28:972–973 (1972).
Wurster DH, Benirschke K: Chromosome studies in some deer, the springbok, and the pronghorn, with notes on placentation in deer. Cytologia 32:273–285 (1967).
Wurster DH, Benirschke K: Indian muntjac, *Muntiacus muntjak*: a deer with a low diploid chromosome number. Science 168:1364–1366 (1970).
Wurster-Hill DH, Gray CW: Giemsa banding patterns in the chromosomes of twelve species of cats (Felidae). Cytogenet Cell Genet 12:377–397 (1973).
Yadav JS, Pillai RK: Somatic karyotypes of two species of Indian frogs (Anura: Amphibia). Cytobios 13:109–115 (1975).
Yang F, Carter NP, Shi L, Ferguson-Smith MA: A comparative study of karyotypes of muntjacs by chromosome painting. Chromosoma 103:642–652 (1995).
Yang F, O'Brien PCM, Milne BS: A complete comparative map for the dog, red fox and human and its integration with canine genetic maps. Genome 62:189–202 (1999).
Yang F, Alkalaeva EZ, Perelman PL, Pardini AT, Harrison WR, et al: Reciprocal chromosome painting among human, aardvark, and elephant (superorder Afrotheria) reveals the likely eutherian ancestral karyotype. Proc Natl Acad Sci USA 100:1062–1066 (2003).
Yokoe M, Sano M, Shibata H, Shibata D, Takayama-Watanabe E, et al: Sperm proteases that may be involved in the initiation of sperm motility in the newt, *Cynops pyrrhogaster*. Int J Mol Sci 15:15210–15224 (2014).
Yong H: Geographic variation in the sex chromosomes of the West Malaysian house shrew *Suncus murinus* (Insectivora, Soricidae). Caryologia 27:65–71 (1974).
Yoshida S: From cyst to tubule: innovations in vertebrate spermatogenesis. Rev Dev Biol 5: 119–131 (2016).
Yoshimoto S, Okada E, Umemoto H, Tamura K, Uno Y, et al: A W-linked DM-domain gene, *DM-W*, participates in primary ovary development in *Xenopus laevis*. Proc Natl Acad Sci USA 105:2469–2474 (2008).
Yosida TH: Sex chromosomes of the tree frog, *Hyla arborea japonica*. J Fac Sci Hokkaido Univ Ser VI Zool 13:352–358 (1957).
Yosida TH: Frequencies of chromosome polymorphism in pairs no. 1, 9, and 13 in three geographical variants of black rats, *Rattus rattus*. Chromosoma 60:391–398 (1977).
Yosida TH, Tsuchiya K, Moriwaki K: Frequency of chromosome polymorphism in *Rattus rattus* collected in Japan. Chromosoma 33: 30–40 (1971).
Yosida TH, Kato H, Tsuchiya K, Moriwaki K, Ochai Y, Monty J: Mauritius type black rats with peculiar karyotypes derived from Robertsonian fission of small metacentrics. Chromosoma 75:51–62 (1979).

Young BE, Lips KR, Reaser JK, Ibáñez R, Salas AW, et al: Population declines and priorities for amphibian conservation in Latin America. Conserv Biol 15:1213–1223 (2001).

Yu S, Graf WD: Telomere capture as a frequent mechanism for stabilization of the terminal chromosomal deletion associated with inverted duplication (inv dup del). Cytogenet Genome Res 129:265–274 (2010).

Yu SL, Lee HY: Comparative karyoplogical analysis of the Korean tree fogs, *Hyla japonica* and *Hyla suweonensis* (Anura, Hylidae). Korean J Zool 33:1–5 (1990).

Yunis JJ, Prakash O: The origin of man: a chromosomal pictorial legacy. Science 215:1525–1530 (1982).

Yunis JJ, Yasmineh WG: Heterochromatin, satellite DNA and cell function. Science 174: 1200–1209 (1971).

Zaborski P: Sur la constance de l'expression de l'antigène H-Y chez le sexe hétérogamètique de quelques Amphibiens et sur la mise en évidence d'un dimorphisme sexuel de l'expression de cet antigène chez l'Amphibien Anoure *Pelodytes puctatus*. C R Acad Sci Ser D 289:1153–1156 (1979).

Zamudio KR, Bell RC, Nali RC, Haddad CFB, Prado CPA: Polyandry, predation, and the evolution of frog reproductive modes. Am Nat 188:S41–S61 (2016).

Zeng Y, Lou S, Liao W, Jehle R: Evolution of sperm morphology in anurans: insights into the roles of mating system and spawning location. BMC Evol Biol 14:1–88 (2014).

Zeyl CW, Green DM: Heteromorphism for a highly repeated sequence in the New Zealand frog *Leiopelma hochstetteri*. Evolution 46:1891–1899 (1992).

Zhdanova NS, Minina JM, Karamisheva TV, Draskovic I, Rubtsov NB, Londoño-Vallejo JA: The very long telomeres in *Sorex granarius* (Soricidae, Eulipothyphla) contain ribosomal DNA. Chromosome Res 15:881–890 (2007).

Ziegler CG, Lamatsch DK, Steinlein C, Engel W, Schartl M, Schmid M: The giant B chromosome of the cyprinid fish *Alburnus alburnus* harbours a retrotransposon-derived repetitive DNA sequence. Chromosome Res 11: 23–35 (2003).

Cytogenet Genome Res 2018;155:249–316
DOI: 10.1159/000490085

Accepted: February 1, 2018
by J. Smith

The Arboranan Frogs:

Appendices

KARGER

E-Mail karger@karger.com
www.karger.com/cgr

Appendix 1. Compilation of all original studies (▶) and review articles (▷) containing cytogenetic data on Arboranae. The species examined and reexamined in the present study are included. The families, subfamilies, genera, and species are listed in alphabetical order. Unnamed species (sp.) are placed at the end of the respective genera. The publications on each species (first column) are listed in chronological order. Remarks on these publications as well as errors or inconsistencies found in them are given in the fourth column. Not all published original studies (▶) could be obtained for the present study.

Author(s) ▶ original data[a] ▷ review articles	Country/ Region[b]	Speci- mens[c]	Locality of sampling, references cited, and remarks[d]	Methods applied[e]	2n/FN[f]
HYLIDAE / Acrisinae / *Acris*					
Acris blanchardi					
▶ Duellman & Cole, 1965[P]	▶ USA	▶ 2♂	▶ 4 miles WNW Lawrence, Douglas County, Kansas State; rep. *Acris crepitans blanchardi*	▶ 1	▶ 22/n.i
▶ Duellman & Cole, 1965[P]	▶ USA	▶ 3♂	▶ ca. 8 road miles S Saffordville, Chase County, Kansas State; rep. *Acris crepitans blanchardi*	▶ 1	▶ 22/n.i.
▶ Duellman & Cole, 1965[P]	▶ USA	▶ 5♂	▶ Lawrence, Douglas County, Kansas State; rep. *Acris crepitans blanchardi*	▶ 1	▶ 22/n.i.
▶ Cole, 1966[P,M,I]	▶ USA	▶ 2♂1♀ 7s.n.i.	▶ Lawrence, Douglas County, Kansas State; rep. *Acris crepitans blanchardi*	▶ 1	▶ 22/44†
▷ Foresti, 1972	▷ n.i.	▷ n.i.	▷ ref. Duellman & Cole, 1965; Cole, 1966; list. *Acris crepitans blanchardi*	▷ n.i.	▷ 22/n.i.
▷ Kuramoto, 1990	▷ N.America	▷ n.i.	▷ ref. Duellman & Cole, 1965; Cole, 1966; ctry.n.sp.; list. *Acris crepitans blanchardi*	▷ 1	▷ 22/n.i.
▷ Catroli & Kasahara, 2009	▷ n.i.	▷ n.i.	▷ ref. Duellman & Cole, 1965; Cole, 1966	▷ n.i.	▷ 22/n.i.
Acris crepitans					
▶ Bushnell et al., 1939[D]	▶ USA	▶ n.i.	▶ Relfoot Lake, Tennessee State	▶ 1	▶ 22‡/n.i.
▶ Nur & Nevo, 1969[P]	▶ USA	▶ 1♂	▶ Knoxville, Knox County, Tennessee State	▶ 1	▶ 22‡/n.i.
▶ Nur & Nevo, 1969[P]	▶ USA	▶ 8♂	▶ Knoxville, Knox County, Tennessee State; 1 to 5 B	▶ 1	▶ 22–27‡/n.i.
▶ Nur & Nevo, 1969[P]	▶ USA	▶ 6♂	▶ Clark Hill Reservoir, McDuffie County, Georgia State	▶ 1	▶ 22‡/n.i.
▶ Nur & Nevo, 1969[P]	▶ USA	▶ 4♂	▶ Murphy, Cherokee County, North Carolina State	▶ 1	▶ 22‡/n.i.
▶ Nur & Nevo, 1969[P]	▶ USA	▶ 1♂	▶ Lynchburg, Campbell County, Virginia State	▶ 1	▶ 22‡/n.i.
▶ Nur & Nevo, 1969[P]	▶ USA	▶ 2♂	▶ Lilypons, Fredrick County, Maryland State	▶ 1	▶ 22‡/n.i.
▶ Nur & Nevo, 1969[P]	▶ USA	▶ 6♂	▶ Binnewater Lake, Ulster County, New York State	▶ 1	▶ 22‡/n.i.
▶ Nur & Nevo, 1969[P]	▶ USA	▶ 1♂	▶ Castalia, Erie County, Ohio State	▶ 1	▶ 22‡/n.i.
▶ Nur & Nevo, 1969[P]	▶ USA	▶ 2♂	▶ Fulton, Whiteside County, Illinois State	▶ 1	▶ 22‡/n.i.
▶ Nur & Nevo, 1969[P]	▶ USA	▶ 3♂	▶ Delafield, Waukesha County, Wisconsin State	▶ 1	▶ 22‡/n.i.
▶ Nur & Nevo, 1969[P]	▶ USA	▶ 2♂	▶ Scotland, Bon Homme County, South Dakota State	▶ 1	▶ 22‡/n.i.
▶ Nur & Nevo, 1969[P]	▶ USA	▶ 10♂	▶ Baldwin Woods, Douglas County, Kansas State	▶ 1	▶ 22‡/n.i.
▶ Bogart, 1970b[P,I]	▶ USA	▶ n.i.	▶ southern USA; loc.n.i.	▶ 1	▶ 22/44†
▷ Duellman, 1970	▷ C.America	▷ n.i.	▷ ref.n.i.; ctry.n.sp.	▷ 1	▷ 22/n.i.
▷ Rabello, 1970	▷ n.i.	▷ n.i.	▷ ref. Bushnell et al., 1939	▷ n.i.	▷ 22/n.i.
▷ Foresti, 1972	▷ n.i.	▷ n.i.	▷ ref. Bushnell et al., 1939	▷ n.i.	▷ 22/n.i.
▶ Bogart, 1973[P]	▶ USA	▶ n.i.	▶ Texas State; loc.u.	▶ 1	▶ 22/44†
▷ King, 1990	▷ N.America	▷ n.i.	▷ ref. Cole, 1966; Morescalchi, 1968a; Nur & Nevo, 1969; Bogart, 1973; ctry.n.sp.	▷ n.i.	▷ 22/44
▷ Kuramoto, 1990	▷ N.America	▷ n.i.	▷ ref. Bushnell et al., 1939; Nur & Nevo, 1969; Bogart, 1970b, 1973; ctry.n.sp.	▷ 1	▷ 22/n.i.
▷ Catroli, 2008	▷ n.i.	▷ n.i.	▷ ref. Bushnell et al., 1939; Duellman & Cole, 1965/err.ref.; Cole, 1966/err.ref.; Nur & Nevo, 1969; Bogart, 1970b, 1973	▷ n.i.	▷ 22/n.i.
▷ Catroli, 2008	▷ n.i.	▷ n.i.	▷ ref. Bushnell et al., 1939; Duellman & Cole, 1965/err.ref.; Cole, 1966/err.ref.; Nur & Nevo, 1969; 1 to 4 B	▷ n.i.	▷ 23–26/n.i.
▷ Catroli & Kasahara, 2009	▷ n.i.	▷ n.i.	▷ ref. Bushnell et al., 1939; Nur & Nevo, 1969; Bogart, 1970b, 1973	▷ n.i.	▷ 22/n.i.
▷ Catroli & Kasahara, 2009	▷ n.i.	▷ n.i.	▷ ref. Bushnell et al., 1939; Nur & Nevo, 1969; Bogart, 1970b, 1973; 1 to 5 B	▷ n.i.	▷ 22–27/n.i.
▶ Present study	▶ USA	▶ 2U	▶ 7 miles N Desoto National Park, Hwt 10, Mississippi State	▶ 1	▶ 22/44
▶ Present study	▶ USA	▶ 2U	▶ Austin, Texas State	▶ 1	▶ 22/44
▶ Present study	▶ USA	▶ 2♂	▶ Palmetto State Park, Texas State	▶ 1	▶ 22/44
Acris gryllus					
▶ Bushnell et al., 1939[D]	▶ USA	▶ n.i.	▶ Athens, Georgia State	▶ 1	▶ 22‡/n.i.
▶ Anderson, 1986[I]	▶ USA	▶ 2♂1♀	▶ Jasper County, South Carolina State; rep. *Acris gryllus gryllus*	▶ 1,2,4,11	▶ 22/44†
▶ Anderson, 1986[I]	▶ USA	▶ 1♀	▶ Orangeburg County, South Carolina State; rep. *Acris gryllus gryllus*	▶ 1,2,4,11	▶ 22/44†
▶ Anderson, 1986[I]	▶ USA	▶ 1♀	▶ Effingham County, Georgia State; rep. *Acris gryllus gryllus*	▶ 1,2,4,11	▶ 22/44†
▶ Anderson, 1986[I]	▶ USA	▶ 2♂	▶ Santa Rosa County, Florida State; rep. *Acris gryllus gryllus*	▶ 1,2,4,11	▶ 22/44†
▷ Rabello, 1970	▷ n.i.	▷ n.i.	▷ ref. Bushnell et al., 1939	▷ n.i.	▷ 22/n.i.
▷ Foresti, 1972	▷ n.i.	▷ n.i.	▷ ref. Bushnell et al., 1939	▷ n.i.	▷ 22/n.i.
▷ King, 1990	▷ N.America	▷ n.i.	▷ ref. Anderson, 1986; ctry.n.sp.	▷ 2,4,11	▷ 22/44
▷ Kuramoto, 1990	▷ N.America	▷ n.i.	▷ ref. Bushnell et al., 1939; ctry.n.sp.	▷ 1	▷ 22/n.i.
▷ Anderson, 1991[I]	▷ Nearctic	▷ n.i.	▷ ref. Anderson, 1986; ctry.n.sp.	▷ 2,4	▷ 22/44†
▷ Catroli, 2008	▷ n.i.	▷ n.i.	▷ ref. Bushnell et al., 1939; Anderson, 1991	▷ n.i.	▷ 22/n.i.
▷ Catroli & Kasahara, 2009	▷ n.i.	▷ n.i.	▷ ref. Bushnell et al., 1939; Anderson, 1991	▷ 17	▷ 22/n.i.
▶ Present study	▶ USA	▶ 1♂	▶ 1 mile N of Edith, 10 miles N Florida State Line, Hwy 441, Georgia State	▶ 1	▶ 22/44
▶ Present study	▶ USA	▶ 4♂6♀	▶ 4 miles N Southport, Brunswick County, North Carolina State	▶ 1	▶ 22/44
▶ Present study	▶ USA	▶ 1♀	▶ NE Florida, loc.u.	▶ 1	▶ 22/44
▶ Present study	▶ USA	▶ 1U	▶ Sam Houston National Forest, Texas State	▶ 1	▶ 22/44

DOI: 10.1159/000490085

Author(s) ▶ original data[a] ▷ review articles	Country/ Region[b]	Speci-mens[c]	Locality of sampling, references cited, and remarks[d]	Methods applied[e]	2n/FN[f]
HYLIDAE / Acrisinae / *Hyliola*					
Hyliola cadaverina					
▶ Maxson & Jameson, 1968[P,M]	▶ USA	▶ 7♂	▶ Trabuco Canyon and Borrego, California State; rep. *Hyla californiae*	▶ 1	▶ 24/48[†]
▶ Anderson, 1986[P,I]	▶ USA	▶ 6♂5♀	▶ Ventura County, California State; rep. *Hyla cadaverina*	▶ 1,2,4,11	▶ 24/48[†]
▶ Anderson, 1986[P,I]	▶ USA	▶ 1♂	▶ Santa Barbara County, California State; rep. *Hyla cadaverina*	▶ 1,2,4,11	▶ 24/48[†]
▷ Foresti, 1972	▷ n.i.	▷ n.i.	▷ ref. Maxson & Jameson, 1968; list. *Hyla californiae*	▷ n.i.	▷ 24/n.i.
▷ Cole, 1974	▷ n.i.	▷ n.i.	▷ ref. Maxson & Jameson, 1968; list. *Hyla cadaverina*	▷ n.i.	▷ 24/n.i.
▷ Morescalchi, 1979	▷ n.i.	▷ n.i.	▷ ref. Bogart, 1973/err.ref.; list. *Hyla cadaverina*	▷ n.i.	▷ 24/n.i.
▷ King, 1990	▷ N.America	▷ n.i.	▷ ref. Anderson, 1986; ctry.n.sp.; list. *Hyla cadaverina*	▷ 2,4,11	▷ 24/48
▷ King, 1990	▷ N.America	▷ n.i.	▷ ref. Maxson & Jameson, 1978/err.ref.; ctry.n.sp.; list. *Hyla californiae*	▷ n.i.	▷ 24/n.i.
▷ Kuramoto, 1990	▷ N.America	▷ n.i.	▷ ref. Maxson & Jameson, 1968; ctry.n.sp.; list. *Hyla cadaverina*	▷ 1	▷ 24/n.i.
▷ Anderson, 1991[I]	▷ Holarctic	▷ n.i.	▷ ref. Anderson, 1986; ctry.n.sp., list. *Hyla cadaverina*	▷ 2,4,11	▷ 24/48[†]
▷ Nunes & Fagundes, 2008b	▷ n.i.	▷ n.i.	▷ ref. Maxson & Jameson, 1968; list. *Pseudacris cadaverina*	▷ n.i.	▷ 24/48
▷ Catroli, 2008	▷ n.i.	▷ n.i.	▷ ref. Maxson & Jameson, 1968; list. *Pseudacris cadaverina*	▷ n.i.	▷ 24/n.i.
▷ Catroli & Kasahara, 2009	▷ n.i.	▷ n.i.	▷ ref. Maxson & Jameson, 1968; Anderson, 1991; list. *Pseudacris cadaverina*	▷ 17	▷ 24/n.i.
▶ Present study	▶ USA	▶ 1♂	▶ California State; loc.u.	▶ 1	▶ 24/48
Hyliola regilla					
▶ Morescalchi, 1965[P,M]	▶ USA	▶ n♂n♀	▶ loc.u., California State	▶ 1	▶ 24/48[†]
▶ Maxson & Jameson, 1968[P,M]	▶ USA	▶ 12♂	▶ La Jolla, Ramona, La Mesa and Carrizo Creek near Jacumba, California State; rep. *Hyla regilla hypochondriaca* and *Hyla regilla deserticola*	▶ 1	▶ 24/48[†]
▶ Bogart, 1970b[P,I]	▶ USA	▶ n.i.	▶ California State; loc.n.i.; rep. *Hyla regilla*	▶ 1	▶ 24/48[†]
▷ Rabello, 1970	▷ n.i.	▷ n.i.	▷ ref. Morescalchi, 1965; list. *Hyla regilla*	▷ n.i.	▷ 24/n.i.
▷ Foresti, 1972	▷ n.i.	▷ n.i.	▷ ref. Morescalchi, 1965; list. *Hyla regilla*	▷ n.i.	▷ 24/n.i.
▷ Cole, 1974	▷ n.i.	▷ n.i.	▷ ref. Morescalchi, 1965; Maxson & Jameson, 1968; list. *Hyla regilla*	▷ n.i.	▷ 24/n.i.
▶ Anderson, 1986[I]	▶ USA	▶ 4♂2♀	▶ Santa Barbara County, California State; rep. *Hyla regilla*	▶ 1,2,4,11	▶ 24/48[†]
▶ Anderson, 1986[I]	▶ USA	▶ 2♂	▶ San Bernardino County, California State; rep. *Hyla regilla*	▶ 1,2,4,11	▶ 24/48[†]
▶ Anderson, 1986[I]	▶ USA	▶ 1♂	▶ Humboldt County, California State; rep. *Hyla regilla*	▶ 1,2,4,11	▶ 24/48[†]
▶ Anderson, 1986[I]	▶ USA	▶ 1♂1♀	▶ Linn County, Oregon State; rep. *Hyla regilla*	▶ 1,2,4,11	▶ 24/48[†]
▷ King, 1990	▷ N.America	▷ n.i.	▷ ref. Morescalchi, 1965; Anderson 1986; ctry.n.sp.; list. *Hyla regilla*	▷ 2,4,11	▷ 24/48
▷ Kuramoto, 1990	▷ N.America	▷ n.i.	▷ ref. Morescalchi, 1965; Maxson & Jameson, 1968; Bogart, 1970b; ctry.n.sp; list. *Hyla regilla*	▷ 1	▷ 24/n.i.
▷ Anderson, 1991[I]	▷ Holarctic	▷ n.i.	▷ ref. Anderson, 1986; ctry.n.sp.; list. *Hyla regilla*	▷ 2,4,11	▷ 24/48[†]
▷ Nunes & Fagundes, 2008b	▷ n.i.	▷ n.i.	▷ ref. Morescalchi, 1965; list. *Pseudacris regilla*	▷ n.i.	▷ 24/48
▷ Catroli, 2008	▷ n.i.	▷ n.i.	▷ ref. Morescalchi, 1965; Maxson & Jameson, 1968; Bogart, 1970b; list. *Pseudacris regilla*	▷ n.i.	▷ 24/n.i.
▷ Catroli & Kasahara, 2009	▷ n.i.	▷ n.i.	▷ ref. Morescalchi, 1965; Maxson & Jameson, 1968; Bogart, 1970b; Anderson, 1991; list. *Pseudacris regilla*	▷ 17	▷ 24/n.i.
▶ Present study	▶ USA	▶ 4n.i	▶ California State; loc.u.	▶ 1	▶ 24/48
HYLIDAE / Acrisinae / *Pseudacris*					
Pseudacris brachyphona					
▶ Duellman & Cole, 1965[N]	▶ USA	▶ 1♂	▶ 4 miles N Murraysville, Westmoreland County, Pennsylvania State	▶ 1	▶ 24/n.i.
▶ Anderson, 1986[I]	▶ USA	▶ 1♂	▶ Tallapoosa County, Alabama State	▶ 1,2,4,11	▶ 24/48[†]
▶ Anderson, 1986[I]	▶ USA	▶ 1♂	▶ Dawson County, Georgia State	▶ 1,2,4,11	▶ 24/48[†]
▷ Rabello, 1970	▷ n.i.	▷ n.i.	▷ ref. Duellman & Cole, 1965	▷ n.i.	▷ 24/n.i.
▷ King, 1990	▷ N.America	▷ n.i.	▷ ref. Duellman & Cole, 1965; Anderson, 1986; ctry.n.sp.	▷ 2,4,11	▷ 24/48
▷ Kuramoto, 1990	▷ N.America	▷ n.i.	▷ ref. Duellman & Cole, 1965; ctry.n.sp.	▷ 1	▷ 24/n.i.
▷ Anderson 1991[I]	▷ Nearctic	▷ n.i.	▷ ref. Anderson, 1986; ctry.n.sp.	▷ 2,4,11	▷ 24/48[†]
▷ Nunes & Fagundes, 2008b	▷ n.i.	▷ n.i.	▷ ref. Anderson, 1991	▷ n.i.	▷ 24/48
▷ Catroli, 2008	▷ n.i.	▷ n.i.	▷ ref. Duellman & Cole, 1965; Anderson, 1991	▷ n.i.	▷ 24/n.i.
▷ Catroli & Kasahara, 2009	▷ n.i.	▷ n.i.	▷ ref. Duellman & Cole, 1965; Anderson, 1991	▷ 17	▷ 24/n.i.
Pseudacris brimleyi					
▶ Wiley, 1982[P,I]	▶ USA	▶ 3♂	▶ Sampson County, North Carolina State	▶ 10	▶ 24/48[†]
▶ Anderson, 1986[I]	▶ USA	▶ 1♂	▶ Jasper County, South Carolina State	▶ 1,2,4,11	▶ 24/48[†]
▷ King, 1990	▷ N.America	▷ n.i.	▷ ref. Wiley, 1982; Anderson, 1986; ctry.n.sp.; err.cit.2n	▷ 2,4,10,11	▷ 26/48
▷ Kuramoto, 1990	▷ N.America	▷ n.i.	▷ ref. Wiley, 1982; ctry.n.sp.	▷ 1	▷ 24/n.i.
▷ Anderson, 1991[I]	▷ Nearctic	▷ n.i.	▷ ref. Anderson, 1986; ctry.n.sp.	▷ 2,4,11	▷ 24/48[†]
▷ Nunes & Fagundes, 2008b	▷ n.i.	▷ n.i.	▷ ref. Anderson, 1991	▷ n.i.	▷ 24/48
▷ Catroli, 2008	▷ n.i.	▷ n.i.	▷ ref. Wiley, 1982; Anderson, 1991; Wiley, 1992/err.ref.	▷ n.i.	▷ 24/n.i.
▷ Catroli & Kasahara, 2009	▷ n.i.	▷ n.i.	▷ ref. Wiley, 1982; Anderson, 1991; Wiley et al., 1992/err.ref.	▷ 17	▷ 24/n.i.
▷ Schmid & Steinlein, 2016a	▷ n.i.	▷ n.i.	▷ ref. Wiley et al., 1992/err.ref.	▷ 14	▷ 24/n.i.
▶ Present study	▶ USA	▶ 1♂	▶ loc.u.	▶ 1	▶ 24/48
▶ Present study	▶ USA	▶ 1n.i.	▶ 1.8 mi SW of Preston, Georgia State	▶ 1	▶ 24/48

Author(s) ▶ original data[a] ▷ review articles	Country/ Region[b]	Speci-mens[c]	Locality of sampling, references cited, and remarks[d]	Methods applied[e]	2n/FN[f]
Pseudacris clarkii					
▶ Bogart, 1973[P]	▶ USA	▶ n.i.	▶ Texas State; rep. *Pseudacris {clarki}*	▶ 1	▶ 24/48†
▷ Morescalchi, 1979	▷ n.i.	▷ n.i.	▷ ref. Bogart, 1973	▷ n.i.	▷ 24/n.i.
▷ Cole, 1974	▷ n.i.	▷ n.i.	▷ ref. Bogart, 1973; list. *Pseudacris {clarki}*	▷ n.i.	▷ 24/n.i.
▷ King, 1990	▷ N.America	▷ n.i.	▷ ref. Bogart, 1973; ctry.n.sp.	▷ n.i.	▷ 24/48
▷ Kuramoto, 1990	▷ N.America	▷ n.i.	▷ ref. Bogart, 1973; ctry.n.sp.	▷ 1	▷ 24/n.i.
▷ Nunes & Fagundes, 2008b	▷ n.i.	▷ n.i.	▷ ref. Bogart, 1973	▷ n.i.	▷ 24/48
▷ Catroli, 2008	▷ n.i.	▷ n.i.	▷ ref. Bogart, 1973	▷ n.i.	▷ 24/n.i.
▷ Catroli & Kasahara, 2009	▷ n.i.	▷ n.i.	▷ ref. Bogart, 1973	▷ n.i.	▷ 24/n.i.
▶ Present study	▶ USA	▶ 1s.n.i.	▶ Texas State; loc.u.	▶ 1	▶ 24/48
▶ Present study	▶ USA	▶ 2s.n.i.	▶ Austin,Texas State	▶ 1	▶ 24/48
Pseudacris crucifer					
▶ Duellman & Cole, 1965[N]	▶ USA	▶ 2♂	▶ Berlin, Camden County, New Jersey State; rep. *Hyla crucifer crucifer*	▶ 1	▶ 24‡/n.i.
▶ Duellman & Cole, 1965[N]	▶ USA	▶ 1♂	▶ NE city limits Chapel Hill, Orange County, North Carolina State; rep. *Hyla crucifer crucifer*	▶ 1	▶ 24‡/n.i.
▶ Wiley, 1982[P,I]	▶ USA	▶ 3♂	▶ Sampson County, North Carolina State; rep. *Hyla crucifer*	▶ 10	▶ 24/48†
▶ Wiley, 1982[P,I]	▶ USA	▶ 2♂	▶ Wake County, North Carolina State; rep. *Hyla crucifer*	▶ 10	▶ 24/48†
▶ Anderson, 1986[P,I]	▶ USA	▶ 3♂	▶ Orange County, New York State; rep. *Hyla crucifer crucifer*	▶ 1,4	▶ 24/48†
▶ Anderson, 1986[P,I]	▶ USA	▶ 2♀	▶ Jasper County, South Carolina State; rep. *Hyla crucifer crucifer*	▶ 1,2,4	▶ 24/48†
▶ Anderson, 1986[P,I]	▶ USA	▶ 2♂	▶ Alachua County, Florida State; rep. *Hyla crucifer bartramiana*	▶ 1,2,4	▶ 24/48†
▷ Rabello, 1970	▷ n.i.	▷ n.i.	▷ ref. Duellman & Cole, 1965; list. *Hyla crucifer crucifer*	▷ n.i.	▷ 24‡/n.i.
▷ Foresti, 1972	▷ n.i.	▷ n.i.	▷ ref. Duellman & Cole, 1965; list. *Hyla crucifer crucifer*	▷ n.i.	▷ 24‡/n.i.
▷ King, 1990	▷ N.America	▷ n.i.	▷ ref. Duellman & Cole, 1965; Wiley, 1982; Anderson, 1986; ctry.n.sp.; list. *Hyla crucifer*	▷ 2,4	▷ 24/48
▷ Kuramoto, 1990	▷ N.America	▷ n.i.	▷ ref. Duellman & Cole, 1965; Wiley, 1982; ctry.n.sp.; list. *Hyla crucifer*	▷ 17	▷ 24/n.i.
▷ Anderson, 1991[I]	▷ Holarctic	▷ n.i.	▷ ref. Anderson, 1986; ctry.n.sp.; list. *Hyla crucifer*	▷ 2,4	▷ 24/48†
▷ Nunes & Fagundes, 2008b	▷ n.i.	▷ n.i.	▷ ref. Anderson, 1991	▷ n.i.	▷ 24/48
▷ Catroli, 2008	▷ n.i.	▷ n.i.	▷ ref. Duellman & Cole, 1965; Wiley, 1982	▷ n.i.	▷ 24/n.i.
▷ Catroli & Kasahara, 2009	▷ n.i.	▷ n.i.	▷ ref. Duellman & Cole, 1965; Wiley, 1982	▷ 17	▷ 24/n.i.
▶ Present study	▶ Canada	▶ 1♂	▶ Whiteshell Provincial Park, Manitoba Province	▶ 1	▶ 24/48
▶ Present study	▶ Canada	▶ 1♂	▶ Guelph, Ontario Province	▶ 1	▶ 24/48
▶ Present study	▶ Canada	▶ 1♂	▶ Guelph, Downy Road, Ontario Province	▶ 1	▶ 24/48
▶ Present study	▶ Canada	▶ 1♂	▶ Camp Nekabong, East of Pembrook, Quebec Province	▶ 1	▶ 24/48
▶ Present study	▶ USA	▶ 1♂	▶ Sanborn Sopchoppy, Florida State	▶ 1	▶ 24/48
▶ Present study	▶ USA	▶ 1s.n.i.	▶ Texas State, loc.u.	▶ 1	▶ 24/48
▶ Present study	▶ USA	▶ 2♂	▶ Sam Houston National Forest, Texas State	▶ 1	▶ 24/48
Pseudacris feriarum					
▶ Wiley, 1982[P,I]	▶ USA	▶ 1♂	▶ Guilford County, North Carolina State; rep. *Pseudacris triseriata feriarum*	▶ 10	▶ 24/48†
▶ Wiley, 1982[P,I]	▶ USA	▶ 4♂	▶ Wake County, North Carolina State; rep. *Pseudacris triseriata feriarum*	▶ 10	▶ 24/48†
▷ Kuramoto, 1990	▷ N.America	▷ n.i.	▷ ref. Wiley, 1982; list. *Pseudacris triseriata feriarum*	▷ 17	▷ 24/n.i.
▷ Catroli, 2008	▷ n.i.	▷ n.i.	▷ ref. Wiley, 1982	▷ n.i.	▷ 24/n.i.
▷ Catroli & Kasahara, 2009	▷ n.i.	▷ n.i.	▷ ref. Duellman & Cole, 1965; Wiley, 1982	▷ 17	▷ 24/n.i.
▶ Present study	▶ USA	▶ 1♂	▶ Georgia State; loc.u.	▶ 1	▶ 24/48
Pseudacris nigrita					
▶ Anderson, 1986[I]	▶ USA	▶ 1♂	▶ Alachua County, Florida State	▶ 1,4,11	▶ 24/48†
▶ Anderson, 1986[I]	▶ USA	▶ 1♂	▶ Jasper County, South Carolina State	▶ 1,2,4,11	▶ 24/48†
▷ King, 1990	▷ N.America	▷ n.i.	▷ ref. Anderson, 1986; ctry.n.sp.	▷ 2,4,11	▷ 24/48
▷ Anderson, 1991[I]	▷ Nearctic	▷ n.i.	▷ ref. Anderson, 1986; ctry.n.sp.	▷ 2,4,11	▷ 24/48†
▷ Nunes & Fagundes, 2008b	▷ n.i.	▷ n.i.	▷ ref. Anderson, 1991	▷ n.i.	▷ 24/48
▷ Catroli, 2008	▷ n.i.	▷ n.i.	▷ ref. Anderson, 1991; Wiley, 1992/err.ref.	▷ n.i.	▷ 24/n.i.
▷ Catroli & Kasahara, 2009	▷ n.i.	▷ n.i.	▷ ref. Anderson, 1991; Wiley et al., 1992	▷ 17	▷ 24/n.i.
▷ Schmid & Steinlein, 2016a	▷ n.i.	▷ n.i.	▷ ref. Wiley et al., 1992	▷ 14	▷ 24/n.i.
Pseudacris ocularis					
▷ Anderson, 1991[I]	▷ Nearctic	▷ n.i.	▷ ref. Anderson, 1986; ctry.n.sp.; list. *Limnaoedus ocularis*	▷ 2,4	▷ 24/48†
▶ Present study	▶ USA	▶ 1♂1♀	▷ Hwy 441, 1 mi N of Edith, 10 mi N Florida State Line, Georgia State	▶ 1	▶ 24/48
Pseudacris ornata					
▶ Schmid, 1978a[P,I]	▶ N.America	▶ 1♀	▶ specimen obtained by animal dealer; loc.n.i.; ctry.n.sp.	▶ 1,2,4,5	▶ 24/48†
▶ Wiley, 1982[P,I]	▶ USA	▶ 3♂	▶ Scotland County, North Carolina State	▶ 10	▶ 24/48†
▶ Anderson, 1986[I]	▶ USA	▶ 1♂	▶ Alachua County, Florida State	▶ 1,4,11	▶ 24/48†
▶ Anderson, 1986[I]	▶ USA	▶ 1♀	▶ Scotland County, North Carolina State	▶ 1,2,4,11	▶ 24/48†
▷ King, 1990	▷ N.America	▷ n.i.	▷ ref. Schmid, 1978a; Wiley, 1982; Anderson, 1986; ctry.n.sp.	▷ 2,4,10, 11,17	▷ 24/48
▷ Kuramoto, 1990	▷ N.America	▷ n.i.	▷ ref. Schmid, 1978a; Wiley, 1982; ctry.n.sp.	▷ 1	▷ 24/n.i.
▷ Anderson, 1991[I]	▷ Nearctic	▷ n.i.	▷ ref. Anderson, 1986; ctry.n.sp.	▷ 2,4,11	▷ 24/48†
▷ Nunes & Fagundes, 2008b	▷ n.i.	▷ n.i.	▷ ref. Anderson, 1991	▷ n.i.	▷ 24/48
▷ Catroli, 2008	▷ n.i.	▷ n.i.	▷ ref. Schmid, 1978a; Wiley, 1982; Anderson, 1991	▷ n.i.	▷ 24/n.i.
▷ Catroli & Kasahara, 2009	▷ n.i.	▷ n.i.	▷ ref. Schmid, 1978a; Wiley, 1982; Anderson, 1991	▷ 17	▷ 24/n.i.
▶ Present study	▶ USA	▶ 1s.n.i.	▶ Savannah Ralin, Georgia State	▶ 1	▶ 24/48
▶ Present study	▶ N.America	▶ 1♀	▶ specimen obtained by animal dealer; loc.n.i.; ctry.n.sp.	▶ 1,2,4,5	▶ 24/48

Author(s) ▶ original data[a] ▷ review articles	Country/ Region[b]	Speci- mens[c]	Locality of sampling, references cited, and remarks[d]	Methods applied[e]	2n/FN[f]
Pseudacris streckeri					
▶ Bogart, 1973[P]	▶ USA	▶ n.i.	▶ Texas State, loc.u.	▶ 1	▶ 24/48†
▷ Cole, 1974	▷ n.i.	▷ n.i.	▷ ref. Bogart, 1973	▷ n.i.	▷ 24/n.i.
▷ Morescalchi, 1979	▷ n.i.	▷ n.i.	▷ ref. Bogart, 1973	▷ n.i.	▷ 24/n.i.
▷ King, 1990	▷ N.America	▷ n.i.	▷ ref. Bogart, 1973; ctry.n.sp.	▷ n.i.	▷ 24/48
▷ Kuramoto, 1990	▷ N.America	▷ n.i.	▷ ref. Bogart, 1973; ctry.n.sp.	▷ 1	▷ 24/n.i.
▷ Nunes & Fagundes, 2008b	▷ n.i.	▷ n.i.	▷ ref. Bogart, 1973	▷ n.i.	▷ 24/48
▷ Catroli, 2008	▷ n.i.	▷ n.i.	▷ ref. Bogart, 1973	▷ n.i.	▷ 24/n.i.
▷ Catroli & Kasahara, 2009	▷ n.i.	▷ n.i.	▷ ref. Bogart, 1973	▷ n.i.	▷ 24/n.i.
▶ Present study	▶ USA	▶ 1s.n.i.	▶ Texas State, loc.u.	▶ 1	▶ 24/48
▶ Present study	▶ USA	▶ 1♂	▶ Travis County, Texas State	▶ 1	▶ 24/48
Pseudacris triseriata					
▶ Duellman & Cole, 1965[N]	▶ USA	▶ 2♂	▶ Berlin, Camden County, New Jersey State; rep. *Pseudacris triseriata kalmi*	▶ 1	▶ 24‡/n.i.
▶ Duellman & Cole, 1965[N]	▶ USA	▶ 5♂	▶ vicinity of Lawrence, Douglas County, Kansas State; rep. *Pseudacris triseriata triseriata*	▶ 1	▶ 24/n.i.
▷ Rabello, 1970	▷ n.i.	▷ n.i.	▷ ref. Duellman & Cole, 1965; list. *Pseudacris tiseriata kalmi*	▷ n.i.	▷ 24‡/n.i.
▷ Rabello, 1970	▷ n.i.	▷ n.i.	▷ ref. Duellman & Cole, 1965; list. *Pseudacris tiseriata triseriata*	▷ n.i.	▷ 24/n.i.
▶ Anderson, 1986[I]	▶ USA	▶ 2♂	▶ Coconino County, Arizona State	▶ 1,2,4,11	▶ 24/48†
▷ King, 1990	▷ N.America	▷ n.i.	▷ ref. Duellman & Cole, 1965; Wiley, 1982; Anderson, 1986; ctry.n.sp.	▷ 2,4,10,11	▷ 24/48
▷ Kuramoto, 1990	▷ N.America	▷ n.i.	▷ ref. Duellman & Cole, 1965; list. *Pseudacris tiseriata kalmi;* ctry.n.sp.	▷ 1	▷ 24/n.i.
▷ Kuramoto, 1990	▷ N.America	▷ n.i.	▷ ref. Duellman & Cole, 1965; list. *Pseudacris triseriata triseriata;* ctry. n. sp.	▷ 1	▷ 24/n.i.
▷ Anderson, 1991[I]	▷ Nearctic	▷ n.i.	▷ ref. Anderson, 1986; ctry.n.sp.	▷ 2,4,11	▷ 24/48†
▷ Nunes & Fagundes, 2008b	▷ n.i.	▷ n.i.	▷ ref. Anderson, 1991	▷ n.i.	▷ 24/48
▷ Catroli, 2008	▷ n.i.	▷ n.i.	▷ ref. Duellman & Cole, 1965; Anderson, 1991	▷ n.i.	▷ 24/n.i.
▷ Catroli, 2008	▷ n.i.	▷ n.i.	▷ ref. Duellman & Cole, 1965; list. *Pseudacris kalmi*	▷ n.i.	▷ 24/n.i.
▷ Catroli & Kasahara, 2009	▷ n.i.	▷ n.i.	▷ ref. Duellman & Cole, 1965; Anderson, 1991	▷ 17	▷ 24/n.i.
▶ Present study	▶ n.i.	▶ n.i.	▶ loc.u.	▶ 1	▶ 24/48
▶ Present study	▶ USA	▶ 1♂	▶ 15 mi E of Ruston, Louisiana State	▶ 1	▶ 24/48
▶ Present study	▶ Canada	▶ 1♂	▶ Arboretum, Guelph, Ontario Province	▶ 1	▶ 24/48
▶ Present study	▶ USA	▶ 1♂	▶ Sam Houston National Forest, Texas State	▶ 1	▶ 24/48
▶ Present study	▶ USA	▶ 1n.i.	▶ Austin, Texas State	▶ 1	▶ 24/48
HYLIDAE / Cophomantinae / *Aplastodiscus*					
Aplastodiscus albofrenatus					
▶ Carvalho et al., 2009b[P,M,I]	▶ Brazil	▶ 7♂1♀	▶ Tijuca Forest, Rio de Janeiro State	▶ 1,2,4,13	▶ 22/44†
Aplastodiscus albosignatus					
▶ Bogart, 1973[P]	▶ Brazil	▶ n.i.	▶ Boracéa, NE of São Paulo; São Paulo State; rep. *Hyla albosignata*	▶ 1	▶ 20/40†
▷ Morescalchi, 1979	▷ n.i.	▷ n.i.	▷ ref. Bogart, 1973; list. *Hyla albosignata*	▷ n.i.	▷ 20/n.i.
▷ King, 1990	▷ S.America	▷ n.i.	▷ ref. Bogart, 1973, ctry.n.sp.; list. *Hyla albosignata*	▷ n.i.	▷ 20/40
▷ Kuramoto, 1990	▷ S.C.America	▷ n.i.	▷ ref. Bogart, 1973, ctry.n.sp.; list. *Hyla albosignata*	▷ 1	▷ 20/n.i.
▷ Nunes & Fagundes, 2008b	▷ n.i.	▷ n.i.	▷ ref. Bogart, 1973	▷ n.i.	▷ 20/40
▷ Catroli, 2008	▷ n.i.	▷ n.i.	▷ ref. Bogart, 1973; Carvalho, 2005	▷ n.i.	▷ 20/n.i.
▶ Carvalho et al., 2009a[P,M,I]	▶ Brazil	▶ 1♂	▶ Piraquara Municipality, Paraná State	▶ 1,2,4,13	▶ 20/40†
▶ Carvalho et al., 2009a[P,M,I]	▶ Brazil	▶ 5♂	▶ São Bento do Sul Municipality, Santa Catarina State	▶ 1,2,4,13	▶ 20/40†
▷ Catroli & Kasahara, 2009	▷ n.i.	▷ n.i.	▷ ref. Bogart, 1973; Carvalho et al., 2009a	▷ 17	▷ 20/n.i.
▶ Present study	▶ Brazil	▶ 1♂	▶ Boracéa, NE of São Paulo; São Paulo State	▶ 1	▶ 20/40
▶ Present study	▶ Brazil	▶ 2U	▶ Teresópolis, W of Rio de Janiero, Rio de Janeiro State	▶ 1	▶ 20/40
Aplastodiscus arildae					
▶ Bogart, 1973[P]	▶ Brazil	▶ n.i.	▶ Boracéa, NE of São Paulo; São Paulo State; rep. *Hyla albofrenata*	▶ 1	▶ 22/44†
▷ King, 1990	▷ S.America	▷ n.i.	▷ ref. Bogart, 1973; ctry.n.sp.; list. *Hyla albofrenata*	▷ n.i.	▷ 22/44
▷ Kuramoto, 1990	▷ S.C.America	▷ n.i.	▷ ref. Bogart, 1973; ctry.n.sp.; list. *Hyla albofrenata*	▷ n.i.	▷ 22/n.i.
▷ Catroli, 2008	▷ n.i.	▷ n.i.	▷ ref. Bogart, 1973; Carvalho, 2005	▷ n.i.	
▷ Nunes & Fagundes, 2008b	▷ n.i.	▷ n.i.	▷ ref. Bogart, 1973; list. *Aplastodiscus albofreanatus*	▷ n.i.	▷ 22/44
▶ Carvalho et al., 2009b[P,M,I]	▶ Brazil	▶ 4♂	▶ Parque Natural Municipal da Serra do Itapety, Mogi das Cruzes, São Paulo State	▶ 1,2,4, 13,14	▶ 22/44†
▶ Carvalho et al., 2009b[P,M,I]	▶ Brazil	▶ 7♂1♀	▶ Tijuca Forest, Rio de Janeiro State; rep. *Aplastodiscus albofrenatus*	▶ 1,2,4, 13,14	▶ 22/44†
▷ Catroli & Kasahara, 2009	▷ n.i.	▷ n.i.	▷ ref. Bogart, 1973; Carvalho et al., 2009b	▷ 17	▷ 22/n.i.
▶ Castro et al., 2012[N]	▶ n.i.	▶ n.i.	▶ n.i.	▶ 1,4	▶ 22/n.i.
▶ Gruber et al., 2012a[P]	▶ Brazil	▶ 1♂	▶ Serra do Japí, Jundiaí, São Paulo State	▶ 1,2,4,6, 9,12–14	▶ 22/44
▶ Gruber et al., 2012a[P]	▶ Brazil	▶ 4♂	▶ Mogi das Cruzes, São Paulo State	▶ 1,2,4,6, 9,12–14	▶ 22/44
▷ Schmid & Steinlein, 2016[a]	▷ n.i.	▷ n.i.	▷ ref. Carvalho et al., 2009b	▷ 14	▷ 22/n.i.
▶ Present study	▶ Brazil	▶ 1♂	▶ Boraceia, São Paulo State	▶ 1,2	▶ 22/44

Author(s) ▶ original data[a] ▷ review articles	Country/ Region[b]	Speci-mens[c]	Locality of sampling, references cited, and remarks[d]	Methods applied[e]	2n/FN[f]
Aplastodiscus callipygius					
▶ Gruber et al., 2012a[P]	▶ Brazil	▶ 7♂	▶ Camanducaia, Minas Gerais State	▶ 1,2,4,6, 9,12–14	▶ 20/40
▶ Castro et al., 2012[N]	▶ n.i.	▶ n.i.	▶ n.i.	▶ 1,4	▶ 20/n.i.
▶ Present study	▶ Brazil	▶ 7♂	▶ Camanducaia, Minas Gerais State	▶ 1,2	▶ 20/40
Aplastodiscus cavicola					
▶ Castro et al., 2012[N]	▶ Brazil	▶ n.i.	▶ loc.n.i.	▶ 1,4	▶ 18/n.i.
Aplastodiscus cochranae					
▷ Catroli, 2008	▷ n.i.	▷ n.i.	▷ ref. Carvalho, 2005	▷ n.i.	▷ 24/n.i.
▶ Carvalho et al., 2009a[P,M,I]	▶ Brazil	▶ 5♂	▶ Rancho Queimado Municipality, Santa Catarina State	▶ 1,2,4,13	▶ 24/48†
▷ Catroli & Kasahara, 2009	▷ n.i.	▷ n.i.	▷ ref. Carvalho et al., 2009a	▷ 17	▷ 24/n.i.
Aplastodiscus ehrhardti					
▷ Catroli, 2008	▷ n.i.	▷ n.i.	▷ ref. Carvalho, 2005	▷ n.i.	▷ 22/n.i.
▶ Carvalho et al., 2009b[P,M,I]	▶ Brazil	▶ 4♂	▶ São Bento do Sul, Santa Catarina State	▶ 1,2,4, 13,14	▶ 22/44†
▷ Catroli & Kasahara, 2009	▷ n.i.	▷ n.i.	▷ ref. Carvalho et al., 2009b	▷ 17	▷ 22/n.i.
Aplastodiscus eugenioi					
▷ Catroli, 2008	▷ n.i.	▷ n.i.	▷ ref. Carvalho, 2005	▷ n.i.	▷ 22/n.i.
▶ Carvalho et al., 2009b[P,M,I]	▶ Brazil	▶ 3♂	▶ Picinguaba, São Paulo State	▶ 1,2,4, 13,14	▶ 22/44†
▷ Catroli & Kasahara, 2009	▷ n.i.	▷ n.i.	▷ ref. Carvalho et al., 2009b	▷ 17	▷ 22/n.i.
▶ Gruber et al., 2012a[P]	▶ Brazil	▶ 1♂1♀	▶ Ubatuba, São Paulo State	▶ 1,2,4,6, 9,12–14	▶ 22/44
▷ Schmid & Steinlein, 2016[a]	▷ n.i.	▷ n.i.	▷ ref. Gruber et al., 2012[a]	▷ 14	▷ 22/n.i.
▶ Present study	▶ Brazil	▶ 1♂1♀	▶ Ubatuba, São Paulo State	▶ 1,2	▶ 22/44
Aplastodiscus ibirapitanga					
▶ Castro et al., 2012[N]	▶ n.i.	▶ n.i.	▶ n.i.	▶ 1,4	▶ 18/n.i.
Aplastodiscus leucopygius					
▶ Bogart, 1973[P]	▶ Brazil	▶ n.i.	▶ Teresópolis, W of Rio de Janiero, Rio de Janeiro State; rep. *Hyla albosignata*	▶ 1	▶ 18/36†
▷ Morescalchi, 1979	▷ n.i.	▷ n.i.	▷ ref. Bogart, 1973; list. *Hyla albosignata*	▷ n.i.	▷ 18/n.i.
▷ King, 1990	▷ S.America	▷ n.i.	▷ ref. Bogart, 1973; ctry.n.sp.; list. *Hyla albosignata*	▷ n.i.	▷ 18/36
▷ Kuramoto, 1990	▷ S.C.America	▷ n.i.	▷ ref. Bogart, 1973; ctry.n.sp.; list. *Hyla albosignata*	▷ 1	▷ 18/n.i.
▷ Nunes & Fagundes, 2008b	▷ n.i.	▷ n.i.	▷ ref. Bogart, 1973; list. *Aplastodiscus albosignatus*	▷ n.i.	▷ 18/36
▷ Catroli, 2008	▷ n.i.	▷ n.i.	▷ ref. Bogart, 1973; Carvalho, 2005	▷ n.i.	▷ 18/n.i.
▶ Carvalho et al., 2009a[P,M,I]	▶ Brazil	▶ 7♂	▶ Mogi das Cruzes Municipality, São Paulo State	▶ 1,2,4,13	▶ 18/36†
▶ Carvalho et al., 2009a[P,M,I]	▶ Brazil	▶ 2♂	▶ Maricá Municipality, Rio de Janeiro State	▶ 1,2,4,13	▶ 18/36†
▷ Catroli & Kasahara, 2009	▷ n.i.	▷ n.i.	▷ ref. Bogart, 1973; Carvalho et al., 2009a	▷ 17	▷ 18/n.i.
▶ Castro et al., 2012[N]	▶ n.i.	▶ n.i.	▶ n.i.	▶ 1,4	▶ 18/n.i.
▶ Gruber et al., 2012a[P]	▶ Brazil	▶ 3♂1♀	▶ Serra do Japí, Jundiaí, São Paulo State	▶ 1,2,4,6, 9,12–14	▶ 18/36
▶ Gruber et al., 2012a[P]	▶ Brazil	▶ 6♂	▶ Mogi das Cruzes, São Paulo State	▶ 1,2,4,6, 9,12–14	▶ 18/36
▷ Schmid & Steinlein, 2016a	▷ n.i.	▷ n.i.	▷ ref. Gruber et al., 2012a	▷ 14	▷ 18/n.i.
▶ Present study	▶ Brazil	▶ n.i.	▶ Teresópolis, W of Rio de Janiero, Rio de Janeiro State	▶ 1	▶ 18/36
Aplastodiscus perviridis					
▶ Feitosa et al., 1995[P]	▶ Brazil	▶ 30♂	▶ Morro do Ferro, Minas Gerais State and Serra da Bocaina, Botucatu, São Paulo State	▶ 1	▶ 24/n.i.
▷ Catroli, 2008	▷ n.i.	▷ n.i.	▷ ref. Carvalho, 2005	▷ n.i.	▷ 24/n.i.
▶ Carvalho et al., 2009a[P,M,I]	▶ Brazil	▶ 6♂1♀	▶ Poços de Caldas Municipality, Minas Gerais State	▶ 1,2,4,13	▶ 24/48†
▶ Carvalho et al., 2009a[P,M,I]	▶ Brazil	▶ 5♂	▶ São Bento do Sul Municipality, Santa Catarina State	▶ 1,2,4,13	▶ 24/48†
▷ Catroli & Kasahara, 2009	▷ n.i.	▷ n.i.	▷ ref. Carvalho et al., 2009a	▷ 17	▷ 24/n.i.
▶ Gruber et al., 2012a[P]	▶ Brazil	▶ 4♂	▶ Camanducaia, Minas Gerais State	▶ 1,2,4,6, 9,12–14	▶ 24/48
▶ Present study	▶ Brazil	▶ 4♂	▶ Camanducaia, Minas Gerais State	▶ 1,2,14	▶ 24/48
Aplastodiscus weygoldti					
▶ Castro et al., 2012[N]	▶ n.i.	▶ n.i.	▶ n.i.	▶ 1,4	▶ 22/n.i.

DOI: 10.1159/000490085

Author(s) ▶ original data[a] ▷ review articles	Country/ Region[b]	Speci- mens[c]	Locality of sampling, references cited, and remarks[d]	Methods applied[e]	2n/FN[f]
HYLIDAE / Cophomantinae / *Boana*					
Boana albomarginata					
▶ Beçak, 1968[P,M]	▶ Brazil	▶ 5♂1♀	▶ Praia Grande and Paranapiacaba, São Paulo State; rep. *Hyla albomarginata*	▶ 1	▶ 24/48†
▷ Rabello, 1970	▷ n.i.	▷ n.i.	▷ ref. Beçak, 1967, 1968, list. *Hyla albomarginata*	▷ n.i.	▷ 24/n.i.
▷ Foresti, 1972	▷ n.i.	▷ n.i.	▷ ref. Beçak, 1967, 1968, list. *Hyla albomarginata*	▷ n.i.	▷ 24/n.i.
▶ Bogart, 1973[P]	▶ Brazil	▶ n.i.	▶ Tijuca Forest; Rio de Janeiro State; rep. *Hyla albofrenata*	▶ 1	▶ 24/48†
▷ Cole, 1974	▷ n.i.	▷ n.i.	▷ ref. Beçak, 1968; list. *Hyla albomarginata*	▷ n.i.	▷ 24/n.i.
▷ Cole, 1974	▷ n.i.	▷ n.i.	▷ ref. Bogart, 1973; list. *Hyla albofrenata*	▷ n.i.	▷ 24/n.i.
▷ Morescalchi, 1979	▷ n.i.	▷ n.i.	▷ ref. Bogart, 1973; list. *Hyla albofrenata*	▷ n.i.	▷ 24/n.i.
▷ King, 1990	▷ S.America	▷ n.i.	▷ ref. Beçak, 1968, ctry.n.sp.; list. *Hyla albomarginata*	▷ n.i.	▷ 24/n.i.
▷ King, 1990	▷ S.America	▷ n.i.	▷ ref. Bogart, 1973; ctry.n.sp.; list. *Hyla albofrenata*	▷ n.i.	▷ 24/48
▷ Kuramoto, 1990	▷ S.C.America	▷ n.i.	▷ ref. Beçak, 1968; ctry.n.sp.; list. *Hyla albomarginata*	▷ 1	▷ 24/n.i.
▷ Kuramoto, 1990	▷ S.C.America	▷ n.i.	▷ ref. Bogart, 1973; ctry.n.sp.; list. *Hyla albofrenata*	▷ 1	▷ 24/n.i.
▶ Nunes & Fagundes, 2008a[P,M]	▶ Brazil	▶ 1♂	▶ Anchieta Municipality, Espírito Santo State; rep. *Hypsiboas albomarginatus*	▶ 1,4	▶ 24/48
▶ Nunes & Fagundes, 2008a[P,M]	▶ Brazil	▶ 2♂	▶ Cariacica Municipality, Espírito Santo State; rep. *Hypsiboas albomarginatus*	▶ 1,4	▶ 24/48
▷ Nunes & Fagundes, 2008b	▷ n.i.	▷ n.i.	▷ ref. Beçak, 1968; list. *Hypsiboas albomarginatus*	▷ n.i.	▷ 24/48
▷ Nunes & Fagundes, 2008b	▷ n.i.	▷ n.i.	▷ ref. Bogart, 1973; list. *Aplastodiscus albofrenatus*	▷ n.i.	▷ 24/48
▷ Catroli, 2008	▷ n.i.	▷ n.i.	▷ ref. Beçak, 1968; Baldissera & Batistic, 1992; Gruber, 2002; Carvalho, 2005; list. *Hypsiboas albomarginatus*	▷ n.i.	▷ 24/n.i.
▷ Catroli, 2008	▷ n.i.	▷ n.i.	▷ Carvalho, 2005; list. *Aplastodiscus albofrenatus*	▷ n.i.	▷ 24/n.i.
▶ Carvalho et al., 2009a[P,M,I]	▶ Brazil	▶ 5♂	▶ Bertioga Municipality, São Paulo State; rep. *Hypsiboas albomarginatus*	▶ 1,2,4,13	▶ 24/48†
▶ Carvalho et al., 2009a[P,M,I]	▶ Brazil	▶ 4♂	▶ Picinguaba Municipality, São Paulo State; rep. *Hypsiboas albomarginatus*	▶ 1,2,4,13	▶ 24/48†
▶ Carvalho et al., 2009a[P,M,I]	▶ Brazil	▶ 2♂	▶ Mogi das Cruzes Municipality, São Paulo State; rep. *Hypsiboas albomarginatus*	▶ 1,2,4,13	▶ 24/48†
▷ Catroli & Kasahara, 2009	▷ n.i.	▷ n.i.	▷ ref. Beçak, 1968; Gruber, 2002; Nunes & Fagundes, 2008a; Carvalho et al., 2009a; list. *Hypsiboas albomarginatus*	▷ 17	▷ 24/n.i.
▷ de Mattos et al., 2014	▷ Brazil	▷ n.i.	▷ ref. Beçak, 1968; Gruber, 2002; Nunes & Fagundes, 2008a; Carvalho et al., 2009a; list. *Hypsiboas albomarginatus*	▷ n.i.	▷ 24/48
▷ Carvalho et al., 2014	▷ Brazil	▷ n.i.	▷ Anchieta and Cariacica, Espírito Santo State; ref. Nunes & Fagundes, 2008a; list. *Hypsiboas albomarginatus*	▷ 4	▷ 24/48
▷ Carvalho et al., 2014	▷ Brazil	▷ n.i.	▷ Bertioga, Mogi das Cruzes and Picinguaba, São Paulo State; ref. Carvalho et al., 2009a; list. *Hypsiboas albomarginatus*	▷ 2,4	▷ 24/48
Boana albopunctata					
▶ Beçak, 1968[P,M]	▶ Brazil	▶ 1♂	▶ São Paulo, São Paulo State; rep. *Hyla albopunctata*	▶ 1	▶ 22/44†
▷ Rabello, 1970	▷ n.i.	▷ n.i.	▷ ref. Beçak, 1967, 1968; list. *Hyla albopunctata*	▷ n.i.	▷ 22/n.i.
▷ Foresti, 1972	▷ n.i.	▷ n.i.	▷ ref. Beçak, 1967, 1968; list. *Hyla albopunctata*	▷ n.i.	▷ 22/n.i.
▶ Bogart, 1973[P]	▶ Brazil	▶ n.i.	▶ southeastern Brazil; loc.n.i.; rep. *Hyla albopunctata*	▶ 1	▶ 22/44†
▷ Morescalchi, 1979	▷ n.i.	▷ n.i.	▷ ref. Bogart, 1973; list. *Hyla albopunctata*	▷ n.i.	▷ 22/n.i.
▷ King, 1990	▷ S.America	▷ n.i.	▷ ref. Beçak, 1968; Bogart, 1973; ctry.n.sp.; list. *Hyla albopunctata*	▷ n.i.	▷ 22/44
▷ Kuramoto, 1990	▷ S.C.America	▷ n.i.	▷ ref. Beçak, 1968; Bogart, 1973; ctry.n.sp.; list. *Hyla albopunctata*	▷ 1	▷ 22/n.i.
▶ Feitosa et al., 1995[P]	▶ Brazil	▶ 28♂	▶ Morro do Ferro, Minas Gerais State; Botucatu, Campinas, São Paulo State; rep. *Hyla albopunctata*	▶ 1	▶ 22/n.i.
▶ Gruber et al., 2007[P]	▶ Brazil	▶ 10♂	▶ Rio Claro, São Paulo State; rep. *Hypsiboas albopunctatus*	▶ 1,2,4, 7,9,12	▶ 22/44†
▶ Gruber et al., 2007[P]	▶ Brazil	▶ 7♂	▶ Rio Claro, São Paulo State; rep. *Hypsiboas albopunctatus*; 1 B	▶ 1,2,4, 7,9,12	▶ 23/46†
▷ Nunes & Fagundes, 2008b	▷ n.i.	▷ n.i.	▷ ref. Beçak, 1968; list. *Hypsiboas albopunctatus*	▷ n.i.	▷ 22/44
▷ Catroli, 2008	▷ n.i.	▷ n.i.	▷ ref. Beçak, 1968; Bogart, 1973; Baldissera & Batistic, 1992; Insaurralde et al., 1999; Gruber et al., 2007; list. *Hypsiboas albopunctatus*	▷ n.i.	▷ 22/n.i.
▷ Catroli, 2008	▷ n.i.	▷ n.i.	▷ ref. Gruber et al., 2007; list. *Hypsiboas albopunctatus*; 1 B	▷ n.i.	▷ 23/n.i.
▷ Catroli & Kasahara, 2009	▷ n.i.	▷ n.i.	▷ ref. Beçak, 1968; Bogart, 1973; Gruber et al., 2007; list. *Hypsiboas albopunctatus*	▷ 17	▷ 22/n.i.
▷ Catroli & Kasahara, 2009	▷ n.i.	▷ n.i.	▷ ref. Gruber et al., 2007; list. *Hypsiboas albopunctatus*; 1 B	▷ 17	▷ 23/n.i.
▶ Ferro et al., 2012[P,M]	▶ Argentina	▶ 3s.n.i.	▶ Santo Domingo, Corrientes Province (27°39′S, 56°08′W); rep. *Hypsiboas albopunctatus*	▶ 1–4,6,9, 13,14	▶ 22/44†
▶ Ferro et al., 2012[P,M]	▶ Argentina	▶ 26s.n.i.	▶ Posadas, Misiones Province (27°26′S, 55°53′W); rep. *Hypsiboas albopunctatus*	▶ 1–4,6,9, 13,14	▶ 22/44†
▶ Ferro et al., 2012[P,M]	▶ Argentina	▶ 5s.n.i.	▶ Posadas, Misiones Province (27°26′S, 55°53′W); rep. *Hypsiboas albopunctatus*; 1 B	▶ 1–4,6,9, 13,14	▶ 23/46†
▶ Ferro et al., 2012[P,M]	▶ Argentina	▶ 6s.n.i.	▶ Itacaruaré, Misiones Province (27°54′S, 55°16′W); rep. *Hypsiboas albopunctatus*	▶ 1–4,6,9, 13,14	▶ 22/44†
▶ Ferro et al., 2012[P,M]	▶ Argentina	▶ 8s.n.i.	▶ Corpus, Misiones Province (27°08′S, 55°27′W); rep. *Hypsiboas albopunctatus*	▶ 1–4,6,9, 13,14	▶ 22/44†
▶ Ferro et al., 2012[P,M]	▶ Argentina	▶ 2s.n.i.	▶ Iguazú, Misiones Province (25°47′S, 54°32′W); rep. *Hypsiboas albopunctatus*	▶ 1–4,6,9, 13,14	▶ 22/44†
▶ Ferro et al., 2012[P,M]	▶ Argentina	▶ 1s.n.i.	▶ Iguazú, Misiones Province (25°47′S, 54°32′W); rep. *Hypsiboas albopunctatus*; 1 B	▶ 1–4,6,9, 13,14	▶ 23/46†

Author(s) ▶ original data[a] ▷ review articles	Country/ Region[b]	Speci- mens[c]	Locality of sampling, references cited, and remarks[d]	Methods applied[e]	2n/FN[f]
▶ Ferro et al., 2012[P,M]	▶ Argentina	▶ 4s.n.i.	▶ Península Andresito, Misiones Province (25°32′S, 54°07′W); rep. *Hypsiboas albopunctatus*	▶ 1–4,6,9, 13,14	▶ 22/44†
▶ Ferro et al., 2012[P,M]	▶ Argentina	▶ 1s.n.i.	▶ Península Andresito, Misiones Province (25°32′S, 54°07′W); rep. *Hypsiboas albopunctatus*; 2 B	▶ 1–4,6,9, 13,14	▶ 24/48†
▶ Ferro et al., 2012[P,M]	▶ Argentina	▶ 18s.n.i.	▶ Puerto Andresito, Misiones Province (25°35′S, 53°59′W); rep. *Hypsiboas albopunctatus*	▶ 1–4,6,9, 13,14	▶ 22/44†
▶ Ferro et al., 2012[P,M]	▶ Argentina	▶ 18s.n.i.	▶ Puerto Andresito, Misiones Province (25°35′S, 53°59′W); rep. *Hypsiboas albopunctatus*; 1 B	▶ 1–4,6,9, 13,14	▶ 23/46†
▶ Ferro et al., 2012[P,M]	▶ Argentina	▶ 3s.n.i.	▶ Puerto Andresito, Misiones Province (25°35′S, 53°59′W); rep. *Hypsiboas albopunctatus*; 2 B	▶ 1–4,6,9, 13,14	▶ 24/48†
▶ Ferro et al., 2012[P,M]	▶ Argentina	▶ 1s.n.i.	▶ Puerto Andresito, Misiones Province (25°35′S, 53°59′W); rep. *Hypsiboas albopunctatus*; 3 B	▶ 1–4,6,9, 13,14	▶ 25/50†
▶ Ferro et al., 2012[P,M]	▶ Argentina	▶ 7s.n.i.	▶ Arroyo Lobo, Misiones Province (25°42′S, 54°05′W); rep. *Hypsiboas albopunctatus*	▶ 1–4,6,9, 13,14	▶ 22/44†
▶ Ferro et al., 2012[P,M]	▶ Argentina	▶ 5s.n.i.	▶ Arroyo Lobo, Misiones Province (25°42′S, 54°05′W); rep. *Hypsiboas albopunctatus*; 1 B	▶ 1–4,6,9, 13,14	▶ 23/46†
▶ De Oliveira et al., 2012[P]	▶ Brazil	▶ 8s.n.i.	▶ Pirenópolis Municipality, Goiás State; rep. *Hypsiboas albopunctatus*	▶ 1	▶ 22/44
▷ de Mattos et al., 2014	▷ Brazil	▷ n.i.	▷ ref. Beçak, 1968; Bogart, 1973; Gruber et al., 2007; De Oliveira et al., 2010/err.ref.; list. *Hypsiboas albopunctatus*	▷ n.i.	▷ 22/44
▷ de Mattos et al., 2014	▷ Brazil	▷ n.i.	▷ ref. Gruber et al., 2007; 1 B; err.cit.FN; list. *Hypsiboas albopunctatus*	▷ n.i.	▷ 23/45
▷ Carvalho et al., 2014	▷ Brazil	▷ n.i.	▷ Rio Claro, São Paulo State; ref. Gruber et al., 2007; list. *Hypsiboas albopunctatus* 1 B	▷ 2,4	▷ 22/n.i.
▶ Gruber et al., 2014[P]	▶ Brazil	▶ 1♂	▶ Rio Claro, São Paulo State (22°25′20″S, 47°34′23″W); rep. *Hypsiboas albopunctatus*	▶ 1	▶ 22/44†
▶ Gruber et al., 2014[P]	▶ Brazil	▶ 3♂	▶ Rio Claro, São Paulo State (22°25′20″S, 47°34′23″W); rep. *Hypsiboas albopunctatus*; 1B	▶ 1,12,18	▶ 23/46†
▶ Present study	▶ Brazil	▶ 4♂	▶ loc.u.	▶ 1	▶ 22/44
▶ Present study	▶ Brazil	▶ 3♂	▶ Campos Elyses, Botucatu, São Paulo State	▶ 1,4,6,8,14	▶ 22/44
▶ Present study	▶ Brazil	▶ 1♂	▶ Fazenda São Joaquim, São Paulo State	▶ 1,4,6,8,14	▶ 22/44
▶ Present study	▶ Brazil	▶ 1♂	▶ A-Ukre, Pará State (07°39′S, 51°21′W), 1B	▶ 1	▶ 22/44
Boana alemani*					
▶ Present study	▶ Venezuela	▶ 3♂,2♀ 1J	▶ Henry Pittier National Park, surroundings of Parque Quamitas, Aragua State	▶ 1,4,5,7	▶ 24/48
Boana atlantica					
▶ Cervini et al., 2012[N]	▶ Brazil	▶ n.i.	▶ Bahia State; loc.n.i.; rep. *Hypsiboas atlanticus*	▶ 1,2,4	▶ 24/48
▶ Carvalho et al., 2014[P,I]	▶ Brazil	▶ n.i.	▶ Atlantic Forest fragment, Itacaré Municipality, Bahia State; rep. *Hypsiboas atlanticus*	▶ 1,2,4	▶ 24/48
Boana bischoffi					
▶ Beçak, 1968[P,M]	▶ Brazil	▶ 1♀	▶ Itapecerica da Serra, São Paulo State; rep. *Hyla multilineata*	▶ 1	▶ 24/48†
▷ Rabello, 1970	▷ n.i.	▷ n.i.	▷ ref. Beçak, 1967, 1968; list. *Hyla multilineata*	▷ n.i.	▷ 24/n.i.
▷ Foresti, 1972	▷ n.i.	▷ n.i.	▷ ref. Beçak, 1967, 1968; list. *Hyla multilineata*	▷ n.i.	▷ 24/n.i.
▶ Foresti, 1972[P,M]	▶ Brazil	▶ 3♂	▶ Rio dos Cedros, Santa Catarina State; rep. *Hyla bischoffi*	▶ 1	▶ 24/48†
▷ Cole, 1974	▷ n.i.	▷ n.i.	▷ ref. Beçak, 1968; list. *Hyla multilineata*	▷ n.i.	▷ 24/n.i.
▷ King, 1990	▷ S.America	▷ n.i.	▷ ref. Beçak, 1968; list. *Hyla multifasciata*/err.sp.n.	▷ n.i.	▷ 24/n.i.
▷ Kuramoto, 1990	▷ S.C.America	▷ n.i.	▷ ref. Beçak, 1968; ctry.n.sp.; list. *Hyla bischoffi multilineata*	▷ 1	▷ 24/n.i.
▶ Raber et al., 2004[P,M,I]	▶ Brazil	▶ 8♂	▶ São Francisco de Paula, Rio Grande do Sul State; rep. *Hyla bischoffi bischoffi*	▶ 1,2,4	▶ 24/48†
▶ Raber et al., 2004[P,M,I]	▶ Brazil	▶ 6♂	▶ Rancho Queimado, Santa Catarina State; rep. *Hyla bischoffi bischoffi*	▶ 1,2,4	▶ 24/48†
▷ Nunes & Fagundes, 2008b	▷ n.i.	▷ n.i.	▷ ref. Beçak, 1968; Raber, 2000; list. *Hypsiboas bischoffi*	▷ n.i.	▷ 24/48
▷ Catroli, 2008	▷ n.i.	▷ n.i.	▷ ref. Beçak, 1968; Foresti, 1972; Raber et al., 2004; list. *Hypsiboas bischoffi*	▷ n.i.	▷ 24/n.i.
▷ Catroli, 2008	▷ n.i.	▷ n.i.	▷ ref. Beçak, 1968; list. *Hypsiboas multifasciatus*/err.sp.n.	▷ n.i.	▷ 24/n.i.
▷ Catroli & Kasahara, 2009	▷ n.i.	▷ n.i.	▷ ref. Foresti, 1972; Raber et al., 2004; list. *Hypsiboas bischoffi*	▷ 17	▷ 24/n.i.
▷ Catroli & Kasahara, 2009	▷ n.i.	▷ n.i.	▷ ref. Beçak, 1968; list. *Hypsiboas multifasciatus*/err.sp.n.	▷ n.i.	▷ 24/n.i.
▷ de Mattos et al., 2014	▷ Brazil	▷ n.i.	▷ ref. Beçak, 1968; Foresti, 1972; Raber et al., 2000/err.ref., list. *Hypsiboas bischoffi*	▷ n.i.	▷ 24/n.i.
▷ Carvalho et al., 2014	▷ Brazil	▷ n.i.	▷ São Francisco de Paula, Rio Grande do Sul State and Rancho Queimado, Santa Catarina State; ref. Raber et al., 2004; list. *Hypsiboas bischoffi*	▷ 2,4	▷ 24/48
Boana boans					
▶ de Mattos et al., 2014[P]	▶ Brazil		▶ São Sebastião do Uatumã, Amazonas State (00°50′ to 01°55′S, 58°50′ to 60°10′W); rep. *Hypsiboas boans*	▶ 1,2,4,14	▶ 24/48
▶ Schmid & Steinlein, 2016a[P]	▶ Venezuela		▶ km 99 on road El Dorado to Santa Elena de Uairén, Bolivar State, ~ 260 m (06°04′S, 61°24′W); rep. *Hypsiboas boans*	▶ 2,4,5,6, 8,14,22	▶ 24/48†
▷ Schmid & Steinlein, 2016a	▷ n.i.		▷ ref. de Mattos et al., 2014; list. *Hypsiboas boans*	▷ 14	▷ 24/n.i.
▶ Schmid & Steinlein, 2016b[P]	▶ Venezuela		▶ km 99 on road El Dorado to Santa Elena de Uairén, Bolivar State, ~ 260 m (06°04′N, 61°24′W); rep. *Hypsiboas boans*	▶ 23	▶ 24/48†
▶ Present study	▶ Venezuela		▶ km 99 on road El Dorado to Santa Elena de Uairén, Bolivar State, ~ 260 m (06°04′N, 61°24′W)	▶ 1,2, 4–8,14	▶ 24/48

Author(s) ▶ original data[a] ▷ review articles	Country/ Region[b]	Speci- mens[c]	Locality of sampling, references cited, and remarks[d]	Methods applied[e]	2n/FN[f]
Boana caingua					
▶ Ananias, 1996[P,M,I]	▶ Brazil	▶ 19♂	▶ Botucatu, São Paulo State; rep. *Hyla caingua*	▶ 1,2,4	▶ 24/48†
▷ Nunes & Fagundes, 2008b	▷ n.i.	▷ n.i.	▷ ref. Ananias, 1996; list. *Hypsiboas caingua*	▷ n.i.	▷ 24/48
▷ Catroli, 2008	▷ n.i.	▷ n.i.	▷ ref. Ananias, 1996; list. *Hypsiboas caingua*	▷ n.i.	▷ 24/n.i.
▷ Catroli & Kasahara, 2009	▷ n.i.	▷ n.i.	▷ ref. Ananias, 1996; list. *Hypsiboas caingua*	▷ 17	▷ 24/n.i.
▷ de Mattos et al., 2014	▷ n.i.	▷ n.i.	▷ ref. Ananias, 1996; list. *Hypsiboas caingua*	▷ n.i.	▷ 24/n.i.
Boana calcarata*					
▶ Present study	▶ Brazil	▶ 1♂	A-Ukre, Pará State (807°39′S, 51°21′W)	▶ 1	▶ 24/48
Boana cinerascens					
▶ Bogart, 1973[P]	▶ Peru	▶ n.i.	▶ loc.n.i.; rep. *Hyla granosa*	▶ 1	▶ 24/48†
▷ Cole, 1974	▷ n.i.	▷ n.i.	▷ ref. Bogart, 1973; list. *Hyla granosa*	▷ n.i.	▷ 24/n.i.
▷ Morescalchi, 1979	▷ n.i.	▷ n.i.	▷ ref. Bogart, 1973; list. *Hyla granosa*	▷ n.i.	▷ 24/n.i.
▷ Kuramoto, 1990	▷ S.C.America	▷ n.i.	▷ ref. Bogart, 1973; ctry.n.sp.; list. *Hyla granosa*	▷ 1	▷ 24/n.i.
▷ Nunes & Fagundes, 2008b	▷ n.i.	▷ n.i.	▷ ref. Bogart, 1973; list. *Hypsiboas cinerascens*	▷ n.i.	▷ 24/48
▷ Catroli, 2008	▷ n.i.	▷ n.i.	▷ ref. Bogart, 1973; list. *Hypsiboas granosus*	▷ n.i.	▷ 24/48
▷ Catroli & Kasahara, 2009	▷ n.i.	▷ n.i.	▷ ref. Bogart, 1973; list. *Hypsiboas cinerascens*	▷ n.i.	▷ 24/n.i.
▶ de Mattos et al., 2012[N]	▶ Brazil	▶ 3♂	▶ campus of Universidade Federal do Amazonas, Manaus, Amazonas State; rep. *Hypsiboas cinerascens*	▶ 1,2,4	▶ 24/n.i.
▷ de Mattos et al., 2014	▷ n.i.	▷ n.i.	▷ ref. Bogart, 1973; list. *Hypsiboas cinerascens*	▷ n.i.	▷ 24/n.i.
▶ de Mattos et al., 2014[P]	▶ Brazil	▶ 4♂	▶ Manaus, Amazonas State (03°04′34″S, 59°57′30″W); rep. *Hypsiboas cinerascens*	▶ 1,2,4,14	▶ 24/48
▶ Present study	▶ Brazil	▶ 1♂	▶ Bosque Nat de Iparia, Huánuco Department	▶ 1	▶ 24/48
Boana cordobae					
▶ Baraquet et al., 2013[P,M,I]	▶ Argentina	▶ 32♂	▶ Achiras (33°09′28.64″S, 64°58′55.13″W), Las Guindas (32°35′35.22″S, 64°42′38.92″W), Pampa de Achala (31°49′41.8″S, 64°51′44.9″W), Los Linderos (32°00′54.05″S, 64°56′42.97″W), Los Tabaquillos 32°23′ 59.75″S, 64°55′33.69″W), La Carolina (32°48′43.94″S, 66°05′48.15″W), Córdoba and San Luis Provinces; rep. *Hypsiboas cordobae*	▶ 1,2	▶ 24/48
▷ de Mattos et al., 2014	▷ Argentina	▷ n.i.	▷ ref. Baraquet et al., 2013; list. *Hypsiboas cordobae*	▷ n.i.	▷ 24/48
Boana crepitans					
▶ Duellman & Cole, 1965[N]	▶ Panama	▶ 1♂	▶ Camp Chagres, Canal Zone; rep. *Hyla crepitans*	▶ 1	▶ 24‡/n.i.
▶ Duellman & Cole, 1965[N]	▶ Panama	▶ 1♂	▶ Finca La Sumbadora, Panamá Province; rep. *Hyla crepitans*	▶ 1	▶ 24‡/n.i.
▷ Duellman, 1970	▷ C.America	▷ n.i.	▷ ref.n.i.; ctry.n.sp.; list. *Hyla crepitans*	▷ 1	▷ 24‡/n.i.
▷ Rabello, 1970	▷ n.i.	▷ n.i.	▷ ref. Duellman & Cole, 1965; list. *Hyla crepitans*	▷ n.i.	▷ 24‡/n.i.
▶ Rabello, 1970[P]	▶ n.i.	▶ 11♂1♀	▶ loc.n.i.; rep. *Hyla crepitans*	▶ 1	▶ 24/48†
▷ Foresti, 1972	▷ n.i.	▷ n.i.	▷ ref. Duellman & Cole, 1965; Rabello, 1969, 1970; list. *Hyla crepitans*	▷ n.i.	▷ 24/n.i.
▶ Bogart, 1973[P]	▶ Colombia	▶ n.i.	▶ loc.n.i.; rep. *Hyla crepitans*	▶ 1	▶ 24/48†
▷ Cole, 1974	▷ n.i.	▷ n.i.	▷ ref. Rabello, 1970; Bogart, 1973; list. *Hyla crepitans*	▷ n.i.	▷ 24/n.i.
▷ Morescalchi, 1979	▷ n.i.	▷ n.i.	▷ ref. Bogart, 1973; list. *Hyla crepitans*	▷ n.i.	▷ 24/n.i.
▷ King, 1990	▷ S.America	▷ n.i.	▷ ref. Rabello, 1970; Bogart, 1973; ctry.n.sp.; list. *Hyla crepitans*	▷ n.i.	▷ 24/48
▷ Kuramoto, 1990	▷ S.C.America	▷ n.i.	▷ ref. Bushnell et al., 1939/err.ref.; Duellman & Cole, 1965, Rabello, 1970; Bogart, 1973; ctry.n.sp.; list. *Hyla crepitans*	▷ 1	▷ 24/n.i.
▶ Gruber et al., 2007[P]	▶ Brazil	▶ 2♂	▶ Piranhas, Alagoas State; rep. *Hypsiboas crepitans*	▶ 1,2,4, 7,9,12	▶ 24/48†
▷ Nunes & Fagundes, 2008b	▷ n.i.	▷ n.i.	▷ ref. Rabello, 1970; list. *Hypsiboas crepitans*	▷ n.i.	▷ 24/48
▷ Catroli, 2008	▷ n.i.	▷ n.i.	▷ ref. Duellman & Cole, 1965; Rabello, 1970; Bogart, 1973; Gruber et al., 2007; list. *Hypsiboas crepitans*	▷ n.i.	▷ 24/n.i.
▷ Catroli & Kasahara, 2009	▷ n.i.	▷ n.i.	▷ ref. Duellman & Cole, 1965; Rabello, 1970; Bogart, 1973; Gruber et al., 2007; Nunes, 2006; list. *Hypsiboas crepitans*	▷ 17	▷ 24/n.i.
▶ Cervini et al., 2012[N]	▶ Brazil	▶ n.i.	▶ Bahia State; loc.n.i.; rep. *Hypsiboas crepitans*	▶ 1,2,4	▶ 24/48
▷ de Mattos, 2014	▷ Brazil	▷ n.i.	▷ ref. Duellman & Cole, 1965; Rabello, 1970; Bogart, 1973; Nunes, 2006; Gruber et al., 2007; list. *Hypsiboas crepitans*	▷ n.i.	▷ 24/48
▷ Carvalho et al., 2014	▷ Brazil	▷ n.i.	▷ Piranhas, Alagoas State; ref. Gruber et al., 2006/err.ref.; list. *Hypsiboas crepitans*	▷ 2,4	▷ 24/48
▶ Carvalho et al., 2014[P,I]	▶ Brazil	▶ n.i.	▶ Estação Ecológica Estadual Wenceslau Guimarães, Wenceslau Guimarães Municipality, Bahia State; rep. *Hypsiboas crepitans*	▶ 1,2,4	▶ 24/48
▶ Present study	▶ Venezuela	▶ 3♂4♀	▶ El Limón, University campus, Aragua State	▶ 1,2,4, 5,7,8	▶ 24/48
▶ Present study	▶ Venezuela	▶ 4♂	▶ Henry Pittier National Park, surroundings of Parque Quamitas, Aragua State	▶ 1,2,4, 5,7,8	▶ 24/48
▶ Present study	▶ Venezuela	▶ 3♂	▶ Henry Pittier National Park, surroundings of the Estación Biológica Rancho Grande, 920–1180 m, Aragua State	▶ 1,2,4, 5,7,8	▶ 24/48
▶ Present study	▶ Colombia	▶ 1♂	▶ Girardot, Cundinamarca Department	▶ 1	▶ 24/48
Boana faber					
▶ Beçak, 1968[P,M]	▶ Brazil	▶ 1♂	▶ Teresópolis, Rio de Janeiro State; rep. *Hyla faber*	▶ 1	▶ 24/48†
▶ Beçak, 1968[P,M]	▶ Brazil	▶ 1♀	▶ São Roque, São Paulo State; rep. *Hyla faber*	▶ 1	▶ 24/48†
▷ Rabello, 1970	▷ n.i.	▷ n.i.	▷ ref. Beçak, 1967, 1968; list. *Hyla faber*	▷ n.i.	▷ 24‡/n.i.
▷ Foresti, 1972	▷ n.i.	▷ n.i.	▷ ref. Beçak, 1967, 1968; list. *Hyla faber*	▷ n.i.	▷ 24/n.i.

Author(s) ▶ original data[a] ▷ review articles	Country/ Region[b]	Speci- mens[c]	Locality of sampling, references cited, and remarks[d]	Methods applied[e]	2n/FN[f]
▷ Cole, 1974	▷ n.i.	▷ n.i.	▷ ref. Beçak, 1968; list. *Hyla faber*	▷ n.i.	▷ 24/n.i.
▷ King, 1990	▷ S.America	▷ n.i.	▷ ref. Beçak, 1968; ctry.n.sp.; list. *Hyla faber*	▷ n.i.	▷ 24/n.i.
▷ Kuramoto, 1990	▷ S.C.America	▷ n.i.	▷ ref. Beçak, 1968; ctry.n.sp.; list. *Hyla faber*	▷ 1	▷ 24/n.i.
▶ Nunes & Fagundes, 2008a[P,M]	▶ Brazil	▶ 2♂	▶ Pedra Azul Municipality, Espírito Santo State; rep. *Hypsiboas faber*	▶ 1,4	▶ 24/48
▷ Nunes & Fagundes, 2008b	▷ n.i.	▷ n.i.	▷ ref. Beçak, 1968; list. *Hypsiboas faber*	▷ n.i.	▷ 24/48
▷ Catroli, 2008	▷ n.i.	▷ n.i.	▷ ref. Beçak, 1968; Carvalho, 2005; Catroli et al., 2006; list. *Hypsiboas faber*	▷ n.i.	▷ 24/n.i.
▶ Carvalho et al., 2009a[P,M,I]	▶ Brazil	▶ 2♂	▶ Mogi das Cruzes Municipality, São Paulo State; rep. *Hypsiboas faber*	▶ 1,2,4,13	▶ 24/48†
▶ Carvalho et al., 2009a[P,M,I]	▶ Brazil	▶ 1♂	▶ Biritiba Mirim Municipality, São Paulo State; rep. *Hypsiboas faber*	▶ 1,2,4,13	▶ 24/48†
▷ Catroli & Kasahara, 2009	▷ n.i.	▷ n.i.	▷ ref. Beçak, 1968; Nunes & Fagundes, 2008a; Carvalho et al., 2009a; list. *Hypsiboas faber*	▷ 17	▷ 24/n.i
▷ de Mattos et al., 2014	▷ Brazil	▷ n.i.	▷ ref. Beçak, 1968; Nunes & Fagundes, 2008a; Carvalho et al., 2009a; list. *Hypsiboas faber*	▷ n.i.	> 24/48
▷ Carvalho et al., 2014	▷ Brazil	▷ n.i.	▷ Pedra Azul, Espírito Santo State; ref. Nunes & Fagundes, 2008a; list. *Hypsiboas faber*	▷ 4	▷ 24/48
▷ Carvalho et al., 2014	▷ Brazil	▷ n.i.	▷ Mogi das Cruzes and Picinguaba, São Paulo State; ref. Carvalho et al., 2009a; list. *Hypsiboas faber*	▷ 4	▷ 24/48
▶ Schmid & Steinlein, 2016a[P]	▶ Brazil	▶ n.i.	▶ near Itanhaém, São Paulo State (24°10′S, 46°47′W); rep. *Hypsoboas faber*	▶ 14	▶ 24/48†
▶ Present study	▶ Brazil	▶ 1♀	▶ near Itanhaém, São Paulo State (24°10′S, 46°47′W); rep. *Hypsoboas faber*	▶ 1,2,4,5,8,14	▶ 24/48
Boana fasciata					
▶ Bogart & Bogart, 1971[N]	▶ Peru	▶ n.i.	▶ Río Pachitea, Bosque Nacional de Iparía, 1200 m, Huanuco Department; rep. *Hyla fasciata*	▶ 1	▶ 24/n.i.
▷ Kuramoto, 1990	▷ S.C.America	▷ n.i.	▷ ref. Bogart & Bogart, 1971; ctry.n.sp.; list. *Hyla fasciata*	▷ 1	▷ 24/n.i.
▷ Nunes & Fagundes, 2008b	▷ n.i.	▷ n.i.	▷ ref. Bogart, 1973/err.ref.; list. *Hypsiboas fasciatus*	▷ n.i.	▷ 24/48*
▶ Present study	▶ Brazil	▶ 1♂	▶ A-Ukre, Pará State (07°39′S, 51°21′W)	▶ 1	▶ 24/48
▶ Present study	▶ Peru	▶ 1♂	▶ Bosque Nacional. de Iparia,Huánuco Department	▶ 1	▶ 24/48
Boana geographica					
▶ Foresti, 1972[P,M]	▶ Brazil	▶ 1♂	▶ Itajibá, Bahia State; rep. *Hyla geographica*	▶ 1	▶ 24/48†
▷ Nunes & Fagundes, 2008b	▷ n.i.	▷ n.i.	▷ ref. Foresti, 1972; list. *Hypsiboas geographicus*	▷ n.i.	▷ 24/48
▷ Catroli, 2008	▷ n.i.	▷ n.i.	▷ ref. Foresti, 1972; list. *Hypsiboas geographicus*	▷ n.i.	▷ 24/n.i.
▷ Catroli & Kasahara, 2009	▷ n.i.	▷ n.i.	▷ ref. Foresti, 1972; list. *Hypsiboas geographicus*	▷ n.i.	▷ 24/n.i.
▷ de Mattos et al., 2014	▷ n.i.	▷ n.i.	▷ ref. Foresti, 1972; list. *Hypsiboas geographicus*	▷ n.i.	▷ 24/n.i.
▶ de Mattos et al., 2014[P]	▶ Brazil	▶ 3♂1♀	▶ Santa Isabel do Rio Negro, Amazonas State (00°24′24″N, 65°01′01″W); rep. *Hypsiboas geographicus*	▶ 1,2,4,14	▶ 24/48
▷ Schmid & Steinlein, 2016a	▷ n.i.	▷ n.i.	▷ ref. de Mattos et al., 2014; list. *Hypsiboas geographicus*	▷ 14	▷ 24/n.i.
Boana goiana					
▷ Catroli, 2008	▷ n.i.	▷ n.i.	▷ ref. Neves et al., 2006; list. *Hypsiboas goianus*	▷ n.i.	▷ 24/n.i.
Boana guentheri					
▶ Raber et al., 2004[P,M,I]	▶ Brazil	▶ 7♂1♀	▶ Terra de Areia Municipality; Rio Grande do Sul State; rep. *Hyla guentheri*	▶ 1,2,4	▶ 24/48†
▷ Nunes & Fagundes, 2008b	▷ n.i.	▷ n.i.	▷ ref. Raber, 2000; list. *Hypsiboas guentheri*	▷ n.i.	▷ 24/48
▷ Catroli, 2008	▷ n.i.	▷ n.i.	▷ ref. Raber et al., 2004; list. *Hypsiboas guentheri*	▷ n.i.	▷ 24/n.i.
▷ Catroli & Kasahara, 2009	▷ n.i.	▷ n.i.	▷ ref. Raber et al., 2004; list. *Hypsiboas guentheri*	▷ 17	▷ 24/n.i.
▷ de Mattos et al., 2014	▷ Brazil	▷ n.i.	▷ ref. Raber et al., 2000/err.ref.; list. *Hypsiboas guentheri*	▷ n.i.	▷ 24/n.i.
▷ Carvalho et al., 2014	▷ Brazil	▷ n.i.	▷ Terra de Areia; Rio Grande do Sul State; ref. Raber et al., 2004; list. *Hypsiboas guentheri*	▷ 2,4	▷ 24/48
Boana joaquini					
▶ Ananias, 1996[P,M,I]	▶ Brazil	▶ 2♂	▶ Urubici, Santa Catarina State; rep. *Hyla pulchella joaquini*	▶ 1,2,4	▶ 24/48†
▶ Ananias et al., 2004[I]	▶ n.i.	▶ n.i.	▶ loc.n.i.; ref. Ananias, 1996; rep. *Hyla pulchella joaquini*	▶ 2,4	▶ 24/48†
▷ Nunes & Fagundes, 2008b	▷ n.i.	▷ n.i.	▷ ref. Ananias, 1996; list. *Hypsiboas joaquini*	▷ n.i.	▷ 24/48
▷ Catroli, 2008	▷ n.i.	▷ n.i.	▷ ref. Ananias, 1996; list. *Hypsiboas joaquini*	▷ n.i.	▷ 24/n.i.
▷ Catroli & Kasahara, 2009	▷ n.i.	▷ n.i.	▷ ref. Ananias, 1996; list. *Hypsiboas joaquini*	▷ 17	▷ 24/n.i.
▷ de Mattos et al., 2014	▷ n.i.	▷ n.i.	▷ ref. Ananias, 1996; list. *Hypsiboas joaquini*	▷ n.i.	▷ 24/n.i.
▷ Carvalho et al., 2014	▷ Brazil	▷ n.i.	▷ ref. Ananias, 2004; list. *Hypsiboas joaquini*	▷ 2,4	▷ 24/48
Boana lanciformis					
▶ de Mattos et al., 2014[P]	▶ Brazil	▶ 3♂	▶ Manaus, Amazonas State (03°04′34″S, 59°57′30″W); rep. *Hypsiboas lanciformis*	▶ 1,2,4,14	▶ 22/44
▶ Present study	▶ Brazil	▶ 22/44	▶ Tabatinga, Amazonas State	▶ 1	▶ 22/44
Boana lundii					
▷ Catroli, 2008	▷ n.i.	▷ n.i.	▷ ref. Catroli et al., 2006; Souza et al., 2007; list. *Hypsiboas lundii*	▷ n.i.	▷ n.i./n.i.
▶ De Oliveira et al., 2012[P]	▶ Brazil	▶ 2s.n.i.	▶ Pirenópolis Municipality, Goiás State; rep. *Hypsiboas lundii*	▶ 1	▶ 24/48
▷ de Mattos et al., 2014	▷ Brazil	▷ n.i.	▷ ref. De Oliveira el al., 2010/err.ref.; list. *Hypsiboas lundii*	▷ n.i.	▷ 24/48

Author(s) ▶ original data[a] ▷ review articles	Country/ Region[b]	Speci- mens[c]	Locality of sampling, references cited, and remarks[d]	Methods applied[e]	2n/FN[f]
Boana marginata					
▶ Ananias et al., 2004[P,M,I]	▶ Brazil	▶ 8♂	▶ São Francisco de Paula, Rio Grande do Sul State; rep. *Hyla marginata*	▶ 1,2,4	▶ 24/48†
▷ Catroli & Kasahara, 2009	▷ n.i.	▷ n.i.	▷ ref. Ananias et al., 2004; list. *Hypsiboas marginatus*	▷ 17	▷ 24/n.i.
▷ Catroli, 2008	▷ n.i.	▷ n.i.	▷ ref. Ananias et al., 2004; list. *Hypsiboas marginatus*	▷ n.i.	▷ 24/n.i.
▷ Hernández-Guzmán et al., 2011	▷ n.i.	▷ n.i.	▷ ref. Ananias et al., 2004; list *Hyla marginata*	▷ n.i.	▷ 24/n.i.
▷ de Mattos et al., 2014	▷ Brazil	▷ n.i.	▷ ref. Ananias et al., 2004; list. *Hypsiboas marginatus*	▷ n.i.	▷ 24/48
▷ Carvalho et al., 2014	▷ Brazil	▷ n.i.	▷ São Francisco de Paula, Rio Grande do Sul State; ref. Ananias et al., 2004; list. *Hypsiboas marginatus*	▷ 2,4	▷ 24/48
Boana multifasciata					
▶ de Oliveira et al., 2012[P]	▶ Brazil	▶ 3s.n.i.	▶ Serranópolis Municipality, Goiás State; rep. *Hypsiboas multifasciatus*	▶ 1	▶ 24/48
▷ de Mattos et al., 2014	▷ Brazil	▷ n.i.	▷ ref. De Oliveira et al., 2010/err.ref.; list. *Hypsiboas multifasciatus*	▷ n.i.	▷ 24/48
▶ de Mattos et al., 2014[P]	▶ Brazil	▶ 1♀	▶ Iranduba, Amazonas State (03°09′47″S, 59°54′29″W); rep. *Hypsiboas multifasciatus*	▶ 1,2,4,14	▶ 24/48
Boana pardalis					
▶ Bogart, 1973[P]	▶ Brazil	▶ n.i.	▶ southeastern Brazil; loc.n.i.; rep. *Hyla pardalis*	▶ 1	▶ 24/48†
▷ Cole, 1974	▷ n.i.	▷ n.i.	▷ ref. Bogart, 1973; list. *Hyla pardalis*	▷ n.i.	▷ 24/n.i.
▷ Morescalchi, 1979	▷ n.i.	▷ n.i.	▷ ref. Bogart, 1973; list. *Hyla pardalis*	▷ n.i.	▷ 24/n.i.
▷ King, 1990	▷ S.America	▷ n.i.	▷ ref. Bogart, 1973; ctry.n.sp.; list. *Hyla pardalis*	▷ n.i.	▷ 24/48
▷ Kuramoto, 1990	▷ S.C.America	▷ n.i.	▷ ref. Bogart, 1973; ctry.n.sp.; list. *Hyla pardalis*	▷ 1	▷ 24/n.i.
▶ Nunes & Fagundes, 2008a[P,M]	▶ Brazil	▶ 4♂	▶ Cariacica Municipality, Espítito Santo State; rep. *Hypsiboas pardalis*	▶ 1,4,13	▶ 24/48
▷ Nunes & Fagundes, 2008b	▷ n.i.	▷ n.i.	▷ ref. Bogart, 1973; list. *Hypsiboas pardalis*	▷ n.i.	▷ 24/48
▷ Catroli, 2008	▷ n.i.	▷ n.i.	▷ ref. Bogart, 1973; Campos et al., 2004; Catroli et al., 2006; list. *Hypsiboas pardalis*	▷ n.i.	▷ 24/n.i.
▷ Catroli & Kasahara, 2009	▷ n.i.	▷ n.i.	▷ ref. Bogart, 1973; Nunes & Fagundes, 2008a; list. *Hypsiboas pardalis*	▷ 17	▷ 24/n.i.
▷ de Mattos et al., 2014	▷ Brazil	▷ n.i.	▷ ref. Bogart, 1973; Nunes & Fagundes, 2008a; list. *Hypsiboas pardalis*	▷ n.i.	▷ 24/48
▷ Carvalho et al., 2014	▷ Brazil	▷ n.i.	▷ Cariacica, Espítito Santo State; ref. Nunes & Fagundes, 2008a; list. *Hypsiboas pardalis*	▷ 4	▷ 24/48
▶ Present study	▶ Brazil	▶ 1♂	▶ Boraceia, São Paulo State	▶ 1	▶ 24/48
Boana polytaenia					
▶ Rabello, 1970[P]	▶ n.i.	▶ 3♂	▶ loc.n.i.; rep. *Hyla polytaenia*	▶ 1	▶ 24/48†
▷ Foresti, 1972	▷ n.i.	▷ n.i.	▷ ref. Rabello, 1969, 1970; list. *Hyla polytaenia*	▷ n.i.	▷ 24/n.i.
▶ Bogart, 1973[P]	▶ Brazil	▶ n.i.	▶ southeastern Brazil; loc.n.i.; rep. *Hyla polytaenia*	▶ 1	▶ 24/48†
▷ Cole, 1974	▷ n.i.	▷ n.i.	▷ ref. Rabello, 1970; Bogart, 1973; list. *Hyla polytaenia*	▷ n.i.	▷ 24/n.i.
▷ Morescalchi, 1979	▷ n.i.	▷ n.i.	▷ ref. Bogart, 1973; list. *Hyla polytaenia*	▷ n.i.	▷ 24/n.i.
▷ King, 1990	▷ S.America	▷ n.i.	▷ ref. Rabello, 1970; Bogart, 1973; ctry.n.sp.; list. *Hyla polytaenia*	▷ n.i.	▷ 24/48
▷ Kuramoto, 1990	▷ S.C.America	▷ n.i.	▷ ref. Rabello, 1970; Bogart, 1973; ctry.n.sp.; list. *Hyla polytaenia*	▷ 1	▷ 24/n.i.
▷ Nunes & Fagundes, 2008b	▷ n.i.	▷ n.i.	▷ ref. Rabello, 1970; list. *Hypsiboas polytaenius*	▷ n.i.	▷ 24/48
▶ Nunes & Fagundes, 2008b[P]	▶ Brazil	▶ 6♂	▶ Parque Estadual da Pedra Azul, Domingos Martins, Espírito Santo State; rep. *Hypsiboas polytaenius*	▶ 1	▶ 24/48
▷ Catroli, 2008	▷ n.i.	▷ n.i.	▷ ref. Rabello, 1970; Rabello et al., 1971; Bogart, 1973; list. *Hypsiboas polytaenius*	▷ n.i.	▷ 24/n.i.
▷ Catroli & Kasahara, 2009	▷ n.i.	▷ n.i.	▷ ref. Rabello, 1970; Rabello et al., 1971; Bogart, 1973; Nunes & Fagundes, 2008b; list. *Hypsiboas polytaenius*	▷ n.i.	▷ 24/n.i.
▷ Hernández-Guzmán et al., 2011	▷ n.i.	▷ n.i.	▷ ref. Nunes & Fagundes, 2008b; list. *Hypsiboas polytaenius*	▷ n.i.	▷ 24/n.i.
▷ de Mattos et al., 2014	▷ Brazil	▷ n.i	▷ ref. Rabello, 1970; Rabello et al., 1971; Bogart, 1973; Nunes & Fagundes, 2008b; list. *Hypsiboas polytaenius*	▷ n.i.	▷ 24/48
▶ Present study	▶ Brazil	▶ 1♂	▶ Boraicea, São Paulo State	▶ 1	▶ 24/48
Boana pombali					
▶ Cervini et al., 2012[N]	▶ Brazil	▶ n.i.	▶ Bahia State; loc.n.i.; rep. *Hypsiboas {pombalis}*	▶ 1,2,4	▶ 24/48
▶ Carvalho et al., 2014[P,I]	▶ Brazil	▶ n.i.	▶ Estação Ecológica Estadual Wenceslau Guimarães, Wenceslau Guimarães Municipality, Bahia State; rep. *Hypsiboas pombali*	▶ 1,2,4	▶ 24/48
Boana prasina					
▶ Beçak, 1968[P,M]	▶ Brazil	▶ 9♂1♀	▶ Campos de Jordão and São Paulo, São Paulo State; rep. *Hyla pulchella prasina*	▶ 1	▶ 24/48†
▷ Rabello, 1970	▷ n.i.	▷ n.i.	▷ ref. Beçak, 1967, 1968; list. *Hyla pulchella prasina*	▷ n.i.	▷ 24/n.i.
▷ Foresti, 1972	▷ n.i.	▷ n.i.	▷ ref. Beçak, 1967, 1968; list. *Hyla pulchella prasina*	▷ n.i.	▷ 24/n.i.
▷ Cole, 1974	▷ n.i.	▷ n.i.	▷ ref. Beçak, 1968; Bogart, 1973/err.ref.; list. *Hyla pulchella*	▷ n.i.	▷ 24/n.i.
▷ Kuramoto, 1990	▷ S.C.America	▷ n.i.	▷ ref. Beçak, 1968; ctry.n.sp.; list. *Hyla prasina*	▷ 1	▷ 24/n.i.
▶ Baldissera et al., 1993[P]	▶ Brazil	▶ 13♂1♀	▶ Serra do Japi, Jundiaí, São Paulo State; rep. *Hyla prasina*	▶ 1,2,4	▶ 24/48†
▶ Ananias, 1996[P,M,I]	▶ Brazil	▶ 5♂	▶ Botucatu, São Paulo State; rep. *Hyla prasina*	▶ 1,2,4	▶ 24/48†
▶ Ananias, 1996[P,M,I]	▶ Brazil	▶ 13♂	▶ Jundiaí, São Paulo State; rep. *Hyla prasina*	▶ 1,2,4	▶ 24/48†
▶ Ananias, 1996[P,M,I]	▶ Brazil	▶ 8♂	▶ São Joaquim; Santa Catarina State; rep. *Hyla prasina*	▶ 1,2,4	▶ 24/48†
▷ Nunes & Fagundes, 2008b	▷ n.i.	▷ n.i.	▷ ref. Beçak, 1968; list. *Hypsiboas prasinus*	▷ n.i.	▷ 24/48
▷ Catroli, 2008	▷ n.i.	▷ n.i.	▷ ref. Beçak, 1968; Baldissera et al., 1993; Ananias, 1996; list. *Hypsiboas prasinus*	▷ n.i.	▷ 24/n.i.
▷ Catroli & Kasahara, 2009	▷ n.i.	▷ n.i.	▷ ref. Beçak, 1968; Baldissera et al., 1993; Ananias, 1996; list. *Hypsiboas prasinus*	▷ 17	▷ 24/n.i.
▷ Hernández-Guzmán et al., 2011	▷ n.i.	▷ n.i.	▷ ref. Baldissera et al., 1993; list. *Hyla prasina*	▷ n.i.	▷ 24/n.i.
▷ de Mattos et al., 2014	▷ Brazil	▷ n.i.	▷ ref. Beçak, 1968; Baldissera et al., 1993; Ananias, 1996; list. *Hypsiboas prasinus*	▷ n.i.	▷ 24/48
▶ Present study	▶ Brazil	▶ n.i.	▶ São Paulo State; loc.u.	▶ 1,2,4	▶ 24/48

Author(s) ▶ original data[a] ▷ review articles	Country/ Region[b]	Speci- mens[c]	Locality of sampling, references cited, and remarks[d]	Methods applied[e]	2n/FN[f]
Boana pulchella					
▶ Saez & Brum, 1960[N]	▶ S.America	▶ n♂	▶ loc.n.i.; ctry.n.sp.; rep. *Hyla raddiana raddiana*	▶ 1	▶ 24/48†
▷ Rabello, 1970	▷ n.i.	▷ n.i.	▷ ref. Saez & Brum, 1960; list. *Hyla pulchella pulchella*	▷ n.i.	▷ 24/n.i.
▷ Foresti, 1972	▷ n.i.	▷ n.i.	▷ ref. Saez & Brum, 1960; list. *Hyla pulchella pulchella*	▷ n.i.	▷ 24/n.i.
▷ Foresti, 1972	▷ n.i.	▷ n.i.	▷ ref. Saez & Brum, 1959, 1960; list. *Hyla raddiana raddiana*	▷ n.i.	▷ 24/n.i.
▶ Bogart, 1973[P]	▶ Argentina	▶ n.i.	▶ loc.n.i.; rep. *Hyla pulchella*	▶ 1	▶ 24/48†
▷ Morescalchi, 1979	▷ n.i.	▷ n.i.	▷ ref. Bogart, 1973; list. *Hyla pulchella*	▷ n.i.	▷ 24/n.i.
▷ King, 1990	▷ S.America	▷ n.i.	▷ ref. Beçak, 1968/err.ref.; Bogart, 1973; ctry.n.sp.; list. *Hyla pulchella*	▷ n.i.	▷ 24/48
▷ King, 1990	▷ S.America	▷ n.i.	▷ ref. Saez & Brum, 1960; list. *Hyla raddiana*	▷ n.i.	▷ 24/n.i.
▷ Kuramoto, 1990	▷ S.C.America	▷ n.i.	▷ ref. Saez & Brum, 1960; Bogart, 1973; ctry.n.sp.; list. *Hyla pulchella pulchella*	▷ 1	▷ 24/n.i.
▶ Ananias, 1996[P,M,I]	▶ Brazil	▶ 16♂	▶ Torres, Rio Grande do Sul State; rep. *Hyla pulchella pulchella*	▶ 1,2,4	▶ 24/48†
▷ Nunes & Fagundes, 2008b	▷ n.i.	▷ n.i.	▷ ref. Beçak, 1968/err.ref.; list. *Hypsiboas pulchellus*	▷ n.i.	▷ 24/48
▷ Catroli & Kasahara, 2009	▷ n.i.	▷ n.i.	▷ ref. Saez & Brum, 1960; Beçak, 1968/err.ref.; Bogart, 1973; Ananias, 1996; list. *Hypsiboas pulchellus*	▷ 17	▷ 24/n.i.
▷ Catroli, 2008	▷ n.i.	▷ n.i.	▷ ref. Saez & Brum, 1960; Beçak, 1968/err.ref.; Bogart, 1973; Ananias, 1996; Insaurralde et al., 1999; list. *Hypsiboas pulchellus*	▷ n.i.	▷ 24/n.i.
▶ Baraquet et al., 2013[P,M,I]	▶ Argentina	▶ 18♂	▶ Río Cuarto (33°06′40.78″S, 64°18′16.88″W), Las Acequias (33°15′26.16″S, 63°55′15.10″W), Alejandro Roca (33°21′06″S, 63°42′10″W), Córdoba Province; rep. *Hypsiboas pulchellus*	▶ 1,2	▶ 24/48
▷ de Mattos et al., 2014	▷ Argentina	▷ n.i.	▷ ref. Saez & Brum, 1960; Beçak, 1968/err.ref.; Bogart, 1973; Ananias, 1996; Baraquet et al., 2013; list. *Hypsiboas pulchellus*	▷ n.i.	▷ 24/48
▶ Present study	▶ Argentina	▶ 1s.n.i.	▶ Río Guasta, Cordoba Province	▶ 1	▶ 24/48
▶ Present study	▶ Argentina	▶ 3s.n.i.	▶ Cordoba Province, loc.u.	▶ 1	▶ 24/48
▶ Present study	▶ Argentina	▶ 1♂	▶ Cordoba Province, loc.u.	▶ 1	▶ 24/48
Boana punctata					
▶ Bogart & Bogart, 1971[N]	▶ Peru	▶ n.i.	▶ Río Pachitea, Bosque Nacional de Iparía, 1200 m, Huanuco Department; rep. *Hyla punctata*	▶ 1	▶ 24/n.i.
▶ Foresti, 1972[P,M]	▶ Brazil	▶ 1♀	▶ Jequié, Bahia State; rep. *Hyla punctata*	▶ 1	▶ 24/48†
▶ Bogart, 1973[P]	▶ Peru	▶ n.i.	▶ loc.n.i.; rep. *Hyla punctata*	▶ 1	▶ 24/48†
▷ Morescalchi, 1979	▷ n.i.	▶ 1♀	▷ ref. Bogart, 1973; list. *Hyla punctata*	▷ n.i.	▷ 24/n.i.
▷ Cole, 1974	▷ n.i.	▷ n.i.	▷ ref. Bogart, 1973; list. *Hyla punctata*	▷ n.i.	▷ 24/n.i.
▶ Anderson, 1986[I]	▶ Paraguay	▷ n.i.	▶ Amamby; rep. *Hyla* cf. *punctata*	▶ 1,2,4	▶ 24/48†
▷ King, 1990	▷ S.America		▷ ref. Bogart, 1973; Anderson, 1986; ctry.n.sp.; list. *Hyla punctata*	▷ 2,4	▷ 24/48
▷ Kuramoto, 1990	▷ S.C.America	▷ n.i.	▷ ref. Bogart & Bogart, 1971; Bogart, 1973; ctry.n.sp.; list. *Hyla punctata*	▷ 1	▷ 24/n.i.
▷ Anderson, 1991[I]	▷ Neotropic	▷ n.i.	▷ ref. Anderson, 1986; ctry.n.sp.; list. *Hyla punctata*	▷ 2,4	▷ 24/48†
▷ Nunes & Fagundes, 2008b	▷ n.i.	▷ n.i.	▷ ref. Bogart, 1973 list. *Hypsiboas punctatus*	▷ n.i.	▷ 24/48
▷ Catroli, 2008	▷ n.i.	▷ n.i.	▷ ref. Bogart & Bogart, 1971; Foresti, 1972; Bogart, 1973; Anderson, 1991; list. *Hypsiboas punctatus*	▷ n.i	▷ 24/n.i.
▷ Catroli & Kasahara, 2009	▷ n.i.	▷ n.i.	▷ ref. Bogart & Bogart, 1971; Foresti, 1972; Bogart, 1973; Anderson, 1991; list. *Hypsiboas punctatus*	▷ 17	▷ 24/n.i.
▷ de Mattos et al., 2014	▷ n.i.	▷ n.i.	▷ ref. Bogart & Bogart, 1971; Foresti, 1972; Bogart, 1973; Anderson, 1991; list. *Hypsiboas punctatus*	▷ n.i.	▷ 24/n.i.
▶ Present study	▶ Brazil	▶ 1♂	▶ Tabatinga, Amazonas State (0415′S, 69°57′W)	▶ 1	▶ 24/48
▶ Present study	▶ Peru	▶ 1♂	▶ Bosque Nacional de Iparia, Huánuco Department	▶ 1	▶ 24/48
Boana raniceps					
▶ Rabello, 1970[P]	▶ n.i.	▶ 5♂1♀	▶ loc.n.i.; rep. *Hyla raniceps*	▶ 1	▶ 24/48†
▷ Foresti, 1972	▷ n.i.	▷ n.i.	▷ ref. Rabello, 1969, 1970; list. *Hyla raniceps*	▷ n.i.	▷ 24/n.i.
▷ Cole, 1974	▷ n.i.	▷ n.i.	▷ ref. Rabello, 1970; list. *Hyla raniceps*	▷ n.i.	▷ 24/n.i.
▷ King, 1990	▷ C.America	▷ n.i.	▷ ref. Rabello, 1970; err.cit.ctry.; list. *Hyla raniceps*	▷ n.i.	▷ 24/n.i.
▷ Kuramoto, 1990	▷ S.C.America	▷ n.i.	▷ ref. Rabello, 1970; ctry.n.sp.; list. *Hyla raniceps*	▷ 1	▷ 24/n.i.
▶ Gruber et al., 2007[P]	▶ Brazil	▶ 1♀	▶ Brasilândia, Mato Grosso do Sul State; rep. *Hypsiboas raniceps*	▶ 1,2,4, 7,9,12	▶ 24/48†
▷ Nunes & Fagundes, 2008b	▷ n.i.	▷ n.i.	▷ ref. Rabello, 1970; list. *Hypsiboas raniceps*	▷ n.i.	▷ 24/48
▷ Catroli, 2008	▷ n.i.	▷ n.i.	▷ ref. Rabello, 1970; Rabello et al., 1971; Insaurralde et al., 1999; Gruber et al., 2007; Souza et al., 2007; list. *Hypsiboas raniceps*	▷ n.i.	▷ 24/n.i.
▷ Catroli & Kasahara, 2009	▷ n.i.	▷ n.i.	▷ ref. Rabello, 1970; Rabello et al., 1971; Gruber et al., 2007; list. *Hypsiboas raniceps*	▷ 17	▷ 24/n.i.
▶ De Oliveira et al., 2012[P]	▶ Brazil	▶ 4s.n.i.	▶ Rio Verde Municipality, Goiás State; rep. *Hypsiboas raniceps*	▶ 1	▶ 24/48
▷ de Mattos et al., 2014	▷ Brazil	▷ n.i.	▷ ref. Rabello, 1970; Rabello et al., 1971; Gruber et al., 2007; Carvalho et al., 2008; list. *Hypsiboas raniceps*	▷ n.i.	▷ 24/48
▶ de Mattos et al., 2014[P]	▶ Brazil	▶ 4♂2♀	▶ Iranduba, Amazonas State (03°09′47″S, 59°54′29″W); rep *Hypsiboas raniceps*	▶ 1,2,4,14	▶ 24/48
▷ Carvalho et al., 2014	▷ Brazil	▷ n.i.	▷ Brasilândia, Mato Grosso do Sul State; ref. Gruber et al., 2006/err.ref.; list. *Hypsiboas raniceps*	▷ 2,4	▷ 24/48
▶ Gruber et al., 2014[P]	▶ Brazil	▶ 1♀	▶ Ilha do Bananal, Goiás State (10°46′07.9″S, 50°00′12.1″W); rep. *Hypsiboas raniceps*	▶ 1,18	▶ 24/48†
▶ Gruber et al., 2014[P]	▶ Brazil	▶ 1♀	▶ Santa Rita do Trivelato, Mato Grosso State (13°47′10.2″S, 55°23′18.0″W); rep. *Hypsiboas raniceps*	▶ 1,18	▶ 24/48†
▶ Present study	▶ Brazil	▶ 1♀	▶ Santa Rita do Trivelato, Mato Grosso State (13°47′10.2″S, 55°23′18.0″W)	▶ 1	▶ 24/48

Author(s) ▶ original data[a] ▷ review articles	Country/ Region[b]	Speci- mens[c]	Locality of sampling, references cited, and remarks[d]	Methods applied[e]	2n/FN[f]
Boana rosenbergi*					
▶ León, 1970[N]	▶ Costa Rica	▶ 1♂	▶ loc.n.i.; rep. *Hyla rosenbergi*	▶ 1	▶ 24‡/n.i.
▷ Foresti, 1972	▷ n.i.	▷ n.i.	▷ ref. León, 1970; list. *Hyla rosenbergi*	▷ n.i.	▷ 24‡/n.i.
▷ Morescalchi, 1979	▷ n.i.	▷ n.i.	▷ ref. León, 1970; list. *Hyla rosenbergi*	▷ n.i.	▷ 24/n.i.
▷ King, 1990	▷ C.America	▷ n.i.	▷ ref. León, 1970; ctry.n.sp.; list. *Hyla rosenbergi*	▷ n.i.	▷ 24/n.i.
▷ Kuramoto, 1990	▷ S.C.America	▷ n.i.	▷ ref. León, 1970; ctry.n.sp.; list. *Hyla rosenbergi*	▷ 1	▷ 24/n.i.
▷ Nunes & Fagundes, 2008b	▷ n.i.	▷ n.i.	▷ ref. León, 1970; list. *Hypsiboas rosenbergi*	▷ n.i.	▷ 24/48*
▷ Catroli, 2008	▷ n.i.	▷ n.i.	▷ ref. León, 1970; list. *Hypsiboas rosenbergi*	▷ n.i.	▷ 24/n.i.
▷ Catroli & Kasahara, 2009	▷ n.i.	▷ n.i.	▷ ref. León, 1970; list. *Hypsiboas rosenbergi*	▷ n.i.	▷ 24/n.i.
▷ de Mattos et al., 2014	▷ n.i.	▷ n.i.	▷ ref. León, 1970; list. *Hypsiboas rosenbergi*	▷ n.i	▷ 24/n.i.
▶ Present study	▶ Costa Rica	▶ 1♀	▶ vicinity of Quepos, Puntarenas Province, 8 m (09°25′N, 84°09′W)	▶ 2,4,5	▶ 24/48
▶ Present study	▶ Costa Rica	▶ 1♀	▶ Tropical Science Center, Rincon de Osa, 20 m, Puntarenas Province (09°35′N, 85°04′W)	▶ 2,4,5	▶ 24/48
Boana rufitela*					
▶ Duellman, 1967[N]	▶ Panama	▶ 2♂	▶ S end Cayo de Agua, Bocas del Toro Province; rep. *Hyla rufitela*	▶ 1	▶ 24‡/n.i.
▶ Duellman, 1967[N]	▶ Panama	▶ 1♂	▶ Bluefields, Península Valientes; Bocas del Toro Province; rep. *Hyla rufitela*	▶ 1	▶ 24‡/n.i.
▷ Duellman, 1970	▷ C.America	▷ n.i.	▷ ref.n.i.; ctry.n.sp.; list. *Hyla rufitela*	▷ 1	▷ 24‡/n.i.
▷ Foresti, 1972	▷ n.i.	▷ n.i.	▷ ref. Duellman, 1967; list. *Hyla rufitela*	▷ n.i.	▷ 24‡/n.i.
▷ Kuramoto, 1990	▷ S.C.America	▷ n.i.	▷ ref. Duellman, 1967; ctry.n.sp.; list. *Hyla rufitela*	▷ 1	▷ 24/n.i.
▷ Nunes & Fagundes, 2008b	▷ n.i.	▷ n.i.	▷ ref. Duellman, 1967; list. *Hypsiboas rufitelus*	▷ n.i.	▷ 24/48*
▷ Catroli, 2008	▷ n.i.	▷ n.i.	▷ ref. Duellman, 1967; list. *Hypsiboas rufitelus*	▷ n.i.	▷ 24/n.i.
▷ Catroli & Kasahara, 2009	▷ n.i.	▷ n.i.	▷ ref. Duellman, 1967; list. *Hypsiboas rufitelus*	▷ n.i.	▷ 24/n.i.
▷ de Mattos et al., 2014	▷ n.i.	▷ n.i.	▷ ref. Duellman, 1967; list. *Hypsiboas rufitelus*	▷ n.i.	▷ 24/n.i.
▶ Present study	▶ Costa Rica	▶ 1♂	▶ Tropical Science Center, Rincon de Osa, 20 m, Puntarenas Province (09°35′N, 85°04′W)	▶ 1,2,4,5, 7,8,14	▶ 24/48
▶ Present study	▶ Costa Rica	▶ 6♀	▶ Cahuita National Park, Limón Province (09°43′N, 82°50′W)	▶ 1,2,4,5, 7,8,14	▶ 24/48
Boana semiguttata					
▶ Foresti, 1996[P,M,I]	▶ Brazil	▶ 1♂	▶ Rio dos Cedros, Santa Catarina State; rep. *Hyla semiguttata*	▶ 1	▶ 24/48†
▶ Ananias, 1996[P,M,I]	▶ Brazil	▶ 4♂	▶ Ponta Grossa, Paraná State; rep. *Hyla semiguttata*	▶ 1,2,4	▶ 24/48†
▶ Ananias et al., 2004[P,M,I]	▶ Brazil	▶ 12♂	▶ Cambará do Sul, Santa Catarina State; rep. *Hyla semiguttata*	▶ 1,2,4	▶ 24/48†
▶ Ananias et al., 2004[P,M,I]	▶ Brazil	▶ 7♂	▶ São Francisco de Paula, Rio Grande do Sul State; rep. *Hyla semiguttata*	▶ 1,2,4	▶ 24/48†
▶ Ananias et al., 2004[P,M,I]	▶ Brazil	▶ 4♂	▶ Piraquara, Paraná State; rep. *Hyla semiguttata*	▶ 1,2,4	▶ 24/48†
▷ Nunes & Fagundes, 2008b	▷ n.i.	▷ n.i.	▷ ref. Foresti, 1972; list. *Hypsiboas semiguttatus*	▷ n.i.	▷ 24/48
▷ Catroli, 2008	▷ n.i.	▷ n.i.	▷ ref. Foresti, 1972; Ananias et al., 2004; list. *Hypsiboas semiguttatus*	▷ n.i.	▷ 24/n.i.
▷ Catroli & Kasahara, 2009	▷ n.i.	▷ n.i.	▷ ref. Foresti, 1972; Ananias et al., 2004; list. *Hypsiboas semiguttatus*	▷ 17	▷ 24/n.i.
▷ Hernández-Guzmán et al., 2011	▷ n.i.	▷ n.i.	▷ ref. Ananias et al., 2004; list. *Hyla semiguttata*	▷ n.i.	▷ 24/n.i.
▷ de Mattos et al., 2014	▷ Brazil	▷ n.i.	▷ ref. Foresti, 1972; Ananias et al., 2004; list. *Hypsiboas semiguttatus*	▷ n.i.	▷ 24/48
▷ Carvalho et al., 2014	▷ Brazil	▷ n.i.	▷ Cambará do Sul, Santa Catarina State, São Francisco de Paula, Rio Grande do Sul State and Piraquara, Paraná State; ref. Ananias et al., 2004; list. *Hypsiboas semiguttatus*	▷ 2,4	▷ 24/48
Boana semilineata					
▶ Nunes & Fagundes, 2008a[P,M]	▶ Brazil	▶ 2♂2U	▶ Santa Teresa Municipality, Espírito Santo State; rep. *Hypsiboas semilineatus*	▶ 1,4,13	▶ 24/48
▷ Catroli & Kasahara, 2009	▷ n.i.	▷ n.i.	▷ ref. Nunes & Fagundes, 2008a; list. *Hypsiboas semilineatus*	▷ 17	▷ 24/n.i.
▶ Cervini et al., 2012[N]	▶ Brazil	▶ n.i.	▶ Bahia State; loc.n.i.; rep. *Hypsiboas semilineatus*	▶ 1,2,4	▶ 24/48
▷ de Mattos et al., 2014	▷ Brazil	▷ n.i.	▷ ref. Nunes & Fagundes, 2008a; list. *Hypsiboas semilineatus*	▷ n.i.	▷ 24/48
▶ Carvalho et al., 2014[P,I]	▶ Brazil	▶ n.i.	▶ Estação Ecológica Estadual Wenceslau Guimarães, Wenceslau Guimarães Municipality, Bahia State; rep. *Hypsiboas semilineatus*	▶ 1,2,4	▶ 24/48
▷ Carvalho et al., 2014	▷ Brazil	▷ n.i.	▷ Santa Teresa, Espírito Santo State; ref. Nunes & Fagundes, 2008a; list. *Hypsiboas semilineatus*	▷ 4	▷ 24/48
Boana wavrini					
▶ de Mattos et al., 2014[P]	▶ Brazil	▶ 1♀	▶ São Sebastião do Uatumã, Amazonas State (00°50′ to 01°55′S, 58°50′ to 60°10′ W); rep. *Hypsiboas wavrini*	▶ 1,2,4,14	▶ 24/48
▶ de Mattos et al., 2014[P]	▶ Brazil	▶ 2♂	▶ Santa Isabel do Rio Negro, Amazonas State (00°24′24″N, 65°01′01″W); rep. *Hypsiboas wavrini*	▶ 1,2,4,14	▶ 24/48
▷ Schmid & Steinlein, 2016a	▷ n.i.	▷ n.i.	▷ ref. De Mattos et al., 2014; list. *Hypsiboas wavrini*	▷ 14	▷ 24/n.i.
***Boana* sp. (aff. *semiguttata*)**					
▶ Ananias et al., 2004[P,M,I]	▶ Argentina	▶ 4♂	▶ Misiones Province; rep. *Hyla* sp. (aff. *semiguttata*)	▶ 1,2,4	▶ 24/48†
▷ Hernández-Guzmán et al., 2011	▷ n.i.	▷ n.i.	▷ ref. Ananias et al., 2004; list. *Hyla* sp. (aff. *semiguttata*)	▷ n.i.	▷ 24/n.i.
▷ Carvalho et al., 2014	▷ Argentina	▷ n.i.	▷ Misiones; ref. Ananias et al., 2004; list. *Hypsiboas* sp. (aff. *semiguttatus*)	▷ 2,4	▷ 24/48
***Boana* sp. (*geographica* group)**					
▶ Bogart & Bogart, 1971[N]	▶ Peru	▶ n.i.	▶ Río Pachitea, Bosque Nacional de Iparía, 1200 m, Huanuco Department; rep. *Hyla geographica* group	▶ 1	▶ 24/n.i.
▷ Kuramoto, 1990	▷ S.C.America	▷ n.i.	▷ ref. Bogart & Bogart, 1971; ctry.n.sp.; list. *Hyla geographica* group	▷ 1	▷ 24/n.i.
▷ Catroli, 2008	▷ n.i.	▷ n.i.	▷ ref. Bogart & Bogart, 1971; list. *Hypsiboas* gr. *geographicus*	▷ n.i.	▷ 24/n.i.
▷ Catroli & Kasahara, 2009	▷ n.i.	▷ n.i.	▷ ref. Bogart & Bogart, 1971; list. *Hypsiboas* gr. *geographica*	▷ n.i.	▷ 24/n.i.
▷ de Mattos et al., 2014	▷ n.i.	▷ n.i.	▷ ref. Bogart & Bogart, 1971; list. *Hypsiboas* gr. *geographicus*	▷ n.i.	▷ 24/n.i.

Author(s) ▶ original data[a] ▷ review articles	Country/ Region[b]	Speci-mens[c]	Locality of sampling, references cited, and remarks[d]	Methods applied[e]	2n/FN[f]
HYLIDAE / Cophomantinae / *Bokermannohyla*					
Bokermannohyla ahenea					
▶ Catroli et al., 2011[P,M]	▶ Brazil	▶ 3♂	▶ São José do Barreiro, São Paulo State	▶ 1,2,4, 6,9,12	▶ 24/48
▶ Present study	▶ Brazil	▶ 3♂	▶ São José do Barreiro, São Paulo State	▶ 1,2	▶ 24/48
Bokermannohyla alvarengai					
▷ Catroli, 2008	▷ n.i.	▷ n.i.	▷ ref. Catroli et al., 2007	▷ n.i.	▷ 24/n.i.
▶ Catroli, 2008[P]	▶ Brazil	▶ 2J	▶ see below Catroli et al., 2011	▶ 1,2,4,12	▶ 24/48
▷ Catroli & Kasahara, 2009	▷ n.i.	▷ n.i.	▷ ref. Catroli, 2008	▷ 17	▷ 24/n.i.
▶ Catroli et al., 2011[P,M]	▶ Brazil	▶ 2J	▶ Cogonhas do Norte, Minas Gerais State	▶ 1,2,4,12	▶ 24/48
▶ Present study	▶ Brazil	▶ 2J	▶ Cogonhas do Norte, Minas Gerais State	▶ 1,2	▶ 24/48
Bokermannohyla circumdata					
▷ Catroli, 2008	▷ n.i.	▷ n.i.	▷ ref. Catroli et al., 2007	▷ n.i.	▷ 24/n.i.
▶ Catroli. 2008[P]	▶ Brazil	▶ 7♂4♀ 1J	▶ see below Catroli et al., 2011	▶ 1,2,4,6, 9,12,13	▶ 24/48
▷ Catroli & Kasahara, 2009	▷ n.i.	▷ n.i.	▷ ref. Catroli, 2008	▷ 17	▷ 24/n.i.
▶ Catroli et al., 2011[P,M]	▶ Brazil	▶ 1♀,1J	▶ Salesópolis, São Paulo State	▶ 1	▶ 24/48
▶ Catroli et al., 2011[P,M]	▶ Brazil	▶ 1♀	▶ Campos do Jordão, São Paulo State	▶ 1,4	▶ 24/48
▶ Catroli et al., 2011[P,M]	▶ Brazil	▶ 7♂2♀	▶ Camanducaia, Minas Gerais State	▶ 1,2,4, 6,9,12	▶ 24/48
▶ Present study	▶ Brazil	▶ 1J	▶ Tijuca Forest; Rio de Janeiro State	▶ 1	▶ 24/48
▶ Present study	▶ Brazil	▶ 1♂	▶ Boracéa, NE of São Paulo, São Paulo State	▶ 1	▶ 24/48
▶ Present study	▶ Brazil	▶ 1s.n.i.	▶ Boracéa, NE of São Paulo, São Paulo State	▶ 1,2	▶ 24/48
Bokermannohyla hylax					
▷ Catroli, 2008	▷ n.i.	▷ n.i.	▷ ref. Campos et al., 2004; Catroli et al., 2007	▷ n.i.	▷ 24/n.i.
▶ Catroli, 2008[P]	▶ Brazil	▶ 1♀1J	▶ see below Catroli et al., 2011	▶ 1,4	▶ 24/48
▷ Catroli & Kasahara, 2009	▷ n.i.	▷ n.i.	▷ ref. Catroli, 2008	▷ 17	▷ 24/n.i.
▶ Catroli et al., 2011[P,M]	▶ Brazil	▶ 1♀1J	▶ Salesópolis, São Paulo State	▶ 1,4	▶ 24/48
▶ Present study	▶ Brazil	▶ 1♀	▶ Campos Elyses, Botucatu, São Paulo State	▶ 1	▶ 24/48
Bokermannohyla ibitiguara					
▷ Catroli, 2008	▷ n.i.	▷ n.i.	▷ ref. Catroli et al., 2007	▷ n.i.	▷ 24/n.i.
▶ Catroli, 2008[P]	▶ Brazil	▶ 2♂	▶ see below Catroli et al., 2011	▶ 1,2,4,6,9	▶ 24/48
▷ Catroli & Kasahara, 2009	▷ n.i.	▷ n.i.	▷ ref. Catroli, 2008	▷ 17	▷ 24/n.i.
▶ Catroli et al., 2011[P,M]	▶ Brazi	▶ 2♂	▶ Furnas, Minas Gerais State	▶ 1,2,4,6,9	▶ 24/48
▶ Present study	▶ Brazil	▶ 2♂	▶ Furnas, Minas Gerais State	▶ 1,2	▶ 24/48
Bokermannohyla izecksohni					
▶ Foresti, 1972[P,M]	▶ Brazil	▶ 1♀	▶ Botucatu, São Paulo State; rep. *Hyla {circundata}*	▶ 1	▶ 24/48[†]
▷ Catroli, 2008	▷ n.i.	▷ n.i.	▷ ref. Foresti, 1972	▷ n.i.	▷ 24/n.i.
▷ Catroli & Kasahara, 2009	▷ n.i.	▷ n.i.	▷ ref. Foresti, 1972	▷ n.i.	▷ 24/n.i.
Bokermannohyla luctuosa					
▶ Baldissera et al., 1991	▶ Brazil	▶ n.i.	▶ loc.n.i	▶ 1	▶ 24/n.i.
▶ Baldissera et al., 1993[P]	▶ Brazil	▶ 2♂	▶ Serra do Japi, Jundiaí, São Paulo State; rep. *Hyla* sp. (aff. *circumdata*)	▶ 1,4	▶ 24/48[†]
▶ Baldissera et al., 1993[P]	▶ Brazil	▶ 1♂1♀	▶ Serra do Japi, Jundiaí, São Paulo State; rep. *Hyla* sp. (aff. *circumdata*); 1 B	▶ 1,4	▶ 25/50[†]
▶ Pombal & Haddad, 1993[N]	▶ Brazil	▶ 2♂	▶ Serra do Japí, Jundiaí Municipality, São Paulo State, 870 m (23°13′S, 46°48′W); rep. *Hyla luctuosa*; ref. *Hyla* sp. in Baldissera et al., 1991	▶ n.i.	▶ 24/48[†]
▶ Pombal & Haddad, 1993[N]	▶ Brazil	▶ 2♂1♀	▶ Serra do Japí, Jundiaí Municipality, São Paulo State, 870 m (23°13′S, 46°48′W); rep. *Hyla luctuosa*; ref. *Hyla* sp. in Baldissera et al., 1991	▶ n.i.	▶ 25/n.i.
▷ Nunes & Fagundes, 2008b	▷ n.i.	▷ n.i.	▷ ref. Baldissera et al., 1993	▷ n.i.	▷ 24/48
▷ Catroli, 2008	▷ n.i.	▷ n.i.	▷ ref. Baldissera et al., 1993	▷ n.i.	▷ 24/n.i.
▷ Catroli, 2008	▷ n.i.	▷ n.i.	▷ ref. Baldissera et al., 1993; 1 B	▷ n.i.	▷ 25/n.i.
▷ Catroli & Kasahara, 2009	▷ n.i.	▷ n.i.	▷ ref. Baldissera et al., 1993	▷ 17	▷ 24/n.i.
▷ Catroli & Kasahara, 2009	▷ n.i.	▷ n.i.	▷ ref. Baldissera et al., 1993; 1 B	▷ 17	▷ 25/n.i.
▷ Hernández-Guzmán et al., 2011	▷ n.i.	▷ n.i.	▷ ref. Baldissera et al., 1993; list. *Hyla* sp. (aff. *circumdata*)	▷ n.i.	▷ 24/n.i.
▷ Hernández-Guzmán et al., 2011	▷ n.i.	▷ n.i.	▷ ref. Baldissera et al., 1993; list. *Hyla* sp. (aff. *circumdata*), 1 B	▷ n.i.	▷ 25/n.i.
▶ Catroli et al., 2011[P,M]	▶ Brazil	▶ 2♂	▶ Jundiaí, São Paulo State	▶ 1,2,4,6,9	▶ 24/48
▶ Present study	▶ Brazil	▶ 2♂	▶ Jundiaí, São Paulo State	▶ 1,2,4	▶ 24/48
Bokermannohyla martinsi					
▶ Catroli et al., 2011[P,M]	▶ Brazil	▶ 3	▶ Nova Lima, Minas Gerais State	▶ 1,2,4, 6,9,12	▶ 24/48
▶ Present study	▶ Brazil	▶ 3♂	▶ Nova Lima, Minas Gerais State	▶ 1,2	▶ 24/48

Author(s) ▶ original data[a] ▷ review articles	Country/ Region[b]	Speci- mens[c]	Locality of sampling, references cited, and remarks[d]	Methods applied[e]	2n/FN[f]
Bokermannohyla saxicola					
▷ Catroli, 2008	▷ n.i.	▷ n.i.	▷ ref. Catroli et al., 2007	▷ n.i.	▷ 24/n.i.
▶ Catroli, 2008[P]	▶ Brazil	▶ 2♂	▶ see below Catroli et al., 2011	▶ 1,2,4,6,9	▶ 24/48
▷ Catroli & Kasahara, 2009	▷ n.i.	▷ n.i.	▷ ref. Catroli, 2008	▷ 17	▷ 24/n.i.
▶ Catroli et al., 2011[P,M]	▶ Brazil	▶ 2♂	▶ Serrá do Cipó, Minas Gerais State	▶ 1,2,4,6,9	▶ 24/48
▶ Present study	▶ Brazil	▶ 2♂	▶ Serrá do Cipó, Minas Gerais State	▶ 1,2	▶ 24/48
***Bokermannohyla* sp. 1**					
▷ Catroli, 2008	▷ n.i.	▷ n.i.	▷ ref. Catroli et al., 2007	▷ n.i.	▷ 24/n.i.
▶ Catroli, 2008[P]	▶ Brazil	▶ 1♀	▶ see below Catroli et al., 2011	▶ 1,4,6,9	▶ 24/48
▷ Catroli & Kasahara, 2009	▷ n.i.	▷ n.i.	▷ ref. Catroli, 2008	▷ 17	▷ 24/n.i.
▶ Catroli et al., 2011[P,M]	▶ Brazil	▶ 1♀	▶ Furnas, Minas Gerais State	▶ 1,4,6,9	▶ 24/48
▶ Present study	▶ Brazil	▶ 1♀	▶ Furnas, Minas Gerais State	▶ 1	▶ 24/48
***Bokermannohyla* sp. 2**					
▶ Catroli et al., 2011[P,M]	▶ Brazil	▶ 1♀	▶ Morretes, Paraná State	▶ 1,2,4,6,9	▶ 24/48
▶ Present study	▶ Brazil	▶ 1♀	▶ Morretes, Paraná State	▶ 1,2	▶ 24/48
***Bokermannohyla* sp. 3**					
▶ Catroli et al., 2011[P,M]	▶ Brazil	▶ 1♂	▶ Nova Lima, Minas Gerais State	▶ 1,2,4,6,9	▶ 24/48
▶ Present study	▶ Brazil	▶ 1♂	▶ Nova Lima, Minas Gerais State	▶ 1,2	▶ 24/48
***Bokermannohyla* sp. 4**					
▶ Catroli et al., 2011[P,M]	▶ Brazil	▶ 1♂1♀	▶ Santa Bárbara, Minas Gerais State	▶ 1,4,6,9	▶ 24/48
▶ Present study	▶ Brazil	▶ 1♂1♀	▶ Santa Bárbara, Minas Gerais State	▶ 1	▶ 24/48
HYLIDAE / Cophomantinae / *Colomascirtus*					
No cytogenetic data					
HYLIDAE / Cophomantinae / *Hyloscirtus*					
Hyloscirtus jahni*					
▶ Present study	▶ Venezuela	▶ 6♂1♀	▶ Monte Zerpa, 2200 m, Mérida State (08°39'S, 71°10'W)	▶ 2,4,6,8	▶ 24/48
HYLIDAE / Cophomantinae / *Myersiohyla*					
No cytogenetic data					
HYLIDAE / Dendropsophinae / *Dendropsophus*					
Dendropsophus anceps					
▶ Foresti, 1972[P]	▶ Brazil	▶ 2♂	▶ Linhares, Espírito Santo State; rep. *Hyla anceps*	▶ 1	▶ 30/56†
▶ Bogart, 1973[P]	▶ Brazil	▶ n.i.	▶ southeastern Brazil; loc.n.i.; rep. *Hyla anceps*	▶ 1	▶ 30/56†
▷ Morescalchi, 1979	▷ n.i.	▷ n.i.	▷ ref. Bogart, 1973; list. *Hyla anceps*	▷ n.i.	▷ 30/n.i.
▷ King, 1990	▷ S.America	▷ n.i.	▷ ref. Bogart, 1973; ctry.n.sp.; list. *Hyla anceps*	▷ n.i.	▷ 30/58
▷ Kuramoto, 1990	▷ S.C.America	▷ n.i.	▷ ref. Bogart, 1973; ctry.n.sp.; list. *Hyla anceps*	▷ 1	▷ 30/n.i.
▷ Skuk & Langone, 1992	▷ n.i.	▷ n.i.	▷ ref. Bogart, 1973; list. *Hyla anceps*	▷ n.i.	▷ 30/58
▷ Nunes & Fagundes, 2008b	▷ n.i.	▷ n.i.	▷ ref. Bogart, 1973; list. *{Dendropsopus} anceps*	▷ n.i.	▷ 30/58
▷ Catroli, 2008	▷ n.i.	▷ n.i.	▷ ref. Foresti, 1972; Bogart, 1973	▷ n.i.	▷ 30/n.i.
▷ Catroli & Kasahara, 2009	▷ n.i.	▷ n.i.	▷ ref. Foresti, 1972; Bogart, 1973	▷ n.i.	▷ 30/n.i.
▶ Present study	▶ Brazil	▶ 2s.n.i.	▶ Rio de Janeiro, Rio de Janeiro State	▶ 1	▶ 30/56
Dendropsophus aperomeus					
▶ Duellman, 1982[N]	▶ Peru	▶ 1♂	▶ 8 km NNE (by road) of Balzapata, 1850 m, Amazonas Department (05°46'S, 74°47'W); rep. *Hyla aperomea*	▶ 1	▶ 30/n.i.
Dendropsophus berthalutzae					
▶ Foresti, 1972[P,M]	▶ Brazil	▶ 1♂	▶ Linhares, Espírito Santo State; rep. *Hyla berthalutzae*; 2 B	▶ 1	▶ 32/56†
▶ Skuk & Langone, 1992[P,M]	▶ Brazil	▶ 2♂2♀	▶ Porto da Cima, Morretes Municipalilty, Paraná State; rep. *Hyla berthalutzae*	▶ 1	▶ 30/54
▶ Gruber et al., 2005[P]	▶ Brazil	▶ 1♂	▶ Guaraqueçaba, Paraná State; rep. *Hyla berthalutzae*	▶ 1,2,4	▶ 30/52
▷ Nunes & Fagundes, 2008b	▷ n.i.	▷ n.i.	▷ ref. Skuk & Langone, 1992; list. *{Dendropsopus} berthalutzae*	▷ n.i.	▷ 30/52,54
▷ Catroli, 2008	▷ n.i.	▷ n.i.	▷ ref. Foresti, 1972; Skuk & Langone, 1992; Gruber et al., 2005	▷ n.i.	▷ 30/n.i.
▷ Catroli & Kasahara, 2009	▷ n.i.	▷ n.i.	▷ ref. Skuk & Langone, 1992; Gruber et al., 2005	▷ 17	▷ 30/n.i.
▷ Catroli & Kasahara, 2009	▷ n.i.	▷ n.i.	▷ ref. Foresti, 1972; err.cit.meth.; 2 B	▷ 17	▷ 32/n.i.
▷ Hernández-Guzmán et al., 2011	▷ n.i.	▷ n.i.	▷ ref. Gruber et al., 2005; list. rep. *Hyla berthalutzae*	▷ n.i.	▷ 30/n.i.
▶ Present study	▶ Brazil	▶ 1♂	▶ Horto Forestal, Rio de Janeiro; Rio de Janeiro State	▶ 1	▶ 30/54
▶ Present study	▶ Brazil	▶ 1♂	▶ Iguape, Sao Paulo State	▶ 1	▶ 30/54

Cytogenet Genome Res 2018;155:249–316
DOI: 10.1159/000490085

Author(s) ▶ original data[a] ▷ review articles	Country/ Region[b]	Speci- mens[c]	Locality of sampling, references cited, and remarks[d]	Methods applied[e]	2n/FN[f]
Dendropsophus bipunctatus					
▶ Foresti, 1973[P,M]	▶ Brazil	▶ 1♂	▶ Linhares, Espírito Santo State; rep. *Hyla bipunctata*	▶ 1	▶ 30/56†
▶ Foresti, 1973[P,M]	▶ Brazil	▶ 1♂	▶ Restinga da Marambaia, (Guanabara) Rio de Janeiro State; rep. *Hyla bipunctata*	▶ 1	▶ 30/56†
▶ Bogart, 1973[P]	▶ Brazil	▶ n.i.	▶ southeastern Brazil; loc.n.i.; rep. *Hyla bipunctata*	▶ 1	▶ 30/58†
▷ Morescalchi, 1979	▷ n.i.	▷ n.i.	▷ ref. Bogart, 1973; list. *Hyla bipunctata*	▷ n.i.	▷ 30/n.i.
▷ King, 1990	▷ S.America	▷ n.i.	▷ ref. Bogart, 1973; ctry.n.sp.; list. *Hyla bipunctata*	▷ n.i.	▷ 30/58
▷ Kuramoto, 1990	▷ S.C.America	▷ n.i.	▷ ref. Bogart, 1973; ctry.n.sp.; list. *Hyla bipunctata*	▷ 1	▷ 30/n.i.
▷ Skuk & Langone, 1992	▷ n.i.	▷ n.i.	▷ ref. Bogart, 1973; list. *Hyla bipunctata*	▷ n.i.	▷ 30/58
▷ Nunes & Fagundes, 2008b	▷ n.i.	▷ n.i.	▷ ref. Bogart, 1973; list. *{Dendropsopus} bipunctatus*	▷ n.i.	▷ 30/58
▷ Catroli, 2008	▷ n.i.	▷ n.i.	▷ ref. Foresti, 1972; Bogart, 1973	▷ n.i.	▷ 30/n.i.
▷ Catroli & Kasahara, 2009	▷ n.i.	▷ n.i.	▷ ref. Foresti, 1972; Bogart, 1973; Nunes, 2006	▷ n.i.	▷ 30/n.i.
▶ Nunes et al., 2011[N]	▶ Brazil	▶ 4s.n.i.	▶ Santa Tereza, Espírito Santo State	▶ 1,4	▶ 30/56
▶ Present study	▶ Brazil	▶ 1♂1♀	▶ Horto Forestal, Rio de Janeiro; Rio de Janeiro State	▶ 1	▶ 30/58
Dendropsophus branneri					
▶ Foresti, 1973[P,M]	▶ Brazil	▶ 1♀	▶ Jequié, Bahia State; rep. *Hyla branneri*	▶ 1	▶ 30/54†
▷ Catroli, 2008	▷ n.i.	▷ n.i.	▷ ref. Foresti, 1972; Bonfim et al., 2007	▷ n.i.	▷ 30/n.i.
▷ Nunes & Fagundes, 2008b	▷ n.i.	▷ n.i.	▷ ref. Foresti, 1972; list. *{Dendropsopus} branneri*	▷ n.i.	▷ 30/54
▷ Catroli & Kasahara, 2009	▷ n.i.	▷ n.i.	▷ ref. Foresti, 1972; Nunes, 2006	▷ 17	▷ 30/n.i.
▶ Nunes et al., 2001[N]	▶ Brazil	▶ 3s.n.i.	▶ Santa Tereza, Espírito Santo State	▶ 1,4	▶ 30/54
Dendropsophus carnifex					
▶ Duellman & Trueb, 1983[N]	▶ Ecuador	▶ n.i.	▶ loc.u.; rep. *Hyla carnifex*	▶ n.i.	▶ 30/n.i.
▷ Catroli, 2008	▷ n.i.	▷ n.i.	▷ ref. Duellman & Trueb, 1983	▷ n.i.	▷ 30/n.i.
▷ Catroli & Kasahara, 2009	▷ n.i.	▷ n.i.	▷ ref. Duellman & Trueb, 1983	▷ n.i.	▷ 30/n.i.
Dendropsophus columbianus					
▶ Duellman & Trueb, 1983[N]	▶ Colombia	▶ n.i.	▶ loc.u.; rep. *Hyla columbiana*	▶ n.i.	▶ 30/n.i.
Dendropsophus cruzi					
▶ Gruber et al., 2005[P]	▶ Brazil	▶ 4♂	▶ Mossâmides, Goiás State; rep. *Hyla cruzi*	▶ 1,2,4	▶ 30/54
▷ Nunes & Fagundes, 2008b	▷ n.i.	▷ n.i.	▷ ref. Gruber, 2002; list. *{Dendropsopus} cruzi*	▷ n.i.	▷ 30/54
▷ Catroli, 2008	▷ n.i.	▷ n.i.	▷ ref. Gruber et al., 2005	▷ n.i.	▷ 30/n.i.
▷ Catroli & Kasahara, 2009	▷ n.i.	▷ n.i.	▷ ref. Gruber et al., 2005	▷ 17	▷ 30/n.i.
▷ Hernández-Guzmán et al., 2011	▷ n.i.	▷ n.i.	▷ ref. Gruber et al., 2005; list. *Hyla cruzi*	▷ n.i.	▷ 30/n.i.
▶ Present study	▶ Brazil	▶ 4♂	▶ Mossâmides, Goiás State	▶ 1	▶ 30/54
Dendropsophus decipiens					
▶ Foresti, 1972[P,M]	▶ Brazil	▶ 1♂	▶ Itaguaí, Rio de Janeiro State; rep. *Hyla decipiens*	▶ 1	▶ 30/56†
▶ Bogart, 1973[P]	▶ Brazil	▶ n.i.	▶ southeastern Brazil; loc.n.i.; rep. *Hyla decipiens*	▶ 1	▶ 30/56†
▷ Morescalchi, 1979	▷ n.i.	▷ n.i.	▷ ref. Bogart, 1973; list. *Hyla decipiens*	▷ n.i.	▷ 30/n.i.
▷ King, 1990	▷ S.America	▷ n.i.	▷ ref. Bogart, 1973; ctry.n.sp.; list. *Hyla decipiens*	▷ n.i.	▷ 30/56
▷ Kuramoto, 1990	▷ S.C.America	▷ n.i.	▷ ref. Bogart, 1973; ctry.n.sp.; list. *Hyla decipiens*	▷ 1	▷ 30/n.i.
▷ Skuk & Langone, 1992	▷ n.i.	▷ n.i.	▷ ref. Bogart, 1973; list. *Hyla decipiens*	▷ n.i.	▷ 30/56
▷ Nunes & Fagundes, 2008b	▷ n.i.	▷ n.i.	▷ ref. Bogart, 1973; list. *{Dendropsopus} decipiens*	▷ n.i.	▷ 30/56
▷ Catroli, 2008	▷ n.i.	▷ n.i.	▷ ref. Foresti, 1972; Bogart, 1973	▷ n.i.	▷ 30/n.i.
▷ Catroli & Kasahara, 2009	▷ n.i.	▷ n.i.	▷ ref. Foresti, 1972; Bogart, 1973	▷ n.i.	▷ 30/n.i.
▶ Present study	▶ Brazil	▶ 1♂	▶ Horto Forestal, Rio de Janeiro; Rio de Janeiro State	▶ 1	▶ 30/56
Dendropsophus ebraccatus*					
▶ Duellman & Cole, 1965[N]	▶ Costa Rica	▶ 2♂	▶ 2 km E Tilaran, Guanacaste Province; rep. *Hyla ebraccata*	▶ 1	▶ 30‡/n.i.
▶ Duellman & Cole, 1965[N]	▶ Mexico	▶ 2♂	▶ 2 km S Valle Nacional, Oaxaca State; rep. *Hyla ebraccata*	▶ 1	▶ 30‡/n.i.
▷ Duellman, 1970	▷ C.America	▷ n.i.	▷ ref.n.i.; ctry.n.sp.; list. *Hyla ebraccata*	▷ 1	▷ 30‡/n.i.
▷ Rabello, 1970	▷ n.i.	▷ n.i.	▷ ref. Duellman & Cole, 1965; list. *Hyla ebraccata*	▷ n.i.	▷ 30‡/n.i.
▷ Foresti, 1972	▷ n.i.	▷ n.i.	▷ ref. Duellman & Cole, 1965; list. *Hyla ebraccata*	▷ n.i.	▷ 30‡/n.i.
▷ King, 1990	▷ S.America	▷ n.i.	▷ ref. Duellman & Cole, 1965; ctry.n.sp.; list. *Hyla ebraccata*	▷ n.i.	▷ 30/n.i.
▷ Kuramoto, 1990	▷ S.C.America	▷ n.i.	▷ ref. Duellman & Cole, 1965; ctry.n.sp.; list. *Hyla ebraccata*	▷ 1	▷ 30/n.i.
▷ Nunes & Fagundes, 2008b	▷ n.i.	▷ n.i.	▷ ref. Kaiser et al., 1996; list. *{Dendropsopus} ebraccatus*	▷ n.i.	▷ 30/58
▷ Catroli, 2008	▷ n.i.	▷ n.i.	▷ ref. Duellman & Cole, 1965; Kaiser et al., 1996	▷ n.i.	▷ 30/n.i.
▷ Catroli & Kasahara, 2009	▷ n.i.	▷ n.i.	▷ ref. Duellman & Cole, 1965; Kaiser et al., 1996	▷ 17	▷ 30/n.i.
▶ Present study	▶ Costa Rica	▶ 1♀	▶ Moravia de Chirripo	▶ 1	▶ 30/58
▶ Present study	▶ Panama	▶ 1♂,1♀	▶ Gamboa, Canal Zone (09°08′N, 79°46′W)	▶ 1	▶ 30/58
▶ Present study	▶ Costa Rica	▶ 1♂	▶ Valle de Talamanca, Limón Province (09°35′N, 83°07′W)	▶ 1,2,4,5, 7,8,14	▶ 30/58
▶ Present study	▶ Costa Rica	▶ 1♂	▶ Valle de la Estrella, Limón Province (09°41′N, 83°00′W)	▶ 1,2,4,5, 7,8,14	▶ 30/58
▶ Present study	▶ Costa Rica	▶ 1♀	▶ Wilson Botanical Garden, San Vito de Java, Puntarenas Province (08°47′N, 82°57′W)	▶ 1,2,4,5, 7,8,14	▶ 30/58
▶ Present study	▶ Costa Rica	▶ 2♂,1♀	▶ vicinity of Quepos, Puntarenas Province, 8 m (09°25′N, 84°09′W)	▶ 1,2,4,5, 7,8,14	▶ 30/58
▶ Present study	▶ Costa Rica	▶ 1♂	▶ Tropical Science Center, Rincon de Osa, 20 m, Puntarenas Province (09°35′N, 85°04′W)	▶ 1,2,4,5, 7,8,14	▶ 30/58
▶ Present study	▶ Costa Rica	▶ 1♀	▶ Hacianda Moravia de Chirripo, Cartago Province	▶ 1	▶ 30/58

DOI: 10.1159/000490085

Author(s) ▶ original data[a] ▷ review articles	Country/ Region[b]	Speci- mens[c]	Locality of sampling, references cited, and remarks[d]	Methods applied[e]	2n/FN[f]
Dendropsophus elegans					
▶ Gruber et al., 2005[P]	▶ Brazil	▶ 3♂	▶ Ubatuba, São Paulo State; rep. *Hyla elegans*	▶ 1,2,4	▶ 30/52
▷ Nunes & Fagundes, 2008b	▷ n.i.	▷ n.i.	▷ ref. Gruber, 2002; list. *{Dendropsopus} elegans*	▷ n.i.	▷ 30/52
▷ Catroli, 2008	▷ n.i.	▷ n.i.	▷ ref. Campos et al., 2004; Gruber et al., 2005	▷ n.i.	▷ 30/n.i.
▷ Catroli & Kasahara, 2009	▷ n.i.	▷ n.i.	▷ ref. Gruber et al., 2005	▷ 17	▷ 30/n.i.
▷ Hernández-Guzmán et al., 2011	▷ n.i.	▷ n.i.	▷ ref. Gruber et al., 2005; list. *Hyla elegans*	▷ n.i.	▷ 30/n.i.
▶ Present study	▶ Brazil	▶ 3♂	▶ Ubatuba, São Paulo State	▶ 1,2	▶ 30/52
Dendropsophus elianeae					
▶ Gruber et al., 2005[P]	▶ Brazil	▶ 5♂	▶ Rio Claro, São Paulo State; rep. *Hyla elianeae*	▶ 1,2,4	▶ 30/54
▷ Nunes & Fagundes, 2008b	▷ n.i.	▷ n.i.	▷ ref. Gruber, 2002; list. *{Dendropsopus} elegans*	▷ n.i.	▷ 30/54
▷ Catroli, 2008	▷ n.i.	▷ n.i.	▷ ref. Gruber et al., 2005	▷ n.i.	▷ 30/n.i.
▷ Catroli & Kasahara, 2009	▷ n.i.	▷ n.i.	▷ ref. Gruber et al., 2005	▷ 17	▷ 30/n.i.
▷ Hernández-Guzmán et al., 2011	▷ n.i.	▷ n.i.	▷ ref. Gruber et al., 2005; list. *Hyla elianeae*	▷ n.i.	▷ 30/n.i.
▶ Medeiros et al., 2013[P]	▶ Brazil	▶ 2♂	▶ Uberlândia, Minas Gerais State	▶ 1,2,4	▶ 30/56
▶ Medeiros et al., 2013[P]	▶ Brazil	▶ 8♂	▶ Botucatu, São Paulo State	▶ 1,2,4	▶ 30/56
▶ Medeiros et al., 2013[P]	▶ Brazil	▶ 4♂	▶ Nova Itapirema, São Paulo State	▶ 1,2,4	▶ 30/56
▶ Medeiros et al., 2013[P]	▶ Brazil	▶ 6♂	▶ Serra da Bodoquena, Matto Grosso do Sul State	▶ 1,2,4	▶ 30/56
▶ Present study	▶ Brazil	▶ 5♂	▶ Serra da Bodoquena, Matto Grosso do Sul State	▶ 1	▶ 30/54
Dendropsophus jimi					
▶ Medeiros et al., 2013[P]	▶ Brazil	▶ 16♂	▶ Uberlândia, Minas Gerais State	▶ 1,2,4,13	▶ 30/50
Dendropsophus labialis					
▶ Bogart, 1973[P]	▶ Colombia	▶ n.i.	▶ loc.n.i.; reported as *Hyla labialis*	▶ 1	▶ 30/50†
▷ Morescalchi, 1979	▷ n.i.	▷ n.i.	▷ ref. Bogart, 1973; list. *Hyla labialis*	▷ n.i.	▷ 30/n.i.
▷ King, 1990	▷ S.America	▷ n.i.	▷ ref. Bogart, 1973; ctry.n.sp.; list. *Hyla labialis*	▷ n.i.	▷ 30/50
▷ Kuramoto, 1990	▷ S.C.America	▷ n.i.	▷ ref. Bogart, 1973; ctry.n.sp.; list. *Hyla labialis*	▷ 1	▷ 30/n.i.
▷ Skuk & Langone, 1992	▷ n.i.	▷ n.i.	▷ ref. Bogart, 1973; list. *Hyla labialis*	▷ n.i.	▷ 30/50
▷ Nunes & Fagundes, 2008b	▷ n.i.	▷ n.i.	▷ ref. Bogart, 1973; list. *{Dendropsopus} labialis*	▷ n.i.	▷ 30/50
▷ Catroli, 2008	▷ n.i.	▷ n.i.	▷ ref. Bogart, 1973	▷ n.i.	▷ 30/n.i.
▷ Catroli & Kasahara, 2009	▷ n.i.	▷ n.i.	▷ ref. Bogart, 1973	▷ n.i.	▷ 30/n.i.
▶ Present study	▶ Colombia	▶ 3♂1♀	▶ city of Bogotá	▶ 1	▶ 30/50
Dendropsophus leali					
▶ Bogart & Bogart, 1971[N]	▶ Peru	▶ n.i.	▶ Río Pachitea, Bosque Nacional de Iparía, 1200 m, Huanuco Department; rep. *Hyla leali*	▶ 1	▶ 30/n.i.
▶ Bogart, 1973[P]	▶ Peru	▶ n.i.	▶ loc.n.i.; rep. *Hyla leali*	▶ 1	▶ 30/60†
▷ Morescalchi, 1979	▷ n.i.	▷ n.i.	▷ ref. Bogart, 1973; list. *Hyla leali*	▷ n.i.	▷ 30/n.i.
▷ King, 1990	▷ S.America	▷ n.i.	▷ ref. Bogart, 1973; ctry.n.sp.; list. *Hyla leali*	▷ n.i.	▷ 30/58
▷ Kuramoto, 1990	▷ S.C.America	▷ n.i.	▷ ref. Bogart & Bogart, 1971; Bogart, 1973; ctry.n.sp.; list. *Hyla leali*	▷ 1	▷ 30/n.i.
▷ Skuk & Langone, 1992	▷ n.i.	▷ n.i.	▷ ref. Bogart, 1973; list. *Hyla leali*	▷ n.i.	▷ 30/60
▷ Nunes & Fagundes, 2008b	▷ n.i.	▷ n.i.	▷ ref. Bogart, 1973; list. *{Dendropsopus} leali*	▷ n.i.	▷ 30/60
▷ Catroli, 2008	▷ n.i.	▷ n.i.	▷ ref. Bogart & Bogart, 1971; Bogart, 1973	▷ n.i.	▷ 30/n.i.
▷ Catroli & Kasahara, 2009	▷ n.i.	▷ n.i.	▷ ref. Bogart & Bogart, 1971; Bogart, 1973	▷ n.i.	▷ 30/n.i.
▶ Present study	▶ Peru	▶ 1♂1♀	▶ Bosque Nacional de Iparia, Huánuco Department	▶ 1	▶ 30/60
Dendropsophus leucophyllatus*					
▶ Bogart, 1970b[P,I]	▶ Peru	▶ n.i.	▶ Rio de Janeiro State; rep. *Hyla leucophyllata*	▶ n.i.	▶ 30/56†
▶ Bogart & Bogart, 1971[N]	▶ Peru	▶ n.i.	▶ Río Pachitea, Bosque Nacional de Iparía, 1200 m, Department; rep. *Hyla leucophyllata*	▶ 1	▶ 30/n.i.
▶ Foresti, 1972[P,M]	▶ Brazil	▶ 1♂	▶ Itaguaí, Rio de Janeiro State; rep. *Hyla leucophyllata*	▶ 1	▶ 30/56†
▶ Bogart, 1973[P]	▶ Peru	▶ n.i.	▶ loc.n.i.; rep. *Hyla leucophyllata*	▶ 1	▶ 30/56†
▷ Morescalchi, 1979	▷ n.i.	▷ n.i.	▷ ref. Bogart, 1973; list. *Hyla leucophyllata*	▷ n.i.	▷ 30/n.i.
▷ King, 1990	▷ S.America	▷ n.i.	▷ ref. Bogart, 1973; ctry.n.sp.; list. *Hyla leucophyllata*	▷ n.i.	▷ 30/52
▷ Kuramoto, 1990	▷ S.C.America	▷ n.i.	▷ ref. Bogart, 1970b, 1973; Bogart & Bogart, 1971; ctry.n.sp.; list. *Hyla leucophyllata*	▷ 1	▷ 30/n.i.
▷ Skuk & Langone, 1992	▷ n.i.	▷ n.i.	▷ ref. Bogart, 1973; list. *Hyla leucophyllata*	▷ n.i.	▷ 30/52
▷ Nunes & Fagundes, 2008b	▷ n.i.	▷ n.i.	▷ ref. Bogart, 1973; list. *{Dendropsopus} leucophyllatus*	▷ n.i.	▷ 30/52
▷ Catroli, 2008	▷ n.i.	▷ n.i.	▷ ref. Bogart, 1970b, 1973; Bogart & Bogart, 1971; Foresti, 1972	▷ n.i.	▷ 30/n.i.
▷ Catroli & Kasahara, 2009	▷ n.i.	▷ n.i.	▷ ref. Bogart, 1970b, 1973; Bogart & Bogart, 1971; Foresti, 1972	▷ n.i.	▷ 30/n.i.
▶ Present study	▶ Peru	▶ 3♂1♀ 1U	▶ Bosque Nat. de Iparia, Huánuco Department	▶ 1	▶ 30/56
▶ Present study	▶ Peru	▶ 2U	▶ Rio de Janeiro,Rio de Janeiro State	▶ 1	▶ 30/56
▶ Present study	▶ Peru	▶ 3♂	▶ A-Ukre, Pará State (07°39'S, 51°21'W)	▶ 1	▶ 30/56
Dendropsophus luteoocellatus					
▶ Present study	▶ Venezuela	▶ 1♀	▶ Los Teques,Urbanización El Barbecho, Miranda State	▶ 1,2,5, 8,14	▶ 30/54
▶ Present study	▶ Venezuela	▶ 9♂	▶ Uria, Juan Crisóstomo Falcón National Park (Sierra de San Luis), 1300 m, Falcón State	▶ 1,2,5, 8,14	▶ 30/54

Author(s) ▶ original data[a] ▷ review articles	Country/ Region[b]	Speci- mens[c]	Locality of sampling, references cited, and remarks[d]	Methods applied[e]	2n/FN[f]
Dendropsophus marmoratus					
▶ Bogart & Bogart, 1971[N]	▶ Peru	▶ n.i.	▶ Río Pachitea, Bosque Nacional de Iparía, 1200 m, Huanuco Department; rep. *Hyla marmorata*	▶ 1	▶ 30/n.i.
▶ Bogart, 1973[P]	▶ Peru	▶ n.i.	▶ loc.n.i.; rep. *Hyla marmorata*	▶ 1	▶ 30/50†
▷ Morescalchi, 1979	▷ n.i.	▷ n.i.	▷ ref. Bogart, 1973; list. *Hyla marmorata*	▷ n.i.	▷ 30/n.i.
▷ King, 1990	▷ S.America	▷ n.i.	▷ ref. Bogart, 1973; ctry.n.sp.; list. *Hyla marmorata*	▷ n.i.	> 30/50
▷ Kuramoto, 1990	▷ S.C.America	▷ n.i.	▷ ref. Bogart & Bogart, 1971; Bogart, 1973; ctry.n.sp.; list. *Hyla marmorata*	▷ 1	▷ 30/n.i.
▷ Skuk & Langone, 1992	▷ n.i.	▷ n.i.	▷ ref. Bogart, 1973; list. *Hyla marmorata*	▷ n.i.	▷ 30/50
▷ Nunes & Fagundes, 2008b	▷ n.i.	▷ n.i.	▷ ref. Bogart, 1973; list. *{Dendropsopus} marmoratus*	▷ n.i.	▷ 30/52
▷ Catroli, 2008	▷ n.i.	▷ n.i.	▷ ref. Bogart & Bogart, 1971; Bogart, 1973	▷ n.i.	▷ 30/n.i.
▷ Catroli & Kasahara, 2009	▷ n.i.	▷ n.i.	▷ ref. Bogart & Bogart, 1971; Bogart, 1973	▷ n.i.	▷ 30/n.i.
▶ Suárez et al., 2013[P,M,I]	▶ Brazil	▶ 3♂1♀	▶ Acampamento Barroso, Juruti Municipality, Pará State	▶ 1–4,6,9, 13,14	▶ 30/50
▶ Suárez et al., 2013[P,M,I]	▶ Brazil	▶ 1♂1♀	▶ Platô Capiranga, Juruti Municipality, Pará State	▶ 1–4,6,9, 13,14	▶ 30/50
▶ Suárez et al., 2013[P,M,I]	▶ Brazil	▶ 1♀	▶ Igarapé Prudêncio, Juruti Municipality, Pará State	▶ 1–4,6,9, 13,14	▶ 30/50
▶ Present study	▶ Peru	▶ 1♂2U	▶ Bosque Natcional de Iparia, Huánuco Department	▶ 1	▶ 30/50
Dendropsophus mathiassoni*					
▶ Present study	▶ Colombia	▶ 1♂	▶ city of Barcelona, Córdoba Department	▶ 1	▶ 30/56
Dendropsophus melanargyreus					
▶ Suárez et al., 2013[P,M,I]	▶ Brazil	▶ 3♂	▶ Ilha de Cotijuba, Belém Municipality, Pará State	▶ 1–4,6,9, 13,14	▶ 30/50
▶ Suárez et al., 2013[P,M,I]	▶ Brazil	▶ 1♀	▶ Serra Sul, Carajás Department, Pará State	▶ 1–4,6,9, 13,14	▶ 30/50
Dendropsophus meridianus					
▶ Foresti, 1972[P,M]	▶ Brazil	▶ 1♂	▶ Itaguaí, Rio de Janeiro State; rep. *Hyla meridiana*	▶ 1	▶ 30/58†
▶ Skuk & Langone, 1986[N]	▶ Brazil	▶ 3♂1♀	▶ Ilha do Fundão, Rio de Janeiro State; rep. *Hyla microcephala meridiana*	▶ 1	▶ 30/58
▷ Kuramoto, 1990	▷ S.C.America	▷ n.i.	▷ ref. Skuk & Langone, 1986; ctry.n.sp.; list. *Hyla microcephala meridiana*	▷ 1	▷ 30/n.i.
▷ Skuk & Langone, 1992	▷ n.i.	▷ n.i.	▷ ref. Skuk & Langone, 1986; list. *Hyla microcephala meridiana*	▷ n.i.	▷ 30/58
▷ Nunes & Fagundes, 2008b	▷ n.i.	▷ n.i.	▷ ref. Skuk & Langone, 1992; list. *{Dendropsopus} meridianus*	▷ n.i.	▷ 30/58
▷ Catroli, 2008	▷ n.i.	▷ n.i.	▷ ref. Foresti, 1972; Skuk & Langone, 1986	▷ n.i.	▷ 30/n.i.
▷ Catroli & Kasahara, 2009	▷ n.i.	▷ n.i.	▷ ref. Foresti, 1972	▷ n.i.	▷ 30/n.i.
▶ Present study	▶ Brazil	▶ 2U	▶ Rio de Janeiro, Rio de Janeiro State	▶ 1	▶ 30/56
Dendropsophus microcephalus					
▶ Duellman & Cole, 1965[N]	▶ Nicaragua	▶ 1♂	▶ Finca Venecia, 7 km N, 16 km E Condega, Esteli Department; rep. *Hyla microcephala martini*	▶ 1	▶ 30‡/n.i.
▶ Duellman & Cole, 1965[N]	▶ Nicaragua	▶ 2♂	▶ Finca Tepeyac, 10.5 km N, 9 km E Matagalpa, Matagalpa State; rep. *Hyla microcephala martini*	▶ 1	▶ 30‡/n.i.
▶ Duellman & Cole, 1965[N]	▶ Mexico	▶ 1♂	▶ 5 km N Chiltepec; Oaxaca State; rep. *Hyla microcephala martini*	▶ 1	▶ 30‡/n.i.
▶ Duellman & Cole, 1965[P]	▶ Panama	▶ 1♂	▶ 3 km W Chepo; Panamá Province; rep. *Hyla microcephala microcephala*	▶ 1	▶ 30‡/n.i.
▶ Duellman & Cole, 1965[P]	▶ Panama	▶ 1♂	▶ El Valle, Coclé Province; rep. *Hyla microcephala microcephala*	▶ 1	▶ 30‡/n.i.
▷ Duellman, 1970	▷ C.America	▷ n.i.	▷ ref.n.i.; ctry.n.sp.; list. *Hyla microcephala*	▷ 1	▷ 30‡/n.i.
▶ León, 1970[N]	▶ Costa Rica	▶ n.i.	▶ loc.n.i.; rep. *Hyla microcephala*	▶ 1	▶ 30‡/n.i.
▷ Rabello, 1970	▷ n.i.	▷ n.i.	▷ ref. Duellman & Cole, 1965; list. *Hyla microcephala microcephala*	▷ n.i.	▷ 30‡/n.i.
▷ Rabello, 1970	▷ n.i.	▷ n.i.	▷ ref. Duellman & Cole, 1965; list. *Hyla microcephala martini*	▷ n.i.	▷ 30‡/n.i.
▷ Foresti, 1972	▷ n.i.	▷ n.i.	▷ ref. Duellman & Cole, 1965; list. *Hyla microcephala martini*	▷ n.i.	▷ 30‡/n.i.
▶ Bogart, 1973[P]	▶ Colombia	▶ n.i.	▶ loc.n.i.; rep. *Hyla microcephala*	▶ 1	▶ 30/60†
▷ Morescalchi, 1979	▷ n.i.	▷ n.i.	▷ ref. Bogart, 1973; list. *Hyla microcephala*	▷ n.i.	▷ 30/n.i.
▶ Anderson, 1986[P,I]	▶ Belize	▶ 3♂1♀	▶ Stann Creek; rep. *Hyla microcephala*	▶ 1,2,4,11	▶ 30/60†
▷ King, 1990	▷ S.America	▷ n.i.	▷ ref. Duellman & Cole, 1965; Bogart, 1973; Anderson, 1986; ctry.n.sp.; list. *Hyla microcephala*	▷ 2,4,11	▷ 30/60
▷ Kuramoto, 1990	▷ S.C.America	▷ n.i.	▷ ref. Duellman & Cole, 1965; León, 1970; Bogart, 1973; ctry.n.sp.; list. *Hyla microcephala microcephala*	▷ 1	▷ 30/n.i.
▷ Kuramoto, 1990	▷ S.C.America	▷ n.i.	▷ ref. Duellman & Cole, 1965; ctry.n.sp.; list. *Hyla microcephala martini*	▷ 1	▷ 30/n.i.
▷ Anderson, 1991[I]	▷ Neotropic	▷ n.i.	▷ ref. Anderson, 1986; ctry.n.sp.; list. *Hyla microcephala*	▷ 2,4	▷ 30/60†
▷ Skuk & Langone, 1992	▷ n.i.	▷ n.i.	▷ ref. Bogart, 1973; list. *Hyla microcephala*	▷ n.i.	▷ 30/58
▷ Nunes & Fagundes, 2008b	▷ n.i.	▷ n.i.	▷ ref. Kaiser et al., 1996; list. *{Dendropsopus} microcephalus*	▷ n.i.	▷ 30/58,60
▷ Catroli, 2008	▷ n.i.	▷ n.i.	▷ ref. Duellman & Cole, 1965; León, 1970; Bogart, 1973; Anderson, 1991; Kaiser et al., 1996	▷ n.i.	▷ 30/n.i.
▷ Catroli & Kasahara, 2009	▷ n.i.	▷ n.i.	▷ ref. Duellman & Cole, 1965; León, 1970; Bogart, 1973; Anderson, 1991; Kaiser et al., 1996	▷ 17	▷ 30/n.i.
▶ Present study	▶ Colombia	▶ 1♂	▶ Giradot	▶ 1	▶ 30/60
▶ Present study	▶ Colombia	▶ 1♂1U	▶ loc.u.	▶ 1	▶ 30/60
▶ Present study	▶ Venezuela	▶ 1♂1♀	▶ El Limón, University campus, Aragua State	▶ 1,2,8	▶ 30/60
▶ Present study	▶ Costa Rica	▶ 2♂1♀	▶ vicinity of Quepos, Puntarenas Province, 8 m (09°25′N, 84°09′W)	▶ 1,2,8	▶ 30/60
▶ Present study	▶ Trinidad	▶ 4♂2U	▶ 6 km SE Talparo, 65 m (10°30′N, 61°16′W)	▶ 1,2	▶ 30/60

Author(s) ▶ original data[a] ▷ review articles	Country/ Region[b]	Speci- mens[c]	Locality of sampling, references cited, and remarks[d]	Methods applied[e]	2n/FN[f]
Dendropsophus microps					
▶ Beçak, 1968[P,M]	▶ Brazil	▶ 3♂1♀	▶ Campos de Jordão, São Paulo State; rep. *Hyla microps*	▶ 1	▶ 30/52†
▷ Rabello, 1970	▷ n.i.	▷ n.i.	▷ ref. Beçak, 1967, 1968; list. *Hyla microps*	▷ n.i.	▷ 30/n.i.
▷ Foresti, 1972	▷ n.i.	▷ n.i.	▷ ref. Beçak, 1967, 1968; list. *Hyla microps*	▷ n.i.	▷ 30/n.i.
▶ Bogart, 1973[P]	▶ Brazil	▶ n.i.	▶ southeastern Brazil; loc.n.i.; rep. *Hyla microps*	▶ 1	▶ 30/52†
▷ Morescalchi, 1979	▷ n.i.	▷ n.i.	▷ ref. Bogart, 1973; list. *Hyla microps*	▷ n.i.	▷ 30/n.i.
▷ King, 1990	▷ S.America	▷ n.i.	▷ ref. Beçak, 1968; Bogart, 1973; ctry.n.sp.; list. *Hyla microps*	▷ n.i.	▷ 30/54
▷ Kuramoto, 1990	▷ S.C.America	▷ n.i.	▷ ref. Beçak, 1968; Bogart, 1973; ctry.n.sp.; list. *Hyla microps*	▷ 1	▷ 30/n.i.
▷ Skuk & Langone, 1992	▷ n.i.	▷ n.i.	▷ ref. Beçak, 1968; Bogart, 1973; list. *Hyla microps*	▷ n.i.	▷ 30/52
▶ Gruber et al., 2005[P]	▶ Brazil	▶ 1♂	▶ Campos do Jordão, São Paulo State; rep. *Hyla microps*	▶ 1,4	▶ 30/52
▶ Gruber et al., 2005[P]	▶ Brazil	▶ 1♂	▶ Ribeirão Branco, São Paulo State; rep. *Hyla microps*	▶ 1	▶ 30/52
▶ Gruber et al., 2005[P]	▶ Brazil	▶ 5♂	▶ São Bento do Sul, Santa Catarina State; rep. *Hyla microps*	▶ 1,2,4	▶ 30/52
▷ Nunes & Fagundes, 2008b	▷ n.i.	▷ n.i.	▷ ref. Beçak, 1968; list. *{Dendropsopus} microps*	▷ n.i.	▷ 30/52
▷ Catroli, 2008	▷ n.i.	▷ n.i.	▷ ref. Beçak, 1968; Bogart, 1973; Gruber et al., 2005	▷ n.i.	▷ 30/n.i.
▷ Catroli & Kasahara, 2009	▷ n.i.	▷ n.i.	▷ ref. Beçak, 1968; Bogart, 1973; Gruber et al., 2005	▷ 17	▷ 30/n.i.
▷ Hernández-Guzmán et al., 2011	▷ n.i.	▷ n.i.	▷ ref. Gruber et al., 2005; list. *Hyla microps*	▷ n.i.	▷ 30/n.i.
▶ Present study	▶ Brazil	▶ 1♂	▶ Boracéa, NE of São Paulo; São Paulo State	▶ 1	▶ 30/52
▶ Present study	▶ Brazil	▶ 1♂	▶ loc.n.i.	▶ 1	▶ 30/52
▶ Present study	▶ Brazil	▶ 1s.n.i.	▶ Boracéa, NE of São Paulo; São Paulo State	▶ 1	▶ 30/52
Dendropsophus minutus					
▶ Rabello, 1970[P]	▶ n.i.	▶ 12♂1♀	▶ loc.n.i.; rep. *Hyla minuta*	▶ 1	▶ 30/60†
▶ Bogart & Bogart, 1971[N]	▶ Peru	▶ n.i.	▶ Río Pachitea, Bosque Nacional de Iparía, 1200 m, Huanuco Department; rep. *Hyla minuta*	▶ 1	▶ 30/n.i.
▷ Foresti, 1972	▷ n.i.	▷ n.i.	▷ ref. Rabello, 1969, 1970; list. *Hyla minuta*	▷ n.i.	▶ 30/n.i.
▶ Bogart, 1973[P]	▶ Brazil	▶ n.i.	▶ southeastern Brazil; loc.n.i.; rep. *Hyla minuta*	▶ 1	▶ 30/60†
▶ Bogart, 1973[P]	▶ Peru	▶ n.i.	▶ loc.n.i.; rep. *Hyla minuta*	▶ 1	▶ 30/60†
▷ Morescalchi, 1979	▷ n.i.	▷ n.i.	▷ ref. Bogart, 1973; list. *Hyla minuta*	▷ n.i.	▷ 30/n.i.
▷ King, 1990	▷ S.America	▷ n.i.	▷ ref. Rabello, 1970; Bogart, 1973; ctry.n.sp; list. *Hyla minuta*	▷ n.i.	▷ 30/60
▷ Kuramoto, 1990	▷ S.C.America	▷ n.i.	▷ ref. Rabello, 1970; Bogart & Bogart, 1971; Bogart, 1973; ctry.n.sp; list. *Hyla minuta*	▷ 1	▷ 30/n.i.
▷ Skuk & Langone, 1992	▷ n.i.	▷ n.i.	▷ ref. Bogart, 1973; list. *Hyla minuta*	▷ n.i.	▷ 30/60
▶ Gruber et al., 2005[P]	▶ Brazil	▶ 3♂	▶ Rio Claro, São Paulo State; rep. *Hyla minuta*	▶ 1,2,4	▶ 30/60
▶ Gruber et al., 2005[P]	▶ Brazil	▶ 1♂	▶ Ubatuba, São Paulo State; rep. *Hyla minuta*	▶ 1,2,4	▶ 30/60
▷ Nunes & Fagundes, 2008b	▷ n.i.	▷ n.i.	▷ ref. Bogart, 1973; list. *{Dendropsopus} minutus*	▷ n.i.	▷ 30/60
▷ Catroli, 2008	▷ n.i.	▷ n.i.	▷ ref. Rabello, 1970; Bogart & Bogart, 1971; Bogart, 1973; Gruber et al., 2005	▷ n.i.	▷ 30/n.i.
▷ Catroli & Kasahara, 2009	▷ n.i.	▷ n.i.	▷ ref. Rabello, 1970; Bogart & Bogart, 1971; Bogart, 1973; Gruber et al., 2005; Nunes, 2006	▷ n.i.	▷ 30/n.i.
▶ Nunes et al., 2001[N]	▶ Brazil	▶ n.i.	▶ Santa Tereza, Espírito Santo State	▶ 1	▶ 30/60
▶ Present study	▶ Brazil	▶ 1♂1U	▶ loc.u.	▶ 1	▶ 30/60
▶ Present study	▶ Brazil	▶ 2♂	▶ Fazenda São Joaquim, São Paulo State	▶ 1	▶ 30/60
▶ Present study	▶ Peru	▶ 1♂1♀	▶ Bosque Nat. de Iparia, Huánuco Department	▶ 1	▶ 30/60
▶ Present study	▶ Venezuela	▶ 4♂	▶ Uria, Juan Crisóstomo Falcón National Park (Sierra de San Luis), 1300 m, Falcón State	▶ 1,2,4,14	▶ 30/60
▶ Present study	▶ Trinidad	▶ 3♂	▶ Simla Research Station, Arima valley, ~ 360 m (10°41′N, 61°17′W)	▶ 1,2,4	▶ 30/60
▶ Present study	▶ Trinidad	▶ 1♀	▶ 6 km SE Talparo, 65 m (10°30′N, 61°16′W)	▶ 1,2,4	▶ 30/60
▶ Present study	▶ Brazil	▶ 1n.i	▶ loc.u.	▶ 1	▶ 30/60
Dendropsophus nahdereri					
▶ Gruber et al., 2005[P]	▶ Brazil	▶ 1♀	▶ Rancho Queimado, Santa Catarina State; rep. *Hyla nahdereri*	▶ 1,2,4	▶ 30/50
▷ Nunes & Fagundes, 2008b	▷ n.i.	▷ n.i.	▷ ref. Gruber, 2002; list. *{Dendropsopus} nahderi*	▷ n.i.	▷ 30/50
▷ Catroli, 2008	▷ n.i.	▷ n.i.	▷ ref. Gruber et al., 2005	▷ n.i.	▷ 30/n.i.
▷ Catroli & Kasahara, 2009	▷ n.i.	▷ n.i.	▷ ref. Gruber et al., 2005	▷ 17	▷ 30/n.i.
▷ Hernández-Guzmán et al., 2011	▷ n.i.	▷ n.i.	▷ ref. Gruber et al., 2005; list. *Hyla nahderi*	▷ n.i.	▷ 30/n.i.
▶ Present study	▶ Brazil	▶ 1♀	▶ Rancho Queimado, Santa Catarina State	▶ 1	▶ 30/50
Dendropsophus nanus					
▶ Rabello, 1970[P]	▶ n.i.	▶ 5♂	▶ loc.n.i.; rep. *Hyla nana*	▶ 1	▶ 30/52†
▷ Foresti, 1972	▷ n.i.	▷ n.i.	▷ ref. Rabello, 1969, 1970; list. *Hyla nana*	▷ n.i.	▷ 30/n.i.
▶ Bogart, 1973[P]	▶ Argentina	▶ n.i.	▶ northern Argentina; loc.n.i.; rep. *Hyla nana*	▶ 1	▶ 30/50†
▷ Morescalchi, 1979	▷ n.i.	▷ n.i.	▷ ref. Bogart, 1973; list. *Hyla nana*	▷ n.i.	▷ 30/n.i.
▷ King, 1990	▷ S.America	▷ n.i.	▷ ref. Rabello, 1970; Bogart, 1973; ctry.n.sp; list. *Hyla nana*	▷ n.i.	▷ 30/50
▷ Kuramoto, 1990	▷ S.C.America	▷ n.i.	▷ ref. Rabello, 1970; Bogart, 1973; ctry.n.sp; list. *Hyla nana*	▷ 1	▷ 30/n.i.
▷ Skuk & Langone, 1992	▷ n.i.	▷ n.i.	▷ ref. Rabello, 1970; Bogart, 1973; list. *Hyla nana*	▷ n.i.	▷ 30/50,52
▶ Skuk & Langone, 1992[P,M]	▶ Argentina	▶ 3♂1♀	▶ km 1141 on road Ruta Nacional 95, Chaco Province; rep. *Hyla nana*	▶ 1	▶ 30/52
▶ Medeiros et al., 2003[P]	▶ Brazil	▶ 40♂3♀	▶ Nova Itapirema, São Paulo State (21°11′S, 49°42′W); rep. *Hyla nana*	▶ 1,2,4,13	▶ 30/52
▶ Gruber et al., 2005[P]	▶ Brazil	▶ 2♂	▶ Rio Claro, São Paulo State; rep. *Hyla nana*	▶ 1,2,4	▶ 30/52
▶ Medeiros et al., 2006[P]	▶ Brazil	▶ 36♂3♀	▶ Nova Aliança, São Paulo State (22°11′S, 49°42′W)	▶ 1,2,4,13	▶ 30/52†
▶ Medeiros et al., 2006[P]	▶ Brazil	▶ 4♂	▶ Nova Aliança, São Paulo State (22°11′S, 49°42′W); 1B	▶ 1,2,4,13	▶ 31/53†
▶ Medeiros et al., 2006[P]	▶ Brazil	▶ 6♂	▶ Botucatu, São Paulo State (22°53′S, 48°26′W)	▶ 1,2,4,13	▶ 30/52†
▶ Medeiros et al., 2006[P]	▶ Brazil	▶ 8♂	▶ Botucatu, São Paulo State (22°53′S, 48°26′W); 1B	▶ 1,2,4,13	▶ 31/53†
▶ Medeiros et al., 2006[P]	▶ Brazil	▶ 1♂	▶ Botucatu, São Paulo State (22°53′S, 48°26′W); 1 to 3 B	▶ 1,2,4,13	▶ 31–33/53–55†

Author(s) ▶ original data[a] ▷ review articles	Country/ Region[b]	Speci- mens[c]	Locality of sampling, references cited, and remarks[d]	Methods applied[e]	2n/FN[f]
▷ Nunes & Fagundes, 2008b	▷ n.i.	▷ n.i.	▷ ref. Rabello, 1970; list. *{Dendropsopus} nanus*	▷ n.i.	▷ 30/52
▷ Catroli, 2008	▷ n.i.	▷ n.i.	▷ ref. Rabello, 1970; Rabello et al., 1971; Bogart, 1973; Skuk & Langone, 1992; Medeiros et al., 2003, 2006; Gruber et al., 2005	▷ n.i.	▷ 30/n.i.
▷ Catroli, 2008	▷ n.i.	▷ n.i.	▷ Medeiros et al., 2006; 1 B	▷ n.i.	▷ 31/n.i.
▷ Catroli & Kasahara, 2009	▷ n.i.	▷ n.i.	▷ ref. Rabello, 1970; Rabello et al., 1971; Bogart, 1973; Skuk & Langone, 1992; Medeiros et al., 2003, 2006; Gruber et al., 2005	▷ 17	▷ 30/n.i.
▷ Catroli & Kasahara, 2009	▷ n.i.	▷ n.i.	▷ Medeiros et al., 2006; 1 B	▷ 17	▷ 31/n.i.
▷ Hernández-Guzmán et al., 2011	▷ n.i.	▷ n.i.	▷ ref. Medeiros et al., 2003; list. *Hyla nana*	▷ n.i.	▷ 30/n.i.
▶ Medeiros et al., 2013[P]	▶ Brazil	▶ 2♂	▶ Botucatu, São Paulo State	▶ 1,2,4,13	▶ 30/52
▶ Medeiros et al., 2013[P]	▶ Brazil	▶ 2♂	▶ Serra da Bodoquena, Mato Grosso do Sul State	▶ 1,2,4,13	▶ 30/52,53
▶ Medeiros et al., 2013[P]	▶ Brazil	▶ 9♂1♀	▶ Telêmaco Borba, Paraná State	▶ 1,2,4,13	▶ 30/52
▶ Present study	▶ Argentina	▶ 1♂	▶ Yuto, Jujuy Province	▶ 1	▶ 30/50
▶ Present study	▶ Argentina	▶ 1U	▶ Concordia, Entra Ríos Province	▶ 1	▶ 30/50
▶ Present study	▶ Brazil	▶ 1♀	▶ A-Ukre, Pará State (07°39′S, 51°21′W)	▶ 1	▶ 30/50
Dendropsophus novaisi					
▶ Teixeira et al., 2016[P]	▶ Brazil	▶ 1♀	▶ Jequié, Bahia State	▶ 1,2,4,14	▶ 30/50†
Dendropsophus oliveirai					
▶ Foresti, 1972[P,N]	▶ Brazil	▶ 1♀	▶ Itajibá, Bahia State, rep. *Hyla oliveirai*	▶ 1	▶ 30/54†
▷ Catroli, 2008	▷ n.i.	▷ n.i.	▷ ref. Foresti, 1972	▷ n.i.	▷ 30/n.i.
▷ Catroli & Kasahara, 2009	▷ n.i.	▷ n.i.	▷ ref. Foresti, 1972	▷ n.i.	▷ 30/n.i.
Dendropsophus parviceps					
▶ Bogart & Bogart, 1971[N]	▶ Peru	▶ n.i.	▶ Río Pachitea, Bosque Nacional de Iparía, 1200 m, Huanuco Department; rep. *Hyla parviceps*	▶ 1	▶ 30/n.i.
▶ Bogart, 1973[P]	▶ Peru	▶ n.i.	▶ loc.n.i.; rep. *Hyla parviceps*	▶ 1	▶ 30/52†
▷ Morescalchi, 1979	▷ n.i.	▷ n.i.	▷ ref. Bogart, 1973; list. *Hyla parviceps*	▷ n.i.	▷ 30/n.i.
▷ King, 1990	▷ S.America	▷ n.i.	▷ ref. Bogart, 1973; ctry.n.sp; list. *Hyla parviceps*	▷ n.i.	▷ 30/52
▷ Kuramoto, 1990	▷ S.C.America	▷ n.i.	▷ ref. Bogart & Bogart, 1971; Bogart, 1973; ctry.n.sp.; list. *Hyla parviceps*	▷ 1	▷ 30/n.i.
▷ Skuk & Langone, 1992	▷ n.i.	▷ n.i.	▷ ref. Bogart, 1973; list. *Hyla parviceps*	▷ n.i.	▷ 30/52
▷ Nunes & Fagundes, 2008b	▷ n.i.	▷ n.i.	▷ ref. Bogart, 1973; list. *{Dendropsopus} parviceps*	▷ n.i.	▷ 30/52,54
▷ Catroli, 2008	▷ n.i.	▷ n.i.	▷ ref. Duellman & Cole, 1965; Bogart & Bogart, 1971; Bogart, 1973	▷ n.i.	▷ 30/n.i.
▷ Catroli & Kasahara, 2009	▷ n.i.	▷ n.i.	▷ ref. Duellman & Cole, 1965; Bogart & Bogart, 1971; Bogart, 1973; err.cit. meth.	▷ 17	▷ 30/n.i.
▶ Present study	▶ Peru	▶ 2♂	▶ Bosque Nat. de Iparia, Huánuco Department	▶ 1	▶ 30/52
▶ Present study	▶ Brazil	▶ 1♂	▶ A-Ukre, Pará State (07°39′S, 51°21′W)	▶ 1	▶ 30/52
Dendropsophus phlebodes*					
▶ Duellman, 1967[P]	▶ Costa Rica	▶ 2♂	▶ 1 km S Puerto Viejo, Heredia Province; rep. *Hyla phlebodes*	▶ 1	▶ 30‡/n.i.
▶ Duellman, 1967[P]	▶ Panama	▶ 1♂	▶ S end Cayo de Agua, Bocas del Toro Province; list. *Hyla phlebodes*	▶ 1	▶ 30‡/n.i.
▷ Duellman, 1970	▷ C.America	▷ n.i.	▷ ref.n.i.; ctry.n.sp.; list. *Hyla phlebodes*	▷ 1	▷ 30‡/n.i.
▷ Foresti, 1972	▷ n.i.	▷ n.i.	▷ ref. Duellman, 1967; list. *Hyla phlebodes*	▷ n.i.	▷ 30‡/n.i.
▷ King, 1990	▷ S.America	▷ n.i.	▷ ref. Duellman & Cole, 1965; ctry.n.sp; list. *Hyla phlebodes*	▷ n.i.	▷ 30/n.i.
▷ Kuramoto, 1990	▷ S.C.America	▷ n.i.	▷ ref. Duellman & Cole, 1965; Duellman, 1967; ctry.n.sp.; list. *Hyla phlebodes*	▷ 1	▷ 30/n.i.
▷ Nunes & Fagundes, 2008b	▷ n.i.	▷ n.i.	▷ ref. Bogart, 1973; list. *{Dendropsopus} phlebodes*	▷ n.i.	▷ 30/56
▷ Catroli, 2008	▷ n.i.	▷ n.i.	▷ ref. Duellman & Cole, 1965; Duellman, 1967; Kaiser et al., 1996	▷ n.i.	▷ 30/n.i.
▷ Catroli & Kasahara, 2009	▷ n.i.	▷ n.i.	▷ ref. Duellman & Cole, 1965; Duellman, 1967; Kaiser et al., 1996	▷ 17	▷ 30/n.i.
▶ Present study	▶ Costa Rica	▶ 1♂1J	▶ Cahuita National Park, Limón Province (09°43′N, 82°50′W)	▶ 1,2,4,5	▶ 30/56
▶ Present study	▶ Costa Rica	▶ 6♂3♀	▶ vicinity of Tanque, Alajuela Province (10°28′N, 84°35′W)	▶ 1,2,4,5	▶ 30/56
Dendropsophus praestans					
▶ Duellman & Trueb, 1983[N]	▶ Colombia	▶ n.i.	loc.u.; rep. *Hyla praestans*	▶ n.i.	▶ 30/n.i.
Dendropsophus rhodopeplus					
▶ Bogart & Bogart, 1971[N]	▶ Peru	▶ n.i.	▶ Río Pachitea, Bosque Nacional de Iparía, 1200 m, Huanuco Department; rep. *Hyla rhodopepla*	▶ 1	▶ 30/n.i.
▶ Bogart, 1973[P]	▶ Peru	▶ n.i.	▶ loc.n.i.; rep. *Hyla rhodopepla*	▶ 1	▶ 30/56†
▷ Morescalchi, 1979	▷ n.i.	▷ n.i.	▷ ref. Bogart, 1973; list. *Hyla rhodopepla*	▷ n.i.	▷ 30/n.i.
▷ King, 1990	▷ S.America	▷ n.i.	▷ ref. Bogart, 1973; ctry.n.sp; list. *Hyla rhodopepla*	▷ n.i.	▷ 30/58
▷ Kuramoto, 1990	▷ S.C.America	▷ n.i.	▷ ref. Bogart & Bogart, 1971; Bogart, 1973; ctry.n.sp; list *Hyla rhodopepla*	▷ 1	▷ 30/n.i.
▷ Skuk & Langone, 1992	▷ n.i.	▷ n.i.	▷ ref. Bogart, 1973; list. *Hyla rhodopepla*	▷ n.i.	▷ 30/58
▷ Nunes & Fagundes, 2008b	▷ n.i.	▷ n.i.	▷ ref. Bogart, 1973; list. *{Dendropsopus} rhodopeplus*	▷ n.i.	▷ 30/58
▶ Present study	▶ Peru	▶ 2♂	▶ Bosque Nat. de Iparia, Huánuco Department	▶ 1	▶ 30/56
Dendropsophus rubicundulus					
▶ Rabello, 1970[P]	▶ n.i.	▶ 1♂	▶ loc.n.i.; rep. *Hyla rubicundula*	▶ 1	▶ 30/54†
▷ Foresti, 1972	▷ n.i.	▷ n.i.	▷ ref. Rabello, 1969, 1970; list. *Hyla rubicundula*	▷ n.i.	▷ 30/n.i.
▷ Foresti, 1972	▷ n.i.	▷ n.i.	▷ ref. Rabello et al., 1968; list. *Hyla elongata*	▷ n.i.	▷ 30/n.i.
▶ Bogart, 1973[P]	▶ Brazil	▶ n.i.	▶ southeastern Brazil; loc.n.i.; rep. *Hyla elongata*	▶ 1	▶ 30/52†
▷ Morescalchi, 1979	▷ n.i.	▷ n.i.	▷ ref. Bogart, 1973; list. *Hyla elongata*	▷ n.i.	▷ 30/n.i.
▷ King, 1990	▷ S.America	▷ n.i.	▷ ref. Rabello, 1970; ctry.n.sp; list. *Hyla rubicundula*	▷ n.i.	▷ 30/n.i.

Author(s) ▶ original data[a] ▷ review articles	Country/ Region[b]	Speci- mens[c]	Locality of sampling, references cited, and remarks[d]	Methods applied[e]	2n/FN[f]
▷ King, 1990	▷ S.America	▷ n.i.	▷ ref. Bogart, 1973; ctry.n.sp; list. *Hyla elongata*	▷ n.i.	▷ 30/58
▷ Kuramoto, 1990	▷ S.C.America	▷ n.i.	▷ ref. Rabello, 1970; Bogart, 1973; ctry.n.sp.; list. *Hyla rubicundula*	▷ 1	▷ 30/n.i.
▷ Skuk & Langone, 1992	▷ n.i.	▷ n.i.	▷ ref. Bogart, 1973; list. *Hyla elongata*	▷ n.i.	▷ 30/58
▷ Skuk & Langone, 1992	▷ n.i.	▷ n.i.	▷ ref. Rabello, 1970; list. *Hyla rubicundula*	▷ n.i.	▷ 30/52
▶ Gruber et al., 2005[P]	▶ Brazil	▶ 2♂	▶ São João d'Aliança, Goiás State; rep. *Hyla rubicundula*	▶ 1,4	▶ 30/54
▷ Nunes & Fagundes, 2008b	▷ n.i.	▷ n.i.	▷ ref. Gruber, 2002; Bogart, 1973; list. *{Dendropsopus} rubicundulus*	▷ n.i.	▷ 30/52,58
▷ Catroli, 2008	▷ n.i.	▷ n.i.	▷ ref. Rabello, 1970; Bogart, 1973; Gruber et al., 2005	▷ n.i.	▷ 30/n.i.
▷ Catroli & Kasahara, 2009	▷ n.i.	▷ n.i.	▷ ref. Rabello, 1970; Bogart, 1973; Gruber et al., 2005	▷ 17	▷ 30/n.i.
▷ Hernández-Guzmán et al., 2011	▷ n.i.	▷ n.i.	▷ ref. Gruber et al., 2005; list. *Hyla rubicundula*	▷ n.i.	▷ 30/n.i.
▶ Present study	▶ Argentina	▶ 1U	▶ Rio Grande, Provincia Tierra del Fuego, Departamento Rio Grande	▶ 1	▶ 30/54
Dendropsophus ruschii					
▷ Nunes & Fagundes, 2008b	▷ n.i.	▷ n.i.	▷ ref. Nunes, 2006; list. *{Dendropsopus} ruschii*	▷ n.i.	▷ 30/56,57
▷ Catroli & Kasahara, 2009	▷ n.i.	▷ n.i.	▷ ref. Nunes, 2006	▷ n.i.	▷ 30/n.i.
Dendropsophus sanborni					
▶ Skuk & Langone, 1992[P,M]	▶ Brazil	▶ 2♂	▶ Parque Estadual Vila Velha, Ponta Grossa Municipality; Paraná State; rep. *Hyla sanborni*	▶ 1	▶ 30/52
▶ Medeiros et al., 2003[P]	▶ Brazil	▶ 27♂	▶ Nova Itapirema, São Paulo State (21°11'S, 49°42'W); rep. *Hyla sanborni*	▶ 1,2,4,13	▶ 30/50
▶ Gruber et al., 2005[P]	▶ Brazil	▶ 6♂	▶ Rio Claro, São Paulo State; rep. *Hyla sanborni*	▶ 1,4	▶ 30/50
▷ Nunes & Fagundes, 2008b	▷ n.i.	▷ n.i.	▷ ref. Skuk & Langone, 1992; list. *{Dendropsopus} sanborni*	▷ n.i.	▷ 30/52
▷ Catroli, 2008	▷ n.i.	▷ n.i.	▷ ref. Skuk & Langone, 1992; Medeiros et al., 2003; Gruber et al., 2005	▷ n.i.	▷ 30/n.i.
▷ Catroli & Kasahara, 2009	▷ n.i.	▷ n.i.	▷ ref. Skuk & Langone, 1992; Medeiros et al., 2003; Gruber et al., 2005	▷ 17	▷ 30/n.i.
▷ Hernández-Guzmán et al., 2011	▷ n.i.	▷ n.i.	▷ ref. Medeiros et al., 2003; list. *Hyla sanborni*	▷ n.i.	▷ 30/n.i.
▶ Medeiros et al., 2013[P]	▶ Brazil	▶ 3♂	▶ Salesópolis, São Paulo State	▶ 1,2,4,13	▶ 30/50
▶ Medeiros et al., 2013[P]	▶ Brazil	▶ 19♂2♀	▶ Botucatu, São Paulo State	▶ 1,2,4,13	▶ 30/50
▶ Medeiros et al., 2013[P]	▶ Brazil	▶ 3♂	▶ Torres, Rio Grande do Sul State	▶ 1,2,4,13	▶ 30/50
▶ Present study	▶ Brazil	▶ 6♂	▶ Rio Claro, São Paulo State	▶ 1	▶ 30/50
Dendropsophus seniculus					
▶ Teixeira et al., 2011[N]	▶ Brazil	▶ 4s.n.i.	▶ Ribeirão Grande, São Paulo State	▶ 1–4	▶ 30/50†
▶ Teixeira et al., 2016[P]	▶ Brazil	▶ 4♂	▶ Ribeirão Grande, São Paulo State	▶ 1,2,4,14	▶ 30/50†
Dendropsophus soaresi					
▷ Catroli, 2008	▷ n.i.	▷ n.i.	▷ ref. Neves et al., 2006	▷ n.i.	▷ 30/n.i.
▶ Teixeira et al., 2016[P]	▶ Brazil	▶ 9♂	▶ Barreiras, Bahia State	▶ 1,2,4,14	▶ 30/50†
Dendropsophus subocularis					
▶ Duellman, 1970[N]	▶ C.America	▶ n♂	▶ loc.n.i.; ctry.n.spec.; rep. *Hyla subocularis*	▶ 1	▶ 30‡/n.i
▷ Kuramoto, 1990	▷ S.C.America	▷ n.i.	▷ ref. Duellman, 1970; ctry.n.sp.; list. *Hyla subocularis*	▷ 1	▷ 30/n.i.
▷ Catroli, 2008	▷ n.i.	▷ n.i.	▷ ref. Duellman, 1970	▷ n.i.	▷ 30/n.i.
▷ Catroli & Kasahara, 2009	▷ n.i.	▷ n.i.	▷ ref. Duellman, 1970	▷ n.i.	▷ 30/n.i.
Dendropsophus walfordi					
▶ Medeiros et al., 2013[P]	▶ Brazil	▶ 16♂	▶ Guajará Mirim, Roraima State	▶ 1,2,4,13	▶ 30/52
Dendropsophus werneri					
▶ Skuk & Langone, 1992[P,M]	▶ Brazil	▶ 3♂1U	▶ Porto da Cima, Morretes Municipality, Paraná State; rep. *Hyla baileyi*	▶ 1	▶ 30/60
▷ Nunes & Fagundes, 2008b	▷ n.i.	▷ n.i.	▷ ref. Skuk & Langone, 1992; list. *{Dendropsopus} werneri*	▷ n.i.	▷ 30/56
***Dendropsophus* sp. (*leucophyllatus* group)**					
▶ Bogart & Bogart, 1971[N]	▶ Peru	▶ n.i.	▶ Río Pachitea, Bosque Nacional de Iparía, 1200 m, Huanuco Department; rep. *Hyla leucophyllata* group	▶ 1	▶ 30/n.i.
▷ Kuramoto, 1990	▷ S.C.America	▷ n.i.	▷ ref. Bogart & Bogart, 1971; ctry.n.sp.	▷ 1	▷ 30/n.i.
▷ Catroli, 2008	▷ n.i.	▷ n.i.	▷ ref. Bogart & Bogart, 1971; list. *Dendropsophus* gr. *leucophyllatus*	▷ n.i.	▷ 30/n.i.
▷ Catroli & Kasahara, 2009	▷ n.i.	▷ n.i.	▷ ref. Bogart & Bogart, 1971; list. *Dendropsophus* gr. *leucophyllatus*	▷ n.i.	▷ 30/n.i.
Dendropsophus* sp. 1					
▶ Present study	▶ Mexico	▶ 1♀	▶ near Palenque, Chiapas Department	▶ 1	▶ 30/56
Dendropsophus* sp. 2					
▶ Present study	▶ Colombia	▶ 1♂	▶ near Popayan, Cauca Department	▶ 1	▶ 30/50
Dendropsophus* sp. 3					
▶ Present study	▶ Peru	▶ 1♂	▶ near Bosque Nat. de Iparia, Huánuco Department	▶ 1	▶ 30/60
Dendropsophus* sp. 4					
▶ Present study	▶ Colombia	▶ 1♂	▶ loc.n.i.	▶ 2	▶ 30/60
Dendropsophus* sp.					
▶ Present study	▶ Colombia	▶ 1♂	▶ loc.n.i.	▶ 4	▶ 30/60

Author(s) ▶ original data[a] ▷ review articles	Country/ Region[b]	Speci-mens[c]	Locality of sampling, references cited, and remarks[d]	Methods applied[e]	2n/FN[f]
HYLIDAE / Dendropsophinae / *Xenohyla*					
Xenohyla truncata					
▶ Suárez et al., 2013 [P,M,I]	▶ Brazil	▶ 1♂2J	▶ Restinga de Maricá, Maricá Municipality, Rio de Janeiro State	▶ 1–4,6, 9,13,14	▶ 24/48
HYLIDAE / Hylinae / *Anotheca*					
Anotheca spinosa					
▶ Duellman & Cole, 1965[P]	▶ Mexico	▶ 1♂	▶ Vista Hermosa, Oaxaca State; rep. *Anotheca coronata*	▶ 1	▶ 24‡/n.i.
▷ Duellman, 1970	▷ C.America	▷ n.i.	▷ ref.n.i.; ctry.n.sp.; list. *Anotheca coronata*	▷ 1	▷ 24‡/n.i.
▷ Foresti, 1972	▷ n.i.	▷ n.i.	▷ ref. Duellman & Cole, 1965; list. *Anotheca coronata*	▷ n.i.	▷ 24‡/n.i.
▶ Sessions, 1978[P,M,I]	▶ Mexico	▶ 1♀	▶ vicinity of Cuautlapan, near Orizaba, Veracruz State	▶ 1	▶ 24/48
▷ Rabello, 1970	▷ n.i.	▷ n.i.	▷ ref. Duellman & Cole, 1965; list. *Anotheca coronata*	▷ n.i.	▷ 24‡/n.i.
▷ King, 1990	▷ S.America	▷ n.i.	▷ ref. Duellman & Cole, 1965; ctry.n.sp.; list. *Anotheca coronata*	▷ n.i.	▷ 24/n.i.
▷ King, 1990	▷ S.America	▷ n.i.	▷ Sessions, 1978; ctry.n.sp.	▷ n.i.	▷ 24/48
▷ Kuramoto, 1990	▷ S.C.America	▷ n.i.	▷ ref. Duellman & Cole, 1965; Sessions, 1978; ctry.n.sp.	▷ 1	▷ 24/n.i.
▷ Catroli, 2008	▷ n.i.	▷ n.i.	▷ ref. Duellman & Cole, 1965; Sessions, 1978; list. *Triprion spinosa*	▷ n.i.	▷ 24/n.i.
▷ Catroli & Kasahara, 2009	▷ n.i.	▷ n.i.	▷ ref. Duellman & Cole, 1965; Sessions, 1978	▷ n.i.	▷ 24/n.i.
▶ Present study	▶ Mexico	▶ 1♂	▶ vicinity of Cuautlapan, Veracruz State; see above Sessions, 1978	▶ 1	▶ 24/48
HYLIDAE / Hylinae / *Bromeliohyla*					
No cytogenetic data					
HYLIDAE / Hylinae / *Charadrahyla*					
Charadrahyla chaneque					
▶ Duellman, 1970[N]	▶ C.America	▶ n♂	▶ loc.n.i.; ctry.n.sp.; rep. *Hyla chaneque*	▶ 1	▶ 24‡/n.i.
▷ Kuramoto, 1990	▷ S.C.America	▷ n.i.	▷ ref. Duellman, 1970; ctry.n.sp; list. *Hyla chaneque*	▷ 1	▷ 24/n.i.
▷ Nunes & Fagundes, 2008b	▷ n.i.	▷ n.i.	▷ ref. Duellman, 1970	▷ n.i.	▷ 24/48*
▷ Catroli, 2008	▷ n.i.	▷ n.i.	▷ ref. Duellman, 1970	▷ n.i.	▷ 24/n.i.
▷ Catroli & Kasahara, 2009	▷ n.i.	▷ n.i.	▷ ref. Duellman, 1970	▷ n.i.	▷ 24/n.i.
HYLIDAE / Hylinae / *Diaglena*					
Diaglena spatulata					
▶ Duellman & Cole, 1965[N]	▶ Mexico	▶ 1♂	▶ Mazatian, Sianola State	▶ 1	▶ 24‡/n.i.
▶ Duellman & Cole, 1965[N]	▶ Mexico	▶ 3♂	▶ 5 km E El Zapote; Guerrero State; rep. *Diaglena reticulata*	▶ 1	▶ 24‡/n.i.
▷ Duellman, 1970	▷ C.America	▷ n.i.	▷ ref.n.i.; ctry.n.sp.; list. *Triprion spatulatus*	▷ n.i.	▷ 24/n.i.
▷ Rabello, 1970	▷ n.i.	▷ n.i.	▷ ref. Duellman & Cole, 1965	▷ n.i.	▷ 24‡/n.i.
▷ Rabello, 1970	▷ n.i.	▷ n.i.	▷ ref. Duellman & Cole, 1965; list. *Diaglena reticulata*	▷ n.i.	▷ 24‡/n.i.
▷ Foresti, 1972	▷ n.i.	▷ n.i.	▷ ref. Duellman & Cole, 1965	▷ n.i.	▷ 24‡/n.i.
▷ Foresti, 1972	▷ n.i.	▷ n.i.	▷ ref. Duellman & Cole, 1965; list. *Diaglena reticulata*	▷ n.i.	▷ 24‡/n.i.
▷ King, 1990	▷ C.America	▷ n.i.	▷ ref. Duellman & Cole, 1965; ctry.n.sp.; list. *Triprion reticulata*	▷ n.i.	▷ 24/n.i.
▷ King, 1990	▷ C.America	▷ n.i.	▷ ref. Duellman & Cole, 1965; ctry.n.sp.; list. *Triprion spatulata*	▷ n.i.	▷ 24/n.i.
▷ Kuramoto, 1990	▷ S.C.America	▷ n.i.	▷ ref. Duellman & Cole, 1965; ctry.n.sp.; list. *Triprion spatulatus reticulatus*	▷ 1	▷ 24/n.i.
▷ Kuramoto, 1990	▷ S.C.America	▷ n.i.	▷ ref. Duellman & Cole, 1965; Duellman, 1970; ctry.n.sp.; list. *Triprion spatulatus*	▷ 1	▷ 24/n.i.
▷ Catroli, 2008	▷ n.i.	▷ n.i.	▷ ref. Duellman & Cole, 1965; Duellman, 1970; list. *Triprion spatulatus*	▷ n.i.	▷ 24/n.i.
▷ Catroli, 2008	▷ n.i.	▷ n.i.	▷ ref. Duellman & Cole, 1965; list. *Triprion spatulatus reticulatus*	▷ n.i.	▷ 24/n.i.
▷ Catroli & Kasahara, 2009	▷ n.i.	▷ n.i.	▷ ref. Duellman & Cole, 1965; Duellman, 1970; list. *Diaglenus spatulatus*	▷ n.i.	▷ 24/n.i.
HYLIDAE / Hylinae / *Dryophytes*					
Dryophytes andersonii					
▶ Duellman & Cole, 1965[N]	▶ USA	▶ 3♂	▶ 2 miles NE Berlin, Burlington County, New Jersey State; rep. *Hyla {andersoni}*	▶ 1	▶ 24‡/n.i.
▷ Rabello, 1970	▷ n.i.	▷ n.i.	▷ ref. Duellman & Cole, 1965; list. *Hyla {andersoni}*	▷ n.i.	▷ 24‡/n.i.
▶ Wasserman, 1970[P]	▶ USA	▶ 1♂	▶ 19 km SW South Toms river, Ocean County, New Jersey State; rep. *Hyla andersoni*	▶ 1	▶ 24/48†
▷ Foresti, 1972	▷ n.i.	▷ n.i.	▷ ref. Duellman & Cole, 1965; Wasserman, 1970; list. *Hyla andersonii*	▷ n.i.	▷ 24/n.i.
▷ Cole, 1974	▷ n.i.	▷ n.i.	▷ ref. Wasserman, 1970; list. *Hyla {andersoni}*	▷ n.i.	▷ 24/n.i.
▶ Wiley, 1982[P,I]	▶ USA	▶ 3♂	▶ Sampson County, North Carolina State; rep. *Hyla {andersoni}*	▶ 10	▶ 24/48†
▶ Anderson, 1986[I]	▶ USA	▶ 4♂2♀	▶ Burlington County, New Jersey State; list. *Hyla andersonii*	▶ 1,4	▶ 24/48†
▶ Anderson, 1986[I]	▶ USA	▶ 1♂	▶ Walton County, Florida State; list. *Hyla andersonii*	▶ 1,4	▶ 24/48†
▶ Anderson, 1986[I]	▶ USA	▶ 1♂1U	▶ Santa Rosa County, Florida State; list. *Hyla andersonii*	▶ 1,2,4	▶ 24/48†
▶ Anderson, 1986[I]	▶ USA	▶ 1♀1U	▶ Okaloosa County, Florida State; list. *Hyla andersonii*	▶ 1,2,4	▶ 24/48†
▶ Anderson, 1986[I]	▶ USA	▶ 2♂	▶ Wayne County, North Carolina State; list. *Hyla andersonii*	▶ 1,4	▶ 24/48†
▶ Anderson, 1986[I]	▶ USA	▶ 1♂	▶ Bladen County, North Carolina State; list. *Hyla andersonii*	▶ 1,2,4	▶ 24/48†
▷ King, 1990	▷ N.America	▷ n.i.	▷ ref. Duellman & Cole, 1965; Wasserman, 1970; Wiley, 1982; Anderson, 1986; list. *Hyla {andersoni}*	▷ 2,4,10	▷ 24/48

DOI: 10.1159/000490085

Author(s) ▶ original data[a] ▷ review articles	Country/ Region[b]	Speci- mens[c]	Locality of sampling, references cited, and remarks[d]	Methods applied[e]	2n/FN[f]
▷ Kuramoto, 1990	▷ N.America	▷ n.i.	▷ ref. Duellman & Cole, 1965; Wasserman, 1970; Wiley, 1982; Anderson & Moler, 1986; ctry.n.sp.; list. *Hyla andersonii*	▷ 17	▷ 24/n.i.
▷ Anderson, 1991[I]	▷ Holarctic	▷ n.i.	▷ ref. Anderson, 1986; ctry.n.sp.; list. *Hyla andersonii*	▷ 2,4	▷ 24/48†
▷ Nunes & Fagundes, 2008b	▷ n.i.	▷ n.i.	▷ ref. Wasserman, 1970; list. *Hyla andersonii*	▷ n.i.	▷ 24/48
▷ Catroli, 2008	▷ n.i.	▷ n.i.	▷ ref. Duellman & Cole, 1965; Wasserman, 1970; Wiley, 1982; Anderson & Moler, 1986; Anderson, 1991; list. *Hyla andersonii*	▷ n.i.	▷ 24/n.i.
▷ Catroli & Kasahara, 2009	▷ n.i.	▷ n.i.	▷ ref. Duellman & Cole, 1965; Wasserman, 1970; Wiley, 1982; Anderson & Moler, 1986; Anderson, 1991; list. *Hyla andersonii*	▷ 17	▷ 24/n.i.
Dryophytes arenicolor					
▶ Duellman & Cole, 1965[N]	▶ Mexico	▶ 1♂	▶ Acahuizotla; Guerrero State; rep. *Hyla arenicolor*	▶ 1	▶ 24‡/n.i.
▷ Duellman, 1970	▷ C.America	▷ n.i.	▷ ref.n.i.; ctry.n.sp.; list. *Hyla arenicolor*	▷ 1	▷ 24‡/n.i.
▶ Anderson, 1986[P,I]	▶ USA	▶ 1♂1♀	▶ Coconino County, Arizona State; rep. *Hyla arenicolor*	▶ 1,4	▶ 24/48†
▶ Anderson, 1986[P,I]	▶ USA	▶ 1♀	▶ Santa Cruz County, Arizona State; rep. *Hyla arenicolor*	▶ 1,2,4	▶ 24/48†
▶ Anderson, 1986[P,I]	▶ Mexico	▶ 1♀	▶ Sonora State; rep. *Hyla arenicolor*	▶ 1,2,4	▶ 24/48†
▷ Rabello, 1970	▷ n.i.	▷ n.i.	▷ ref. Duellman & Cole, 1965; list. *Hyla arenicolor*	▷ n.i.	▷ 24‡/n.i.
▷ Foresti, 1972	▷ n.i.	▷ n.i.	▷ ref. Duellman & Cole, 1965; list. *Hyla arenicolor*	▷ n.i.	▷ 24‡/n.i.
▷ King, 1990	▷ N.America	▷ n.i.	▷ ref. Duellman & Cole, 1965; Wasserman, 1970; Anderson, 1986; ctry.n.sp.; list. *Hyla arenicolor*	▷ 2,4	▷ 24/48
▷ Kuramoto, 1990	▷ S.C.America	▷ n.i.	▷ ref. Duellman & Cole, 1965; ctry.n.sp.; list. *Hyla arenicolor*	▷ 1	▷ 24/n.i.
▷ Anderson, 1991[I]	▷ Holarctic	▷ n.i.	▷ ref. Anderson, 1986; ctry.n.sp.; list. *Hyla arenicolor*	▷ 2,4	▷ 24/48†
▷ Nunes & Fagundes, 2008b	▷ n.i.	▷ n.i.	▷ ref. Anderson, 1991; list. *Hyla arenicolor*	▷ n.i.	▷ 24/48
▷ Catroli, 2008	▷ n.i.	▷ n.i.	▷ ref. Duellman & Cole, 1965; Wasserman, 1970; Anderson, 1991; list. *Hyla arenicolor*	▷ n.i.	▷ 24/n.i.
▷ Catroli & Kasahara, 2009	▷ n.i.	▷ n.i.	▷ ref. Duellman & Cole, 1965; Wasserman, 1970; Anderson, 1991; list. *Hyla arenicolor*	▷ 17	▷ 24/n.i.
Dryophytes avivocus					
▶ Bushnell et al., 1939[D]	▶ USA	▶ n.i.	▶ Relfoot Lake, Tennessee State; rep. *Hyla avivoca*	▶ 1	▶ 24‡/n.i.
▷ Rabello, 1970	▷ n.i.	▷ n.i.	▷ ref. Bushnell et al., 1939; list. *Hyla avivoca*	▷ n.i.	▷ 24/n.i.
▷ Foresti, 1972	▷ n.i.	▷ n.i.	▷ ref. Bushnell et al., 1939; list. *Hyla avivoca*	▷ n.i.	▷ 24/n.i.
▶ Wiley, 1982[P,I]	▶ USA	▶ 3♂	▶ Relfoot Lake, Tennessee State; rep. *Hyla avivoca*	▶ 10	▶ 24/48†
▶ Anderson, 1986[I]	▶ USA	▶ 5♂	▶ Jasper County, South Carolina State; rep. *Hyla avivoca ogechiensis*	▶ 1,2,4,11	▶ 24/48†
▶ Anderson, 1986[I]	▶ USA	▶ 1♂	▶ Washington County, Florida State; rep. *Hyla avivoca avivoca*	▶ 1,2,4,11	▶ 24/48†
▷ King, 1990	▷ N.America	▷ n.i.	▷ ref. Bushnell et al., 1939; Wiley, 1982; Anderson, 1986; ctry.n.sp.; list. *Hyla avivoca*	▷ 2,4,10,11	▷ 24/48
▷ Kuramoto, 1990	▷ N.America	▷ n.i.	▷ ref. Bushnell et al., 1939; Wiley, 1982; ctry.n.sp.; list. *Hyla avivoca*	▷ 17	▷ 24/n.i.
▷ Anderson, 1991[I]	▷ Holarctic	▷ n.i.	▷ ref. Anderson, 1986; ctry.n.sp.; list. *Hyla avivoca*	▷ 2,4,11	▷ 24/48†
▷ Nunes & Fagundes, 2008b	▷ n.i.	▷ n.i.	▷ ref. Wiley, 1982; list. *Hyla avivoca*	▷ n.i.	▷ 24/48
▷ Catroli, 2008	▷ n.i.	▷ n.i.	▷ ref. Bushnell et al., 1939; Wiley, 1982; Anderson, 1991; list. *Hyla avivoca*	▷ n.i.	▷ 24/n.i.
▷ Catroli & Kasahara, 2009	▷ n.i.	▷ n.i.	▷ ref. Bushnell et al., 1939; Wiley, 1982; Anderson, 1991; list. *Hyla avivoca*	▷ 17	▷ 24/n.i.
▶ Present study	▶ USA	▶ 1♂	▶ Hickman County, Kentucky State	▶ 1	▶ 24/48
Dryophytes chrysoscelis					
▶ Bogart & Wasserman, 1972[P,M,I]	▶ USA	▶ n.i.	▶ Texas, South Carolina, New Jersey, Louisiana, Mississippi, Indiana and Illinois States	▶ 1	▶ 24/48†
▷ Foresti, 1972	▷ n.i.	▷ n.i.	▷ ref. Wasserman, 1970; list. *Hyla chrysoscelis*	▷ n.i.	▷ 24/n.i.
▶ Bogart, 1973[P]	▶ USA	▶ n.i.	▶ Texas State; list. *Hyla chrysoscelis*	▶ 1	▶ 24/48†
▷ Cole, 1974	▷ n.i.	▷ n.i.	▷ ref. Bogart & Wasserman, 1972; Bogart, 1973; list. *Hyla chrysoscelis*	▷ n.i.	▷ 24/n.i.
▷ Morescalchi, 1979	▷ n.i.	▷ n.i.	▷ ref. Bogart, 1973; list. *Hyla chrysoscelis*	▷ n.i.	▷ 24/n.i.
▶ Wiley, 1982[P,I]	▶ USA	▶ 1♂	▶ Jackson County, North Carolina State; chr.6 morph; rep. *Hyla chrysoscelis*	▶ 10	▶ 24/48†
▶ Wiley, 1982[P,I]	▶ USA	▶ 3♂	▶ Wilson County, Tennessee State; chr.6 morph; rep. *Hyla chrysoscelis*	▶ 10	▶ 24/48†
▶ Wiley, 1982[P,I]	▶ USA	▶ 1♂	▶ Cumberland County, Tennessee State; chr.6 morph; rep. *Hyla chrysoscelis*	▶ 10	▶ 24/48†
▶ Wiley, 1982[P,I]	▶ USA	▶ 1♂	▶ St. James Parish, Louisiana State; chr.6 morph; rep. *Hyla chrysoscelis*	▶ 10	▶ 24/48†
▶ Wiley, 1982[P,I]	▶ USA	▶ 1♂	▶ St. Charles Parish, Louisiana State; chr.6 morph; rep. *Hyla chrysoscelis*	▶ 10	▶ 24/48†
▶ Wiley, 1982[P,I]	▶ USA	▶ 4♂	▶ Wake County, North Carolina State; chr.8 morph; rep. *Hyla chrysoscelis*	▶ 10	▶ 24/48†
▶ Wiley, 1982[P,I]	▶ USA	▶ 1♂	▶ Burke County, North Carolina State; chr.8 morph; rep. *Hyla chrysoscelis*	▶ 10	▶ 24/48†
▶ Wiley, 1982[P,I]	▶ USA	▶ 1♂	▶ Bladen County, North Carolina State; chr.8 morph; rep. *Hyla chrysoscelis*	▶ 10	▶ 24/48†
▶ Wiley, 1982[P,I]	▶ USA	▶ 1♂	▶ Rockingham County, North Carolina State; chr.8 morph; rep. *Hyla chrysoscelis*	▶ 10	▶ 24/48†
▶ Wiley, 1982[P,I]	▶ USA	▶ 1♂	▶ Dare County, North Carolina State; chr.8 morph; rep. *Hyla chrysoscelis*	▶ 10	▶ 24/48†
▶ Wiley, 2003[P]	▶ USA	▶ 2♂2♀	▶ Pitt County, North Carolina State; rep. *Hyla chrysoscelis*	▶ 12	▶ 24†/48†
▶ Anderson, 1986[I]	▶ USA	▶ 1♂	▶ Cobb County, Georgia State; chr.6 morph; rep. *Hyla chrysoscelis*	▶ 1,4,11	▶ 24/48†
▶ Anderson, 1986[I]	▶ USA	▶ 1♀	▶ Portage County, Wisconsin State; chr.6 morph; rep. *Hyla chrysoscelis*	▶ 1,2,4,11	▶ 24/48†
▶ Anderson, 1986[I]	▶ USA	▶ 1♂	▶ Talladega County, Alabama State; chr.6 morph; rep. *Hyla chrysoscelis*	▶ 1,2,4,11	▶ 24/48†
▶ Anderson, 1986[I]	▶ USA	▶ 1♀	▶ Okaloosa County, Florida State; chr.6 morph; rep. *Hyla chrysoscelis*	▶ 1,2,4,11	▶ 24/48†
▶ Anderson, 1986[I]	▶ USA	▶ 3♂2♀	▶ Jasper County, South Carolina State; chr.8 morph; rep. *Hyla chrysoscelis*	▶ 1,2,4,11	▶ 24/48†
▶ Anderson, 1986[I]	▶ USA	▶ 1♂	▶ Dorchester County, Maryland State; chr.8 morph; rep. *Hyla chrysoscelis*	▶ 1,4,11	▶ 24/48†
▶ Anderson, 1986[I]	▶ USA	▶ 4♂	▶ Alachua County, Florida State; chr.11 morph; rep. *Hyla chrysoscelis*	▶ 1,2,4,11	▶ 24/48†
▶ Wiley et al., 1989[P]	▶ USA	▶ 3s.n.i.	▶ Todd County, Kentucky State; rep. *Hyla chrysoscelis*	▶ 4,12,13	▶ 24†/48†
▶ Wiley et al., 1989[P]	▶ USA	▶ 2s.n.i.	▶ Meade County, Kentucky State; rep. *Hyla chrysoscelis*	▶ 4,12,13	▶ 24†/48†
▶ Wiley et al., 1989[P]	▶ USA	▶ 6s.n.i.	▶ McCreary County, Kentucky State; rep. *Hyla chrysoscelis*	▶ 4,12,13	▶ 24†/48†

Author(s) ▶ original data[a] ▷ review articles	Country/ Region[b]	Speci-mens[c]	Locality of sampling, references cited, and remarks[d]	Methods applied[e]	2n/FN[f]
▶ Wiley et al., 1989[P]	▶ USA	▶ 15s.n.i.	▶ Lee County, Alabama State; rep. *Hyla chrysoscelis*	▶ 4,12,13	▶ 24†/48†
▶ Wiley et al., 1989[P]	▶ USA	▶ 6s.n.i.	▶ Red River Parish, Louisiana State; rep. *Hyla chrysoscelis*	▶ 4,12,13	▶ 24†/48†
▶ Wiley et al., 1989[P]	▶ USA	▶ 6s.n.i.	▶ Tangipahoa Parish, Louisiana State; rep. *Hyla chrysoscelis*	▶ 4,12,13	▶ 24†/48†
▶ Wiley et al., 1989[P]	▶ USA	▶ 20s.n.i.	▶ Wayne County, West Virginia State; rep. *Hyla chrysoscelis*	▶ 4,12,13	▶ 24†/48†
▶ Wiley et al., 1989[P]	▶ USA	▶ 3s.n.i.	▶ Tucker County, West Virginia State; rep. *Hyla chrysoscelis*	▶ 4,12,13	▶ 24†/48†
▶ Wiley et al., 1989[P]	▶ USA	▶ 13s.n.i.	▶ Summers County, West Virginia State; rep. *Hyla chrysoscelis*	▶ 4,12,13	▶ 24†/48†
▶ Wiley et al., 1989[P]	▶ USA	▶ 8s.n.i.	▶ Dane County, Wisconsin State; rep. *Hyla chrysoscelis*	▶ 4,12,13	▶ 24†/48†
▶ Wiley et al., 1989[P]	▶ USA	▶ 10s.n.i.	▶ Juneau County, Wisconsin State; rep. *Hyla chrysoscelis*	▶ 4,12,13	▶ 24†/48†
▶ Wiley et al., 1989[P]	▶ USA	▶ 2s.n.i.	▶ Dare County, North Carolina State; rep. *Hyla chrysoscelis*	▶ 4,12,13	▶ 24†/48†
▶ Wiley et al., 1989[P]	▶ USA	▶ 1s.n.i.	▶ McDowell County, North Carolina State; rep. *Hyla chrysoscelis*	▶ 4,12,13	▶ 24†/48†
▶ Wiley et al., 1989[P]	▶ USA	▶ 7s.n.i.	▶ Warren County, North Carolina State; rep. *Hyla chrysoscelis*	▶ 4,12,13	▶ 24†/48†
▶ Wiley et al., 1989[P]	▶ USA	▶ 4s.n.i.	▶ Vance County, North Carolina State; rep. *Hyla chrysoscelis*	▶ 4,12,13	▶ 24†/48†
▶ Wiley et al., 1989[P]	▶ USA	▶ 6s.n.i.	▶ Polk County, North Carolina State; rep. *Hyla chrysoscelis*	▶ 4,12,13	▶ 24†/48†
▶ Wiley et al., 1989[P]	▶ USA	▶ 5s.n.i	▶ Watauga County, North Carolina State; rep. *Hyla chrysoscelis*	▶ 4,12,13	▶ 24†/48†
▶ Wiley et al., 1989[P]	▶ USA	▶ 6s.n.i.	▶ Mitchell County, North Carolina State; rep. *Hyla chrysoscelis*	▶ 4,12,13	▶ 24†/48†
▶ Wiley et al., 1989[P]	▶ USA	▶ 7s.n.i.	▶ Transylvania County, North Carolina State; rep. *Hyla chrysoscelis*	▶ 4,12,13	▶ 24†/48†
▶ Wiley et al., 1989[P]	▶ USA	▶ 1s.n.i.	▶ Brunswick County, Virginia State; rep. *Hyla chrysoscelis*	▶ 4,12,13	▶ 24†/48†
▶ Wiley et al., 1989[P]	▶ USA	▶ 6s.n.i.	▶ Mecklemburg County, Virginia State; rep. *Hyla chrysoscelis*	▶ 4,12,13	▶ 24†/48†
▶ Wiley et al., 1989[P]	▶ USA	▶ 4s.n.i.	▶ Aiken County, South Carolina State; rep. *Hyla chrysoscelis*	▶ 4,12,13	▶ 24†/48†
▶ Wiley et al., 1989[P]	▶ USA	▶ 10s.n.i.	▶ Alachua County, Florida State; rep. *Hyla chrysoscelis*	▶ 4,12,13	▶ 24†/48†
▶ Wiley et al., 1989[P]	▶ USA	▶ 63s.n.i.	▶ Leon County, Florida State; rep. *Hyla chrysoscelis*	▶ 4,12,13	▶ 24†/48†
▶ Wiley et al., 1989[P]	▶ USA	▶ 3s.n.i.	▶ Okaloosa County, Florida State; rep. *Hyla chrysoscelis*	▶ 4,12,13	▶ 24†/48†
▶ Wiley et al., 1989[P]	▶ USA	▶ 1s.n.i.	▶ Liberty County, Florida State; rep. *Hyla chrysoscelis*	▶ 4,12,13	▶ 24†/48†
▶ Wiley et al., 1989[P]	▶ USA	▶ 4s.n.i.	▶ Pulasky County, Arkansas State; rep. *Hyla chrysoscelis*	▶ 4,12,13	▶ 24†/48†
▶ Wiley et al., 1989[P]	▶ USA	▶ 4s.n.i.	▶ Payne County, Oklahoma State; rep. *Hyla chrysoscelis*	▶ 4,12,13	▶ 24†/48†
▶ Wiley et al., 1989[P]	▶ USA	▶ 4s.n.i.	▶ Ottawa County, Oklahoma State; rep. *Hyla chrysoscelis*	▶ 4,12,13	▶ 24†/48†
▶ Wiley et al., 1989[P]	▶ USA	▶ 2s.n.i.	▶ McCurtain County, Oklahoma State; rep. *Hyla chrysoscelis*	▶ 4,12,13	▶ 24†/48†
▶ Wiley et al., 1989[P]	▶ USA	▶ 11s.n.i.	▶ Howell County, Missouri State; rep. *Hyla chrysoscelis*	▶ 4,12,13	▶ 24†/48†
▶ Wiley et al., 1989[P]	▶ USA	▶ 4s.n.i.	▶ Williamson County, Illinois State; rep. *Hyla chrysoscelis*	▶ 4,12,13	▶ 24†/48†
▶ Wiley et al., 1989[P]	▶ USA	▶ 3s.n.i.	▶ Bastrop County, Texas State; rep. *Hyla chrysoscelis*	▶ 4,12,13	▶ 24†/48†
▶ Wiley et al., 1989[P]	▶ USA	▶ 3s.n.i.	▶ Montgomery County, Kansas State; rep. *Hyla chrysoscelis*	▶ 4,12,13	▶ 24†/48†
▷ King, 1990	▷ N.America	▷ n.i.	▷ ref. Bogart & Wasserman, 1972; Bogart, 1973; Wiley, 1982, 1983; Anderson, 1986; ctry.n.sp.; list. *Hyla chrysoscelis*	▷ 2,4,10,11	▷ 24/48
▷ Kuramoto, 1990	▷ N.America	▷ n.i.	▷ ref. Bogart & Wasserman, 1972; Bogart, 1973; Wiley, 1982, 1983; ctry.n.sp.; list. *Hyla chrysoscelis*	▷ 17	▷ 24/n.i.
▷ Anderson, 1991[I]	▷ Holarctic	▷ n.i.	▷ ref. Anderson, 1986; ctry.n.sp. list. *Hyla chrysoscelis*	▷ 2,4,11	▷ 24/48†
▶ Wiley & Little, 2000[P]	▶ USA	▶ 3♂	▶ Summers County, West Virginia State; rep. *Hyla chrysoscelis*	▶ 12	▶ 24†/48†
▶ Wiley & Little, 2000[P]	▶ USA	▶ 2♂2♀	▶ Pitt County, North Carolina State; rep. *Hyla chrysoscelis*	▶ 12	▶ 24†/48†
▶ Wiley & Little, 2000[P]	▶ USA	▶ 2♂	▶ Houston County, Georgia State; rep. *Hyla chrysoscelis*	▶ 12	▶ 24†/48†
▶ Wiley, 2003[P]	▶ USA	▶ 2♂	▶ Summers County, West Virginia State; rep. *Hyla chrysoscelis*	▶ 12	▶ 24†/48†
▶ Wiley, 2003[P]	▶ USA	▶ 2♂2♀	▶ Pitt County, North Carolina State; rep. *Hyla chrysoscelis*	▶ 12	▶ 24†/48†
▷ Nunes & Fagundes, 2008b	▷ n.i.	▷ n.i.	▷ ref. Bogart, 1973; list. *Hyla chrysoscelis*	▷ n.i.	▷ 24/48
▷ Catroli, 2008	▷ n.i.	▷ n.i.	▷ ref. Bogart & Wasserman, 1972; Bogart, 1973; Wiley, 1982, 1983, 1992, 2003; Wiley et al., 1989; Anderson, 1991; Wiley & Little, 2000; list. *Hyla chrysoscelis*	▷ n.i.	▷ 24/n.i.
▷ Catroli & Kasahara, 2009	▷ n.i.	▷ n.i.	▷ ref. Bogart & Wasserman, 1972; Bogart, 1973; Wiley, 1982, 1983, 2003; Wiley et al., 1989, 1992; Anderson, 1991; Wiley & Little, 2000; list. *Hyla chrysoscelis*	▷ 17	▷ 24/n.i.
▷ Schmid & Steinlein, 2016a	▷ n.i.	▷ n.i.	▷ ref. Wiley et al., 1992	▷ 14	▷ 24/n.i.
▶ Present study	▶ USA	▶ 2♂	▶ Caney Creek, Sam Houstan National Park, Texas State	▶ 1	▶ 24/48
▶ Present study	▶ USA	▶ 1U	▶ Elgin, Texas State	▶ 1	▶ 24/48
▶ Present study	▶ USA	▶ 1♂	▶ Hwy 969, Texas State	▶ 1	▶ 24/48
▶ Present study	▶ USA	▶ 2♂1♀	▶ Texas State	▶ 1	▶ 24/48
▶ Present study	▶ USA	▶ 1♂1U	▶ Tillman, South Carolina State	▶ 1	▶ 24/48
▶ Present study	▶ USA	▶ 1s.n.i.	▶ loc.u.	▶ 1	▶ 24/48
▶ Present study	▶ USA	▶ 1♂	▶ loc.u.	▶ 1	▶ 24/48
▶ Present study	▶ USA	▶ 1♂	▶ Caney Creek, Sam Houston National Forest, Texas State	▶ 1	▶ 24/48
Dryophytes cinereus					
▶ Bushnell et al., 1939[D]	▶ USA	▶ n.i.	▶ Relfoot Lake, Tennessee State; rep. *Hyla cinerea cinerea*	▶ 1	▶ 24‡/n.i.
▷ Rabello, 1970	▷ n.i.	▷ n.i.	▷ ref. Bushnell et al., 1939; list. *Hyla cinerea*	▷ n.i.	▷ 24/n.i.
▷ Foresti, 1972	▷ n.i.	▷ n.i.	▷ ref. Bushnell et al., 1939; list. *Hyla cinerea*	▷ n.i.	▷ 24/n.i.
▶ Schmid, 1978a[P,I]	▶ N.America	▶ 2♂1♀	▶ specimens obtained by animal dealer; loc.n.i.; list. *Hyla cinerea*	▶ 1,2,4,5	▶ 24/48†
▶ Wiley, 1982[P,I]	▶ USA	▶ 3♂	▶ Rayne, Louisiana State; rep. *Hyla cinerea*	▶ 10	▶ 24/48†
▶ Wiley, 1982[P,I]	▶ USA	▶ 2♂	▶ Berkeley County, South Carolina State; rep. *Hyla cinerea*	▶ 10	▶ 24/48†
▶ Anderson, 1986[I]	▶ USA	▶ 1♂	▶ unknown county, Florida State; list. *Hyla cinerea*	▶ 1,4	▶ 24/48†
▶ Anderson, 1986[I]	▶ USA	▶ 2♂	▶ Gulf County, Florida State; list. *Hyla cinerea*	▶ 1,4	▶ 24/48†
▶ Anderson, 1986[I]	▶ USA	▶ 2♂	▶ Alachua County, Florida State; list. *Hyla cinerea*	▶ 1,4	▶ 24/48†
▶ Anderson, 1986[I]	▶ USA	▶ 4♂1♀	▶ Jasper County, South Carolina State; list. *Hyla cinerea*	▶ 1,2,4	▶ 24/48†
▶ Anderson, 1986[I]	▶ USA	▶ 3♂1♀	▶ Dorchester County, Maryland State; list. *Hyla cinerea*	▶ 1,4	▶ 24/48†
▶ Anderson, 1986[I]	▶ USA	▶ 1♂1♀	▶ Lonoke County, Arkansas State; list. *Hyla cinerea*	▶ 1,4	▶ 24/48†
▶ Anderson, 1986[I]	▶ USA	▶ 4♂	▶ Orleans County, Louisiana State; list. *Hyla cinerea*	▶ 1,4	▶ 24/48†
▶ Anderson, 1986[I]	▶ USA	▶ 1♂	▶ Hunting Island, Georgia State; list. *Hyla cinerea*	▶ 1,4	▶ 24/48†

Author(s) ▶ original data[a] ▷ review articles	Country/ Region[b]	Speci-mens[c]	Locality of sampling, references cited, and remarks[d]	Methods applied[e]	2n/FN[f]
▷ King, 1990	▷ N.America	▷ n.i.	▷ ref. Bushnell et al., 1939; Schmid, 1978a, 1980/err.ref.; Wiley, 1982; ctry. n.sp.; list. *Hyla cinerea*	▷ 2,4,10, 17	▷ 24/48
▷ Kuramoto, 1990	▷ N.America	▷ n.i.	▷ ref. Bushnell et al., 1939; Schmid, 1978a, 1980/err.ref.; Wiley, 1982; Anderson & Moler, 1986; ctry.n.sp.; list. *Hyla cinerea*	▷ 17	▷ 24/n.i.
▷ Anderson, 1991[I]	▷ Holarctic	▷ n.i.	▷ ref. Anderson, 1986; ctry.n.sp.; list. *Hyla arenicolor*	▷ 2,4	▷ 24/48†
▷ Nunes & Fagundes, 2008b	▷ n.i.	▷ n.i.	▷ ref. Anderson, 1991; list. *Hyla cinerea*	▷ n.i.	▷ 24/48
▷ Catroli, 2008	▷ n.i.	▷ n.i.	▷ ref. Bushnell et al., 1939, Schmid 1978a, 1980/err.ref.; Wiley, 1982; Anderson & Moler, 1986; Anderson, 1991; Wiley et al., 1992; list. *Hyla cinerea*	▷ n.i.	▷ 24/n.i.
▷ Catroli & Kasahara, 2009	▷ n.i.	▷ n.i.	▷ ref. Bushnell et al., 1939, Schmid 1978a, 1980/err.ref.; Wiley, 1982; Anderson & Moler, 1986; Anderson, 1991; Wiley et al., 1992; list. *Hyla cinerea*	▷ 17	▷ 24/n.i.
▷ Schmid & Steinlein, 2016a	▷ n.i.	▷ n.i.	▷ ref. Wiley et al., 1992; list. *Hyla cinerea*	▷ 14	▷ 24/n.i.
▶ Present study	▶ USA	▶ 1J1U	▶ Palmeto Sate Park, Texas State	▶ 1	▶ 24/48
▶ Present study	▶ USA	▶ 1U	▶ Sam Houston National Park, Texas State	▶ 1	▶ 24/48
▶ Present study	▶ USA	▶ 1♀	▶ 24 miles S Georgetown, Hwy 17, South Carolina State	▶ 1	▶ 24/48
▶ Present study	▶ USA	▶ 1U	▶ Goose Island, Georgia State	▶ 1	▶ 24/48
▶ Present study	▶ USA	▶ 1U	▶ loc.n.i.	▶ 1,2,4,5	▶ 24/48
Dryophytes euphorbiaceus					
▶ Duellman & Cole, 1965[N]	▶ Mexico	▶ 1♂	▶ Llano de las Flores, Oaxaca State; rep. *Hyla euphorbiacea*	▶ 1	▶ 24‡/n.i.
▶ Anderson, 1986[I]	▶ Mexico	▶ 1♂	▶ Puerto de Aire, Veracruz State; rep. *Hyla* cf. *euphorbiacea*	▶ 1,2,4	▶ 24/48†
▷ Duellman, 1970	▷ C.America	▷ n.i.	▷ ref.n.i.; ctry.n.sp.; list. *Hyla euphorbiacea*	▷ 1	▷ 24‡/n.i.
▷ Rabello, 1970	▷ n.i.	▷ n.i.	▷ ref. Duellman & Cole, 1965; list. *Hyla euphorbiacea*	▷ n.i.	▷ 24‡/n.i.
▷ Foresti, 1972	▷ n.i.	▷ n.i.	▷ ref. Duellman & Cole, 1965; list. *Hyla euphorbiacea*	▷ n.i.	▷ 24‡/n.i.
▷ King, 1990	▷ C.America	▷ n.i.	▷ ref. Duellman & Cole, 1965; Anderson, 1986; ctry.n.sp.; list. *Hyla euphorbiacea*	▷ 2,4	▷ 24/48
▷ Kuramoto, 1990	▷ S.C.America	▷ n.i.	▷ ref. Duellman & Cole, 1965; ctry.n.sp.; list. *Hyla euphorbiacea*	▷ 1	▷ 24/n.i.
▷ Anderson, 1991[I]	▷ Holarctic	▷ n.i.	▷ ref. Anderson, 1986; ctry.n.sp.; list. *Hyla euphorbiacea*	▷ 2,4	▷ 24/48†
▷ Nunes & Fagundes, 2008b	▷ n.i.	▷ n.i.	▷ ref. Duellman & Cole, 1965; list. *Hyla euphorbiacea*	▷ n.i.	▷ 24/48
▷ Catroli, 2008	▷ n.i.	▷ n.i.	▷ ref. Duellman & Cole, 1965; Anderson, 1991; list. *Hyla euphorbiacea*	▷ n.i.	▷ 24/n.i.
▷ Catroli & Kasahara, 2009	▷ n.i.	▷ n.i.	▷ ref. Duellman & Cole, 1965; Anderson, 1991; list. *Hyla euphorbiacea*	▷ 17	▷ 24/n.i.
▶ Present study	▶ Mexico	▶ 1♂	▶ Iz. De Matamoras	▶ 1	▶ 24/48
▶ Present study	▶ Mexico	▶ 1♂	▶ 2 mi S of Xoxa, S of Oaxaca	▶ 1	▶ 24/48
Dryophytes eximius					
▶ Duellman & Cole, 1965[N]	▶ Mexico	▶ 3♂	▶ 3.5 km W Cuautlixco; Morelos State; rep. *Hyla eximia*	▶ 1	▶ 24‡/n.i.
▷ Duellman, 1970	▷ C.America	▷ n.i.	▷ ref.n.i.; ctry.n.sp.; list. *Hyla eximia*	▷ 1	▷ 24‡/n.i.
▶ Anderson, 1986[I]	▶ Mexico	▶ 1♀	▶ El Chico National Park; Hidalgo State; rep. *Hyla eximia*	▶ 1,4	▶ 24/48†
▶ Anderson, 1986[I]	▶ Mexico	▶ 1♂	▶ Valle de Bravo, Mexico State; rep. *Hyla eximia*	▶ 1,4	▶ 24/48†
▷ King, 1990	▷ C.America	▷ n.i.	▷ ref. Duellman & Cole, 1965; Anderson, 1986; ctry.n.sp.; list. *Hyla eximia*	▷ 2,4	▷ 24/48
▷ Kuramoto, 1990	▷ N.America	▷ n.i.	▷ ref. Duellman & Cole, 1965; ctry.n.sp.; list. *Hyla eximia*	▷ 1	▷ 24/n.i.
▷ Anderson, 1991[I]	▷ Holarctic	▷ n.i.	▷ ref. Anderson, 1986; ctry.n.sp.; list. *Hyla eximia*	▷ 2,4	▷ 24/48†
▷ Nunes & Fagundes, 2008b	▷ n.i.	▷ n.i.	▷ ref. Duellman & Cole, 1965; list. *Hyla eximia*	▷ n.i.	▷ 24/48
▷ Catroli, 2008	▷ n.i.	▷ n.i.	▷ ref. Duellman & Cole, 1965; Anderson, 1991; list. *Hyla eximia*	▷ n.i.	▷ 24/n.i.
▷ Catroli & Kasahara, 2009	▷ n.i.	▷ n.i.	▷ ref. Duellman & Cole, 1965; Anderson, 1991; list. *Hyla eximia*	▷ 17	▷ 24/n.i.
▶ Present study	▶ Mexico	▶ 2♂	▶ Patzcuaro, Michoacán State	▶ 1	▶ 24/48
▶ Present study	▶ Mexico	▶ 1♂	▶ Uruapan, Michoacán State	▶ 1	▶ 24/48
▶ Present study	▶ Mexico	▶ 4♂	▶ 20 miles N Toluca, Mexico State	▶ 1	▶ 24/48
▶ Present study	▶ Mexico	▶ 1♂	▶ 2 miles S Ixtapan de la Sur, Mexico State	▶ 1	▶ 24/48
▶ Present study	▶ USA	▶ 1♂	▶ Southern Arizona State, loc.u.	▶ 1	▶ 24/48
Dryophytes femoralis					
▶ Wiley, 1982[P,I]	▶ USA	▶ 3♂	▶ Bladen County, North Carolina State; rep. *Hyla femoralis*	▶ 10	▶ 24/48†
▶ Wiley, 1982[P,I]	▶ USA	▶ 1♂	▶ Sampson County, North Carolina State; rep. *Hyla femoralis*	▶ 10	▶ 24/48†
▶ Wiley, 1982[P,I]	▶ USA	▶ 1♂	▶ Dare County, North Carolina State; rep. *Hyla femoralis*	▶ 10	▶ 24/48†
▶ Anderson & Moler, 1986[P]	▶ USA	▶ 1♀	▶ Blackwater River State Forest, 10 km NNW Baker, Okaloosa County, Florida State; rep. *Hyla femoralis*	▶ 1	▶ 24†/48†
▶ Anderson, 1986[P,I]	▶ USA	▶ 8♂	▶ Alachua County, Florida State; rep. *Hyla femoralis*	▶ 1,2,4	▶ 24/48†
▶ Anderson, 1986[P,I]	▶ USA	▶ 3♂	▶ Okaloosa County, Florida State; rep. *Hyla femoralis*	▶ 1,2,4	▶ 24/48†
▶ Anderson, 1986[P,I]	▶ USA	▶ 1♀	▶ Putnam County, Florida State; rep. *Hyla femoralis*	▶ 1,2,4	▶ 24/48†
▶ Anderson, 1986[P,I]	▶ USA	▶ 1♂	▶ Santa Rosa County, Florida State; rep. *Hyla femoralis*	▶ 1,2,4	▶ 24/48†
▶ Anderson, 1986[P,I]	▶ USA	▶ 3♀	▶ La Fayette County, Florida State; rep. *Hyla femoralis*	▶ 1,2,4	▶ 24/48†
▶ Anderson, 1986[P,I]	▶ USA	▶ 2♂1♀	▶ Effingham County, Georgia State; rep. *Hyla femoralis*	▶ 1,2,4	▶ 24/48†
▶ Anderson, 1986[P,I]	▶ USA	▶ 6♂5♀	▶ Jasper County, South Carolina State; rep. *Hyla femoralis*	▶ 1,2,4	▶ 24/48†
▶ Anderson, 1991[P,I]	▶ N.America	▶ 35♂17♀	▶ loc.n.i.; ctry.n.sp.; rep. *Hyla femoralis*	▶ 4	▶ 24/48†
▷ King, 1990	▷ N.America	▷ n.i.	▷ ref. Wiley, 1982; Anderson, 1986; ctry.n.sp.; list. *Hyla femoralis*	▷ 2,4,10	▷ 24/48
▷ Kuramoto, 1990	▷ N.America	▷ n.i.	▷ ref. Wiley, 1982; Anderson & Moler, 1986; ctry.n.sp.; list. *Hyla femoralis*	▷ 17	▷ 24/n.i.
▷ Anderson, 1991[I]	▷ Holarctic	▷ n.i.	▷ ref. Anderson, 1986; ctry.n.sp.; list. *Hyla femoralis*	▷ 2,4	▷ 24/48†
▶ Wiley, 2003[P]	▶ USA	▶ 2♂2♀	▶ Pitt County, North Carolina State; list. *Hyla femoralis*	▶ 12,13	▶ 24†/48†
▶ Schmid & Steinlein, 2003[P,M]	▶ N.America	▶ 2♂2♀	▶ specimens obtained by animal dealer; loc.n.i.; ctry.n.sp.; list. *Hyla femoralis*	▶ 4,5,6, 12,13	▶ 24/48†
▷ Nunes & Fagundes, 2008b	▷ n.i.	▷ n.i.	▷ ref. Anderson, 1991; list. *Hyla femoralis*	▷ n.i.	▷ 24/48

Author(s) ▶ original data[a] ▷ review articles	Country/ Region[b]	Speci-mens[c]	Locality of sampling, references cited, and remarks[d]	Methods applied[e]	2n/FN[f]
▷ Catroli, 2008	▷ n.i.	▷ n.i.	▷ ref. Wiley, 1982, 1992, 2003; Anderson & Moler, 1986; Anderson, 1991; Schmid & Steinlein, 2003; list. *Hyla femoralis*	▷ n.i.	▷ 24/n.i.
▷ Catroli & Kasahara, 2009	▷ n.i.	▷ n.i.	▷ ref. Wiley, 1982, 2003; Anderson & Moler, 1986; Anderson, 1991; Wiley et al., 1992; Schmid & Steinlein, 2003; list. *Hyla femoralis*	▷ 17	▷ 24/n.i.
▷ Schmid & Steinlein, 2016	▷ n.i.	▷ n.i.	▷ list. Wiley et al. 1992; Schmid et al., unpubl.; list. *Hyla femoralis*	▷ 14	▷ 24/n.i.
▶ Present study	▶ USA	▶ 1♂	▶ Georgia State, loc.u.	▶ 1	▶ 24/48
▶ Present study	▶ USA	▶ 1J	▶ 45 mi W of Conway, Hwy 378 Lake City Bypass, South Carolina State	▶ 1	▶ 24/48
▶ Present study	▶ USA	▶10♂10♀	▶ Leon County, Florida State	▶ 1,5,8,14	▶ 24/48
Dryophytes gratiosus					
▶ Wiley, 1982[P,I]	▶ USA	▶ 2♂	▶ Wake County, North Carolina State; rep. *Hyla gratiosa*	▶ 10	▶ 24/48†
▶ Wiley, 1982[P,I]	▶ USA	▶ 2♂	▶ New Hanover County, North Carolina State; rep. *Hyla gratiosa*	▶ 10	▶ 24/48†
▶ Wiley, 1982[P,I]	▶ USA	▶ 1♂	▶ Tallahassee, Florida State; rep. *Hyla gratiosa*	▶ 10	▶ 24/48†
▶ Anderson, 1986[I]	▶ USA	▶ 4♂	▶ Jasper County, South Carolina State; rep. *Hyla gratiosa*	▶ 1,4	▶ 24/48†
▶ Anderson, 1986[I]	▶ USA	▶ 1♀	▶ Caroline County, Maryland State; rep. *Hyla gratiosa*	▶ 1,4	▶ 24/48†
▶ Anderson, 1986[I]	▶ USA	▶ 1♀	▶ Effingham County, Georgia State; rep. *Hyla gratiosa*	▶ 1,4	▶ 24/48†
▶ Anderson, 1986[I]	▶ USA	▶ 1♂	▶ Alachua County, Florida State; rep. *Hyla gratiosa*	▶ 1,2,4	▶ 24/48†
▷ King, 1990	▷ S.America	▷ n.i.	▷ ref. Bogart, 1973; Wiley, 1982; Anderson, 1986; err.cit.ctry.; list. *Hyla gratiosa*	▷ 2,4,10	▷ 24/48
▷ Kuramoto, 1990	▷ N.America	▷ n.i.	▷ Wiley, 1982; ctry.n.sp.; list. *Hyla gratiosa*	▷ 17	▷ 24/n.i.
▷ Anderson, 1991[I]	▷ Holarctic	▷ n.i.	▷ ref. Anderson, 1986; ctry.n.sp.; list. *Hyla gratiosa*	▷ 2,4	▷ 24/48†
▷ Nunes & Fagundes, 2008b	▷ n.i.	▷ n.i.	▷ ref. Bogart, 1973; list. *Hyla gratiosa*	▷ n.i.	▷ 24/48
▷ Catroli, 2008	▷ n.i.	▷ n.i.	▷ ref. Bogart, 1973; Wiley, 1982; Anderson, 1991; list. *Hyla gratiosa*	▷ n.i.	▷ 24/n.i.
▷ Catroli & Kasahara, 2009	▷ n.i.	▷ n.i.	▷ ref. Bogart, 1973; Wiley, 1982; Anderson, 1991; list. *Hyla gratiosa*	▷ 17	▷ 24/n.i.
Dryophytes immaculatus					
▷ Kuramoto, 1990	▷ Asia	▷ n.i.	▷ ref. Tang et al., 1984; ctry.n.sp.; list. *Hyla arborea immaculata*	▷ 1	▷ 24/n.i.
▷ Catroli, 2008	▷ n.i.	▷ n.i.	▷ ref. Tang et al., 1984; list. *Hyla immaculata*	▷ n.i.	▷ 24/n.i.
▷ Catroli & Kasahara, 2009	▷ n.i.	▷ n.i.	▷ ref. Tang et al., 1984; list. *Hyla immaculata*	▷ 17	▷ 24/n.i.
Dryophytes japonicus					
▷ Oguma & Makino, 1937[N]	▷ n.i.	▷ n.i.	▷ ref. Iriki, 1930, 1932a,b; list. *Hyla arborea japonica*	▷ n.i.	▷ 24/n.i.
▶ Yosida, 1957[P,D,M,I]	▶ Japan	▶ n♂	▶ suburbs of Misima; rep. *Hyla arborea japonica*	▶ 1	▶ 24/48†
▶ Seto & Makino, 1964[P]	▶ Japan	▶ n.i.	▶ campus of Hokkaido University and suburbs of Sapporo; rep. *Hyla arborea japonica*	▶ 1	▶ 24/48†
▶ Seto, 1964[P]	▶ Japan	▶ n.i.	▶ campus of Hokkaido University and suburbs of Sapporo; rep. *Hyla arborea japonica*	▶ 1	▶ 24/48†
▶ Anderson, 1986[P,I]	▶ Japan	▶ 1♂1♀	▶ Shizuoka; rep. *Hyla japonica*	▶ 1,4,11	▶ 24/48†
▶ Anderson, 1986[P,I]	▶ Japan	▶ 4♂	▶ Kyoto; rep. *Hyla japonica*	▶ 1,2,4,11	▶ 24/48†
▷ Rabello, 1970	▷ n.i.	▷ n.i.	▷ ref. Iriki, 1930; list. *Hyla arborea japonica*	▷ n.i.	▷ 24/n.i.
▷ Foresti, 1972	▷ n.i.	▷ n.i.	▷ ref. Iriki, 1930; list. *Hyla arborea japonica*	▷ n.i.	▷ 24/n.i.
▷ King, 1990	▷ Japan	▷ n.i.	▷ ref. Andengwrson, 1986; list. *Hyla japonica*	▷ 2,4,11	▷ 24/48
▶ Yu & Lee, 1990[P,I]	▶ Korea	▶ 2♂8♀	▶ Kangchon, Kangwon-do	▶ 2	▶ 24/48†
▶ Yu & Lee, 1990[P,I]	▶ Korea	▶ 15♂3♀	▶ Suweon, Kyonggi-do	▶ 2	▶ 24/48†
▶ Yu & Lee, 1990[P,I]	▶ Korea	▶ 18♂6♀	▶ Sanchong, Kyongsangnam-do	▶ 2	▶ 24/48†
▶ Yu & Lee, 1990[P,I]	▶ Korea	▶ 1♀	▶ Incheon	▶ 2	▶ 24/48†
▶ Yu & Lee, 1990[P,I]	▶ Korea	▶ 5♂10♀	▶ Namhae, Kyongsangnam-do	▶ 2	▶ 24/48†
▷ Kuramoto, 1990	▷ Asia	▷ n.i.	▷ ref. Yosida, 1957; Matsuda, 1963; Seto, 1964; Seto & Makino, 1964; Kuramoto et al., 1974; ctry.n.sp.; list. *Hyla japonica*	▷ 2	▷ 24/n.i.
▷ Anderson, 1991[I]	▷ Holarctic	▷ n.i.	▷ ref. Anderson, 1986; ctry.n.sp.; list. *Hyla japonica*	▷ 2,4,11	▷ 24/48†
▶ Miura, 1995[P]	▶ Japan	▶ 1♂2♀	▶ Hiroshima and Higashihiroshima	▶ 12	▶ 24/48†
▷ Nunes & Fagundes, 2008b	▷ n.i.	▷ n.i.	▷ ref. Anderson, 1991; list. *Hyla japonica*	▷ n.i.	▷ 24/48
▷ Catroli, 2008	▷ n.i.	▷ n.i.	▷ ref. Yosida, 1957; Matsuda, 1963; Seto, 1964; Seto & Makino, 1964; Kuramoto et al., 1974; Anderson, 1991; Miura, 1995; list. *Hyla japonica*	▷ n.i.	▷ 24/n.i.
▷ Catroli & Kasahara, 2009	▷ n.i.	▷ n.i.	▷ ref. Yosida, 1957; Matsuda, 1963; Seto, 1964; Seto & Makino, 1964; Kuramoto et al., 1974; Anderson, 1991; Miura, 1995; list. *Hyla japonica*	▷ 17	▷ 24/n.i.
▷ Hernández-Guzmán et al., 2011	▷ n.i.	▷ n.i.	▷ ref. Seto, 1964; list. *Hyla arborea japonica*	▷ n.i.	▷ 24/n.i.
▶ Present study	▶ Japan	▶ 1♂2♀	▶ Hiroshima and Higashihiroshima	▶ 1,2	▶ 24/48
Dryophytes plicatus					
▶ Duellman & Cole, 1965[N]	▶ Mexico	▶ 1♂	▶ El Chico National Park; Hidalgo State; *rep. Hyla lafrentzi*	▶ 1	▶ 24‡/n.i.
▷ Duellman, 1970	▷ C.America	▷ n.i	▷ ref. Duellman & Cole, 1965; ctry.n.sp.; list. *Hyla plicata*	▷ 1	▷ 24‡/n.i.
▷ Rabello, 1970	▷ n.i.	▷ n.i.	▷ ref. Duellman & Cole, 1965; list. *Hyla lafrentzi*	▷ n.i.	▷ 24‡/n.i.
▷ Foresti, 1972	▷ n.i.	▷ n.i.	▷ ref. Duellman & Cole, 1965; list. *Hyla lafrentzi*	▷ n.i.	▷ 24‡/n.i.
▷ King, 1990	▷ C.America	▷ n.i.	▷ ref. Duellman & Cole, 1965; ctry.n.sp.; list. *Hyla lafrentzi*	▷ n.i.	▷ 24/n.i.
▷ Kuramoto, 1990	▷ S.C.America	▷ n.i.	▷ ref. Duellman & Cole, 1965; ctry.n.sp.; list. *Hyla plicata*	▷ 1	▷ 24/n.i.
▷ Catroli, 2008	▷ n.i.	▷ n.i.	▷ ref. Duellman & Cole, 1965; list. *Hyla plicata*	▷ n.i.	▷ 24/n.i.
▷ Catroli & Kasahara, 2009	▷ n.i.	▷ n.i.	▷ ref. Duellman & Cole, 1965; list. *Hyla plicata*	▷ n.i.	▷ 24/n.i.
Dryophytes squirellus					
▶ Wiley, 1982[P,I]	▶ USA	▶1♂	▶ Wake County, North Carolina State; rep. *Hyla squirella*	▶ 10	▶ 24/48†
▶ Wiley, 1982[P,I]	▶ USA	▶1♂	▶ Sampson County, North Carolina State; rep. *Hyla squirella*	▶ 10	▶ 24/48†
▶ Wiley, 1982[P,I]	▶ USA	▶2♂	▶ Columbus County, North Carolina State; rep. *Hyla squirella*	▶ 10	▶ 24/48†
▶ Anderson, 1986[I]	▶ USA	▶1♂3♀	▶ Boone County, North Carolina State; rep. *Hyla squirella*	▶ 1,2,4,11	▶ 24/48†

DOI: 10.1159/000490085

Author(s) ▶ original data[a] ▷ review articles	Country/ Region[b]	Speci- mens[c]	Locality of sampling, references cited, and remarks[d]	Methods applied[e]	2n/FN[f]
▶ Anderson, 1986[I]	▶ USA	▶2♂1♀	▶ Jasper County, South Carolina State; rep. *Hyla squirella*	▶ 1,2,4,11	▶ 24/48†
▶ Anderson, 1986[I]	▶ USA	▶1♂	▶ Charlton County, Georgia State; rep. *Hyla squirella*	▶ 1,4,11	▶ 24/48†
▶ Anderson, 1986[I]	▶ USA	▶1♀	▶ Putnam County, Florida State; rep. *Hyla squirella*	▶ 1,4,11	▶ 24/48†
▶ Bogart, 1973[P]	▶ USA	▶ n.i.	▶ Texas State; list. *Hyla squirella*	▶ 1	▶ 24/48†
▶ Anderson, 1991[P,I]	▶ N.America	▶ n♂6♀	▶ loc.n.i.; ctry.n.sp.; rep. *Hyla squirella*	▶ 2,11	▶ 24/48†
▷ Cole, 1974	▷ n.i.	▷ n.i.	▷ ref. Bogart, 1973; list. *Hyla squirella*	▷ n.i.	▷ 24/n.i.
▷ Morescalchi, 1979	▷ n.i.	▷ n.i.	▷ ref. Bogart, 1973; list. *Hyla squirella*	▷ n.i.	▷ 24/n.i.
▷ King, 1990	▷ N.America	▷ n.i.	▷ ref. Bogart, 1973; Wiley, 1982; Anderson, 1986; ctry.n.sp.; list. *Hyla squirella*	▷ 2,4,10, 11	▷ 24/48
▷ Kuramoto, 1990	▷ N.America	▷ n.i.	▷ ref. Bogart, 1973; Wiley, 1982; ctry.n.sp.; list. *Hyla squirella*	▷ 17	▷ 24/n.i.
▷ Anderson, 1991[I]	▷ Holarctic	▷ n.i.	▷ ref. Anderson, 1986; ctry.n.sp.; list. *Hyla squirella*	▷ n.i.	▷ 24/48†
▷ Nunes & Fagundes, 2008b	▷ n.i.	▷ n.i.	▷ ref. Bogart, 1973; list. *Hyla squirella*	▷ n.i.	▷ 24/48
▷ Catroli, 2008	▷ n.i.	▷ n.i.	▷ ref. Bogart, 1973; Wiley, 1982, 1992; Anderson, 1991; list. *Hyla squirella*	▷ n.i.	▷ 24/n.i.
▷ Catroli & Kasahara, 2009	▷ n.i.	▷ n.i.	▷ ref. Bogart, 1973; Wiley, 1982; Anderson, 1991; Wiley et al., 1992; list. *Hyla squirella*	▷ 17	▷ 24/n.i.
▷ Schmid & Steinlein, 2016a	▷ n.i.	▷ n.i.	▷ ref. Wiley, 1992; list. *Hyla squirella*	▷ 14	▷ 24/n.i.
▶ Present study	▶ USA	▶2♂	▶ North Carolina State; loc.u.	▶ 1	▶ 24/48
▶ Present study	▶ USA	▶1♂	▶ loc.u.	▶ 1	▶ 24/48
▶ Present study	▶ USA	▶1♂	▶ South Carolina State, loc.u.	▶ 1	▶ 24/48
Dryophytes suweonensis					
▶ Yu & Lee, 1990[P,I]	▶ Korea	▶ 20♂5♀	▶ Suweon, Kangwon-do; rep. *Hyla suweonensis*	▶ 2	▶ 24/48
▶ Yu & Lee, 1990[P,I]	▶ Korea	▶ 9♂2♀	▶ Incheon; rep. *Hyla suweonensis*	▶ 2	▶ 24/48
▷ King, 1990	▷ Korea	▷ n.i.	▷ ref. Kuramoto, 1980; list. *Hyla suweonensis*	▷ 1	▷ 24/48
▷ Kuramoto, 1990	▷ Asia	▷ n.i.	▷ ref. Kuramoto, 1980; ctry.n.sp.; list. *Hyla suweonensis*	▷ 1	▷ 24/n.i.
▷ Nunes & Fagundes, 2008b	▷ n.i.	▷ n.i.	▷ ref. Kuramoto, 1990; list. *Hyla suweonensis*	▷ n.i.	▷ 24/48
▷ Catroli, 2008	▷ n.i.	▷ n.i.	▷ ref. Kuramoto, 1980; list. *Hyla suweonensis*	▷ n.i.	▷ 24/n.i.
▷ Catroli & Kasahara, 2009	▷ n.i.	▷ n.i.	▷ ref. Kuramoto, 1980; list. *Hyla suweonensis*	▷ n.i.	▷ 24/n.i.
Dryophytes versicolor					
▶ Bushnell et al., 1939[D]	▶ USA	▶ n.i.	▶ Relfoot Lake, Tennessee State; err.id.spec; rep. *Hyla versicolor versicolor*	▶ 1	▶ 24‡/n.i.
▶ Wasserman, 1970[P]	▶ USA	▶ 10♂2♀ 19tp	▶ Alpine, Bergen County, New Jersey State; rep. *Hyla versicolor*	▶ 1	▶ 48/96†
▶ Wasserman, 1970[N]	▶ USA	▶ n.i.	▶ ref. Bogart, pers.comm.; Bastrop, Bastrop County; Texas Sate;	▶ 1	▶ 48/n.i.
▷ Foresti, 1972	▷ n.i.	▷ n.i.	▷ ref. Bushnell et al. 1939; list. *Hyla versicolor*	▷ n.i.	▷ 24/n.i.
▷ Foresti, 1972	▷ n.i.	▷ n.i.	▷ ref. Wasserman, 1970; list. *Hyla versicolor*	▷ n.i.	▷ 48/n.i.
▶ Bogart & Wasserman, 1972[P,M,I]	▶ USA	▶ n.i.	▶ Texas, South Carolina, New Jersey, Louisiana, Mississippi, Indiana and Illinois States	▶ 1	▶ 48/96†
▶ Wiley, 1982[P]	▶ USA	▶7♂	▶ Hampshire County, Massachusetts State; rep. *Hyla versicolor*	▶ 10	▶ 48/96†
▶ Anderson, 1986[I]	▶ USA	▶2♂	▶ Page County, Virginia State; rep. *Hyla versicolor*	▶ 1,2,4	▶ 48/96†
▶ Anderson, 1986[I]	▶ USA	▶1♂	▶ Portage County, Wisconsin State; rep. *Hyla versicolor*	▶ 1,2,4	▶ 48/96†
▷ Rabello, 1970	▷ n.i.	▷ n.i.	▷ ref. Bushnell et al., 1939; list. *Hyla versicolor*	▷ n.i.	▷ 24/n.i.
▶ Wiley et al., 1989[P]	▶ USA	▶15s.n.i.	▶ Ross County, Ohio State; rep. *Hyla versicolor*	▶ 4,12,13	▶ 48†/96†
▶ Wiley et al., 1989[P]	▶ USA	▶1s.n.i.	▶ Meade County, Kentucky State; rep. *Hyla versicolor*	▶ 4,12,13	▶ 48†/96†
▶ Wiley et al., 1989[P]	▶ USA	▶1s.n.i.	▶ Butler County, Pennsylvania State; rep. *Hyla versicolor*	▶ 4,12,13	▶ 48†/96†
▶ Wiley et al., 1989[P]	▶ USA	▶5s.n.i.	▶ Greenbriar County, West Virginia State; rep. *Hyla versicolor*	▶ 4,12,13	▶ 48†/96†
▶ Wiley et al., 1989[P]	▶ USA	▶2s.n.i.	▶ Brunswick County, Virginia State; rep. *Hyla versicolor*	▶ 4,12,13	▶ 48†/96†
▶ Wiley et al., 1989[P]	▶ USA	▶15s.n.i.	▶ Mecklenburg County, Virginia State; rep. *Hyla versicolor*	▶ 4,12,13	▶ 48†/96†
▶ Wiley et al., 1989[P]	▶ USA	▶10s.n.i.	▶ South Hampton County, Massachusetts State; rep. *Hyla versicolor*	▶ 4,12,13	▶ 48†/96†
▶ Wiley et al., 1989[P]	▶ USA	▶8s.n.i.	▶ Cherokee County, Oklahoma State; rep. *Hyla versicolor*	▶ 4,12,13	▶ 48†/96†
▶ Wiley et al., 1989[P]	▶ USA	▶2s.n.i.	▶ McCurtain County, Oklahoma State; rep. *Hyla versicolor*	▶ 4,12,13	▶ 48†/96†
▶ Wiley et al., 1989[P]	▶ USA	▶3s.n.i.	▶ Ottawa County, Oklahoma State; rep. *Hyla versicolor*	▶ 4,12,13	▶ 48†/96†
▶ Wiley et al., 1989[P]	▶ USA	▶3s.n.i.	▶ Barry County, Missouri State; rep. *Hyla versicolor*	▶ 4,12,13	▶ 48†/96†
▶ Wiley et al., 1989[P]	▶ USA	▶4s.n.i.	▶ McDonough County, Illinois State; rep. *Hyla versicolor*	▶ 4,12,13	▶ 48†/96†
▶ Wiley et al., 1989[P]	▶ USA	▶15s.n.i.	▶ Juneau County, Wisconsin State; rep. *Hyla versicolor*	▶ 4,12,13	▶ 48†/96†
▶ Wiley et al., 1989[P]	▶ USA	▶14s.n.i.	▶ Bastrop County, Texas State; rep. *Hyla versicolor*	▶ 4,12,13	▶ 48†/96†
▶ Wiley et al., 1989[P]	▶ USA	▶2s.n.i.	▶ Warren County, North Carolina State; rep. *Hyla versicolor*	▶ 4,12,13	▶ 48†/96†
▷ King, 1990	▷ N.America	▷ n.i.	▷ ref. Bushnell et al., 1939; Wasserman, 1970; León, 1970/err.ref.; Bogart & Wasserman, 1972; Wiley, 1982; Anderson, 1986; ctry.n.sp.; list. *Hyla versicolor*	▷ 2,4,10	▷ 48/96
▷ Kuramoto, 1990	▷ N.America	▷ n.i.	▷ ref. Wasserman, 1970; Bogart & Wasserman, 1972; Wiley, 1982; ctry.n.sp.; list. *Hyla versicolor*	▷ 17	▷ 48/n.i.
▷ Anderson, 1991[I]	▷ Holarctic	▷ n.i.	▷ ref. Anderson, 1986; ctry.n.sp.; rep. *Hyla versicolor*	▷ 2,4	▷ 48/96†
▶ Wiley & Little, 2000[P]	▶ USA	▶2♂1♀	▶ Mecklenburg County, Virginia State; rep. *Hyla versicolor*	▶ 12	▶ 48†/96†
▷ Nunes & Fagundes, 2008b	▷ n.i.	▷ n.i.	▷ ref. Wasserman, 1970; list. *Hyla versicolor*	▷ n.i.	▷ 48/n.i.
▷ Catroli, 2008	▷ n.i.	▷ n.i.	▷ Bushnell et al., 1939; Wasserman, 1970; León, 1970/err.ref.; Bogart & Wasserman, 1972; Wiley, 1982, 1992; Wiley et al., 1989; Anderson, 1991; Wiley & Little, 2000; list. *Hyla versicolor*	▷ n.i.	▷ 48/n.i.
▷ Catroli & Kasahara, 2009	▷ n.i.	▷ n.i.	▷ Bushnell et al., 1939; Wasserman, 1970; León, 1970/err.ref.; Bogart & Wasserman, 1972; Wiley, 1982; Wiley et al., 1989, 1992; Anderson, 1991; Wiley & Little, 2000; list. *Hyla versicolor*	▷ 17	▷ 48/n.i.
▷ Schmid & Steinlein, 2016a	▷ n.i.	▷ n.i.	▷ Wiley et al. 1992; list. *Hyla versicolor*	▷ 14	▷ 48/n.i.
▶ Present study	▶ USA	▶1s.n.i	▶ Bastrop State Park, Texas State	▶ 1	▶ 48/48

Author(s) ▶ original data[a] ▷ review articles	Country/ Region[b]	Speci-mens[c]	Locality of sampling, references cited, and remarks[d]	Methods applied[e]	2n/FN[f]
▶ Present study	▶ USA	▶2♂	▶ Bastrop State Park, Texas State	▶ 1	▶ 48/48
▶ Present study	▶ USA	▶1s.n.i	▶ Bastrop State Park, Texas State	▶ 1	▶ 48/48
▶ Present study	▶ USA	▶2♂,5J	▶ Bastrop State Park, Texas State	▶ 1	▶ 48/48
HYLIDAE / Hylinae / *Duellmanohyla*					
Duellmanohyla ignicolor					
▶ Duellman & Cole, 1965[N]	▶ Mexico	▶ 2♂	▶ 4.2 km S Vista Hermosa, Oaxaca State; rep. *Ptychohyla ignicolor*	▶ 1	▶ 24‡/n.i.
▷ Duellman, 1970	▷ C.America	▷ n.i.	▷ ref.n.i.; ctry.n.sp.; list. *Ptychohyla ignicolor*	▷ 1	▷ 24/n.i.
▷ Rabello, 1970	▷ n.i.	▷ n.i.	▷ ref. Duellman & Cole, 1965; list. *Ptychohyla ignicolor*	▷ n.i.	▷ 24‡/n.i.
▷ King, 1990	▷ C.America	▷ n.i.	▷ ref. Duellman & Cole, 1965; ctry.n.sp.; list. *Ptychohyla ignicolor*	▷ n.i.	▷ 24/n.i.
▷ Kuramoto, 1990	▷ S.C.America	▷ n.i.	▷ ref. Duellman & Cole, 1965; ctry.n.sp.; list. *Ptychohyla ignicolor*	▷ 1	▷ 24/n.i.
▷ Catroli, 2008	▷ n.i.	▷ n.i.	▷ ref. Duellman & Cole, 1965	▷ n.i.	▷ 24/n.i.
▷ Catroli & Kasahara, 2009	▷ n.i.	▷ n.i.	▷ ref. Duellman & Cole, 1965	▷ n.i.	▷ 24/n.i.
Duellmanohyla rufioculis					
▶ Duellman, 1967[N]	▶ Costa Rica	▶ 3♂	▶ 15 km SW San Isidro del General, San José Province; rep. *Hyla rufioculis*	▶ 1	▶ 24‡/n.i.
▷ Duellman, 1970	▷ C.America	▷ n.i.	▷ ref.n.i.; ctry.n.sp.; list. *Hyla rufioculis*	▷ 1	▷ 24‡/n.i.
▷ Foresti, 1972	▷ n.i.	▷ n.i.	▷ ref. Duellman, 1967; list. *Hyla rufioculis*	▷ n.i.	▷ 24‡/n.i.
▷ Kuramoto, 1990	▷ S.C.America	▷ n.i.	▷ ref. Duellman, 1967; ctry.n.sp.; list. *Hyla rufioculis*	▷ 1	▷ 24/n.i.
▷ Catroli, 2008	▷ n.i.	▷ n.i.	▷ ref. Duellman, 1967	▷ n.i.	▷ 24/n.i.
▷ Catroli & Kasahara, 2009	▷ n.i.	▷ n.i.	▷ ref. Duellman, 1967	▷ n.i.	▷ 24/n.i.
Duellmanohyla uranochroa					
▶ León, 1970[N]	▶ Costa Rica	▶ 1♂	▶ loc.n.i.; rep. *Hyla uranochroa*	▶ 1	▶ 24‡/n.i.
▷ Foresti, 1972	▷ n.i.	▷ n.i.	▷ ref. León, 1970; list. *Hyla uranochroa*	▷ n.i.	▷ 24‡/n.i.
▷ Morescalchi, 1979	▷ n.i.	▷ n.i.	▷ ref. León, 1970; list. *Hyla uranochroa*	▷ n.i.	▷ 24/n.i.
▷ King, 1990	▷ C.America	▷ n.i.	▷ ref. León, 1970; ctry.n.sp.; list. *Hyla uranochroa*	▷ n.i.	▷ 24/n.i.
▷ Kuramoto, 1990	▷ S.C.America	▷ n.i.	▷ ref. León, 1970; ctry.n.sp.; list. *Hyla uranochroa*	▷ 1	▷ 24/n.i.
▷ Catroli, 2008	▷ n.i.	▷ n.i.	▷ ref. León, 1970	▷ n.i.	▷ 24/n.i.
▷ Catroli & Kasahara, 2009	▷ n.i.	▷ n.i.	▷ ref. León, 1970	▷ n.i.	▷ 24/n.i.
HYLIDAE / Hylinae / *Ecnomiohyla*					
No cytogenetic data					
HYLIDAE / Hylinae / *Exerodonta*					
Exerodonta smaragdina					
▶ Duellman, 1970[N]	▶ C.America	▶ n♂	▶ loc.n.i.; ctry.n.sp.; list. *Hyla smaragdina*	▶ 1	▶ 24‡/n.i.
▷ Kuramoto, 1990	▷ S.C.America	▷ n.i.	▷ ref. Duellman, 1970; list. *Hyla smaragdina*	▷ 1	▷ 24/n.i.
▷ Nunes & Fagundes, 2008b	▷ n.i.	▷ n.i.	▷ ref. Duellman, 1970	▷ n.i.	▷ 24/48*
▷ Catroli, 2008	▷ n.i.	▷ n.i.	▷ ref. Duellman, 1970	▷ n.i.	▷ 24/n.i.
▷ Catroli & Kasahara, 2009	▷ n.i.	▷ n.i.	▷ ref. Duellman, 1970	▷ n.i.	▷ 24/n.i.
HYLIDAE / Hylinae / *Hyla*					
Hyla annectans					
▶ Li et al, 1981[I]	▶ Europe	▶ 1♂1♀	▶ western China; loc.u.	▶ 1	▶ 24/48†
▷ Kuramoto, 1990	▷ Asia	▷ n.i.	▷ ref. Li et al., 1981; ctry.n.sp.	▷ 1	▷ 24/n.i.
▷ Nunes & Fagundes, 2008b	▷ n.i.	▷ n.i.	▷ ref. Li et al., 1981r	▷ n.i.	▷ 24/48
▷ Catroli, 2008	▷ n.i.	▷ n.i.	▷ ref. Li et al., 1981	▷ n.i.	▷ 24/n.i.
▷ Catroli & Kasahara, 2009	▷ n.i.	▷ n.i.	▷ ref. Li et al., 1981	▷ 17	▷ 24/n.i.
Hyla arborea					
▷ Oguma & Makino, 1937[N]	▷ n.i.	▷ n.i.	▷ ref. Galgano, 1933	▷ n.i.	▷ 24/n.i.
▶ Morescalchi, 1965[P,M,I]	▶ Italy	▶ n♂n♀	▶ surroundings of Naples	▶ 1	▶ 24/48†
▶ Anderson, 1986[I]	▶ Italy	▶ 1♂	▶ loc.u.	▶ 1	▶ 24/48†
▷ Rabello, 1970	▷ n.i.	▷ n.i.	▷ ref. Galgano, 1933	▷ n.i.	▷ 24/n.i.
▶ Ullerich, 1970[P]	▶ Europe	▶ n♂	▶ specimens obtained by animal dealer; loc.n.i.; ctry.n.sp.	▶ 1	▶ 24/48†
▷ Foresti, 1972	▷ n.i.	▷ n.i.	▷ ref. Galgano, 1933	▷ n.i.	▷ 24/n.i.
▷ Foresti, 1972	▷ n.i.	▷ n.i.	▷ ref. Morescalchi, 1965; list. *Hyla arborea arborea*	▷ n.i.	▷ 24/n.i.
▶ Bogart, 1973[P]	▶ Israel	▶ n.i.	▶ loc.n.i.	▶ 1	▶ 24/48†
▷ Cole, 1974	▷ n.i.	▷ n.i.	▷ ref. Seto, 1964; Morescalchi, 1965; Bogart, 1973	▷ n.i.	▷ 24/n.i.
▶ Schmid, 1978a[N]	▶ Europe	▶ 1♂2♀	▶ specimens obtained by animal dealer; loc.n.i.; ctry.n.sp.	▶ 1	▶ 24/n.i.
▷ King, 1990	▷ Europe	▷ n.i.	▷ ref. Seto, 1964; Morescalchi, 1965; Ullerich, 1970; Bogart, 1973; Schmid, 1978a; Anderson, 1986; ctry.n.sp.	▷ 2,4,17	▷ 24/48
▷ Kuramoto, 1990	▷ Europe	▷ n.i.	▷ ref. Morescalchi, 1965; Ullerich, 1970; Schmid, 1978a; ctry.n.sp.	▷ 17	▷ 24/n.i.
▷ Anderson, 1991[I]	▷ Holarctic	▷ n.i.	▷ ref. Anderson, 1986; ctry.n.sp.	▷ 2,4	▷ 24/48†
▶ Martirosyan & Stepanyan, 2007[P]	▶ Armenia	▶ n♂	▶ Alaverdi; rep. *Hyla arborea schelkownikovi*	▶ 1,4	▶ 24/48
▷ Nunes & Fagundes, 2008b	▷ n.i.	▷ n.i.	▷ ref. Morescalchi, 1965	▷ n.i.	▷ 24/48
▷ Catroli, 2008	▷ n.i.	▷ n.i.	▷ ref. Seto, 1964; Morescalchi, 1965; Ullerich, 1970; Bogart, 1973; Schmid, 1978a; Anderson, 1991	▷ n.i.	▷ 24/n.i.

Author(s) ▶ original data[a] ▷ review articles	Country/ Region[b]	Speci-mens[c]	Locality of sampling, references cited, and remarks[d]	Methods applied[e]	2n/FN[f]
▷ Catroli & Kasahara, 2009	▷ n.i.	▷ n.i.	▷ ref. Morescalchi, 1965; Ullerich, 1970; Bogart, 1973; Schmid, 1978a; Anderson, 1991; Martirosyan & Stepanyan, 2007	▷ 17	▷ 24/n.i.
▷ Hernández-Guzmán et al., 2011	▷ n.i.	▷ n.i.	▷ ref. Martirosyan & Stepanyan, 2007; list. *Hyla arborea schelkownikovi*	▷ n.i.	▷ 24/n.i.
▶ Present study	▶ Israel	▶ 1♂1U	▶ Yakum	▶ 1	▶ 24/48
▶ Present study	▶ Serbia	▶ 2J	▶ Belgrade	▶ 1	▶ 24/48
▶ Present study	▶ Croatia	▶ 1s.n.i.1J	▶ island of Korcula	▶ 1	▶ 24/48
Hyla chinensis					
▶ Anderson, 1986[P,I]	▶ Taiwan	▶ 5♂1♀	▶ Sun-ken	▶ 1,2,4,11	▶ 24/48†
▷ King, 1990	▷ Taiwan	▷ n.i.	▷ ref. Anderson, 1986	▷ 2,4,11	▷ 24/48
▶ Kuramoto, 1990	▶ Asia	▷ n.i.	▶ ref. Kuramoto, unpubl.; ctry.n.sp.	▶ n.i.	▶ 24/n.i.
▷ Anderson, 1991[I]	▷ Holarctic	▷ n.i.	▷ ref. Anderson, 1986; ctry.n.sp.	▷ 2,4,11	▷ 24/48†
▷ Nunes & Fagundes, 2008b	▷ n.i.	▷ n.i.	▷ ref. Anderson, 1991	▷ n.i.	▷ 24/48
▷ Catroli, 2008	▷ n.i.	▷ n.i.	▷ ref. Anderson, 1991	▷ n.i.	▷ 24/n.i.
▷ Catroli & Kasahara, 2009	▷ n.i.	▷ n.i.	▷ ref. Anderson, 1991	▷ 17	▷ 24/n.i.
Hyla hallowellii					
▷ Anderson, 1991[I]	▷ Ryukyu Islands	▷ n.i.	▷ ref. Kuramoto et al., 1973/err.ref.; rep. *Hyla {hallowelli}*	▷ 1	▷ 24/48†
▷ King, 1990		▷ n.i.	▷ ref. Kuramoto et al., 1973/err.ref.; Anderson, 1986; rep. *Hyla {hallowelli}*	▷ n.i.	▷ 24/48
▷ Kuramoto, 1990	▷ Japan	▷ n.i.	▷ ref. Kuramoto et al., 1974; ctry.n.sp.	▷ 1	▷ 24/n.i.
▷ Anderson, 1991[I]	▷ Asia	▷ n.i.	▷ ref. Anderson, 1986; ctry.n.sp.	▷ 2,4	▷ 24/48†
▷ Nunes & Fagundes, 2008b	▷ Holarctic	▷ n.i.	▷ ref. Kuramoto, 1974/err.ref.	▷ n.i.	▷ 24/48
▷ Catroli, 2008	▷ n.i.	▷ n.i.	▷ ref. Kuramoto et al., 1974; Anderson 1991	▷ n.i.	▷ 24/n.i.
▷ Catroli & Kasahara, 2009	▷ n.i.	▷ n.i.	▷ ref. Kuramoto et al., 1974; Anderson 1991	▷ 17	▷ 24/n.i.
	▷ n.i.				
Hyla meridionalis					
▶ Anderson, 1986[I]	▶ Canary Islands	▶ 2♂	▶ La Gomera	▶ 1	▶ 24/48†
▶ Anderson, 1986[I]	▶ France	▶ 2♂1♀	▶ St. Remy	▶ 1	▶ 24/48†
▶ Anderson, 1986[I]	▶ France	▶ 1♂	▶ Pyrénées-Orientales Department	▶ 1,2	▶ 24/48†
▷ King, 1990	▷ Europe	▷ n.i.	▷ ref. Anderson, 1986; ctry.n.sp.	▷ 2	▷ 24/48
▷ Anderson, 1991[I]	▷ Holarctic	▷ n.i.	▷ ref. Anderson, 1986; ctry.n.sp.	▷ 2	▷ 24/48†
▷ Nunes & Fagundes, 2008b	▷ n.i.	▷ n.i.	▷ ref. Anderson, 1991	▷ n.i.	▷ 24/48
▷ Catroli, 2008	▷ n.i.	▷ n.i.	▷ ref. Duellman & Cole, 1965/err.ref.; Anderson, 1991	▷ n.i.	▷ 24/n.i.
▷ Catroli & Kasahara, 2009	▷ n.i.	▷ n.i.	▷ ref. Anderson, 1991	▷ 17	▷ 24/n.i.
Hyla savignyi					
▶ Anderson, 1986[I]	▶ Turkey	▶ 1♀	▶ Adana	▶ 1,2,4	▶ 24/48†
▷ King, 1990	▷ Turkey	▷ n.i.	▷ ref. Anderson, 1986	▷ 2,4	▷ 24/48
▷ Kuramoto, 1990	▷ Europe	▷ n.i.	▷ ref. Bogart, 1973	▷ 1	▷ 24/n.i.
▷ Anderson, 1991[I]	▷ Holarctic	▷ n.i.	▷ ref. Anderson, 1986; ctry.n.sp.	▷ 2,4	▷ 24/48†
▶ Al-Shehri & Al-Saleh, 2005[P,M,I]	▶ Saudi Arabia	▶ 10♂ 10♀	▶ Abha region, Aseer Province; err.FN (44) in text	▶ 1	▶ 24/48†
▶ Martirosyan & Stepanyan, 2007[P]	▶ Armenia	▶ 2♂2♀	▶ vicinity of Urtsadzor, Ararat region, 1,100 m (39°55′N, 44°50′E)	▶ 1,2,4	▶ 24/48
▶ Martirosyan & Stepanyan, 2007[P]	▶ Armenia	▶ 4♂4♀	▶ vicinity of Ejmiatsin, Armavyr region, 870 m (40°09′N, 44°17′E)	▶ 1,2,4	▶ 24/48
▷ Nunes & Fagundes, 2008b	▷ n.i.	▷ n.i.	▷ ref. Anderson, 1991	▷ n.i.	▷ 24/48
▷ Catroli, 2008	▷ n.i.	▷ n.i.	▷ ref. Bogart, 1973; Anderson, 1991	▷ n.i.	▷ 24/n.i.
▷ Catroli & Kasahara, 2009	▷ n.i.	▷ n.i.	▷ ref. Bogart, 1973; Anderson, 1991; Al-Shehri & Al-Saleh, 2005; Martirosyan & Stepanyan, 2007	▷ 17	▷ 24/n.i.
▷ Hernández-Guzmán et al., 2011	▷ n.i.	▷ n.i.	▷ ref. Martirosyan & Stepanyan, 2007	▷ n.i.	▷ 24/n.i.
HYLIDAE / Hylinae / *Isthmohyla*					
Isthmohyla angustilineata					
▶ León, 1970[P]	▶ Costa Rica	▶ 1♂	▶ loc.n.i.; rep. *Hyla angustilineata*	▶ 1	▶ 24‡/n.i.
▷ Foresti, 1972	▷ n.i.	▷ n.i.	▷ ref. León, 1970; list. *Hyla angustilineata*	▷ n.i.	▷ 24‡/n.i.
▷ Morescalchi, 1979	▷ n.i.	▷ n.i.	▷ ref. León, 1970; list. *Hyla angustilineata*	▷ n.i.	▷ 24/n.i.
▷ King, 1990	▷ C.America	▷ n.i.	▷ ref. León, 1970; ctry.n.sp.; list. *Hyla angustilineata*	▷ n.i.	▷ 24/n.i.
▷ Kuramoto, 1990	▷ S.C.America	▷ n.i.	▷ ref. León, 1970; ctry.n.sp.; list. *Hyla angustilineata*	▷ 1	▷ 24/n.i.
▷ Nunes & Fagundes, 2008b	▷ n.i.	▷ n.i.	▷ ref. León, 1970	▷ n.i.	▷ 24/48*
▷ Catroli, 2008	▷ n.i.	▷ n.i.	▷ ref. León, 1970	▷ n.i.	▷ 24/n.i.
▷ Catroli & Kasahara, 2009	▷ n.i.	▷ n.i.	▷ ref. León, 1970	▷ n.i.	▷ 24/n.i.
Isthmohyla lancasteri					
▶ Anderson, 1986[I]	▶ Costa Rica	▶ 1♂	▶ Pavones; rep. *Hyla lancasteri*	▶ 1,2,4	▶ 24/48†
▷ King, 1990	▷ Costa Rica	▷ n.i.	▷ ref. Anderson, 1986; list. *Hyla lancasteri*	▷ 2,4	▷ 24/48
▷ Anderson, 1991[I]	▷ Neotropic	▷ n.i.	▷ ref. Anderson, 1986; ctry.n.sp.; list. *Hyla lancasteri*	▷ 2,4	▷ 24/48†
▷ Nunes & Fagundes, 2008b	▷ n.i.	▷ n.i.	▷ ref. Anderson, 1991	▷ n.i.	▷ 24/48
▷ Catroli, 2008	▷ n.i.	▷ n.i.	▷ ref. Anderson, 1991	▷ n.i.	▷ 24/n.i.
▷ Catroli & Kasahara, 2009	▷ n.i.	▷ n.i.	▷ ref. Anderson, 1991	▷ 17	▷ 24/n.i.

Author(s) ▶ original data[a] ▷ review articles	Country/ Region[b]	Speci- mens[c]	Locality of sampling, references cited, and remarks[d]	Methods applied[e]	2n/FN[f]
Isthmohyla pictipes					
▶ Duellman, 1970[N]	▶ C.America	▶ n♂	▶ loc.n.i.; ctry.n.sp.; rep. *Hyla pictipes*	▶ 1	▶ 24‡/n.i.
▷ Kuramoto, 1990	▷ S.C.America	▷ n.i.	▷ ref. Duellman, 1970; ctry.n.sp.; list. *Hyla pictipes*	▷ 1	▷ 24/n.i.
▷ Nunes & Fagundes, 2008b	▷ n.i.	▷ n.i.	▷ ref. Duellman, 1970	▷ n.i.	▷ 24/48*
▷ Catroli, 2008	▷ n.i.	▷ n.i.	▷ ref. Duellman, 1970	▷ n.i.	▷ 24/n.i.
▷ Catroli & Kasahara, 2009	▷ n.i.	▷ n.i.	▷ ref. Duellman, 1970	▷ n.i.	▷ 24/n.i.
Isthmohyla pseudopuma					
▶ León, 1970[P]	▶ Costa Rica	▶ 1♂	▶ loc.n.i.; rep. *Hyla pseudopuma*	▶ 1	▶ 24‡/n.i.
▷ Duellman, 1970	▷ C.America	▷ n.i.	▷ ref.n.i.; ctry.n.sp.; list. *Hyla pseudopuma*	▷ 1	▷ 24‡/n.i.
▷ Foresti, 1972	▷ n.i.	▷ n.i.	▷ ref. León, 1970; list. *Hyla pseudopuma*	▷ n.i.	▷ 24‡/n.i.
▷ Morescalchi, 1979	▷ n.i.	▷ n.i.	▷ ref. León, 1970; list. *Hyla pseudopuma*	▷ n.i.	▷ 24/n.i.
▷ Kuramoto, 1990	▷ S.C.America	▷ n.i.	▷ ref. León, 1970; ctry.n.sp.; list. *Hyla pseudopuma*	▷ 1	▷ 24/n.i.
▷ Nunes & Fagundes, 2008b	▷ n.i.	▷ n.i.	▷ ref. León, 1970	▷ n.i.	▷ 24/48*
▷ Catroli, 2008	▷ n.i.	▷ n.i.	▷ ref. León, 1970	▷ n.i.	▷ 24/n.i.
▷ Catroli & Kasahara, 2009	▷ n.i.	▷ n.i	▷ ref. León, 1970	▷ n.i.	▷ 24/n.i.
Isthmohyla rivularis					
▶ León, 1970[P]	▶ Costa Rica	▶ 1♂	▶ loc.n.i.; rep. *Hyla rivularis*	▶ 1	▶ 24/48†
▷ Duellman, 1970	▷ C.America	▷ n.i.	▷ ref.n.i.; ctry.n.sp.; list. *Hyla rivularis*	▷ 1	▷ 24‡/n.i.
▷ Foresti, 1972	▷ n.i.	▷ n.i.	▷ ref. León, 1970; list. *Hyla rivularis*	▷ n.i.	▷ 24‡/n.i.
▷ Morescalchi, 1979	▷ n.i.	▷ n.i.	▷ ref. León, 1970; list. *Hyla rivularis*	▷ n.i.	▷ 24/n.i.
▷ King, 1990	▷ C.America	▷ n.i.	▷ ref. León, 1970; ctry.n.sp.; list. *Hyla rivularis*	▷ n.i.	▷ 24/n.i.
▷ Kuramoto, 1990	▷ S.C.America	▷ n.i.	▷ ref. León, 1970; ctry.n.sp.; list. *Hyla rivularis*	▷ 1	▷ 24/n.i.
▷ Nunes & Fagundes, 2008b	▷ n.i.	▷ n.i.	▷ ref. León, 1970	▷ n.i.	▷ 24/48
▷ Catroli, 2008	▷ n.i.	▷ n.i.	▷ ref. León, 1970	▷ n.i.	▷ 24/n.i.
▷ Catroli & Kasahara, 2009	▷ n.i.	▷ n.i.	▷ ref. León, 1970	▷ n.i.	▷ 24/n.i.
Isthmohyla tica					
▶ León, 1970[N]	▶ Costa Rica	▶ 1♂	▶ loc.n.i.; rep. *Hyla tica*	▶ 1	▶ 24‡/n.i.
▷ Duellman, 1970	▷ C.America	▷ n.i.	▷ ref.n.i.; ctry.n.sp.; list. *Hyla tica*	▷ 1	▷ 24‡/n.i.
▷ Foresti, 1972	▷ n.i.	▷ n.i.	▷ ref. León, 1970; list. *Hyla tica*	▷ n.i.	▷ 24‡/n.i.
▷ Morescalchi, 1979	▷ n.i.	▷ n.i.	▷ ref. León, 1970; list. *Hyla tica*	▷ n.i.	▷ 24/n.i.
▷ King, 1990	▷ C.America	▷ n.i.	▷ ref. León, 1970; ctry.n.sp.; list. *Hyla tica*	▷ n.i.	▷ 24/n.i.
▷ Kuramoto, 1990	▷ S.C.America	▷ n.i.	▷ ref. León, 1970; ctry.n.sp.; list. *Hyla tica*	▷ 1	▷ 24/n.i.
▷ Nunes & Fagundes, 2008b	▷ n.i.	▷ n.i.	▷ ref. León, 1970	▷ n.i.	▷ 24/48*
▷ Catroli, 2008	▷ n.i.	▷ n.i.	▷ ref. León, 1970	▷ n.i.	▷ 24/n.i.
▷ Catroli & Kasahara, 2009	▷ n.i.	▷ n.i.	▷ ref. León, 1970	▷ n.i.	▷ 24/n.i.
Isthmohyla zeteki*					
▶ Present study	▶ Costa Rica	▶ 1♀	▶ Poas Volcano National Park, Alajuela Province (10°12′N, 84°13′W)	▶ 2,4	▶ 24/48
HYLIDAE / Hylinae / *Megastomatohyla*					
No cytogenetic data					
HYLIDAE / Hylinae / *Plectrohyla*					
Plectrohyla ixil					
▶ Duellman, 1970[N]	▶ C.America	▶ n♂	▶ loc.n.i.; ctry.n.sp.	▶ 1	▶ 24‡/n.i.
▷ Kuramoto, 1990	▷ S.C.America	▷ n.i.	▷ ref. Duellman, 1970; ctry.n.sp. list. *{Plectohyla} ixil*	▷ 1	▷ 24/n.i.
▷ Catroli, 2008	▷ n.i.	▷ n.i.	▷ ref. Duellman, 1970	▷ n.i.	▷ 24/n.i.
▷ Catroli & Kasahara, 2009	▷ n.i.	▷ n.i.	▷ ref. Duellman, 1970	▷ n.i.	▷ 24/n.i.
Plectrohyla sagorum					
▶ Duellman, 1970[N]	▶ C.America	▶ n♂	▶ loc.n.i.; ctry.n.sp.	▶ 1	▶ 24‡/n.i.
▷ Kuramoto, 1990	▷ S.C.America	▷ n.i.	▷ ref. Duellman, 1970; ctry.n.sp.; list. *{Plectohyla} sagorum*	▷ 1	▷ 24/n.i.
▷ Catroli, 2008	▷ n.i.	▷ n.i.	▷ ref. Duellman, 1970	▷ n.i.	▷ 24/n.i.
▷ Catroli & Kasahara, 2009	▷ n.i.	▷ n.i.	▷ ref. Duellman, 1970	▷ n.i.	▷ 24/n.i.
HYLIDAE / Hylinae / *Ptychohyla*					
Ptychohyla achrochorda					
▶ Duellman, 1967[N]	▶ Mexico	▶ 3♂	▶ 5 km S Yetia, Oaxaca State; rep. *Hyla erythromma*	▶ 1	▶ 24‡/n.i.
▷ Duellman, 1970	▷ C.America	▷ n.i.	▷ ref.n.i.; ctry.n.sp.; list. *Hyla erythromma*	▷ 1	▷ 24‡/n.i.
▷ Foresti, 1972	▷ n.i.	▷ n.i.	▷ ref. Duellman, 1967; list. *Hyla erythromma*	▷ n.i.	▷ 24‡/n.i.
▷ Kuramoto, 1990	▷ S.C.America	▷ n.i.	▷ ref. Duellman, 1967; ctry.n.sp.; list. *Hyla erythromma*	▷ 1	▷ 24/n.i.
▷ Catroli, 2008	▷ n.i.	▷ n.i.	▷ ref. Duellman, 1967; list. *Ptychohyla erythromma*	▷ n.i.	▷ 24/n.i.
▷ Catroli & Kasahara, 2009	▷ n.i.	▷ n.i.	▷ ref. Duellman, 1967; list. *Ptychohyla erythromma*	▷ n.i.	▷ 24/n.i.

Author(s) ▶ original data[a] ▷ review articles	Country/ Region[b]	Speci-mens[c]	Locality of sampling, references cited, and remarks[d]	Methods applied[e]	2n/FN[f]
Ptychohyla legleri*					
▶ Duellman, 1967[N]	▶ Costa Rica	▶ 3♂	▶ 14 km N San Isidro del General, San José Province; rep. *Hyla legleri*	▶ 1	▶ 24[‡]/n.i.
▷ Duellman, 1970	▷ C.America	▷ n.i.	▷ ref.n.i.; ctry.n.sp.; list. *Hyla legleri*	▷ 1	▷ 24[‡]/n.i.
▷ Foresti, 1972	▷ n.i.	▷ n.i.	▷ ref. Duellman, 1967; list. *Hyla legleri*	▷ n.i.	▷ 24[‡]/n.i.
▷ Kuramoto, 1990	▷ S.C.America	▷ n.i.	▷ ref. Duellman, 1967; ctry.n.sp.; list. *Hyla legleri*	▷ 1	▷ 24/n.i.
▷ Catroli, 2008	▷ n.i.	▷ n.i.	▷ ref. Duellman, 1967	▷ n.i.	▷ 24/n.i.
▷ Catroli & Kasahara, 2009	▷ n.i.	▷ n.i.	▷ ref. Duellman, 1967	▷ n.i.	▷ 24/n.i.
▶ Present study	▶ Costa Rica	▶ 1♀	▶ Wilson Botanical Garden, San Vito de Java, Puntarenas Province (08°47′N, 82°57′W)	▶ 1,4,5, 8, 14	▶ 24/48
Ptychohyla leonhardschultzei					
▶ Duellman, 1970	▶ C.America	▶ n♂	▶ loc.n.i.; ctry.n.sp.	▶ 1	▶ 24/n.i.
▷ Kuramoto, 1990	▷ S.C.America	▷ n.i.	▷ ref. Duellman, 1970; ctry.n.sp.	▷ 1	▷ 24/n.i.
▷ Catroli, 2008	▷ n.i.	▷ n.i.	▷ ref. Duellman, 1970	▷ n.i.	▷ 24/n.i.
▷ Catroli & Kasahara, 2009	▷ n.i.	▷ n.i.	▷ ref. Duellman, 1970	▷ n.i.	▷ 24/n.i.
HYLIDAE / Hylinae / *Rheohyla*					
No cytogenetic data					
HYLIDAE / Hylinae / *Sarcohyla*					
Sarcohyla arborescandens					
▶ Duellman & Cole, 1965[N]	▶ Mexico	▶ 3♂	▶ 4.2 km S Vista Hermosa; Oaxaca State; rep. *Hyla hazelae*	▶ 1	▶ 24/n.i.
▷ Duellman, 1970	▷ C.America	▷ n.i.	▷ ref. Duellman & Cole, 1965; ctry.n.sp.; list. *Hyla arborescandens*	▷ n.i.	▷ 24/n.i.
▷ Rabello, 1970	▷ n.i.	▷ n.i.	▷ ref. Duellman & Cole, 1965; list. *Hyla hazelae*	▷ n.i.	▷ 24/n.i.
▷ Foresti, 1972	▷ n.i.	▷ n.i.	▷ ref. Duellman & Cole, 1965; list. *Hyla hazelae*	▷ n.i.	▷ 24/n.i.
▷ King, 1990	▷ C.America	▷ n.i.	▷ ref. Duellman & Cole, 1965; ctry.n.sp.; list. *Hyla hazelae*	▷ n.i.	▷ 24/n.i.
▷ Kuramoto, 1990	▷ S.C.America	▷ n.i.	▷ ref. Duellman & Cole, 1965; ctry.n.sp.; list. *Hyla arborescandens*	▷ 1	▷ 24/n.i.
▷ Nunes & Fagundes, 2008b	▷ n.i.	▷ n.i.	▷ ref. Duellman & Cole, 1965; list. *Plectrohyla arborescandens*	▷ n.i.	▷ 24/48*
▷ Catroli, 2008	▷ n.i.	▷ n.i.	▷ ref. Duellman & Cole, 1965; list. *Plectrohyla arborescandens*	▷ n.i.	▷ 24/n.i.
▷ Catroli, 2008	▷ n.i.	▷ n.i.	▷ ref. Duellman & Cole, 1965; list. *Plectrohyla hazelae*	▷ n.i.	▷ 24/n.i.
▷ Catroli & Kasahara, 2009	▷ n.i.	▷ n.i.	▷ ref. Duellman & Cole, 1965; Duellman, 1970; list. *Plectrohyla arborescandens*	▷ n.i.	▷ 24/n.i.
Sarcohyla pentheter					
▶ Duellman & Cole, 1965[N]	▶ Mexico	▶ 1♂	▶ 29 km SSE Juchatengo, Oaxaca State; rep. *Hyla bistincta*	▶ 1	▶ 24[‡]/n.i.
▷ Duellman, 1970	▷ C.America	▷ n.i.	▷ ref. Duellman Cole, 1965; ctry.n.sp.; list. *Hyla pentheter*	▷ n.i.	▷ 24[‡]/n.i.
▷ Rabello, 1970	▷ n.i.	▷ n.i.	▷ ref. Duellman & Cole, 1965; list. *Hyla bistincta*	▷ n.i.	▷ 24[‡]/n.i.
▷ Foresti, 1972	▷ n.i.	▷ n.i.	▷ ref. Duellman & Cole, 1965; list. *Hyla {bidistincta}*	▷ n.i.	▷ 24[‡]/n.i.
▷ Foresti, 1972	▷ n.i.	▷ n.i.	▷ ref. Duellman, 1967; list. *Hyla pentheter*	▷ n.i.	▷ 24[‡]/n.i.
▷ King, 1990	▷ C.America	▷ n.i.	▷ ref. Duellman & Cole, 1965; ctry.n.sp.; list. *Hyla bistincta*	▷ n.i.	▷ 24/n.i.
▷ King, 1990	▷ C.America	▷ n.i.	▷ ref. Duellman, 1967; ctry.n.sp.; list. *Hyla pentheter*	▷ n.i.	▷ 24/n.i.
▷ Kuramoto, 1990	▷ S.C.America	▷ n.i.	▷ ref. Duellman & Cole, 1965; ctry.n.sp.; list. *Hyla pentheter*	▷ n.i.	▷ 24/n.i.
▷ Catroli, 2008	▷ n.i.	▷ n.i.	▷ ref. Duellman & Cole, 1965; list. *Plectrohyla bistincta*	▷ n.i.	▷ 24/n.i
▷ Catroli, 2008	▷ n.i.	▷ n.i.	▷ ref. Duellman & Cole, 1965; Duellman, 1967; list. *Plectrohyla pentheter*	▷ n.i.	▷ 24/n.i.
▷ Nunes & Fagundes, 2008b	▷ n.i.	▷ n.i.	▷ ref. Duellman & Cole, 1965; list. *Plectrohyla bistincta*	▷ n.i.	▷ 24/48*
▷ Catroli & Kasahara, 2009	▷ n.i.	▷ n.i.	▷ ref. Duellman & Cole, 1965; Duellman, 1967, 1970; list. *Plectrohyla pentheter*	▷ n.i.	▷ 24/n.i.
Sarcohyla robertsorum					
▶ Duellman & Cole, 1965[N]	▶ Mexico	▶ 1♂	▶ El Chico National Park, Hidalgo State; rep. *Hyla robertsorum*	▶ 1	▶ 24[‡]/n.i.
▷ Duellman, 1970	▷ C.America	▷ n.i.	▷ ref.n.i.; ctry.n.sp.; list. *Hyla robertsorum*	▷ 1	▷ 24[‡]/n.i.
▷ Rabello, 1970	▷ n.i.	▷ n.i.	▷ ref. Duellman & Cole, 1965; list. *Hyla robertsorum*	▷ n.i.	▷ 24[‡]/n.i.
▷ Foresti, 1972	▷ n.i.	▷ n.i.	▷ ref. Duellman & Cole, 1965; list. *Hyla robertsorum*	▷ n.i.	▷ 24[‡]/n.i.
▷ King, 1990	▷ C.America	▷ n.i.	▷ ref. Duellman & Cole, 1965; ctry.n.sp.; list. *Hyla robertsorum*	▷ n.i.	▷ 24/n.i.
▷ Kuramoto, 1990	▷ S.C.America	▷ n.i.	▷ ref. Duellman & Cole, 1965; ctry.n.sp.; list. *Hyla robertsorum*	▷ 1	▷ 24/n.i.
▷ Nunes & Fagundes, 2008b	▷ n.i.	▷ n.i.	▷ ref. Duellman & Cole, 1965	▷ n.i.	▷ 24/28*
▷ Catroli, 2008	▷ n.i.	▷ n.i.	▷ ref. Duellman & Cole, 1965; list. *Plectrohyla robertsorum*	▷ n.i.	▷ 24/n.i.
▷ Catroli & Kasahara, 2009	▷ n.i.	▷ n.i.	▷ ref. Duellman & Cole, 1965; list. *Plectrohyla robertsorum*	▷ n.i.	▷ 24/n.i.
HYLIDAE / Hylinae / *Smilisca*					
Smilisca baudinii					
▶ Duellman & Cole, 1965[N]	▶ Mexico	▶ 1♂	▶ 7.3 km SW Matatan, Sinaloa State; rep. *Smilisca {baudini}*	▶ 1	▶ 24[‡]/n.i.
▶ Duellman & Cole, 1965[N]	▶ Mexico	▶ 1♂	▶ 5.6 km S San Andres de la Cruz, Guerrero State; rep. *Smilisca {baudini}*	▶ 1	▶ 24[‡]/n.i.
▶ Duellman & Cole, 1965[N]	▶ Mexico	▶ 3♂	▶ 3 km S Acahuizotla, Guerrero State; rep. *Smilisca {baudini}*	▶ 1	▶ 24[‡]/n.i.
▶ Duellman & Trueb, 1966[N]	▶ n.i.	▶ n.i.	▶ loc.n.i.; rep. *Smilisca {baudini}*	▶ 1	▶ 24/n.i.
▶ León, 1970[N]	▶ Costa Rica	▶ n.i.	▶ loc.n.i.; rep. *Smilisca {baudini}*	▶ 1	▶ 24[‡]/n.i.
▷ Duellman, 1970	▷ C.America	▷ n.i.	▷ ref.n.i.; ctry.n.sp.	▷ 1	▷ 24/n.i.
▷ Rabello, 1970	▷ n.i.	▷ n.i.	▷ ref. Duellman & Cole, 1965; list. *Smilisca {baudini}*	▷ n.i.	▷ 24[‡]/n.i.
▶ Bogart, 1973[P]	▶ Mexico	▶ n.i.	▶ loc.n.i.; rep. *Hyla {baudini}*	▶ 1	▶ 24/48[†]
▷ Cole, 1974	▷ n.i.	▷ n.i.	▷ ref. Bogart, 1973; list. *Smilisca {baudini}*	▷ n.i.	▷ 24/n.i.

Cytogenet Genome Res 2018;155:249–316
DOI: 10.1159/000490085

Author(s) ▶ original data[a] ▷ review articles	Country/ Region[b]	Speci-mens[c]	Locality of sampling, references cited, and remarks[d]	Methods applied[e]	2n/FN[f]
▶ Cole, 1974[P]	▶ Mexico	▶ 1♂	▶ 35.6 miles by highway 85 S ciudad de Valles, San Luis Potosí State; rep. *Smilisca {baudini}*	▶ 1	▶ 24/48[†]
▶ Cole, 1974[P]	▶ Mexico	▶ 1♂	▶ 21.9 miles by highway 140 SE Jalapa, Veracruz State; rep. *Smilisca {baudini}*	▶ 1	▶ 24/48[†]
▷ Morescalchi, 1979	▷ n.i.	▷ n.i.	▷ ref. Bogart, 1973; Cole, 1974	▷ n.i.	▷ 24/n.i.
▶ Anderson, 1986[I]	▶ Mexico	▶ 1♀	▶ Colima, State; loc.u	▶ 1,2,4	▶ 24/48[†]
▶ Anderson, 1986[I]	▶ Belize	▶ 1J	▶ Stann Creek	▶ 1,2,4	▶ 24/48[†]
▷ King, 1990	▷ C.America	▷ n.i.	▷ ref. Duellman & Cole, 1965; Bogart, 1973; Cole, 1974; Anderson, 1986; ctry.n.sp.; list. *Smilisca {baudini}*	▷ 2,4	▷ 24/48
▷ Kuramoto, 1990	▷ S.C.America	▷ n.i.	▷ ref. Duellman & Cole, 1965; León, 1970; Cole, 1974; Bogart, 1973; ctry.n.sp.	▷ 1	▷ 24/n.i.
▷ Anderson, 1991[I]	▷ Neotropic	▷ n.i.	▷ ref. Anderson, 1986; ctry.n.sp.	▷ 2,4	▷ 24/48[†]
▷ Nunes & Fagundes, 2008b	▷ n.i.	▷ n.i.	▷ ref. Bogart, 1973; list. *Smilisca {baudini}*	▷ n.i.	▷ 24/48
▷ Catroli, 2008	▷ n.i.	▷ n.i.	▷ ref. Duellman & Cole, 1965; León, 1970; Bogart, 1973; Cole, 1974; Anderson, 1991	▷ n.i.	▷ 24/n.i.
▷ Catroli & Kasahara, 2009	▷ n.i.	▷ n.i.	▷ ref. Duellman & Cole, 1965; Duellman & Trueb, 1966; León, 1970; Bogart, 1973; Cole, 1974; Anderson, 1991	▷ 17	▷ 24/n.i.
▶ Hernández-Guzmán et al., 2011[P,M]	▶ Mexico	▶ 1♂1♀	▶ urban zone of Villa Luís Gil Pérez, Tabasco State (17°52′43.05″N, 93°04′25.03″W)	▶ 1	▶ 24[‡]/48[†]
▶ Present study	▶ Costa Rica	▶ 3♂	▶ Cahuita National Park, Limón Province (09°43′N, 82°50′W)	▶ 1,2,4, 5,7,8	▶ 24/48
▶ Present study	▶ Colombia	▶ 1♂	▶ loc.u.	▶ 1	▶ 24/48
▶ Present study	▶ Mexico	▶ 1J	▶ Cuatlapan	▶ 1	▶ 24/48
▶ Present study	▶ Mexico	▶ 1s.n.i.	▶ near Lake Catemaco, Vera Cruz Department	▶ 1	▶ 24/48
Smilisca cyanosticta					
▶ Duellman & Cole, 1965[N]	▶ Mexico	▶ 2♂	▶ 11 km N Vista Hermosa, Oaxaca State	▶ 1	▶ 24[‡]/n.i.
▶ Duellman & Cole, 1965[N]	▶ Mexico	▶ 1♂	▶ 8 km S San Mateo Yetla, Oaxaca State	▶ 1	▶ 24[‡]/n.i.
▶ Duellman & Trueb, 1966[N]	▶ n.i.	▶ n.i.	▶ loc.n.i.	▶ 1	▶ 24/n.i.
▷ Duellman, 1970	▷ C.America	▷ n.i.	▷ ref.n.i.; ctry.n.sp.	▷ 1	▷ 24/n.i.
▷ Rabello, 1970	▷ n.i.	▷ n.i.	▷ ref. Duellman & Cole, 1965; list. *Smilisca {cyanostica}*	▷ n.i.	▷ 24[‡]/n.i.
▷ King, 1990	▷ C.America	▷ n.i.	▷ ref. Duellman & Cole, 1965; ctry.n.sp.; list. *Smilisca {cyanostricta}*	▷ n.i.	▷ 24/n.i.
▷ Kuramoto, 1990	▷ S.C.America	▷ n.i.	▷ ref. Duellman & Cole, 1965; Duellman & Trueb, 1966; ctry.n.sp.	▷ 1	▷ 24/n.i.
▷ Catroli, 2008	▷ n.i.	▷ n.i.	▷ ref. Duellman & Cole, 1965; Duellman & Trueb, 1966	▷ n.i.	▷ 24/n.i.
▷ Catroli & Kasahara, 2009	▷ n.i.	▷ n.i.	▷ ref. Duellman & Cole, 1965; Duellman & Trueb, 1966	▷ n.i.	▷ 24/n.i.
Smilisca fodiens					
▶ Duellman & Cole, 1965[N]	▶ Mexico	▶ 2♂	▶ 3.7 km E Villa Union, Sinaloa State; rep. *Pternohyla fodiens*	▶ 1	▶ 24[‡]/n.i.
▶ Duellman & Cole, 1965[N]	▶ Mexico	▶ 1♂	▶ 1 km N Villa Union, Sinaloa State; rep. *Pternohyla fodiens*	▶ 1	▶ 24[‡]/n.i.
▷ Duellman, 1970	▷ C.America	▷ n.i.	▷ ref.n.i.; ctry.n.sp.; list. *Pternohyla fodiens*	▷ 1	▷ 24[‡]/n.i.
▷ Rabello, 1970	▷ n.i.	▷ n.i.	▷ ref. Duellman & Cole, 1965; list. *Pternohyla fodiens*	▷ n.i.	▷ 24[‡]/n.i.
▶ Cole, 1974[P]	▶ Mexico	▶ 4♂	▶ 1 km by road S center of Navojoa, Sonora State: rep. *Pternohyla fodiens*	▶ 1	▶ 24/48[†]
▷ Morescalchi, 1979	▷ n.i.	▷ n.i.	▷ ref. Cole, 1974; list. *Pternohyla fodiens*	▷ n.i.	▷ 24/n.i.
▷ King, 1990	▷ C.America	▷ n.i.	▷ ref. Duellman & Cole, 1965; Cole, 1974; ctry.n.sp.; list. *Pternohyla fodiens*	▷ n.i.	▷ 24/48
▷ Kuramoto, 1990	▷ S.C.America	▷ n.i.	▷ ref. Duellman & Cole, 1965; Cole, 1974; ctry.n.sp.; list. *Pternohyla fodiens*	▷ 1	▷ 24/n.i.
▷ Nunes & Fagundes, 2008b	▷ n.i.	▷ n.i.	▷ ref. Cole, 1974	▷ n.i.	▷ 24/48
▷ Catroli, 2008	▷ n.i.	▷ n.i.	▷ ref. Duellman & Cole, 1965; Cole, 1974	▷ n.i.	▷ 24/n.i.
▷ Catroli & Kasahara, 2009	▷ n.i.	▷ n.i	▷ ref. Duellman & Cole, 1965; Cole, 1974	▷ n.i.	▷ 24/n.i.
▶ Present study	▶ USA	▶ 1♂	▶ southern Arizona State, loc.u.	▶ 1	▶ 24/48
Smilisca phaeota					
▶ Duellman & Trueb, 1966[N]	▶ n.i.	▶ n.i.	▶ loc.n.i.	▶ 1	▶ 24/n.i.
▷ Duellman, 1970	▷ C.America	▷ n.i.	▷ ref.n.i.; ctry.n.sp.	▷ 1	▶ 24/n.i.
▷ Kuramoto, 1990	▷ S.C.America	▷ n.i.	▷ ref. Duellman & Trueb, 1966; ctry.n.sp.	▷ n.i.	▷ 24/n.i.
▷ Catroli, 2008	▷ n.i.	▷ n.i.	▷ ref. Duellman & Trueb, 1966	▷ n.i.	▷ 24/n.i.
▷ Catroli & Kasahara, 2009	▷ n.i.	▷ n.i.	▷ ref. Duellman & Trueb, 1966	▷ n.i.	▷ 24/n.i.
▶ Present study	▶ Costa Rica	▶ 2♂	▶ Cahuita National Park, Limón Province (09°43′N, 82°50′W)	▶ 1,2,4–8	▶ 24/48
▶ Present study	▶ Costa Rica	▶ 1♀	▶ Wilson Botanical Garden, San Vito de Java, Puntarenas Province (08°47′N, 82°57′W)	▶ 1,2,4–8	▶ 24/48
▶ Present study	▶ Costa Rica	▶ 2♂1♀	▶ Tropical Science Center, Rincon de Osa, 20 m, Puntarenas Province (09°35′N, 85°04′W)	▶ 1,2,4–8	▶ 24/48
▶ Present study	▶ Costa Rica	▶ 1♀	▶ vicinity of Barra del Colorado, Limón province (10°41′N, 83°38′W)	▶ 1,2,4–8	▶ 24/48
Smilisca puma					
▶ Duellman & Trueb, 1966[N]	▶ n.i.	▶ n.i.	▶ loc.n.i.	▶ 1	▶ 24/n.i.
▷ Duellman, 1970	▷ C.America	▷ n.i.	▷ ref.n.i.; ctry.n.sp.	▷ 1	▷ 24/n.i.
▷ Kuramoto, 1990	▷ S.C.America	▷ n.i.	▷ ref. Duellman & Trueb, 1966; ctry.n.sp.	▷ 1	▷ 24/n.i.
▷ Catroli, 2008	▷ n.i.	▷ n.i.	▷ ref. Duellman & Trueb, 1966	▷ n.i.	▷ 24/n.i.
▷ Catroli & Kasahara, 2009	▷ n.i.	▷ n.i.	▷ ref. Duellman & Trueb, 1966	▷ n.i.	▷ 24/n.i.

Author(s) ▶ original data[a] ▷ review articles	Country/ Region[b]	Speci-mens[c]	Locality of sampling, references cited, and remarks[d]	Methods applied[e]	2n/FN[f]
Smilisca sila					
▶ Duellman & Trueb, 1966[N]	▶ n.i.	▶ n.i.	▶ loc.n.i.	▶ 1	▶ 24‡/n.i.
▷ Duellman, 1970	▷ C.America	▷ n.i.	▷ ref.n.i.; ctry.n.sp.	▷ 1	▷ 24‡/n.i.
▷ Kuramoto, 1990	▷ S.C.America	▷ n.i.	▷ ref. Duellman & Trueb, 1966; ctry.n.sp.	▷ 1	▷ 24/n.i.
▷ Catroli, 2008	▷ n.i.	▷ n.i.	▷ ref. Duellman & Trueb, 1966	▷ n.i.	▷ 24/n.i.
▷ Catroli & Kasahara, 2009	▷ n.i.	▷ n.i.	▷ ref. Duellman & Trueb, 1966	▷ n.i.	▷ 24/n.i.
Smilisca sordida					
▶ Duellman & Trueb, 1966[N]	▶ n.i.	▶ n.i.	▶ loc.n.i.	▶ 1	▶ 24/n.i.
▷ Duellman, 1970	▷ C.America	▷ n.i.	▷ ref.n.i.; ctry.n.sp.	▷ 1	▷ 24/n.i.
▶ Bogart, 1973[P]	▶ Panama	▶ n.i.	▶ loc.n.i.; rep. *Hyla sordida*	▶ 1	▶ 24/48†
▷ Cole, 1974	▷ n.i.	▷ n.i.	▷ ref. Bogart, 1973	▷ n.i.	▷ 24/n.i.
▷ Morescalchi, 1979	▷ n.i.	▷ n.i.	▷ ref. Bogart, 1973	▷ n.i.	▷ 24/n.i.
▷ King, 1990	▷ C.America	▷ n.i.	▷ ref. Bogart, 1973; ctry.n.sp.	▷ n.i.	▷ 24/48
▷ Kuramoto, 1990	▷ S.C.America	▷ n.i.	▷ ref. Duellman & Trueb, 1966; Bogart, 1973; ctry.n.sp.	▷ 1	▷ 24/n.i.
▷ Nunes & Fagundes, 2008	▷ n.i.	▷ n.i.	▷ ref. Bogart, 1973	▷ n.i.	▷ 24/48
▷ Catroli, 2008	▷ n.i.	▷ n.i.	▷ ref. Duellman & Trueb, 1966; Bogart, 1973	▷ n.i.	▷ 24/n.i.
▷ Catroli & Kasahara, 2009	▷ n.i.	▷ n.i.	▷ ref. Duellman & Trueb, 1966; Bogart, 1973	▷ n.i.	▷ 24/n.i.
▶ Present study	▶ Costa Rica	▶ 6♂1♀	▶ Valle de la Estrella, Limón Province (09°41′N, 83°00′W)	▶ 1,2,4–8	▶ 24/48
▶ Present study	▶ Costa Rica	▶ 2♂2♀	▶ Valle de Orosi, Cartago Province0 (9°48′N, 83°51′W)	▶ 1,2,4–8	▶ 24/48
▶ Present study	▶ Costa Rica	▶ 2♂	▶ Tropical Science Center, Rincon de Osa, 20 m, Puntarenas Province (09°35′N, 85°04′W)	▶ 1,2,4–8	▶ 24/48
▶ Present study	▶ Costa Rica	▶ 1s.n.i.	▶ loc.u.	▶ 1	▶ 24/48
HYLIDAE / Hylinae / *Tlalocohyla*					
Tlalocohyla loquax*					
▶ Duellman, 1970[N]	▶ C.America	▶ n♂	▶ loc.n.i; ctry.n.sp.; list. *Hyla loquax*	▶ 1	▶ 24‡/n.i.
▷ Kuramoto, 1990	▷ S.C.America	▷ n.i.	▷ ref. Duellman, 1970; ctry.n.sp.; list. *Hyla loquax*	▷ 1	▷ 24/n.i.
▷ Nunes & Fagundes, 2008b	▷ n.i.	▷ n.i.	▷ ref. Duellman, 1970	▷ n.i.	▷ 24/48*
▷ Catroli, 2008	▷ n.i.	▷ n.i.	▷ ref. Duellman, 1970	▷ n.i.	▷ 24/n.i.
▷ Catroli & Kasahara, 2009	▷ n.i.	▷ n.i.	▷ ref. Duellman, 1970	▷ n.i.	▷ 24/n.i.
▶ Present study	▶ C.America	▶ n♂	▶ specimen obtained by animal dealer; loc.u	▶ 2	▶ 24/48
Tlalocohyla smithii					
▶ Duellman & Cole, 1965[N]	▶ Mexico	▶ 1♂	▶ 3.7 km E Villa Union, Sinaloa State; rep. *Hyla {smithi}*	▶ 1	▶ 24‡/n.i.
▶ Duellman & Cole, 1965[N]	▶ Mexico	▶ 1♂	▶ 7.3 km SW Matatan, Sinaloa State; rep. *Hyla {smithi}*	▶ 1	▶ 24‡/n.i.
▶ Duellman & Cole, 1965[N]	▶ Mexico	▶ 1♂	▶ San Vicente, Guerrero State; rep. *Hyla {smithi}*	▶ 1	▶ 24‡/n.i.
▷ Duellman, 1970	▷ C.America	▷ n.i.	▷ ref.n.i.; ctry.n.sp.; list. *Hyla smithii*	▷ 1	▷ 24‡/n.i.
▷ Foresti, 1972	▷ n.i.	▷ n.i.	▷ ref. Duellman & Cole, 1965; list. *Hyla {smithi}*	▷ n.i.	▷ 24‡/n.i.
▶ Anderson, 1986[I]	▶ Mexico	▶ 2♀	▶ Sinaloa; rep. *Hyla smithii*	▶ 1,4	▶ 24/48†
▷ Rabello, 1970	▷ n.i.	▷ n.i.	▷ ref. Duellman & Cole, 1965; list. *Hyla {smithi}*	▷ n.i.	▷ 24‡/n.i.
▷ King, 1990	▷ C.America	▷ n.i.	▷ ref. Duellman & Cole, 1965; Anderson, 1986; ctry.n.sp.; list. *Hyla {smithi}*	▷ 2	▷ 24/48
▷ Kuramoto, 1990	▷ S.C.America	▷ n.i.	▷ ref. Duellman & Cole, 1965; ctry.n.sp.; list. *Hyla smithii*	▷ 1	▷ 24/n.i.
▷ Anderson, 1991[I]	▷ Neotropic	▷ n.i.	▷ ref. Anderson, 1986; ctry.n.sp.; list. *Hyla smithii*	▷ 2,4	▷ 24/48†
▷ Catroli, 2008	▷ n.i.	▷ n.i.	▷ ref. Duellman & Cole, 1965; Anderson, 1991	▷ n.i.	▷ 24/n.i.
▷ Nunes & Fagundes, 2008b	▷ n.i.	▷ n.i.	▷ ref. Wiley, 1982/err.ref.	▷ n.i.	▷ 24/48
▷ Catroli & Kasahara, 2009	▷ n.i.	▷ n.i.	▷ ref. Duellman & Cole, 1965; Anderson, 1991	▷ 17	▷ 24/n.i.
HYLIDAE / Hylinae / *Triprion*					
Triprion petasatus					
▶ Duellman & Cole, 1965[N]	▶ Mexico	▶ 1♂	▶ 12 km E Chichen-Itza, Yucatán Department	▶ 1	▶ 24‡/n.i.
▶ Duellman & Cole, 1965[N]	▶ Mexico	▶ 2♂	▶ Dzibilchaltun, Yucatán Department	▶ 1	▶ 24‡/n.i.
▷ Duellman, 1970	▷ C.America	▷ n.i.	▷ ref.n.i.; ctry.n.sp.	▷ 1	▷ 24/n.i.
▷ Rabello, 1970	▷ n.i.	▷ n.i.	▷ ref. Duellman & Cole, 1965	▷ n.i.	▷ 24‡/n.i.
▶ Cole, 1974[P]	▶ Mexico	▶ 3♂1♀	▶ E edge Mérida on highway 180, Yucatán State	▶ 1	▶ 24/48†
▷ Morescalchi, 1979	▷ n.i.	▷ n.i.	▷ ref. Cole, 1974	▷ n.i.	▷ 24/n.i.
▷ King, 1990	▷ C.America	▷ n.i.	▷ ref. Duellman & Cole, 1965; Cole, 1974; ctry.n.sp.	▷ n.i.	▷ 24/48
▷ Kuramoto, 1990	▷ S.C.America	▷ n.i.	▷ ref. Duellman & Cole, 1965; Cole, 1974	▷ 1	▷ 24/n.i.
▷ Nunes & Fagundes, 2008b	▷ n.i.	▷ n.i.	▷ ref. Cole, 1974	▷ n.i.	▷ 24/48
▷ Catroli, 2008	▷ n.i.	▷ n.i.	▷ ref. Duellman & Cole, 1965; Cole, 1974	▷ n.i.	▷ 24/n.i.
▷ Catroli & Kasahara, 2009	▷ n.i.	▷ n.i.	▷ ref. Duellman & Cole, 1965; Cole, 1974	▷ n.i.	▷ 24/n.i.
HYLIDAE / Lophyohylinae / *Aparasphenodon*					
Aparasphenodon bokermanni					
▶ Gruber et al., 2012b[P,M]	▶ Brazil	▶ 1♂	▶ Cananéia, São Paulo State (25°01′19″S, 47°55′41″W)	▶ 1,2,4,6, 9,13,14	▶ 24/48
▶ Present study	▶ Brazil	▶ 1♂	▶ Cananéia, São Paulo State (25°01′19″S, 47°55′41″W)	▶ 1,2	▶ 24/48

Author(s) ▶ original data[a] ▷ review articles	Country/ Region[b]	Speci-mens[c]	Locality of sampling, references cited, and remarks[d]	Methods applied[e]	2n/FN[f]
Aparasphenodon brunoi					
▶ Foresti, 1972[P,M]	▶ Brazil	▶ 1♂	▶ Grumari (Guanabara) Rio de Janeiro State	▶ 1	▶ 24/48†
▶ Bogart, 1973[P]	▶ Brazil	▶ n.i.	▶ southeastern Brazil; loc.n.i.	▶ 1	▶ 24/48†
▷ Cole, 1974	▷ n.i.	▷ n.i.	▷ ref. Bogart, 1973	▷ n.i.	▷ 24/n.i.
▷ Morescalchi, 1979	▷ n.i.	▷ n.i.	▷ ref. Bogart, 1973	▷ n.i.	▷ 24/n.i.
▷ King, 1990	▷ S.America	▷ n.i.	▷ ref. Bogart, 1973; ctry.n.sp.	▷ n.i.	▷ 24/48
▷ Kuramoto, 1990	▷ S.C.America	▷ n.i.	▷ ref. Bogart, 1973; ctry.n.sp.	▷ 1	▷ 24/n.i.
▶ Kasahara et al., 1998[P]	▶ Brazil	▶ n.i.	▶ loc.n.i.	▶ 12	▶ 24/48†
▶ Kasahara et al., 2003[P,M]	▶ Brazil	▶ 2♂	▶ Aracruz, Espírito Santo State	▶ 1,2,4,7, 9,12,13	▶ 24/48†
▷ Nunes & Fagundes, 2008b	▷ n.i.	▷ n.i.	▷ ref. Foresti, 1972	▷ n.i.	▷ 24/48
▶ Nunes & Fagundes, 2008b[P]	▶ Brazil	▶ 1♂	▶ Parque Estadual Paulo César Vinha, Guarapari, Espírito Santo State	▶ 1	▶ 24/48
▷ Catroli, 2008	▷ n.i.	▷ n.i.	▷ ref. Foresti, 1972; Bogart, 1973; Kasahara et al., 2003	▷ n.i.	▷ 24/n.i.
▷ Catroli & Kasahara, 2009	▷ n.i.	▷ n.i.	▷ ref. Foresti, 1972; Bogart, 1973; Kasahara et al., 2003; Nunes & Fagundes, 2008b	▷ 17	▷ 24/n.i.
▷ Hernández-Guzmán et al., 2011	▷ n.i.	▷ n.i.	▷ ref. Nunes & Fagundes, 2008b	▷ n.i.	▷ 24/n.i.
▶ Present study	▶ Brazil	▶ 2♀	▶ Recreio, São Paulo State	▶ 1,2,7	▶ 24/48
HYLIDAE / Lophyohylinae /*Argenteohyla*					
Argenteohyla siemersi					
▷ Catroli, 2008	▷ n.i.	▷ n.i.	▷ ref. Morand & Hernando, 1996; list. *Argenteohyla siemersi pederseni*	▷ n.i.	▷ 24/n.i.
▷ Catroli & Kasahara, 2009	▷ n.i.	▷ n.i.	▷ ref. Morand & Hernando, 1996	▷ 17	▷ 24/n.i.
HYLIDAE / Lophyohylinae / *Corythomantis*					
Corythomantis greeningi					
▶ Kasahara et al., 2003[P,M]	▶ Brazil	▶ 2♀	▶ Piranhas, Alagoas State	▶ 1,2,4,7, 9,12,13	▶ 24/48†
▷ Catroli, 2008	▷ n.i.	▷ n.i.	▷ ref. Kasahara et al., 2003	▷ n.i.	▷ 24/n.i.
▷ Catroli & Kasahara, 2009	▷ n.i.	▷ n.i	▷ ref. Kasahara et al., 2003	▷ 17	▷ 24/n.i.
▶ Present study	▶ Brazil	▶ 2♀	▶ Piranhas, Alagoas State	▶ 1,2	▶ 24/48
HYLIDAE / Lophyohylinae / *Dryaderces*					
No cytogenetic data					
HYLIDAE / Lophyohylinae / *Itapotihyla*					
Itapotihyla langsdorffii					
▶ Foresti, 1972[P,M]	▶ Brazil	▶ 1♂	▶ Grumari, (Guanabara) Rio de Janeiro State; rep. *Hyla langsdorffii*	▶ 1	▶ 24/48†
▶ Kasahara et al., 2003[P,M]	▶ Brazil	▶ 1♂	▶ Rio Claro, São Paulo State; rep. *Osteocephalus langsdorffii*	▶ 1,2,4,7, 9,12,13	▶ 24/48†
▶ Kasahara et al., 2003[P,M]	▶ Brazil	▶ 1♂	▶ Ubatuba, São Paulo State; rep. *Osteocephalus langsdorffii*	▶ 1,2,4,7, 9,12,13	▶ 24/48†
▷ Nunes & Fagundes, 2008b	▷ n.i.	▷ n.i.	▷ ref. Foresti, 1972	▷ n.i.	▷ 24/48
▶ Nunes & Fagundes, 2008b[P]	▶ Brazil	▶ 2♂	▶ Estação Biológica de Santa Lúcia, Santa Teresa, Espírito Santo State	▶ 1,4	▶ 24/48
▷ Catroli, 2008	▷ n.i.	▷ n.i.	▷ ref. Foresti, 1972; Kasahara et al., 2003	▷ n.i.	▷ 24/n.i.
▷ Catroli & Kasahara, 2009	▷ n.i.	▷ n.i.	▷ ref. Foresti, 1972; Kasahara et al., 2003; Nunes & Fagundes, 2008b	▷ 17	▷ 24/n.i.
▷ Hernández-Guzmán et al., 2011	▷ n.i.	▷ n.i.	▷ ref. Nunes & Fagundes, 2008b	▷ n.i.	▷ 24/n.i.
▶ Gruber et al., 2012b[P,M]	▶ Brazil	▶ 2♂	▶ Ilhéus, Bahia State (14°47′29″S, 39°02′41″W)	▶ 1,2,4,6, 9,13,14	▶ 24/48
▶ Gruber et al., 2012b[P,M]	▶ Brazil	▶ 1♀	▶ Rio Claro, São Paulo State (22°25′20″S, 47°34′23″W)	▶ 1,2,4,6, 9,13,14	▶ 24/48
▷ Schmid & Steinlein, 2016a	▷ n.i.	▷ n.i.	▷ ref. Gruber et al., 2012b	▷ 14	▷ 24/n.i.
▶ Present study	▶ Brazil	▶ 1s.n.i.	▶ Rio Claro, São Paulo State (22°25′20″S, 47°34′23″W)	▶ 1,2	▶ 24/48
HYLIDAE / Lophyohylinae / *Nyctimantis*					
No cytogenetic data					
HYLIDAE / Lophyohylinae / *Osteocephalus*					
Osteocephalus taurinus					
▶ Anderson, 1996[P,I]	▶ F. Guiana	▶ 1♀	▶ Sinnamary; err.FN (24) in text	▶ 1,2,4	▶ 24/48†
▷ Catroli, 2008	▷ n.i.	▷ n.i.	▷ ref. Anderson, 1996	▷ n.i.	▷ 24/n.i.
▷ Catroli & Kasahara, 2009	▷ n.i.	▷ n.i.	▷ ref. Anderson, 1996	▷ 17	▷ 24/n.i.
▶ Present study	▶ Venezuela	▶ 1♀	▶ loc.u.	▶ 1,5,6,8	▶ 24/48

Author(s) ▶ original data[a] ▷ review articles	Country/ Region[b]	Speci-mens[c]	Locality of sampling, references cited, and remarks[d]	Methods applied[e]	2n/FN[f]
HYLIDAE / Lophyohylinae / *Osteopilus*					
Osteopilus dominicensis					
▶ Anderson, 1996[P,I]	▶ Dom.Rep.	▶ 1♀	▶ La Vega Province; err.FN (24) in text	▶ 1,2,4	▶ 24/48†
▶ Anderson, 1996[P,I]	▶ Dom.Rep.	▶ 1♂	▶ Santo Domingo; err.FN (24) in text	▶ 1,2,4	▶ 24/48†
▷ Catroli, 2008	▷ n.i.	▷ n.i.	▷ ref. Anderson, 1996	▷ n.i.	▷ 24/n.i.
▷ Catroli & Kasahara, 2009	▷ n.i.	▷ n.i.	▷ ref. Anderson, 1996	▷ 17	▷ 24/n.i.
▶ Present study	▶ Dom.Rep.	▶ 1♂,1♀	▶ Bavaro, Punta Cana, Golf Course	▶ 1	▶ 24/48
Osteopilus marianae					
▶ Anderson, 1996[P,I]	▶ Jamaica	▶ 1♀	▶ Trewlany Parish; err.FN (24) in text	▶ 1	▶ 24/48†
▷ Catroli, 2008	▷ n.i.	▷ n.i.	▷ ref. Anderson, 1996	▷ n.i.	▷ 24/n.i.
▷ Catroli & Kasahara, 2009	▷ n.i.	▷ n.i.	▷ ref. Anderson, 1996	▷ 17	▷ 24/n.i.
▶ Present study	▶ Jamaica	▶ 1♀	▶ loc.u.	▶ 1	▶ 24/48
Osteopilus ocellatus					
▶ Cole, 1974[P]	▶ Jamaica	▶ 1♂2♀	▶ Barbecue Bottom, ca. 9 miles S Duncans,Trewlany Parish; list. *Hyla brunnea*	▶ 1	▶ 34/48†
▷ Morescalchi, 1979	▷ n.i.	▷ n.i.	▷ ref. Cole, 1974; list. *Hyla brunnea*	▷ n.i.	▷ 34/n.i.
▷ King, 1990	▷ C.America	▷ n.i.	▷ ref. Cole, 1974; ctry.n.sp.; list. *Hyla brunnea*	▷ n.i.	▷ 34/48
▷ Kuramoto, 1990	▷ S.C.America	▷ n.i.	▷ ref. Cole, 1974; ctry.n.sp.; list. *Osteopilus brunneus*	▷ 1	▷ 34/n.i.
▶ Anderson, 1996[P,I]	▶ Jamaica	▶ 2♂2♀	▶ Trewlany Parish; err.FN (24) in text; list. *Osteopilus brunneus*	▶ 1,2,4,14	▶ 34/48†
▶ Anderson, 1996[P,I]	▶ Jamaica	▶ 2♂	▶ St. Andrew Parish; err.FN (24) in text; list. *Osteopilus brunneus*	▶ 1,2,4,14	▶ 34/48†
▷ Nunes & Fagundes, 2008b	▷ n.i.	▷ n.i.	▷ ref. Cole, 1974; err.cit.2n; list. *Osteopilus brunneus*	▷ n.i.	▷ 24/48
▷ Catroli, 2008	▷ n.i.	▷ n.i.	▷ ref. Cole, 1974; Anderson, 1996; list. *Osteopilus brunneus*	▷ n.i.	▷ 34/n.i.
▷ Catroli & Kasahara, 2009	▷ n.i.	▷ n.i.	▷ ref. Cole, 1974; Anderson, 1996; list. *Osteopilus brunneus*	▷ 17	▷ 34/n.i.
▶ Present study	▶ Jamaica	▶ 1♂2♀ 1J	▶ Vicinity of Quick Step, 395 m, Trelawny Parish (18°15′N, 77°43′W)	▶ 1,2,4,5,8	▶ 34/48
▶ Present study	▶ Jamaica	▶ 1n.i.	▶ loc.u.	▶ 1	▶ 34/48
▶ Present study	▶ Jamaica	▶ 1♂	▶ loc.u.	▶ 1	▶ 34/48
Osteopilus septentrionalis					
▶ Duellman & Cole, 1965[N]	▶ USA	▶ 3♂	▶ Miami-Dade County, Florida State; rep. *Hyla septentrionalis*	▶ 1	▶ 24‡/n.i.
▷ Rabello, 1970	▷ n.i.	▷ n.i.	▷ ref. Duellman & Cole, 1965; list. *Hyla septentrionalis*	▷ n.i.	▷ 24‡/n.i.
▷ Foresti, 1972	▷ n.i.	▷ n.i.	▷ ref. Duellman & Cole, 1965; list. *Hyla septentrionalis*	▷ n.i.	▷ 24‡/n.i.
▶ Cole, 1974[P]	▶ USA	▶ 2♂5♀	▶ Naples, Collier County and Key West, Monroe County, Florida State	▶ 1	▶ 24/48†
▷ Morescalchi, 1979	▷ n.i.	▷ n.i.	▷ ref. Cole, 1974; list. *Hyla septentrionalis*	▷ n.i.	▷ 24/n.i.
▶ Schmid, 1978a[P,I]	▶ Bahamas	▶ 1♂	▶ specimen obtained by animal dealer; loc.n.i.; rep. *Hyla septentrionalis*	▶ 1,2,4,5	▶ 24/48†
▶ Schmid, 1978a[P,I]	▶ Cuba	▶ 1♀	▶ specimen obtained by animal dealer; loc.n.i.; rep. *Hyla septentrionalis*	▶ 1,2,4,5	▶ 24/48†
▷ King, 1990	▷ C.America	▷ n.i.	▷ ref. Duellman & Cole, 1965; Bogart, 1973; Cole, 1974; Schmid, 1978a; err.cit.ctry.; list. *Hyla septentrionalis*	▷ 2,4,17	▷ 24/48
▷ Kuramoto, 1990	▷ S.C.America	▷ n.i.	▷ ref. Duellman & Cole, 1965; Cole, 1974; Schmid, 1978a; 1980a/err.ref.; ctry.n.sp.	▷ 17	▷ 24/n.i.
▶ Anderson, 1996[P,I]	▶ Bahamas	▶ 1♂2♀	▶ Bimini; err.FN (24) in text	▶ 1,2,4,14	▶ 24/48†
▶ Anderson, 1996[P,I]	▶ Cuba	▶ 1♂	▶ Guantánamo Province; err.FN (24) in text	▶ 1,2,4,14	▶ 24/48†
▶ Anderson, 1996[P,I]	▶ USA	▶ 4♂1♀	▶ Miami-Dade County, Florida State; err.FN (24) in text	▶ 1,2,3,14	▶ 24/48†
▶ Anderson, 1996[P,I]	▶ USA	▶ 1♀	▶ St. Lucie County, Florida State; err.FN (24) in text	▶ 1,2,4,14	▶ 24/48†
▷ Catroli, 2008	▷ n.i.	▷ n.i.	▷ ref. Duellman & Cole, 1965; Bogart, 1973; Cole, 1974; Schmid, 1978a, 1980/err.ref.; Anderson, 1996	▷ n.i.	▷ 24/n.i.
▷ Catroli & Kasahara, 2009	▷ n.i.	▷ n.i.	▷ ref. Duellman & Cole, 1965; Bogart, 1973; Cole, 1974; Schmid, 1978a, 1980/err.ref.; Anderson, 1996	▷ 17	▷ 24/n.i.
▶ Schmid & Steinlein, 2016b[P]	▶ Cuba	▶ 1♀	▶ Reserva Ecológica Rancho Azucarero, Artemisa, La Habana Province (22°50′N, 82°49′W)	▶ 23	▶ 24/48†
▶ Present study	▶ Cuba	▶ 1J	▶ loc.u.	▶ 1	▶ 24/48
▶ Present study	▶ USA	▶ 1♂	▶ Florida State, loc.u.	▶ 1	▶ 24/48
▶ Present study	▶ USA	▶ 1♂	▶ Everglades National Park, Florida State	▶ 1	▶ 24/48
▶ Present study	▶ Cuba	▶ 1♀	▶ Reserva Ecológica Rancho Azucarero, Artemisa, La Habana Province (22°50′N, 82°49′W)	▶ 1,2,4, 5,14	▶ 24/48
Osteopilus wilderi					
▶ Anderson, 1996[P,I]	▶ Jamaica	▶ 1♀	▶ Trelawney Parish; err.FN (26) in text	▶ 1,2,4	▶ 28/52†
▶ Anderson, 1996[P,I]	▶ Jamaica	▶ 1♀	▶ Quick Step; err.FN (26) in text	▶ 1,2,4	▶ 28/52†
▷ Catroli, 2008	▷ n.i.	▷ n.i.	▷ ref. Anderson, 1996	▷ n.i.	▷ 28/n.i.
▷ Catroli & Kasahara, 2009	▷ n.i.	▷ n.i.	▷ ref. Anderson, 1996	▷ 17	▷ 28/n.i.
▶ Present study	▶ Jamaica	▶ 1s.n.i.	▶ loc.u.	▶ 1	▶ 28/52
HYLIDAE / Lophyohylinae / *Phyllodytes*					
Phyllodytes edelmoi					
▶ Gruber et al., 2012b[P,M]	▶ Brazil	▶ 1♂2♀	▶ Maceió, Alagoas State (09°40′06″S, 35°43′59″W)	▶ 1,2,4,6, 9,13,14	▶ 22/44
▷ Schmid & Steinlein, 2016a	▷ n.i.	▷ n.i.	▷ ref. Gruber et al, 2012b	▷ 14	▷ 22/n.i.
▶ Present study	▶ Brazil	▶ 1♂2♀	▶ Maceió, Alagoas State (09°40′06″S, 35°43′59″W)	▶ 1	▶ 22/44

Author(s) ▶ original data[a] ▷ review articles	Country/ Region[b]	Speci-mens[c]	Locality of sampling, references cited, and remarks[d]	Methods applied[e]	2n/FN[f]
Phyllodytes luteolus					
▶ Gruber et al., 2012b[P,M]	▶ Brazil	▶ 2♂	▶ Guaraparí, Espírito Santo State (20°39′01″S, 40°29′10″W)	▶ 1,2,4,6, 9,13,14	▶ 22/44
▶ Present study	▶ Brazil	▶ 2♂	▶ Guaraparí, Espírito Santo State (20°39′01″S, 40°29′10″W)	▶ 1	▶ 22/44
HYLIDAE / Lophyohylinae / *Phytotriades*					
No cytogenetic data					
HYLIDAE / Lophyohylinae / *Tepuihyla*					
No cytogenetic data					
HYLIDAE / Lophyohylinae / *Trachycephalus*					
Trachycephalus mesophaeus					
▶ Foresti, 1972[P,M]	▶ Brazil	▶ 1♂	▶ Itajibá, Bahia State; rep. *Phrynohyas mesophaea*	▶ 1	▶ 24/48†
▷ Catroli, 2008	▷ n.i.	▷ n.i.	▷ ref. Foresti, 1972	▷ n.i.	▷ 24/n.i.
▷ Catroli & Kasahara, 2009	▷ n.i.	▷ n.i.	▷ ref. Foresti, 1972	▷ n.i.	▷ 24/n.i.
▶ Gruber et al., 2012b[P,M]	▶ Brazil	▶ 3♂2♀	▶ Ilhéus, Bahia State (14°47′29″S, 39°02′41″W)	▶ 1,2,4,6, 9,13,14	▶ 24/48
▶ Gruber et al., 2012b[P,M]	▶ Brazil	▶ 1J	▶ Ubatuba, São Paulo State (23°26′19″S, 45°05′25″W)	▶ 1,2,4,6, 9,13,14	▶ 24/48
▶ Gruber et al., 2012b[P,M]	▶ Brazil	▶ 1♂	▶ Biritiba Mirim, São Paulo State (23°34′17″S, 46°02′15″W)	▶ 1,2,4,6, 9,13,14	▶ 24/48
▶ Present study	▶ Brazil	▶ 1♂	▶ Biritiba Mirim, São Paulo State (23°34′17″S, 46°02′15″W)	▶ 1,2	▶ 24/48
Trachycephalus typhonius					
▶ Duellman & Cole, 1965[N]	▶ Mexico	▶ 1♂	▶ 4 km NNE Felipe Carrillo Puerto, Quintana Roo State; rep. *Phrynohyas spilomma*	▶ 1	▶ 24‡/n.i.
▶ Duellman & Cole, 1965[N]	▶ Nicaragua	▶ 1♂	▶ Finca San Isidro, 10 km S Chinandega; Chinandega Department; rep. *Phrynohyas spilloma*	▶ 1	▶ 24‡/n.i.
▶ Rabello, 1970[P]	▶ n.i.	▶ 1♂2♀	▶ loc.n.i.; rep. *Phrynohyas venulosa*	▶ 1	▶ 24/48†
▷ Rabello, 1970	▷ n.i.	▷ n.i.	▷ ref. Duellman & Cole, 1965; list. *Phrynohyas spilomma*	▷ n.i.	▷ 24‡/n.i.
▶ Bogart & Bogart, 1971[N]	▶ Costa Rica	▶ n.i.	▶ loc.n.i.; rep. *Phrynohyas venulosa*	▶ 1	▶ 24/n.i.
▷ Foresti, 1972	▷ n.i.	▷ n.i.	▷ ref. Rabello et al., 1968; list. *Phrynohyas hebes*	▷ n.i.	▷ 24/n.i.
▷ Foresti, 1972	▷ n.i.	▷ n.i.	▷ ref. Duellman & Cole, 1965; list. *Phrynohyas spilomma*	▷ n.i.	▷ 24‡/n.i.
▷ Foresti, 1972	▷ n.i.	▷ n.i.	▷ ref. Rabello, 1969, 1970; list. *Phrynohyas venulosa*	▷ n.i.	▷ 24/n.i.
▶ Bogart, 1973[P]	▶ Colombia	▶ n.i.	▶ loc.n.i.; rep. *Phrynohyas venulosa*	▶ 1	▶ 24/48†
▷ Cole, 1974	▷ n.i.	▷ n.i.	▷ ref. Rabello, 1970; Bogart, 1973; list. *Phrynohyas venulosa*	▷ n.i.	▷ 24/n.i.
▷ Morescalchi, 1979	▷ n.i.	▷ n.i.	▷ ref. Bogart, 1973; list. *Phrynohyas venulosa*	▷ n.i.	▷ 24/n.i.
▷ King, 1990	▷ S.America	▷ n.i.	▷ ref. Duellman & Cole, 1965; ctry.n.sp.; list. *Phrynohyas spilomma*	▷ n.i.	▷ 24/n.i.
▷ King, 1990	▷ S.America	▷ n.i.	▷ ref. Duellman & Cole, 1965; Bogart, 1973; ctry.n.sp.; list. *Phrynohyas venulosa*	▷ n.i.	▷ 24/48
▷ Kuramoto, 1990	▷ S.C.America	▷ n.i.	▷ ref. Duellman & Cole, 1965; Rabello, 1970; Bogart & Bogart, 1971; Bogart, 1973; ctry.n.sp.; list. *Phrynohyas venulosa*	▷ 1	▷ 24/n.i.
▷ Catroli, 2008	▷ n.i.	▷ n.i.	▷ ref. Duellman & Cole, 1965; Rabello, 1970; Bogart & Bogart, 1971; Bogart, 1973; Insaurralde et al., 1999; Souza et al., 2007; list. *Trachycephalus venulosus*	▷ n.i.	▷ 24/n.i.
▷ Catroli & Kasahara, 2009	▷ n.i.	▷ n.i.	▷ ref. Duellman & Cole, 1965; Rabello, 1970; Bogart & Bogart, 1971; Bogart, 1973; list. *Trachycephalus venulosus*	▷ n.i.	▷ 24/n.i.
▶ Gruber et al., 2012b[P,M]	▶ Brazil	▶ 1♀	▶ Porto Primavera, Mato Grosso do Sul State (22°26′01″S, 52°58′11″W)	▶ 1,2,4,6, 9,13,14	▶ 24/48
▶ Gruber et al., 2012b[P,M]	▶ Brazil	▶ 1♂	▶ Rio Claro, São Paulo State (22°25′20″S, 47°34′23″W)	▶ 1,2,4,6, 9,13,14	▶ 24/48
▶ Present study	▶ Venezuela	▶ 1♀	▶ km 100 on road El Dorado to Santa Elena de Uairén, Bolivar State, ~ 260 m (06°04′N, 61°24′W)	▶ 1,2	▶ 24/48
▶ Present study	▶ Colombia	▶ 1♂	▶ loc.u.	▶ 1	▶ 24/48
***Trachycephalus* sp.**					
▷ Kuramoto, 1990	▷ S.C.America	▷ n.i.	▷ ref. Bogart, 1973; ctry.n.sp.	▷ 1	▷ 24/n.i.
▷ Catroli, 2008	▷ n.i.	▷ n.i.	▷ ref. Bogart, 1973	▷ n.i.	▷ 24/n.i.
▷ Catroli & Kasahara, 2009	▷ n.i.	▷ n.i.	▷ ref. Bogart, 1973	▷ n.i.	▷ 24/n.i.
***Trachycephalus* sp.**					
▶ Gruber et al., 2012b[P,M]	▶ Brazil	▶ 1♂	▶ Paranaíta, Mato Grosso State	▶ 1,2,4,6, 9,13,14	▶ 24/48
▶ Present study	▶ Brazil	▶ 1♂	▶ Paranaíta, Mato Grosso State	▶ 1,2	▶ 24/48

Author(s) ▶ original data[a] ▷ review articles	Country/ Region[b]	Speci- mens[c]	Locality of sampling, references cited, and remarks[d]	Methods applied[e]	2n/FN[f]
HYLIDAE / Pseudinae / *Lysapsus*					
Lysapsus bolivianus					
▶ Busin et al., 2006[P,M]	▶ Brazil	▶ 15♂ 12♀	▶ Santarém, Pará State; rep. *Lysapsus limellus bolivianus*	▶ 1,2,4,13	▶ 24/48[†]
▶ Busin et al., 2006[P,M]	▶ Brazil	▶ 12♂ 10♀	▶ Manaus, Amazonas State; rep. *Lysapsus limellus bolivianus*	▶ 1,2,4,13	▶ 24/48[†]
▶ Busin et al., 2006[P,M]	▶ Brazil	▶ 12♂ 14♀	▶ Guajará-Mirim, Rondônia State; rep. *Lysapsus limellus bolivianus*	▶ 1,2,4,13	▶ 24/48[†]
▷ Catroli, 2008	▷ n.i.	▷ n.i.	▷ ref. Busin et al., 2006; list. *Pseudis bolivianus*	▷ n.i.	▷ 24/n.i.
▷ Catroli & Kasahara, 2009	▷ n.i.	▷ n.i.	▷ ref. Busin et al., 2006; list. *Pseudis boliviana*	▷ 17	▷ 24/n.i.
Lysapsus caraya					
▶ Busin et al., 2006[P,M]	▶ Brazil	▶ 12♂8♀	▶ Santa Terezinha, Mato Grosso State	▶ 1,2,4,13	▶ 24/48[†]
▷ Catroli, 2008	▷ n.i.	▷ n.i.	▷ ref. Busin et al., 2006; list. *Pseudis caraya*	▷ n.i.	▷ 24/n.i.
▷ Catroli & Kasahara, 2009	▷ n.i.	▷ n.i.	▷ ref. Busin et al., 2006; list. *Pseudis caraya*	▷ 17	▷ 24/n.i.
Lysapsus laevis					
▶ Suárez et al., 2013[P,M,I]	▶ Brazil	▶ 11♂	▶ Ilha de Marajo, Soure, Soure Municipality, Pará State	▶ 1–4,6, 9,13,14	▶ 24/48
Lysapsus limellum					
▶ Barrio & Pistol de Rubel, 1970[P]	▶ Argentina	▶ 18♂2♀	▶ Santo Tomé, Santa Fe Province; rep. *Lysapsus limellus limellus*	▶ 1	▶ 24/48[†]
▷ Kuramoto, 1990	▷ S.C.America	▷ n.i.	▷ ref. Barrio & Pistol de Rubel, 1970; ctry.n.sp.; list. *Lysapsus limellus*	▷ 1	▷ 24/n.i.
▶ Busin et al., 2006[P,M]	▶ Brazil	▶ 9♂6♀	▶ Corumbá, Mato Grosso do Sul State; rep. *Lysapsus limellus limellus*	▶ 1,2,4,13	▶ 24/48[†]
▶ Busin et al., 2006[P,M]	▶ Brazil	▶ 7♂5♀	▶ Nossa Senhora do Livramento, Mato Grosso State; rep. *Lysapsus limellus limellus*	▶ 1,2,4,13	▶ 24/48[†]
▷ Catroli, 2008	▷ n.i.	▷ n.i.	▷ ref. Barrio & Pistol de Rubel, 1970; list. *Pseudis limellum*	▷ n.i.	▷ 24/n.i.
▷ Catroli, 2008	▷ n.i.	▷ n.i.	▷ ref. Busin et al., 2006; list. *Pseudis limellum limellus*	▷ n.i.	▷ 24/n.i.
▷ Catroli & Kasahara, 2009	▷ n.i.	▷ n.i.	▷ ref. Barrio & Pistol de Rubel, 1970; Busin et al., 2006; list. *Pseudis limellum*	▷ 17	▷ 24/n.i.
HYLIDAE / Pseudinae / *Pseudis*					
Pseudis bolbodactyla					
▷ Catroli, 2008	▷ n.i.	▷ n.i.	▷ ref. Busin et al., unpubl.	▷ n.i.	▷ 24/n.i.
▶ Busin et al., 2008[P,I]	▶ Brazil	▶ 15♂ 10♀	▶ Quirinópolis, Goiás State (18°30′00″S, 50°24′00″W)	▶ 1–4,13	▶ 24/48[†]
▷ Catroli & Kasahara, 2009	▷ n.i.	▷ n.i.	▷ ref. Busin et al., 2008	▷ 17	▷ 24/n.i.
▶ De Oliveira et al., 2012[P]	▶ Brazil	▶ 3s.n.i.	▶ Pirenópolis Municipality, Goiás State; err.FN(46) in text	▶ 1	▶ 24/48[†]
Pseudis cardosoi					
▶ Busin et al., 2001[P,M,I]	▶ Brazil	▶ 8♂5♀	▶ Tainhas, Contendas River, Rio Grande do Sul State (29°16′53″S, 50°15′28″W); list. *Pseudis* sp. (aff. *minuta*)	▶ 1,2,4	▶ 28/48[†]
▷ Catroli, 2001	▷ n.i.	▷ n.i.	▷ ref. Busin et al., 2001	▷ n.i.	▷ 28/n.i.
▷ Catroli & Kasahara, 2009	▷ n.i.	▷ n.i.	▷ ref. Busin et al., 2001	▷ 17	▷ 28/n.i.
Pseudis fusca					
▷ Catroli, 2008	▷ n.i.	▷ n.i.	▷ ref. Busin et al., unpubl.	▷ n.i.	▷ 24/n.i.
▶ Busin et al., 2008[P,I]	▶ Brazil	▶ 6♂	▶ Coronel Murta, Minas Gerais State (16°36′00″S, 42°12′00″W)	▶ 1–4,13	▶ 24/48[†]
▶ Busin et al., 2008[P,I]	▶ Brazil	▶ 1♂1♀	▶ Salinas, Minas Gerais State (16°06′00″S, 42°12′00″W)	▶ 1–4,13	▶ 24/48[†]
▷ Catroli & Kasahara, 2009	▷ n.i.	▷ n.i.	▷ ref. Busin et al., 2008	▷ 17	▷ 24/n.i.
Pseudis minuta					
▶ Barrio & Pistol de Rubel, 1970[P]	▶ Argentina	▶ 3♂3♀	▶ Buenos Aires Province; loc.n.i.; rep. *Lysapsus mantidactylus*	▶ 1	▶ 24/48[†]
▶ Barrio & Pistol de Rube, 1970I[P]	▶ Uruguay	▶ 2♂2♀	▶ Santiago Vázquez; rep. *Lysapsus mantidactylus*	▶ 1	▶ 24/48[†]
▶ Barrio & Pistol de Rubel, 1970[P]	▶ Uruguay	▶ 2♂	▶ Solís; rep. *Lysapsus mantidactylus*	▶ 1	▶ 24/48[†]
▷ Kuramoto, 1990	▷ S.C.America	▷ n.i.	▷ ref. Barrio & Pistol de Rubel, 1970; ctry.n.sp.; list. *Lysapsus mantidactylus*	▷ 1	▷ 24/n.i.
▶ Busin et al., 2001[P,M,I]	▶ Brazil	▶ 4♂5♀	▶ Eldorado do Sul, Guaíba mine, Rio Grande do Sul State (30°00′00″S, 51°25′00″W)	▶ 1,2,4	▶ 24/48[†]
▶ Busin et al., 2001[P,M,I]	▶ Brazil	▶ 2♂2♀ 3J	▶ São Jerônimo, Rio Grande do Sul State (30°27′30″S, 52°08′52″W)	▶ 1,2,4	▶ 24/48[†]
▷ Catroli, 2008	▷ n.i.	▷ n.i.	▷ ref. Busin et al., 2001	▷ n.i.	▷ 24/n.i.
▷ Catroli & Kasahara, 2009	▷ n.i.	▷ n.i.	▷ ref. Busin et al., 2001	▷ 17	▷ 24/n.i.
Pseudis paradoxa					
▶ Saez & Brum, 1960[N]	▶ S.America	▶ n♂	▶ loc.n.i.; ctry.n.sp.; err.2n	▶ 1	▶ 22/44[†]
▶ Bogart, 1973[P]	▶ Brazil	▶ n.i.	▶ southeastern Brazil; loc.n.i.	▶ 1	▶ 24/48[†]
▷ Kuramoto, 1990	▷ S.C.America	▷ n.i.	▷ ref. Saez & Brum, 1960; ctry.n.sp., err.2n	▷ 1	▷ 22/n.i.

Author(s) ▶ original data[a] ▷ review articles	Country/ Region[b]	Speci- mens[c]	Locality of sampling, references cited, and remarks[d]	Methods applied[e]	2n/FN[f]
▷ Kuramoto, 1990	▷ S.C.America	▷ n.i.	▷ ref. Bogart, 1973; ctry.n.sp.	▷ 1	▷ 24/n.i.
▶ Busin et al., 2008[P,I]	▶ Brazil	▶ 3♂4♀	▶ Bacabal, Maranhão State (04°12′00″S, 44°48′00″W); rep. *Pseudis paradoxa paradoxa*	▶ 1–4,13	▶ 24/48[†]
▷ Catroli, 2008	▷ n.i.	▷ n.i.	▷ ref. Barrio & Pistol de Rubel, 1970/err.ref.	▷ n.i.	▷ 24/n.i.
▷ Catroli & Kasahara, 2009	▷ n.i.	▷ n.i.	▷ ref. Barrio & Pistol de Rubel, 1970/err.ref.	▷ n.i.	▷ 24/n.i.
Pseudis platensis					
▶ Beçak, 1968[P,M]	▶ Brazil	▶ 3♂4♀	▶ São José do Rio Preto, São Paulo State; rep. *Pseudis paradoxus platensis*	▶ 1	▶ 24/48[†]
▷ Rabello, 1970	▷ n.i.	▷ n.i.	▷ ref. Beçak, 1967, 1968; list. *Pseudis paradoxus platensis*	▷ n.i.	▷ 24/n.i.
▶ Barrio & Pistol de Rubel, 1970[P]	▶ Argentina	▶ 4♂2♀	▶ vicinity of Santa Fe; rep. *Pseudis paradoxus platensis*	▶ 1	▶ 24/48[†]
▷ Kuramoto, 1990	▷ S.C.America	▷ n.i.	▷ ref. Beçak, 1968; Barrio & Pistol de Rubel, 1970; ctry.n.sp.; list. *Pseudis {parasoxa} platensis*	▷ n.i.	▷ 24/n.i.
▷ Catroli, 2008	▷ n.i.	▷ n.i.	▷ ref. Busin et al., unpubl.	▷ n.i.	▷ 24/n.i.
▶ Busin et al., 2008[P,I]	▶ Brazil	▶ 8♂5♀	▶ Corumbá, Mato Grosso do Sul State (18°54′00″S, 56°48′00″W); rep. *Pseudis paradoxa platensis*	▶ 1–4,13	▶ 24/48[†]
▷ Catroli & Kasahara, 2009	▷ n.i.	▷ n.i.	▷ ref. Busin et al., 2008	▷ 17	▷ 24/n.i.
Pseudis tocantins					
▷ Catroli, 2008	▷ n.i.	▷ n.i.	▷ ref. Busin et al., unpubl.	▷ n.i.	▷ 24/n.i.
▶ Busin et al., 2008[P,I]	▶ Brazil	▶ 4♂7♀	▶ Porto Nacional, Tocantins State (10°42′00″S, 48°24′00″W)	▶ 1–4,13	▶ 24/48[†]
▷ Catroli & Kasahara, 2009	▷ n.i.	▷ n.i.	▷ ref. Busin et al., 2008	▷ 17	▷ 24/n.i.
▶ Gatto et al., 2016[P]	▶ Brazil	▶ 4♂7♀	▶ Porto Nacional, Tocantins State (10°42′00″S, 48°24′00″W); see above Busin et al., 2008	▶ 19–21	▶ 24[†]/48[†]
HYLIDAE / Pseudinae / *Scarthyla*					
Scarthyla goinorum					
▶ Suárez et al., 2013[P,M,I]	▶ Brazil	▶ 2♂1♀	▶ Comunidade Praia Grande, Cruzeiro do Sul Municipality, Acre State	▶ 1–4,6, 9,13,14	▶ 22/44
▷ Schmid & Steinlein, 2016a	▷ n.i.	▷ n.i.	▷ ref. Suárez et al., 2013	▷ 14	▷ 22/n.i.
HYLIDAE / Scinaxinae / *Julianus*					
Julianus uruguayus					
▶ Cardozo et al., 2011[P,M]	▶ Uruguay	▶ 3♂	▶ Quebrada de los Cuervos, Treinta y Tres Department; rep. *Scinax uruguayus*	▶ 1,4	▶ 24/48
▷ Peixoto et al., 2015	▷ Uruguay	▷ n.i.	▷ Treinta y Tres Department; ref. Cardozo et al., 2011; err.cit.meth.; list. *Scinax uruguayus*	▷ 2,4	▷ 24/48
HYLIDAE / Scinaxinae / *Ololygon*					
Ololygon albicans					
▶ Cardozo et al., 2011[P,M]	▶ Brazil	▶ 4♂	▶ Petrópolis, Rio de Janeiro State; rep. *Scinax albicans*	▶ 1,2,4	▶ 24/48
▷ Peixoto et al., 2015	▷ Brazil	▷ n.i.	▷ Petrópolis, Rio de Janeiro State; ref. Cardozo et al., 2011; list. *Scinax albicans*	▷ 2,4	▷ 24/48
▶ Gruber et al., 2017[P,M]	▶ Brazil	▶ 1♂	▶ Petrópolis, Rio de Janeiro State (22°30′17″S, 43°10′56″W)	▶ 1,2,4,6, 9,12–14	▶ 24/48[†]
▶ Present study	▶ Brazil	▶ 1♂	▶ Petrópolis, Rio de Janeiro State (22°30′17″S, 43°10′56″W)	▶ 1	▶ 24/48
▶ Present study	▶ Brazil	▶ 1♂	▶ Petrópolis, Rio de Janeiro State (22°30′17″S, 43°10′56″W)	▶ 1,2	▶ 24/48
Ololygon arduoa					
▶ Peixoto et al., 2015[P]	▶ Brazil	▶ 3♂1♀	▶ Santa Teresa, Espírito Santo State (19°43′21″S, 40°42′48″W); rep. *Scinax arduous*	▶ 1,2,4, 6,9,15	▶ 24/48
Ololygon argyreornata					
▶ Nunes & Fagundes, 2008b[P]	▶ Brazil	▶ 2♂	▶ Parque Estadual Paulo César Vinha, Guarapari, Espírito Santo State; rep. *Scinax argyreornatus*	▶ 1	▶ 24/48
▷ Catroli & Kasahara, 2009	▷ n.i.	▷ n.i.	▷ ref. Nunes & Fagundes, 2008b; list. *Scinax argyreornatus*	▷ n.i.	▷ 24/n.i.
▷ Hernández-Guzmán et al., 2011	▷ n.i.	▷ n.i.	▷ ref. Nunes & Fagundes, 2008b; list. *Scinax argyreornatus*	▷ n.i.	▷ 24/n.i.
▶ Cardozo et al., 2011[P,M]	▶ Brazil	▶ 3♂2J	▶ Ilha do Cardoso, São Paulo State; rep. *Scinax argyreornatus*	▶ 1,2,4	▶ 24/48
▷ Peixoto et al., 2015	▷ Brazil	▷ n.i.	▷ Guarapari, Espírito Santo State; ref. Nunes & Fagundes, 2008b; list. *Scinax argyreornatus*	▷ n.i.	▷ 24/48
▷ Peixoto et al., 2015	▷ Brazil	▷ n.i.	▷ Ilha do Cardoso, São Paulo State; ref. Cardozo et al., 2011; list. *Scinax argyreornatus*	▷ 2,4	▷ 24/48
▶ Gruber et al., 2017[P,M]	▶ Brazil	▶ 2♂	▶ Ilha do Cardoso, São Paulo State (25°07′31″S, 47°58′06″W)	▶ 1,2,4,6, 9,12–14	▶ 24/48[†]
▶ Present study	▶ Brazil	▶ 2♂	▶ Ilha do Cardoso, São Paulo State (25°07′31″S, 47°58′06″W)	▶ 1,2	▶ 24/48
Ololygon ariadne					
▶ Cardozo et al., 2011[P,M]	▶ Brazil	▶ 1♀	▶ São José do Barreiro, São Paulo State; rep. *Scinax ariadne*	▶ 1,2,4	▶ 24/48
▷ Peixoto et al., 2015	▷ Brazil	▷ n.i.	▷ São José do Barreiro, São Paulo State; ref. Cardozo et al., 2011; list. *Scinax Ariadne*	▷ 2,4	▷ 24/48
▶ Present study	▶ Brazil	▶ 1♀	▶ São José do Barreiro, São Paulo State	▶ 1,2	▶ 24/48

Author(s) ▶ original data[a] ▷ review articles	Country/ Region[b]	Speci-mens[c]	Locality of sampling, references cited, and remarks[d]	Methods applied[e]	2n/FN[f]
Ololygon aromothyella					
▶ Souza et al., 2011[N]	▶ Brazil	▶ n.i.	▶ Agudo, Rio Grande do Sul State and Concórdia, Santa Catarina State; rep. *Scinax* cf. *aromothyella*	▶ 1,2,4	▶ 24/48†
▶ Cardozo et al., 2011[P,M]	▶ Argentina	▶ 1♂	▶ Balneario del Arroyo Cuña Pirú, Aristóbulo del Valle, Cainguas Department, Misiones Province; rep. *Scinax aromothyella*	▶ 1,2,4	▶ 24/48
▶ Cardozo et al., 2011[P,M]	▶ Argentina	▶ 1♂	▶ Parque Provincial Salto Encantado, Cainguas Department, Misiones Province; rep. *Scinax aromothyella*	▶ 1,2,4	▶ 24/48
▶ Cardozo et al., 2011[P,M]	▶ Uruguay	▶ 3♂	▶ Quebrada de los Cuervos, Treinta y Tres Department; rep. *Scinax aromothyella*	▶ 1,2,4	▶ 24/48
▷ Peixoto et al., 2015	▷ Argentina	▷ n.i.	▷ Misiones Province; ref. Cardozo et al., 2011; list. *Scinax aromothyella*	▷ 2,4	▷ 24/48
▷ Peixoto et al., 2015	▷ Uruguay	▷ n.i.	▷ Treinta y Tres Department; ref. Cardozo et al., 2011; list. *Scinax aromothyella*	▷ 2,4	▷ 24/48
Ololygon aurata					
▶ Nogueira et al., 2015a[P,M]	▶ Brazil	▶ 5♀	▶ Jequié, Bahia State (13°51′04″S, 40°04′52″W); rep. *Scinax auratus*	▶ 1,2,4, 7,9,13	▶ 24/48
▷ Peixoto et al., 2015	▷ Brazil	▷ n.i.	▷ Jequié, Bahia State; ref. Nogueira et al., 2015a; list. *Scinax auratus*	▷ 2,4	▷ 24/48
Ololygon belloni					
▶ Peixoto et al., 2015[P]	▶ Brazil	▶ 4♂1♀	▶ Castelo, Espírito Santo State (20°33′25″S, 41°16′44″W); rep. *Scinax belloni*	▶ 1,2,4, 6,9,15	▶ 24/48
Ololygon berthae					
▶ Cardozo et al., 2011[P,M]	▶ Argentina	▶ 3♂	▶ Gdor. Virasoro, Santo Tomé Department, Corrientes Province; rep. *Scinax berthae*	▶ 1,2,4	▶ 24/48
▶ Cardozo et al., 2011[P,M]	▶ Argentina	▶ 1♂	▶ Fachinal, Candelaria Department, Misiones Province; rep. *Scinax berthae*	▶ 1,2,4	▶ 24/48
▶ Cardozo et al., 2011[P,M]	▶ Argentina	▶ 1♂	▶ Profundidad, Candelaria Department, Misiones Province; rep. *Scinax berthae*	▶ 1,2,4	▶ 24/48
▶ Cardozo et al., 2011[P,M]	▶ Argentina	▶ 1♂	▶ Campus UNaM, Villa Lanús, Capital Department, Misiones Province; rep. *Scinax berthae*	▶ 1,2,4	▶ 24/48
▶ Cardozo et al., 2011[P,M]	▶ Argentina	▶ 5♂	▶ Barrio Villa Blosset, Posadas, Capital Department, Misiones Province; rep. *Scinax berthae*	▶ 1,2,4	▶ 24/48
▶ Cardozo et al., 2011[P,M]	▶ Argentina	▶ 2♂	▶ Pto. Iguazú, Iguazú Department, Misiones Province; rep. *Scinax berthae*	▶ 1,2,4	▶ 24/48
▶ Cardozo et al., 2011[P,M]	▶ Brazil	▶ 1♂	▶ Itirapina, São Paulo State; rep. *Scinax berthae*	▶ 1,2,4	▶ 24/48
▶ Cardozo et al., 2011[P,M]	▶ Uruguay	▶ 2♂	▶ Bañado de Santa Teresa, Rocha Department; rep. *Scinax berthae*	▶ 1,2,4	▶ 24/48
▷ Peixoto et al., 2015	▷ Argentina	▷ n.i.	▷ Provinces of Corrientes and Misiones; ref. Cardozo et al., 2011; list. *Scinax berthae*	▷ 2,4	▷ 24/48
▷ Peixoto et al., 2015	▷ Brazil	▷ n.i.	▷ Itirapina, São Paulo State; ref. Cardozo et al., 2011; list. *Scinax berthae*	▷ 2,4	▷ 24/48
▷ Peixoto et al., 2015	▷ Uruguay	▷ n.i.	▷ Rocha Department; ref. Cardozo et al., 2011; list. *Scinax berthae*	▷ 2,4	▷ 24/48
Ololygon brieni					
▶ Bogart, 1973[P]	▶ Brazil	▶ n.i.	▶ southeastern Brazil; loc.n.i.; rep. *Hyla brieni*	▶ 1	▶ 24/48†
▷ Cole, 1974	▷ n.i.	▷ n.i.	▷ ref. Bogart, 1973; list. *Hyla brieni*	▷ n.i.	▷ 24/n.i.
▷ King, 1990	▷ S.America	▷ n.i.	▷ ref. Bogart, 1973; ctry.n.sp.; list. *Hyla brieni*	▷ n.i.	▷ 24/48
▷ Kuramoto, 1990	▷ S.C.America	▷ n.i.	▷ ref. Bogart, 1973; ctry.n.sp.; list. *Ololygon brieni*	▷ 1	▷ 24/n.i.
▷ Nunes & Fagundes, 2008	▷ n.i.	▷ n.i.	▷ ref. Bogart, 1973; list. *Scinax brieni*	▷ n.i.	▷ 24/48
▷ Catroli, 2008	▷ n.i.	▷ n.i.	▷ ref. Bogart, 1973; list. *Scinax brieni*	▷ n.i.	▷ 24/n.i.
▷ Catroli & Kasahara, 2009	▷ n.i.	▷ n.i.	▷ ref. Bogart, 1973; list. *Scinax brieni*	▷ n.i.	▷ 24/n.i.
▷ Peixoto et al., 2015	▷ Brazil	▷ n.i.	▷ southeastern Brazil; ref. Bogart, 1973; list. *Scinax brieni*	▷ n.i.	▷ 24/48
▶ Present study[P]	▶ Brazil	▶ 1♂	▶ Boraceia, São Paulo State	▶ 1	▶ 24/48
Ololygon canastrensis					
▶ Cardozo et al., 2011[P,M]	▶ Brazil	▶ 2♂1♀	▶ Furnas, Minas Gerais State; rep. *Scinax canastrensis*	▶ 1,2,4	▶ 24/48
▷ Peixoto et al., 2015	▷ Brazil	▷ n.i.	▷ Furnas, Minas Gerais State; ref. Cardozo et al., 2011; list. *Scinax canastrensis*	▷ 2,4	▷ 24/48
▶ Present study	▶ Brazil	▶ 2♂1♀	▶ Furnas, Minas Gerais State	▶ 1,2	▶ 24/48
Ololygon catharinae					
▶ Bogart, 1973[P]	▶ Brazil	▶ n.i.	▶ southeastern Brazil; loc.n.i.; rep. *Hyla catharinae*	▶ 1	▶ 24/48†
▷ Cole, 1974	▷ n.i.	▷ n.i.	▷ ref. Bogart, 1973; list. *Hyla catharinae*	▷ n.i.	▷ 24/n.i.
▷ Morescalchi, 1979	▷ n.i.	▷ n.i.	▷ ref. Bogart, 1973; list. *Hyla catharinae*	▷ n.i.	▷ 24/n.i.
▷ King, 1990	▷ S.America	▷ n.i.	▷ ref. Bogart, 1973; ctry.n.sp.; list. *Hyla catharinae*	▷ n.i.	▷ 24/48
▷ Kuramoto, 1990	▷ S.C.America	▷ n.i.	▷ ref. Bogart, 1973; ctry.n.sp.; list. *Ololygon catharinae brieni*	▷ 1	▷ 24/n.i.
▷ Nunes & Fagundes, 2008b	▷ n.i.	▷ n.i.	▷ ref. Bogart, 1973; list. *Scinax catharinae*	▷ n.i.	▷ 24/48
▷ Catroli, 2008	▷ n.i.	▷ n.i.	▷ ref. Bogart, 1973; list. *Scinax catharinae*	▷ n.i.	▷ 24/n.i.
▷ Catroli & Kasahara, 2009	▷ n.i.	▷ n.i.	▷ ref. Bogart, 1973; list. *Scinax catharinae*	▷ n.i.	▷ 24/n.i.
▷ Peixoto et al., 2015	▷ Brazil	▷ n.i.	▷ southeastern Brazil; ref. Bogart, 1973; list. *Scinax catharinae*	▷ n.i.	▷ 24/48
▶ Present study	▶ Brazil	▶ 1s.n.i.	▶ southeastern Brazil; loc.n.i.	▶ 1	▶ 24/48
Ololygon centralis					
▷ Catroli, 2008	▷ n.i.	▷ n.i.	▷ ref. Ribeiro et al., 2006; list. *Scinax centralis*	▷ n.i.	▷ 24/n.i.

Author(s) ▶ original data[a] ▷ review articles	Country/ Region[b]	Speci-mens[c]	Locality of sampling, references cited, and remarks[d]	Methods applied[e]	2n/FN[f]
Ololygon cosenzai					
▶ Peixoto et al., 2015[P]	▶ Brazil	▶ 2♂2♀	▶ Ervália, Minas Gerais State (20°50′55″S, 43°30′26″W); rep. *Scinax cosenzai*	▶ 1,2,4, 6,9,15	▶ 24/48
▶ Peixoto et al., 2015[P]	▶ Brazil	▶ 2♂	▶ Araponga, Minas Gerais State (20°43′03″S, 42°28′48″W); rep *Scinax cosenzai*	▶ 1,2,4, 6,9,15	▶ 24/48
Ololygon hiemalis					
▶ Cardozo et al., 2011[P,M]	▶ Brazil	▶ 2♂	▶ Jundiai, São Paulo State; rep. *Scinax hiemalis*	▶ 1,2,4	▶ 24/48
▷ Peixoto et al., 2015	▷ Brazil	▷ n.i.	▷ Jundiai, São Paulo State; ref. Cardozo et al., 2011; list. *Scinax hiemalis*	▷ 2,4	▷ 24/48
▶ Gruber et al., 2017[P,M]	▶ Brazil	▶ 3♂	▶ Mogi Guaçu, São Paulo State (22°21′37″S, 57°07′07″W)	▶ 1,2,4,6, 9,12–14	▶ 24/48†
▶ Present study	▶ Brazil	▶ 3♂	▶ Mogi Guaçu, São Paulo State (22°21′37″S, 57°07′07″W)	▶ 1,2	▶ 24/48
Ololygon littoralis					
▶ Cardozo et al., 2011[P,M]	▶ Brazil	▶ 3♂	▶ Ubatuba, São Paulo State; rep. *Scinax littoralis*	▶ 1,2,4	▶ 24/48
▷ Peixoto et al., 2015	▷ Brazil	▷ n.i.	▷ Ubatuba, São Paulo State; ref. Cardozo et al., 2011; list. *Scinax littoralis*	▷ 2,4	▷ 24/48
▶ Gruber et al., 2017[P,M]	▶ Brazil	▶ 1♂	▶ Bertioga, São Paulo State (23°56′27″S, 45°19′48″W)	▶ 1,2,4,6, 9,12–14	▶ 24/48†
▶ Gruber et al., 2017[P,M]	▶ Brazil	▶ 1♂	▶ Ubatuba, São Paulo State (23°26′19″S, 45°05′25″W)	▶ 1,2,4,6, 9,12–14	▶ 24/48†
▶ Present study	▶ Brazil	▶ 1♂	▶ Ubatuba, São Paulo State (23°26′19″S, 45°05′25″W	▶ 1,2	▶ 24/48
Ololygon longilinea					
▶ Cardozo et al., 2011[P,M]	▶ Brazil	▶ 2♂	▶ Poços de Caldas, Minas Gerais State; rep. *Scinax longilineus*	▶ 1,2,4	▶ 24/48
▷ Peixoto et al., 2015	▷ Brazil	▷ n.i.	▷ Poços de Caldas, Minas Gerais State; ref. Cardozo et al., 2011; list. *Scinax longilineus*	▷ 2,4	▷ 24/48
▶ Present study	▶ Brazil	▶ 2♂	▶ Poços de Caldas, Minas Gerais State	▶ 1,2	▶ 24/48
Ololygon obtriangulata					
▶ Cardozo et al., 2011[P,M]	▶ Brazil	▶ 1♂	▶ São José do Barreiro, São Paulo State; rep. *Scinax obtriangulatus*	▶ 1,2,4	▶ 24/48
▷ Peixoto et al., 2015	▷ Brazil	▷ n.i.	▷ São José do Barreiro, São Paulo State; ref. Cardozo et al., 2011; list. *Scinax obtriangulatus*	▷ 2,4	▷ 24/48
▶ Gruber et al., 2017[P,M]	▶ Brazil	▶ 1♀	▶ Biritiba Mirim, São Paulo State (23°34′17″S, 46°02′15″W)	▶ 1,2,4,6, 9,12–14	▶ 24/48†
▶ Present study	▶ Brazil	▶ 1♀	▶ Biritiba Mirim, São Paulo State (23°34′17″S, 46°02′15″W)	▶ 1,2	▶ 24/48
Ololygon perpusilla					
▶ Bogart, 1973[P]	▶ Brazil	▶ n.i.	▶ southeastern Brazil; loc.n.i.; rep. *Hyla perpusilla*	▶ 1	▶ 24/48†
▷ Cole, 1974	▷ n.i.	▷ n.i.	▷ ref. Bogart, 1973; list. *Hyla perpusilla*	▷ n.i.	▷ 24/n.i.
▷ Morescalchi, 1979	▷ n.i.	▷ n.i.	▷ ref. Bogart, 1973; list. *Hyla perpusilla*	▷ n.i.	▷ 24/n.i.
▷ King, 1990	▷ S.America	▷ n.i.	▷ ref. Bogart, 1973; ctry.n.sp.; list. *Hyla perpusilla*	▷ n.i.	▷ 24/48
▷ Kuramoto, 1990	▷ S.C.America	▷ n.i.	▷ ref. Bogart, 1973; ctry.n.sp.	▷ 1	▷ 24/n.i.
▷ Nunes & Fagundes, 2008b	▷ n.i.	▷ n.i.	▷ ref. Bogart, 1973; list. *Scinax perpusillus*	▷ n.i.	▷ 24/48
▷ Catroli, 2008	▷ n.i.	▷ n.i.	▷ ref. Bogart, 1973; list. *Scinax perpusillus*	▷ n.i.	▷ 24/n.i.
▷ Catroli & Kasahara, 2009	▷ n.i.	▷ n.i.	▷ ref. Bogart, 1973; list. *Scinax perpusillus*	▷ n.i.	▷ 24/n.i.
▷ Peixoto et al., 2015	▷ Brazil	▷ n.i.	▷ southeastern Brazil; ref. Bogart, 1973; list. *Scinax perpusillus*	▷ n.i.	▷ 24/48
▶ Present study	▶ Brazil	▶ n.i.	▶ southeastern Brazil; loc.u.	▶ 1	▶ 24/48
▶ Present study	▶ Brazil	▶ 1♀, 1s.n.i.	▶ Recreio, São Paulo State	▶ 1	▶ 24/48
Ololygon rizibilis					
▷ Catroli, 2008	▷ n.i.	▷ n.i.	▷ ref. Leme et al., 2004; list. *Scinax rizibilis*	▷ n.i.	▷ 24/n.i.
▶ Cardozo et al., 2011[P,M]	▶ Brazil	▶ 5♂3♀	▶ Riberão Branco, São Paulo State; rep. *Scinax rizibilis*	▶ 1,2,4	▶ 24/48
▷ Peixoto et al., 2015	▷ Brazil	▷ n.i.	▷ Riberão Branco, São Paulo State; ref. Cardozo et al., 2011; list. *Scinax rizibilis*	▷ 2,4	▷ 24/48
▶ Present study	▶ Brazil	▶ 5♂3♀	▶ Riberão Branco, São Paulo State	▶ 1,2	▶ 24/48
Ololygon trapicheiroi					
▷ Catroli, 2008	▷ n.i.	▷ n.i.	▷ ref. Leme et al., 2004; list. *Scinax trapicheiroi*	▷ n.i.	▷ 24/n.i.
▶ Cardozo et al., 2011[P,M]	▶ Brazil	▶ 1♀	▶ Ubatuba, São Paulo State; rep. *Scinax trapicheiroi*	▶ 1	▶ 24/48
▷ Peixoto et al., 2015	▷ Brazil	▷ n.i.	▷ Ubatuba, São Paulo State; ref. Cardozo et al., 2011; err.cit.meth.; list. *Scinax trapicheiroi*	▷ 4	▷ 24/48
▶ Present study	▶ Brazil	▶ 1♀	▶ Ubatuba, São Paulo State;	▶ 1	▶ 24/48
Ololygon tripui					
▶ Peixoto et al., 2016[P,M]	▶ Brazil	▶ 9♂3♀	▶ Parque Estadual Serra do Brigadeiro, Ervália Municipality, Minas Gerais State (20°51′52″S, 42°31′17″W)	▶ 1,2,4, 13,15	▶ 24/48
▶ Peixoto et al., 2016[P,M]	▶ Brazil	▶ 6♂2♀	▶ Mata do Sossego, Simonésia Municipality, Minas Gerais State (20°04′22″S, 42°04′12″W)	▶ 1,2,4, 13,15	▶ 24/48
▶ Peixoto et al., 2016[P,M]	▶ Brazil	▶ 6♂3♀	▶ Usina da Fumaça, Muriaé Municipality, Minas Gerais State (21°00′58″S, 42°26′36″W)	▶ 1,2,4, 13,15	▶ 24/48
▶ Peixoto et al., 2016[P,M]	▶ Brazil	▶ 3♂	▶ Estaçao Ecológica do Tripuí, Ouro Preto Municipality, Minas Gerais State (20°23′22″S, 43°32′20″W)	▶ 1,2,4, 13,15	▶ 24/48

Author(s) ▶ original data[a] ▷ review articles	Country/ Region[b]	Speci-mens[c]	Locality of sampling, references cited, and remarks[d]	Methods applied[e]	2n/FN[f]
Ololygon v-signata					
▶ Peixoto et al., 2015[P]	▶ Brazil	▶ 3♂	▶ São Pedro da Serra, Rio de Janeiro State (22°18′26″S, 42°18′16″W); rep. *Scinax v-signatus*	▶ 1,2,4, 6,9,15	▶ 24/48
***Ololygon* sp. 1 (group *perpusilla*)**					
▶ Cardozo et al., 2011[P,M]	▶ Brazil	▶ 1♂1♀	▶ Paratí, Rio de Janeiro State; rep. *Scinax* sp. 1 (*perpusillus* group)	▶ 1,4	▶ 24/48
▷ Peixoto et al., 2015	▷ Brazil	▷ n.i.	▷ Paratí, Rio de Janeiro State; ref. Cardozo et al., 2011; err.cit.meth.; list. *Scinax* sp. 1	▷ 4	▷ 24/48
▶ Present study	▶ Brazil	▶ 1♂1♀	▶ Paratí, Rio de Janeiro State	▶ 1	▶ 24/48
***Ololygon* sp. 2 (group *perpusilla*)**					
▶ Cardozo et al., 2011[P,M]	▶ Brazil	▶ 1♂1♀ 1J	▶ Ilha do Cardoso, São Paulo State; rep. *Scinax* sp. 2 (*perpusillus* group)	▶ 1,2,4	▶ 24/48
▷ Peixoto et al., 2015	▷ Brazil	▷ n.i.	▷ Ilha do Cardoso, São Paulo State; ref. Cardozo et al., 2011; list. *Scinax* sp. 2	▷ 2,4	▷ 24/48
▶ Present study	▶ Brazil	▶ 1♂1♀	▶ Ilha do Cardoso, São Paulo State	▶ 1,2	▶ 24/48
***Ololygon* sp. (group *perpusilla*)**					
▶ Peixoto et al., 2015[P]	▶ Brazil	▶ 10♂1♀	▶ Pocrane, Minas Gerais State (19°38′31″S, 41°41′24″W); rep. *Scinax* sp. (gr. *perpusillus*)	▶ 1,2,4, 6,9,15	▶ 24/48
▶ Peixoto et al., 2015[P]	▶ Brazil	▶ 1♂	▶ Caratinga, Minas Gerais State (19°43′21″S, 41°48′47″W); rep. *Scinax* sp. (gr. *perpusillus*)	▶ 1,2,4, 6,9,15	▶ 24/48
HYLIDAE / Scinaxinae / *Scinax*					
Scinax acuminatus					
▷ Catroli, 2008	▷ n.i.	▷ n.i.	▷ ref. Cardozo & Baldo, 2006	▷ n.i.	▷ 24/n.i.
▶ Cardozo et al., 2011[P,M]	▶ Argentina	▶ 2♂	▶ Charata, Chacabuco Department, Chaco Province	▶ 1,2,4	▶ 24/48
▶ Cardozo et al., 2011[P,M]	▶ Argentina	▶ 3♂3♀	▶ El Oscuro, Curuzú Cuatiá Department, Corrientes Province	▶ 1,2,4	▶ 24/48
▷ Peixoto et al., 2015	▷ Argentina	▷ n.i.	▷ Provinces of Chaco and Corrientes; ref. Cardozo et al., 2011	▷ 2,4	▷ 24/48
Scinax alter					
▶ Nunes & Fagundes, 2008b[P]	▶ Brazil	▶ 2♂	▶ Estação Biológica de Santa Lúcia, Santa Teresa, Espírito Santo State	▶ 1	▶ 24/48
▶ Nunes & Fagundes, 2008b[P]	▶ Brazil	▶ 1♂	▶ Parque Estadual Paulo César Vinha, Guarapari, Espírito Santo State	▶ 1	▶ 24/48
▶ Nunes & Fagundes, 2008b[P]	▶ Brazil	▶ 1♂	▶ Lagoa de Samarco, Anchieta, Espírito Santo State	▶ 1	▶ 24/48
▷ Catroli, 2008	▷ n.i.	▷ n.i.	▷ ref. Leme et al., 2004	▷ n.i.	▷ 24/n.i.
▷ Catroli & Kasahara, 2009	▷ n.i.	▷ n.i.	▷ ref. Nunes & Fagundes, 2008b	▷ 17	▷ 24/n.i.
▷ Hernández-Guzmán et al., 2011	▷ n.i.	▷ n.i.	▷ ref. Nunes & Fagundes, 2008b	▷ n.i.	▷ 24/n.i.
▶ Cardozo et al., 2011[P,M]	▶ Brazil	▶ 1♂	▶ Ubatuba, São Paulo State	▶ 1,2,4	▶ 24/48
▶ Cardozo et al., 2011[P,M]	▶ Brazil	▶ 1♀	▶ Rancho Queimado, Santa Catarina State	▶ 1,2,4	▶ 24/48
▷ Peixoto et al., 2015	▷ Brazil	▷ n.i.	▷ Santa Teresa, Guarapari and Anchieta, Espírito Santo State, ref. Nunes & Fagundes, 2008b	▷ n.i.	▷ 24/48
▷ Peixoto et al., 2015	▷ Brazil	▷ n.i.	▷ Ubatuba, São Paulo State and Rancho Queimado, Santa Catarina State; ref. Cardozo et al., 2011	▷ 2,4	▷ 24/48
▶ Present study	▶ Brazil	▶ 1♀	▶ Rancho Queimado, Santa Catarina State	▶ 1,2	▶ 24/48
Scinax baumgardneri*					
▶ Present study	▶ Peru	▶ 1n.i.	▶ Bosque Nacional de Iparia, Huánuco Department	▶ 1	▶ 24/48
Scinax boesemani					
▶ Nogueira et al., 2015b[P,M]	▶ Brazil	▶ 1♂	▶ Bragança, Pará State	▶ 1,2,4,6,9	▶ 24/48
▷ Peixoto et al., 2015	▷ Brazil	▷ n.i.	▷ Bragança, Pará State; ref. Nogueira et al., 2015b	▷ 2,4	▷ 24/48
▶ Present study	▶ Brazil	▶ 1♂	▶ Camp Coloso, Manaus Project	▶ 1	▶ 24/48
Scinax boulengeri*					
▶ Duellman, 1967[N]	▶ Panama	▶ 2♂	▶ El Real, Darien Province; rep. *Hyla boulengeri*	▶ 1	▶ 24‡/n.i.
▶ Duellman, 1967[N]	▶ Costa Rica	▶ 1♂	▶ 1 km S Puerto Viejo; rep. *Hyla boulengeri*	▶ 1	▶ 24‡/n.i.
▷ Duellman, 1970	▷ C.America	▷ n.i.	▷ ref.n.i.; ctry.n.sp.; list. *Hyla boulengeri*	▷ 1	▷ 24‡/n.i.
▷ Foresti, 1972	▷ n.i.	▷ n.i.	▷ ref. Duellman, 1967; list. *Hyla boulengeri*	▷ n.i.	▷ 24‡/n.i.
▷ Kuramoto, 1990	▷ S.C.America	▷ n.i.	▷ ref. Duellman, 1967; ctry.n.sp.; list. *Ololygon boulengeri*	▷ 1	▷ 24/n.i.
▷ Catroli, 2008	▷ n.i.	▷ n.i.	▷ ref. Duellman, 1967	▷ n.i.	▷ 24/n.i.
▷ Catroli & Kasahara, 2009	▷ n.i.	▷ n.i.	▷ ref. Duellman, 1967	▷ n.i.	▷ 24/n.i.
▷ Peixoto et al., 2015	▷ Panama	▷ n.i.	▷ Darien Province; ref. Duellman, 1967	▷ n.i.	▷ 24/n.i.
▷ Peixoto et al., 2015	▷ Costa Rica	▷ n.i.	▷ Heredia Province; ref. Duellman, 1967	▷ n.i.	▷ 24/n.i.
▶ Present study	▶ Costa Rica	▶ 2♂	▶ Valle de Talamanca, Limón Province (09°35′N, 83°07′W)	▶ 1,2,4,5,7	▶ 24/48
▶ Present study	▶ Costa Rica	▶ 3♂	▶ vicinity of Quepos, Puntarenas Province, 8 m (09°25′N, 84°09′W)	▶ 1,2,4,5,7	▶ 24/48
Scinax caissara					
▶ Lourenço et al., 2016[P,M]	▶ Brazil	▶ 2♂	▶ Cananéia Municipality, São Paulo State	▶ 1,4	▶ 24/48
Scinax caldarum					
▶ Gruber et al., 2017[P,M]	▶ Brazil	▶ 1♂	▶ Poços de Caldas, Minas Gerais State (21°47′18″S, 46°33′45″W)	▶ 1,2,4,6, 9,12–14	▶ 24/48†

Author(s) ▶ original data[a] ▷ review articles	Country/ Region[b]	Speci- mens[c]	Locality of sampling, references cited, and remarks[d]	Methods applied[e]	2n/FN[f]
Scinax camposseabrai					
▶ Nogueira et al., 2015b[P,M]	▶ Brazil	▶ 3♂1♀	▶ Maracás, Bahia State	▶ 1,2,4,6,9	▶ 24/48
▷ Peixoto et al., 2015	▷ Brazil	▷ n.i.	▷ Maracás, Bahia State; ref. Nogueira et al., 2015b	▷ 2,4	▷ 24/48
Scinax constrictus					
▶ Sousa, 2007	▶ Brazil	▶ not known	▶ not known	▶ 1	▶ 22/44
▶ De Oliveira et al., 2012[P]	▶ Brazil	▶ 9s.n.i.	▶ Barrio Alto and Nazário Municipalities, Goiás State	▶ 1	▶ 22/44
Scinax crospedospilus					
▶ Gruber et al., 2017[P,M]	▶ Brazil	▶ 2♂	▶ Mogi das Cruzes, São Paulo State (23°31'29"S, 46°11'14"W)	▶ 1,2,4,6, 9,12–14	▶ 24/48†
Scinax curicica					
▶ Cardozo et al., 2011[P,M]	▶ Brazil	▶ 1♂	▶ Serra do Cipó, Minas Gerais State	▶ 1,2,4	▶ 24/48
▷ Peixoto et al., 2015	▷ Brazil	▷ n.i.	▷ Serra do Cipó, Minas Gerais State; ref. Cardozo et al., 2011	▷ 2,4	▷ 24/48
▶ Present study	▶ Brazil	▶ 1♂	▶ Serra do Cipó, Minas Gerais State	▶ 1,2	▶ 24/48
Scinax cuspidatus					
▶ Foresti, 1972[P,M]	▶ Brazil	▶ 1♀	▶ Grumari, (Guanabara) Rio de Janeiro State; rep. *Hyla cuspidata*	▶ 1	▶ 24/48†
▷ Catroli, 2008	▷ n.i.	▷ n.i.	▷ ref. Foresti, 1972	▷ n.i.	▷ 24/n.i.
▷ Catroli & Kasahara, 2009	▷ n.i.	▷ n.i.	▷ ref. Foresti, 1972	▷ n.i.	▷ 24/n.i.
▶ Present study	▶ Brazil	▶ 1♂	▶ Tijuca, Rio de Janeiro State	▶ 1	▶ 24/48
Scinax duartei					
▶ Foresti, 1972[P,M]	▶ Brazil	▶ 1♀	▶ Itatiaia, Rio de Janeiro State; rep. *Hyla duartei*	▶ 1	▶ 24/48†
▷ Catroli, 2008	▷ n.i.	▷ n.i.	▷ ref. Foresti, 1972	▷ n.i.	▷ 24/n.i.
▷ Catroli & Kasahara, 2009	▷ n.i.	▷ n.i.	▷ ref. Foresti, 1972	▷ n.i.	▷ 24/n.i.
▶ Cardozo et al., 2011[P,M]	▶ Brazil	▶ 1♂	▶ Itamonte, Minas Gerais State	▶ 1,2,4	▶ 24/48
▶ Cardozo et al., 2011[P,M]	▶ Brazil	▶ 1♂	▶ Petrópolis, Rio de Janeiro State	▶ 1,2,4	▶ 24/48
▷ Peixoto et al., 2015	▷ Brazil	▷ n.i.	▷ Itamonte, Minas Gerais State and Petrópolis, Rio de Janeiro State; ref. Cardozo et al., 2011	▷ 2,4	▷ 24/48
▶ Present study	▶ Brazil	▶ 1♂	▶ Petrópolis, Rio de Janeiro State	▶ 1,2	▶ 24/48
Scinax elaeochrous					
▶ Duellman, 1967[N]	▶ Costa Rica	▶ 3♂	▶ 1 km S Puerto Viejo, Heredia Province; rep. *Hyla elaeochroa*	▶ 1	▶ 24‡/n.i.
▷ Duellman, 1970	▷ C.America	▷ n.i.	▷ ref.n.i.; ctry.n.sp.; list. *Hyla elaeochroa*	▷ 1	▷ 24‡/n.i.
▷ Foresti, 1972	▷ n.i.	▷ n.i.	▷ ref. Duellman & Cole, 1965/err.ref.; list. *Hyla elaeochroa*	▷ n.i.	▷ 24‡/n.i.
▶ Anderson, 1986[I]	▶ Costa Rica	▶ 2♀	▶ Pavones; rep. *Ololygon elaeochroa*	▶ 1,2,4	▶ 24/48†
▷ King, 1990	▷ C.America	▷ n.i.	▷ ref. Anderson, 1986; ctry.n.sp.; list. *Ololygon elaeochroa*	▷ 2,4	▷ 24/48
▷ Kuramoto, 1990	▷ S.C.America	▷ n.i.	▷ ref. Duellman, 1967; ctry.n.sp.; list. *Ololygon elaeochroa*	▷ 1	▷ 24/n.i.
▷ Anderson, 1991[I]	▷ Neotropic	▷ n.i.	▷ ref. Anderson, 1986; ctry.n.sp.; list. *Ololygon elaeochroa*	▷ 2,4	▷ 24/48†
▷ Catroli, 2008	▷ n.i.	▷ n.i.	▷ ref. Duellman, 1967; Anderson, 1991	▷ n.i.	▷ 24/n.i.
▷ Catroli & Kasahara, 2009	▷ n.i.	▷ n.i.	▷ ref. Duellman, 1967; Anderson, 1991	▷ 17	▷ 24/n.i.
▷ Peixoto et al., 2015	▷ Costa Rica	▷ n.i.	▷ Heredia Province; ref. Duellman, 1967	▷ n.i.	▷ 24/n.i.
▶ Present study	▶ Costa Rica	▶ 4♂2♀ 10J	▶ Cahuita National Park, Limón Province (09°43'N, 82°50'W)	▶ 1,2,4, 5,7,8	▶ 24/48
▶ Present study	▶ Costa Rica	▶ 4♂1♀ 1J	▶ vicinity of Tanque, Alajuela Province (10°28'N, 84°35'W)	▶ 1,2,4, 5,7,8	▶ 24/48
Scinax eurydice					
▷ Catroli, 2008	▷ n.i.	▷ n.i.	▷ ref. Leme et al., 2004	▷ n.i.	▷ 24/n.i.
▶ Cardozo et al., 2011[P,M]	▶ Brazil	▶ 1♂	▶ Jundiaí, São Paulo State	▶ 1,2,4	▶ 24/48
▶ Cardozo et al., 2011[P,M]	▶ Brazil	▶ 1♂2♀	▶ São Jose dos Campos, São Paulo State	▶ 1,2,4	▶ 24/48
▶ Nogueira et al., 2015a[P,M]	▶ Brazil	▶ 3♂2J	▶ Jequié, Bahia State (13°51'04"S, 40°04'52"W)	▶ 1,2,4, 7,9,13	▶ 24/48
▷ Peixoto et al., 2015	▷ Brazil	▷ n.i.	▷ Jundiaí and São Jose dos Campos, São Paulo State; ref. Cardozo et al., 2011	▷ 2,4	▷ 24/48
▷ Peixoto et al., 2015	▷ Brazil	▷ n.i.	▷ Jequié, Bahia State; ref. Nogueira et al., 2015[a]	▷ 2,4	▷ 24/48
▶ Gruber et al., 2017[P,M]	▶ Brazil	▶ 1♂	▶ Serra do Japi, Jundiaí, São Paulo State (37°25'81"S, 46°05'36"W)	▶ 1,2,4,6, 9,12–14	▶ 24/48†
▶ Present study	▶ Brazil	▶ 1♂	▶ Serra do Japi, Jundiaí, São Paulo State (37°25'81"S, 46°05'36"W)	▶ 1,2	▶ 24/48
Scinax funereus					
▶ Bogart & Bogart, 1971[N]	▶ Peru	▶ n.i.	▶ Río Pachitea, Bosque Nacional de Iparía, 1200 m, Huánuco Department; rep. *Hyla funerea*	▶ 1	▶ 24/n.i.
▷ Kuramoto, 1990	▷ S.C.America	▷ n.i.	▷ ref. Bogart & Bogart, 1971; list. *Ololygon funerea*	▷ 1	▷ 24/n.i.
▷ Catroli, 2008	▷ n.i.	▷ n.i.	▷ ref. Bogart & Bogart, 1971	▷ n.i.	▷ 24/n.i.
▷ Catroli & Kasahara, 2009	▷ n.i.	▷ n.i.	▷ ref. Bogart & Bogart, 1971	▷ n.i.	▷ 24/n.i.
▷ Peixoto et al., 2015	▷ Peru	▷ n.i.	▷ Huánuco Department; ref. Duellman, 1967/err.ref.	▷ n.i.	▷ 24/n.i.
Scinax fuscomarginatus					
▶ Rabello, 1970[P]	▶ n.i.	▶ 6♂	▶ loc.n.i.; rep. *Hyla fuscomarginata*	▶ 1	▶ 24/48†
▷ Foresti, 1972	▷ n.i.	▷ n.i.	▷ ref. Rabello, 1969, 1970; list. *Hyla fuscomarginata*	▷ n.i.	▷ 24/n.i.
▷ Foresti, 1972	▷ n.i.	▷ n.i.	▷ ref. Beçak et al., 1970; list. *Hyla parkeri*	▷ n.i.	▷ 24/n.i.

Author(s) ▶ original data[a] ▷ review articles	Country/ Region[b]	Speci-mens[c]	Locality of sampling, references cited, and remarks[d]	Methods applied[e]	2n/FN[f]
▷ Cole, 1974	▷ n.i.	▷ n.i.	▷ ref. Rabello, 1970; list. *Hyla fuscomarginata*	▷ n.i.	▷ 24/n.i.
▷ King, 1990	▷ S.America	▷ n.i.	▷ ref. Rabello, 1970; ctry.n.sp.; list. *Hyla fuscomarginata*	▷ n.i.	▷ 24/n.i.
▷ Kuramoto, 1990	▷ S.C.America	▷ n.i.	▷ ref. Beçak et al., 1970b; Rabello, 1970; ctry.n.sp.; list. *Ololygon fuscomarginata*	▷ 1	▷ 24/n.i.
▷ Nunes & Fagundes, 2008b	▷ n.i.	▷ n.i.	▷ ref. Rabello, 1970	▷ n.i.	▷ 24/48
▷ Catroli, 2008	▷ n.i.	▷ n.i.	▷ ref. Rabello, 1970; Rabello et al., 1971	▷ n.i.	▷ 24/n.i.
▷ Catroli & Kasahara, 2009	▷ n.i.	▷ n.i.	▷ ref. Rabello, 1970; Rabello et al., 1971	▷ n.i.	▷ 24/n.i.
▶ Cardozo et al., 2011[P,M]	▶ Argentina	▶ 9♂2♀	▶ Reserva Santa Maria, Ituzaingó, Ituzaingó Department, Corrientes Province	▶ 1,4	▶ 24/48
▷ Peixoto et al., 2015	▷ n.i.	▷ n.i.	▷ ref. Rabello, 1970	▷ n.i.	▷ 24/n.i.
▷ Peixoto et al., 2015	▷ Argentina	▷ n.i.	▷ Corrientes Province; ref. Cardozo et al., 2011; err.cit.meth.	▷ 2,4	▷ 24/48
Scinax fuscovarius					
▶ Rabello, 1970[P]	▶ n.i.	▶ 5♂1♀	▶ loc.n.i.; rep. *Hyla trachythorax*	▶ 1	▶ 24/48†
▶ Rabello, 1970[P]	▶ n.i.	▶ 1♂	▶ loc.n.i.; rep. *Hyla trachythorax*; 1 extra chromosome in a gonad	▶ 1	▶ 25/50†
▷ Foresti, 1972	▷ n.i.	▷ n.i.	▷ ref. Rabello, 1969, 1970; list. *Hyla {trachytorax}*	▷ n.i.	▷ 24/n.i.
▶ Anderson, 1986[I]	▶ Paraguay	▶ 1♂	▶ Itapua; rep. *Ololygon fuscovaria*	▶ 1,2,4	▶ 24/48†
▷ King, 1990	▷ C.America	▷ n.i.	▷ ref. Anderson, 1986; ctry.n.sp.; list. *Ololygon fuscovaria*	▷ 2,4	▷ 24/48
▷ King, 1990	▷ S.America	▷ n.i.	▷ ref. Rabello, 1970; ctry.n.sp.; list. *Hyla trachythorax*	▷ n.i.	▷ 24/n.i.
▷ Kuramoto, 1990	▷ S.C.America	▷ n.i.	▷ ref. Rabello, 1970; ctry.n.sp.; list. *Ololygon trachythorax*	▷ 1	▷ 24/n.i.
▷ Anderson, 1991[I]	▷ Neotropic	▷ n.i.	▷ ref. Anderson, 1986; ctry.n.sp.; list. *Ololygon fuscovaria*	▷ 2,4	▷ 24/48†
▶ Baldissera et al., 1993[P]	▶ Brazil	▶ 3♂	▶ Rio Casca, Minas Gerais State; rep. *Hyla fuscovaria*	▶ 1,4	▶ 24/48†
▶ Baldissera et al., 1993[P]	▶ Brazil	▶ 1♀	▶ Rio Claro, São Paulo State; rep. *Hyla fuscovaria*	▶ 1,4	▶ 24/48†
▶ Kasahara et al., 2003[P,M]	▶ Brazil	▶ 4♂4♀ 1J	▶ Rio Claro, São Paulo State	▶ 1,2,4,7, 9,12,13	▶ 24/48†
▶ Kasahara et al., 2003[P,M]	▶ Brazil	▶ 1♂	▶ Santa Maria da Serra, São Paulo State	▶ 1,2,4,7, 9,12,13	▶ 24/48†
▶ Kasahara et al., 2003[P,M]	▶ Brazil	▶ 1J	▶ Itatiba, São Paulo State	▶ 1,2,4,7, 9,12,13	▶ 24/48†
▷ Nunes & Fagundes, 2008b	▷ n.i.	▷ n.i.	▷ ref. Anderson, 1991	▷ n.i.	▷ 24/48
▷ Catroli, 2008	▷ n.i.	▷ n.i.	▷ ref. Rabello, 1970; Anderson, 1991; Baldissera & Batistic 1992; Leme et al., 2004; Baldissera et al., 1993; Kasahara et al., 2003; Cardozo & Baldo, 2006; Trentin et al., 2007	▷ n.i.	▷ 24/n.i.
▷ Catroli, 2008	▷ n.i.	▷ n.i.	▷ ref. Rabello, 1970; 1 extra chromosome	▷ n.i.	▷ 25/n.i.
▷ Catroli & Kasahara, 2009	▷ n.i.	▷ n.i.	▷ ref. Rabello, 1970; Anderson, 1991; Baldissera et al., 1993; Kasahara et al., 2003	▷ 17	▷ 24/n.i.
▷ Catroli & Kasahara, 2009	▷ n.i.	▷ n.i.	▷ ref. Rabello, 1970; err.cit.meth.; 1 extra chromosome	▷ 17	▷ 25/n.i.
▶ Neri et al., 2009[N]	▶ Brazil	▶ n.i.	▶ Jequié Municipality, Bahia State	▶ 2,4	▶ 24/48
▷ Hernández-Guzmán et al., 2011	▷ n.i.	▷ n.i.	▷ ref. Baldissera et al., 1993; list. *Hyla fuscovaria*	▷ n.i.	▷ 24/n.i.
▶ Cardozo et al., 2011[P,M]	▶ Argentina	▶ 1♀	▶ El Alto, Sierras Department, Catamarca Province	▶ 1,2,4	▶ 24/48
▶ Cardozo et al., 2011[P,M]	▶ Argentina	▶ 1♂	▶ Laguna Rodeo, Parque Provincial Yala, Manuel Belgrano Department, Jujuy Province	▶ 1,2,4	▶ 24/48
▶ Cardozo et al., 2011[P,M]	▶ Argentina	▶ 1♂1♀	▶ Balneario del Arroyo Cuña Pirú, Aristóbulo del Valle, Cainguas Department, Misiones Province	▶ 1,2,4	▶ 24/48
▶ Cardozo et al., 2011[P,M]	▶ Argentina	▶ 1♂	▶ Barrio Villa Sarita, Posadas, Capital Department, Misiones Province	▶ 1,2,4	▶ 24/48
▶ Cardozo et al., 2011[P,M]	▶ Argentina	▶ 8♂1♀	▶ Campus UNaM, Villa Lanús, Capital Department, Misiones Province	▶ 1,2,4	▶ 24/48
▶ Cardozo et al., 2011[P,M]	▶ Argentina	▶ 2♂	▶ Posadas, Capital Department, Misiones Province	▶ 1,2,4	▶ 24/48
▶ Cardozo et al., 2011[P,M]	▶ Argentina	▶ 1♀	▶ Parque Provincial Cametti, Andresito, Gral. Manuel Belgrano Department, Misiones Province	▶ 1,2,4	▶ 24/48
▶ Cardozo et al., 2011[P,M]	▶ Argentina	▶ 1♀	▶ Refugio Tangará, El Soberbio, Guaraní Department, Misiones Province	▶ 1,2,4	▶ 24/48
▶ Cardozo et al., 2011[P,M]	▶ Argentina	▶ 2♂1♀	▶ Encón Grande, Rosario de Lerma Department, Salta Province	▶ 1,2,4	▶ 24/48
▶ Cardozo et al., 2011[P,M]	▶ Brazil	▶ 1♂	▶ Condoi, Paraná State	▶ 1,2,4	▶ 24/48
▶ Cardozo et al., 2011[P,M]	▶ Brazil	▶ 2♀1J	▶ Rio Claro, São Paulo State	▶ 1,2,4	▶ 24/48
▶ Cardozo et al., 2011[P,M]	▶ Brazil	▶ 1♂	▶ Santa Maria da Serra, São Paulo State	▶ 1,2,4	▶ 24/48
▶ Cardozo et al., 2011[P,M]	▶ Brazil	▶ 1♂	▶ Socorro, São Paulo State	▶ 1,2,4	▶ 24/48
▷ Peixoto et al., 2015	▷ n.i.	▷ n.i.	▷ ref. Rabello, 1970; 1 extra chromosome	▷ n.i.	▷ 25/n.i.
▷ Peixoto et al., 2015	▷ n.i.	▷ n.i.	▷ ref. King, 1990; Anderson, 1991	▷ n.i.	▷ 24/n.i.
▷ Peixoto et al., 2015	▷ Brazil	▷ n.i.	▷ Rio Casca, Minas Gerais State and Rio Claro, São Paulo State; ref. Baldissera et al., 1993	▷ n.i.	▷ 24/n.i.
▷ Peixoto et al., 2015	▷ Brazil	▷ n.i.	▷ Rio Claro, Santa Maria da Serra and Itatiba, São Paulo State; ref. Kasahara et al., 2003	▷ 2,4	▷ 24/48
▷ Peixoto et al., 2015	▷ Argentina	▷ n.i.	▷ Provinces of Catamarca, Jujuy, Misiones and Salta; ref. Cardozo et al., 2011	▷ 2,4	▷ 24/48
▷ Peixoto et al., 2015	▷ Brazil	▷ n.i.	▷ Condoi, Paraná State and Rio Claro, Santa Maria da Serra and Socorro, São Paulo State; ref. Cardozo et al., 2011	▷ 2,4	▷ 24/48
▶ Gruber et al., 2017[P,M]	▶ Brazil	▶ 1♀	▶ Biritiba Mirim, São Paulo State (23°34′17″S, 46°02′15″W)	▶ 1,2,4,6, 9,12–14	▶ 24/48†
▶ Gruber et al., 2017[P,M]	▶ Brazil	▶ 1♂	▶ Socorro, São Paulo State (22°35′29″S, 46°31′44″W)	▶ 1,2,4,6, 9,12–14	▶ 24/48†
▶ Present study	▶ Brazil	▶ 1♀	▶ Caraguatatuba, São Paulo State	▶ 1	▶ 24/48
▶ Present study	▶ Brazil	▶ 5♂1♀	▶ Campos Elyses, Botucatu, São Paulo State	▶ 1	▶ 24/48
▶ Present study	▶ Brazil	▶ 1♂1♀	▶ Pedra Branca, Grota do Berro, Botucatu, São Paulo State	▶ 1	▶ 24/48
▶ Present study	▶ Brazil	▶ 13♂	▶ Fazenda São Joaquim, São Paulo State	▶ 1,2,5, 8,14	▶ 24/48

Author(s) ▶ original data[a] ▷ review articles	Country/ Region[b]	Speci- mens[c]	Locality of sampling, references cited, and remarks[d]	Methods applied[e]	2n/FN[f]
Scinax garbei					
▶ Bogart & Bogart, 1971[N]	▶ Peru	▶ n.i.	▶ Río Pachitea, Bosque Nacional de Iparía, 1200 m, Huánuco Department; rep. *Hyla garbei*	▶ 1	▶ 24/n.i.
▷ Kuramoto, 1990	▷ S.C.America	▷ n.i.	▷ ref. Bogart & Bogart, 1971; ctry.n.sp.; list. *Ololygon garbei*	▷ 1	▷ 24/n.i.
▷ Catroli, 2008	▷ n.i.	▷ n.i.	▷ ref. Bogart & Bogart, 1971	▷ n.i.	▷ 24/n.i.
▷ Catroli & Kasahara, 2009	▷ n.i.	▷ n.i.	▷ ref. Bogart & Bogart, 1971	▷ n.i.	▷ 24/n.i.
▶ Nogueira et al., 2015b[P,M]	▶ Brazil	▶ 1♂3♀	▶ Bragança, Pará State	▶ 1,2,4,6,9	▶ 24/48
▷ Peixoto et al., 2015	▷ Peru	▷ n.i.	▷ Huánuco Department; ref. Bogart & Bogart, 1971	▷ n.i.	▷ 24/n.i.
▷ Peixoto et al., 2015	▷ Brazil	▷ n.i.	▷ Bragança, Pará State; ref. Nogueira et al., 2015b	▷ 2,4	▷ 24/48
▶ Present study	▶ Colombia	▶ 1n.i.	▶ Teresita	▶ 1	▶ 24/48
▶ Present study	▶ Peru	▶ 1♂	▶ Bosque Nacional de Iparia, Huánuco Department	▶ 1	▶ 24/48
Scinax granulatus					
▷ Catroli, 2008	▷ n.i.	▷ n.i.	▷ ref. Cardozo & Baldo, 2006	▷ n.i.	▷ 24/n.i.
▶ Souza et al., 2011[N]	▶ Brazil	▶ n.i.	▶ Passo Fundo, Rio Grande do Sul State; rep. *Scinax* cf. *granulatus*	▶ 1,2,4	▶ 24/48[†]
▶ Cardozo et al., 2011[P,M]	▶ Argentina	▶ 3♂	▶ Parque Provincial Moconá, Guaraní Department, Misiones Province	▶ 1,2,4	▶ 24/48
▶ Cardozo et al., 2011[P,M]	▶ Uruguay	▶ 1♂	▶ Playa Penino, San José Department	▶ 1,2,4	▶ 24/48
▷ Peixoto et al., 2015	▷ Argentina	▷ n.i.	▷ Misiones Province; ref. Cardozo et al., 2011	▷ 2,4	▷ 24/48
▷ Peixoto et al., 2015	▷ Uruguay	▷ n.i.	▷ San José Department; ref. Cardozo et al., 2011	▷ 2,4	▷ 24/48
Scinax hayii					
▶ Beçak, 1968[P,M]	▶ Brazil	▶ 3♂5♀	▶ Campos de Jordão and Itapecerica de Serra, São Paulo State; rep. *Hyla hayi*	▶ 1	▶ 24/48[†]
▷ Rabello, 1970	▷ n.i.	▷ n.i.	▷ ref. Beçak, 1967, 1968; list. *Hyla {hayi}*	▷ n.i.	▷ 24/n.i.
▷ Foresti, 1972	▷ n.i.	▷ n.i.	▷ ref. Beçak, 1967, 1968; list. *Hyla {hayi}*	▷ n.i.	▷ 24/n.i.
▷ Cole, 1974	▷ n.i.	▷ n.i.	▷ ref. Beçak, 1968; list. *Hyla {hayi}*	▷ n.i.	▷ 24/n.i.
▷ King, 1990	▷ S.America	▷ n.i.	▷ ref. Beçak, 1968; ctry.n.sp.; list. *Hyla hayii*	▷ n.i.	▷ 24/n.i.
▷ Kuramoto, 1990	▷ S.C.America	▷ n.i.	▷ ref. Beçak, 1968; ctry.n.sp.; list. *Ololygon hayii*	▷ 1	▷ 24/n.i.
▶ Baldissera et al., 1993[P]	▶ Brazil	▶ 2♂	▶ Serra do Japi, Jundiaí, São Paulo State; rep. *Hyla hayii*	▶ 1,4	▶ 24/48[†]
▷ Nunes & Fagundes, 2008b	▷ n.i.	▷ n.i.	▷ ref. Beçak, 1968	▷ n.i.	▷ 24/48
▷ Catroli, 2008	▷ n.i.	▷ n.i.	▷ ref. Beçak, 1968; Baldissera et al., 1993; Leme et al., 2004	▷ n.i.	▷ 24/n.i.
▷ Catroli & Kasahara, 2009	▷ n.i.	▷ n.i.	▷ ref. Beçak, 1968; Baldissera et al., 1993	▷ 17	▷ 24/n.i.
▷ Hernández-Guzmán et al., 2011	▷ n.i.	▷ n.i.	▷ ref. Baldissera et al., 1993; list. *Hyla hayii*	▷ n.i.	▷ 24/n.i.
▶ Cardozo et al., 2011[P,M]	▶ Brazil	▶ 3♂1♀	▶ Camanducaia, Minas Gerais State	▶ 1,2,4	▶ 24/48
▶ Cardozo et al., 2011[P,M]	▶ Brazil	▶ 1♂	▶ Campos do Jordão, São Paulo State	▶ 1,2,4	▶ 24/48
▷ Peixoto et al., 2015	▷ Brazil	▷ n.i.	▷ Campos de Jordão and Itapecerica de Serra, São Paulo State; ref. Beçak, 1968	▷ n.i.	▷ 24/n.i.
▷ Peixoto et al., 2015	▷ Brazil	▷ n.i.	▷ Serra do Japi, Jundiaí, São Paulo State; ref. Baldissera et al., 1993	▷ 4	▷ 24/n.i.
▷ Peixoto et al., 2015	▷ Brazil	▷ n.i.	▷ Camanducaia, Minas Gerais State and Campos do Jordão, São Paulo State, ref. Cardozo et al., 2011	▷ 2,4	▷ 24/48
▶ Gruber et al., 2017[P,M]	▶ Brazil	▶ 1♂	▶ Biritiba Mirim, São Paulo State (23°34′17″S, 46°02′15″W)	▶ 1,2,4,6, 9,12–14	▶ 24/48[†]
▶ Gruber et al., 2017[P,M]	▶ Brazil	▶ 1♂	▶ Mogi das Cruzes, São Paulo State (23°31′29″S, 46°11′14″W)	▶ 1,2,4,6, 9,12–14	▶ 24/48[†]
▶ Present study	▶ Brazil	▶ 1♂	▶ Mogi das Cruzes, São Paulo State (23°31′29″S, 46°11′14″W)	▶ 1,2,4	▶ 24/48
Scinax kautskyi					
▶ Neri et al., 2009[N]	▶ Brazil	▶ n.i.	▶ Jequié Municipality, Bahia State	▶ 2,4	▶ 24/48
Scinax nasicus					
▷ Catroli, 2008	▷ n.i.	▷ n.i.	▷ ref. Cardozo & Baldo, 2006	▷ n.i.	▷ 24/n.i.
▶ Cardozo et al., 2011[P,M]	▶ Argentina	▶ 7♂2♀	▶ Campus UNaM. Villa Lanús, Capital Department, Misiones Province	▶ 1,2,4	▶ 24/48
▶ Cardozo et al., 2011[P,M]	▶ Argentina	▶ 1♂	▶ Ruta Provincial No. 5 and San Francisco River, Pichanal, Orán Department, Salta Province	▶ 1,2,4	▶ 24/48
▶ Cardozo et al., 2011[P,M]	▶ Argentina	▶ 1♂	▶ Vera, Vera Department, Santa Fe Province	▶ 1,2,4	▶ 24/48
▶ Cardozo et al., 2011[P,M]	▶ Argentina	▶ 1♂	▶ Añatuya, Gral. Taboada Department, Santiago del Estero Province	▶ 1,2,4	▶ 24/48
▷ Peixoto et al., 2015	▷ Argentina	▷ n.i.	▷ Provinces of Misiones, Salta, Santa Fe and Santiago del Estero; ref. Cardozo et al., 2011	▷ 2,4	▷ 24/48
Scinax pachycrus					
▶ Foresti, 1972[P,M]	▶ Brazil	▶ 1♂	▶ Jequié, Bahia State; rep. *Hyla {pachychrus}*	▶ 1	▶ 24/48[†]
▶ Nogueira et al., 2015b[P,M]	▶ Brazil	▶ 2♂1♀ 1U	▶ Jequié, Bahia State	▶ 1,2,4,6,9	▶ 24/48
▷ Peixoto et al., 2015	▷ Brazil	▷ n.i.	▷ Jequié, Bahia State; ref. Nogueira et al., 2015b	▷ 2,4	▷ 24/48
Scinax perereca					
▶ Pombal et al., 1995	▶ Brazil	▶ 2♂	▶ Fazendinha São Luis, Ribeirão Branco Municipality, 800 m, São Paulo State (24°13′S, 48°46′W)	▶ 1,4	▶ 24/48[†]
▷ Catroli, 2008	▷ n.i.	▷ n.i.	▷ ref. Leme et al., 2004; Cardozo & Baldo, 2006	▷ n.i.	▷ 24/n.i.
▷ Catroli & Kasahara, 2009	▷ n.i.	▷ n.i.	▷ ref. Pombal et al., 1995	▷ 17	▷ 24/n.i.
▷ Hernández-Guzmán et al., 2011	▷ n.i.	▷ n.i.	▷ ref. Pombal et al., 1995	▷ n.i.	▷ 24/n.i.
▶ Cardozo et al., 2011[P,M]	▶ Argentina	▶ 5♂	▶ Parque Provincial Moconá, Guaraní Department, Misiones Province	▶ 1,2,4	▶ 24/48
▶ Cardozo et al., 2011[P,M]	▶ Argentina	▶ 1♂	▶ Colonia Victoria, San Pedro Department, Misiones Province	▶ 1,2,4	▶ 24/48
▶ Cardozo et al., 2011[P,M]	▶ Argentina	▶ 1♂	▶ Arroyo 3 vueltas, Colonia Victoria, San Pedro Department, Misiones Province	▶ 1,2,4	▶ 24/48

Author(s) ▶ original data[a] ▷ review articles	Country/ Region[b]	Speci-mens[c]	Locality of sampling, references cited, and remarks[d]	Methods applied[e]	2n/FN[f]
▶ Cardozo et al., 2011[P,M]	▶ Brazil	▶ 4♂	▶ Riberão Branco, São Paulo State	▶ 1,2,4	▶ 24/48
▷ Peixoto et al., 2015	▷ Brazil	▷ n.i.	▷ Riberão Branco, São Paulo State; ref. Pombal et al., 1995	▷ 4	▷ 24/48
▷ Peixoto et al., 2015	▷ Argentina	▷ n.i.	▷ Misiones Province; ref. Cardozo et al., 2011	▷ 2,4	▷ 24/48
▷ Peixoto et al., 2015	▷ Brazil	▷ n.i.	▷ Riberão Branco, São Paulo State; ref. Cardozo et al., 2011	▷ 2,4	▷ 24/48
Scinax rostratus					
▶ Bogart, 1973[P]	▶ Peru	▶ n.i.	▶ loc.n.i.; rep. *Hyla rostrata*	▶ 1	▶ 24/48†
▷ Cole, 1974	▷ n.i.	▷ n.i.	▷ ref. Bogart, 1973; list. *Hyla rostrata*	▷ n.i.	▷ 24/n.i.
▷ Morescalchi, 1979	▷ n.i.	▷ n.i.	▷ ref. Bogart, 1973; list. *Hyla rostrata*	▷ n.i.	▷ 24/n.i.
▷ King, 1990	▷ S.America	▷ n.i.	▷ ref. Bogart, 1973; ctry.n.sp.; list. *Hyla rostrata*	▷ n.i.	▷ 24/48
▷ Kuramoto, 1990	▷ S.C.America	▷ n.i.	▷ ref. Bogart, 1973; ctry.n.sp.; list. *Ololygon rostrata*	▷ 1	▷ 24/n.i.
▷ Nunes & Fagundes, 2008b	▷ n.i.	▷ n.i.	▷ ref. Bogart, 1973	▷ n.i.	▷ 24/48
▷ Catroli, 2008	▷ n.i.	▷ n.i.	▷ ref. Bogart, 1973	▷ n.i.	▷ 24/n.i.
▷ Catroli & Kasahara, 2009	▷ n.i.	▷ n.i.	▷ ref. Bogart, 1973	▷ n.i.	▷ 24/n.i.
▷ Peixoto et al., 2015	▷ Peru	▷ n.i.	▷ ref. Bogart, 1973	▷ n.i.	▷ 24/48†
▶ Present study	▶ Colombia	▶ n.i.	▶ Teresita	▶ 1	▶ 24/48
▶ Present study	▶ Venezuela	▶ 1♀	▶ Palmichal Forest Reserve, Carabobo State (10°18′N, 68°13′W)	▶ 1	▶ 24/48
Scinax ruber					
▶ Bogart & Bogart, 1971[N]	▶ Peru	▶ n.i.	▶ Río Pachitea, Bosque Nacional de Iparía, 1200 m, Huanuco Department; rep. *Hyla rubra*	▶ 1	▶ 24/n.i.
▶ Bogart, 1973[P]	▶ Colombia	▶ n.i.	▶ loc.n.i.; rep. *Hyla rubra*	▶ 1	▶ 24/48†
▷ Cole, 1974	▷ n.i.	▷ n.i.	▷ ref. Bogart, 1973; list. *Hyla rubra*	▷ n.i.	▷ 24/n.i.
▷ Morescalchi, 1979	▷ n.i.	▷ n.i.	▷ ref. Bogart, 1973; list. *Hyla rubra*	▷ n.i.	▷ 24/n.i.
▷ King, 1990	▷ S.America	▷ n.i.	▷ ref. Bogart, 1973; ctry.n.sp.; list. *Hyla rubra*	▷ n.i.	▷ 24/48
▷ Kuramoto, 1990	▷ S.C.America	▷ n.i.	▷ ref. Bogart & Bogart, 1971; Bogart, 1973; ctry.n.sp.; list. *Ololygon rubra*	▷ 1	▷ 24/n.i.
▷ Catroli & Kasahara, 2009	▷ n.i.	▷ n.i.	▷ ref. Bogart, 1973	▷ n.i.	▷ 24/n.i.
▶ Present study	▶ Venezuela	▶ 1♀	▶ Jusepín, Monagas State	▶ 1,2,5	▶ 24/48
▶ Present study	▶ Venezuela	▶ 2♂	▶ Campamento Río Grande, Sierra de Imataca, ~ 300 m, Bolivar State (8°03′S, 61°38′W)	▶ 1,2,5	▶ 24/48
▶ Present study	▶ Trinidad	▶ 1♂	▶ 6 km SE Talparo, 65 m (10°30′N, 61°16′W)	▶ 1	▶ 24/48
▶ Present study	▶ Colombia	▶ n.i.	▶ Florencia, Caquetá Department	▶ 1	▶ 24/48
▶ Present study	▶ Colombia	▶ 1♂	▶ Florencia, Caquetá Department	▶ 1	▶ 24/48
▶ Present study	▶ Peru	▶ 1♂	▶ Bosque Nacional de Iparia. Huánuco Department	▶ 1	▶ 24/48
Scinax similis					
▷ Catroli, 2008	▷ n.i.	▷ n.i.	▷ ref. Leme et al., 2004	▷ n.i.	▷ 24/n.i.
▶ Cardozo et al., 2011[P,M]	▶ Argentina	▶ 1♂	▶ Reserva Santa Maria, Ituzaingó, Ituzaingó Department, Corrientes Province	▶ 1,2,4	▶ 24/48
▶ Cardozo et al., 2011[P,M]	▶ Argentina	▶ 1♂	▶ Campo San Juan, Candelaria Department, Misiones Province	▶ 1,2,4	▶ 24/48
▶ Cardozo et al., 2011[P,M]	▶ Argentina	▶ 4♂	▶ Campus UNaM, Villa Lanús, Capital Department, Misiones Province	▶ 1,2,4	▶ 24/48
▶ Cardozo et al., 2011[P,M]	▶ Argentina	▶ 1♂	▶ Ruta Nacional No. 12, Km 1329, Posadas, Capital Department, Misiones Province	▶ 1,2,4	▶ 24/48
▶ Cardozo et al., 2011[P,M]	▶ Brazil	▶ 2♀	▶ Três Marias, Minas Gerais State	▶ 1,2,4	▶ 24/48
▶ Cardozo et al., 2011[P,M]	▶ Brazil	▶ 2♂	▶ Itirapina, São Paulo State	▶ 1,2,4	▶ 24/48
▶ Cardozo et al., 2011[P,M]	▶ Brazil	▶ 1♂	▶ Rio Claro, São Paulo State	▶ 1,2,4	▶ 24/48
▶ De Oliveira et al., 2012[P]	▶ Brazil	▶ 4s.n.i.	▶ Pirenópolis Municipality, Goiás State	▶ 1	▶ 24/48
▷ Peixoto et al., 2015	▷ Argentina	▷ n.i.	▷ Provinces of Corrientes and Misiones; ref. Cardozo et al., 2011	▷ 2,4	▷ 24/48
▷ Peixoto et al., 2015	▷ Brazil	▷ n.i.	▷ Três Marias, Minas Gerais State and Itirapina, São Paulo State; ref. Cardozo et al., 2011	▷ 2,4	▷ 24/48
▶ Gruber et al., 2017[P,M]	▶ Brazil	▶ 2♀	▶ Três Marias, Minas Gerais State (18°12′43″S, 45°13′19″W)	▶ 1,2,4,6, 9,12–14	▶ 24/48†
▶ Present study	▶ Brazil	▶ 1n.i.	▶ loc.u.	▶ 1,2	▶ 24/48
Scinax squalirostris					
▶ Barrio & Pistol de Rubel, 1970[P]	▶ Argentina	▶ 1♂	▶ Paraná de las Palmas and Canal 6, Paraná Delta, Buenos Aires Province; rep. *Hyla squalirostris*	▶ 1	▶ 24/48†
▷ Foresti, 1972	▷ n.i.	▷ n.i.	▷ ref. Barrio & Pistol de Rubel, 1970; list. *Hyla squalirostris*	▷ n.i.	▷ 24/n.i.
▶ Foresti, 1972[P,M]	▶ Brazil	▶ 2♂	▶ Itatinga, São Paulo State; rep. *Hyla squalirostris*	▶ 1	▶ 24/48†
▷ Cole, 1974	▷ n.i.	▷ n.i.	▷ ref. Barrio & Pistol de Rubel, 1970; list. *Hyla squalirostris*	▷ n.i.	▷ 24/n.i.
▷ King, 1990	▷ S.America	▷ n.i.	▷ ref. Barrio & Pistol de Rubel, 1970; ctry.n.sp.; list. *Hyla squalirostris*	▷ n.i.	▷ 24/n.i.
▷ Kuramoto, 1990	▷ S.C.America	▷ n.i.	▷ ref. Barrio & Pistol de Rubel, 1970; list. *Ololygon squalirostris*	▷ n.i.	▷ 24/n.i.
▷ Nunes & Fagundes, 2008b	▷ n.i.	▷ n.i.	▷ ref. Barrio & Pistol de Rubel, 1970	▷ n.i.	▷ 24/48
▷ Catroli, 2008	▷ n.i.	▷ n.i.	▷ ref. Barrio & Pistol de Rubel, 1970; Foresti, 1972; Trentin et al., 2007	▷ n.i.	▷ 24/n.i.
▷ Catroli & Kasahara, 2009	▷ n.i.	▷ n.i.	▷ ref. Barrio & Pistol de Rubel, 1970; Foresti, 1972; Trentin et al., 2007	▷ n.i.	▷ 24/n.i.
▶ Cardozo et al., 2011[P,M]	▶ Argentina	▶ 8♂	▶ Barrio Villa Blosset, Posadas, Capital Department, Misiones Province	▶ 1,2,4	▶ 24/48
▶ Cardozo et al., 2011[P,M]	▶ Argentina	▶ 1♂	▶ Campus UNaM, Villa Lanús, Capital Department, Misiones Province	▶ 1,2,4	▶ 24/48
▶ Cardozo et al., 2011[P,M]	▶ Brazil	▶ 1♂	▶ Itirapina, São Paulo State	▶ 1,2,4	▶ 24/48
▶ Cardozo et al., 2011[P,M]	▶ Uruguay	▶ 1♂	▶ Delta del Tigre, San José Department	▶ 1,2,4	▶ 24/48
▷ Peixoto et al., 2015	▷ n.i.	▷ n.i.	▷ ref. Barrio & Pistol de Rubel, 1970	▷ n.i.	▷ 24/48
▷ Peixoto et al., 2015	▷ Argentina	▷ n.i.	▷ Misiones Province; ref. Cardozo et al., 2011	▷ 2,4	▷ 24/48
▷ Peixoto et al., 2015	▷ Brazil	▷ n.i.	▷ Itirapina, São Paulo State; ref. Cardozo et al., 2011	▷ 2,4	▷ 24/48
▷ Peixoto et al., 2015	▷ Uruguay	▷ n.i.	▷ San José Department; ref. Cardozo et al., 2011	▷ 2,4	▷ 24/48

Author(s) ▶ original data[a] ▷ review articles	Country/ Region[b]	Speci- mens[c]	Locality of sampling, references cited, and remarks[d]	Methods applied[e]	2n/FN[f]
Scinax staufferi					
▶ Duellman, 1967[N]	▶ Nicaragua	▶ 2♂	▶ 16.5 km NE San Juan del Sur, Rivas Department; rep. *Hyla staufferi*	▶ 1	▶ 24‡/n.i.
▶ Duellman, 1967[N]	▶ Nicaragua	▶ 1♂	▶ 7.7 km NE San Juan del Sur, Rivas Department; rep. *Hyla staufferi*	▶ 1	▶ 24‡/n.i.
▶ Anderson, 1986[I]	▶ Belize	▶ 2♀	▶ Stann Creek; rep. *Ololygon staufferi staufferi*	▶ 1,2,4	▶ 24/48†
▷ Duellman, 1970	▷ C.America	▷ n.i.	▷ ref.n.i.; ctry.n.sp.; list. *Hyla staufferi*	▷ 1	▷ 24‡/n.i.
▷ Foresti, 1972	▷ n.i.	▷ n.i.	▷ ref. Duellman, 1967; list. *Hyla staufferi*	▷ n.i.	▷ 24‡/n.i.
▷ King, 1990	▷ C.America	▷ n.i.	▷ ref. Anderson, 1986; ctry.n.sp.; list. *Ololygon staufferi*	▷ 2,4	▷ 24/48
▷ Kuramoto, 1990	▷ S.C.America	▷ n.i.	▷ ref. Duellman, 1967; ctry.n.sp.; list. *Ololygon staufferi*	▷ 1	▷ 24/n.i.
▷ Anderson, 1991[I]	▷ Neotropic	▷ n.i.	▷ ref. Anderson, 1986; ctry.n.sp.; list. *Ololygon staufferi*	▷ 2,4	▷ 24/48†
▷ Catroli, 2008	▷ n.i.	▷ n.i.	▷ ref. Duellman, 1967; Anderson, 1991	▷ n.i.	▷ 24/n.i.
▷ Catroli & Kasahara, 2009	▷ n.i.	▷ n.i.	▷ ref. Duellman, 1967; Anderson, 1991	▷ 17	▷ 24/n.i.
▷ Peixoto et al., 2015	▷ Nicaragua	▷ n.i.	▷ Rivas Department; ref. Duellman, 1967	▷ n.i.	▷ 24/n.i.
▶ Present study	▶ Mexico	▶ 1♂	▶ 2 mi S of Ixtapan de la Sal	▶ 1	▶ 24/48
Scinax trilineatus					
▶ Nogueira et al., 2015b[P,M]	▶ Brazil	▶ 1♂	▶ Bragança, Pará State	▶ 1,2,4,6,9	▶ 24/48
Scinax x-signatus					
▶ Neri et al., 2009[N]	▶ Brazil	▶ n.i.	▶ Jequié Municipality, Bahia State	▶ 2,4	▶ 24/48
▶ Nogueira et al., 2015b[P,M]	▶ Brazil	▶ 1♂1♀ 1U	▶ Jequié, Bahia State	▶ 1,2,4,6,9	▶ 24/48
▶ Nogueira et al., 2015b[P,M]	▶ Brazil	▶ 1♂3♀	▶ Bragança, Pará State	▶ 1,2,4,6,9	▶ 24/48
▷ Peixoto et al., 2015	▷ Brazil	▷ n.i.	▷ Jequié, Bahia State and Bragança, Pará State; ref. Nogueira et al., 2015b	▷ 2,4	▷ 24/48
▶ Present study	▶ Venezuela	▶ 3♂	▶ El Limón, University campus, Aragua State	▶ 1,2,4, 5,7,8	▶ 24/48
***Scinax* sp. (*ruber* group)**					
▷ Catroli & Kasahara, 2009	▷ n.i.	▷ n.i.	▷ ref. Nunes, 2006; Nunes & Fagundes, 2008b	▷ n.i.	▷ 24/n.i.
***Scinax* sp.**					
▶ Nunes & Fagundes, 2008b[P]	▶ Brazil	▶ 2♂	▶ Lagoa de Samarco, Anchieta, Espírito Santo State	▶ 1	▶ 24/48
▷ Peixoto et al., 2015	▷ Brazil	▷ n.i.	▷ Anchieta, Espírito Santo State; ref. Nunes & Fagundes, 2008b	▷ n.i.	▷ 24/48
Scinax* sp.					
▶ Present study	▶ n.i.	▶ n.i.	▶ loc.u.	▶ 1	▶ 24/48
HYLIDAE / Scinaxinae / *Sphaenorhynchus*					
Sphaenorhynchus caramaschii					
▶ Suárez et al., 2013[P,M,I]	▶ Brazil	▶ 3♂1J	▶ Fazenda Intermontes, Ribeirão Grande Municipality, São Paulo State	▶ 1–4,6, 9,13,14	▶ 24/48
Sphaenorhynchus carneus					
▶ Suárez et al., 2013[P,M,I]	▶ Brazil	▶ 2♂	▶ Comunidade Praia Grande, Cruzeiro do Sul Municipality, Acre State	▶ 1–4,6, 9,13,14	▶ 26/52
Sphaenorhynchus dorisae					
▶ Suárez et al., 2013[P,M,I]	▶ Brazil	▶ 1♂	▶ Comunidade Praia Grande, Cruzeiro do Sul Municipality, Acre State	▶ 1–4,6, 9,13,14	▶ 24/48
▶ Suárez et al., 2013[P,M,I]	▶ Brazil	▶ 1♂	▶ Comunidade Praia Grande, Cruzeiro do Sul Municipality, Acre State; 1 B	▶ 1–4,6, 9,13,14	▶ 25/49
▶ Suárez et al., 2013[P,M,I]	▶ Brazil	▶ 2♂	▶ Comunidade Praia Grande, Cruzeiro do Sul Municipality, Acre State; 2 B	▶ 1–4,6, 9,13,14	▶ 26/50
▷ Schmid & Steinlein, 2016a	▷ n.i.	▷ n.i.	▷ ref. Suárez et al., 2013	▷ 14	▷ 24/n.i.
Sphaenorhynchus lacteus					
▶ Suárez et al., 2013[P,M,I]	▶ Brazil	▶ 2♂	▶ Levylândia, Ananindeua Municipality, Pará State	▶ 1–4,6, 9,13,14	▶ 24/48
▶ Suárez et al., 2013[P,M,I]	▶ Brazil	▶ 1♂	▶ Santa Barbara, Santa Barbara Municipality, Pará State	▶ 1–4,6,$ 9,13,14	▶ 24/48
▶ Suárez et al., 2013[P,M,I]	▶ Brazil	▶ 2♂	▶ Marabá, Marabá Municipality, Pará State	▶ 1–4,6, 9,13,14	▶ 24/48
▷ Schmid & Steinlein, 2016°	▷ n.i.	▷ n.i.	▷ ref. Suárez et al., 2013	▷ 14	▷ 24/n.i.
▶ Present study	▶ Guyana	▶ 1♀	▶ Georgetown	▶ 1	▶ 24/48
Sphaenorhynchus planicola					
▶ Foresti, 1972[P,M]	▶ Brazil	▶ 2♂	▶ Linhares, Espírito Santo State; rep. *{Sphaenorhyncus} planicola*	▶ 1	▶ 24/48†
▷ Catroli, 2008	▷ n.i.	▷ n.i.	▷ ref. Foresti, 1972	▷ n.i.	▷ 24/n.i.
▷ Catroli & Kasahara, 2009	▷ n.i.	▷ n.i.	▷ ref. Foresti, 1972	▷ n.i.	▷ 24/n.i.

Author(s) ▶ original data[a] ▷ review articles	Country/ Region[b]	Speci- mens[c]	Locality of sampling, references cited, and remarks[d]	Methods applied[e]	2n/FN[f]
PELODRYADIDAE / Litoriinae / *Litoria*					
Litoria adelaidensis					
▶ King, 1980[P,M,I]	▶ Australia	▶ 3s.n.i.	▶ Forrestfield, Perth, Western Australia State	▶ 1,2,4	▶ 26/52†
▷ King, 1990	▷ Australia	▷ n.i.	▷ ref. King, 1980	▷ 2	▷ 26/52
▷ Kuramoto, 1990	▷ Aus./NG.	▷ n.i.	▷ ref. King, 1980; ctry.n.sp.	▷ 17	▷ 26/n.i.
▷ Catroli, 2008	▷ n.i.	▷ n.i.	▷ ref. King, 1980	▷ n.i.	▷ 26/n.i.
▷ Catroli & Kasahara, 2009	▷ n.i.	▷ n.i.	▷ ref. King, 1980	▷ 17	▷ 26/n.i.
▶ Present study	▶ Australia	▶ 1U	▶ Western Australia State, loc.u.	▶ 1	▶ 26/52
Litoria amboinensis					
▶ Menzies & Tippet, 1976[N]	▶ New Guinea	▶ 5♂	▶ Brown River, Central Province; rep. *Litoria {amboinensis}*	▶ 1	▶ 26‡/n.i.
▷ Morescalchi, 1979	▷ n.i.	▷ n.i.	▷ ref. Menzies & Tippett, 1976; list. *Litoria {amboinensis}*	▷ n.i.	▷ 26/n.i.
▷ King, 1990	▷ New Guinea	▷ n.i.	▷ ref. Menzies & Tippett, 1976; list. *Litoria {amboienensis}*	▷ n.i.	▷ 26/n.i.
▷ Kuramoto, 1990	▷ Aus./NG.	▷ n.i.	▷ ref. Menzies & Tippett, 1976; ctry.n.sp.; list. *Litoria {amboinensis}*	▷ 1	▷ 26/n.i.
▷ Catroli, 2008	▷ n.i.	▷ n.i.	▷ ref. Menzies & Tippett, 1976; list. *Litoria {amboinesis}*	▷ n.i.	▷ 26/n.i.
▷ Catroli & Kasahara, 2009	▷ n.i.	▷ n.i.	▷ ref. Menzies & Tippett, 1976; list. *Litoria {amboinensis}*	▷ n.i.	▷ 26/n.i.
Litoria angiana					
▶ Duellman, 1967[P]	▶ New Guinea	▶ 1♂	▶ Edie Creek, Morobe Province; rep. *Hyla angiana*; err.2n	▶ 1	▶ 30‡/n.i.
▷ Foresti, 1972	▷ n.i.	▷ n.i.	▷ ref. Duellman, 1967; list. *Hyla angiana*; err.2n	▷ n.i.	▷ 30‡/n.i.
▷ Morescalchi, 1973	▷ n.i.	▷ n.i.	▷ ref. Duellman, 1967; list. *Hyla (Litoria) angiana*; err.2n	▷ n.i.	▷ 30‡/n.i.
▶ Menzies & Tippet, 1976[P]	▶ New Guinea	▶ 4♂	▶ Schrader Mountains, central highlands	▶ 1	▶ 26‡/n.i.
▶ Menzies & Tippet, 1976[P]	▶ New Guinea	▶ 2♂	▶ Kubor Mountains, central highlands	▶ 1	▶ 26‡/n.i.
▶ Menzies & Tippet, 1976[P]	▶ New Guinea	▶ 1♂	▶ Mount Elimbari, central highlands	▶ 1	▶ 26‡/n.i.
▶ Menzies & Tippet, 1976[P]	▶ New Guinea	▶ 1♂	▶ Mount Kaindi, central highlands	▶ 1	▶ 26‡/n.i.
▷ Morescalchi, 1979	▷ n.i.	▷ n.i.	▷ ref. Menzies & Tippett, 1976	▷ n.i.	▷ 26/n.i.
▷ King, 1990	▷ New Guinea	▷ n.i.	▷ ref. Menzies & Tippett, 1976	▷ n.i.	▷ 26/n.i.
▷ Kuramoto, 1990	▷ Aus./NG.	▷ n.i.	▷ ref. Duellman, 1967; ctry.n.sp.; err.2n	▷ 1	▷ 30/n.i.
▷ Kuramoto, 1990	▷ Aus./NG.	▷ n.i.	▷ ref. Menzies & Tippett, 1976, ctry.n.sp.	▷ 1	▷ 26/n.i.
▷ Catroli, 2008	▷ n.i.	▷ n.i.	▷ ref. Menzies & Tippett, 1976	▷ n.i.	▷ 26/n.i.
▷ Catroli & Kasahara, 2009	▷ n.i.	▷ n.i.	▷ ref. Menzies & Tippett, 1976	▷ n.i.	▷ 26/n.i.
▷ Catroli & Kasahara, 2009	▷ n.i.	▷ n.i.	▷ ref. Duellman, 1967; err.2n	▷ n.i.	▷ 30/n.i.
Litoria arfakiana					
▶ Duellman, 1967[P]	▶ New Guinea	▶ 1♂	▶ Morobe Province; rep. *Hyla arfakiana*	▶ 1	▶ 26‡/n.i.
▷ Foresti, 1972	▷ n.i.	▷ n.i.	▷ ref. Duellman, 1967; list. *Hyla arfakiana*	▷ n.i.	▷ 26‡/n.i.
▷ Morescalchi, 1973	▷ n.i.	▷ n.i.	▷ ref. Duellman, 1967; list. *Hyla (Litoria) arfakiana*	▷ n.i.	▷ 26/n.i.
▶ Menzies & Tippet, 1976[N]	▶ New Guinea	▶ 1♂	▶ Waitabe, Central Province	▶ 1	▶ 26‡/n.i.
▷ Morescalchi, 1979	▷ n.i.	▷ n.i.	▷ ref. Menzies & Tippett, 1976	▷ n.i.	▷ 26/n.i.
▷ King, 1990	▷ New Guinea	▷ n.i.	▷ ref. Menzies & Tippett, 1976	▷ n.i.	▷ 26/n.i.
▷ Kuramoto, 1990	▷ Aus./NG.	▷ n.i.	▷ ref. Menzies & Tippett, 1976; Kuramoto & Allison; unpubl.; ctry.n.sp.	▷ 1	▷ 26/n.i.
▶ Kuramoto & Allison, 1991[P,M]	▶ New Guinea	▶ 3♂	▶ vicinity of Wau, Morobe Province	▶ 1	▶ 26/52†
▷ Catroli, 2008	▷ n.i.	▷ n.i.	▷ ref. Menzies & Tippett, 1976; Kuramoto & Allison, 1991	▷ n.i.	▷ 26/n.i.
▷ Catroli & Kasahara, 2009	▷ n.i.	▷ n.i.	▷ ref. Duellman, 1967; Menzies & Tippett, 1976; Kuramoto & Allison, 1991	▷ n.i.	▷ 26/n.i.
Litoria bicolor					
▶ Menzies & Tippet, 1976[N]	▶ New Guinea	▶ 2♂1tp	▶ Brown River, Central Province; list. *Litoria bicolor 'A'*	▶ 1	▶ 26/n.i.
▶ Menzies & Tippet, 1976[N]	▶ New Guinea	▶ 2♂	▶ Weam, Western Province; list. *Litoria bicolor 'F'*	▶ 1	▶ 26‡/n.i.
▷ Morescalchi, 1979	▷ n.i.	▷ n.i.	▷ ref. Menzies & Tippett, 1976	▷ n.i.	▷ 26/n.i.
▷ King, 1990	▷ Australia	▷ n.i.	▷ ref. Morescalchi, 1968a; Menzies & Tippett, 1976; King, 1981	▷ 2	▷ 26/52
▷ Kuramoto, 1990	▷ Aus./NG.	▷ n.i.	▷ ref. Morescalchi, 1970; Menzies & Tippett, 1976; ctry.n.sp.	▷ 1	▷ 26/n.i.
▷ Catroli, 2008	▷ n.i.	▷ n.i.	▷ ref. Morescalchi, 1968; Menzies & Tippett, 1976; King, 1980/err.ref.	▷ n.i.	▷ 26/n.i.
▷ Catroli & Kasahara, 2009	▷ n.i.	▷ n.i.	▷ ref. Menzies & Tippett, 1976; King, 1980/err.ref.	▷ 17	▷ 26/n.i.
▶ Present study	▶ Australia	▶ 1♂	▶ Finniss River, Northern Territory (12°54′S, 130°26′E)	▶ 1,2,4,14	▶ 26/52
▶ Present study	▶ Australia	▶ 11♂ 12♀	▶ Jabiru, Kakadu National Park, Northern Territory (12°40′S, 132°50′E)	▶ 1,2,4,14	▶ 26/52
▶ Present study	▶ Australia	▶ 2♂4♀ 2U	▶ Ja Ja, Kakadu National Park, Nothern Territory (12°34′S, 132°53′E)	▶ 1,2,4,14	▶ 26/52
Litoria congenita					
▶ Menzies & Tippet, 1976[N]	▶ New Guinea	▶ 4♂	▶ Port Moresby, Central Province	▶ 1	▶ 26‡/n.i.
▷ Morescalchi, 1979	▷ n.i.	▷ n.i.	▷ ref. Menzies & Tippett, 1976	▷ n.i.	▷ 26/n.i.
▷ King, 1990	▷ New Guinea	▷ n.i.	▷ ref. Menzies & Tippett, 1976	▷ n.i.	▷ 26/n.i.
▷ Kuramoto, 1990	▷ Aus./NG	▷ n.i.	▷ ref. Menzies & Tippett, 1976; ctry.n.sp.	▷ 1	▷ 26/n.i.
▷ Catroli, 2008	▷ n.i.	▷ n.i.	▷ ref. Menzies & Tippett, 1976	▷ n.i.	▷ 26/n.i.
▷ Catroli & Kasahara, 2009	▷ n.i.	▷ n.i.	▷ ref. Menzies & Tippett, 1976	▷ n.i.	▷ 26/n.i.
Litoria contrastens					
▶ Menzies & Tippet, 1976[N]	▶ New Guinea	▶ 2♂	▶ Wau, Morobe Province	▶ 1	▶ 26‡/n.i.
▷ Morescalchi, 1979	▷ n.i.	▷ n.i.	▷ ref. Menzies & Tippett, 1976; list. *Litoria {contrastans}*	▷ n.i.	▷ 26/n.i.
▷ King, 1990	▷ New Guinea	▷ n.i.	▷ ref. Menzies & Tippett, 1976	▷ n.i.	▷ 26/n.i.
▷ Kuramoto, 1990	▷ Aus./NG.	▷ n.i.	▷ ref. Menzies & Tippett, 1976; ctry.n.sp.	▷ 1	▷ 26/n.i.
▷ Catroli, 2008	▷ n.i.	▷ n.i.	▷ ref. Menzies & Tippett, 1976	▷ n.i.	▷ 26/n.i.
▷ Catroli & Kasahara, 2009	▷ n.i.	▷ n.i.	▷ ref. Menzies & Tippett, 1976	▷ n.i.	▷ 26/n.i.

Author(s) ▶ original data[a] ▷ review articles	Country/ Region[b]	Speci- mens[c]	Locality of sampling, references cited, and remarks[d]	Methods applied[e]	2n/FN[f]
Litoria cooloolensis					
▶ King, 1980[P,M,I]	▶ Australia	▶ 2s.n.i.	▶ Cooloola National Park, Queensland State	▶ 1,2,4	▶ 26/52[†]
▷ King, 1990	▷ Australia	▷ n.i.	▷ ref. King, 1980	▷ 2	▷ 26/52
▶ King, 1990[P]	▶ n.i.	▶ n.i.	▶ loc.n.i.	▶ 2	▶ 26/52[†]
▷ Kuramoto, 1990	▷ Aus./NG.	▷ n.i.	▷ ref. King 1980; ctry.n.sp.	▷ 17	▷ 26/n.i.
▷ Catroli, 2008	▷ n.i.	▷ n.i.	▷ ref. King, 1980	▷ n.i.	▷ 26/n.i.
▷ Catroli & Kasahara, 2009	▷ n.i.	▷ n.i.	▷ ref. King, 1980	▷ 17	▷ 26/n.i.
Litoria coplandi					
▶ Morescalchi, 1979[N]	▶ n.i.	▶ n.i.	▶ ref. Morescalchi & Ingram, unpubl.	▶ n.i.	▶ 26/n.i.
▶ King et al., 1990[P]	▶ Australia	▶ 1s.n.i.	▶ Hayes Creek, Northern Territory	▶ 13	▶ 26/52[†]
▷ King, 1990	▷ Australia	▷ n.i.	▷ ref. King, 1981	▷ 2,10,13	▷ 26/52
▶ King, 1990[P]	▶ n.i.	▶ n.i.	▶ loc.n.i.	▶ 1,2,10,13	▶ 26/52[†]
▷ Kuramoto, 1990	▷ Aus./NG.	▷ n.i.	▷ ref. Morescalchi, 1979; ctry.n.sp.	▷ 1	▷ 26/n.i.
▶ King, 1991[P]	▶ n.i.	▶ n.i.	▶ loc.n.i.	▶ 2,10	▶ 26/52[†]
▷ Catroli, 2008	▷ n.i.	▷ n.i.	▷ ref. King, 1980/err.ref., 1990	▷ n.i.	▷ 26/n.i.
▷ Catroli & Kasahara, 2009	▷ n.i.	▷ n.i.	▷ ref. Morescalchi, 1979; King, 1980/err.ref., 1990	▷ 17	▷ 26/n.i.
▶ Present study	▶ Australia	▶ 10♂ 13♀	▶ Bowerbird, Kakadu National Park, Northern Territory (12°25′S, 132°40′E)	▶ 1,2,4, 8,14	▶ 26/52
Litoria darlingtoni					
▶ Duellman, 1967[N]	▶ New Guinea	▶ 2♂	▶ Eastern Highlands Province; rep. *Hyla darlingtoni*	▶ 1	▶ 26[‡]/n.i.
▷ Foresti, 1972	▷ n.i.	▷ n.i.	▷ ref. Duellman, 1967; list. *Hyla darlingtoni*	▷ n.i.	▷ 26[‡]/n.i.
▷ Morescalchi, 1973	▷ n.i.	▷ n.i.	▷ ref. Duellman, 1967; list. *Hyla (Litoria) darlingtoni*	▷ n.i.	▷ 26/n.i.
▶ Menzies & Tippet, 1976[N]	▶ New Guinea	▶ 1♂	▶ Elimbari, (Chimbu) Simbu Province	▶ 1	▶ 26[‡]/n.i.
▷ Morescalchi, 1979	▷ n.i.	▷ n.i.	▷ ref. Menzies & Tippett, 1976	▷ n.i.	▷ 26/n.i.
▷ King, 1990	▷ New Guinea	▷ n.i.	▷ ref. Menzies & Tippett, 1976	▷ n.i.	▷ 26/n.i.
▷ Kuramoto, 1990	▷ Aus./NG.	▷ n.i.	▷ ref. Duellman, 1967; Menzies & Tippett, 1976; ctry.n.sp.	▷ 1	▷ 26/n.i.
▷ Catroli, 2008	▷ n.i.	▷ n.i.	▷ ref. Menzies & Tippett, 1976	▷ n.i.	▷ 26/n.i.
▷ Catroli & Kasahara, 2009	▷ n.i.	▷ n.i.	▷ ref. Duellman, 1967; Menzies & Tippett, 1976	▷ n.i.	▷ 26/n.i.
▶ Present study	▶ New Guinea	▶ n.i.	▶ loc.u.	▶ 1	▶ 26/52
Litoria dentata					
▶ Morescalchi, 1979[N]	▶ n.i.	▶ n.i.	▶ ref. Morescalchi & Ingram, unpubl.	▶ n.i.	▶ 26/n.i.
▷ King, 1990	▷ Australia	▷ n.i.	▷ ref. King, 1981	▷ n.i.	▷ 26/52
▶ King, 1990[P]	▶ n.i.	▶ n.i.	▶ loc.n.i.	▶ 1	▶ 26/52
▷ Kuramoto, 1990	▷ Aus./NG.	▷ n.i.	▷ ref. Morescalchi, 1979; ctry.n.sp.	▷ n.i.	▷ 26/n.i.
▷ Catroli, 2008	▷ n.i.	▷ n.i.	▷ ref. Morescalchi, 1979; King, 1980/err.ref., 1990	▷ n.i.	▷ 26/n.i.
▷ Catroli & Kasahara, 2009	▷ n.i.	▷ n.i.	▷ ref. Morescalchi, 1979; King, 1980/err.ref., 1990	▷ n.i.	▷ 26/n.i.
▶ Present study	▶ Australia	▶ 1U	▶ Southeastern Queensland, loc.u.	▶ 1,2	▶ 26/52
Litoria ewingii					
▷ King, 1990	▷ Australia	▷ n.i.	▷ ref. King, 1981	▷ n.i.	▷ 26/52
▷ Kuramoto, 1990	▷ Aus./NG.	▷ n.i.	▷ ref. Woodruff, 1972; ctry.n.sp.	▷ 1	▷ 26/n.i.
▷ Catroli, 2008	▷ n.i.	▷ n.i.	▷ ref. King, 1980/err.ref.	▷ n.i.	▷ 26/n.i.
▷ Catroli & Kasahara, 2009	▷ n.i.	▷ n.i.	▷ ref. Woodruff, 1972; King, 1980/err.ref.	▷ n.i.	▷ 26/n.i.
▶ Present study	▶ Australia	▶ 1U	▶ Southeastern Australia, loc.u.	▶ 1	▶ 26/52
Litoria fallax					
▶ Morescalchi, 1979[N]	▶ n.i.	▶ n.i.	▶ ref. Morescalchi & Ingram, unpubl.; rep. *Litoria glauerti*	▶ n.i.	▶ 26/n.i.
▷ King, 1990	▷ Australia	▷ n.i.	▷ ref. King, 1981	▷ n.i.	▷ 26/52
▷ Kuramoto, 1990	▷ Aus./NG.	▷ n.i.	▷ ref. Morescalchi, 1979; ctry.n.sp.	▷ 1	▷ 26/n.i.
▷ Catroli, 2008	▷ n.i.	▷ n.i.	▷ ref. King, 1980/err.ref.	▷ n.i.	▷ 26/n.i.
▷ Catroli & Kasahara, 2009	▷ n.i.	▷ n.i.	▷ ref. Morescalchi, 1979; King, 1980/err.ref.	▷ n.i.	▷ 26/n.i.
▷ Schmid & Steinlein, 2016a	▷ n.i.	▷ n.i	▷ ref. Schmid et al., unpubl.	▷ 14	▷ 26/n.i.
▶ Present study	▶ Australia	▶ 12♂7♀	▶ Conondale, Queensland State (26°44′S, 152°44′E)	▶ 1,2,4,14	▶ 26/52
Litoria freycineti					
▷ King, 1990	▷ Australia	▷ n.i.	▷ ref. King, 1981	▷ n.i.	▷ 26/52
▷ Catroli, 2008	▷ n.i.	▷ n.i.	▷ ref. King, 1980/err.ref.	▷ n.i.	▷ 26/n.i.
▷ Catroli & Kasahara, 2009	▷ n.i.	▷ n.i.	▷ ref. King, 1980/err.ref.	▷ n.i.	▷ 26/n.i.
▶ Present study	▶ Australia	▶ 1U	▶ Southern Queensland, loc.u.	▶ 1	▶ 26/52
Litoria inermis					
▷ King, 1990	▷ Australia	▷ n.i.	▷ ref. King, 1981 and unpubl.	▷ 2	▷ 26/52
▷ Catroli, 2008	▷ n.i.	▷ n.i.	▷ ref. King, 1980/err.ref.	▷ n.i.	▷ 26/n.i.
▷ Catroli & Kasahara, 2009	▷ n.i.	▷ n.i.	▷ ref. King, 1980/err.ref., 1990	▷ 17	▷ 26/n.i.
▶ Present study	▶ Australia	▶ 3♂	▶ Millstream, Queensland State (17°39′S, 145°51′E)	▶ 1,2,4,14	▶ 26/52
Litoria jervisiensis					
▷ King, 1990	▷ Australia	▷ n.i.	▷ ref. King, 1981	▷ n.i.	▷ 26/52
▷ Catroli, 2008	▷ n.i.	▷ n.i.	▷ ref. King, 1980/err.ref.	▷ n.i.	▷ 26/n.i.
▷ Catroli & Kasahara, 2009	▷ n.i.	▷ n.i.	▷ ref. King, 1980/err.ref.	▷ n.i.	▷ 26/n.i.
▶ Present study	▶ Australia	▶ 1U	▶ New South Wales State, loc.u.	▶ 1	▶ 26/52

Author(s) ▶ original data[a] ▷ review articles	Country/ Region[b]	Specimens[c]	Locality of sampling, references cited, and remarks[d]	Methods applied[e]	2n/FN[f]
Litoria latopalmata					
▶ Bogart, 1973[P]	▶ Australia	▶ n.i.	▶ eastern Australia; loc.n.i.; rep. *Hyla latopalmata*	▶ 1	▶ 26/52[†]
▷ Morescalchi, 1979	▷ n.i.	▷ n.i.	▷ ref. Bogart, 1973; list. *Hyla latopalmata*	▷ n.i.	▷ 26/n.i.
▷ King, 1990	▷ Australia	▷ n.i.	▷ ref. Bogart, 1973; King, 1981	▷ n.i.	▷ 26/52
▷ Kuramoto, 1990	▷ Aus./NG.	▷ n.i.	▷ ref. Bogart, 1973; ctry.n.sp.	▷ 1	▷ 26/n.i.
▷ Catroli, 2008	▷ n.i.	▷ n.i.	▷ ref. Bogart, 1973; King, 1980/err.ref.	▷ n.i.	▷ 26/n.i.
▷ Catroli & Kasahara, 2009	▷ n.i.	▷ n.i.	▷ ref. Bogart, 1973; King, 1980/err.ref.	▷ n.i.	▷ 26/n.i.
▶ Present study	▶ Australia	▶1♂1♀	▶ Indooroopilly	▶ 1	▶ 26/52
▶ Present study	▶ Australia	▶3♂	▶ Millstream, Queensland State (17°39′S, 145°51′E)	▶ 1,4,14	▶ 26/52
Litoria meiriana					
▶ King, 1980[P,M,I]	▶ Australia	▶ 8s.n.i.	▶ Mitchell Plateau, Western Australia State; Canon Hill, Northern Territory	▶ 1,2,4	▶ 26/52[†]
▷ King, 1990	▷ Australia	▷ n.i.	▷ ref. King, 1980	▷ 2	▷ 26/52
▶ King, 1990[P]	▶ n.i.	▶ n.i.	▶ loc.n.i.	▶ 1,2	▶ 26/52[†]
▷ Kuramoto, 1990	▷ Aus./NG.	▷ n.i.	▷ ref. King, 1980; ctry.n.sp.	▷ 17	▷ 26/n.i.
▷ Catroli, 2008	▷ n.i.	▷ n.i.	▷ ref. King, 1980	▷ n.i.	▷ 26/n.i.
▷ Catroli & Kasahara, 2009	▷ n.i.	▷ n.i.	▷ ref. King, 1980	▷ 17	▷ 26/n.i.
▶ Present study	▶ Australia	▶8♂9♀ 1U	▶ Bowerbird, Kakadu National Park, Northern Territory (12°25′S, 132°40′E)	▶ 1,2,4,14	▶ 26/52
▶ Present study	▶ Australia	▶2♂3♀	▶ Ja Ja, Kakadu National Park, Nothern Territory (12°34′S, 132°53′E)	▶ 1,2,4,14	▶ 26/52
Litoria microbelos					
▷ King, 1990	▷ Australia	▷ n.i.	▷ ref. King, 1981	▷ n.i.	▷ 26/52
▷ Catroli, 2008	▷ n.i.	▷ n.i.	▷ ref. King, 1980/err.ref.	▷ n.i.	▷ 26/n.i.
▷ Catroli & Kasahara, 2009	▷ n.i.	▷ n.i.	▷ ref. King, 1980/err.ref.	▷ n.i.	▷ 26/n.i.
▶ Present study	▶ Australia	▶ 2U	▶ Scott Creek, Northern Territory (14°50′S, 132°08′E)	▶ 1	▶ 26/52
Litoria micromembrana					
▶ Duellman, 1967[P]	▶ New Guinea	▶ 1♂	▶ Morobe Province; rep. *Hyla becki*	▶ 1	▶ 26[‡]/n.i.
▷ Foresti, 1972	▷ n.i.	▷ n.i.	▷ ref. Duellman, 1967; list. *Hyla becki*	▷ n.i.	▷ 26[‡]/n.i.
▷ Morescalchi, 1973	▷ n.i.	▷ n.i.	▷ ref. Duellman, 1967; list. *Hyla (Litoria) becki*	▷ n.i.	▷ 26/n.i.
▷ Kuramoto, 1990	▷ Aus./NG.	▷ n.i.	▷ ref. Duellman, 1967; ctry.n.sp.; list. *Hyla becki*	▷ 1	▷ 26/n.i.
▷ Catroli & Kasahara, 2009	▷ n.i.	▷ n.i.	▷ ref. Duellman, 1967; list. *Litoria becki*	▷ n.i.	▷ 26/n.i.
Litoria modica					
▶ Menzies & Tippet, 1976[N]	▶ New Guinea	▶ 6♂	▶ Kaironk, Madang Province; Gumine, (Chimbu) Simbu Province; Tambul, Western Highlands Province; Waitape, Central Province	▶ 1	▶ 26[‡]/n.i.
▷ Morescalchi, 1979	▷ n.i.	▷ n.i.	▷ ref. Menzies & Tippett, 1976	▷ n.i.	▷ 26/n.i.
▷ King, 1990	▷ New Guinea	▷ n.i.	▷ ref. Menzies & Tippett, 1976	▷ n.i.	▷ 26/n.i.
▷ Kuramoto, 1990	▷ Aus./NG.	▷ n.i.	▷ ref. Menzies & Tippett, 1976; ctry.n.sp.	▷ n.i.	▷ 26/n.i.
▷ Catroli, 2008	▷ n.i.	▷ n.i.	▷ ref. Menzies & Tippett, 1976	▷ n.i.	▷ 26/n.i.
▷ Catroli & Kasahara, 2009	▷ n.i.	▷ n.i.	▷ ref. Menzies & Tippett, 1976	▷ n.i.	▷ 26/n.i.
Litoria nasuta					
▶ Menzies & Tippet, 1976[N]	▶ New Guinea	▶ 2♂	▶ Port Moresby, Central Province	▶ 1	▶ 26[‡]/n.i.
▷ Morescalchi, 1979	▷ n.i.	▷ n.i.	▷ ref. Morescalchi & Ingram, 1974; Menzies & Tippett, 1976	▷ n.i.	▷ 26/n.i.
▷ King, 1990	▷ Australia	▷ n.i.	▷ ref. Morescalchi & Ingram, 1974; Menzies & Tippett, 1976; King, 1981	▷ n.i.	▷ 26/52
▷ Kuramoto, 1990	▷ Aus./NG.	▷ n.i.	▷ ref. Morescalchi & Ingram, 1974; Menzies & Tippett, 1976; ctry.n.sp.	▷ 1	▷ 26/n.i.
▷ Catroli, 2008	▷ n.i.	▷ n.i.	▷ ref. Morescalchi & Ingram, 1974; Menzies & Tippett, 1976; King, 1980/err. ref.	▷ n.i.	▷ 26/n.i.
▷ Catroli & Kasahara, 2009	▷ n.i.	▷ n.i.	▷ ref. Morescalchi & Ingram, 1974; Menzies & Tippett, 1976; King, 1980/err. ref.	▷ n.i.	▷ 26/n.i.
▶ Present study	▶Australia	▶ 1♂4♀ 2U	▶ Scott Creek, Northern Territory (14°50′S, 132°08′E)	▶ 1,2,4, 8,14	▶ 26/52
▶ Present study	▶Australia	▶ 2♀	▶ Jabiru, Kakadu National Park, Northern Territory (12°40′S, 132°50′E)	▶ 1,2,4, 8,14	▶ 26/52
Litoria nigrofrenata					
▷ King, 1990	▷ Aus./NG.	▷ n.i.	▷ ref. King, 1981	▷ n.i.	▷ 26/52
▷ Catroli, 2008	▷ n.i.	▷ n.i.	▷ ref. King, 1980/err.ref.	▷ n.i.	▷ 26/n.i.
▷ Catroli & Kasahara, 2009	▷ n.i.	▷ n.i.	▷ ref. King, 1980/err.ref.	▷ n.i.	▷ 26/n.i.
▶ Present study	▶Australia	▶ 2U	▶ Cape York peninsula, Queensland State; loc.u.	▶ 1	▶ 26/52
Litoria nigropunctata					
▶ Menzies & Tippet, 1976[N]	▶ New Guinea	▶ 1♂	▶ Lae, Morobe Province	▶ 1	▶ 26[‡]/n.i.
▷ Morescalchi, 1979	▷ n.i.	▷ n.i.	▷ Menzies & Tippett, 1976	▷ n.i.	▷ 26/n.i.
▷ King, 1990	▷ New Guinea	▷ n.i.	▷ Menzies & Tippett, 1976	▷ n.i.	▷ 26/n.i.
▷ Kuramoto, 1990	▷ Aus./NG.	▷ n.i.	▷ Menzies & Tippett, 1976; ctry.n.sp.	▷ 1	▷ 26/n.i.
▷ Catroli, 2008	▷ n.i.	▷ n.i.	▷ Menzies & Tippett, 1976	▷ n.i.	▷ 26/n.i.
▷ Catroli & Kasahara, 2009	▷ n.i.	▷ n.i.	▷ Menzies & Tippett, 1976	▷ n.i.	▷ 26/n.i.

Author(s) ▶ original data[a] ▷ review articles	Country/ Region[b]	Speci- mens[c]	Locality of sampling, references cited, and remarks[d]	Methods applied[e]	2n/FN[f]
Litoria olongburensis					
▶ King, 1980[P,M,I]	▶ Australia	▶ 2s.n.i.	▶ Cooloola National Park, Queensland State	▶ 1,2,4	▶ 26/52†
▷ King, 1990	▷ New Guinea	▷ n.i.	▷ ref. King, 1980; err.cit.ctry.	▷ 2	▷ 26/52
▷ Kuramoto, 1990	▷ Aus./NG.	▷ n.i.	▷ ref. King, 1980; ctry.n.sp.	▷ 17	▷ 26/n.i.
▷ Catroli, 2008	▷ n.i.	▷ n.i.	▷ ref. King, 1980	▷ n.i.	▷ 26/n.i.
▷ Catroli & Kasahara, 2009	▷ n.i.	▷ n.i.	▷ ref. King, 1980	▷ 17	▷ 26/n.i.
Litoria pallida*					
▶ Present study	▶ Australia	▶ 8♂8♀ 7U	▶ Scott Creek, Northern Territory (14°50′S, 132°08′E)	▶ 1,2,4, 5,8,14	▶ 26/52
▶ Present study	▶ Australia	▶ 2♂1U	▶ Millstream, Queensland State (17°39′S, 145°51′E)	▶ 1,2,4, 5,8,14	▶ 26/52
Litoria peronii					
▶ Morescalchi, 1979[N]	▶ n.i.	▶ n.i.	▶ ref. Morescalchi & Ingram, unpubl.; list. *{Litoria peroni}*	▶ n.i.	▶ 26/n.i.
▶ King, 1980[P,M,I]	▶ Australia	▶ 5s.n.i.	▶ Balranald and Griffith, New South Wales State; rep. *{Litoria peroni}*	▶ 1,2,4	▶ 26/52†
▶ King et al., 1990[P]	▶ Australia	▶ 1s.n.i.	▶ Kialoa, New South Wales State	▶ 13	▶ 26/52†
▷ King, 1990	▷ Australia	▷ n.i.	▷ ref. King, 1980	▷ 2,4, 10,13	▷ 26/52
▶ King, 1990[P]	▶ n.i.	▶ n.i.	▶ loc.n.i.	▶ 2	▶ 26/52†
▷ Kuramoto, 1990	▷ Aus./NG.	▷ n.i.	▷ ref. Morescalchi, 1979; King, 1980; ctry.n.sp.	▷ 17	▷ 26/n.i.
▷ Catroli, 2008	▷ n.i.	▷ n.i.	▷ ref. King, 1980	▷ n.i.	▷ 26/n.i.
▷ Catroli & Kasahara, 2009	▷ n.i.	▷ n.i.	▷ ref. Morescalchi, 1979; King, 1980	▷ 17	▷ 26/n.i.
▶ Present study	▶ Australia	▶ 3♂3♀	▶ Balnarald, New South Wales State (34°38′S, 143°33′E)	▶ 1,2,4,5, 7,8,14	▶ 26/52
▶ Present study	▶ Australia	▶ 1♂	▶ Kroombit, Queensland State (24°22′S, 150°55′E)	▶ 1,2	▶ 26/52
▶ Present study	▶ Australia	▶ 2♂1♀	▶ Watagan State Forest, New South Wales State (33°00′S, 151°23′E)	▶ 1,2,4	▶ 26/52
Litoria personata*					
▶ Present study	▶ Australia	▶ 3♂2♀	▶ Bowerbird, Kakadu National Park, Northern Territory (12°25′S, 132°40′E)	▶ 1,2,4,14	▶ 26/52
Litoria pygmaea					
▶ Menzies & Tippet, 1976[N]	▶ New Guinea	▶ 1♂	▶ Brown River, Central Province	▶ 1	▶ 26‡/n.i.
▷ Morescalchi, 1979	▷ n.i.	▷ n.i.	▷ ref. Menzies & Tippett, 1976	▷ n.i.	▷ 26/n.i.
▷ King, 1990	▷ New Guinea	▷ n.i.	▷ ref. Menzies & Tippett, 1976	▷ n.i.	▷ 26/n.i.
▷ Kuramoto, 1990	▷ Aus./NG.	▷ n.i.	▷ ref. Menzies & Tippett, 1976; ctry.n.sp.	▷ 1	▷ 26/n.i.
▷ Catroli, 2008	▷ n.i.	▷ n.i.	▷ ref. Menzies & Tippett, 1976	▷ n.i.	▷ 26/n.i.
▷ Catroli & Kasahara, 2009	▷ n.i.	▷ n.i.	▷ ref. Menzies & Tippett, 1976	▷ n.i.	▷ 26/n.i.
Litoria rothii					
▶ King et al., 1990[P]	▶ Australia	▶ 1s.n.i.	▶ Hayes Creek, Northern Territory	▶ 13	▶ 26/52†
▷ King, 1990	▷ Aus./NG.	▷ n.i.	▷ ref. King, 1981, 1985	▷ 2,10,13	▷ 26/52
▷ Catroli, 2008	▷ n.i.	▷ n.i.	▷ ref. King, 1980/err.ref.	▷ n.i.	▷ 26/n.i.
▷ Catroli & Kasahara, 2009	▷ n.i.	▷ n.i.	▷ ref. King, 1980/err.ref., 1985	▷ 17	▷ 26/n.i.
▶ Present study	▶ Australia	▶ 7♂10♀ 3U	▶ Scott Creek, Northern Territory (14°50′S, 132°08′E)	▶ 1,2,4,5, 7,8,14	▶ 26/52
▶ Present study	▶ Australia	▶ 2♂1♀	▶ Garden of Northern Territory Museum, Fannie Bay, Darwin, Northern Territory (12°26′S, 130°50′E)	▶ 1,2,4,5, 7,8,14	▶ 26/52
Litoria rubella					
▶ Menzies & Tippet, 1976[N]	▶ New Guinea	▶ 4♂	▶ Weam, Western Province	▶ 1	▶ 26‡/n.i.
▷ Morescalchi, 1979	▷ n.i.	▷ n.i.	▷ ref. Menzies & Tippett, 1976; Morescalchi & Ingram, unpubl.	▷ n.i.	▷ 26/n.i.
▶ Morescalchi, 1979[N]	▶ n.i.	▶ n.i.	▶ ref. Morescalchi & Ingram, unpubl.	▶ n.i.	▶ 26/n.i.
▷ King, 1990	▷ Australia	▷ n.i.	▷ ref. Menzies & Tippett, 1976; King, 1981	▷ 2,10	▷ 26/52
▷ Kuramoto, 1990	▷ Aus./NG.	▷ n.i.	▷ ref. Menzies & Tippett, 1976; Morescalchi, 1979; ctry.n.sp.	▷ 1	▷ 26/n.i.
▶ King, 1991[P]	▶ n.i.	▶ n.i.	▶ Australia	▶ 1,10	▶ 26/52†
▷ Catroli, 2008	▷ n.i.	▷ n.i.	▷ ref. Menzies & Tippett, 1976; King, 1980/err.ref.	▷ n.i.	▷ 26/n.i.
▷ Catroli & Kasahara, 2009	▷ n.i.	▷ n.i.	▷ ref. Menzies & Tippett, 1976; Morescalchi, 1979; King, 1980/err.ref.	▷ 17	▷ 26/n.i.
▶ Present study	▶ Australia	▶ 1♂	▶ Mannum, on River Murray	▶ 1	▶ 26/52
▶ Present study	▶ Australia	▶ 1♂	▶ Gladstone, Queensland state	▶ 1	▶ 26/52
▶ Present study	▶ Australia	▶ 3♂7♀ 1U	▶ vicinity of Howard Springs, Northern Territory (12°29′S, 131°03′E)	▶ 1,2,4	▶ 26/52
Litoria thesaurensis					
▶ Menzies & Tippet, 1976[N]	▶ New Guinea	▶ 2♂	▶ Lae, Morobe Province	▶ 1	▶ 26‡/n.i.
▷ Morescalchi, 1979	▷ n.i.	▷ n.i.	▷ ref. Menzies & Tippett, 1976	▷ n.i.	▷ 26/n.i.
▷ King, 1990	▷ New Guinea	▷ n.i.	▷ ref. Menzies & Tippett, 1976	▷ n.i.	▷ 26/n.i.
▷ Kuramoto, 1990	▷ Aus./NG.	▷ n.i.	▷ ref. Menzies & Tippett, 1976; Kuramoto & Allison, unpubl.; ctry.n.sp.	▷ 1	▷ 26/n.i.
▶ Kuramoto & Allison, 1991[P,M]	▶ New Guinea	▶ 1♂	▶ vicinity of Wau, Morobe Province	▶ 1	▶ 26/52†
▷ Catroli, 2008	▷ n.i.	▷ n.i.	▷ ref. Menzies & Tippett, 1976; Kuramoto & Allison, 1991	▷ n.i.	▷ 26/n.i.
▷ Catroli & Kasahara, 2009	▷ n.i.	▷ n.i.	▷ ref. Menzies & Tippett, 1976; Kuramoto & Allison, 1991	▷ n.i.	▷ 26/n.i.

DOI: 10.1159/000490085

Author(s) ▶ original data[a] ▷ review articles	Country/ Region[b]	Speci-mens[c]	Locality of sampling, references cited, and remarks[d]	Methods applied[e]	2n/FN[f]
Litoria timida					
▶ Menzies & Tippet, 1976[N]	▶ New Guinea	▶ 2♂	▶ Brown River, Central Province	▶ 1	▶ 26‡/n.i.
▷ Morescalchi, 1979	▷ n.i.	▷ n.i.	▷ ref. Menzies & Tippett, 1976	▷ n.i.	▷ 26/n.i.
▷ King, 1990	▷ New Guinea	▷ n.i.	▷ ref. Menzies & Tippett, 1976	▷ n.i.	▷ 26/n.i.
▷ Kuramoto, 1990	▷ Aus./NG.	▷ n.i.	▷ ref. Menzies & Tippett, 1976; ctry.n.sp.	▷ 1	▷ 26/n.i.
▷ Catroli, 2008	▷ n.i.	▷ n.i.	▷ ref. Menzies & Tippett, 1976	▷ n.i.	▷ 26/n.i.
▷ Catroli & Kasahara, 2009	▷ n.i.	▷ n.i.	▷ ref. Menzies & Tippett, 1976	▷ n.i.	▷ 26/n.i.
Litoria tornieri					
▷ King, 1990	▷ Australia	▷ n.i.	▷ ref. King, 1981	▷ n.i.	▷ 26/52
▷ Catroli, 2008	▷ n.i.	▷ n.i.	▷ ref. King, 1980/err.ref.	▷ n.i.	▷ 26/n.i.
▷ Catroli & Kasahara, 2009	▷ n.i.	▷ n.i.	▷ ref. King, 1980/err.ref.	▷ n.i.	▷ 26/n.i.
▶ Present study	▶ Australia	▶ 1♂	▶ Scott Creek, Northern Territory (14°50′S, 132°08′E)	▶ 1,2,4,14	▶ 26/52
▶ Present study	▶ Australia	▶ 14♂5♀	▶ Ja Ja, Kakadu National Park, Northern Territory (12°34′S, 132°53′E)	▶ 1,2,4,14	▶ 26/52
Litoria tyleri*					
▶ Present study	▶ Australia	▶ 1U	▶ Southeastern Queensland, loc.u.	▶ 1	▶ 26/52
Litoria verreauxii					
▷ King, 1990	▷ Australia	▷ n.i.	▷ ref. King, 1981	▷ n.i.	▷ 26/52
▷ Catroli, 2008	▷ n.i.	▷ n.i.	▷ ref. King, 1980/err.ref.	▷ n.i.	▷ 26/n.i.
▷ Catroli & Kasahara, 2009	▷ n.i.	▷ n.i.	▷ ref. King, 1980/err.ref.	▷ n.i.	▷ 26/n.i.
▷ Schmid & Steinlein, 2016a	▷ n.i.	▷ n.i.	▷ ref. Schmid et al., unpubl.	▷ n.i.	▷ 26/n.i.
▶ Present study	▶ Australia	▶ 1♂1♀	▶ Southeastern Queensland State, loc.u.	▶ 1,2,4, 5,14	▶ 26/52
Litoria vocivincens					
▶ Menzies & Tippet, 1976[N]	▶ New Guinea	▶ 1tp	▶ Brown River, Central Province	▶ 1	▶ 26/n.i.
▷ Morescalchi, 1979	▷ n.i.	▷ n.i.	▷ ref. Menzies & Tippett, 1976	▷ n.i.	▷ 26/n.i.
▷ King, 1990	▷ New Guinea	▷ n.i.	▷ ref. Menzies & Tippett, 1976	▷ n.i.	▷ 26/n.i.
▷ Kuramoto, 1990	▷ Aus./NG.	▷ n.i.	▷ ref. Menzies & Tippett, 1976; ctry.n.sp.	▷ 1	▷ 26/n.i.
▷ Catroli, 2008	▷ n.i.	▷ n.i.	▷ ref. Menzies & Tippett, 1976	▷ n.i.	▷ 26/n.i.
▷ Catroli & Kasahara, 2009	▷ n.i.	▷ n.i.	▷ ref. Menzies & Tippett, 1976	▷ n.i.	▷ 26/n.i.
Litoria watjulumensis					
▷ King, 1990	▷ Australia	▷ n.i.	▷ ref. King, 1981; list. *Litoria {wotjulumensis}*	▷ n.i.	▷ 26/52
▷ Catroli, 2008	▷ n.i.	▷ n.i.	▷ ref. King, 1980/err.ref; list. *Litoria {wotjulumensis}*	▷ n.i.	▷ 26/n.i.
▷ Catroli & Kasahara, 2009	▷ n.i.	▷ n.i.	▷ ref. King, 1980/err.ref.	▷ n.i.	▷ 26/n.i.
▶ Present study	▶ Australia	▶ 1♂	▶ Scott Creek, Northern Territory (14°50′S, 132°08′E)	▶ 1,2,4,5, 7,8,14	▶ 26/52
▶ Present study	▶ Australia	▶ 8♂6♀ 1U	▶ Bowerbird, Kakadu National Park, Northern Territory (12°25′S, 132°40′E)	▶ 1,2,4,5, 7,8,14	▶ 26/52
Litoria wollastoni					
▶ Menzies & Tippet, 1976[N]	▶ New Guinea	▶ 4♂	▶ Baiyer River, Enga Province; Mau River, Milne Bay Province; Waitape, Central Province	▶ 1	▶ 26‡/n.i.
▷ Morescalchi, 1979	▷ n.i.	▷ n.i.	▷ ref. Menzies & Tippett, 1976	▷ n.i.	▷ 26/n.i.
▷ King, 1990	▷ New Guinea	▷ n.i.	▷ ref. Menzies & Tippett, 1976	▷ n.i.	▷ 26/n.i.
▷ Kuramoto, 1990	▷ Aus./NG.	▷ n.i.	▷ ref. Duellman, 1967; Menzies & Zweifel, 1974; Menzies & Tippett, 1976; Kuramoto & Allison, unpubl.; ctry.n.sp.	▷ 1	▷ 26/n.i.
▶ Kuramoto & Allison, 1991[P,M]	▶ New Guinea	▶ 4♂	▶ vicinity of Wau, Morobe Province	▶ 1	▶ 26/52†
▷ Catroli, 2008	▷ n.i.	▷ n.i.	▷ ref. Menzies & Tippett, 1976; Kuramoto & Allison, 1991	▷ n.i.	▷ 26/n.i.
▷ Catroli & Kasahara, 2009	▷ n.i.	▷ n.i.	▷ ref. Menzies & Tippett, 1976; Kuramoto & Allison, 1991	▷ n.i.	▷ 26/n.i.
***Litoria* sp. (*bicolor* group)**					
▷ Kuramoto, 1990	▷ Aus./NG.	▷ n.i.	▷ ref. Menzies & Tippett, 1976; ctry.n.sp.	▷ 1	▷ 26/n.i.
PELODRYADIDAE / Pelodryadinae / *Nyctimystes*					
Nyctimystes cheesmani					
▶ Menzies & Tippet, 1976[N]	▶ New Guinea	▶ 5♂	▶ Baiyer River, Enga Province	▶ 1	▶ 26‡/n.i.
▷ Morescalchi, 1979	▷ n.i.	▷ n.i.	▷ ref. Menzies & Tippett, 1976	▷ n.i.	▷ 26/n.i.
▷ King, 1990	▷ New Guinea	▷ n.i.	▷ ref. Menzies & Tippett, 1976	▷ n.i.	▷ 26/n.i.
▷ Kuramoto, 1990	▷ Aus./NG.	▷ n.i.	▷ ref. Menzies & Tippett, 1976; ctry.n.sp.	▷ 17	▷ 26/n.i.
▷ Catroli, 2008	▷ n.i.	▷ n.i.	▷ ref. Menzies & Tippett, 1976; list. *Litoria {cheesmanae}*	▷ n.i.	▷ 26/n.i.
▷ Catroli & Kasahara, 2009	▷ n.i.	▷ n.i.	▷ ref. Menzies & Tippett, 1976; list. *Litoria {cheesmanae}*	▷ n.i.	▷ 26/n.i.
Nyctimystes daymani					
▶ Menzies & Tippet, 1976[N]	▶ New Guinea	▶ 2♂	▶ Baiyer River, Enga Province	▶ 1	▶ 26‡/n.i.
▷ Morescalchi, 1979	▷ n.i.	▷ n.i.	▷ ref. Menzies & Tippett, 1976	▷ n.i.	▷ 26/n.i.
▷ King, 1990	▷ New Guinea	▷ n.i.	▷ ref. Menzies & Tippett, 1976	▷ n.i.	▷ 26/n.i.
▷ Kuramoto, 1990	▷ Aus./NG.	▷ n.i.	▷ ref. Menzies & Tippett, 1976; ctry.n.sp.	▷ 17	▷ 26/n.i.
▷ Catroli, 2008	▷ n.i.	▷ n.i.	▷ ref. Menzies & Tippett, 1976; *list. Litoria daymani*	▷ n.i.	▷ 26/n.i
▷ Catroli & Kasahara, 2009	▷ n.i.	▷ n.i.	▷ ref. Menzies & Tippett, 1976; *list. Litoria daymani*	▷ n.i.	▷ 26/n.i.

Author(s) ▶ original data[a] ▷ review articles	Country/ Region[b]	Speci- mens[c]	Locality of sampling, references cited, and remarks[d]	Methods applied[e]	2n/FN[f]
Nyctimystes disruptus					
▶ Duellman, 1967[N]	▶ New Guinea	▶ 2♂	▶ Morobe Province; rep. *Nyctimystes papua*; err.id.sp.	▶ 1	▶ 26‡/n.i.
▷ Foresti, 1972	▷ n.i.	▷ n.i.	▷ ref. Duellman, 1967; list. *Nyctimystes papua*	▷ n.i.	▷ 26‡/n.i
▷ Kuramoto, 1990	▷ Aus./NG.	▷ n.i.	▷ ref. Duellman, 1967; ctry.n.sp.	▷ 1	▷ 26/n.i.
▷ Catroli, 2008	▷ n.i.	▷ n.i.	▷ ref. Duellman, 1967; list. *Litoria papua*	▷ n.i.	▷ 26/n.i.
▷ Catroli & Kasahara, 2009	▷ n.i.	▷ n.i.	▷ ref. Duellman, 1967; list. *Litoria papua*	▷ n.i.	▷ 26/n.i.
Nyctimystes foricula					
▶ Duellman, 1967[N]	▶ New Guinea	▶ 2♂	▶ Morobe Province	▶ 1	▶ 26‡/n.i.
▷ Foresti, 1972	▷ n.i.	▷ n.i.	▷ ref. Duellman, 1967	▷ n.i.	▷ 26‡/n.i.
▷ King, 1990	▷ New Guinea	▷ n.i.	▷ ref. Duellman, 1967	▷ n.i.	▷ 26/n.i.
▷ Kuramoto, 1990	▷ Aus./NG.	▷ n.i.	▷ ref. Duellman, 1967; Kuramoto & Allison, unpubl.; ctry.n.sp.	▷ 1	▷ 26/n.i.
▶ Kuramoto & Allison, 1991[P,M]	▶ New Guinea	▶ 2♂	▶ vicinity of Wau, Morobe Province	▶ 1	▶ 26/52†
▷ Catroli, 2008	▷ n.i.	▷ n.i.	▷ ref. Duellman, 1967; Kuramoto & Allison, 1991; list. *Litoria foricula*	▷ n.i.	▷ 26/n.i.
▷ Catroli & Kasahara, 2009	▷ n.i.	▷ n.i.	▷ ref. Duellman, 1967; Kuramoto & Allison, 1991; list. *Litoria foricula*	▷ n.i.	▷ 26/n.i.
Nyctimystes humeralis					
▶ Menzies & Tippet, 1976[N]	▶ New Guinea	▶ 2♂	▶ Baiyer River, Enga Province	▶ 1	▶ 26‡/n.i.
▷ Morescalchi, 1979	▷ n.i.	▷ n.i.	▷ ref. Menzies & Tippett, 1976	▷ n.i.	▷ 26/n.i.
▷ King, 1990	▷ New Guinea	▷ n.i.	▷ ref. Menzies & Tippett, 1976	▷ n.i.	▷ 26/n.i.
▷ Kuramoto, 1990	▷ Aus./NG.	▷ n.i.	▷ ref. Menzies & Tippett, 1976; ctry.n.sp.	▷ 1	▷ 26/n.i.
▷ Catroli, 2008	▷ n.i.	▷ n.i.	▷ ref. Menzies & Tippett, 1976; list. *Litoria humeralis*	▷ n.i.	▷ 26/n.i.
▷ Catroli & Kasahara, 2009	▷ n.i.	▷ n.i.	▷ ref. Menzies & Tippett, 1976; list. *Litoria humeralis*	▷ n.i.	▷ 26/n.i.
Nyctimystes infrafrenatus					
▶ Menzies & Tippet, 1976[P,M]	▶ New Guinea	▶ 8s.n.i.	▶ University campus, Port Moresby, Central Province; rep. *Litoria infrafrenata*	▶ 1	▶ 24/48†
▶ Menzies & Tippet, 1976[P]	▶ New Guinea	▶ 1♂	▶ Lae, Morobe Province; rep. *Litoria infrafrenata*	▶ 1	▶ 24/48†
▶ Menzies & Tippet, 1976[P]	▶ New Guinea	▶ 1s.n.i.	▶ Daru, Western Province; rep. *Litoria infrafrenata*	▶ 1	▶ 24/48†
▶ Menzies & Tippet, 1976[P]	▶ New Guinea	▶ 1♂	▶ Goodenough Island; rep. *Litoria infrafrenata*	▶ 1	▶ 24/48†
▶ Menzies & Tippet, 1976[P]	▶ New Guinea	▶ 1♂	▶ Hoskins, New Britain Island; rep. *Litoria infrafrenata*	▶ 1	▶ 24/48†
▷ Morescalchi, 1979	▷ n.i.	▷ n.i.	▷ ref. Menzies & Tippett, 1976; list. *Litoria infrafrenata*	▷ n.i.	▷ 24/n.i.
▶ King, 1980[P,M,I]	▶ Australia	▶ 4s.n.i.	▶ Tully and Edward River, Queensland State; rep. *Litoria infrafrenata*	▶ 1,2,4	▶ 24/48†
▷ King, 1990	▷ Aus./NG.	▷ n.i.	▷ ref. Menzies & Tippett, 1976; King, 1980; ctry.n.sp.; list. *Litoria infrafrenata*	▷ 2,4	▷ 24/48
▷ Kuramoto, 1990	▷ Aus./NG.	▷ n.i.	▷ ref. Menzies & Tippett, 1976; King, 1980; ctry.n.sp.; list. *Litoria infrafrenata*	▷ 17	▷ 24/n.i.
▶ Schmid et al., 1995[P]	▶ Australia	▶ 1♂	▶ specimens obtained by animal dealer; loc.n.i.; rep. *Litoria infrafrenata*	▶ 2,6,11	▶ 24/48†
▷ Catroli, 2008	▷ n.i.	▷ n.i.	▷ ref. Menzies & Tippett, 1976; King, 1980; list. *Litoria infrafrenata*	▷ n.i.	▷ 24/n.i.
▷ Catroli & Kasahara, 2009	▷ n.i.	▷ n.i.	▷ ref. Menzies & Tippett, 1976; King, 1980; list. *Litoria infrafrenata*	▷ 17	▷ 24/n.i.
▶ Present study	▶ Australia	▶ 1♂	▶ specimens obtained by animal dealer; loc.n.i.; rep. *Litoria infrafrenata*	▶ 1,2,4,5, 7,8,14	▶ 24/48
Nyctimystes kubori					
▶ Duellman, 1967[N]	▶ New Guinea	▶ 3♂	▶ Morobe Province	▶ 1	▶ 26‡/n.i.
▶ Duellman, 1967[N]	▶ New Guinea	▶ 2♂	▶ Eastern Highlands Province	▶ 1	▶ 26‡/n.i.
▷ Foresti, 1972	▷ n.i.	▷ n.i.	▷ ref. Duellman, 1967	▷ n.i.	▷ 26‡/n.i.
▶ Menzies & Tippet, 1976[N]	▶ New Guinea	▶ 4♂	▶ Kaironk, Morobe Province; Gumine, (Chimbu) Simbu Province; Kubor, Western Highlands Province	▶ 1	▶ 26‡/n.i.
▷ Morescalchi, 1979	▷ n.i.	▷ n.i.	▷ ref. Menzies & Tippett, 1976	▷ n.i.	▷ 26/n.i.
▷ King, 1990	▷ New Guinea	▷ n.i.	▷ ref. Duellman, 1967; Menzies & Tippett, 1976	▷ n.i.	▷ 26/n.i.
▷ Kuramoto, 1990	▷ Aus./NG.	▷ n.i.	▷ ref. Duellman, 1967; Menzies & Tippett, 1976; ctry.n.sp.	▷ 1	▷ 26/n.i.
▷ Catroli, 2008	▷ n.i.	▷ n.i.	▷ ref. Duellman, 1967; Menzies & Tippett, 1976; list. *Litoria kubori*	▷ n.i.	▷ 26/n.i.
▷ Catroli & Kasahara, 2009	▷ n.i.	▷ n.i.	▷ ref. Duellman, 1967; Menzies & Tippett, 1976; list. *Litoria kubori*	▷ n.i.	▷ 26/n.i.
Nyctimystes papua					
▶ Menzies & Tippet, 1976[N]	▶ New Guinea	▶ 1♂	▶ Waitape, Central Province	▶ 1	▶ 26‡/n.i.
▷ Morescalchi, 1979	▷ n.i.	▷ n.i.	▷ ref. Menzies & Tippett, 1976	▷ n.i.	▷ 26/n.i.
▷ King, 1990	▷ New Guinea	▷ n.i.	▷ ref. Menzies & Tippett, 1976	▷ n.i.	▷ 26/n.i.
▷ Kuramoto, 1990	▷ Aus./NG.	▷ n.i.	▷ ref. Menzies & Tippett, 1976; ctry.n.sp.	▷ 1	▷ 26/n.i.
▷ Catroli, 2008	▷ n.i.	▷ n.i.	▷ ref. Menzies & Tippett, 1976; list. *Litoria papua*	▷ n.i.	▷ 26/n.i.
▷ Catroli & Kasahara, 2009	▷ n.i.	▷ n.i.	▷ ref. Menzies & Tippett, 1976; list. *Litoria papua*	▷ n.i.	▷ 26/n.i.
Nyctimystes pulcher					
▶ Kuramoto, 1990[N]	▶ Aus./NG.	▶ n.i.	▶ ref. Kuramoto & Allison, unpubl.; ctry.n.sp.; list. *Nyctimystes pulchra*	▶ 1	▶ 26/n.i.
▶ Kuramoto & Allison, 1991[P,M]	▶ New Guinea	▶ 1♂	▶ vicinity of Wau, Morobe Province; rep. *Nyctimystes pulchra*	▶ 1	▶ 26/52†
▷ Catroli, 2008	▷ n.i.	▷ n.i.	▷ ref. Kuramoto & Allison, 1991; list. *Litoria pulchra*	▷ n.i.	▷ 26/n.i.
▷ Catroli & Kasahara, 2009	▷ n.i.	▷ n.i.	▷ ref. Kuramoto & Allison, 1991; list. *Litoria pulchra*	▷ n.i.	▷ 26/n.i.
Nyctimystes semipalmatus					
▶ Duellman, 1967[N]	▶ New Guinea	▶ 1♂	▶ Oro (Northern) Province; rep. *Nyctimystes semipalmata*	▶ 1	▶ 26‡/n.i.
▷ Foresti, 1972	▷ n.i.	▷ n.i.	▷ ref. Duellman, 1967; list. *Nyctimystes semipalmata*	▷ n.i.	▷ 26‡/n.i.
▷ King, 1990	▷ Australia	▷ n.i.	▷ ref. Duellman, 1967; err.cit.ctry.; list. *Nyctimystes semipalmata*	▷ n.i.	▷ 26/n.i.
▷ Kuramoto, 1990	▷ Aus./NG.	▷ n.i.	▷ ref. Duellman, 1967; ctry.n.sp.; list. *Nyctimystes semipalmata*	▷ 1	▷ 26/n.i.
▷ Catroli, 2008	▷ n.i.	▷ n.i.	▷ ref. Duellman, 1967; list. *Litoria semipalmata*	▷ n.i.	▷ 26/n.i.
▷ Catroli & Kasahara, 2009	▷ n.i.	▷ n.i.	▷ ref. Duellman, 1967; list. *Litoria semipalmata*	▷ n.i.	▷ 26/n.i.

Author(s) ▶ original data[a] ▷ review articles	Country/ Region[b]	Speci- mens[c]	Locality of sampling, references cited, and remarks[d]	Methods applied[e]	2n/FN[f]
***Nyctimystes* sp. (*montana* group)**					
▶ Menzies & Tippet, 1976[N]	▶ New Guinea	▶ 4♂	▶ Ei Creek and Waitape, Central Province	▶ 1	▶ 26‡/n.i.
▷ Morescalchi, 1979	▷ n.i.	▷ n.i.	▷ ref. Menzies & Tippett, 1976; list. *Nyctimystes montana*/err.cit.sp.	▷ n.i.	▷ 26/n.i.
▷ King, 1990	▷ New Guinea	▷ n.i.	▷ ref. Menzies & Tippett, 1976; list. *Nyctimystes montana*/err.cit.sp.	▷ n.i.	▷ 26/n.i.
▷ Kuramoto, 1990	▷ n.i.	▷ n.i.	▷ ref. Menzies & Tippett, 1976; ctry.n.sp.; list. *Nyctimystes montana*/ err.cit.sp.	▷ 1	▷ 26/n.i.
▷ Catroli, 2008	▷ n.i.	▷ n.i.	▷ ref. Menzies & Tippett, 1976; list. *Litoria* gr. *montana*	▷ n.i.	▷ 26/n.i.
▷ Catroli, 2008	▷ n.i.	▷ n.i.	▷ ref. Menzies & Tippett, 1976; list. *Litoria montana*/err.cit.sp.	▷ n.i.	▷ 26/n.i.
▷ Catroli & Kasahara, 2009	▷ n.i.	▷ n.i.	▷ ref. Menzies & Tippett, 1976; list. *Litoria* gr. *montana*	▷ n.i.	▷ 26/n.i.
▷ Catroli & Kasahara, 2009	▷ n.i.	▷ n.i.	▷ ref. Menzies & Tippett, 1976; list. *Litoria montana*/err.cit.sp.	▷ n.i.	▷ 26/n.i.
PELODRYADIDAE / Pelodryadinae / *Ranoidea*					
Ranoidea alboguttata					
▷ King, 1990	▷ Australia	▷ n.i.	▷ ref. Morescalchi & Ingram, 1974; King, 1981; list. *Litoria alboguttata*	▷ n.i.	▷ 26/52
▷ Kuramoto, 1990	▷ Aus./NG.	▷ n.i.	▷ ref. Morescalchi & Ingram, 1974; ctry.n.sp.; list. *Litoria alboguttata*	▷ 1	▷ 26/n.i.
▷ Catroli, 2008	▷ n.i.	▷ n.i.	▷ ref. Morescalchi & Ingram, 1974; King, 1980/err.ref.; list. *Litoria alboguttata*	▷ n.i.	▷ 26/n.i.
▷ Catroli & Kasahara, 2009	▷ n.i.	▷ n.i.	▷ ref. King, 1980/err.ref.; list. *Litoria alboguttata*	▷ n.i.	▷ 26/n.i.
▶ Present study	▶ Australia	▶ 2U	▶ Queensland State, loc.u.	▶ 1	▶ 26/52
Ranoidea aurea					
▷ King, 1990	▷ Australia	▷ n.i.	▷ ref. King, 1981; list. *Litoria aurea*	▷ n.i.	▷ 26/52
▷ Catroli, 2008	▷ n.i.	▷ n.i.	▷ ref. King, 1980/err.ref.; list. *Litoria aurea*	▷ n.i.	▷ 26/n.i.
▷ Catroli & Kasahara, 2009	▷ n.i.	▷ n.i.	▷ ref. King, 1980/err.ref.; list. *Litoria aurea*	▷ n.i.	▷ 26/n.i.
Ranoidea australis					
▶ King et al, 1979[P,M]	▶ Australia	▶ 3s.n.i.	▶ Cannon Hill, Northern Territory; rep. *Cyclorana australis*	▶ 1	▶ 26/52†
▷ King, 1990	▷ Australia	▷ n.i.	▷ ref. King et al., 1979 and unpubl.; list. *Cyclorana australis*	▷ 2,10	▷ 26/52
▶ King, 1990[I]	▶ Australia	▶ n.i.	▶ loc.n.i.; rep. *Cyclorana australis*	▶ 2,4	▶ 26/52
▷ Kuramoto, 1990	▷ Aus./NG.	▷ n.i.	▷ ref. King et al., 1979; Morescalchi, 1979; ctry.n.sp.; list. *Cyclorana australis*	▷ 1	▷ 26/n.i.
▶ King, 1991[P]	▶ n.i.	▶ n.i.	▶ loc.n.i.; rep. *Cyclorana australis*	▶ 2	▶ 26/52†
▷ Catroli, 2008	▷ n.i.	▷ n.i.	▷ ref. King et al., 1979; Morescalchi, 1979; list. *Litoria australis*	▷ n.i.	▷ 26/n.i.
▷ Catroli & Kasahara, 2009	▷ n.i.	▷ n.i.	▷ ref. King et al., 1979; list. *Litoria australis*	▷ 17	▷ 26/n.i.
▶ Schmid & Steinlein, 2016a[P]	▶ Australia	▶ n.i.	▶ Adelaide River, Nothern Territory (12°39′S, 131°19′E); rep. *Dryopsophus australis*	▶ 2,4,5, 8,14	▶ 26/52†
▶ Present study	▶ Australia	▶ n.i.	▶ Adelaide River, Northern Territory (12°39′S, 131°19′E)	▶ 1,2,4, 5,8,14	▶ 26/52
Ranoidea barringtonensis*					
▶ Present study	▶ Australia	▶ 10♂ 1♀	▶ Kroombit, Queensland State (24°22′S, 150°55′E)	▶ 1,2,4, 5,14	▶ 26/52
▶ Present study	▶ Australia	▶ 7♂1♀	▶ Conondale, Queensland State (26°44′S, 152°44′E)	▶ 1,2,4, 5,14	▶ 26/52
Ranoidea booroolongensis					
▶ Donnellan & Mahony, 2004[P]	▶ Australia	▶ 10s.n.i.	▶ Abercrombie River, New South Wales State (34°12′S, 149°40′E); rep. *Litoria booroolongensis*	▶ 1,2	▶ 26/52†
▶ Donnellan & Mahony, 2004[P]	▶ Australia	▶ 2s.n.i.	▶ Sofala, New South Wales State (33°05′S, 149°41′E); rep. *Litoria booroolongensis*	▶ 1,2	▶ 26/52†
▷ Catroli, 2008	▷ n.i.	▷ n.i.	▷ ref. Donnellan & Mahony, 2004; list. *Litoria booroolongensis*	▷ n.i.	▷ 26/n.i.
▷ Catroli & Kasahara, 2009	▷ n.i.	▷ n.i.	▷ ref. Donnellan & Mahony, 2004; list. *Litoria booroolongensis*	▷ 17	▷ 26/n.i.
▶ Present study	▶ Australia	▶ 1U	▶ Abercrombie River, New South Wales State (34°12′S, 149°40′E)	▶ 4	▶ 26/52
Ranoidea brevipes					
▶ King et al., 1979[P,M]	▶ Australia	▶ 1s.n.i.	▶ Mon Repos, Bundaberg, Queensland State; rep. *Cyclorana brevipes*	▶ 1	▶ 26/52†
▷ King, 1990	▷ Australia	▷ n.i.	▷ ref. King et al., 1979; list. *Cyclorana brevipes*	▷ n.i.	▷ 26/52
▷ Kuramoto, 1990	▷ Aus./NG.	▷ n.i.	▷ ref. King et al., 1979; Morescalchi, 1979; ctry.n.sp.; list. *Cyclorana brevipes*	▷ 1	▷ 26/n.i.
▷ Catroli, 2008	▷ n.i.	▷ n.i.	▷ ref. King et al., 1979; Morescalchi, 1979; list. *Litoria brevipes*	▷ n.i.	▷ 26/n.i.
▷ Catroli & Kasahara, 2009	▷ n.i.	▷ n.i.	▷ ref. King et al., 1979; list. *Litoria brevipes*	▷ n.i.	▷ 26/n.i.
Ranoidea caerulea					
▶ Stephenson & Stephenson, 1970[P,M,I]	▶ Australia	▶ 3♂1♀	▶ loc.n.i.; rep. *Hyla caerulea*	▶ 1	▶ 26/52†
▷ Foresti, 1972	▷ n.i	▷ n.i	▷ ref. Stephenson & Stephenson, 1970; list. *Hyla caerulea*	▷ n.i	▷ 26/n.i.
▶ Menzies & Tippet, 1976[N]	▶ New Guinea	▶ 1s.n.i.	▶ Port Moresby, Central Province; rep. *Litoria caerulea*	▶ 1	▶ 26/n.i.
▷ Morescalchi, 1979	▷ n.i.	▷ n.i.	▷ ref. Menzies & Tippet, 1976; list. *Litoria caerulea*	▷ n.i.	▷ 26/n.i.
▶ King et al., 1990[N]	▶ Australia	▶ 1s.n.i.	▶ Hayes Creek, Northern Territoty; rep. *Litoria caerulea*	▶ 13	▶ 26/n.i.
▷ King, 1990	▷ Australia	▷ n.i.	▷ ref. Stephenson & Stephenson, 1970; Menzies & Tippett, 1976; King, 1981; list. *Litoria caerulea*	▷ 13	▷ 26/52

Author(s) ▶ original data[a] ▷ review articles	Country/ Region[b]	Speci- mens[c]	Locality of sampling, references cited, and remarks[d]	Methods applied[e]	2n/FN[f]
▶ King, 1990[P]	▶ n.i.	▶ n.i.	▶ loc.n.i.; rep. *Litoria caerulea*	▶ 1	▶ 26/52
▷ Kuramoto, 1990	▷ Aus./NG.	▷ n.i.	▷ ref. Stephenson & Stephenson, 1970; Menzies & Tippet, 1976; ctry.n.sp.; list. *Litoria caerulea*	▷ 1	▷ 26/n.i.
▷ Catroli, 2008	▷ n.i.	▷ n.i.	▷ ref. Stephenson & Stephenson, 1970; Menzies & Tippet, 1976; King et al., 1979; King, 1980/err.ref., 1990; list. *Litoria caerulea*	▷ n.i.	▷ 26/n.i.
▷ Catroli & Kasahara, 2009	▷ n.i.	▷ n.i.	▷ ref. Stephenson & Stephenson, 1970; Menzies & Tippet, 1976; King, 1980/err.ref., 1990; list. *Litoria caerulea*	▷ 17	▷ 26/n.i.
▶ Present study	▶ Australia	▶ 5♂9♀ 3U	▶ Garden of Norhern Territory Museum, Fannie Bay, Darwin, Northern Territory (12°26′S, 130°50′E)	▶ 1,2,4, 5,8,14	▶ 26/52
Ranoidea chloris					
▷ Morescalchi, 1979	▷ n.i.	▷ n.i.	▷ ref. Morescalchi & Ingram, 1974; list. *Litoria chloris*	▷ n.i.	▷ 26/n.i.
▶ King, 1980[P,M,I]	▶ Australia	▶ 2s.n.i.	▶ Warrie National Park, Queensland State; rep. *Litoria chloris*	▶ 1,2,4	▶ 26/52[†]
▷ King, 1990	▷ Australia	▷ n.i.	▷ ref. Morescalchi & Ingram, 1974; King 1980; list. *Litoria chloris*	▷ 2,4	▷ 26/52
▶ King, 1990[P]	▶ n.i.	▶ n.i.	▶ loc.n.i.; rep. *Litoria chloris*	▶ 2	▶ 26/52[†]
▷ Kuramoto, 1990	▷ Aus./NG.	▷ n.i.	▷ ref. Morescalchi & Ingram, 1974; King 1980; ctry.n.sp.; list. *Litoria chloris*	▷ 17	▷ 26/n.i.
▷ Catroli, 2008	▷ n.i.	▷ n.i.	▷ ref. Morescalchi & Ingram, 1974; King 1980; list. *Litoria chloris*	▷ n.i.	▷ 26/n.i.
▷ Catroli & Kasahara, 2009	▷ n.i.	▷ n.i.	▷ ref. Morescalchi & Ingram, 1974; King 1980; list. *Litoria chloris*	▷ 17	▷ 26/n.i.
▶ Present study	▶ Australia	▶ 1U	▶ Queensland State, loc.u.	▶ 1	▶ 26/52
Ranoidea citropa					
▶ King, 1990[P]	▶ n.i.	▶ n.i.	▶ loc.n.i.; rep. *Litoria citropa*	▶ 1	▶ 26/52
▷ Catroli, 2008	▷ n.i.	▷ n.i.	▷ ref. King, 1980/err.ref.; 1990; list. *Litoria citropa*	▷ n.i.	▷ 26/n.i.
▷ Catroli & Kasahara, 2009	▷ n.i.	▷ n.i.	▷ ref. King, 1990; list. *Litoria citropa*	▷ n.i.	▷ 26/n.i.
▶ Present study	▶ Australia	▶ 1U	▶ New South Wales State, loc.u.	▶ 1	▶ 26/52
Ranoidea cryptotis					
▶ King et al., 1979[P,M]	▶ Australia	▶ 1s.n.i.	▶ Kununurra, Western Australia State; rep. *Cyclorana cryptotis*	▶ 1	▶ 26/52[†]
▷ King, 1990	▷ Australia	▷ n.i.	▷ ref. King et al., 1979 and unpubl.; list. *Cyclorana cryptotis*	▷ 2	▷ 26/52
▷ Kuramoto, 1990	▷ Aus./NG.	▷ n.i.	▷ ref. King et al., 1979; ctry.n.sp.; list. *Cyclorana cryptotis*	▷ 1	▷ 26/n.i.
▷ Catroli, 2008	▷ n.i.	▷ n.i.	▷ ref. King et al., 1979; list. *Litoria cryptotis*	▷ n.i.	▷ 26/n.i.
▷ Catroli & Kasahara, 2009	▷ n.i.	▷ n.i.	▷ ref. King et al., 1979; list. *Litoria cryptotis*	▷ 17	▷ 26/n.i.
Ranoidea cyclorhynchus					
▷ King, 1990	▷ Australia	▷ n.i.	▷ ref. King, 1981; list. *Litoria cyclorhynchus*	▷ n.i.	▷ 26/52
▷ Catroli, 2008	▷ n.i.	▷ n.i.	▷ ref. King, 1980/err.ref.; list. *Litoria cyclorhynchus*	▷ n.i.	▷ 26/n.i.
▷ Catroli & Kasahara, 2009	▷ n.i.	▷ n.i.	▷ ref. King, 1980/err.ref.; list. *Litoria cyclorhyncha*	▷ n.i	▷ 26/n.i
▶ Present study	▶ Australia	▶ 1♀	▶ Western Australia, loc.u.	▶ 1,2,4, 5,7,8	▶ 26/52
Ranoidea dahlii					
▷ King, 1990	▷ Australia	▷ n.i.	▷ ref. Tyler et al., 1978; King, 1981 and unpubl.; list. *Litoria dahlii*	▷ 2	▷ 26/52
▷ Kuramoto, 1990	▷ Aus./NG.	▷ n.i.	▷ ref. Tyler et al., 1978; ctry.n.sp.; list. *Litoria dahlii*	▷ 1	▷ 26/n.i.
▷ Catroli, 2008	▷ n.i.	▷ n.i.	▷ ref. Tyler et al., 1978; King, 1980/err.ref.; list. *Litoria dahlii*	▷ n.i.	▷ 26/n.i.
▷ Catroli & Kasahara, 2009	▷ n.i.	▷ n.i.	▷ ref. Tyler et al., 1978; King, 1980/err.ref., 1990; list. *Litoria dahlii*	▷ 17	▷ 26/n.i.
▶ Present study	▶ Australia	▶ n.i.	▶ Adelaide River floodplain, Arnhem Highway, Northern Territory	▶ 1,2,4, 5,7,8	▶ 26/52
Ranoidea dayi					
▶ King, 1990[P]	▶ n.i.	▶ n.i.	▶ loc.n.i.; rep. *Nyctimystes hosmeri*	▶ 1	▶ 26/52
▷ Catroli & Kasahara, 2009	▷ n.i.	▷ n.i.	▷ ref. King, 1990; list. *Litoria dayi*	▷ n.i.	▷ 26/n.i.
▶ Present study	▶ Australia	▶ 8♂	▶ Pilgrim Sands, Queensland State (16°04′S, 145°27′E)	▶ 2,14	▶ 26/52
Ranoidea eucnemis					
▷ King, 1990	▷ Aus./NG.	▷ n.i.	▷ ref. King, 1981; ctry.n.sp.; list. *Litoria eucnemis*	▷ n.i.	▷ 26/52
▶ King, 1990[P]	▶ n.i.	▶ n.i.	▶ loc.n.i.; rep. *Litoria serrata*	▶ 1	▶ 26/52
▷ Catroli, 2008	▷ n.i.	▷ n.i.	▷ ref. King, 1980/err.ref., 1990; list. *Litoria eucnemis*	▷ n.i.	▷ 26/n.i.
▷ Catroli & Kasahara, 2009	▷ n.i.	▷ n.i.	▷ ref. King, 1980/err.ref., 1990; list. *Litoria eucnemis*	▷ n.i.	▷ 26/n.i.
▷Schmid & Steinlein, 2016a	▷ n.i.	▷ n.i.	▷ ref. Schmid et al., unpubl.; list. *Dryopsophus eucnemis*	▷ 14	▷ 26/n.i.
▶ Present study	▶ Australia	▶ 2♂	▶ vicinity of Rockhampton, Queensland State (23°23′S, 150°30′E)	▶ 1,2,4,14	▶ 26/52
Ranoidea genimaculata					
▶ Menzies & Tippet, 1976[N]	▶ New Guinea	▶ 1♂	▶ Lae, Morobe Province; rep. *Litoria genimaculata*	▶ 1	▶ 26[‡]/n.i.
▷ Morescalchi, 1979	▷ n.i.	▷ n.i.	▷ ref. Menzies & Tippett, 1976; list. *Litoria genimaculata*	▷ n.i.	▷ 26/ n.i.
▷ King, 1990	▷ New Guinea	▷ n.i.	▷ ref. Menzies & Tippett, 1976; King, 1981; err.cit.2n; list. *Litoria genimaculata*	▷ n.i.	▷ 24/ n.i.
▷ Kuramoto, 1990	▷ Aus./NG.	▷ n.i.	▷ ref. Menzies & Tippett, 1976; ctry.n.sp.; list. *Litoria genimaculata*	▷ 1	▷ 26/n.i.
▷ Catroli, 2008	▷ n.i.	▷ n.i.	▷ ref. Menzies & Tippett, 1976; King, 1980/err.ref.; err.cit.2n; list. *Litoria genimaculata*	▷ n.i.	▷ 24/ n.i.
▷ Catroli & Kasahara, 2009	▷ n.i.	▷ n.i.	▷ ref. Menzies & Tippett, 1976; King, 1980/err.ref.; err.cit.2n; list. *Litoria genimaculata*	▷ n.i.	▷ 24/ n.i.
▶ Kakampuy et al., 2013[P,M,I]	▶ Australia	▶ 19♂	▶ Mount Lewis, Queensland State; rep. *Litoria genimaculata*	▶ 1,2,4	▶ 26/52

DOI: 10.1159/000490085

Author(s) ▶ original data[a] ▷ review articles	Country/ Region[b]	Speci-mens[c]	Locality of sampling, references cited, and remarks[d]	Methods applied[e]	2n/FN[f]
▶ Schmid & Steinlein, 2016a[P]	▶ Australia	▶ n.i.	▶ Mount Lewis, Queensland State (16°34′S, 145°16′E); rep. *Dryopsophus ge-nimaculatus*	▶ 2,4,5, 8,14	▶ 26/52†
▶ Present study	▶ Australia	▶ 19♂1♀	▶ Mount Lewis, Queensland State (16°34′S, 145°16′E)	▶ 1,2,4,5, 7,8,14	▶ 26/52
Ranoidea gracilenta*					
▶ Menzies & Tippet, 1976[N]	▶ New Guinea	▶ 6♂	▶ Daru, Western Province; rep. *Litoria gracilenta*	▶ 1	▶ 26‡/n.i.
▷ Morescalchi, 1979	▷ n.i.	▷ n.i.	▷ ref. Menzies & Tippett, 1976; list. *Litoria gracilenta*	▷ n.i.	▷ 26/n.i.
▷ King, 1990	▷ Aus./NG.	▷ n.i.	▷ ref. Menzies & Tippett, 1976; King, 1981; ctry.n.sp.; list. *Litoria gracilenta*	▷ n.i.	▷ 26/52
▷ Kuramoto, 1990	▷ Aus./NG.	▷ n.i.	▷ ref. Menzies & Tippett, 1976; ctry.n.sp.; list. *Litoria gracilenta*	▷ 1	▷ 26/n.i.
▷ Catroli, 2008	▷ n.i.	▷ n.i.	▷ ref. Menzies & Tippett, 1976; list. *Litoria gracilenta*	▷ n.i.	▷ 26/n.i.
▷ Catroli & Kasahara, 2009	▷ n.i.	▷ n.i.	▷ ref. Menzies & Tippett, 1976; King, 1980/err.ref.; list. *Litoria gracilenta*	▷ n.i.	▷ 26/n.i.
▶ Present study	▶ Australia	▶ 1U	▶ Queensland State, loc.u.	▶ 1,4	▶ 26/52
Ranoidea impura					
▶ Menzies & Tippet, 1976[N]	▶ New Guinea	▶ 4♂	▶ Brown River, Central Province; rep. *Litoria impura*	▶ 1	▶ 26‡/n.i.
▷ Morescalchi, 1979	▷ n.i.	▷ n.i.	▷ ref. Menzies & Tippett, 1976; list. *Litoria impura*	▷ n.i.	▷ 26/n.i.
▷ King, 1990	▷ New Guinea	▷ n.i.	▷ ref. Menzies & Tippett, 1976; list. *Litoria impura*	▷ n.i.	▷ 26/n.i.
▷ Kuramoto, 1990	▷ Aus./NG.	▷ n.i.	▷ ref. Menzies & Tippett, 1976; ctry.n.sp.; list. *Litoria impura*	▷ 1	▷ 26/n.i.
▷ Catroli, 2008	▷ n.i.	▷ n.i.	▷ ref. Menzies & Tippett, 1976; list. *Litoria impura*	▷ n.i.	▷ 26/n.i.
▷ Catroli & Kasahara, 2009	▷ n.i.	▷ n.i.	▷ ref. Menzies & Tippett, 1976; list. *Litoria impura*	▷ n.i.	▷ 26/n.i.
Ranoidea jungguy					
▶ Donnellan & Mahony, 2004[P]	▶ Australia	▶ 2s.n.i.	▶ Mission Beach, Queensland State (17°55′S, 146°06′E); rep. *Litoria jungguy*	▶ 1	▶ 26/52†
▶ Donnellan & Mahony, 2004[P]	▶ Australia	▶ 2s.n.i.	▶ Tully River, Queensland State (17°48′S, 145°41′E); rep. *Litoria jungguy*	▶ 1	▶ 26/52†
▶ Donnellan & Mahony, 2004[P]	▶ Australia	▶ 3s.n.i.	▶ Johnston River SF, Queensland State (17°32′S, 146°03′E); rep. *Litoria jungguy*	▶ 1	▶ 26/52†
▶ Donnellan & Mahony, 2004[P]	▶ Australia	▶ 4s.n.i.	▶ Millaa Millaa Falls, Queensland State (17°30′S, 145°37′E); rep. *Litoria jungguy*	▶ 1,2	▶ 26/52†
▶ Donnellan & Mahony, 2004[P]	▶ Australia	▶ 8s.n.i.	▶ Upper Barron, Queensland State (17°23′S, 145°31′E); rep. *Litoria jungguy*	▶ 1,2	▶ 26/52†
▶ Donnellan & Mahony, 2004[P]	▶ Australia	▶ 1s.n.i.	▶ Kauri Creek, Queensland State (17°11′S, 145°34′E); rep. *Litoria jungguy*	▶ 1	▶ 26/52†
▶ Donnellan & Mahony, 2004[P]	▶ Australia	▶ 9s.n.i.	▶ Tinaroo Falls Dam, Queensland State (17°11′S, 145°34′E); rep. *Litoria jungguy*	▶ 1,2	▶ 26/52†
▶ Donnellan & Mahony, 2004[P]	▶ Australia	▶ 2s.n.i.	▶ Granite Gorge, Queensland State (17°04′S, 145°21′E); rep. *Litoria jungguy*	▶ 1	▶ 26/52†
▶ Donnellan & Mahony, 2004[P]	▶ Australia	▶ 10s.n.i.	▶ Crystal Cascades, Queensland State (16°56′S, 145°42′E); rep. *Litoria jungguy*	▶ 1	▶ 26/52†
▶ Donnellan & Mahony, 2004[P]	▶ Australia	▶ 2s.n.i.	▶ Kuranda, Queensland State (16°49′S, 145°38′E); rep. *Litoria jungguy*	▶ 1	▶ 26/52†
▶ Donnellan & Mahony, 2004[P]	▶ Australia	▶ 18s.n.i.	▶ Barron River, Queensland State (16°38′S, 145°33′E); rep. *Litoria jungguy*	▶ 1	▶ 26/52†
▶ Donnellan & Mahony, 2004[P]	▶ Australia	▶ 3s.n.i.	▶ Mount Lewis, Queensland State (16°35′S, 145°17′E); rep. *Litoria jungguy*	▶ 1	▶ 26/52†
▶ Donnellan & Mahony, 2004[P]	▶ Australia	▶ 4s.n.i.	▶ Mount Windsor, Queensland State (16°15′S, 145°05′E); rep. *Litoria jungguy*	▶ 1	▶ 26/52†
▷ Catroli, 2008	▷ n.i.	▷ n.i.	▷ ref. Donnellan & Mahony, 2004; list. *Litoria jungguy*	▷ n.i.	▷ 26/n.i.
▷ Catroli & Kasahara, 2009	▷ n.i.	▷ n.i.	▷ ref. Donnellan & Mahony, 2004; list. *Litoria jungguy*	▷ 17	▷ 26/n.i.
▶ Present study	▶ Australia	▶ 6♂2♀	▶ Barron River, Queensland State (16°38′S, 145°33′E)	▶ 1,2,4, 7,14	▶ 26/52
Ranoidea lesueurii					
▶ King, 1980[P,M,I]	▶ Australia	▶ 8s.n.i.	▶ Cairns, Queensland State; Gudgenby, Australian Capital Territory; rep. *Litoria {lesueuri}*	▶ 1,2,4	▶ 26/52†
▶ King, 1988[P]	▶ Australia	▶ n.i.	▶ loc.n.i.; rep. *Litoria {lesueuri}*	▶ 1,2, 10,13	▶ 26/52†
▶ King et al., 1990[P]	▶ Australia	▶ 1s.n.i.	▶ Kuranda, Queensland State; rep. *Litoria {lesueuri} coger*	▶ 13	▶ 26/52†
▷ King, 1990	▷ Australia	▷ n.i.	▷ ref. King, 1980; list. *Litoria {lesueuri}*	▷ 2,4, 10,13	▷ 26/52
▶ King, 1990[P]	▶ n.i.	▶ n.i.	▶ loc.n.i.; rep. *Litoria {lesueuri}*	▶ 2,13	▶ 26/52†
▷ Kuramoto, 1990	▷ Aus./NG.	▷ n.i.	▷ ref. King, 1980; ctry.n.sp.; list. *Litoria lesueurii*	▷ 17	▷ 26/n.i.
▶ Donnellan & Mahony, 2004[P]	▶ Australia	▶ 5s.n.i.	▶ Cann River, Victoria State (37°31′S, 149°10′E); rep. *Litoria {lesueuri}*	▶ 1	▶ 26/52†
▶ Donnellan & Mahony, 2004[P]	▶ Australia	▶ 1s.n.i.	▶ Maddens Plains, New South Wales State (34°17′S, 150°55′E); rep. *Litoria {lesueuri}*	▶ 1	▶ 26/52†
▶ Donnellan & Mahony, 2004[P]	▶ Australia	▶ 12s.n.i.	▶ Hacking River, New South Wales State (34°08′S, 151°04′E); rep. *Litoria {lesueuri}*	▶ 1,2	▶ 26/52†
▷ Catroli, 2008	▷ n.i.	▷ n.i.	▷ ref. King, 1980; Donnellan & Mahony, 2004; list. *Litoria lesueurii*	▷ n.i.	▷ 26/n.i.
▷ Catroli & Kasahara, 2009	▷ n.i.	▷ n.i.	▷ ref. King, 1980; Donnellan & Mahony, 2004; list. *Litoria lesueurii*	▷ 17	▷ 26/n.i.
Ranoidea longipes					
▶ King et al., 1979[P,M]	▶ Australia	▶ 2s.n.i.	▶ Kununurra, Western Australia State; rep. *Cyclorana longipes*	▶ 1	▶ 26/52†
▷ King, 1990	▷ Australia	▷ n.i.	▷ ref. King et al., 1979; list. *Cyclorana longipes*	▷ n.i.	▷ 26/52
▷ Kuramoto, 1990	▷ Aus./NG.	▷ n.i.	▷ ref. King et al., 1979; ctry.n.sp.; list. *Cyclorana longipes*	▷ 1	▷ 26/n.i.
▷ Catroli, 2008	▷ n.i.	▷ n.i.	▷ ref. King et al., 1979; list. *Litoria longipes*	▷ n.i.	▷ 26/n.i.
▷ Catroli & Kasahara, 2009	▷ n.i.	▷ n.i.	▷ ref. King et al., 1979; list. *Litoria longipes*	▷ n.i.	▷ 26/n.i.
▶ Present study	▶ Australia	▶ 2U	▶ Boroloola, Northern Territory (16°04′S, 136°18′E)	▶ 5,8	▶ 26/52

Author(s) ▶ original data[a] ▷ review articles	Country/ Region[b]	Speci- mens[c]	Locality of sampling, references cited, and remarks[d]	Methods applied[e]	2n/FN[f]
Ranoidea maini					
▶ King et al, 1979[P,M]	▶ Australia	▶ 1s.n.i.	▶ Alice Springs, Northern Territory; rep. *Cyclorana maini*	▶ 1	▶ 26/52[†]
▶ King et al, 1979[P,M]	▶ Australia	▶ 1s.n.i.	▶ Whitecliffs, Western Australia State; rep. *Cyclorana maini*	▶ 1	▶ 26/52[†]
▶ King et al, 1979[P,M]	▶ Australia	▶ 1s.n.i.	▶ Wiluna, Western Australia State; rep. *Cyclorana maini*	▶ 1	▶ 26/52[†]
▷ King, 1990	▷ Australia	▷ n.i.	▷ ref. King et al., 1979; list. *Cyclorana maini*	▷ n.i.	▷ 26/52
▷ Kuramoto, 1990	▷ Aus./NG.	▷ n.i.	▷ ref. King et al., 1979; ctry.n.sp.; list. *Cyclorana maini*	▷ 1	▷ 26/n.i.
▷ Catroli, 2008	▷ n.i.	▷ n.i.	▷ ref. King et al., 1979; list. *Litoria maini*	▷ n.i.	▷ 26/n.i.
▷ Catroli & Kasahara, 2009	▷ n.i.	▷ n.i.	▷ ref. King et al., 1979; list. *Litoria maini*	▷ n.i.	▷ 26/n.i.
▶ Present study	▶ Australia	▶ 2U	▶ Western Australia State; loc.u.	▶ 1,14	▶ 26/52
Ranoidea moorei					
▶ King, 1980[P,M,I]	▶ Australia	▶ 10s.n.i.	▶ Chidlows, Forrestfield, Mount Barker, and Perth, Western Australia State; rep. *Litoria moorei*	▶ 1,2,4	▶ 26/52[†]
▶ King, 1988[P]	▶ Australia	▶ n.i.	▶ loc.n.i.; rep. *Litoria moorei*	▶ 13	▶ 26/52[†]
▶ King et al., 1990[P]	▶ Australia	▶ 1s.n.i.	▶ Chidlows, Western Australia State; rep. *Litoria moorei*	▶ 13	▶ 26/52[†]
▷ King, 1990	▷ Australia	▷ n.i.	▷ ref. King, 1980; list. *Litoria moorei*	▷ 2,4,13	▷ 26/52
▶ King, 1990[P]	▶ n.i.	▶ n.i.	▶ loc.n.i.; rep. *Litoria moorei*	▶ 13	▶ 26/52[†]
▷ Kuramoto, 1990	▷ Aus./NG.	▷ n.i.	▷ ref. King, 1980; ctry.n.sp.; list. *Litoria moorei*	▷ 17	▷ 26/n.i.
▷ Catroli, 2008	▷ n.i.	▷ n.i.	▷ ref. King, 1980; list. *Litoria moorei*	▷ n.i.	▷ 26/n.i.
▷ Catroli & Kasahara, 2009	▷ n.i.	▷ n.i.	▷ ref. King, 1980; King et al., 1990; list. *Litoria moorei*	▷ 17	▷ 26/n.i.
▶ Present study	▶ Australia	▶ 1♀	▶ vicinity of Perth, Western Australia State	▶ 2,4,7	▶ 26/52
Ranoidea nannotis					
▶ King, 1988[P]	▶ Australia	▶ n.i.	▶ loc.n.i; rep. *Litoria nannotis*	▶ 13	▶ 26/52[†]
▶ King et al., 1990[P]	▶ Australia	▶ 1s.n.i.	▶ Kuranda, Queensland State; rep. *Litoria nannotis*	▶ 13	▶ 26/52[†]
▷ King, 1990	▷ Australia	▷ n.i.	▷ ref. King, 1987; list. *Litoria nannotis*	▷ 2,13	▷ 26/52
▶ King, 1990[P]	▶ n.i.	▶ n.i.	▶ loc.n.i; rep. *Litoria nannotis*	▶ 13	▶ 26/52[†]
▷ Catroli, 2008	▷ n.i.	▷ n.i.	▷ ref. King, 1987; list. *Litoria nannotis*	▷ n.i.	▷ 26/n.i.
▷ Catroli & Kasahara, 2009	▷ n.i.	▷ n.i.	▷ ref. King, 1987; list. *Litoria nannotis*	▷ 17	▷ 26/n.i.
▶ Present study	▶ Australia	▶ 2♂	▶ Cape Tripulation, Queensland State (16°05′S, 145°27′E)	▶ 1,2,4, 8,14	▶ 26/52
▶ Present study	▶ Australia	▶ 3♂2♀	▶ Paluma, Queensland State (19°00′S, 146°11′E)	▶ 1,2,4, 8,14	▶ 26/52
Ranoidea novaehollandiae					
▶ King et al., 1990[P]	▶ Australia	▶ 1s.n.i.	▶ Goodooga, Queensland State; rep. *Cyclorana novaehollandiae*	▶ 13	▶ 26/52[†]
▷ King, 1990	▷ Australia	▷ n.i.	▷ ref. King, 1990 and unpubl.; list. *Cyclorana novaehollandiae*	▷ 2,10,13	▷ 26/52
▶ King, 1990[I]	▶ n.i.	▶ n.i.	▶ loc.n.i; rep. *Cyclorana novaehollandiae*	▶ 2,4	▶ 26/52
▷ Catroli, 2008	▷ n.i.	▷ n.i.	▷ ref. King et al., 1979; King, 1990; list. *Litoria novaehollandiae*	▷ n.i.	▷ 26/n.i.
▷ Catroli & Kasahara, 2009	▷ n.i.	▷ n.i.	▷ ref. King, 1990; list. *Litoria novaehollandiae*	▷ 17	▷ 26/n.i.
▶ Present study	▶ Australia	▶ 1♀	▶ Goodooga, Queensland State (29°06′S, 147°28′E)	▶ 1,2,4,8	▶ 26/52
Ranoidea nyakalensis					
▶ Kakampuy et al., 2013[P,M,I]	▶ Australia	▶ 4♂	▶ Paluma, Queensland State; rep. *Litoria nyakalensis*	▶ 1,2,4	▶ 26/52
▶ Present study	▶ Australia	▶ 4♂	▶ Paluma, Queensland State (19°00′S, 146°11′E); rep. *Litoria nyakalensis*	▶ 1,2,4, 5,7,14	▶ 26/52
Ranoidea pearsoniana					
▶ King, 1980[P,M,I]	▶ Australia	▶ 2s.n.i.	▶ Canondale Range, Queensland State; rep. *Litoria {pearsoni}*	▶ 1,2,4	▶ 26/52[†]
▷ King, 1990	▷ Australia	▷ n.i.	▷ ref. King, 1980; list. *Litoria pearsoniana*	▷ 2	▷ 26/52
▶ King, 1990[P]	▶ n.i.	▶ n.i.	▶ loc.n.i.; rep. *Litoria pearsoniana*	▶ 2	▶ 26/52[†]
▷ Kuramoto, 1990	▷ Aus./NG.	▷ n.i.	▷ ref. King, 1980; ctry.n.sp.; list. *Litoria pearsoniana*	▷ 17	▷ 26/n.i.
▷ Catroli, 2008	▷ n.i.	▷ n.i.	▷ ref. King, 1980; list. *Litoria pearsoniana*	▷ n.i.	▷ 26/n.i.
▷ Catroli & Kasahara, 2009	▷ n.i.	▷ n.i.	▷ ref. King, 1980; list. *Litoria pearsoniana*	▷ 17	▷ 26/n.i.
Ranoidea phyllochroa					
▶ Stephenson & Stephenson, 1970[P,M,I]	▶ Australia	▶ 2♂	▶ loc.n.i.; rep. *Hyla phyllochroa*; err.id.sp., see King, 1980	▶ 1	▶ 26/52[†]
▷ Foresti, 1972	▷ n.i.	▷ n.i.	▷ ref. Stephenson et al., 1970/err.ref; list. *Hyla phyllochroa*	▷ n.i.	▷ 26/n.i.
▶ King, 1980[P,M,I]	▶ Australia	▶ 2s.n.i.	▶ Bellingen, New South Wales State; rep. *Litoria phyllochroa*	▶ 1,2,4	▶ 26/52[†]
▷ King, 1990	▷ Australia	▷ n.i.	▷ ref. King, 1980; list. *Litoria phyllochroa*	▷ 2	▷ 26/52
▶ King, 1990[P]	▶ Australia	▶ n.i.	▶ loc.n.i.; rep. *Litoria phyllochroa*	▶ 2	▶ 26/52[†]
▷ Kuramoto, 1990	▷ Aus./NG.	▷ n.i.	▷ ref. Stephenson & Stephenson, 1970; King, 1980; ctry.n.sp.; list. *Litoria phyllochroa*	▷ 17	▷ 26/n.i.
▷ Catroli, 2008	▷ n.i.	▷ n.i.	▷ ref. King, 1980; list. *Litoria phyllochroa*	▷ n.i.	▷ 26/n.i.
▷ Catroli & Kasahara, 2009	▷ n.i.	▷ n.i.	▷ ref. Stephenson & Stephenson, 1970; King, 1980; list. *Litoria phyllochroa*	▷ 17	▷ 26/n.i.
▶ Present study	▶ Australia	▶ 1U	▶ New South Wales State, loc.u.	▶ 4	▶ 26/52

Author(s) ▶ original data[a] ▷ review articles	Country/ Region[b]	Speci- mens[c]	Locality of sampling, references cited, and remarks[d]	Methods applied[e]	2n/FN[f]
Ranoidea platycephala					
▶ King et al., 1979[P,M]	▶ Australia	▶ 2s.n.i.	▶ Palmerston, Western Australia State; rep. *Cyclorana platycephalus*	▶ 1	▶ 26/52†
▷ King, 1990	▷ Australia	▷ n.i.	▷ ref. King et al., 1979, 1985 and unpubl.; list. *Cyclorana platycephalus*	▷ 2,10	▷ 26/52
▶ King, 1990[I]	▶ n.i.	▶ n.i.	▶ loc.n.i.; rep. *Cyclorana platycephalus*	▶ 2,4	▶ 26/52
▷ Kuramoto, 1990	▷ Aus./NG.	▷ n.i.	▷ ref. King et al., 1979; ctry.n.sp.; list. *Cyclorana platycephala*	▷ 1	▷ 26/n.i.
▷ Catroli, 2008	▷ n.i.	▷ n.i.	▷ ref. King et al., 1979, 1985; list. *Litoria platycephala*	▷ n.i.	▷ 26/n.i.
▷ Catroli & Kasahara, 2009	▷ n.i.	▷ n.i.	▷ ref. King et al., 1979, 1985, 1990; list. *Litoria platycephala*	▷ 17	▷ 26/n.i.
▶ Present study	▶ Australia	▶ 1♂	▶ Rocklands, Queensland State (23°25′S, 150°36′E)	▶ 1,4,5,8	▶ 26/52
Ranoidea raniformis					
▶ King, 1980[P,M,I]	▶ Australia	▶ 9s.n.i.	▶ Devenport, Tasmania State; Griffith, New South Wales State; Millel, South Australia State; rep. *Litoria raniformis*	▶ 1,2,4	▶ 26/52†
▶ King et al., 1990[P]	▶ Australia	▶ 2s.n.i.	▶ Chiltern, Victoria State; rep. *Litoria raniformis*	▶ 13	▶ 26/52†
▷ King, 1990	▷ Australia	▷ n.i.	▷ ref. King, 1980; list. *Litoria raniformis*	▷ 2,4, 10,13	▷ 26/52
▶ King, 1990[P]	▶ n.i.	▶ n.i.	▶ loc.n.i.; rep. *Litoria raniformis*	▶ 2,10	▶ 26/52†
▷ Kuramoto, 1990	▷ Aus./NG.	▷ n.i.	▷ ref. Tyler et al., 1978; King, 1980; ctry.n.sp.; list. *Litoria raniformis*	▷ 17	▷ 26/n.i.
▷ Catroli, 2008	▷ n.i.	▷ n.i.	▷ ref. King, 1980; list. *Litoria raniformis*	▷ n.i.	▷ 26/n.i.
▷ Catroli & Kasahara, 2009	▷ n.i.	▷ n.i.	▷ ref. Tyler et al., 1978; King, 1980; King et al., 1990; list. *Litoria raniformis*	▷ 17	▷ 26/n.i.
▶ Present study	▶ Australia	▶ 1U	▶ South Australia State, loc.u.	▶ 1,4	▶ 26/52
Ranoidea rheocola					
▷ King, 1990	▷ Australia	▷ n.i.	▷ ref. King, 1981; list. *Litoria rheocolus*	▷ n.i.	▷ 26/52
▷ Catroli, 2008	▷ n.i.	▷ n.i.	▷ ref. King, 1980/err.ref.; list. *Litoria rheocola*	▷ n.i.	▷ 26/n.i.
▷ Catroli & Kasahara, 2009	▷ n.i.	▷ n.i.	▷ ref. King, 1980/err.ref.; list. *Litoria rheocola*	▷ n.i.	▷ 26/n.i.
▶ Present study	▶ Australia	▶ 3♂	▶ vicinity of Rockhampton, Queensland State (23°23′S, 150°30′E)	▶ 1,2,4, 5,8,14	▶ 26/52
▶ Present study	▶ Australia	▶ 10♂1♀	▶ Mount Lewis, Queensland State (16°34′S, 145°16′E)	▶ 1,2	▶ 26/52
Ranoidea serrata*					
▶ Present study	▶ Australia	▶ 1U	Paluma, Queensland State (19°00′S, 146°11′E)	▶ 1,4	▶ 26/52
Ranoidea splendida					
▷ King, 1990	▷ Australia	▷ n.i.	▷ ref. King, 1981; list. *Litoria splendida*	▷ 2	▷ 26/52
▷ Catroli, 2008	▷ n.i.	▷ n.i.	▷ ref. King, 1980/err.ref.; list. *Litoria splendida*	▷ n.i.	▷ 26/n.i.
▷ Catroli & Kasahara, 2009	▷ n.i.	▷ n.i.	▷ ref. King, 1980/err.ref.; list. *Litoria splendida*	▷ 17	▷ 26/n.i.
Ranoidea subglandulosa					
▷ King, 1990	▷ Australia	▷ n.i.	▷ ref. King, 1981; list. *Litoria glandulosa*	▷ n.i.	▷ 26/52
▷ Catroli & Kasahara, 2009	▷ n.i.	▷ n.i.	▷ ref. King, 1980/err.ref.; list. *Litoria subglandulosa*	▷ n.i.	▷ 26/n.i.
Ranoidea vagitus*					
▶ Present study	▶ Australia	▶ 1U	▶ Western Australia State, loc.u.	▶ 2	▶ 26/52
Ranoidea verrucosa					
▶ King et al., 1979[N,M]	▶ Australia	▶ 1s.n.i.	▶ Narrabri, New South Wales State; rep. *Cyclorana verrucosus*	▶ 1	▶ 26/52†
▷ King, 1990	▷ Australia	▷ n.i.	▷ ref. King et al., 1979; list. *Cyclorana verrucosus*	▷ n.i.	▷ 26/52
▷ Kuramoto, 1990	▷ Aus./NG.	▷ n.i.	▷ ref. King et al., 1979; ctry.n.sp.; list. *Cyclorana verrucosa*	▷ 1	▷ 26/n.i.
▷ Catroli, 2008	▷ n.i.	▷ n.i.	▷ ref. King et al., 1979; list. *Litoria verrucosa*	▷ n.i.	▷ 26/n.i.
▷ Catroli & Kasahara, 2009	▷ n.i.	▷ n.i.	▷ ref. King et al., 1979; list. *Litoria verrucosa*	▷ n.i.	▷ 26/n.i.
▶ Present study	▶ Australia	▶ n.i.	▶ loc.n.i.	▶ 1	▶ 26/52
Ranoidea wilcoxii					
▶ Donnellan & Mahony, 2004[P]	▶ Australia	▶ 7s.n.i.	▶ Ourimbah, New South Wales State (33°20′S, 151°19′E); rep. *Litoria wilcoxii*	▶ 1,2	▶ 26/52†
▶ Donnellan & Mahony, 2004[P]	▶ Australia	▶ 7s.n.i.	▶ Olney SF, New South Wales State (33°06′S, 151°13′E); rep. *Litoria wilcoxii*	▶ 1,2	▶ 26/52†
▶ Donnellan & Mahony, 2004[P]	▶ Australia	▶ 1s.n.i.	▶ Ellenborough Falls, New South Wales State (31°37′S, 152°18′E); rep. *Litoria wilcoxii*	▶ 1	▶ 26/52†
▶ Donnellan & Mahony, 2004[P]	▶ Australia	▶ 6s.n.i.	▶ Thora, New South Wales State (30°25′S, 152°02′E); rep. *Litoria wilcoxii*	▶ 1,2	▶ 26/52†
▶ Donnellan & Mahony, 2004[P]	▶ Australia	▶ 1s.n.i.	▶ Wiangarie, New South Wales State (28°23′S, 153°06′E); rep. *Litoria wilcoxii*	▶ 1	▶ 26/52†
▶ Donnellan & Mahony, 2004[P]	▶ Australia	▶ 1s.n.i.	▶ Connondale Range, Queensland State (26°45′S, 152°37′E); rep. *Litoria wilcoxii*	▶ 1	▶ 26/52†
▶ Donnellan & Mahony, 2004[P]	▶ Australia	▶ 1s.n.i.	▶ Byfield, Queensland State (22°50′S, 150°38′E); rep. *Litoria wilcoxii*	▶ 1	▶ 26/52†
▶ Donnellan & Mahony, 2004[P]	▶ Australia	▶ 1s.n.i.	▶ Lolworth Creek, Queensland State (19°56′S, 145°44′E); rep. *Litoria wilcoxii*	▶ 1	▶ 26/52†
▶ Donnellan & Mahony, 2004[P]	▶ Australia	▶ 3s.n.i.	▶ Hervey Range, Queensland State (19°25′S, 146°47′E); rep. *Litoria wilcoxii*	▶ 1	▶ 26/52†
▶ Donnellan & Mahony, 2004[P]	▶ Australia	▶ 4s.n.i.	▶ Ollera Creek, Queensland State (18°59′S, 146°20′E); rep. *Litoria wilcoxii*	▶ 1,2	▶ 26/52†
▶ Donnellan & Mahony, 2004[P]	▶ Australia	▶ 2s.n.i.	▶ Wallaman Falls, Queensland State (18°36′S, 145°48′E); rep. *Litoria wilcoxii*	▶ 1	▶ 26/52†

Author(s) ▶ original data[a] ▷ review articles	Country/ Region[b]	Speci- mens[c]	Locality of sampling, references cited, and remarks[d]	Methods applied[e]	2n/FN[f]
▶ Donnellan & Mahony, 2004[P]	▶ Australia	▶ 1s.n.i.	▶ Granite Gorge, Queensland State (17°04′S, 145°21′E); rep. *Litoria wilcoxii*	▶ 1	▶ 26/52†
▷ Catroli, 2008	▷ n.i.	▷ n.i.	▷ ref. Donnellan & Mahony, 2004; list. *Litoria wilcoxii*	▷ n.i.	▷ 26/n.i.
▷ Catroli & Kasahara, 2009	▷ n.i.	▷ n.i.	▷ ref. Donnellan & Mahony, 2004; list. *Litoria wilcoxii*	▷ 17	▷ 26/n.i.
▶ Present study	▶ Australia	▶ 4♂1U	▶ Millstream, Queensland State (17°39′S, 145°51′E)	▶ 2,4,5, 8,14	▶ 26/52
PHYLLOMEDUSIDAE / *Agalychnis*					
Agalychnis annae*					
▶ León, 1970[N]	▶ Costa Rica	▶ 1♂	▶ loc.n.i.	▶ 1	▶ 26‡/n.i.
▷ Foresti, 1972	▷ n.i.	▷ n.i.	▷ ref. León, 1970	▷ n.i.	▷ 26‡/n.i.
▷ Morescalchi, 1979	▷ n.i.	▷ n.i.	▷ ref. León, 1970; err.cit.2n	▷ n.i.	▷ 24/n.i.
▷ King, 1990	▷ C.America	▷ n.i.	▷ ref. León, 1970; ctry.n.sp.; err.cit.2n	▷ n.i.	▷ 24/n.i.
▷ Kuramoto, 1990	▷ S.C.America	▷ n.i.	▷ ref. León, 1970; ctry.n.sp.	▷ 1	▷ 26/n.i.
▷ Catroli, 2008	▷ n.i.	▷ n.i.	▷ ref. León, 1970	▷ n.i.	▷ 26/n.i.
▷ Catroli & Kasahara, 2009	▷ n.i.	▷ n.i.	▷ ref. León, 1970	▷ n.i.	▷ 26/n.i.
▶ Present study	▶ Costa Rica	▶ n.i.	▶ loc.n.i.	▶ 2,5	▶ 26/52
Agalychnis callidryas					
▶ Duellman & Cole, 1965[N]	▶ Panama	▶ 1♂	▶ S slope Cerro La Campana; Panamá Province, list. *Phyllomedusa callidryas callidryas*	▶ 1	▶ 26/n.i.
▶ Duellman & Cole, 1965[N]	▶ Panama	▶ 3♂	▶ Barro Colorado Island, Canal Zone; list. *Phyllomedusa callidryas callydrias*	▶ 1	▶ 26/n.i.
▶ Duellman & Cole, 1965[P]	▶ Nicaragua	▶ 2♂	▶ Casa Colorada, 22 km S Managua; Masaya Department; list. *Phyllomedusa helenae*	▶ 1	▶ 26/n.i.
▶ Duellman & Cole, 1965[P]	▶ Nicaragua	▶ 2♂	▶ Finca Tepeyac, 10.5 km N, 9 km E Matagalpa, Matagalpa Department; list. *Phyllomedusa helenae*	▶ 1	▶ 26/n.i.
▶ León, 1970[N]	▶ Costa Rica	▶ n.i.	▶ loc.n.i.	▶ 1	▶ 26‡/n.i.
▷ Duellman, 1970	▷ C.America	▷ n.i.	▷ ref.n.i.; ctry.n.sp.	▷ 1	▷ 26/n.i.
▷ Rabello, 1970	▷ n.i.	▷ n.i.	▷ ref. Duellman & Cole, 1965; list. *Phyllomedusa callidryas callidryas*	▷ n.i.	▷ 26/n.i.
▷ Rabello, 1970	▷ n.i.	▷ n.i.	▷ Duellman & Cole, 1965; list. *Phyllomedusa helenae*	▷ n.i.	▷ 26/n.i.
▷ Foresti, 1972	▷ n.i.	▷ n.i.	▷ ref. Duellman & Cole, 1965; list. *Phyllomedusa callidryas callidryas*	▷ n.i.	▷ 26/n.i.
▷ Foresti, 1972	▷ n.i.	▷ n.i.	▷ ref. Duellman & Cole, 1965; list. *Phyllomedusa helenae*	▷ n.i.	▷ 26/n.i.
▷ King, 1990	▷ S.America	▷ n.i.	▷ ref. Duellman & Cole, 1965; err.cit.ctry.; list. *Phyllomedusa callidryas*	▷ n.i.	▷ 26/n.i.
▷ King, 1990	▷ S.America	▷ n.i.	▷ ref. Duellman & Cole, 1965; err.cit.ctry.; list. *Phyllomedusa helenae*	▷ n.i.	▷ 26/n.i.
▷ Kuramoto, 1990	▷ S.C.America	▷ n.i.	▷ ref. Duellman & Cole, 1965; León 1970; ctry.n.sp.	▷ 1	▷ 26/n.i.
▶ Schmid et al., 1995[P]	▶ Costa Rica	▶ 12♂4♀	▶ Cahuita, Limón Province (09°47′N, 82°53′W)	▶ 1,2, 4–6,12	▶ 26/52†
▶ Schmid et al., 1995[P]	▶ Costa Rica	▶ 2♂	▶ Valle de las Rosas, Limón Province (09°40′N, 82°58′W)	▶ 1,2, 4–6,12	▶ 26/52†
▶ Schmid et al., 1995[P]	▶ Costa Rica	▶ 1♂	▶ Valle de Talamanca, Limón Province (09°35′N, 82°56′W)	▶ 1,2, 4–6,12	▶ 26/52†
▶ Schmid et al., 1995[P]	▶ Costa Rica	▶ 2♂	▶ Quepos, Puntarenas Province (09°25′N, 84°10′W)	▶ 1,2, 4–6,12	▶ 26/52†
▶ Schmid et al., 1995[P]	▶ Costa Rica	▶ 5♂	▶ Rincón, Guanacaste Province (08°42′N, 83°30′W)	▶ 1,2, 4–6,12	▶ 26/52†
▷ Nunes & Fagundes, 2008b	▷ n.i.	▷ n.i.	▷ ref. Duellman & Cole, 1965; list. *Agalychnis callidryas*	▷ n.i.	▷ 26/52
▷ Nunes & Fagundes, 2008b	▷ n.i.	▷ n.i.	▷ ref. Duellman & Cole, 1965; list. *Agalychnis helenae*	▷ n.i.	▷ 26/52
▷ Catroli, 2008	▷ n.i.	▷ n.i.	▷ ref. Duellman & Cole, 1965; León, 1970; Schmid et al., 1995	▷ n.i.	▷ 26/n.i.
▷ Catroli & Kasahara, 2009	▷ n.i.	▷ n.i.	▷ ref. Duellman & Cole, 1965; León, 1970; Schmid et al., 1995	▷ 17	▷ 26/n.i.
▶ Present study	▶ n.i.	▶ n.J	▶ loc.u.	▶ 1	▶ 26/52
▶ Present study	▶ Panama	▶ 1♀	▶ loc.u.	▶ 1	▶ 26/52
▶ Present study	▶ Costa Rica	▶ 12♂4♀	▶ Cahuita, Limón Province (09°47′N, 82°53′W)	▶ 2,5,7, 8,14	▶ 26/52
▶ Present study	▶ Costa Rica	▶ 2♂	▶ Valle de las Rosas, Limón Province (09°40′N, 82°58′W)	▶ 2,5,7, 8,14	▶ 26/52
▶ Present study	▶ Costa Rica	▶ 1♂	▶ Valle de Talamanca, Limón Province (09°35′N, 82°56′W)	▶ 2,5,7, 8,14	▶ 26/52
▶ Present study	▶ Costa Rica	▶ 2♂	▶ Quepos, Puntarenas Province (09°25′N, 84°10′W)	▶ 2,5,7, 8,14	▶ 26/52
▶ Present study	▶ Costa Rica	▶ 5♂	▶ Rincón, Guanacaste Province (08°42′N, 83°30′W)	▶ 2,5,7, 8,14	▶ 26/52
Agalychnis dacnicolor					
▶ Duellman & Cole, 1965[N]	▶ Mexico	▶ 1♂	▶ Villa Unión, Sinaloa Department; rep. *Phyllomedusa dacnicolor*	▶ 1	▶ 26‡/n.i.
▶ Duellman & Cole, 1965[N]	▶ Mexico	▶ 1♂	▶ 3.5 km W Cuautlixco, Morelos Department; rep. *Phyllomedusa dacnicolor*	▶ 1	▶ 26‡/n.i.
▶ Duellman & Cole, 1965[N]	▶ Mexico	▶ 1♂	▶ Acahuizotla, Guerrero State, rep. *Phyllomedusa dacnicolor*	▶ 1	▶ 26‡/n.i.
▷ Duellman, 1970	▷ C.America	▷ n.i.	▷ ref.n.i.; ctry.n.sp.; list. *Pachymedusa dacnicolor*	▷ 1	▷ 26‡/n.i.
▶ Cole, 1971[P]	▶ Mexico	▶ 1s.n.i.	▶ Acapulco, Guerrero State; rep. *Pachymedusa dacnicolor*	▶ 1	▶ 26/52†
▶ Cole, 1971[P]	▶ Mexico	▶ 3s.n.i.	▶ northern outskirts of Alamos, Sonora State; rep. *Pachymedusa dacnicolor*	▶ 1	▶ 26/52†
▶ Cole, 1971[P]	▶ Mexico	▶ 1s.n.i.	▶ Quintera mine, Sierra de Promontorios, ca. 9 miles (by road) W Alamos, Sonora State, rep. *Pachymedusa dacnicolor*	▶ 1	▶ 26/52†

Author(s) ▶ original data[a] ▷ review articles	Country/ Region[b]	Speci-mens[c]	Locality of sampling, references cited, and remarks[d]	Methods applied[e]	2n/FN[f]
▷ Foresti, 1972	▷ n.i.	▷ n.i.	▷ ref. Duellman & Cole, 1965; Cole, 1971; list. *Phyllomedusa dacnicolor*	▷ n.i.	▷ 26/n.i.
▷ King, 1990	▷ S.America	▷ n.i.	▷ ref. Duellman & Cole, 1965; Cole, 1971; Schmid, 1980a; err.cit.ctry; list. *Pachymedusa dacnicolor*	▷ 2,4,17	▷ 26/52
▷ Kuramoto, 1990	▷ S.C.America	▷ n.i.	▷ ref. Duellman & Cole, 1965; Cole, 1971; Schmid, 1980a; ctry.n.sp.; list. *Pachymedusa dacnicolor*	▷ 17	▷ 26/n.i.
▷ Catroli, 2008	▷ n.i.	▷ n.i.	▷ ref. Duellman & Cole, 1965; Cole, 1971; Schmid, 1980a; list. *Pachymedusa dacnicolor*	▷ n.i.	▷ 26/n.i.
▷ Catroli & Kasahara, 2009	▷ n.i.	▷ n.i.	▷ ref. Duellman & Cole, 1965; Cole, 1971; Schmid, 1980a; list. *Pachymedusa dacnicolor*	▷ 17	▷ 26/n.i.
▶ Present study	▶ Mexico	▶ 1s.n.i.	▶ loc.u.; animal supplied by animal dealer	▶ 2,5	▶ 26/52
Agalychnis lemur					
▶ León, 1970[N]	▶ Costa Rica	▶ 1♂	▶ loc.n.i.; rep. *Phyllomedusa lemur*	▶ 1	▶ 26‡/n.i.
▷ Duellman, 1970	▷ C.America	▷ n.i.	▷ ref.n.i.; ctry.n.sp.; list. *Phyllomedusa lemur*	▷ 1	▷ 26‡/n.i.
▷ Foresti, 1972	▷ n.i.	▷ n.i.	▷ ref. León, 1970; list. *Phyllomedusa lemur*	▷ n.i.	▷ 26‡/n.i.
▷ Morescalchi, 1979	▷ n.i.	▷ n.i.	▷ ref. León, 1970; list. *Phyllomedusa lemur*	▷ n.i.	▷ 26/n.i.
▷ King, 1990	▷ C.America	▷ n.i.	▷ ref. León, 1970; ctry.n.sp.; list. *Phyllomedusa lemur*	▷ n.i.	▷ 26/n.i.
▷ Kuramoto, 1990	▷ S.C.America	▷ n.i.	▷ ref. León, 1970; ctry.n.sp.; list. *Phyllomedusa lemur*	▷ 1	▷ 26/n.i.
▷ Catroli, 2008	▷ n.i.	▷ n.i.	▷ ref. León, 1970; list. *Hylomantis lemur*	▷ n.i.	▷ 26/n.i.
▷ Catroli & Kasahara, 2009	▷ n.i.	▷ n.i.	▷ ref. León, 1970; list. *Hylomantis lemur*	▷ n.i.	▷ 26/n.i.
PHYLLOMEDUSIDAE / *Callimedusa*					
Callimedusa tomopterna					
▶ Batistic, 1989[P,I]	▶ Brazil	▶ 2♂	▶ Manaus, Amazonas State; rep. *Phyllomedusa tomopterna*	▶ 1	▶ 26/52†
▷ Catroli, 2008	▷ n.i.	▷ n.i.	▷ ref. Batistic, 1989; list. *Phyllomedusa tomopterna*	▷ n.i.	▷ 26/n.i.
▷ Catroli & Kasahara, 2009	▷ n.i.	▷ n.i.	▷ ref. Batistic, 1989; list. *Phyllomedusa tomopterna*	▷ n.i.	▷ 26/n.i.
PHYLLOMEDUSIDAE / *Cruziohyla*					
Cruziohyla calcarifer					
▶ Duellman & Cole, 1965[N]	▶ Panama	▶ 1♂	▶ Laguna, Darien Province; rep. *Phyllomedusa calcarifer*	▶ 1	▶ 26‡/n.i.
▷ Duellman, 1970	▷ C.America	▷ n.i.	▷ ref.n.i.; ctry.n.sp.; list. *Agalychnis calcarifer*	▷ 1	▷ 26‡/n.i.
▷ Rabello, 1970	▷ n.i.	▷ n.i.	▷ ref. Duellman & Cole, 1965; list. *Phyllomedusa calcarifer*	▷ n.i.	▷ 26‡/n.i.
▷ Foresti, 1972	▷ n.i.	▷ n.i.	▷ ref. Duellman & Cole, 1965; list. *Phyllomedusa calcarifer*	▷ n.i.	▷ 26‡/n.i.
▷ King, 1990	▷ S.America	▷ n.i.	▷ ref. Duellman & Cole, 1965; ctry.n.sp.; list. *Phyllomedusa calcarifer*	▷ n.i.	▷ 26/n.i.
▷ Kuramoto, 1990	▷ S.C.America	▷ n.i.	▷ ref. Duellman & Cole, 1965; ctry.n.sp.; list. *Agalychnis calcarifer*	▷ 1	▷ 26/n.i.
▷ Nunes & Fagundes, 2008b	▷ n.i.	▷ n.i.	▷ ref. Duellman & Cole, 1965	▷ n.i.	▷ 26/52*
▷ Catroli, 2008	▷ n.i.	▷ n.i.	▷ ref. Duellman & Cole, 1965	▷ n.i.	▷ 26/n.i.
▷ Catroli & Kasahara, 2009	▷ n.i.	▷ n.i.	▷ ref. Duellman & Cole, 1965	▷ n.i.	▷ 26/n.i.
PHYLLOMEDUSIDAE / *Hylomantis*					
No cytogenetic data					
PHYLLOMEDUSIDAE / *Phasmahyla*					
Phasmahyla cochranae*					
▶ Present study	▶ Brazil	▶ 1juv	▶ City of São Paulo, São Paulo State	▶ 1	▶ 26/52
Phasmahyla exilis					
▶ Nunes & Fagundes, 2008b[P]	▶ Brazil	▶ 3♂	▶ Reserva Biológica de Duas Bocas, Cariacica, Espírito Santo State	▶ 1	▶ 26/52
▷ Catroli & Kasahara, 2009	▷ n.i.	▷ n.i.	▷ ref. Nunes & Fagundes, 2008b	▷ n.i.	▷ 26/n.i.
▷ Winkelmann and Noleto, 2015	▷ n.i.	▷ n.i.	▷ ref. Nunes & Fagundes, 2008b; list. *Phyllomedusa exilis*	▷ n.i.	▷ 26/52
Phasmahyla spectabilis					
▶ Barth et al., 2013[P]	▶ Brazil	▶ 3s.n.i.	▶ Camacan, Bahia State	▶ 1,4,13	▶ 26/52†
▶ Barth et al., 2014[P]	▶ Brazil	▶ 2s.n.i.	▶ Camacan, Bahia State	▶ 2,6,9,14	▶ 26/52†
PHYLLOMEDUSIDAE / *Phrynomedusa*					
No cytogenetic data					
PHYLLOMEDUSIDAE / *Phyllomedusa*					
Phyllomedusa bahiana					
▶ Foresti, 1972[P,M]	▶ Brazil	▶ 1♂	▶ Jequié, Bahia State	▶ 1	▶ 26/52†
▷ King, 1990	▷ C.America	▷ n.i.	▷ ref. León, 1970; ctry.n.sp.; list. *Phyllomedusa {bahania}*	▷ n.i.	▷ 26/n.i.
▶ Barth et al., 2013[P]	▶ Brazil	▶ 12s.n.i.	▶ Feira de Santana, Bahia State	▶ 1,4,13	▶ 26/52†
▶ Barth et al., 2013[P]	▶ Brazil	▶ 1s.n.i.	▶ Vitória da Conquista, Bahia State	▶ 1,4,13	▶ 26/52†
▶ Barth et al., 2013[P]	▶ Brazil	▶ 2s.n.i.	▶ Ituberá, Bahia State	▶ 1,4,13	▶ 26/52†
▶ Barth et al., 2013[P]	▶ Brazil	▶ 7s.n.i.	▶ Camacan, Bahia State	▶ 1,4,13	▶ 26/52†
▶ Barth et al., 2013[P]	▶ Brazil	▶ 10s.n.i.	▶ Ilhéus, Bahia State	▶ 1,4,13	▶ 26/52†

Author(s) ▶ original data[a] ▷ review articles	Country/ Region[b]	Speci- mens[c]	Locality of sampling, references cited, and remarks[d]	Methods applied[e]	2n/FN[f]
▶ Barth et al., 2014[P]	▶ Brazil	▶ 12s.n.i.	▶ Feira de Santana, Bahia State	▶ 2,6,9,14	▶ 26/52†
▶ Barth et al., 2014[P]	▶ Brazil	▶ 1s.n.i.	▶ Vitória da Conquista, Bahia State	▶ 2,6,9,14	▶ 26/52†
▶ Barth et al., 2014[P]	▶ Brazil	▶ 2s.n.i.	▶ Ituberá, Bahia State	▶ 2,6,9,14	▶ 26/52†
▶ Barth et al., 2014[P]	▶ Brazil	▶ 7s.n.i.	▶ Camacan, Bahia State	▶ 2,6,9,14	▶ 26/52†
▶ Barth et al., 2014[P]	▶ Brazil	▶ 10s.n.i.	▶ Ilhéus, Bahia State	▶ 2,6,9,14	▶ 26/52†
▶ Barth et al., 2014[P]	▶ Brazil	▶ 1s.n.i.	▶ Areia Branca, Sergipe State	▶ 2,6,9,14	▶ 26/52†
▶ Bruschi et al., 2014a[P]	▶ Brazil	▶ 8s.n.i.	▶ Alagoinhas, Bahia State	▶ 1–4, 13,14	▶ 26/52†
▷ Winkelmann and Noleto, 2015	▷ n.i.	▷ n.i.	▷ ref. Barth et al., 2013; Bruschi et al., 2014a	▷ n.i.	▷ n.i.
▷ Schmid & Steinlein, 2016a	▷ n.i.	▷ n.i.	▷ ref. Barth et al., 2014; Bruschi et al., 2014a	▷ 14	▷ 26/n.i.
Phyllomedusa boliviana					
▶ Barrio, 1976[P]	▶ Argentina	▶ 6♂	▶ vicinity of Yuto, Jujuy Province; rep. *Phyllomedusa pailona*	▶ 1	▶ 26/52†
▶ Barrio, 1976[P]	▶ Argentina	▶ 5♂1♀	▶ Campamento de YPF, Río Pescado, Salta Province; rep. *Phyllomedusa pailona*	▶ 1	▶ 26/52†
▷ Kuramoto, 1990	▷ S.C.America	▷ n.i.	▷ ref. Barrio, 1976; ctry.n.sp.	▷ 1	▷ 26/n.i.
▷ Catroli, 2008	▷ n.i.	▷ n.i.	▷ ref. Barrio, 1976	▷ n.i.	▷ 26/n.i.
▷ Catroli & Kasahara, 2009	▷ n.i.	▷ n.i.	▷ ref. Barrio, 1976	▷ n.i.	▷ 26/n.i.
Phyllomedusa burmeisteri					
▶ Batistic et al., 1975[P,M]	▶ Brazil	▶ 8♂	▶ Sacra Família, Rio de Janeiro State	▶ 1	▶ 26/52†
▶ Batistic, 1989[P,I]	▶ Brazil	▶ 10♂	▶ Sacra Família, Rio de Janeiro State	▶ 1,2,4	▶ 26/52†
▶ Batistic, 1989[P,I]	▶ Brazil	▶ 1♂	▶ Linhares, Espírito Santo State	▶ 1,2,4	▶ 26/52†
▶ Batistic, 1989[P,I]	▶ Brazil	▶ 1♂	▶ Viçosa, Minas Gerais Santo State	▶ 1,2,4	▶ 26/52†
▶ Batistic, 1989[P,I]	▶ Brazil	▶ 1♂	▶ Itatiba, São Paulo State	▶ 1,2,4	▶ 26/52†
▷ King, 1990	▷ S.America	▷ n.i.	▷ Batistic et al., 1975; ctry.n.sp.	▷ n.i.	▷ 26/52
▷ Kuramoto, 1990	▷ S.C.America	▷ n.i.	▷ ref. Batistic et al., 1975; ctry.n.sp.	▷ 1	▷ 26/n.i.
▷ Nunes & Fagundes, 2008b	▷ n.i.	▷ n.i.	▷ ref. Beçak & Denaro, 1970/err.ref.; list. *Phasmahyla burmeisteri*	▷ n.i.	▷ 26/52
▷ Catroli, 2008	▷ n.i.	▷ n.i.	▷ ref. Batistic, 1989	▷ n.i.	▷ 26/n.i.
▷ Catroli & Kasahara, 2009	▷ n.i.	▷ n.i.	▷ ref. Batistic et al., 1975; Batistic, 1989	▷ 17	▷ 26/n.i.
▷ Paiva et al., 2010	▷ n.i.	▷ n.i.	▷ ref. Batistic, 1989	▷ 4	▷ 26/n.i.
Phyllomedusa camba					
▶ Paiva et al., 2010[P]	▶ Brazil	▶ 2 s.n.i.	▶ Ministro Andreazza, Rôndonia State (11°04′27″S, 61°31′01″W)	▶ 1,4	▶ 26/46†
▷ Catroli & Kasahara, 2009	▷ n.i.	▷ n.i.	▷ ref. Paiva et al., 2010	▷ 17	▷ 26/n.i.
▷ Winkelmann and Noleto, 2015	▷ n.i.	▷ n.i.	▷ ref. Paiva et al., 2010	▷ n.i.	▷ 26/46
Phyllomedusa distincta					
▶ Batistic, 1989[P,I]	▶ Brazil	▶ 4♂2U	▶ Apiaí, São Paulo State	▶ 1,2,4	▶ 26/52†
▶ Batistic, 1989[P,I]	▶ Brazil	▶ 3♂	▶ Guapiara, São Paulo State	▶ 1,2,4	▶ 26/52†
▶ Batistic, 1989[P,I]	▶ Brazil	▶ 1♂	▶ Iporanga, São Paulo State	▶ 1,2,4	▶ 26/52†
▶ Batistic, 1989[P,I]	▶ Brazil	▶ 1♂	▶ Juréia, São Paulo State	▶ 1,2,4	▶ 26/52†
▶ Batistic, 1989[P,I]	▶ Brazil	▶ 4♂	▶ Ribeirão Branco, São Paulo State	▶ 1,2,4	▶ 26/52†
▶ Batistic, 1989[P,I]	▶ Brazil	▶ 1♂	▶ Joinville, Santa Catarina State	▶ 1,2,4	▶ 26/52†
▶ Batistic, 1989[P,I]	▶ Brazil	▶ 3♂	▶ Barra Velha, Santa Catarina State	▶ 1,2,4	▶ 26/52†
▶ Batistic, 1989[P,I]	▶ Brazil	▶ 3♂	▶ Rio Vermelho, São Bento do Sul, Santa Catarina State	▶ 1,2,4	▶ 26/52†
▶ Batistic, 1989[P,I]	▶ Brazil	▶ 6♂	▶ Florianópolis, Santa Catarina State	▶ 1,2,4	▶ 26/52†
▶ Kasahara et al., 1998[P]	▶ Brazil	▶ n.i.	▶ loc.n.i.	▶ 12	▶ 26/52†
▷ Nunes & Fagundes, 2008b	▷ n.i.	▷ n.i.	▷ ref. Kasahara et al. 1998; list. *Phasmahyla distincta*	▷ n.i.	▷ 26/52
▷ Catroli, 2008	▷ n.i.	▷ n.i.	▷ ref. Batistic, 1989; Kasahara et al., 2007	▷ n.i.	▷ 26/n.i.
▷ Catroli & Kasahara, 2009	▷ n.i.	▷ n.i.	▷ ref. Batistic, 1989; Haddad et al., 1994	▷ 17	▷ 26/n.i.
▷ Paiva et al., 2010	▷ n.i.	▷ n.i.	▷ ref. Batistic, 1989; Kasahara et al., 2007	▷ 4	▷ 26/n.i.
▶ Gruber et al., 2013[P]	▶ Brazil	▶ 1♂	▶ Guaraqueçaba, Paraná State	▶ 1,2,4,6, 9,12–14	▶ 26/52†
▶ Gruber et al., 2013[P]	▶ Brazil	▶ 5♂1♀	▶ Riberão Branco, São Paulo State	▶ 1,2,4,6, 9,12–14	▶ 26/52†
▶ Bruschi et al., 2014a[P]	▶ Brazil	▶ 3s.n.i.	▶ Iporanga, São Paulo State	▶ 1–4, 13,14	▶ 26/52†
▷ Winkelmann and Noleto, 2015	▷ n.i.	▷ n.i.	▷ ref. Pombal & Haddad, 1992; Gruber et al., 2013; Bruschi et al., 2014a	▷ n.i.	▷ 26/52
▷ Schmid & Steinlein, 2016a	▷ n.i.	▷ n.i.	▷ ref. Gruber et al., 2013; Bruschi et al., 2014a	▷ 14	▷ 26/n.i.
▶ Present study	▶ Brazil	▶ 5♂1♀	▶ Riberão Branco, São Paulo State	▶ 1,2	▶ 26/52
Phyllomedusa iheringii					
▶ Batistic, 1989[P,I]	▶ Brazil	▶ 4♂	▶ Bagé, Rio Grande do Sul State	▶ 1,2,4	▶ 26/52†
▷ Catroli, 2008	▷ n.i.	▷ n.i.	▷ ref. Batistic, 1989	▷ n.i.	▷ 26/n.i.
▷ Catroli & Kasahara, 2009	▷ n.i.	▷ n.i.	▷ ref. Batistic, 1989	▷ 17	▷ 26/n.i.
▷ Paiva et al., 2010	▷ n.i.	▷ n.i.	▷ ref. Batistic, 1989	▷ 4	▷ 26/n.i.
Phyllomedusa rustica					
▶ Bruschi et al., 2014b[P]	▶ Brazil	▶ 10♂	▶ Água Doce Municipality, Santa Catarina State, 1330 m (26°35′59.9″S, 51°34′39.4″W)	▶ 1-4,13	▶ 26/52†
▷ Winkelmann and Noleto, 2015	▷ n.i.	▷ n.i.	▷ ref. Bruschi et al., 2014b	▷ n.i.	▷ 26/52

Author(s) ▶ original data[a] ▷ review articles	Country/ Region[b]	Speci- mens[c]	Locality of sampling, references cited, and remarks[d]	Methods applied[e]	2n/FN[f]
Phyllomedusa sauvagii					
▶ Barrio, 1976[P]	▶ Argentina	▶ 3♂1♀	▶ Frías, Santiago del Estero Province	▶ 1	▶ 26/52†
▶ Barrio, 1976[P]	▶ Argentina	▶ 3♂	▶ Parque Nacional El Rey, Salta Province	▶ 1	▶ 26/52†
▶ Batistic, 1989[P,I]	▶ Brazil	▶ 1U	▶ Corumbá, Mato Grosso do Sul State	▶ 1,2,4	▶ 26/52†
▶ Batistic, 1989[P,I]	▶ Paraguay	▶ 2♂	▶ Transchaco	▶ 1,2,4	▶ 26/52†
▶ Batistic, 1989[P,I]	▶ Paraguay	▶ 2♂	▶ Pozo Colorado	▶ 1,2,4	▶ 26/52†
▶ Batistic, 1989[P,I]	▶ Paraguay	▶ 1♂	▶ Chaco Paraguayo	▶ 1,2,4	▶ 26/52†
▶ Batistic, 1989[P,I]	▶ Argentina	▶ 2♂1♀	▶ Chamical, La Rioja Province	▶ 1,2,4	▶ 26/52†
▷ Kuramoto, 1990	▷ S.C.America	▷ n.i.	▷ ref. Barrio, 1976; ctry.n.sp.	▷ 1	▷ 26/n.i.
▷ Catroli, 2008	▷ n.i.	▷ n.i.	▷ ref. Barrio, 1976	▷ n.i.	▷ 26/n.i.
▷ Catroli & Kasahara, 2009	▷ n.i.	▷ n.i.	▷ ref. Barrio, 1976; Morand & Hernando, 1997	▷ 17	▷ 26/n.i.
▷ Paiva et al., 2010	▷ n.i.	▷ n.i.	▷ ref. Batistic, 1989; Morand & Hernando, 1997	▷ 4	▷ 26/n.i.
▶ Present study	▶ Argentina	▶ 1♂	▶ Chamical, La Rioja Province	▶ 1	▶ 26/52
Phyllomedusa tarsius					
▶ Bogart, 1973[P]	▶ Peru	▶ n.i.	▶ loc.n.i.	▶ 1	▶ 26/50†
▷ Morescalchi, 1979	▷ n.i.	▷ n.i.	▷ ref. Bogart, 1973	▷ n.i.	▷ 26/n.i.
▶ Batistic, 1989[P,I]	▶ Brazil	▶ 1♂	▶ Manaus, Amazonas State	▶ 1,2	▶ 26/46†
▷ King, 1990	▷ S.America	▷ n.i.	▷ ref. Bogart, 1973; ctry.n.sp.;	▷ n.i.	▷ 26/52
▷ Kuramoto, 1990	▷ S.C.America	▷ n.i.	▷ ref. Bogart, 1973; ctry.n.sp.	▷ 1	▷ 26/n.i.
▷ Nunes & Fagundes, 2008b	▷ n.i.	▷ n.i.	▷ ref. Bogart, 1973; list. *Phasmahyla tarsius*	▷ n.i.	▷ 26/52
▷ Catroli, 2008	▷ n.i.	▷ n.i.	▷ ref. Bogart, 1973	▷ n.i.	▷ 26/n.i.
▷ Catroli & Kasahara, 2009	▷ n.i.	▷ n.i.	▷ ref. Bogart, 1973	▷ n.i.	▷ 26/n.i.
▷ Paiva et al., 2010	▷ n.i.	▷ n.i.	▷ ref. Batistic, 1989	▷ 4	▷ 26/n.i.
▶ Bruschi et al., 2014a[P]	▶ Brazil	▶ 5s.n.i.	▶ Reserva Ducke, Manaus, Amazonas State	▶ 1–4, 13,14	▶ 26/46†
▶ Bruschi et al., 2014a[P]	▶ Ecuador	▶ 5s.n.i.	▶ Yasuní, Orellana Province	▶ 1–4, 13,14	▶ 26/46†
▷ Winkelmann and Noleto, 2015	▷ n.i.	▷ n.i.	▷ ref. Bogart, 1973	▷ n.i.	▷ 26/52
▷ Winkelmann and Noleto, 2015	▷ n.i.	▷ n.i.	▷ ref. Bruschi et al., 2014a	▷ n.i.	▷ 26/46
Phyllomedusa tetraploidea					
▶ Beçak et al., 1970a[P,M]	▶ Brazil	▶ 2♂	▶ São Paulo State; loc.n.i.; rep. *Phyllomedusa burmeisteri*	▶ 1	▶ 52/104†
▷ Rabello, 1970	▷ n.i.	▷ n.i.	▷ ref. Beçak et al., in press; list. *Phyllomedusa burmeisteri*	▷ n.i.	▷ 52/n.i.
▷ Foresti, 1972	▷ n.i.	▷ n.i.	▷ ref. Beçak et al., 1970a; list. *Phyllomedusa burmeisteri*	▷ n.i.	▷ 52/n.i.
▶ Batistic et al., 1975[M]	▶ Brazil	▶ 2♂	▶ Botucatu, São Paulo State; rep. *Phyllomedusa burmeisteri*	▶ 1	▶ 52/n.i.
▶ Barrio, 1976[P]	▶ Argentina	▶ 2♂	▶ vicinity of Bernardo de Irigoyen, Misiones Province; rep. *Phyllomedusa burmeisteri*	▶ 1	▶ 52/104†
▶ Barrio, 1976[P]	▶ Argentina	▶ 1♂1♀	▶ vicinity of 2 de Mayo, Misiones Province; rep. *Phyllomedusa burmeisteri*	▶ 1	▶ 52/104†
▶ Batistic, 1989[P,I]	▶ Brazil	▶ 10♂2♀ 3U	▶ Holambra II, Paranapanema, São Paulo State; rep. *Phyllomedusa* sp. 4n	▶ 1,2,4	▶ 52/104†
▶ Batistic, 1989[P,I] & pers.comm.	▶ Brazil	▶ 4♂	▶ Ribeirão Branco, São Paulo State; rep. *Phyllomedusa* sp. 4n	▶ 1,2,4	▶ 52/104†
▷ King, 1990	▷ S.America	▷ n.i.	▷ ref. Beçak et al., 1970a; ctry.n.sp.; list. *Phyllomedusa burmeisteri*	▷ n.i.	▷ 52/104
▷ Kuramoto, 1990	▷ S.C.America	▷ n.i.	▷ ref. Beçak et al., 1970a; Barrio, 1976; ctry.n.i.; list. *Phyllomedusa burmeisteri*	▷ 1	▷ 52/n.i.
▷ Nunes & Fagundes, 2008b	▷ n.i.	▷ n.i.	▷ ref. Beçak et al., 1970a; Pombal & Haddad, 1992; list. *Phasmahyla tetraploidea*	▷ n.i.	▷ 52/104
▷ Catroli, 2008	▷ n.i.	▷ n.i.	▷ ref. Beçak, 1970; Batistic et al., 1975; Barrio, 1976; Batistic, 1989; Kasahara et al., 2007	▷ n.i.	▷ 52/n.i.
▷ Catroli & Kasahara, 2009	▷ n.i.	▷ n.i.	▷ ref. Beçak et al., 1970a, Barrio, 1976; Batistic, 1989; Haddad et al., 1994	▷ 17	▷ 52/n.i.
▶ Barth et al., 2009[P]	▶ Brazil	▶ 2♂	▶ São Paulo State; loc.n.i.	▶ 1	▶ 52/104†
▷ Paiva et al., 2010	▷ n.i.	▷ n.i.	▷ ref. Batistic, 1989; Kasahara et al., 2007; list. *Phyllomedusa* sp. 4n	▷ 4	▷ 52/n.i.
▶ Gruber et al., 2013[P]	▶ Brazil	▶ 8♂1♀	▶ Riberão Branco, São Paulo State	▶ 1,2,4,6, 9,12–14	▶ 52/104†
▶ Gruber et al., 2013[P]	▶ Brazil	▶ 1♀	▶ Tatuí, São Paulo State	▶ 1,2,4,6, 9,12–14	▶ 52/104†
▷ Winkelmann and Noleto, 2015	▷ n.i.	▷ n.i.	▷ ref. Beçak et al., 1970a; Pombal & Haddad, 1992; Gruber et al., 2013	▷ n.i.	▷ 52/104
▶ Winkelmann and Noleto, 2015	▶ Brazil	▶ 2♂	▶ vicinity of União da Vitória Municipality, Paraná State (26°13′49″S, 51°05′11″W)	▶ 1,2	▶ 52/104
▷ Schmid & Steinlein, 2016a	▷ n.i.	▷ n.i.	▷ ref. Gruber et al., 2013	▷ 14	▷ 52/n.i.
▶ Present study	▶ Brazil	▶ 8♂1♀	▶ Riberão Branco, São Paulo State	▶ 1,2	▶ 52/104
Phyllomedusa vaillantii					
▶ Bruschi et al., 2014a[P]	▶ Brazil	▶ 1s.n.i.	▶ Jacareacanga, Pará State	▶ 1–4, 13,14	▶ 26/52†
▶ Bruschi et al., 2014a[P]	▶ Brazil	▶ 4s.n.i.	▶ Porto Velho, Rondônia State	▶ 1–4, 13,14	▶ 26/52†
▶ Bruschi et al., 2014a[P]	▶ Ecuador	▶ 5s.n.i.	▶ Yasuní, Orellana Province	▶ 1–4, 13,14	▶ 26/52†
▷ Winkelmann and Noleto, 2015	▷ n.i.	▷ n.i.	▷ ref. Bruschi et al., 2014a	▷ n.i.	▷ 26/52
▷ Schmid & Steinlein, 2016a	▷ n.i.	▷ n.i.	▷ ref. Bruschi et al., 2014a	▷ 14	▷ 26/n.i.

Author(s) ▶ original data[a] ▷ review articles	Country/ Region[b]	Speci- mens[c]	Locality of sampling, references cited, and remarks[d]	Methods applied[e]	2n/FN[f]
Phyllomedusa venusta					
▶ Duellman, 1970[N]	▶ C.America	▶ n♂	▶ loc.n.i.; ctry.n.sp.	▶ 1	▶ 26‡/n.i.
▷ Kuramoto, 1990	▷ S.C.America	▷ n.i.	▷ ref. Duellman, 1970; ctry.n.sp.	▷ 1	▷ 26/n.i.
▷ Catroli, 2008	▷ n.i.	▷ n.i.	▷ ref. Duellman, 1970	▷ n.i.	▷ 26/n.i.
▷ Catroli & Kasahara, 2009	▷ n.i.	▷ n.i.	▷ ref. Duellman, 1970	▷ n.i.	▷ 26/n.i.
PHYLLOMEDUSIDAE / *Pithecopus*					
Pithecopus ayeaye					
▶ Batistic, 1989[P,I]	▶ Brazil	▶ 4♂	▶ Morro do Ferro, Poços de Caldas, Minas Gerais State; rep. *Phyllomedusa ayeaye*	▶ 1,2,4	▶ 26/52†
▷ Catroli, 2008	▷ n.i.	▷ n.i.	▷ ref. Batistic, 1989; list. *Phyllomedusa ayeaye*	▷ n.i.	▷ 26/n.i
▷ Catroli & Kasahara, 2009	▷ n.i.	▷ n.i.	▷ ref. Batistic, 1989; list. *Phyllomedusa ayeaye*	▷ 17	▷ 26/n.i.
▷ Paiva et al., 2010	▷ n.i.	▷ n.i.	▷ ref. Batistic, 1989; list. *Phyllomedusa ayeaye*	▷ 4	▷ 26/n.i.
Pithecopus azureus					
▷ Catroli, 2008	▷ n.i.	▷ n.i.	▷ ref. Barrio, 1976; Souza et al., 2007; list. *Phyllomedusa azurea*	▷ n.i.	▷ 26/n.i.
▷ Catroli & Kasahara, 2009	▷ n.i.	▷ n.i.	▷ ref. Barrio, 1976; list. *Phyllomedusa azurea*	▷ 17	▷ 26/n.i.
Pithecopus hypochondrialis					
▶ Foresti, 1972[P,M]	▶ Brazil	▶ 1♂	▶ Jequié, Bahia State; rep. *Phyllomedusa hypochondrialis*	▶ 1	▶ 26/52†
▶ Barrio, 1976[P]	▶ Argentina	▶ 3♂	▶ Villa Ocampo, Santa Fe Province; rep. *Phyllomedusa hypochondrialis*	▶ 1	▶ 26/52†
▶ Barrio, 1976[P]	▶ Argentina	▶ 1♂1♀	▶ vicinity of Resistencia, Chaco Province; rep. *Phyllomedusa hypochondrialis*	▶ 1	▶ 26/52†
▶ Batistic, 1989[P,I]	▶ Paraguay	▶ 1♂	▶ Transchaco, km 222; rep. *Phyllomedusa hypochondrialis*	▶ 1,2	▶ 26/52†
▶ Batistic, 1989[P,I]	▶ Paraguay	▶ 2♂	▶ Pozo Colorado; rep. *Phyllomedusa hypochondrialis*	▶ 1,2	▶ 26/52†
▶ Batistic, 1989[P,I]	▶ Brazil	▶ 2♂	▶ Bela Vista, Mato Grosso do Sul State; rep. *Phyllomedusa hypochondrialis*	▶ 1,2	▶ 26/52†
▶ Batistic, 1989[P,I]	▶ Brazil	▶ 2♂	▶ Tucuruí, Pará State; rep. *Phyllomedusa hypochondrialis*	▶ 1,2,4	▶ 26/52†
▷ King, 1990	▷ C.America	▷ n.i.	▷ ref. León, 1970/err.ref.; err.cit.ctry; list. *Phyllomedusa hypochondrialis*	▷ n.i.	▷ 26/n.i.
▷ Kuramoto, 1990	▷ S.C.America	▷ n.i.	▷ ref. Barrio, 1976; ctry.n.sp.; list. *Phyllomedusa hypochondrialis*	▷ 1	▷ 26/n.i.
▷ Catroli, 2008	▷ n.i.	▷ n.i.	▷ ref. Foresti, 1972; Batistic, 1989; list. *Phyllomedusa hypochondrialis*	▷ n.i.	▷ 26/n.i.
▷ Catroli & Kasahara, 2009	▷ n.i.	▷ n.i.	▷ ref. Foresti, 1972; Batistic, 1989; list. *Phyllomedusa hypochondrialis*	▷ 17	▷ 26/n.i.
▷ Paiva et al., 2010	▷ n.i.	▷ n.i.	▷ ref. Batistic, 1989; Morand & Hernando, 1997; list. *Phyllomedusa hypochondrialis*	▷ 4	▷ 26/n.i.
▶ Bruschi et al., 2013[P]	▶ Brazil	▶ 6s.n.i.	▶ Laranjal do Jari, Amapá State; rep. *Phyllomedusa hypochondrialis*	▶ 1,2,4	▶ 26/52†
▶ Bruschi et al., 2013[P]	▶ Brazil	▶ 6s.n.i.	▶ Prainha, Pará State; rep. *Phyllomedusa hypochondrialis*	▶ 1,2,4	▶ 26/52†
▶ Bruschi et al., 2013[P]	▶ Brazil	▶ 6s.n.i.	▶ Monte Alegre, Pará State; rep. *Phyllomedusa hypochondrialis*	▶ 1,2,4	▶ 26/52†
▶ Bruschi et al., 2013[P]	▶ Brazil	▶ 11s.n.i.	▶ Alenquer, Pará State; rep. *Phyllomedusa hypochondrialis*	▶ 1,2,4	▶ 26/52†
▶ Bruschi et al., 2013[P]	▶ Brazil	▶ 5s.n.i.	▶ Oriximiná, Pará State; rep. *Phyllomedusa hypochondrialis*	▶ 1,2,4	▶ 26/52†
▶ Bruschi et al., 2013[P]	▶ Brazil	▶ 5s.n.i.	▶ Óbidos, Pará State; rep. *Phyllomedusa hypochondrialis*	▶ 1,2,4	▶ 26/52†
▶ Bruschi et al., 2013[P]	▶ Brazil	▶ 6s.n.i.	▶ Belterra, Pará State; rep. *Phyllomedusa hypochondrialis*	▶ 1–4	▶ 26/52†
▶ Bruschi et al., 2013[P]	▶ Brazil	▶ 1s.n.i.	▶ Alta Floresta, Mato Grosso State; rep. *Phyllomedusa* cf. *hypochondrialis*	▶ 1–4	▶ 26/52†
▶ Bruschi et al., 2013[P]	▶ Brazil	▶ 6s.n.i.	▶ Chapada dos Guimarães, Mato Grosso State; rep. *Phyllomedusa* cf. *hypochondrialis*	▶ 1,2,4	▶ 26/52†
▶ Bruschi et al., 2013[P]	▶ Brazil	▶ 1s.n.i.	▶ Santa Terezinha, Mato Grosso State; rep. *Phyllomedusa* cf. *hypochondrialis*	▶ 1,2,4	▶ 26/52†
▶ Present study	▶ Brazil	▶ 1♂	▶ loc.u.	▶ 1	▶ 26/52
▶ Present study	▶ Brazil	▶ 1s.n.i.	▶ A-Ukre, Pará State (07°39′S, 51°21′W)	▶ 1	▶ 26/52
Pithecopus nordestinus					
▷ Catroli, 2008	▷ n.i.	▷ n.i.	▷ ref. Barth et al., 2007; list. *Phyllomedusa nordestina*	▷ n.i.	▷ 26/n.i.
▶ Barth et al., 2009[N]	▶ n.i.	▶ n.i.	▶ loc.n.i.	▶ n.i.	▶ 26/n.i.
▶ Bruschi et al., 2012[P]	▶ Brazil	▶ 9♂,1J	▶ Vitória da Conquista, Bahia State	▶ 1,2,4,13	▶ 26/52†
▶ Bruschi et al., 2012[P]	▶ Brazil	▶ 6♂	▶ Jequié, Bahia State	▶ 1,2,4,13	▶ 26/52†
▶ Barth et al., 2013[P]	▶ Brazil	▶ 5s.n.i.	▶ Vitória da Conquista, Bahia State	▶ 1,4,13	▶ 26/52†
▶ Barth et al., 2013[P]	▶ Brazil	▶ 1s.n.i.	▶ Camacan, Bahia State	▶ 1,4,13	▶ 26/52†
▶ Barth et al., 2013[P]	▶ Brazil	▶ 5s.n.i.	▶ Ituberá, Bahia State	▶ 1,4,13	▶ 26/52†
▶ Barth et al., 2013[P]	▶ Brazil	▶ 1s.n.i.	▶ Feira da Santana, Bahia State	▶ 1,4,13	▶ 26/52†
▶ Barth et al., 2014[P]	▶ Brazil	▶ 5s.n.i.	▶ Vitória de Conquista, Bahia State; rep. *Phyllomedusa nordestina*	▶ 2,6,9,14	▶ 26/52†
▶ Barth et al., 2014[P]	▶ Brazil	▶ 1s.n.i.	▶ Camacan, Bahia State; rep. *Phyllomedusa nordestina*	▶ 2,6,9,14	▶ 26/52†
▶ Barth et al., 2014[P]	▶ Brazil	▶ 5s.n.i.	▶ Ituberá, Bahia State; rep. *Phyllomedusa nordestina*	▶ 2,6,9,14	▶ 26/52†
▶ Barth et al., 2014[P]	▶ Brazil	▶ 1s.n.i.	▶ Feira de Santana, Bahia State; rep. *Phyllomedusa nordestina*	▶ 2,6,9,14	▶ 26/52†
▷ Winkelmann and Noleto, 2015	▷ n.i.	▷ n.i.	▷ ref. Bruschi et al., 2012, 2013/err.ref; Barth et al., 2013; list. *Phyllomedusa nordestina*	▷ n.i.	▷ 26/52
▶ present study	▶ Brazil	▶.n.i.	▶ loc.u.	▶ 1	▶ 26/52
Pithecopus palliatus					
▶ Bogart, 1970b[P,I]	▶ Peru	▶ n.i.	▶ loc.n.i.; rep. *Phyllomedusa* sp.	▶ 1	▶ 26/52†
▶ Bogart, 1973[P]	▶ Peru	▶ n.i.	▶ loc.n.i.; rep. *Phyllomedusa palliata*	▶ 1	▶ 26/52†
▷ Morescalchi, 1979	▷ n.i.	▷ n.i.	▷ ref. Bogart, 1973; list. *Phyllomedusa palliata*	▷ n.i.	▷ 26/n.i.
▷ King, 1990	▷ S.America	▷ n.i.	▷ ref. Bogart, 1973; ctry.n.sp.; list. *Phyllomedusa palliata*	▷ n.i.	▷ 26/52
▷ Kuramoto, 1990	▷ S.C.America	▷ n.i.	▷ ref. Bogart, 1970a/err.ref., 1970b, 1973; ctry.n.sp.; list. *Phyllomedusa palliata*	▷ 1	▷ 26/n.i.

DOI: 10.1159/000490085

Author(s) ▶ original data[a] ▷ review articles	Country/ Region[b]	Speci- mens[c]	Locality of sampling, references cited, and remarks[d]	Methods applied[e]	2n/FN[f]
▷ Nunes & Fagundes, 2008b	▷ n.i.	▷ n.i.	▷ ref. Bogart, 1973; list. *Phasmahyla palliata*	▷ n.i.	▷ 26/52
▷ Catroli, 2008	▷ n.i.	▷ n.i.	▷ ref. Bogart, 1970a/err.ref., 1970b, 1973; list. *Phyllomedusa palliata*	▷ n.i.	▷ 26/n.i.
▷ Catroli & Kasahara, 2009	▷ n.i.	▷ n.i.	▷ ref. Bogart, 1973; list. *Phyllomedusa palliata*	▷ n.i.	▷ 26/n.i.
▷ Winkelmann and Noleto, 2015	▷ n.i.	▷ n.i.	▷ ref. Bogart, 1973; list. *Phyllomedusa palliata*	▷ n.i.	▷ 26/52
▶ Present study	▶ Peru	▶ 1J	▶ Bosque Nacional de Iparia, Huánuco Department	▶ 1	▶ 26/52
Pithecopus rohdei					
▶ Batistic, 1989[P,I]	▶ Brazil	▶ 3♂	▶ Horto Florestal de Santa Cruz, Rio de Janeiro State; rep. *Phyllomedusa rohdei*	▶ 1,2,4	▶ 26/52†
▶ Batistic, 1989[P,I]	▶ Brazil	▶ 6♂2♀	▶ Itaguaí, Rio de Janeiro State; rep. *Phyllomedusa rohdei*	▶ 1,2,4	▶ 26/52†
▶ Batistic1989[P,I],	▶ Brazil	▶ 1♂	▶ Biritiba-Mirim, São Paulo State; rep. *Phyllomedusa rohdei*	▶ 1,2,4	▶ 26/52†
▶ Batistic, 1989[P,I]	▶ Brazil	▶ 1U	▶ Paranapiacaba, São Paulo State; rep. *Phyllomedusa rohdei*	▶ 1,2,4	▶ 26/52†
▶ Batistic, 1989[P,I]	▶ Brazil	▶ 2♂1♀	▶ Linhares, Espírito Santo State; rep. *Phyllomedusa rohdei*	▶ 1,2,4	▶ 26/52†
▶ Barth et al., 2009[P]	▶ Brazil	▶ 11♂1♀	▶ Ilhéus Municipality (14°47'S, 39°10'W) and Ituberá Municipality (13°49'S, 39°10'W), Bahia State; rep. *Phyllomedusa rohdei*	▶ 1,4	▶ 26/52†
▷ Nunes & Fagundes, 2008b	▷ n.i.	▷ n.i.	▷ ref. Batistic, 1989; list. *Phasmahyla rohdei*	▷ n.i.	▷ 26/52
▶ Nunes & Fagundes, 2008b[P]	▶ Brazil	▶ 1♂	▶ Linhares, Espírito Santo State; rep. *Phyllomedusa rohdei*	▶ 1	▶ 40/80
▷ Catroli, 2008	▷ n.i.	▷ n.i.	▷ ref. Batistic, 1989; list. *Phyllomedusa rohdei*	▷ n.i.	▷ 26/n.i.
▷ Catroli & Kasahara, 2009	▷ n.i.	▷ n.i.	▷ ref. Batistic, 1989; Nunes & Fagundes, 2008b; Moran & Hernando, 1997; Paiva et al., 2010; list. *Phyllomedusa rohdei*	▷ 17	▷ 26/n.i.
▷ Catroli & Kasahara, 2009	▷ n.i.	▷ n.i.	▷ ref. Nunes & Fagundes, 2008b; Morand & Hernando, 1997; list. *Phyllomedusa rohdei*	▷ 17	▷ 40/n.i.
▷ Paiva et al., 2010	▷ n.i.	▷ n.i.	▷ ref. Batistic, 1989; list. *Phyllomedusa rohdei*	▷ 4	▷ 26/n.i.
▶ Paiva et al., 2010[P]	▶ Brazil	▶ 7s.n.i.	▶ Biritiba-Mirim, São Paulo State (23°34'21"S, 46°02'19"W); rep. *Phyllomedusa rohdei*	▶ 1,4	▶ 26/46†
▶ Bruschi et al., 2012[P]	▶ Brazil	▶ 10♂	▶ Ilhéus, Bahia State; rep. *Phyllomedusa rohdei*	▶ 1,2,4,13	▶ 26/52†
▶ Barth et al., 2013[P]	▶ Brazil	▶ 6s.n.i.	▶ Ituberá, Bahia State; rep. *Phyllomedusa rohdei*	▶ 1,4,13	▶ 26/52†
▶ Barth et al., 2013[P]	▶ Brazil	▶ 8s.n.i.	▶ Ilhéus, Bahia State; rep. *Phyllomedusa rohdei*	▶ 1,4,13	▶ 26/52†
▶ Barth et al., 2014[P]	▶ Brazil	▶ 6s.n.i.	▶ Ituberá, Bahia State; rep. *Phyllomedusa rohdei*	▶ 2,6,9,14	▶ 26/52†
▶ Barth et al., 2014[P]	▶ Brazil	▶ 8s.n.i.	▶ Ilhéus, Bahia State; rep. *Phyllomedusa rohdei*	▶ 2,6,9,14	▶ 26/52†
▷ Winkelmann and Noleto, 2015	▷ n.i.	▷ n.i.	▷ ref. Barth et al., 2009; Paiva et al., 2010; Bruschi et al., 2012; list. *Phyllomedusa rohdei*	▷ n.i.	▷ 26/52
▷ Winkelmann and Noleto, 2015	▷ n.i.	▷ n.i.	▷ ref. Nunes & Fagundes, 2008b; list. *Phyllomedusa rohdei*	▷ n.i.	▷ 40/80
▷ Schmid & Steinlein, 2016a	▷ n.i.	▷ n.i.	▷ ref. Barth et al., 2014; list. *Phyllomedusa rohdei*	▷ 14	▷ 26/n.i.
Pithecopus trinitatis*					
▶ Present study	▶ Trinidad	▶ 2♂1♀	▶ Simla Research Station, Arima valley, ~ 360 m (10°41'N, 61°17'W)	▶ 14	▶ 26/48
***Pithecopus* sp. (aff. *hypochondrialis*)**					
▶ Bruschi et al., 2013[P]	▶ Brazil	▶ 5s.n.i.	▶ São Luis, Maranhão State; rep. *Phyllomedusa* (sp. aff. *hypochondrialis*)	▶ 1,2,4	▶ 26/52†
▶ Bruschi et al., 2013[P]	▶ Brazil	▶ 8s.n.i.	▶ Bacabeira, Maranhão State; rep. *Phyllomedusa* (sp. aff. *hypochondrialis*)	▶ 1,2,4	▶ 26/52†
▶ Bruschi et al., 2013[P]	▶ Brazil	▶ 8s.n.i.	▶ Urbano Santos, Maranhão State; rep. *Phyllomedusa* (sp. aff. *hypochondrialis*)	▶ 1,2,4	▶ 26/52†
▶ Bruschi et al., 2013[P]	▶ Brazil	▶ 13s.n.i.	▶ Porto Nacional, Tocantins State; rep. *Phyllomedusa* (sp. aff. *hypochondrialis*)	▶ 1,2,4	▶ 26/52†
▶ Bruschi et al., 2013[P]	▶ Brazil	▶ 18s.n.i.	▶ Barreiras, Bahia State; rep. *Phyllomedusa* (sp. aff. *hypochondrialis*)	▶ 1,2,4	▶ 26/52†
▶ Bruschi et al., 2013[P]	▶ Brazil	▶ 8s.n.i.	▶ Uberlândia, Minas Gerais State; rep. *Phyllomedusa* (sp. aff. *hypochondrialis*)	▶ 1,2,4	▶ 26/52†

[a] **N** Diploid or haploid chromosome number (2n or n). **FN** Fundamental number (number of chromosome arms, FN). **D** Drawings of mitotic and/or meiotic chromosomes (with or without camera lucida). **P** Photographs of mitotic and/or meiotic chromosomes. **I** Idiogram. **M** Chromosome length measurements. **pers.comm.:** personal communication. The mitotic chromosomes of the species labelled with * are published here for the first time.

[b] **Argent.:** Argentina. **Dom.Rep.:** Dominican Republic. **F.Guiana:** French Guiana. **C.America:** Central America. **S.America:** South America. **S.C.America.:** South and Central America. **Aus./NG.:** Australia/New Guinea. **n.i.:** not indicated.

[c] **J**: juvenile. **U**: sex unknown. **n.i.:** not indicated. **s.n.i.:** sex not indicated. **tp:** tadpole.

[d] Locality of sampling according to original publication, personal communication of authors, and further information obtained from museums where the cytogenetically examined specimens are deposited. Apparent spelling errors in the names of the sampling localities found in previous publications have been corrected. **anim.:** animals. **ctry.n.sp.:** country not specified. **err.leg.:** erroneous legend. **err.ref.:** erroneous reference. **err.cit.ctry.:** erroneous citation of country. **err.cit.2n:** erroneous citation of diploid chromosome number. **err.cit.FN:** erroneous citation of fundamental number. **ex.ref.miss.:** exact reference missing in the reference list of the corresponding publication. **list.:** listed as. **loc.n.i.:** sampling locality not indicated. **loc.u.:** sampling locality unknown. **m:** meters above sea level. **pers.comm.:** personal communication. **ref.:** refers to. **ref.n.i.:** reference not indicated. **rep.:** reported as. **unpubl.:** unpublished. *{ }*: erroneous or deviant spelling of species name.

[e] **1:** Conventional staining (haematoxylin, aceto orcein, Giemsa stain). **2:** C-banding (followed by Giemsa staining). **3:** C-banding (followed by fluorochrome staining). **4:** NOR silver staining. **5:** Quinacrine staining. **6:** DAPI staining. **7:** Distamycin A/DAPI counterstaining. **8:** Mithramycin staining or distamycin A/mithramycin counterstaining. **9:** Chromomycin A_3 staining or distamycin A/chromomycin A_3 counterstaining. **10**: G-banding. **11:** Restriction endonuclease banding. **12:** BrdU-replication banding. **13:** FISH with 18S + 28S or 45S rDNA probes. **14:** FISH with telomeric DNA probes. **15:** FISH with DNA microsatellites. **16:** Fluorochrome staining (not specified). **17:** Differential staining (not specified). **18:** Chromosome microdissection and painting. **19:** FISH with 5S rDNA probes. **20:** FISH with repetitive DNA probes. **21:** Comparative genomic hybridization. **22:** Hoechst 33258 staining. **23:** Immunofluorescence against repetitive 5-methylcytosine-rich DNA.

[f] **2n†/FN†:** the 2n or/and FN are not explicitly mentioned in the text of the corresponding publication but inferred from the chromosome illustration, idiogram or text. Small arms in the chromosomes were neglected if their total length does not exceed twice the width of a chromatid in the long arm of the same chromosome (see page chapter 3.1). **2n‡:** inferred from chromosome counts in meiosis. **n.i.:** not indicated.

Appendix 2.

Materials and Methods section in the publication of Duellman et al. [2016] used to produce the timetree for inferring the historical biogeography of the Arboranae (Fig. 2). References in this Appendix 2 are listed in the general list of References (pages 223–247).

Duellman et al. [2016] began by assembling all available data in Genbank for 2 mitochondrial genes which were the best-represented, the large (16S) and small (12S) rRNA subunits. They reviewed both gene alignments in MEGA 5.2 [Tamura et al., 2011] and trimmed them to include only sequences of Hylidae (s.l.), as well as those of the chosen 4 outgroup taxa – viz., *Ceuthomantis smaragdinus*, *Dendrobates auratus*, *Haddadus binotatus*, and *Rhinoderma darwinii*. They excluded taxa that have since been synonymized with recognized taxa according to Frost [2015].

They also built alignments for an additional 10 genes that were well-represented: cytochrome b (*cytb*), chemokine receptor 4 (*CXCR4*), histone 3a (*H3A*), sodium-calcium exchanger 1 (*NCX1*), pro-opiomelanocortin (*POMC*), recombination-activating gene 1 (*RAG1*), rhodopsin (*Rho*), seventh-in-absentia (*SIA*), solute carrier family 8 member 3 (*SLC8a3*), and tyrosinase precursor (*Tyr*). To do this, they again queried GenBank for all hylid sequences for each of the 10 genes and examined all available sequences of Hylidae (s.l.). Again, they used the latest taxonomy [Frost 2015] and, for each gene, examined all available sequences for each recognized species and chose the longest and most complete (or, if all else equal the first sequence). This approach was similar, but not identical, to that of Pyron & Wiens [2011]. However, they added 7 additional genes for which there were at least 25 distinct hylid taxa: 28S rRNA (*28S*), cellular myelocytomatosis exon 2 (*cmyc2*), cellular myelocytomatosis exon 3 (*cmyc3*), cytochrome oxidase subunit I (*COI*), tensin 3 (*TNS3*), NADH dehydrogenase subunit 1 (*ND1*), and NADH dehydrogenase subunit 2 (*ND2*).

All genes were aligned independently in MEGA 5.2, using MUSCLE 3.6 under default parameters [Edgar, 2004], and rarely with adjustments made by eye. Protein translations were reviewed for protein-coding genes to ensure correct alignment with respect to reading frame. Individual gene trees were constructed as an additional check on data quality, and a few obviously misplaced sequences (either mislabeled in Genbank, or containing sequence errors) were removed. All 19 genes were concatenated for a total of 503 ingroup taxa (species) and 4 outgroup taxa, and 16,128 aligned sites. For comparison, Pyron & Wiens [2011] included 1,824 sequences of Hylidae (s.l.) across the 12 genes in common with the data of Duellman et al. [2016] set whereas Duellman et al. [2016] used 2,306 sequences across those same genes. They included 7 additional genes for a total of 3,069 sequences of Hylidae (s.l.), which corresponds to 68% more sequences (overall) than in Pyron & Wiens [2011]. GenBank accession numbers for all genes of all taxa included in the study of Duellman et al. [2016] are given in their Appendix 1 along with a summary of sequence coverage per gene.

Maximum likelihood analyses were performed on the concatenated dataset. As with phylogenetic analyses of other large data sets [Padial et al., 2014; Pyron et al., 2011; Pyron et al., 2013], Bayesian methods and standard bootstrap analysis were computationally intractable. RAxML 8.0.24 [Stamatakis, 2014] was implemented on the CIPRES Science Gateway [Miller et al., 2010]. The dataset was partitioned by gene (i.e., 19 partitions), and analyzed using the GTRGAMMA model for all partitions (the maximized available model in RAxML). All parameters for the ML analyses were estimated by the program during the run. Branch support in the trees was provided by rapid bootstrap analysis (1000 replicates). The rapid bootstrap method has been shown to produce almost identical values as the standard bootstrap method (r = 0.99), but it is many times faster [Stamatakis et al., 2008]. As a test of the consistency of the rapid bootstrap method, they made 5 separate runs and determined that 82% of the nodes were identical or varied within 5% and nearly all (99%) of the well-supported nodes (>70% support value) were in this category. This was consistent with the typical pattern of bootstrap support values, where higher support values have smaller variances [Hedges, 1992]. Alternate partitioning schemes were considered, including the best scheme under the BIC according to PartitionFinder 1.1.1 [Lanfear et al., 2012, 2014]. There were no significant changes in topology in the resulting trees between the 2 schemes.

Divergence times were estimated using RelTime as implemented in MEGA7 [Tamura et al., 2013]. The full dataset (16,128 aligned sites) was loaded and all sites were analyzed. The topology from the ML analysis was used as the starting tree, with the 4 outgroup taxa defined manually. Local clocks were used, with a clock stringency of "few clocks" set (clock rates merged on two StdErrors). The GTR+I+G model was implemented with 5 discrete gamma categories.

Two minimum and 2 maximum constraints were used as calibrations. The minimum divergence time between Hylinae and Acridinae was set at 16 million years ago (Mya), based on remains of *Hyla* from the Miocene of Austria [Sanchiz, 1998]. The divergence time between Phyllomedusidae and Pelodryadidae was constrained between 35–70 Mya, based on the timeframe when Australian pelodryadid arboranans (represented by *Litoria*) could disperse from South America through Antarctica [Li & Powell, 2001; Sanmartin & Ronquist, 2004; Springer et al., 1998; Woodburne & Case, 1996]. The maximum age for the Caribbean genus *Osteopilus* was set at 37.2 Mya, the date at which the Greater and Lesser Antilles became permanently subaerial, based on geologic facts [Iturralde-Vinent & MacPhee, 1999]. RelTime analyses were also performed with various combinations of minimum and maximum calibrations in order to gauge the effects of each calibration on the inferred divergence times.

Finally, as a check on RelTime estimates, they also estimated divergence times using an uncorrelated relaxed clock model in BEAST v1.8.1 [Drummond & Rambaut, 2007] on their maximum likelihood tree with the topology constrained, using default optimization settings for priors and operators and 2 independent runs. they set uniform priors on the 3 calibrated nodes, as for the RelTime analyses, then ran the search for 100 million generations, unpartitioned, and discarded the first 10 million generations as the burn-in. They used Tree Annotator v1.8.0 [Drummond & Rambaut, 2007] to visualize the inferred phylogeny and the node age estimates. Because the time estimates from BEAST were nearly identical to those of RelTime (r = 0.96, slope = 1.06), they presented only those from RelTime because they have more conservative confidence intervals.

Divergence times were similar across all RelTime analyses calibrated with different sets of maximums and minimums. There were 4 sets of identical results: (1) those analyses that included the minimum calibration of 35 Mya placed on the split between Phyllomedusidae and Pelodryadidae and maximum of 70 Mya placed on that same split, regardless of whether the other minimum and/or maximum were included; (2) those analyses that included the minimum calibration of 16 Mya placed on the split between Hylinae and Acridinae and the maximum of 70 Mya placed on the split between Phyllomedusidae and Pelodryadidae, regardless of the presence of the 37.2 Mya *Osteopilus* maximum; (3) those analyses that included the minimum calibration of 35 Mya placed on the split between Phyllomedusidae and Pelodryadidae and maximum of 37.2 Mya on *Osteopilus*, regardless of the presence of the 16 Mya minimum placed on the split between Hylinae and Acridinae; and (4) the analysis with the minimum calibration of 16 Mya placed on the split between Hylinae and Acridinae and the maximum of 37.2 Mya placed at the base of *Osteopilus*. Presented in their timetree (Fig. 17) are the divergence times estimated from an analysis with all 4 calibrations, which corresponded to the results of set 1. The date estimates of analyses in set 2 were ~11% younger than those of set 1. The date estimates of analyses in set 3 were ~17% older than those of set 1. The date estimates of the analysis in set 4 were ~7% older than those of set 1. These results showed that no single calibration had a major (e.g., >50%) impact on the resulting time estimates.

Appendix 3. Compilation of previous publications from which figures have been taken for the present monograph

Reference	Journal[a]	Figure[b]	Page	Species[c]
Sessions, 1990	Herpetologica 34:70–73	1	71	*Anotheca spinosa*
Miura, 1995	Chromosoma 103:567–574	4	572	*Rana* species, *Dryophytes japonicus*, *Bufo j. japonicus*
Gruber et al., 2013	BMC Genetics 14:75	1	2	*Phyllomedusa distincta, P. tetraploidea,* and hybrid
Gruber et al., 2013	BMC Genetics 14:75	2	3	*Phyllomedusa distincta, P. tetraploidea,* and hybrid
Gruber et al., 2013	BMC Genetics 14:75	3	4	*Phyllomedusa distincta, P. tetraploidea,* and hybrid
Gruber et al., 2013	BMC Genetics 14:75	5	5	*Phyllomedusa distincta, P. tetraploidea,* and hybrid
Cardozo et al., 2011	Copeia 2011:251–263	1	255	*Scinax albicans*
Cardozo et al., 2011	Copeia 2011:251–263	1	255	*Scinax argyreornatus*
Cardozo et al., 2011	Copeia 2011:251–263	1	255	*Scinax ariadne*
Cardozo et al,. 2011	Copeia 2011:251–263	1	255	*Scinax aromothyella*
Cardozo et al., 2011	Copeia 2011:251–263	1	255	*Scinax berthae*
Cardozo et al., 2011	Copeia 2011:251–263	1	255	*Scinax canastrensis*
Cardozo et al., 2011	Copeia 2011:251–263	1	255	*Scinax hiemalis*
Cardozo et al., 2011	Copeia 2011:251–263	1	255	*Scinax littoralis*
Cardozo et al., 2011	Copeia 2011:251–263	2	256	*Scinax longilineus*
Cardozo et al., 2011	Copeia 2011:251–263	2	256	*Scinax obtriangulatus*
Cardozo et al., 2011	Copeia 2011:251–263	2	256	*Scinax rizibilis*
Cardozo et al., 2011	Copeia 2011:251–263	2	256	*Scinax trapicheiroi*
Cardozo et al., 2011	Copeia 2011:251–263	2	256	*Scinax* sp. 1 (group perpusillus)
Cardozo et al., 2011	Copeia 2011:251–263	2	256	*Scinax* sp. 2 (group perpusillus)
Cardozo et al., 2011	Copeia 2011:251–263	3	257	*Scinax acuminatus*
Cardozo et al., 2011	Copeia 2011:251–263	3	257	*Scinax alter*
Cardozo et al., 2011	Copeia 2011:251–263	3	257	*Scinax curicica*
Cardozo et al., 2011	Copeia 2011:251–263	3	257	*Scinax duartei*
Cardozo et al., 2011	Copeia 2011:251–263	3	257	*Scinax eurydice*
Cardozo et al., 2011	Copeia 2011:251–263	3	257	*Scinax fuscomarginatus*
Cardozo et al., 2011	Copeia 2011:251–263	3	257	*Scinax fuscovarius*
Cardozo et al., 2011	Copeia 2011:251–263	3	257	*Scinax granulatus*
Cardozo et al., 2011	Copeia 2011:251–263	4	258	*Scinax hayii*
Cardozo et al., 2011	Copeia 2011:251–263	4	258	*Scinax nasicus*
Cardozo et al., 2011	Copeia 2011:251–263	4	258	*Scinax perereca*
Cardozo et al., 2011	Copeia 2011:251–263	4	258	*Scinax similis*
Cardozo et al., 2011	Copeia 2011:251–263	4	258	*Scinax squalirostris*
Cardozo et al., 2011	Copeia 2011:251–263	4	258	*Julianus uruguayus*
Cardozo et al., 2011	Copeia 2011:251–263	5	259	*Scinax albicans*
Cardozo et al., 2011	Copeia 2011:251–263	5	259	*Scinax argyreornatus*
Cardozo et al., 2011	Copeia 2011:251–263	5	259	*Scinax ariadne*
Cardozo et al., 2011	Copeia 2011:251–263	5	259	*Scinax aromothyella*
Cardozo et al., 2011	Copeia 2011:251–263	5	259	*Scinax berthae*
Cardozo et al., 2011	Copeia 2011:251–263	5	259	*Scinax canastrensis*
Cardozo et al., 2011	Copeia 2011:251–263	5	259	*Scinax hiemalis*
Cardozo et al., 2011	Copeia 2011:251–263	5	259	*Scinax littoralis*
Cardozo et al., 2011	Copeia 2011:251–263	6	259	*Scinax longilineus*
Cardozo et al., 2011	Copeia 2011:251–263	6	259	*Scinax obtriangulatus*
Cardozo et al., 2011	Copeia 2011:251–263	6	259	*Scinax rizibilis*
Cardozo et al., 2011	Copeia 2011:251–263	6	259	*Scinax* sp. 2 (gr. *perpusillus*)
Cardozo et al., 2011	Copeia 2011:251–263	7	260	*Scinax acuminatus*
Cardozo et al., 2011	Copeia 2011:251–263	7	260	*Scinax alter*
Cardozo et al., 2011	Copeia 2011:251–263	7	260	*Scinax curicica*
Cardozo et al., 2011	Copeia 2011:251–263	7	260	*Scinax duartei*
Cardozo et al., 2011	Copeia 2011:251–263	7	260	*Scinax eurydice*
Cardozo et al., 2011	Copeia 2011:251–263	7	260	*Scinax fuscovarius*
Cardozo et al., 2011	Copeia 2011:251–263	8	260	*Scinax granulatus*
Cardozo et al., 2011	Copeia 2011:251–263	8	260	*Scinax hayii*
Cardozo et al., 2011	Copeia 2011:251–263	8	260	*Scinax nasicus*
Cardozo et al., 2011	Copeia 2011:251–263	8	260	*Scinax perereca*
Cardozo et al., 2011	Copeia 2011:251–263	8	260	*Scinax similis*
Cardozo et al., 2011	Copeia 2011:251–263	8	260	*Scinax squalirostris*
Kasahara et al., 2003	Cytogenet Genome Res 103:155–162	1	156	*Aparasphenodon brunoi*
Kasahara et al., 2003	Cytogenet Genome Res 103:155–162	1	156	*Corythomantis greeningi*
Kasahara et al., 2003	Cytogenet Genome Res 103:155–162	1	156	*Osteocephalus langsdorffii*

Reference	Journal[a]	Figure[b]	Page	Species[c]
Kasahara et al., 2003	Cytogenet Genome Res 103:155–162	1	156	*Scinax fuscovarius*
Kasahara et al., 2003	Cytogenet Genome Res 103:155–162	5	160	*Aparasphenodon brunoi*
Catroli et al., 2011	J Herpetol 45:120–128	1	123	*Bokermannohyla ahenea*
Catroli et al., 2011	J Herpetol 45:120–128	1	123	*Bokermannohyla circumdata*
Catroli et al., 2011	J Herpetol 45:120–128	1	123	*Bokermannohyla hylax*
Catroli et al., 2011	J Herpetol 45:120–128	1	123	*Bokermannohyla luctuosa*
Catroli et al., 2011	J Herpetol 45:120–128	1	123	*Bokermannohyla* sp. 1
Catroli et al., 2011	J Herpetol 45:120–128	1	123	*Bokermannohyla* sp. 2
Catroli et al., 2011	J Herpetol 45:120–128	1	123	*Bokermannohyla* sp. 3
Catroli et al., 2011	J Herpetol 45:120–128	1	123	*Bokermannohyla* sp. 4
Catroli et al., 2011	J Herpetol 45:120–128	2	124	*Bokermannohyla alverangai*
Catroli et al., 2011	J Herpetol 45:120–128	2	124	*Bokermannohyla ibitiguara*
Catroli et al., 2011	J Herpetol 45:120–128	2	124	*Bokermannohyla saxicola*
Catroli et al., 2011	J Herpetol 45:120–128	2	124	*Bokermannohyla martinsi*
Catroli et al., 2011	J Herpetol 45:120–128	4	125	*Bokarmannohyla ahenea*
Catroli et al., 2011	J Herpetol 45:120–128	4	125	*Bokarmannohyla circumdata*
Catroli et al., 2011	J Herpetol 45:120–128	4	125	*Bokarmannohyla luctuosa*
Catroli et al., 2011	J Herpetol 45:120–128	4	125	*Bokarmannohyla* sp. 2
Catroli et al., 2011	J Herpetol 45:120–128	4	125	*Bokarmannohyla* sp. 3
Catroli et al., 2011	J Herpetol 45:120–128	4	125	*Bokarmannohyla alvarengai*
Catroli et al., 2011	J Herpetol 45:120–128	4	125	*Bokarmannohyla ibitiguara*
Catroli et al., 2011	J Herpetol 45:120–128	4	125	*Bokarmannohyla saxicola*
Catroli et al., 2011	J Herpetol 45:120–128	4	125	*Bokarmannohyla martinsi*
Baldissera et al., 1993	Rev Bras Genet 16:335–345	1	338	*Boana prasina*
Baldissera et al., 1993	Rev Bras Genet 16:335	1	338	*Boana hayii*
Baldissera et al., 1993	Rev Bras Genet 16:335	1	338	*Scinax fuscovarius*
Baldissera et al., 1993	Rev Bras Genet 16:335	2	339	*Bokermannohyla luctuosa*
Baldissera et al., 1993	Rev Bras Genet 16:335	2	339	*Bokermannohyla luctuosa*
Baldissera et al., 1993	Rev Bras Genet 16:335	5	342	*Boana prasina*
Gruber et al., 2005	Folia Biologica 51:68–75	1	70	*Dendropsophus berthalutzae*
Gruber et al., 2005	Folia Biologica 51:68–75	1	70	*Dendropsophus cruzi*
Gruber et al., 2005	Folia Biologica 51:68–75	1	70	*Dendropsophus elegans*
Gruber et al., 2005	Folia Biologica 51:68–75	1	70	*Dendropsophus elianae*
Gruber et al., 2005	Folia Biologica 51:68–75	2	71	*Dendropsophus microps*
Gruber et al., 2005	Folia Biologica 51:68–75	2	71	*Dendropsophus microps*
Gruber et al., 2005	Folia Biologica 51:68–75	2	71	*Dendropsophus minutus*
Gruber et al., 2005	Folia Biologica 51:68–75	2	71	*Dendropsophus minutus*
Gruber et al., 2005	Folia Biologica 51:68–75	3	72	*Dendropsophus nahdereri*
Gruber et al., 2005	Folia Biologica 51:68–75	3	72	*Dendropsophus nanus*
Gruber et al., 2005	Folia Biologica 51:68–75	3	72	*Dendropsophus rubicundulus*
Gruber et al., 2005	Folia Biologica 51:68–75	3	72	*Dendropsophu sanborni*
Gruber et al., 2005	Folia Biologica 51:68–75	4	73	*Dendropsophus cruzi*
Gruber et al., 2005	Folia Biologica 51:68–75	4	73	*Dendropsophus elegans*
Gruber et al., 2005	Folia Biologica 51:68–75	4	73	*Dendropsophus minutus*
Gruber et al., 2007	Genetica 130:51:281–291	1	283	*Boana albopunctata*
Gruber et al., 2007	Genetica 130:51:281–291	1	283	*Boana albopunctata*
Gruber et al., 2007	Genetica 130:51:281–291	1	283	*Boana raniceps*
Gruber et al., 2007	Genetica 130:51:281–291	1	283	*Boana crepitans*
Gruber et al., 2007	Genetica 130:51:281–291	1	283	*Boana albopunctata*
Gruber et al., 2007	Genetica 130:51:281–291	1	283	*Boana raniceps*
Gruber et al., 2007	Genetica 130:51:281–291	1	283	*Boana crepitans*
Gruber et al., 2012	Comp Cytogenet 6:409–423	1	413	*Aparasphenodon bokermanni*
Gruber et al., 2012	Comp Cytogenet 6:409–423	1	413	*Itapotihyla langsdorffii*
Gruber et al., 2012	Comp Cytogenet 6:409–423	1	413	*Trachycephalus* sp.
Gruber et al., 2012	Comp Cytogenet 6:409–423	1	413	*Trachycephalus mesophaeus*
Gruber et al., 2012	Comp Cytogenet 6:409–423	1	413	*Trachycephalus typhonius*
Gruber et al., 2012	Comp Cytogenet 6:409–423	1	413	*Phyllodytes edelmoi*
Gruber et al., 2012	Comp Cytogenet 6:409–423	1	413	*Phyllodytes luteolus*
Gruber et al., 2012	Comp Cytogenet 6:409–423	2	414	*Aparasphenodon bokermanni*
Gruber et al., 2012	Comp Cytogenet 6:409–423	2	414	*Itapotihyla langsdorffii*
Gruber et al., 2012	Comp Cytogenet 6:409–423	2	414	*Trachycephalus* sp.
Gruber et al., 2012	Comp Cytogenet 6:409–423	2	414	*Trachycephalus mesophaeus*
Gruber et al., 2012	Comp Cytogenet 6:409–423	2	414	*Trachycephalus typhonius*
Gruber et al., 2012	BMC Genet 13:28	1	3	*Aplastodiscus perviridis*
Gruber et al., 2012	BMC Genet 13:28	1	3	*Aplastodiscus arildae*
Gruber et al., 2012	BMC Genet 13:28	1	3	*Aplastodiscus eugenioi*
Gruber et al., 2012	BMC Genet 13:28	1	3	*Aplastodiscus callipygius*

Reference	Journal[a]	Figure[b]	Page	Species[c]
Gruber et al., 2012	BMC Genet 13:28	1	3	*Aplastodiscus leucopygius*
Gruber et al., 2012	BMC Genet 13:28	3	4	*Aplastodiscus perviridis*
Gruber et al., 2012	BMC Genet 13:28	3	4	*Aplastodiscus arildae*
Gruber et al., 2012	BMC Genet 13:28	3	4	*Aplastodiscus eugenioi*
Gruber et al., 2012	BMC Genet 13:28	3	4	*Aplastodiscus callipygius*
Gruber et al., 2012	BMC Genet 13:28	3	4	*Aplastodiscus leucopygius*

[a] The complete reference can be found in the Reference section.

[b] The original figures were modified for the present mongraph. The order of the chromosomes in the karyotypes of the original publications and in the present monograph differs. Schematic drawings of chromomes and idiograms were redesigned.

[c] Species names are according to Duellman et al. [2016] and Frost [2017].

Cytogenet Genome Res 2018;155:317–324
DOI:10.1159/000492201

The Arboranan Frogs:
Species Index

KARGER

E-Mail karger@karger.com
www.karger.com/cgr

Arabic numerals indicate positions of the genera and species in the main text, numbers in italics positions in the tables, and numbers in bold positions in the figures.

Amphibia – Arboranae

DOI:10.1159/000492201

DOI:10.1159/000492201

Amphibia – Other Taxa

Cytogenet Genome Res 2018;155:325
DOI: 10.1159/000493810

Dedication

Michael Schmid

We dedicate this publication to Prof. Michael Schmid who passed away shortly after the publisher received the final approval for publication. Michael will be sorely missed. He was globally recognized as a leader in cytogenetics and chromosome research and had an enormously productive career. His passion for frogs combined with his methodological expertise have revealed a fascinating history of chromosome mutation in anurans that will serve to guide future research in genomics. As a colleague, mentor, and friend, his enthusiasm was infectious. We feel fortunate to have been included in his quest to karyotype all the frogs in the world and as editors in this and previous (The Chromosomes of Terraranan Frogs; The Hemiphractid Frogs) major cytogenetic publications.

James P. (Jim) Bogart
S. Blair Hedges

KARGER

E-Mail karger@karger.com
www.karger.com/cgr